iF yearbook communication 2006

Herausgeber ▪ Editor
iF International Forum Design GmbH, Messegelände, 30521 Hannover, Germany
Phone +49(0)511-8932400, Fax: +49(0)511-8932401, www.ifdesign.de, info@ifdesign.de
iF Team
Geschäftsführung, Managing Director: Ralph Wiegmann
Anja-Martina Kirschning, Charity Sala Lirazan, Gabriele Bertemann, Korinna Gramsch,
Andrea Schewior, Heike Meier, Dirk Bartelsmeier, Anne-Kathrin Aumann, Sabine Böhmer,
Rainer Schwarz, Birgit Schultz, Frank Zierenberg
iF Press Office
Claudia Neumann Communication I NeumanLuz Cooperation, Eigelstein 103–113, 50668 Köln, Germany
Phone +49(0)221-913949-0, Fax +49(0)221-913949-19, design@neumann-luz.de
Produktion ▪ Production
Birkhäuser – Publishers for Architecture, Marion Plassmann
Projektmanagement Birkhäuser ▪ Project Management
Elena Dinter
Gestaltung aktuelle Ausgabe ▪ Design of Current Edition
Beate Wichmann, Frankfurt
Satz und Lithografie ▪ Typesetting and Lithography
Raasch & Partner GmbH, Dreieich
Druck ▪ Print
Aumüller Druck KG, Regensburg
Fotografie Jury ▪ Jury Photography
Dirk Meußling, Isernhagen KB
Übersetzung ▪ Translation
Kevin Stiers, Hamburg
Textredaktion ▪ Copy Editing
Andrea Schlitzer, Köln
Corporate Design iF communication design award
Helke Brandt und Partner, Hannover
Bibliographic information published by Die Deutsche Bibliothek
Die deutsche Bibliothek lists this publication in the Deutsche Nationalbibliografie;
detailed bibliographic data is available on the Internet at http://dnb.ddb.de.

Verlag ▪ Publisher
Birkhäuser – Publishers for Architecture
P.O.Box 133, CH-4010 Basel, Switzerland
Member of Springer Science+Business Media
www.birkhauser.ch
Printed in Germany
ISBN-10: 3-7643-7620-1
ISBN-13: 978-3-7643-7620-8

iF yearbook communication 2006

Birkhäuser – Verlag für Architektur
Birkhäuser – Publishers for Architecture
Basel . Boston . Berlin

inhalt ▪ contents

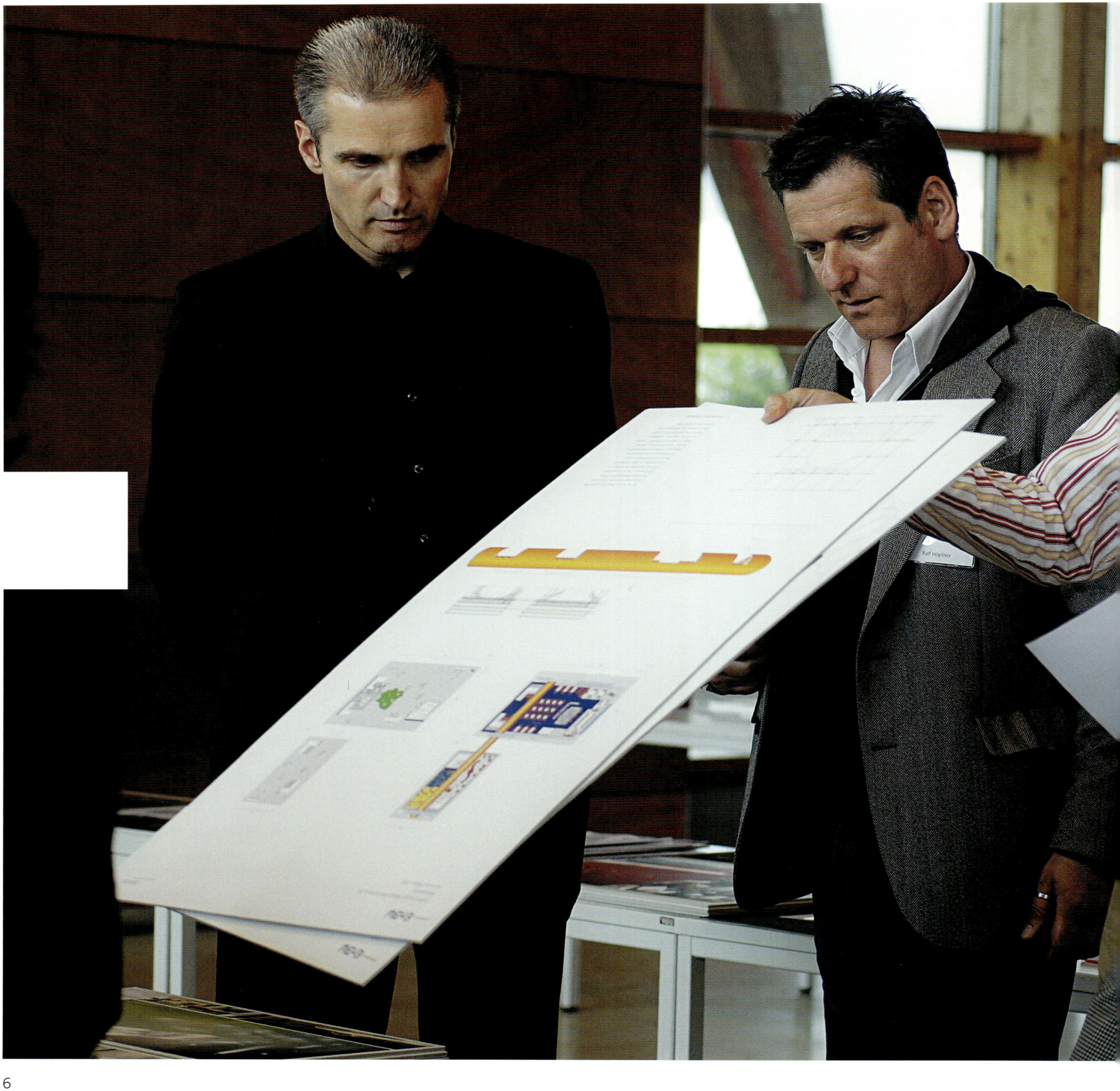
Ralf Höpfner

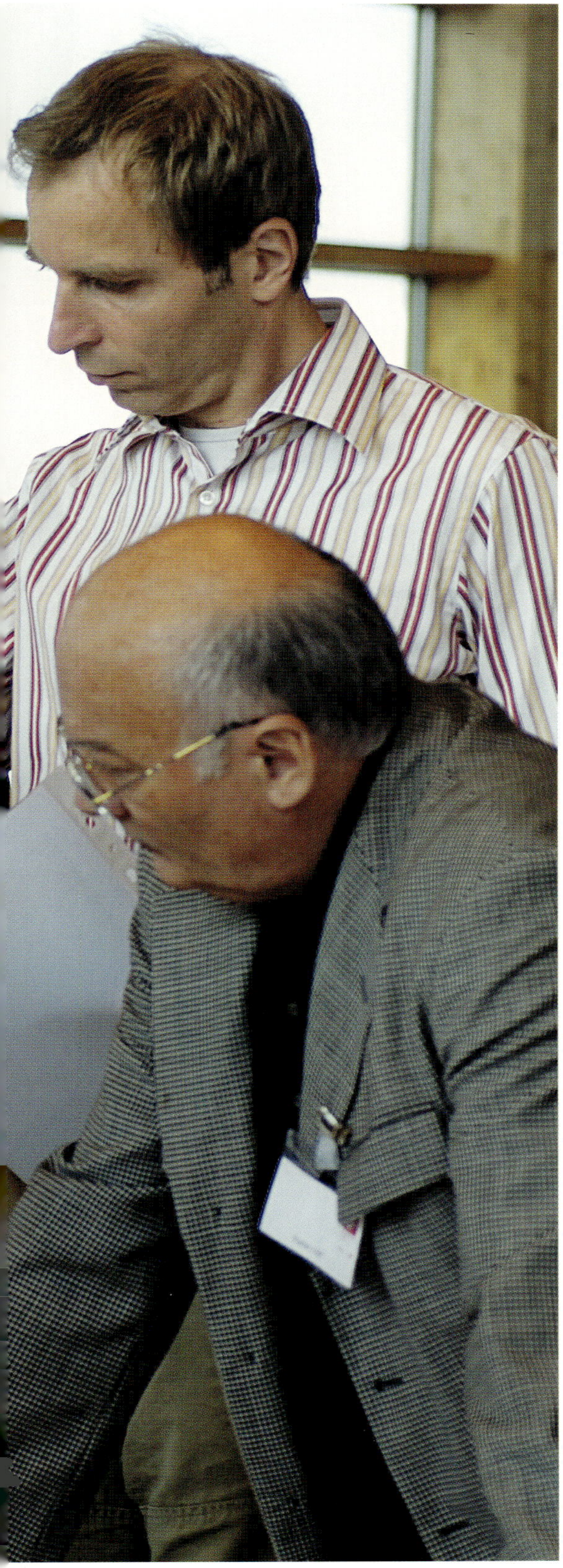

Weltmeisterliches Design – der iF communication design award 2006

Der iF communication design award geht in sein drittes Jahr – ein Jahr, in dem nicht nur Deutschland an einer gigantischen Kommunikationswelle partizipierte, ausgelöst von zwei kleinen großen Buchstaben: WM. Was als sportlicher Fußballhöhepunkt die Zuschauer erfreuen soll, handelt auf anderen Ebenen von Marken, Markenrechten, Markenschutz, Vermarktung, Lizenzen, gutem Logo/schlechtem Logo, lustigem Maskottchen/lächerlichem Maskottchen und vielen Dingen mehr, in jedem Fall aber von Kommunikation. Diese scheint, geht es ums Geschäft und die Öffentlichkeit, manches Mal nicht einfach zu sein, ist registriert, limitiert und unter ständiger Beobachtung – sei es von Veranstalter- oder Unternehmerseite.

Fußball(kommunikation) verbindet und will in die Welt hinaus, mehr als jedes andere sportliche Ereignis sonst. Und König Fußball hat mehr als die meisten anderen Ereignisse das Thema Emotion für sich gepachtet. Ein Thema, das, glaubt man den vorausschauenden Trendexperten, wieder zunehmend Entscheidungen beeinflusst und die reine „Herz-Ansprache" weiter in den Hintergrund drängt. Kopf und Bauch sollen in dem magischen Entscheidungs-Dreieck die Vorherrschaft haben.

Die WM fegte also über uns hinweg, eine globale Kommunikationsmaschinerie, die keinen Zweifel aufkommen ließ, dass die Verantwortlichen bei der FIFA ihre unternehmerischen Hausaufgaben gemacht haben. Das Ergebnis waren nicht nur lokale, verkaufsfördernde Maßnahmen wie WM-Würstchen, WM-Berliner und andere Backwaren („11 Brötchen sollt ihr sein!"), sondern ein über allem thronendes, umstrittenes Logo. Einige Stimmen aus der Designszene sahen gar das von Otl Aicher erstmals 1972 zu den Olympischen Spielen eingesetzte Corporate Design, ein Aushängeschild gelungenen Designs aus Deutschland, beschädigt. Längst hat der Sport sein einstmals „prolliges" Image abgelegt und buhlen Lifestylelabel um die Gunst, Spieler ausrüsten zu können. Fußball besitzt mittlerweile eigene Fashion-Accessoires, ist in den Fokus der Trends und des Designs gelangt und eine Plattform geworden, um auf ihr zu kommunizieren.

Ein Ereignis wie die WM hat gezeigt, wie wichtig Design ist – und wie viel wichtiger intelligent durchdachte Designlösungen im Vorfeld eines Projekts sind, damit die Kommunikation der Botschaft nicht in der Oberfläche der Gestaltung untergeht. Bei einer WM gilt nicht nur die Darstellung einer stimmigen Einheit, eines Auftritts des gesamten Wettbewerbs, sondern auch die Präsentation der „Untereinheiten", also der Mannschaften und Länder – im Design bedeutet dies u.a. die Gestaltung einer ganzheitlichen Corporate Identity.

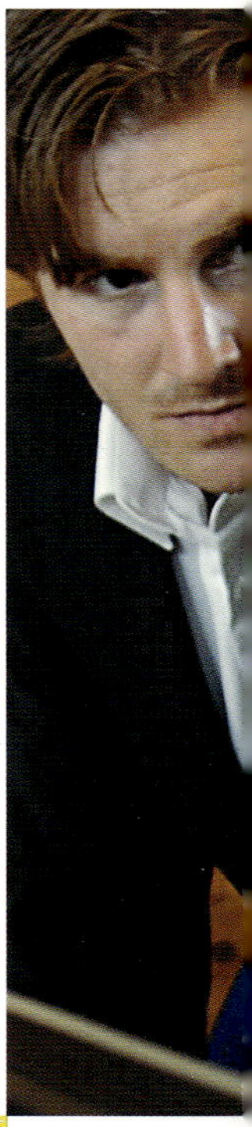

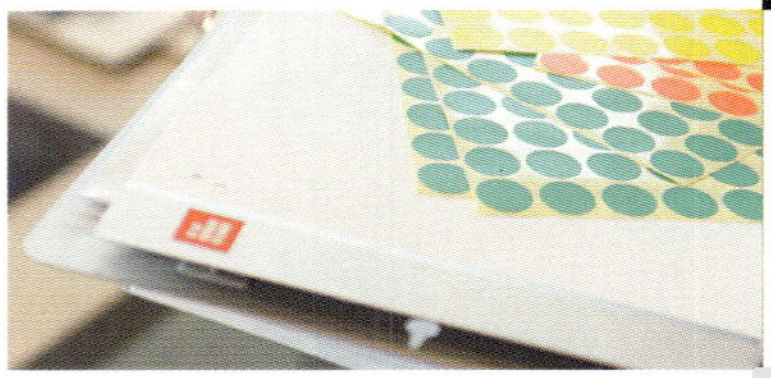

Im iF communication design award 2006 sah die Jury nicht nur WM-geprägte Einreichungen, viele andere Themen standen im Mittelpunkt der Beiträge: Produktpräsentationen, Messeauftritte, Corporate Communication, Gimmicks, Websites, Kulturvermittlung, Protestaktionen und vieles mehr wurde von den Experten begutachtet. Wie immer keine leichte Aufgabe, Profis und Studierende zu bewerten – eher jede Menge Diskussionspotenzial. Am Ende jedoch fielen Entscheidungen, und 247 Arbeiten erhielten insgesamt das iF Label für ausgezeichnete Kommunikationsleistungen. Die Juroren haben als Team ihren vollen Einsatz gegeben, manches Mal als Einzelspieler fungiert und dennoch am Ende harmoniert, und für ihr Engagement danken wir sehr herzlich.

Zu beobachten war in diesem Jahr, dass sich auf den ersten Blick die Einreichungen nicht unterschieden, das Besondere sehr schwer zu entdecken war. Die Welt rückt zusammen, längst ist die Globalisierung auch in der Gestaltung angekommen und wird es immer schwieriger (aber durchaus nicht unmöglich), sich zu unterscheiden, das Einzigartige nach außen zu kehren und für sich sprechen zu lassen. Wer für andere Märkte gestaltet, muss sich nicht nur mit den Eigenheiten der ungewohnten Umgebung, sondern auch mit den Gegenspielern vertraut machen. Erst im Wettbewerb stellt sich heraus, wer der Bessere ist, wer die Nase vorn hat – und damit das erste und vielleicht entscheidende Tor schießt. Und aus Sicht der Jury ist dann knapp daneben eben immer noch vorbei. Es reicht nicht aus, die bunteren und somit aussagekräftigeren Trikots zu tragen oder die verrücktesten Schuhe auf den Rasen zu bringen, die sich mit Goldanstrich gleich selbst feiern. Wer im iF communication design award einen der 30 iF gold awards gewinnen will, der muss zu den Besten gehören und sich im Finale durchsetzen. Und wie beim Fußball gilt frei nach Sepp Herberger: „Nach dem iF award ist vor dem iF award".

Allen Gewinnern des diesjährigen iF communication design award gilt unser Glückwunsch, Sie haben als Team oder Einzelplayer ein großartiges Spiel gemacht!

Ralph Wiegmann, iF Geschäftsführer

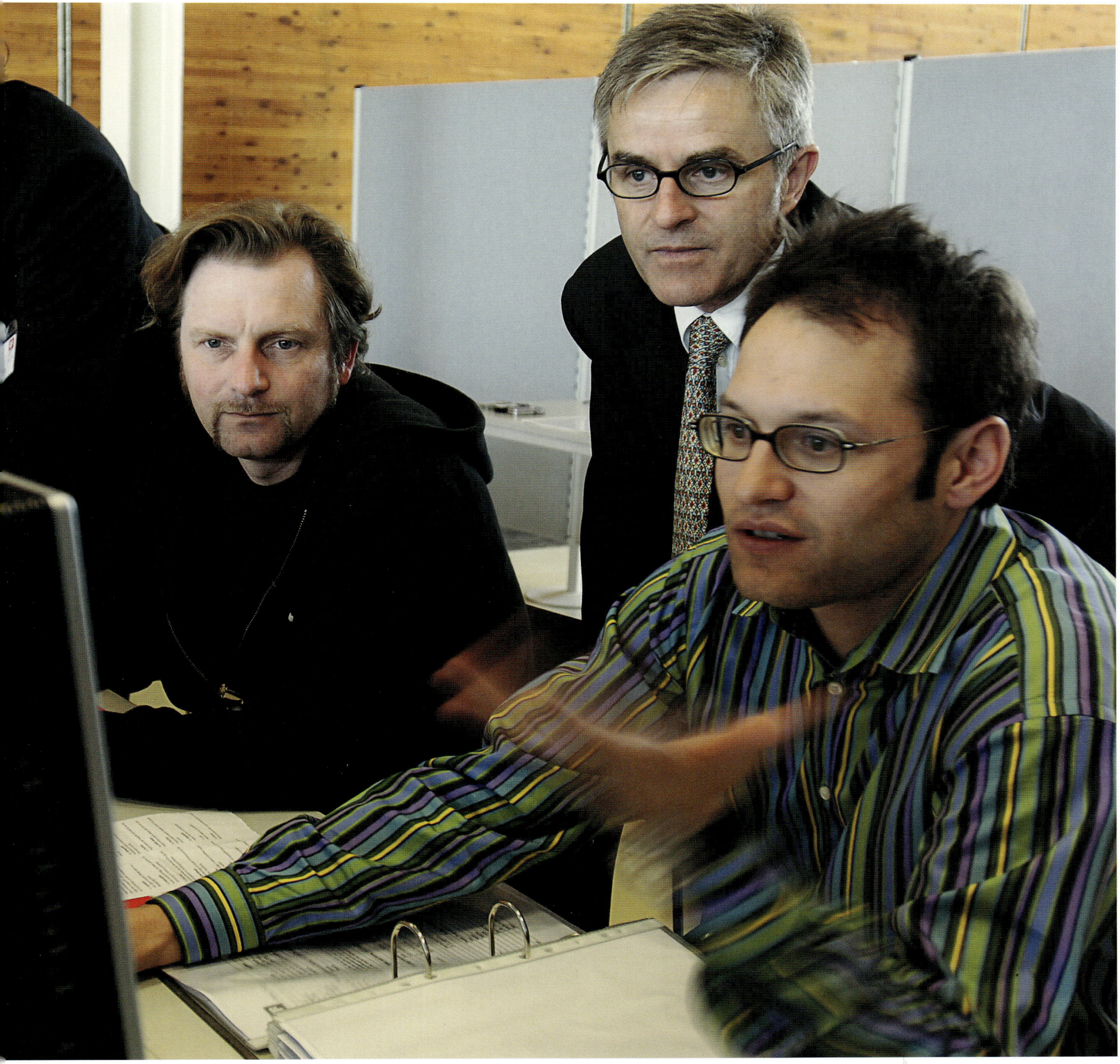

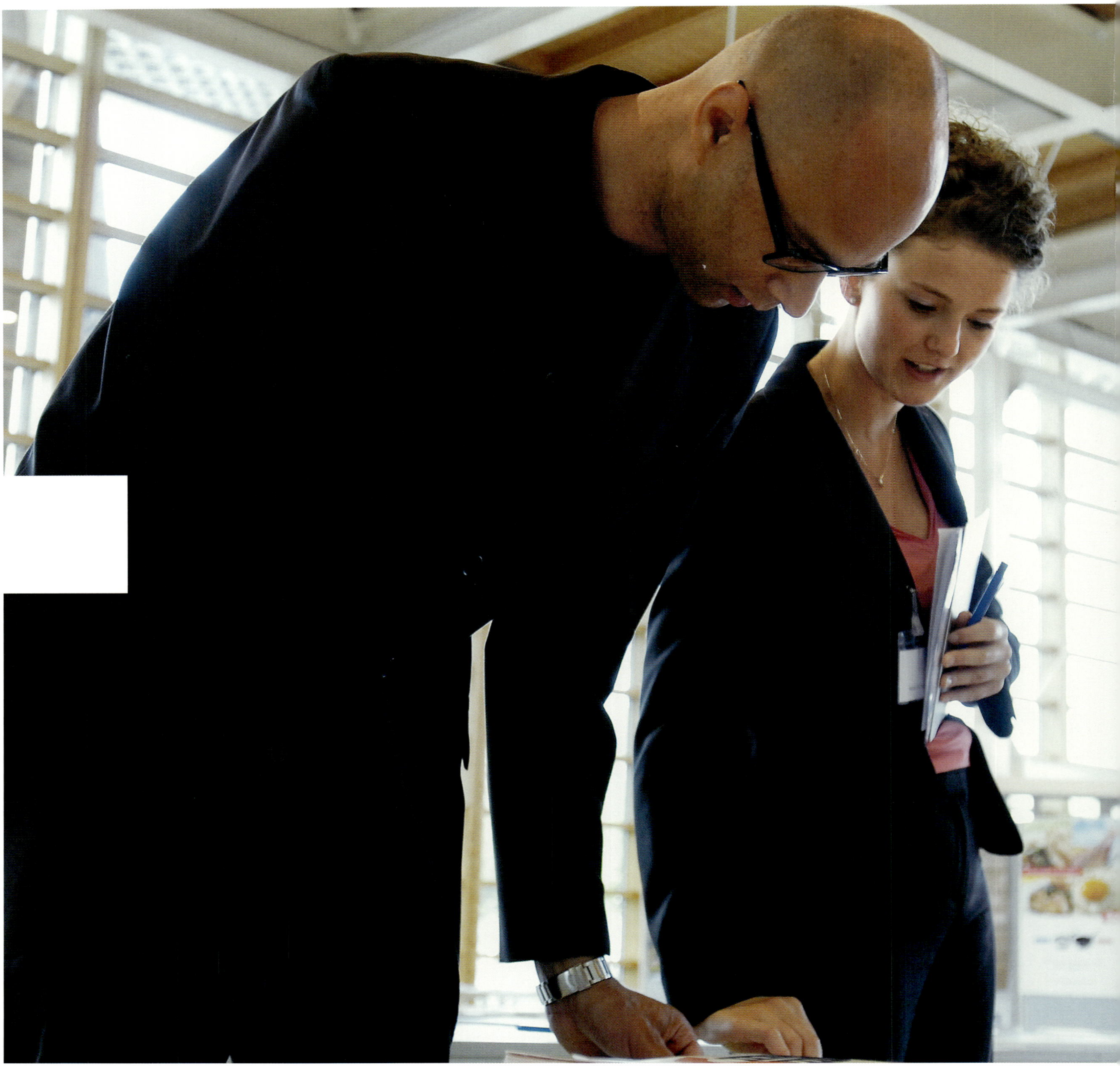

Design Cup runneth over – the iF communication design award 2006

The iF communication design award is now approaching its third year of existence – a year in which Germany and many other nations are riding a gigantic communication wave spawned by two little words: World Cup. What at one level represents the penultimate football highlight for spectators everywhere, serves at another level as a gigantic stage for brands, branding rights, brand protection, marketing, licensing, good logo/bad logo, funny mascot/ridiculous mascot, etc. – in all cases, however, involving communication. And communication in this context does not appear to be all that easy, since the minute it is a question of money and the public, it becomes registered, restricted and continually monitored – be it on the part of the organizers or the relevant enterprises.

Football (communication) brings people together and embraces the great, wide world more than any other sporting event. And football has a monopoly on the topic of emotion more than almost any other event – a topic which, according to the trend forecasters, is increasingly influencing our decisions and displacing pure "appeals to the heart". The intellect and the gut are to play the dominant role in the magical decision-making triangle in the future.

So we are being steamrollered by the World Cup, a global communications machinery which leaves no room for doubt that the FIFA functionaries have done their entrepreneurial homework. The results have not been limited to just local, sales-promoting measures like World Cup frankfurters and World Cup baked goods, but also comprise the event's ubiquitous, controversial logo. Several members of the design community are even of the opinion that the new logo has done permanent damage to the corporate design first deployed by Otl Aicher for the 1972 Olympics in Germany – a calling card for successful German design. Sports has long since dispensed with its "commoner" image, as lifestyle labels line up for the privilege of outfitting players. Nowadays football even has its own fashion accessories, and has become a focal point for trends and design developments, serving as a veritable communication platform.

An event like the World Cup has proven how important design is, and how important intelligent, well-thought-out design solutions can be during the run-up to a project, to insure the message being communicated doesn't get lost at the surface level of the design. During a World Cup, not only is the presentation of a coherent whole – a united appearance for the overall tournament – important, but also the way the "sub-units", i.e. the teams and nations themselves, are presented. In the sphere of design, this means designing a holistic corporate identity.

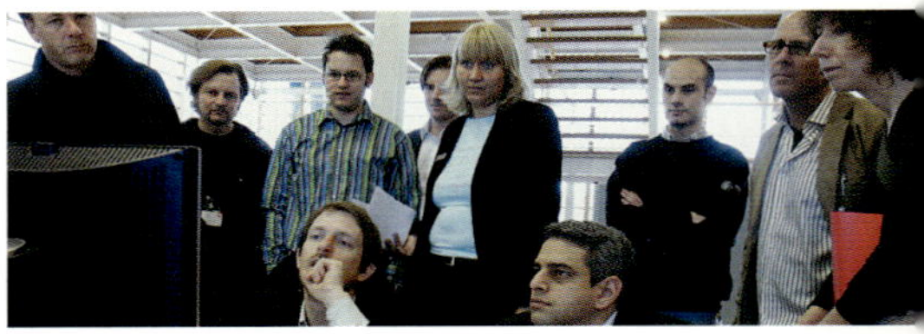

For the iF communication design award 2006 competition, the jury was treated not only to projects with a World Cup twist, but also to many other project types: product presentations, tradeshow appearances, corporate communication, gimmicks, websites, arts & culture promotion, protest actions and much more. As always, it was no easy task judging the work of the design pros and students, and there was ample potential for discussion. At the end of the day however, decisions were made, and 247 projects were awarded an iF seal of communication design excellence. The jurors gave their best as a team, sometimes as individualists but always harmonizing in the end, and we thank them for their dedication and commitment.

This year there was a certain superficial sameness to the projects which made it difficult to discover what was special about each one. The world is getting smaller, and globalization has long since conquered the domain of design. As a result, it is becoming more difficult (although not impossible) to differentiate oneself, to emphasize the uniqueness of a project and develop a distinctive voice. Whoever is designing something for a different market not only has to learn the peculiarities of that unfamiliar environment, but also needs to study the opposing players. Not until the match begins can it be determined who is better, who has an edge over his or her opponent, and who will score that first, and perhaps decisive, goal. From the jury's point of view, missing the goal by inches is the same as missing it by a mile. It is not enough to wear the most colorful and therefore most expressive jersey or to dress players in the gaudiest of shoes with a bright gold finish that celebrates itself. Anybody who wants to win one of the 30 iF gold awards has to be one of the very best and be capable of winning the final match. Or, to borrow a phrase from Germany's legendary national trainer Sepp Herberger: "After the iF awards is before the iF awards."

Our sincere congratulations to the winners of this year's iF communication design awards – as a team or as individualists, you played a great match!

Ralph Wiegmann, iF Managing Director

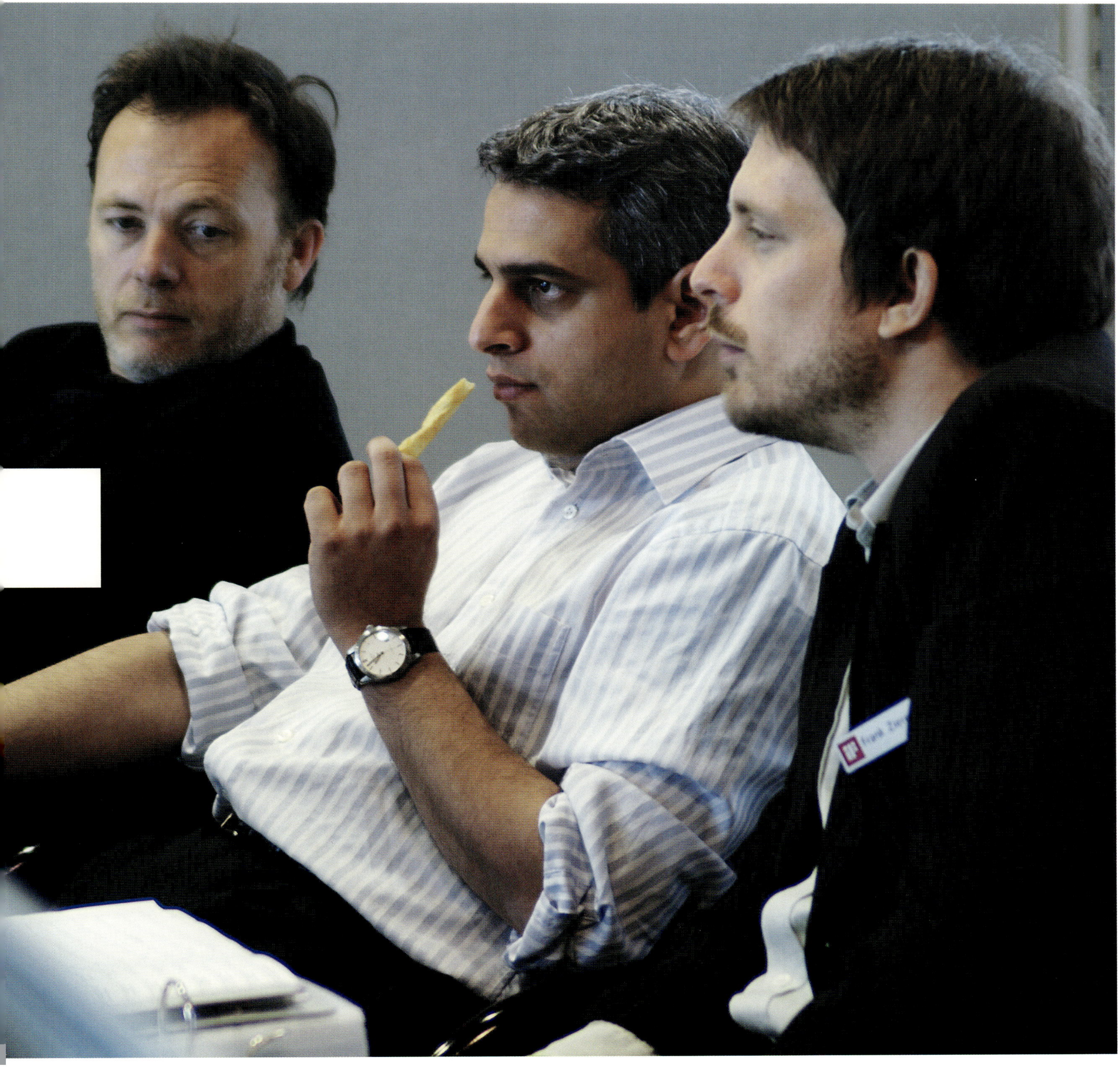

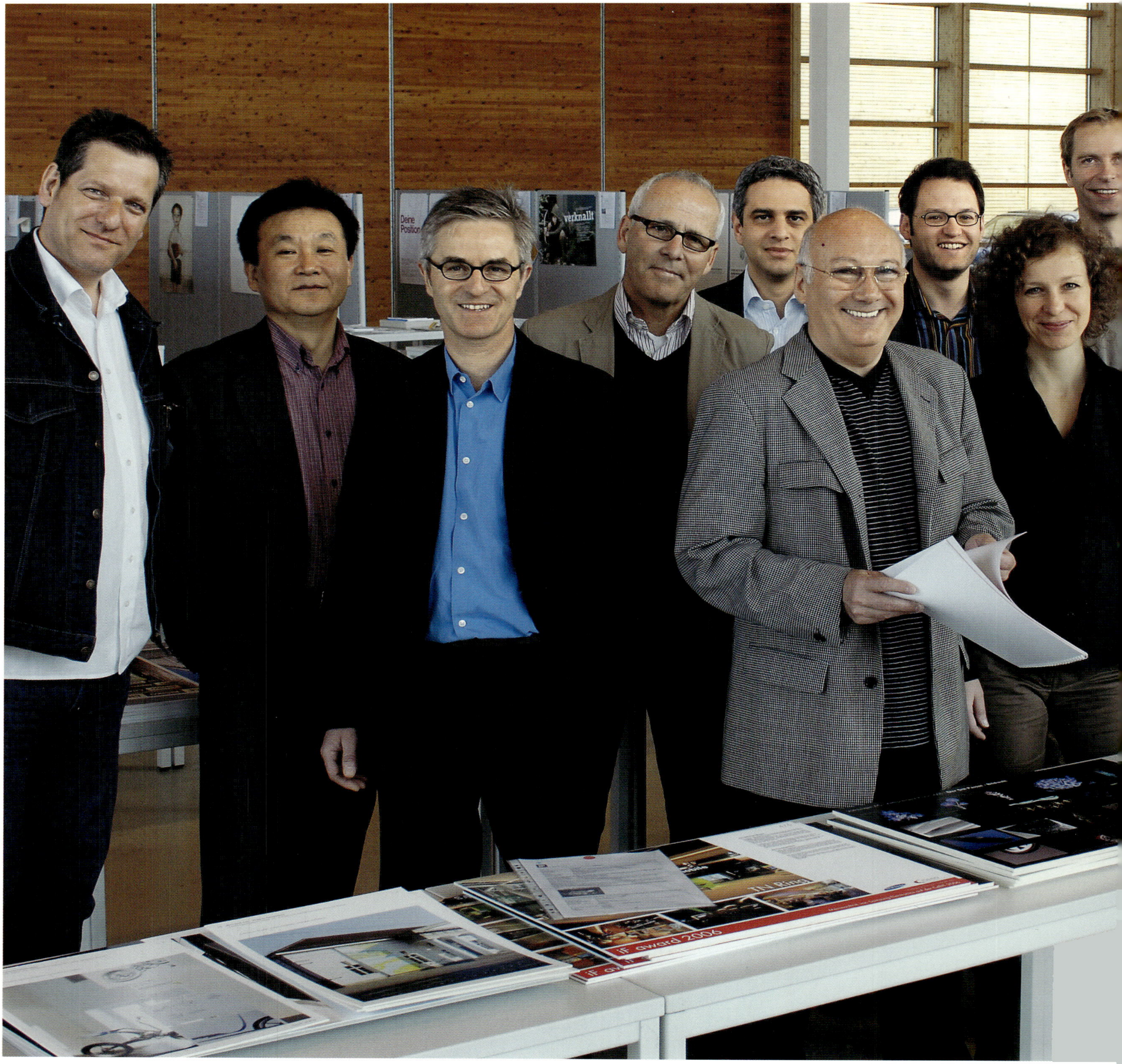
Deine
Position
verknallt
iF award 2006

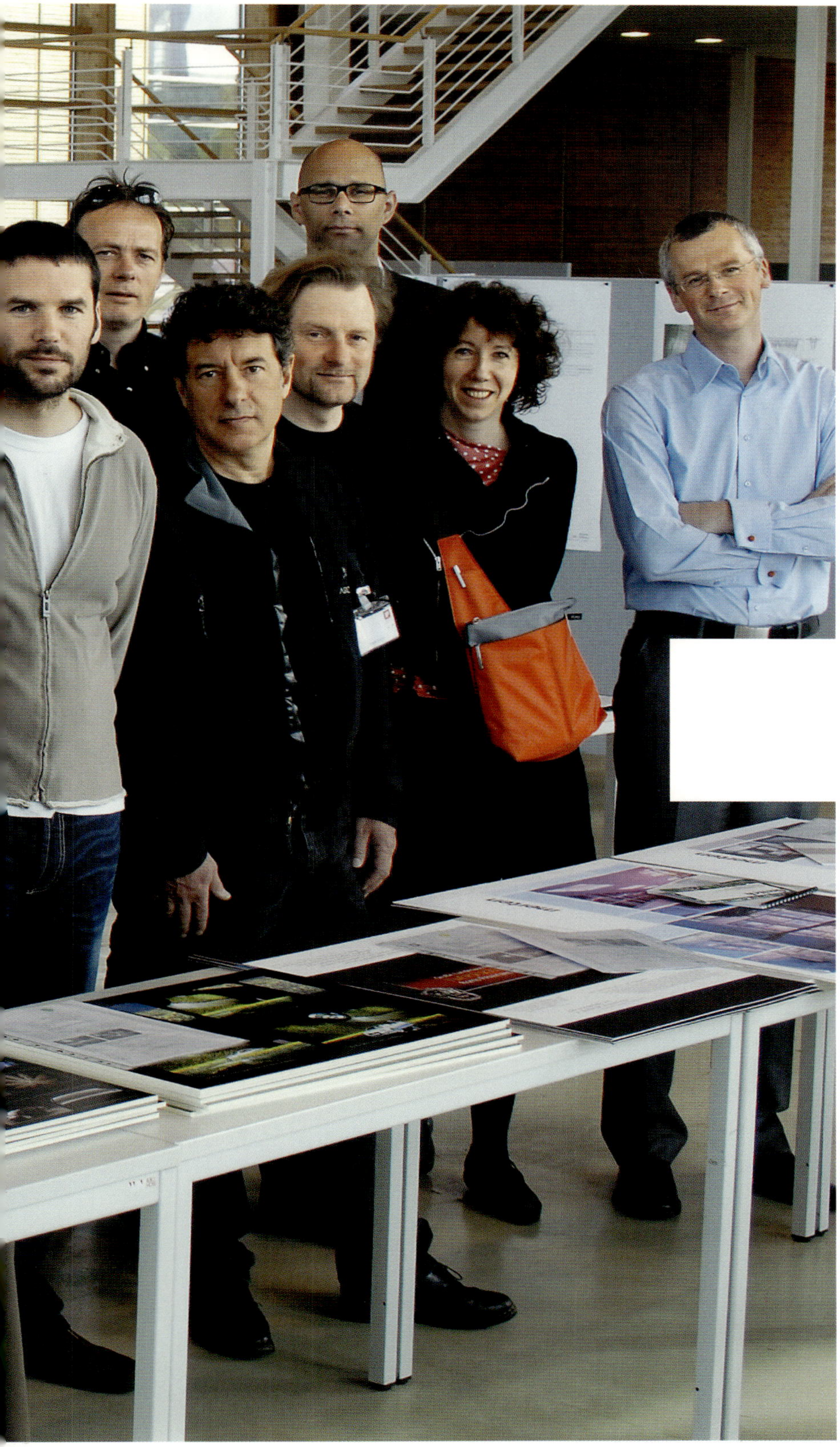

The jury in 2006

1 Ralf Höpfner
2 Jae-Jin Shim
3 Ruedi Alexander Müller-Beyeler
4 Prof. Manfred Becker
5 Ravin Mehta
6 Rainer Hilf
7 Pascal Urscheler
8 Manu Lange
9 Michael Volkmer
10 James Auger
11 Prof. Dr. Gerhard Buurman
12 Pippo Lionni
13 Marcus Korell
14 Daniel Thoma
15 Fiona Raby
16 Prof. Detlef Rahe

Chairman of the jury 2006. Ruedi Alexander Müller-Beyeler

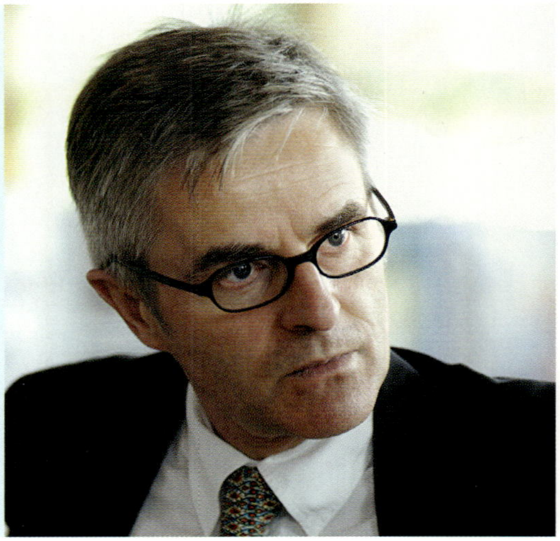

■ Ruedi Alexander Müller-Beyeler schloss sein Studium an der Eidgenössischen Technischen Hochschule in Zürich (ETH) als Ingenieur ab und ist seit Januar 2006 Partner von Pleon Schweiz in Zürich, verantwortlich für Markenstrategie und Design Management. Zuvor gründete er Nose Design Intelligence, Zürich, (1990) und leitete dieses Unternehmen 15 Jahre lang als CEO. Davor war er u. a. als Berater für Corporate und Brand Identity bei Interbrand Zintzmeyer & Lux tätig. Ruedi Alexander Müller-Beyeler ist Präsident des Schulrates der Hochschule für Gestaltung und Kunst in Zürich (hgkz).

■ Ruedi Alexander Müller-Beyeler graduated in Engineering from the Federal Institute of Technology in Zurich in 1978. Since January 2006 he has been a partner at Pleon Switzerland in Zurich, responsible for brand strategy and design management. Before that he established Nose Design Intelligence, Zurich (1990) and headed up the company for 15 years as its CEO. Previously he was active as a consultant for corporate and brand identity at Interbrand Zintzmeyer & Lux. Ruedi Alexander Müller-Beyeler is President of the Advisory Board of the University of Applied Sciences and Arts in Zurich (hgkz).

iF communication design award – der Name ist Programm

■ communication design – der Name dieses jüngsten und sehr erfolgreichen iF Preises ist Programm. Alles, was wir in diesem Buch zur Betrachtung vorlegen, ist gestaltet, weil Designer gezielt kommunizieren wollten: eine Ausstellung, ein Produkt, ein Unternehmen oder eine Geschichte, einen Bericht, eine Struktur. Gemeinsam ist den Beiträgen des iF communication design award, dass die Botschaft zuerst war – vorliegend als Briefing, Skript, Positionierung oder als Strategie auf der Suche nach der geeigneten Form. Geeignet ist die Form, wenn die Botschaft beim Zielpublikum ankommt und von einer kritischen Mehrheit angenommen wird.

Design kann auch andersherum gedacht werden: Aus dem gestalterischen Tun heraus entsteht erst das, was sich durch seine Form ausdrückt. Die Botschaft ist jetzt nachgelagert. „Das Formen der Gedanken während des Design" könnte man diese Annäherung frei nach Kleist nennen. Am Anfang ist nur eine vage Idee von dem, was entstehen soll – ein Gefühl von etwas, das in der Luft liegt.

Bei genauerer Betrachtung ist die zweite Herangehensweise an Design natürlich in der ersten immer auch enthalten. Das macht Design so spannend und wichtig für unsere Zeit. Denn der Prozess des erfolgreichen Design resultiert in Innovationen. Und das ist, was die Märkte suchen.

Für die Jury des iF communication design award ging es einmal mehr darum, diejenigen Beiträge zu finden, die einerseits eine kommunikative Absicht erkennen lassen, andererseits dieses schwierig zu benennende Etwas ausstrahlen, was den Zeitgeist trifft, den Trend setzt.

Dieses Etwas zu erkennen war nicht immer so schwierig wie heute. Noch vor 15 bis 20 Jahren galt zumindest in Europa ein Designdogma, das klar machte, was als Design anzuerkennen sei und was nicht. Bauhaus und die Ulmer Schule hatten Setzungen gemacht, die lange Zeit als allein gültig akzeptiert waren. Die Jurys hatten lediglich darüber zu wachen, dass die Regeln des guten Design von denjenigen befolgt wurden, die sich Designer nannten.

Heute haben sich diese Dogmen verflüchtigt. Der Kunst folgend bedient sich Design einer stetig wachsenden Vielfalt von Medien und Techniken, manifestieren sich Ästhetiken scheinbar nach Belieben. Scheinbar – wenn da nicht dieses Etwas wäre, was die Preisträger voraushaben …

Ruedi Alexander Müller-Beyeler, Jury-Vorsitzender

iF communication design award – the name is the mission

■ communication design – the name of the newest, already highly successful iF award is at the same time its mission. Everything presented in this book was designed because designers wanted to communicate in a targeted fashion – an exhibition, a product, an enterprise, a story, a report, a structure. A shared characteristic of the winning projects for this iF communication design award was that the message came first – for example in the guise of a briefing, script, positioning or strategy that was seeking the right form. The form is appropriate if the message reaches its target group and is accepted by a critical majority.

But design can also be approached from a different angle: that which the form expresses can only emerge through the designer's own activity. In this case, the message is downstream. "Shaping your thoughts while designing" would be one way of describing this approximation, borrowing freely from Kleist. At the beginning only a vague notion of what is to be created is present – a feeling of something in the air.

Upon closer examination the second approach is naturally always intrinsic to the first one. This is what makes design so exciting and so important to our age. Because the process of successful design results in innovations, which is what markets today are looking for.

The job of the iF communication design award jury was once again to detect those projects that had a clear communicative intent on the one hand, but which radiated a special something – a trendsetting, zeitgeist component – on the other.

Recognizing the latter has never been more difficult than it is today. 15 or 20 years ago there was a design dogma, at least in Europe, which made clear what could be considered good design and what couldn't. Bauhaus and the Ulm School developed rules that for a long time were accepted as authoritative. Juries simply had to monitor whether the people who called themselves designers were adhering to the rules of design.

Today these dogma have evaporated. Following the example of art, design is making use of a steadily growing variety of media and technologies, and new esthetics appear to be popping up at random. Appear to be – if there were not that special something, which the prizewinners can call their own ...

Ruedi Alexander Müller-Beyeler, Jury Chairman

CeBIT usability award 2006

Der CeBIT usability award: Ein Preis für herausragende Leistung im Bereich digital media

Die Deutsche Messe AG hat in Kooperation mit iF International Forum Design GmbH in diesem Jahr erstmalig unter allen eingesandten Beiträgen in der Disziplin „digital media" den CeBIT usability award vergeben.

Bei diesem Sonderpreis wurde der Beitrag prämiert, bei dem die Benutzerfreundlichkeit und Bedienbarkeit bestmöglich umgesetzt wurde. Dabei stehen insbesondere eine intuitive, zielgruppengerechte Navigation sowie die Gebrauchstauglichkeit im Vordergrund.

Ganz im Sinne der CeBIT geht es bei diesem Sonderpreis darum, besonders innovative Lösungskonzepte zu prämieren, die „Digital Solutions for Work&Life" repräsentieren und die Initiative der Unternehmen fördern, sich aktiv für die Alltagstauglichkeit ihrer angebotenen Technik zu engagieren.

Der Gewinner des CeBIT usability award wird in die Presse- und Öffentlichkeitsarbeit der CeBIT 2007 eingebunden. Die namentliche Integrierung des Gewinners in die Besucherkampagne der CeBIT, in Pressemitteilungen zum award und auf der CeBIT Homepage entspricht einem Mediavolumen von 250.000 Euro.

The CeBIT usability award: A prize for outstanding achievement in the digital media sector

For the first time in 2006, Deutsche Messe AG – in cooperation with iF International Forum Design GmbH – has selected a project from the digital media category as the recipient of a CeBIT usability award.

This special prize goes to the most usable and user-friendly project, with the emphasis on intuitive navigation that is ideally geared to the target audience.

In the spirit of CeBIT, this award is reserved for especially innovative concepts in the sphere of "Digital Solutions for Work & Life". This award is intended to encourage the active commitment of enterprises to developing technology which is perfectly geared to everyday use.

The winner of the CeBIT usability award will be included in the press publicity and public relations campaign for CeBIT 2007. The inclusion of the winner's name in the CeBIT visitor advertising campaign as well as in press releases covering the award and finally, on the CeBIT homepage, corresponds to a media volume in the amount of 250,000 euros.

CeBIT usability award 250,000 euros

Auftraggeber Client
Eastman Kodak Company
Rochester, NY, USA

Projekt Project
Kodak Picture Kiosk
Bildterminal
Picture terminal

Design
Jo Gray
Tim Wood
Hye-Jin Nae
Eastman Kodak Company
Rochester, NY, USA

Die Kodak Picture Kiosk G4 Hardwarelösung bietet eine schnellere Performance und ein neues, zeitgemäßes Styling, um den Erlebniswert auf Seiten des Kunden beim Ausdrucken seiner Bilder zu steigern und den Handel auf dem schnell wachsenden Digital Printing Markt zu unterstützen.

- Mit dem optimierten Warenkorb-Workflow kann der Kunde seinen Einkauf ohne Unterbrechung tätigen und verschiedene Aufträge zu einer Bestellung zusammenfassen.
- Bei der neuen Benutzeroberfläche wurde die Grafik überarbeitet und gleichzeitig der Workflow verbessert.
- Für die Bildersuche auf großen Speicherkarten mit mehreren Hundert Fotos wurden die Navigationsfunktionen erweitert.
- Mit dem Ziel, das Interesse auf Kundenseite zu stimulieren und die Verkaufszahlen zu erhöhen, wurde die Unterstützung durch Werbe- und Promotionsmaßnahmen verbessert, um die individuelle Absatzförderung der Einzelhändler zu steigern.
- Die Anlage ermöglicht die Einrichtung eines Multi Order Station Systems, wodurch der Händler seine Investition optimal auslasten kann.
- Die Kodak Perfect Touch Technologie gewährleistet höchste Bildqualität.

The Kodak Picture Kiosk G4 hardware solution offers faster performance and new contemporary styling designed to elevate the consumer printing experience while helping retailers compete in the rapidly expanding digital printing market.

- Shopping Cart workflow that enables different products to be added to an order without multiple checkouts
- New User Interface with updated graphics and improved workflows
- Enhanced navigation controls for searching through hundreds of pictures on large memory cards
- Improved advertising and promotional support to drive consumer trial and product up-sells as well as the ability to support specific retailer advertising and promotional activities
- Enables multi Order Station system configurations, allowing retailers to better utilize their capital assets
- Kodak Perfect Touch Technology for high quality images

CeBIT
Join the vision

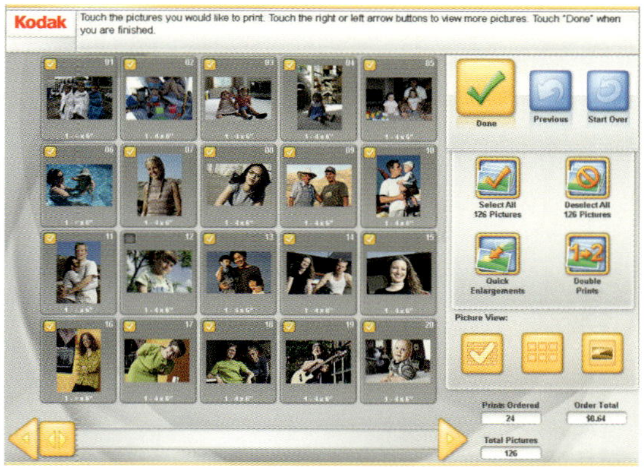

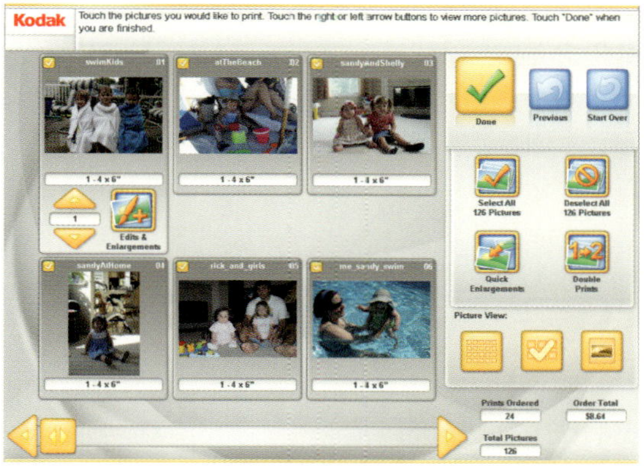

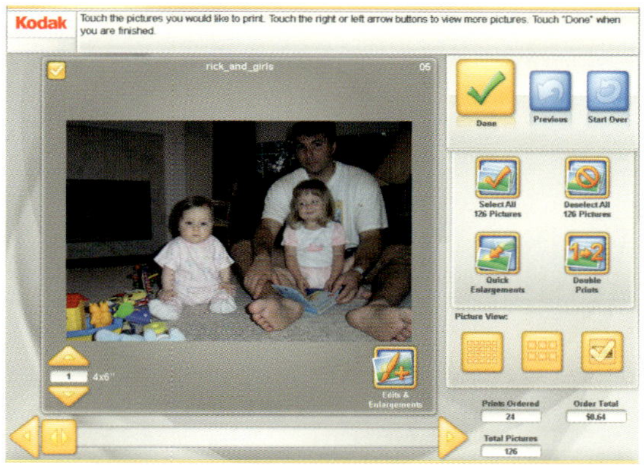

special media partner award 2006

Der special media partner award 2006: Ein viel versprechender Start!

Die Bedeutung der Medien als Filter für die Wogen an Informationen, die uns tagtäglich erreichen, wächst von Jahr zu Jahr. Die Redaktionen der Tageszeitungen und Magazine übernehmen es für uns, die wichtigere von den unwichtigeren Informationen zu unterscheiden, sie bewerten sie – kurz: ihr Einfluss ist in unserem Informationszeitalter kaum zu überschätzen, ihre Orientierungshilfe unerlässlich. Dies gilt in ganz besonderem Maße für eine kompetente Fachpresse im Bereich Design, Architektur und Marketing – also Medien, die sich an der Schnittstelle bewegen, die den iF communication design award so spannend macht.

Dieser Funktion und Bedeutung wurden drei ausgewählte Medien gerecht, die den iF communication design award schon länger als Medienpartner begleiten. In diesem Jahr bildeten die Redakteure dieser hochkarätigen Magazine eine Jury, die den erstmals ausgelobten special media partner award verlieh: AXIS, das bedeutende japanische Designmagazin, das führende deutsche Magazin für Design und Publishing PAGE und W&V, das Wochenmagazin für Marketing, Werbung, Medien und E-Business. Die Jury hat ihre absoluten Favoriten aus allen Kategorien benannt und zeigte sich beeindruckt von der Qualität vieler Entwürfe: Vorhang auf für die Entwürfe, die in den Augen unserer Medienpartner voll überzeugten!

The special media partner award 2006: An auspicious beginning!

The media's role in filtering the waves of information that reach us every day continues to grow from year to year. Newspaper and magazine editors have taken over the task for us of deciding which information is important and unimportant. In today's information age, their influence can scarcely be overestimated, and their function as an orientation aid is indispensable. This is particularly true as regards a competent trade press in the realm of design, architecture and marketing – i.e. media that are active at the very interface which makes the iF communication design award so exciting.

This important role was exercised this year by three select media that have been active as media partners to the iF communication design award for a number of years. This year the editors of these high-class magazines formed a jury to present the first-ever "special media partner award": AXIS, a major Japanese design magazine, PAGE, Germany's leading design and publishing magazine and W&V, the weekly magazine for marketing, advertising, media and e-business. The jury selected its absolute favorites from all the various categories and reported being impressed by the quality of a good many projects: Hats off to the projects that succeeded in completely persuading our media partners!

AXIS PAGE W&V

www.axisinc.co.jp www.page-online.de www.wuv.de

special media partner award 2006

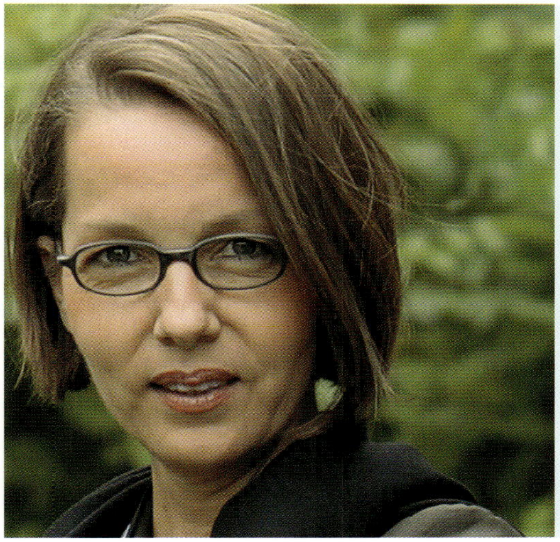

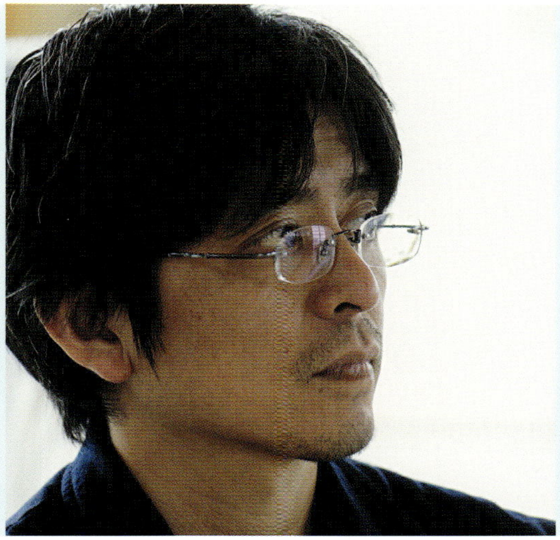

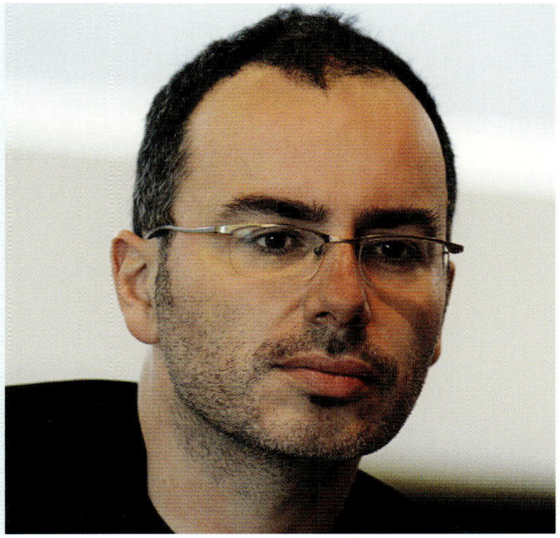

■ Gabriele Günder studierte Visuelle Kommunikation in Mainz. Ihre Diplomarbeit über computergestütztes Gestalten wurde 1988 als Buch verlegt. Als freie Art-Direktorin war sie für MACup, Design Report, Screen Multimedia, Mediale und andere Magazine tätig. Seit 1997 ist sie Chefredakteurin von PAGE, dem führenden Magazin für Design, Medienproduktion und Trends, das im Verlag redtec publishing, einem Unternehmen der Mediengruppe Süddeutscher Verlag GmbH, erscheint.

■ Gabriele Günder studied Visual Communications in Mainz; her thesis on computer-aided design was published as a book in 1988. As a freelance art director she has worked for MACup, Design Report, Screen Multimedia, Mediale and a number of other magazines. Since 1997 she has been Managing Editor of PAGE, the leading magazine for design, media production and trends, published by redtec publishing, part of the Süddeutscher Verlag GmbH media group.

■ Katsutoshi Ishibashi wurde 1965 in Osaka geboren. An der Osaka University of Foreign Studies erwarb er einen Abschluss am Institut für Arabische Sprache. Seit 1998 ist er Chefredakteur des Designmagazins AXIS. Seit 1981 berichtet AXIS zweisprachig über designaffine Entwicklungen, sowohl im gegenständlichen als auch im übertragenen Sinne. Die Themenpalette umfasst neue Designgegenstände, Architektur, Städte, Inneneinrichtungen, Grafisches, Mode und Multimedia sowie zukunftsgerichtete Wissenschaft, Konstruktion und Medizin.

■ Katsutoshi Ishibashi was born in Osaka in 1965 and graduated from the Department of Arabic language at the Osaka University of Foreign Studies. He has been Editor-in-Chief of the AXIS design magazine since 1998. AXIS is a bilingual design magazine launched in 1981 that provides information about various phenomena and objects discovered throughout the world, with a particular focus on design. The magazine covers subjects such as products, architecture, cities, interiors, graphics, fashion, and multimedia, as well as cutting-edge science, engineering, and medicine.

■ Frank Zimmer studierte Geschichte, Philosophie und Rechtswissenschaft an den Universitäten Köln und Mainz und promovierte zum Dr. phil. Er war von 1999 bis 2000 für das Design-Buchprogramm des Verlages form in Frankfurt verantwortlich. 2000 wechselte er zum Europa Fachpresse Verlag nach München, wo er für das Marketing-Fachmagazin „W&V" arbeitet. 2004 wurde er Chef vom Dienst von „W&V Online", zusätzlich betreut er seit 2006 die Online-Auftritte der Zeitschriften „Kontakter", „Media & Marketing" und „Marketing-Journal".

■ Frank Zimmer studied History, Philosophy and Law at the University of Cologne and the University of Mainz, ultimately obtaining a Ph.D. in Philosophy. Between 1999 and 2000 he was responsible for the design book series put out by "form" publishers in Frankfurt, before signing on at Munich-based Europa Fachpresse Verlag publishers in 2000, where he now works for the marketing magazine "W&V". In 2004 he became Managing Editor of the "W&V Online" service and, since 2006, has also been responsible for the online versions of "Kontakter", "Media & Marketing" and "Marketing-Journal" magazines.

01 digital media

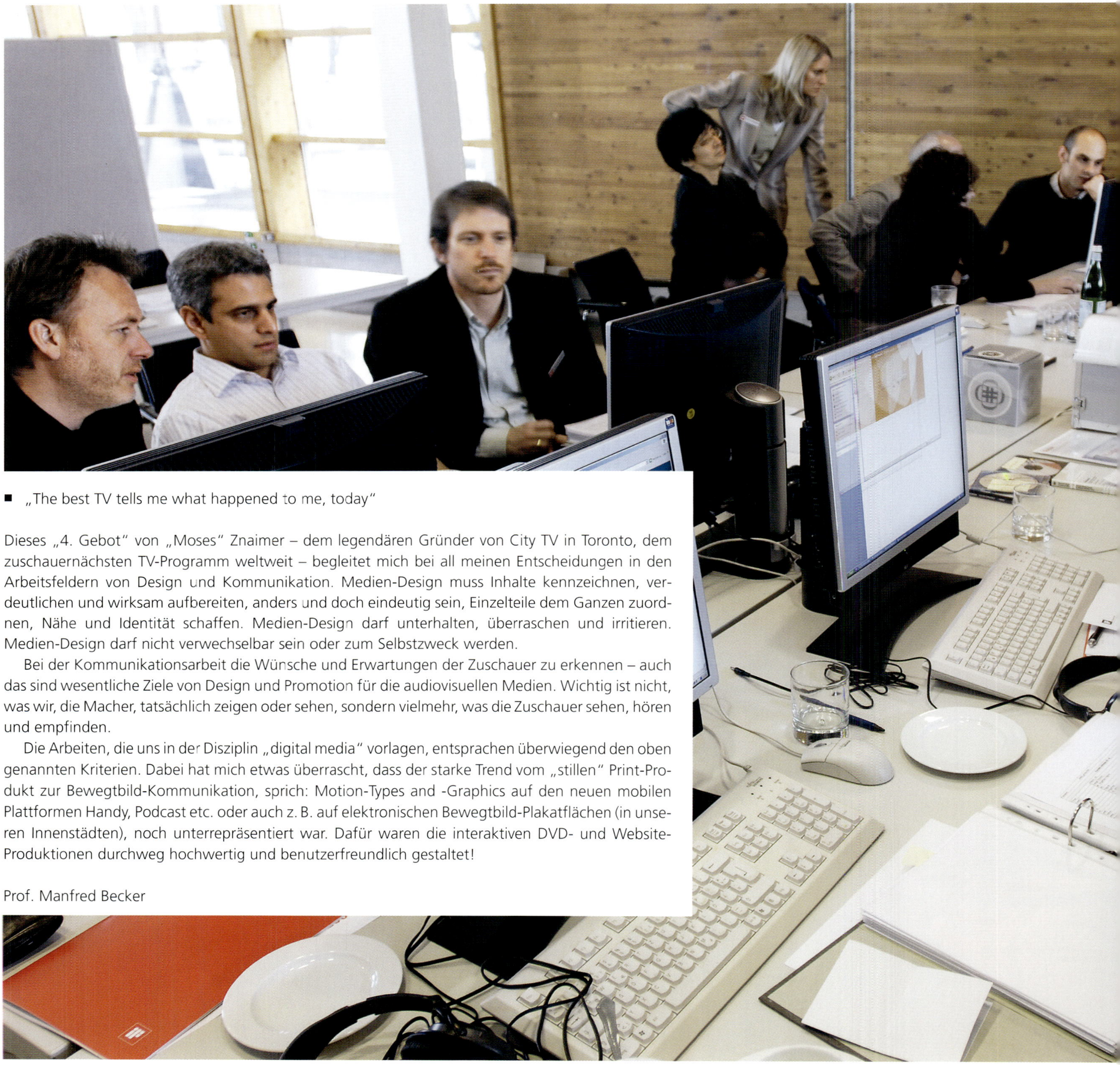

■ „The best TV tells me what happened to me, today"

Dieses „4. Gebot" von „Moses" Znaimer – dem legendären Gründer von City TV in Toronto, dem zuschauernächsten TV-Programm weltweit – begleitet mich bei all meinen Entscheidungen in den Arbeitsfeldern von Design und Kommunikation. Medien-Design muss Inhalte kennzeichnen, verdeutlichen und wirksam aufbereiten, anders und doch eindeutig sein, Einzelteile dem Ganzen zuordnen, Nähe und Identität schaffen. Medien-Design darf unterhalten, überraschen und irritieren. Medien-Design darf nicht verwechselbar sein oder zum Selbstzweck werden.

Bei der Kommunikationsarbeit die Wünsche und Erwartungen der Zuschauer zu erkennen – auch das sind wesentliche Ziele von Design und Promotion für die audiovisuellen Medien. Wichtig ist nicht, was wir, die Macher, tatsächlich zeigen oder sehen, sondern vielmehr, was die Zuschauer sehen, hören und empfinden.

Die Arbeiten, die uns in der Disziplin „digital media" vorlagen, entsprachen überwiegend den oben genannten Kriterien. Dabei hat mich etwas überrascht, dass der starke Trend vom „stillen" Print-Produkt zur Bewegtbild-Kommunikation, sprich: Motion-Types and -Graphics auf den neuen mobilen Plattformen Handy, Podcast etc. oder auch z. B. auf elektronischen Bewegtbild-Plakatflächen (in unseren Innenstädten), noch unterrepräsentiert war. Dafür waren die interaktiven DVD- und Website-Produktionen durchweg hochwertig und benutzerfreundlich gestaltet!

Prof. Manfred Becker

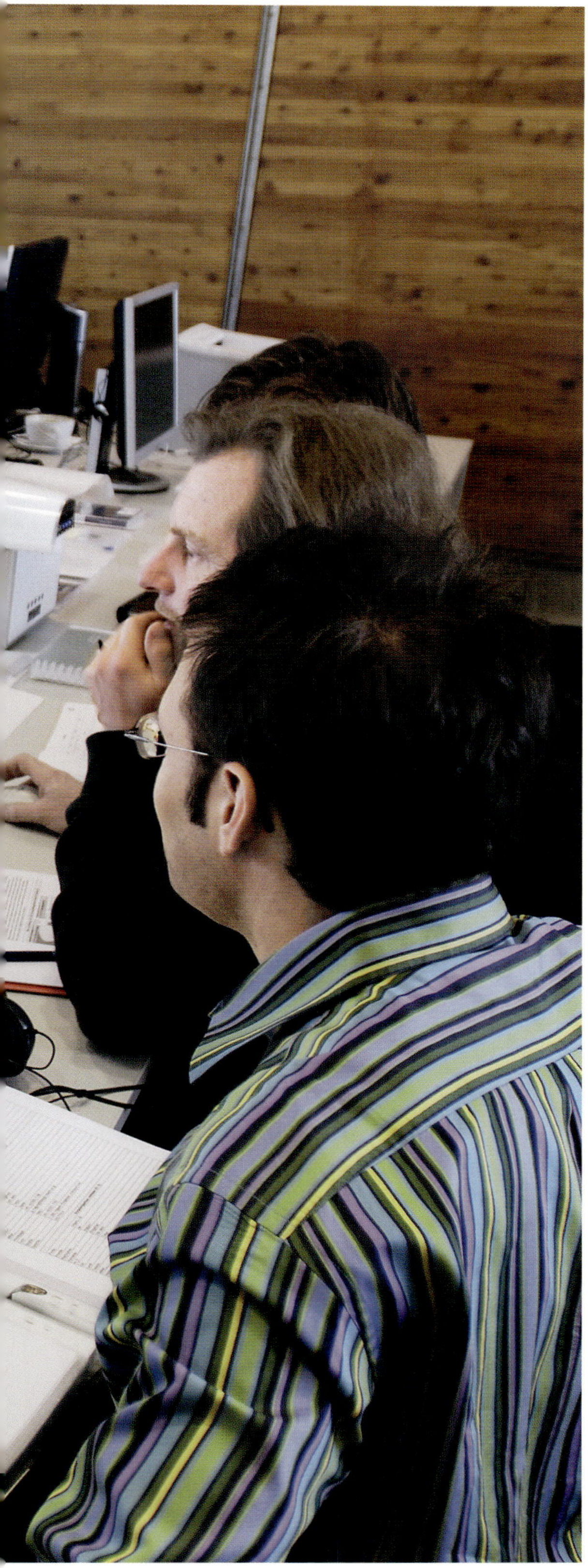

digital media

■ "The best TV tells me what happened to me, today"

This "4th Law" of "Moses" Znaimer – the legendary founder of City TV in Toronto, the world's most popular TV channel – informs all the decisions I make in my design and communication work. Media design must reflect, clarify and effectively convey content, be unorthodox yet clearly defined, relate individual parts to the whole and at the same time establish identity and rapport. Media design can entertain, surprise and irritate, but should not be indistinguishable nor an end in itself.

Relating communication work to the desires and expectations of viewers is also an essential aim of any design or promotion intended for audiovisual media. The important thing is not what we, the makers, actually convey or see, but rather what our viewers ultimately see, hear and experience.

The majority of projects submitted to us in the "digital media" category met these important criteria. Yet I was somewhat surprised that the strong trend away from "still" print products to communication via moving pictures, i.e. motion types and graphics on new mobile platforms like cell phones and pod casts or electronic movie screen images (in downtown areas) was underrepresented. Nevertheless, the interactive DVD and website productions did stand out for their uniformly high standards and user-friendly design.

Prof. Manfred Becker

■ Ravin Mehta (*1967), studierter Informatiker, begann seine Karriere 1995 bei Pixelpark im Bereich Technische Konzeption und baute dort ab 1998 das Intranet-Kompetenz-Center auf. Von Sommer 2000 bis Sommer 2005 leitete er als Managing Director den Kölner Standort der Pixelpark AG. Seit 2005 verantwortet er das Agenturgeschäft der Pixelpark AG mit Standorten in Köln und Berlin.

■ Ravin Mehta (*1967) studied computer science before beginning his career in the Technical Conception department of Pixelpark in 1995. Beginning in 1998 he built up the Intranet Competence Center there. From summer 2000 to summer 2005 he served as Managing Director of the Cologne branch of Pixelpark AG. Since 2005 he has been responsible for the agency business of Pixelpark AG, with offices in Cologne and Berlin.

■ Marcus Korell (*1965) studierte Visuelle Kommunikation an der Fachhochschule Dortmund. Von 1992 bis 1998 war er für Heuer & Sachse Werbeagentur in Hamburg tätig und ist seit 1998 bei Scholz & Friends, Hamburg. Hier war er Senior Art Director sowie Creative Director und ist derzeit Family Leiter in der Betreuung unterschiedlicher Etats. Marcus Korell gewann nationale wie internationale Auszeichnungen, u.a. beim New York Festival, den London International Awards, beim ADC Deutschland oder dem Golden Award of Montreux.

■ Marcus Korell (*1965) studied Visual Communications at the Dortmund University of Applied Sciences. From 1992 to 1998 he worked for the advertising agency Heuer & Sachse in Hamburg, before joining Scholz & Friends, likewise Hamburg-based. He has served there as Senior Art Director and Creative Director and is currently responsible for several different accounts. Marcus Korell has won national and international awards, including at the New York Festival, the London International Awards, ADC Deutschland and the Golden Award of Montreux.

■ Fiona Raby ist Mitinhaberin des 1994 gegründeten Designstudios Dunne & Raby in London. Dem Studio dienen seine Produkte und Dienstleistungen als Medium, um unter Designern sowie Industrie und Öffentlichkeit Gesprächs- und Denkprozesse über kulturelle und ethische Implikationen neuer Technologien anzuregen. Bei vielen ihrer Projekte arbeitet sie mit industriellen Forschungsinstituten, Bildungs- oder Kultureinrichtungen zusammen. Ihre Werke finden sich unter anderem in der Dauerausstellung des MOMA in New York und dem Victoria & Albert Museum in London.

■ Fiona Raby is a partner in the design practice Dunne & Raby (London) set up in 1994. They use products and services as a medium to stimulate discussion and debate amongst designers, industry and the public about the social, cultural and ethical implications of emerging technologies. Many of their projects are collaborative, working with industrial research labs, academia and cultural institutions. They have work in the permanent collection, MOMA, New York, and at the Victorian & Albert Museum, London.

Ravin Mehta

Marcus Korell

Fiona Raby

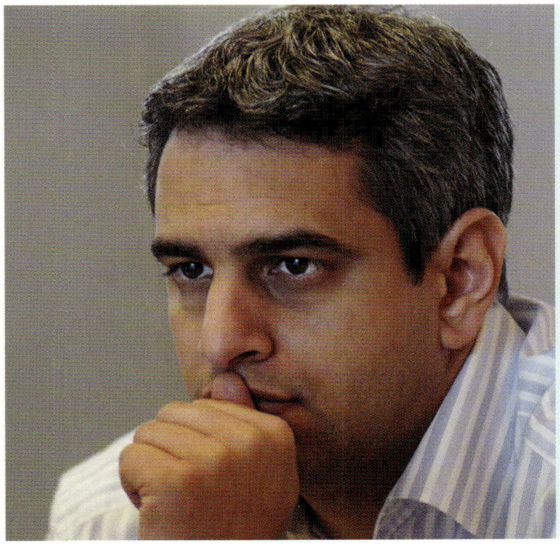

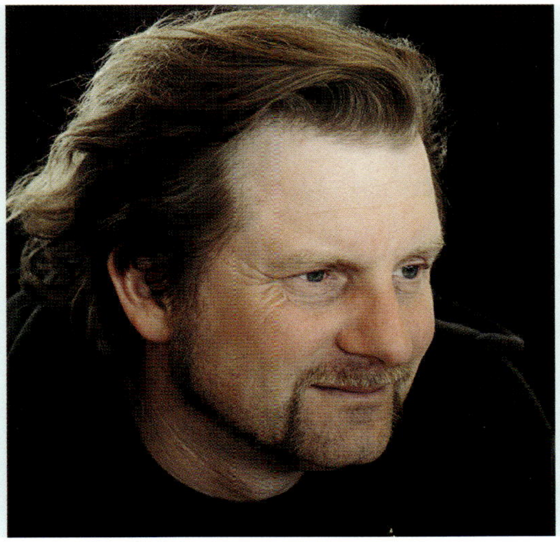

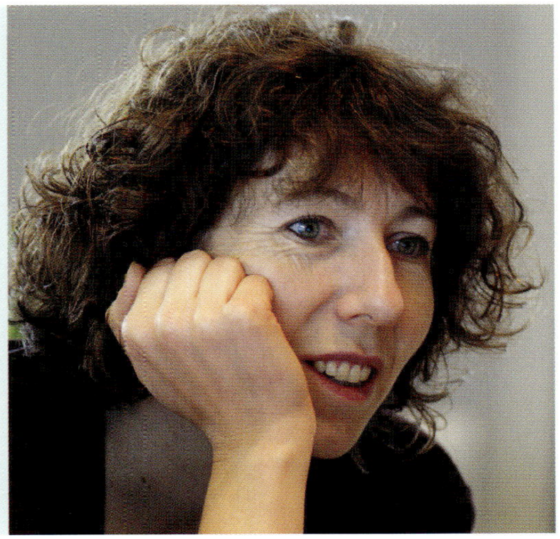

Prof. Manfred Becker

Prof. Dr. Gerhard Buurman

Pascal Urscheler

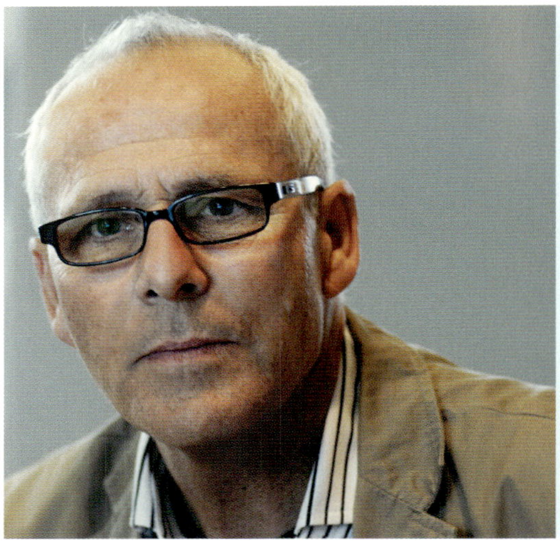

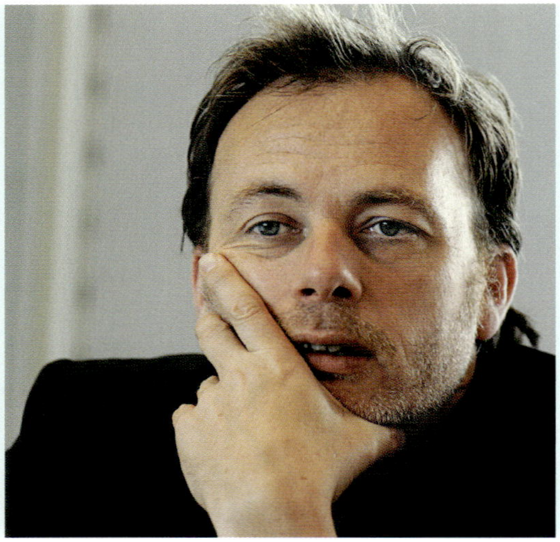

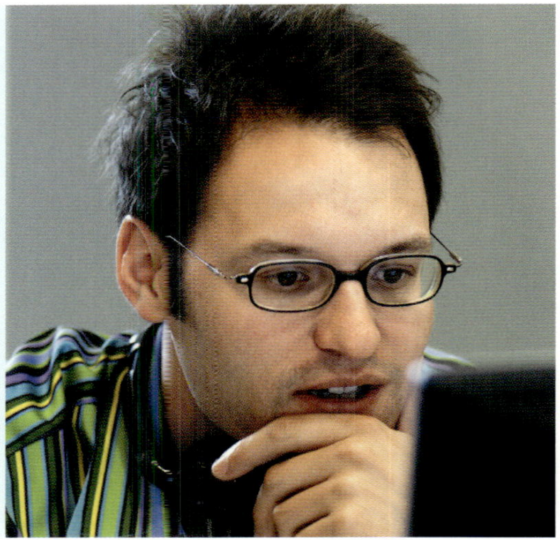

■ Prof. Manfred Becker (*1946) studierte Visuelle Kommunikation an der FH Wuppertal. Ab 1988 war er für RTL TELEVISION u.a. als Art Director, Creative Director, Vorsitzender der Geschäftsführung sowie Creative Managing Director von RTL CREATION tätig. Seit 2003 ist er Creative Consultant für die RTL GROUP-Unternehmen. Er ist Professor für Film & TV Design / Promotion an der Filmakademie Baden-Württemberg, „Mentor" und Dozent für das Fortbildungsprogramm Media Design Producing an der ifs – internationale filmschule köln und seit 1996 Präsident des Verbands EYES & EARS of EUROPE.

■ Prof. Manfred Becker (*1946) studied Visual Communications at the Univ. of Applied Sciences in Wuppertal. He joined RTL TELEVISION in 1988 and has worked there as Art Director, Creative Director and Chairman of the Board, as well as Creative Managing Director of RTL CREATION. He has been a Creative Consultant for the RTL GROUP since 2003. He is professor of Film & TV Design / Promotion at the Baden-Württemberg Film Academy, mentor and lecturer in the Media Design Producing further education program at Cologne's International Film School (ifs), and since 1996 has been president of the EYES & EARS of EUROPE association.

■ Prof. Dr. Gerhard Buurman studierte Industrial Design an der Universität Essen und promovierte zum Thema „Online-Technologien im Automobildesign". Seit 1999 ist er Dozent an der Hochschule für Gestaltung und Kunst in Zürich und übernahm 2001 die Gründung und den Aufbau des Studienbereichs Interaction Design und Game Design. Seit 2006 arbeitet er im Rahmen der Hochschularbeit an der Gründung eines nationalen Forschungszentrums für Interaction Design im Bereich „Finance and Banking" und ist Partner bei PLASMADESIGN ZÜRICH.

■ Prof. Dr. Gerhard Buurman studied Industrial Design at the University of Essen, writing his doctoral thesis on "Online Technologies in Automotive Design". He has been a lecturer at the Zurich School of Art and Design (hgkz) since 1999 and in 2001 took on responsibility for setting up and developing courses in Interaction Design and Game Design. Since 2006 his work for the hgkz has also involved setting up a national research center for Interaction Design in the financial services and banking sector. He is a partner of PLASMADESIGN ZÜRICH.

■ Pascal Urscheler ist Geschäftsführer und Teilhaber der .nemos GmbH. Zuvor war er Geschäftsführer der Carat Interactive AG. Neben seiner beruflichen Laufbahn absolvierte er Weiterbildungen zum Marketingplaner und Marketingleiter, ist Mitglied im Kernteam, das die Zürcher Kreativwirtschaft in Zusammenarbeit mit den Wirtschaftsförderungen von Kanton und Stadt Zürich international neu positioniert, und engagiert sich als aktives Mitglied in der Jury des Crossmedia Award, der jährlich durch die Goldbach Media Gruppe vergeben wird.

■ Pascal Urscheler is Managing Director and partner of .nemos GmbH. Prior to that he was Managing Director of Carat Interactive AG. Parallel to his employment he completed continuing education programs in marketing planning and marketing management. He is a member of the core team that works together with business and trade promotion agencies in the canton and city of Zurich to internationally reposition the Zurich's creative economy. Urscheler is actively involved as a jury member of the annual Crossmedia Award from the Goldbach Media Group.

gold award: digital media atmosphere

Auftraggeber Client
adidas AG
Herzogenaurach, Germany

Projekt Project
adidas Sport Style Y-3 Cubes
Website

Design
Jörg Waldschütz
Antje Thomsen
Peter Kirsch
André Bourguignon
NEUE DIGITALE
Frankfurt am Main, Germany

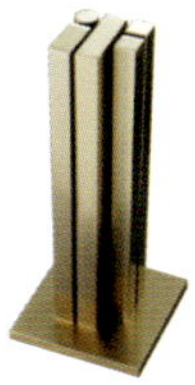

Der User sieht zunächst ein Model in einem Kubus. Ein Klick – und das Model stößt sich von der Wand ab, dreht sich um die eigene Achse und springt in eine neue akrobatische Pose. Ein weiterer Klick – und der gesamte Raum dreht sich um 90 Grad. Und mit ihm das Model, das die aktuelle Y-3 Kollektion trägt und so in vielfältigen Ansichten präsentiert. Neben den in Kuben inszenierten Styles präsentiert sich im Produktbereich die gesamte Y-3 Kollektion.
Y-3 ist eine High-End-Fashionmarke, die aus einer Kooperation zwischen adidas und dem japanischen Designer Yohij Yamamoto entstanden ist.

The Y-3 autumn/winter collection 2005/2006 is presented in rotating cubes. The styles are displayed from lots of different views in a new multimedia space dimension. Models in acrobatic poses jump from wall to wall, seemingly defying the principle of gravity within the cubes. In addition to the styles demonstrated in the cubes the entire Y-3 collection is displayed in the product area.
Y-3 is a high end fashion brand originating from a cooperation between adidas and the Japanese designer Yohij Yamamoto.

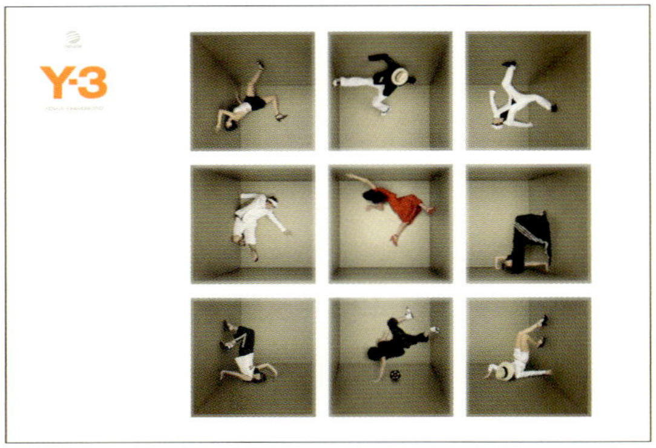

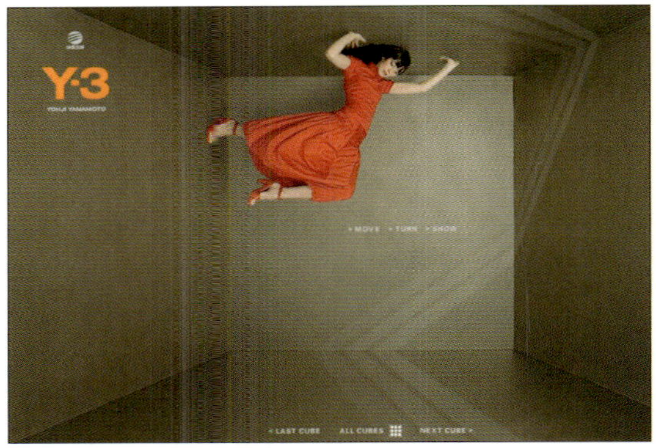

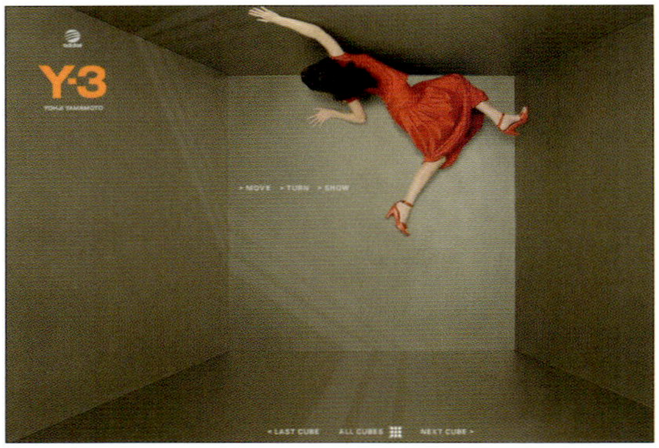

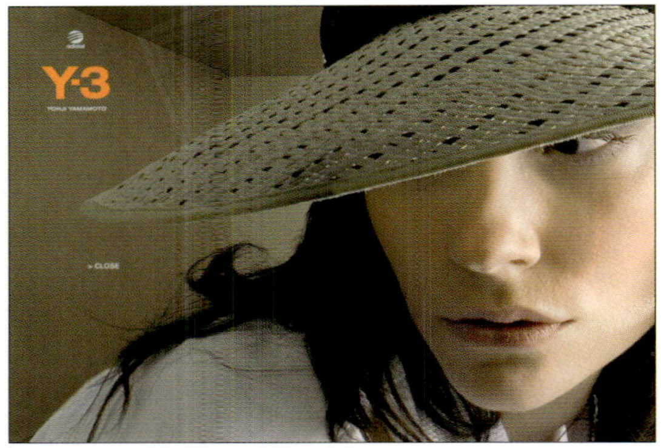

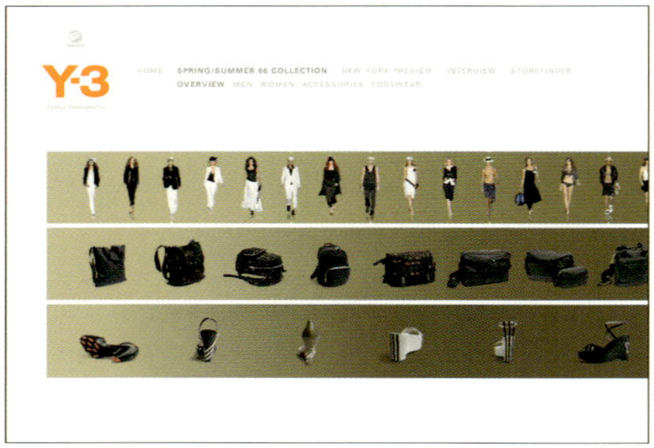

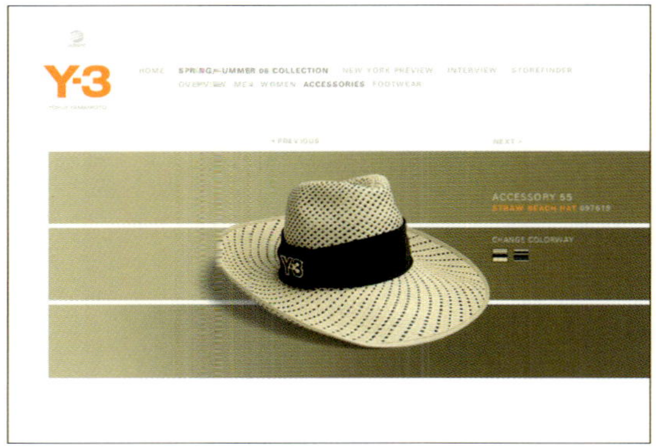

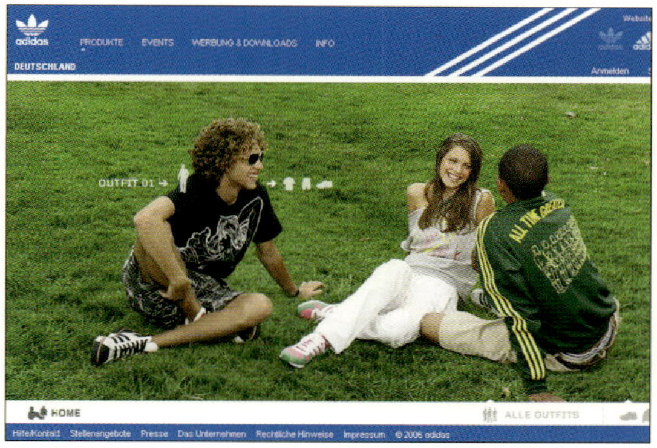

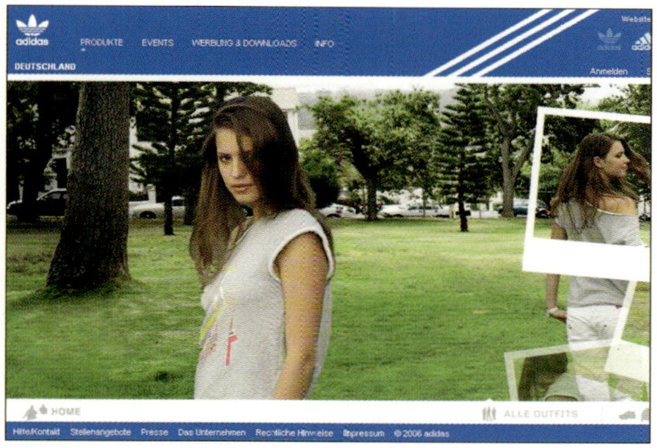

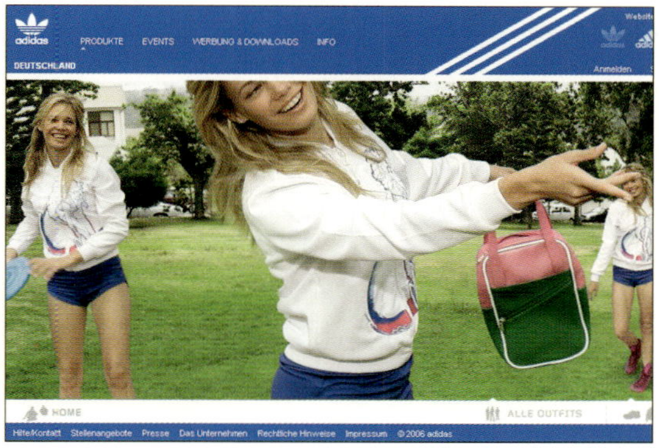

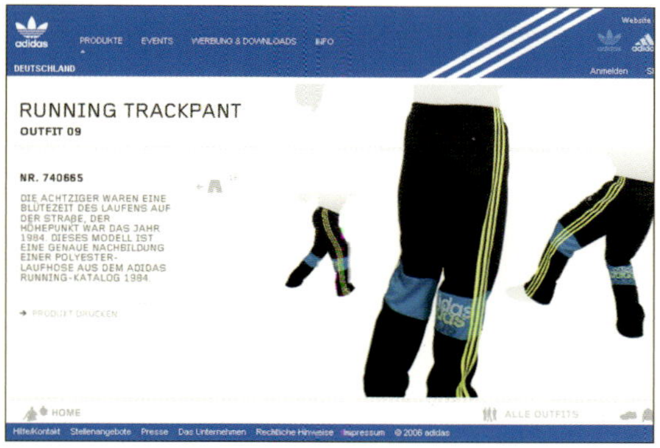

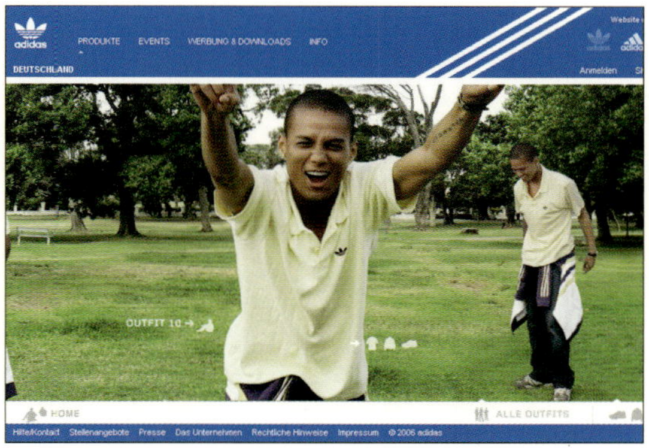

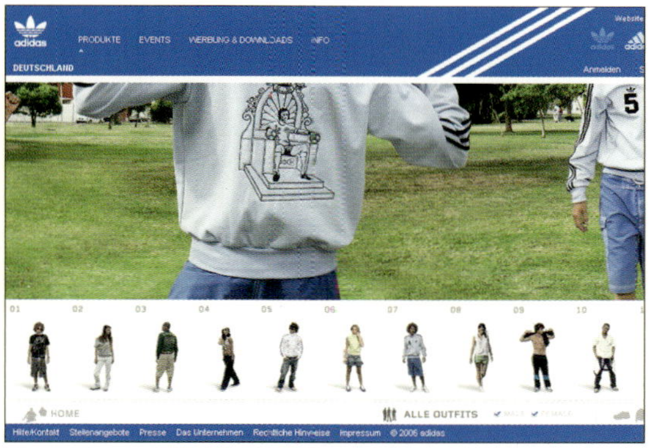

Auftraggeber Client
adidas AG
Herzogenaurach, Germany

Projekt Project
adidas Originals Centralpark
Website

Design
Jörg Waldschütz
André Bourguignon
Ozan Halici
Peter Kirsch
NEUE DIGITALE
Frankfurt am Main, Germany

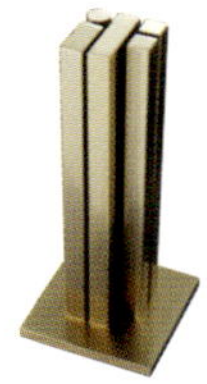

Die Website für adidas Originals ist an die Offline-Kampagne angelehnt. Die Motive und Art der Fotografie sprechen eine einfache Sprache: Freundschaft, junge Menschen im Park. Der User erhält den Eindruck, selbst Teil der Gruppe zu sein. Mit Einstieg in die Seite landet er im Park View, in dem er sich im 360°-Rundumblick durch den Park bewegt. Dort kann er – per Klick – zu unterschiedlichen Ansichten der einzelnen Outfits gelangen, aber auch direkt die Detailseite zu einem Produkt aufrufen. Über die zuschaltbaren zusätzlichen Module „Outfit Viewer" und „Product Viewer" navigiert er zu einem Outfit oder einer Produktdetailseite.

The Web special for the adidas Originals is based on the global marketing campaign. The themes and character of the photographs speak a simple language: friendship, young people in the park. Users have the impression that they are part of the group themselves. On entering the site users find themselves in the Park View by which they can take a 360° look around. Then they can click to see various views of the outfits, or directly go to the product detail page. Users can activate the modules "Outfit Viewer" and "Product Viewer" at any time to navigate to a certain outfit or product.

gold award: digital media atmosphere

Auftraggeber Client
OBI Bau- und Heimwerkermärkte GmbH & Co.
Franchise Center KG
Wermelskirchen, Germany

Projekt Project
OBI Jubelorchester
Microsite

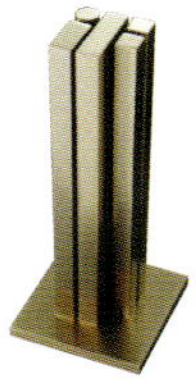

Design
denkwerk
Köln, Germany

Als Nationaler Förderer der FIFA WM 2006™ präsentiert sich OBI seinen Kunden als begeisterter Unterstützer des Fußball-Ereignisses. Auf einer Microsite wird eine eigene OBI Fankultur inszeniert. Das passende Design transportiert diese Stimmung mal schrill, mal augenzwinkernd und sorgt mit dem „Jubelorchester" für jede Menge Spaß: Bei der E-Card-Funktion können die Nutzer den OBI Song „OOOBI ist das schön" mit verschiedenen OBI Fans – ausgestattet mit Torfsäcken, Kreissäge, Gießkanne etc. – instrumentieren und die Komposition als E-Card versenden. Beim Abspielen werden die Filmloops des ausgewählten Jubelorchesters gezeigt.

As a national main promoter of the FIFA World Cup 2006™, OBI presents itself as an enthusiastic supporter of this football event. On the microsite, an own OBI fan culture is staged. The proper design takes this temper once hilarious, once with "a twinkle-in-the-eye" and creates a lot of fun with the "Jubelorchester – cheering orchestra". The user can arrange the OBI song "OOOBI ist das schön" with the help of different OBI fans, equipped with peat bags, circular saw, watering can etc. and send this new composition as an e-card. When playing the card, the film loops and the chosen orchestra will be seen.

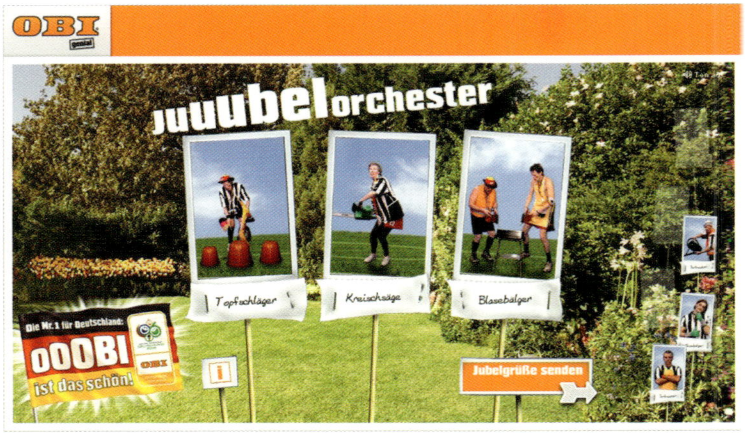

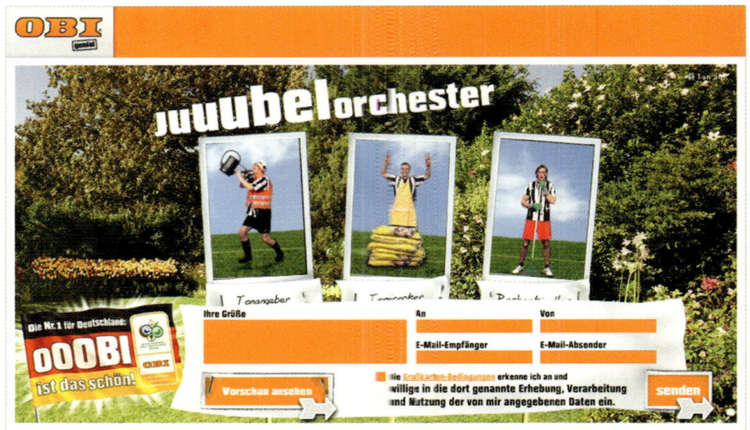

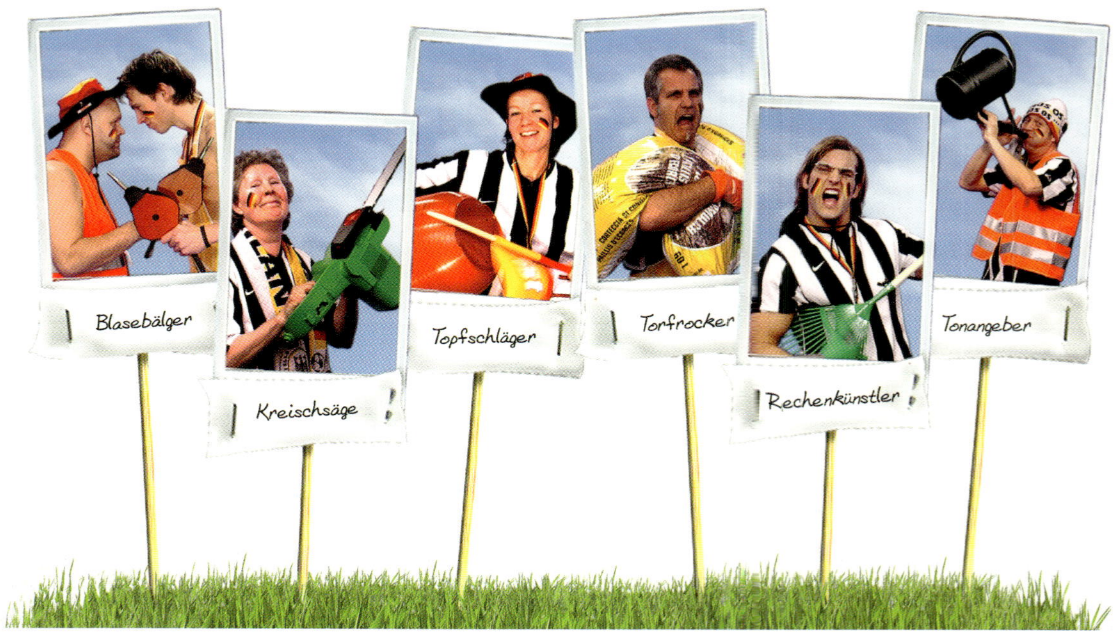

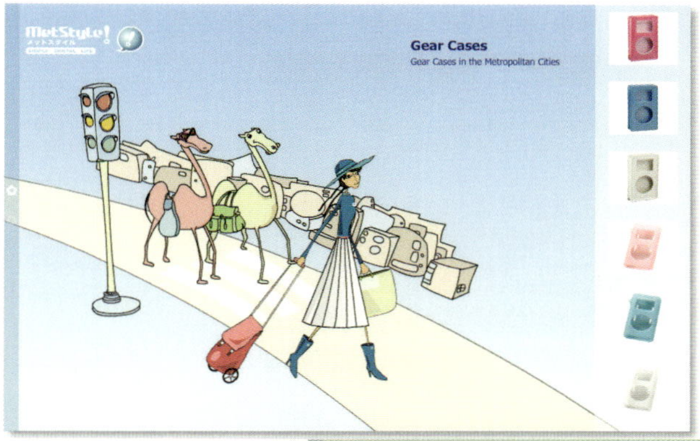

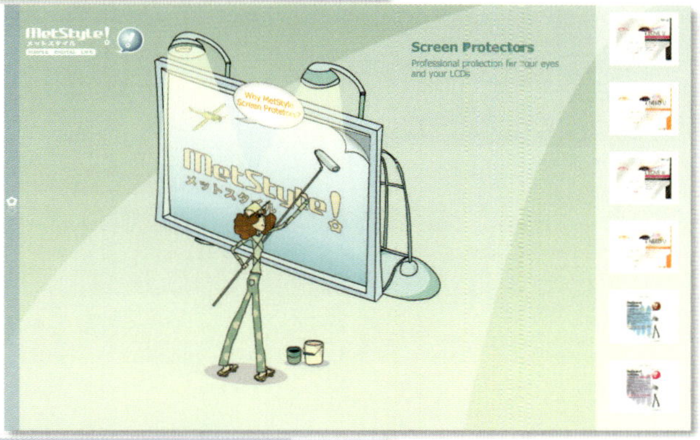

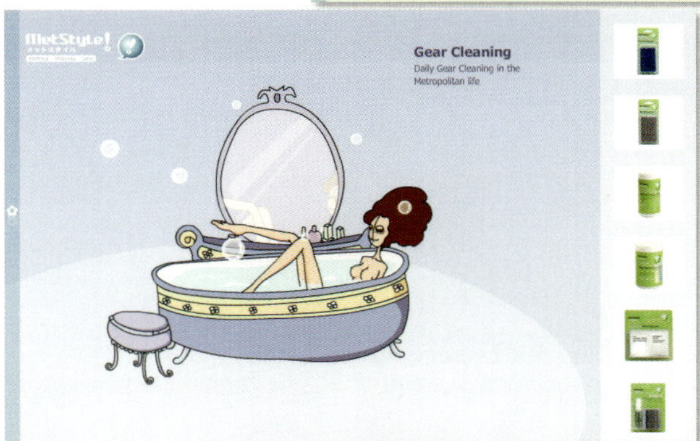

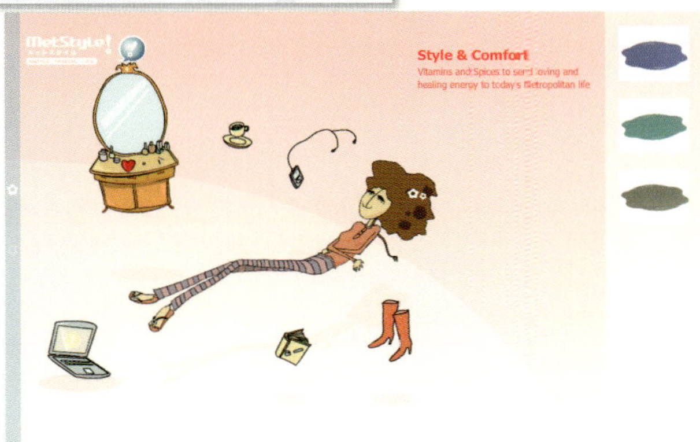

Auftraggeber Client
Celesta inc
Taipei, Taiwan

Projekt Project
www.metstyle.com
Website

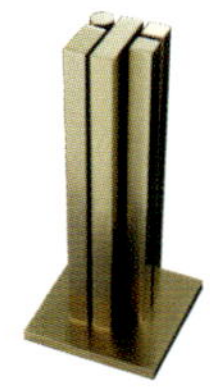

Design
Comprador International
Marketing Services
Taipei, Taiwan

Anders als konventionelle produktorientierte Websites bietet die Metstyle Website für den Zugriff eine situationsbezogene Benutzeroberfläche. Anhand der Darstellung einer modernistischen Metropole lenkt die Metstyle Website den Blick auf spezielle Aspekte urbanen Lebens und entführt den Anwender auf Entdeckungsreise in eine wundersame digitale Welt. Der Betrachter ortet die diversen Metstyle Produktgruppen, indem er mit der Maus über die auf der Homepage abgebildeten Personen und Objekte fährt. Per Mausklick öffnen sich die Produktinformationen. Die Metstyle Website ist eine Kombination von Benutzerfreundlichkeit und Design auf höchstem Niveau.

In contrast to conventional product-oriented websites, the Metstyle website offers a situation-based user interface to grant website access. Depicting a modernized metropolitan setting, the Metstyle site reflects real urban life, taking the user into an amazing, fun-to-explore digital world. The user locates the company's various products by riding the mouse over people and objects on the homepage – one click and the user can then access the product information. The Metstyle website skillfully combines the elements of usability and design.

gold award: digital media screendesign

Auftraggeber Client
MTV Networks GmbH & Co. OHG
On Air
Berlin, Germany

Projekt Project
On Air Design 2005
Channel Design

Design
Dinko Lacic
Hagen Biewer
Peter Gotzoll
Benjamin Brettschneider
Tolga Peters
Tim Goode
Thomas Sabel
MTV Networks GmbH & Co. OHG
On Air
Berlin, Germany

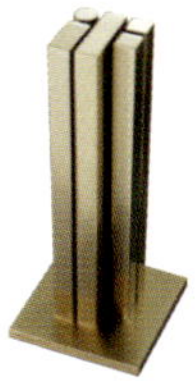

Die Welt der Sticker verselbstständigt sich und greift unmerklich in unseren Alltag ein.

The world of stickers becomes independent and influences our day-to-day life.

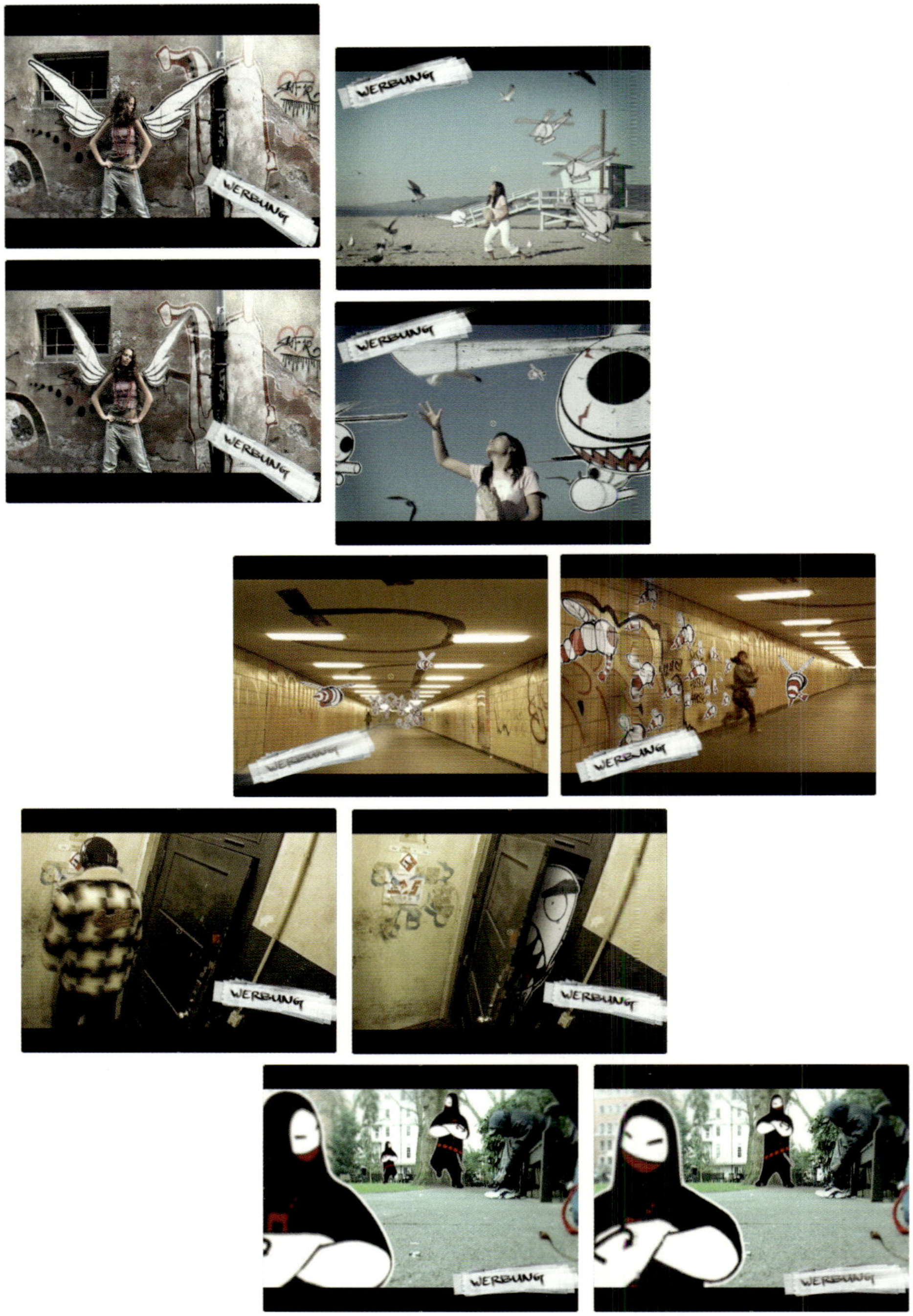

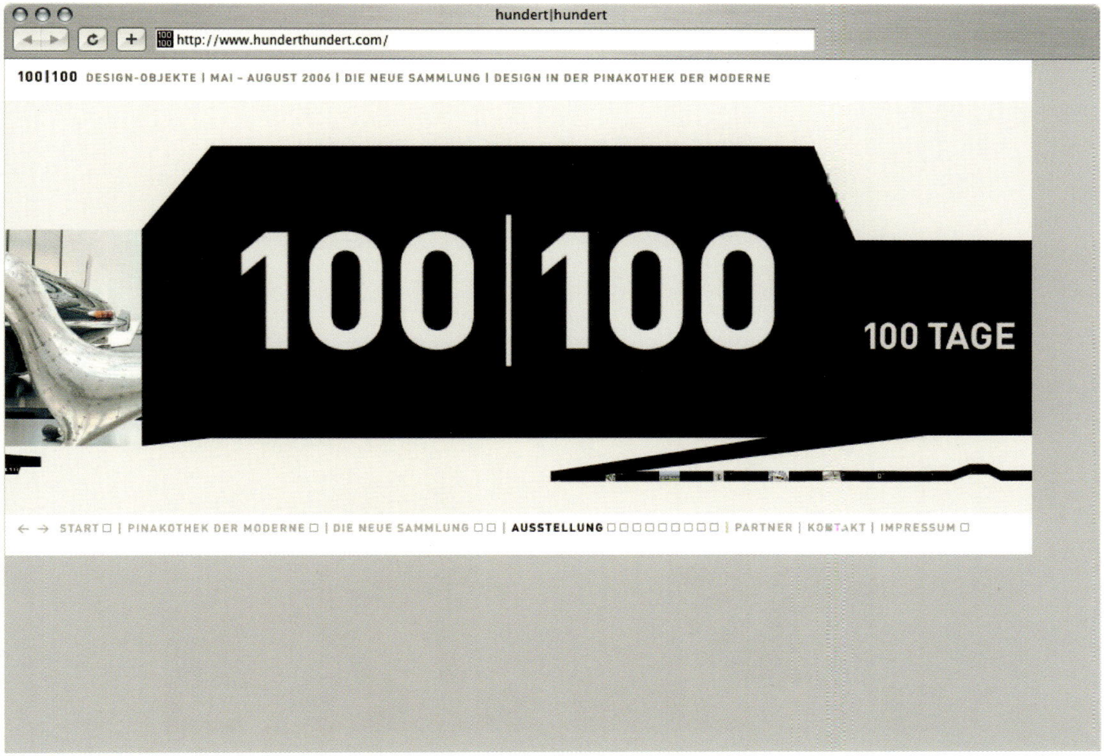

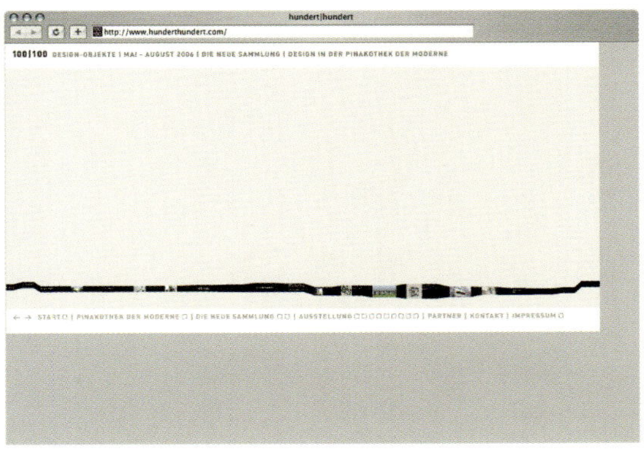

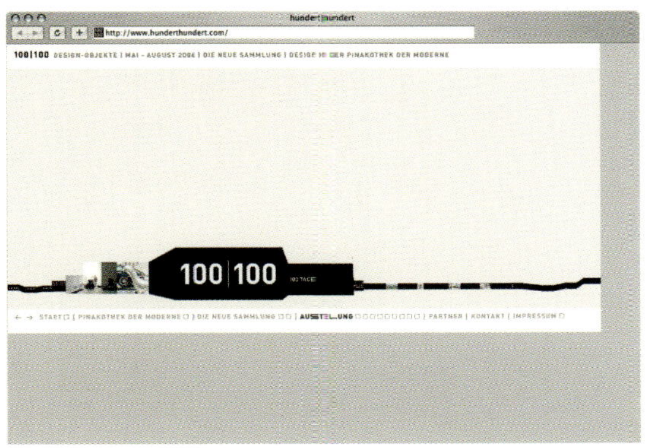

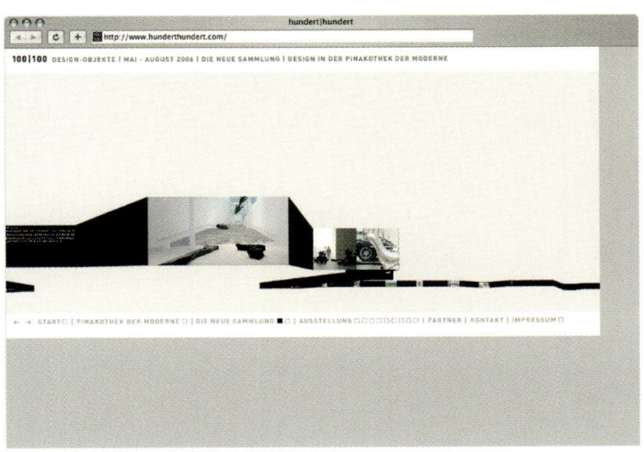

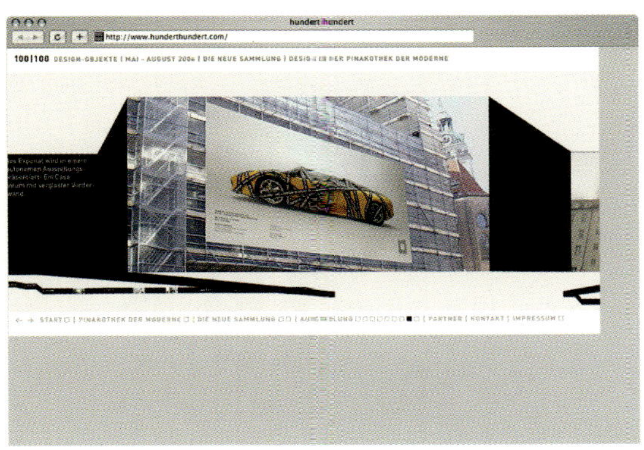

Auftraggeber Client
hundertlhundert GmbH
München, Germany

Projekt Project
100l100
Website

Design
Knut Maierhofer
Michael Keller
Bruno Marek
Patrick Märki
KMS Team GmbH
München, Germany

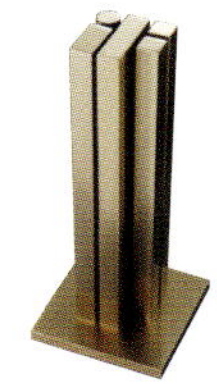

Der Internetauftritt der Ausstellung 100l100 soll die ihr zugrunde liegende Idee – eine Containerreihe quer durch die Pinakothek – auf den ersten Blick transportieren. Ein horizontal fortlaufendes Band fungiert als Navigationsmenü, die einzelnen Inhalte sind darauf abgebildet. Der Cursor beeinflusst die Bewegung des Bands, wird ein Inhalt ausgewählt, zoomt sich der entsprechende Ausschnitt heran.

The website of the exhibition 100l100 took up the underlying conceptual idea of the project – a row of containers extending diagonally across the Pinakothek. A continuous horizontally moving strip functions as a navigational menu, displaying the individual contents in succession. The cursor controls the movement of the strip. When a particular content is selected, the view automatically zooms in on the relevant detail.

Auftraggeber Client
Rechtsanwälte Haftner, Schobel + Simoncic
St. Pölten, Austria

Projekt Project
Fallweise die Besten.
Website

Design
Thomas Wagner
Matthias Steinperl
Berthold Kremser
Monika Gerersdorfer
wagner steinperl.
St. Pölten, Austria

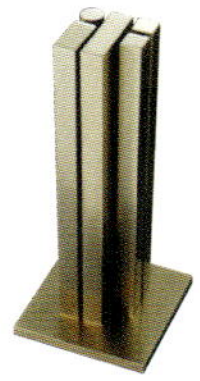

Die Corporate Website der Rechtsanwälte Haftner, Schobel + Simoncic zeigt die „trockene" Materie der Juristerei in einer vollkommen neuen und originellen Art und Weise. Bereits im Intro wird dem Betrachter klar, dass das, was folgen wird, nicht dem herkömmlichen Klischee des Rechtsanwalts entspricht. Auf der Website selbst werden die Partner in jeweils eigenständigen Sujets (im Zusammenhang mit ihrer Kompetenz) präsentiert. Durch die Art der Inszenierung wird beim Betrachter Sympathie durch ein Augenzwinkern hergestellt und im zweiten Schritt – durch die Aufzählung der jeweiligen Schwerpunkte – die Kompetenz der Anwälte dargestellt.

The corporate website of the lawyer's office "Haftner, Schobel + Simoncic" presents the rather factual topic of jurisprudence in an absolutely new and humorous way. Watching the intro, the user already notices that what follows will not show the common cliché of a lawyer. On the website each partner presents himself in a picture related to his specific competence. Through this humorous presentation the user sympathizes with the lawyers, and in a second step gets to know their different competencies.

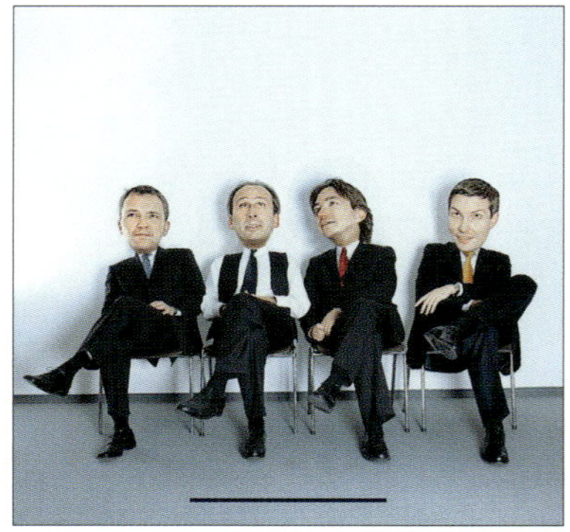

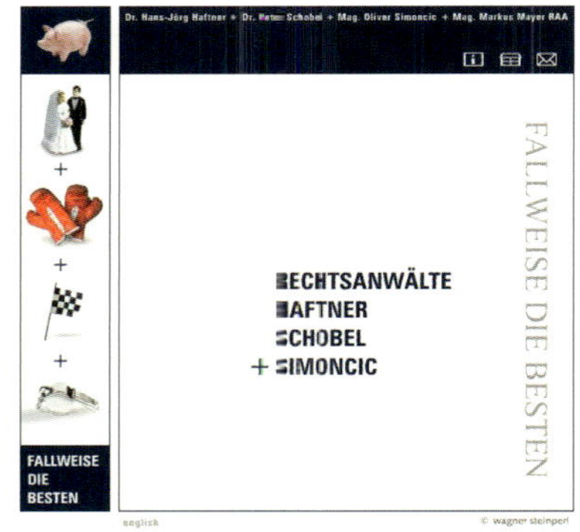

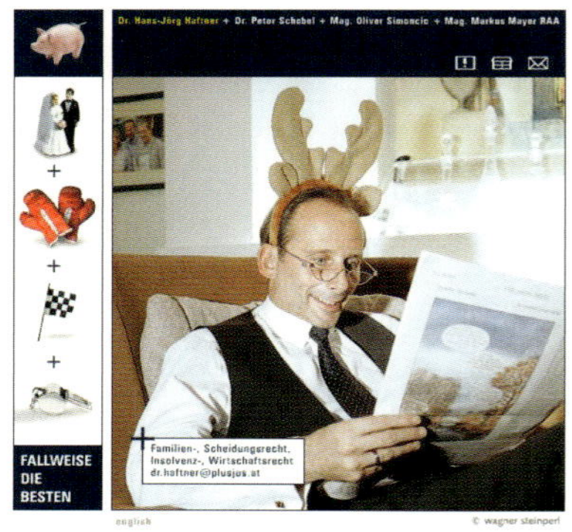

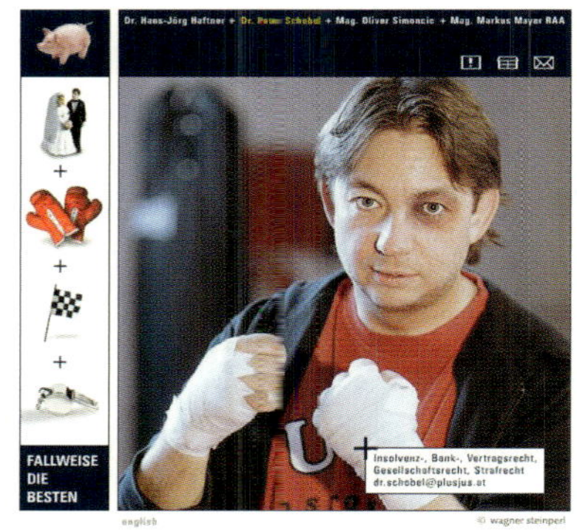

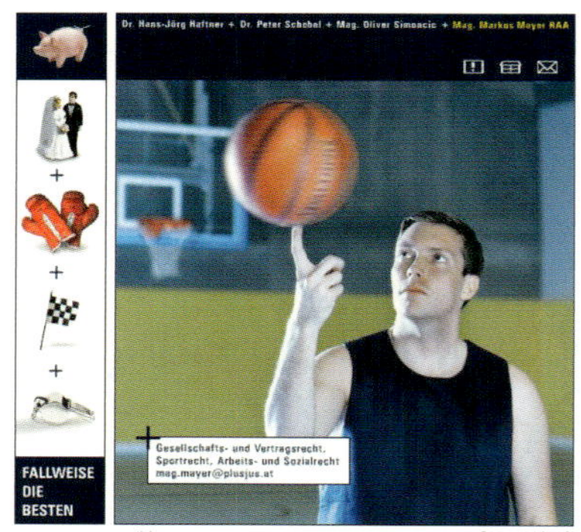

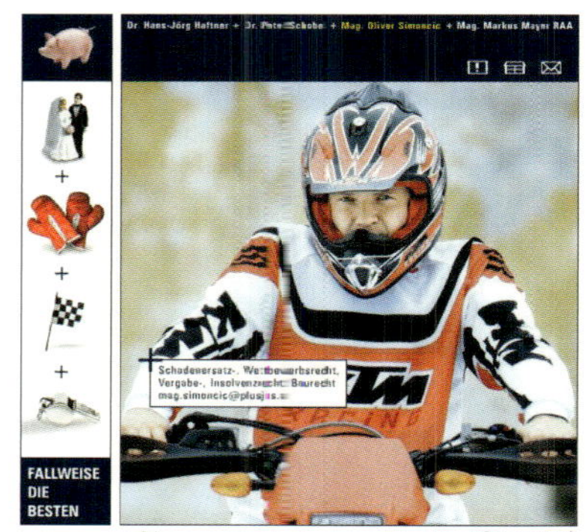

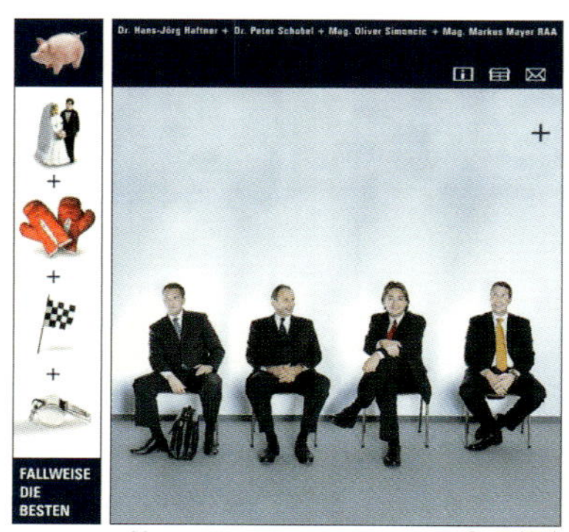

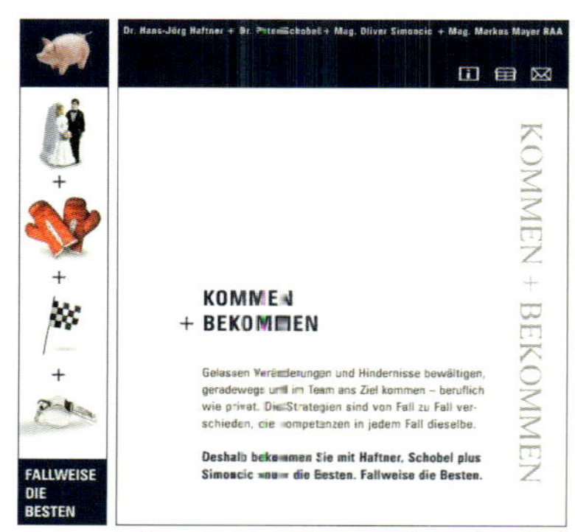

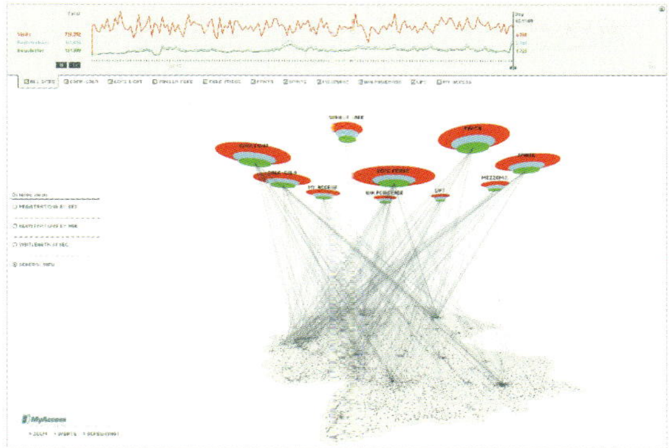

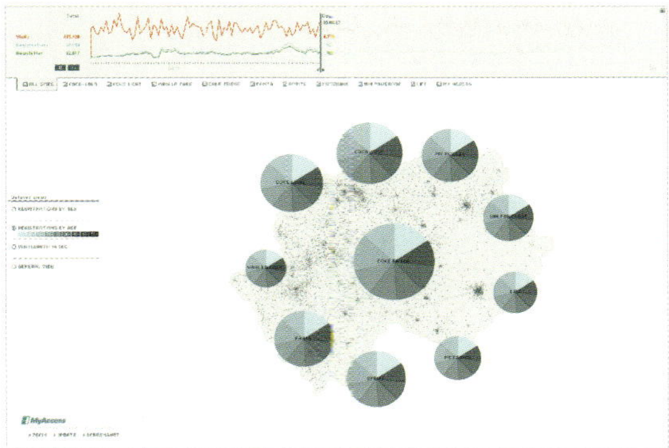

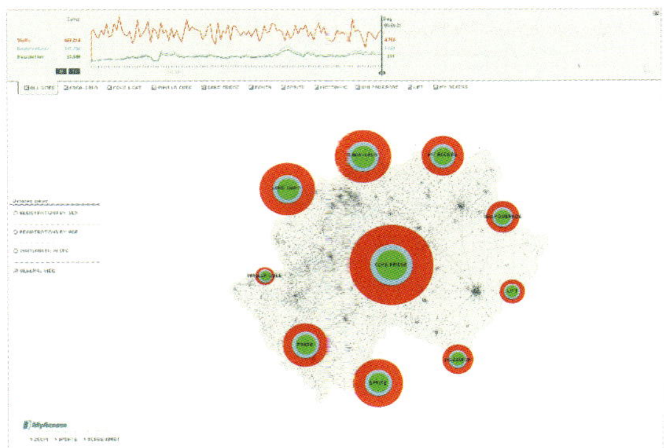

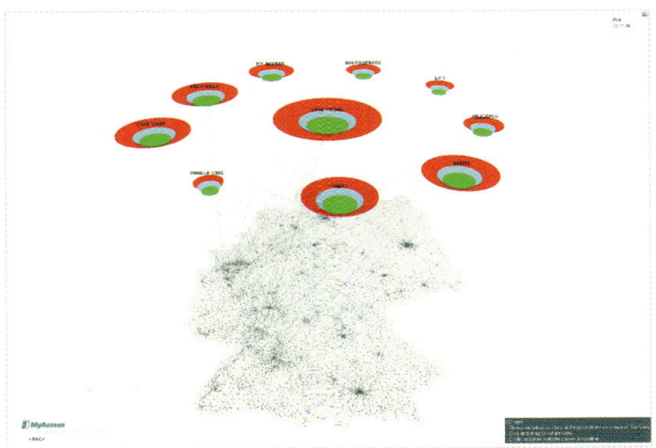

Auftraggeber Client
Coca-Cola GmbH
Berlin, Germany

Projekt Project
Coca-Cola Datenvisualisierung
Online-Webanalyse-Tool
Online web analytics tool

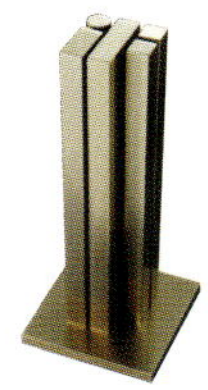

Design
Scholz & Volkmer GmbH
Wiesbaden, Germany

Online-Tool zur Visualisierung des Nutzerverhaltens auf den Coca-Cola Markenwebsites. Ziel ist es, Informationen aus Logfiletabellen und deren komplexe Zusammenhänge nachvollziehbar darzustellen, um Effektivität und Effizienz von Online-Marketingmaßnahmen zu steigern. Die Vergleichbarkeit und Analyse der Websites im Verhältnis zueinander wird verdeutlicht. Auf nur einem Screen zeigt die Applikation Anzahl und Aktivität der User in Deutschland im zeitlichen Ablauf.

Online tool to visualise user behaviour on the different brand websites of Coca-Cola. The objective is to be able to compare and analyse the websites. Information from log-file tables and their complex interrelationships are presented on only one screen in a clear and understandable manner. Internal Controlling and Marketing gets the benefit of a new level of experience: compression of data to knowledge, in order to create new interrelationships. Added value for Controlling and Marketing: the effectiveness and efficiency of online marketing activities is increased.

Auftraggeber Client
decodeunicode
Fachhochschule Mainz
Mainz, Germany

Projekt Project
www.decodeunicode.org – the basic multilingual plane
Website

Design
Johannes Bergerhausen
Siri Poarangan
Wenzel Spingler
Mathias Wollin
decodeunicode
Mainz, Germany

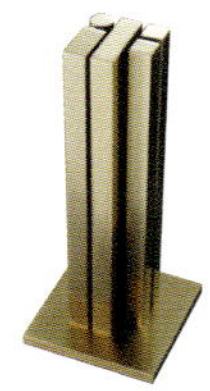

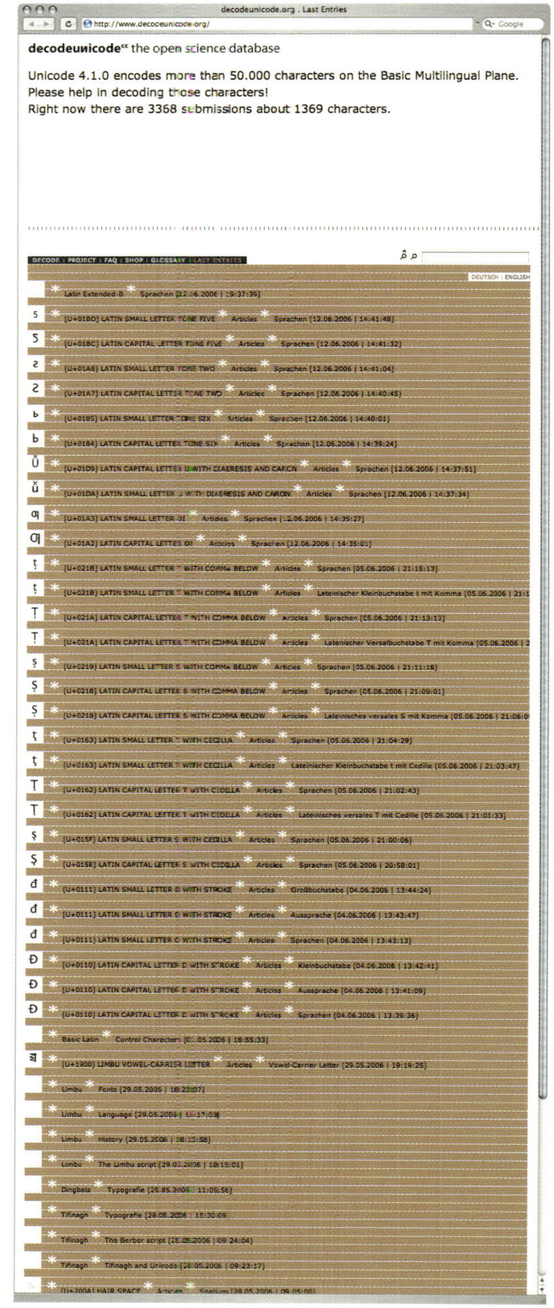

decodeunicode ist eine unabhängige Online-Plattform für digitale Schriftkultur und präsentiert alle heute auf dem Rechner zugänglichen Schriftzeichen der Welt. Mehr als 50.000 Unicode-Zeichen sind seit Mai 2005 intuitiv zugänglich. Jedes Zeichen wird plattformunabhängig in drei Größen angezeigt. Über die Wikifunktion werden Informationen zu jedem einzelnen Zeichen gesammelt. Testen Sie die Volltextsuche, beispielsweise mit dem Wort „star"!
Das Projekt wurde mit insgesamt sechs internationalen Designpreisen ausgezeichnet. Die Website nutzen mehr als 700 Besucher täglich aus über 100 Ländern. Im ersten Jahr wurden mehr als neun Millionen Seiten aufgerufen. Das Projekt wurde vom Bundesministerium für Bildung und Forschung (BMBF) unterstützt.

decodeunicode is an independent online-platform for digital type culture and presents all characters that are accessible today on a personal computer. Since May 2005, more than 50,000 Unicode-characters have been easily accessible. Each character is rendered in three different sizes. With a wiki-function, we are collecting information about each and every character. Just try the full-text search with the word "star"!
The project was awarded six international design awards. The website is used by more than 700 visitors a day, coming from more than 100 countries, having viewed nine million pages in the first year. The project was supported by the Federal Ministry of Education and Research (BMBF).

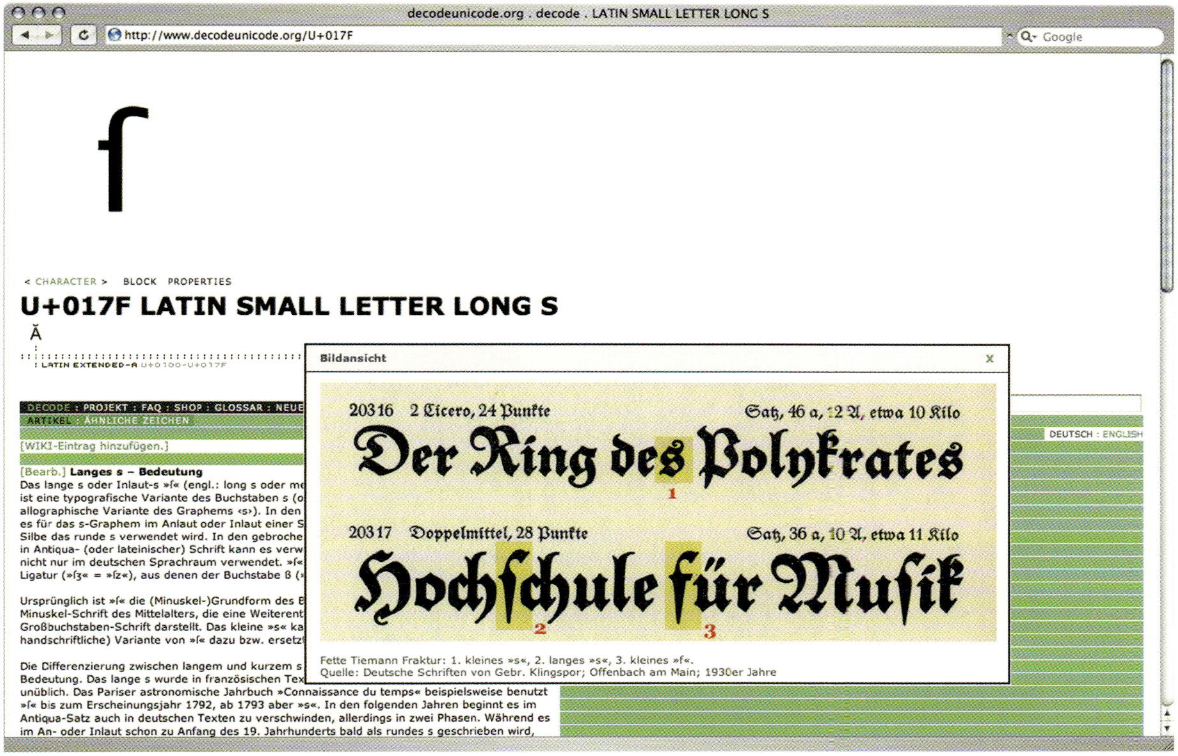
decodeunicode.org . decode . LATIN SMALL LETTER LONG S
http://www.decodeunicode.org/U+017F
Google
ſ
< CHARACTER > BLOCK PROPERTIES
U+017F LATIN SMALL LETTER LONG S
Ă
LATIN EXTENDED-A U+0100-U+017F
DECODE : PROJEKT : FAQ : SHOP : GLOSSAR : NEUE
ARTIKEL : ÄHNLICHE ZEICHEN
DEUTSCH : ENGLISH
[WIKI-Eintrag hinzufügen.]
[Bearb.] Langes s – Bedeutung
Das lange s oder Inlaut-s »ſ« (engl.: long s oder me
ist eine typografische Variante des Buchstaben s (o
allographische Variante des Graphems ‹s›). In den
es für das s-Graphem im Anlaut oder Inlaut einer S
Silbe das runde s verwendet wird. In den gebroche
in Antiqua- (oder lateinischer) Schrift kann es verw
nicht nur im deutschen Sprachraum verwendet. »ſ«
Ligatur (»ſʒ« = »ſz«), aus denen der Buchstabe ß (»
Ursprünglich ist »ſ« die (Minuskel-)Grundform des B
Minuskel-Schrift des Mittelalters, die eine Weiterent
Großbuchstaben-Schrift darstellt. Das kleine »s« ka
handschriftliche) Variante von »ſ« dazu bzw. ersetz
Die Differenzierung zwischen langem und kurzem s
Bedeutung. Das lange s wurde in französischen Tex
unüblich. Das Pariser astronomische Jahrbuch »Connaissance du temps« beispielsweise benutzt
»ſ« bis zum Erscheinungsjahr 1792, ab 1793 aber »s«. In den folgenden Jahren beginnt es im
Antiqua-Satz auch in deutschen Texten zu verschwinden, allerdings in zwei Phasen. Während es
im An- oder Inlaut schon zu Anfang des 19. Jahrhunderts bald als rundes s geschrieben wird,
Bildansicht
x
20316 2 Cicero, 24 Punkte
Satz, 46 a, 12 A, etwa 10 Kilo
Der Ring des Polykrates
1
20317 Doppelmittel, 28 Punkte
Satz, 36 a, 10 A, etwa 11 Kilo
Hochſchule für Muſik
2
3
Fette Tiemann Fraktur: 1. kleines »s«, 2. langes »s«, 3. kleines »f«.
Quelle: Deutsche Schriften von Gebr. Klingspor; Offenbach am Main; 1930er Jahre

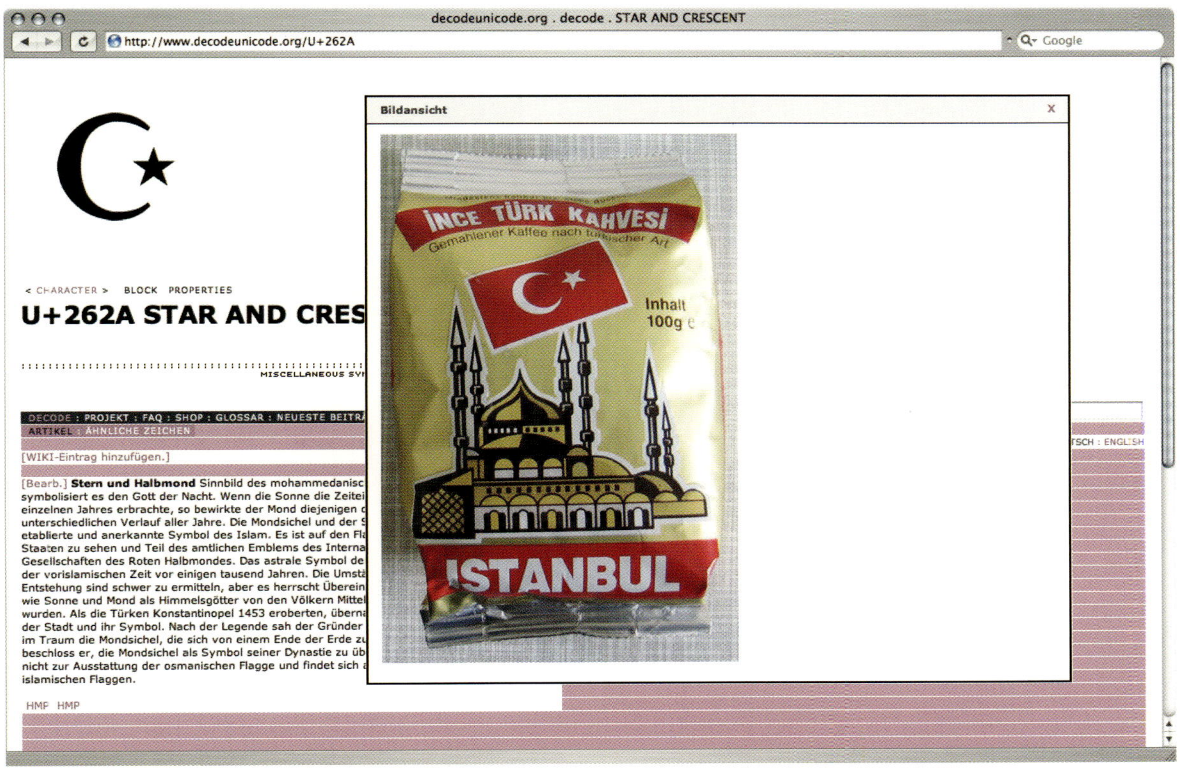
decodeunicode.org . decode . STAR AND CRESCENT
http://www.decodeunicode.org/U+262A
Google
☪
< CHARACTER > BLOCK PROPERTIES
U+262A STAR AND CRES
MISCELLANEOUS SY
DECODE : PROJEKT : FAQ : SHOP : GLOSSAR : NEUESTE BEITRÄ
ARTIKEL : ÄHNLICHE ZEICHEN
[WIKI-Eintrag hinzufügen.]
[Bearb.] Stern und Halbmond Sinnbild des mohammedanisc
symbolisiert es den Gott der Nacht. Wenn die Sonne die Zeitei
einzelnen Jahres erbrachte, so bewirkte der Mond diejenigen d
unterschiedlichen Verlauf aller Jahre. Die Mondsichel und der S
etablierte und anerkannte Symbol des Islam. Es ist auf den Fla
Staaten zu sehen und Teil des amtlichen Emblems des Interna
Gesellschaften des Roten Halbmondes. Das astrale Symbol de
der vorislamischen Zeit vor einigen tausend Jahren. Die Umstä
Entstehung sind schwer zu ermitteln, aber es herrscht Überein
wie Sonne und Mond als Himmelsgötter von den Völkern Mittel
wurden. Als die Türken Konstantinopel 1453 eroberten, überna
der Stadt und ihr Symbol. Nach der Legende sah der Gründer
im Traum die Mondsichel, die sich von einem Ende der Erde zu
beschloss er, die Mondsichel als Symbol seiner Dynastie zu üb
nicht zur Ausstattung der osmanischen Flagge und findet sich a
islamischen Flaggen.
HMP HMP
TSCH : ENGLISH
Bildansicht
x
İNCE TÜRK KAHVESİ
Gemahlener Kaffee nach türkischer Art
Inhalt
100g
İSTANBUL

gold award: digital media typography

Auftraggeber Client
MTV Networks GmbH & Co. OHG
On Air
Berlin, Germany

Projekt Project
MTV Noise
Programm-Package
Programme package

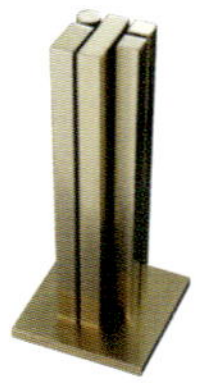

Design
Dinko Lacic
MTV Networks GmbH & Co. OHG
On Air
Berlin, Germany

Die Musik dekonstruiert das Logo der Musikgenre-Sendung „Noise"!

Music deconstructs the logo of the music-genre show "Noise"!

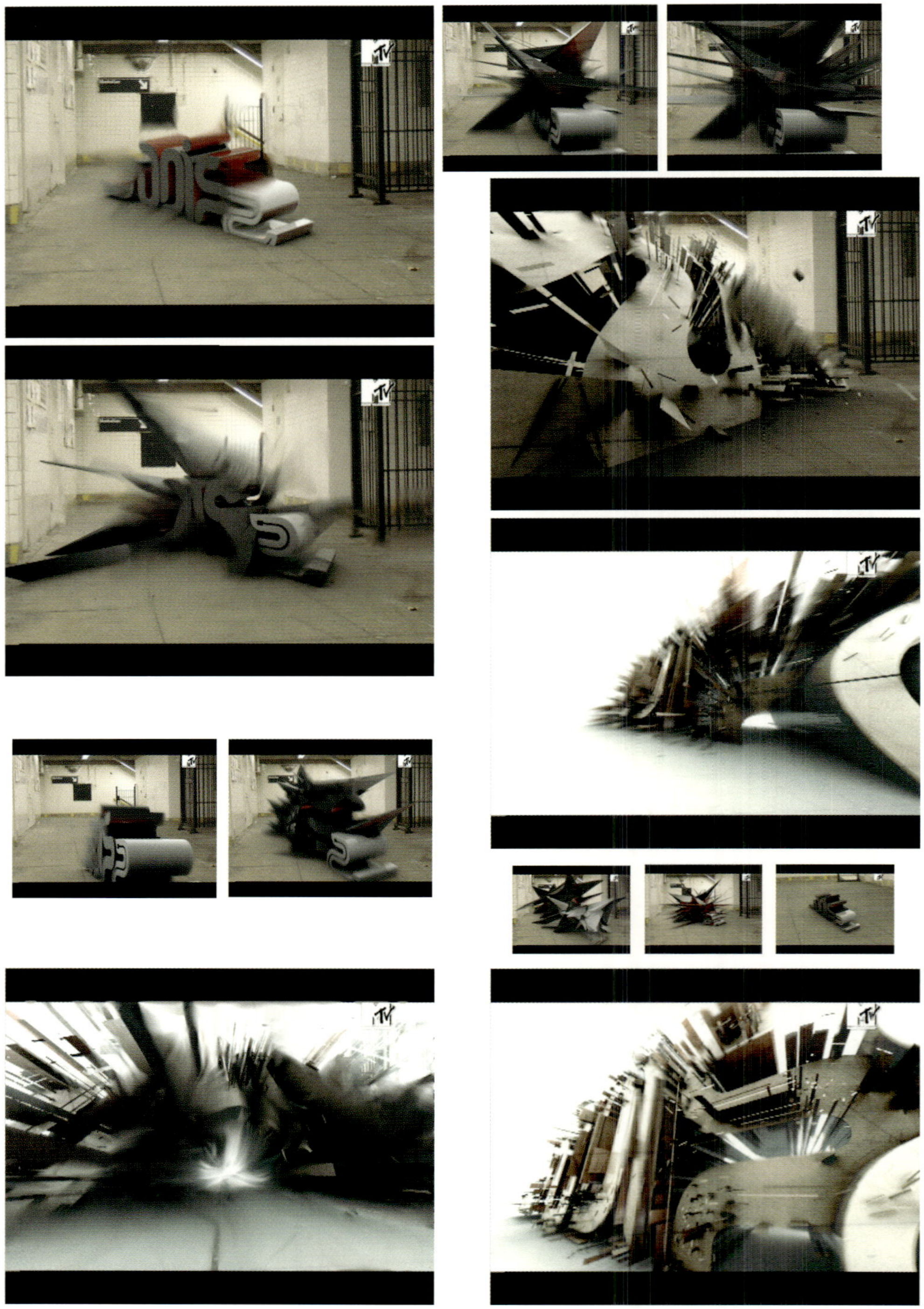

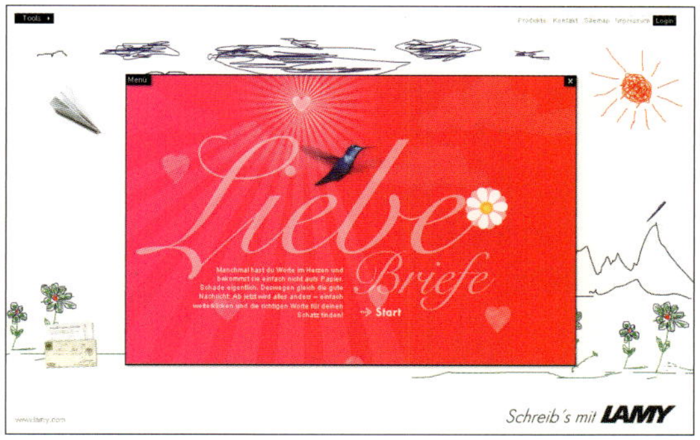

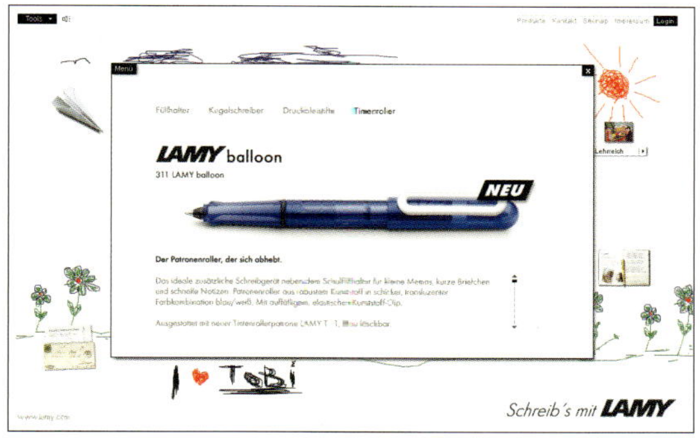

Auftraggeber Client
C. Josef Lamy GmbH
Heidelberg, Germany

Projekt Project
Lamy – Website für Jugendliche
Website

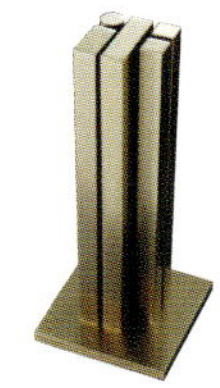

Design
Markus Beige
Markus Gimenez
Berger Baader Hermes
München, Germany

„Schreib's mit Lamy" – Das Coole an der Schule. Die Ansprache erfolgt sehr emotional und intuitiv über Bildwelten. In unterschiedlichen Themenwelten wird auf Bedürfnisse der Schüler eingegangen. Liebe, Berufe, Spiele und Kommunikation werden gleichzeitig unterhaltend und informativ dargestellt. Um einen Dialog mit der Zielgruppe zu erreichen und gleichzeitig einen viralen Effekt auszulösen, wird der Mauszeiger auf dem weißen Desktop zum Füllhalter. Egal ob Spruch, Telefonnummer oder einfach nur Herzen, der Jugendliche kann seinen Desktop selbst scribblen und an Freunde weiterschicken.

"Write with Lamy" – and be cool in school. This website approaches users at a highly emotional and intuitive level through its deployment of imagery. The various needs and interests of school students are explored in various topic areas, with romance, jobs, games and communication presented in an entertaining and informative way. In order to achieve a dialog with the target group and at the same time achieve a viral effect, the mouse pointer turns into a fountain pen on the white desktop. Whether they want to write a message or a phone number or simply draw hearts, young people can doodle their own desktop designs and send them to friends.

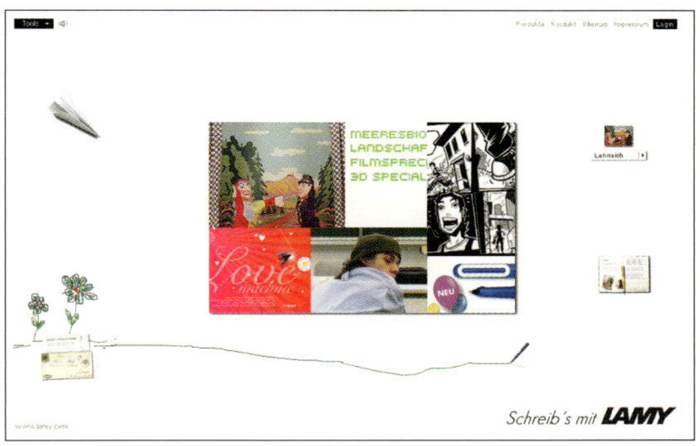

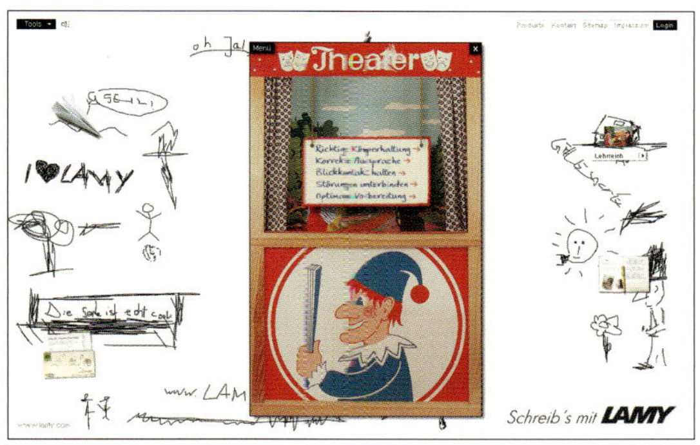

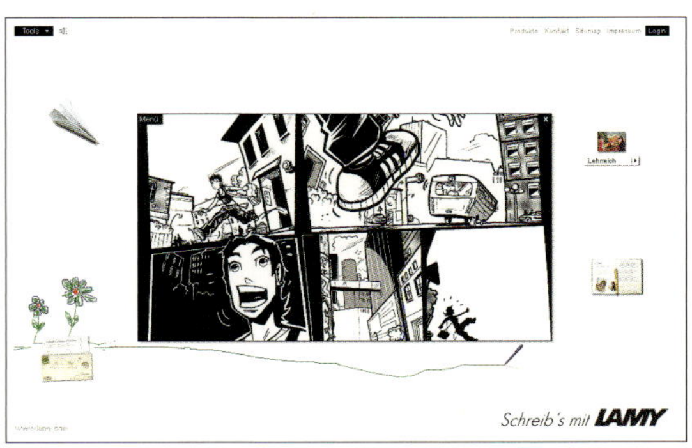

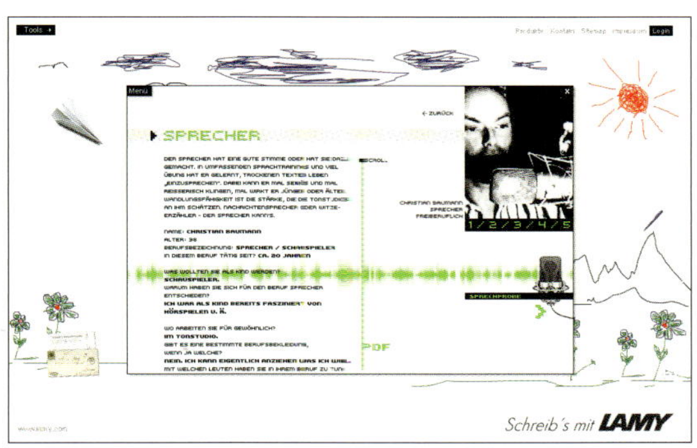

gold award: digital media interface   multiple winner: digital media screendesign

Auftraggeber Client
Carpe Diem GmbH & Co KG
Fuschl am See, Austria

Projekt Project
CARPE DIEM AT WORK
Website

Design
Rob Nikowitsch (Creative Director)
Stephanie Wolfsteiner (Designer)
Jan Massberg,
Paul Stevens (Flash)
Oliver Kochs (Camera, OKAY Film Munich)
Bernd Leinfelder (Technical Director)
Kriz Eideloth (Programmer)
Robert Schlittenbauer (Account Supervisor)
Markus Nikowitsch (Sounddesign)
H2OMEDIA AG
München, Germany

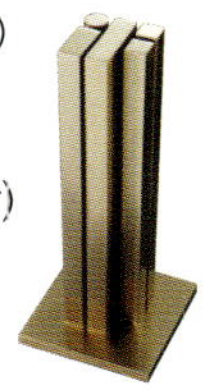

Virale Aktion zur deutschen Markteinführung der Carpe Diem fresh Getränke:
Auf der Website carpediematwork.com können die kreativsten Büros Carpe Diem Getränke fürs ganze Office gewinnen.

This viral campaign accompanies the market launch of Carpe Diem fresh beverages:
On carpediematwork.com, the most creative companies have the opportunity to win Carpe Diem beverages for the entire office.

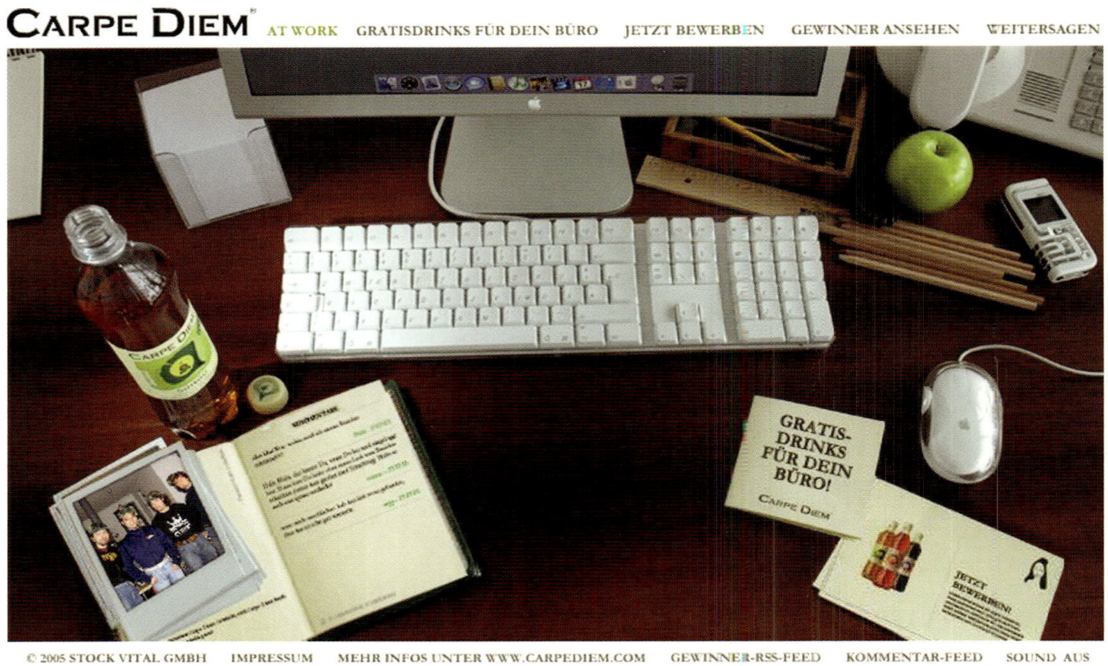

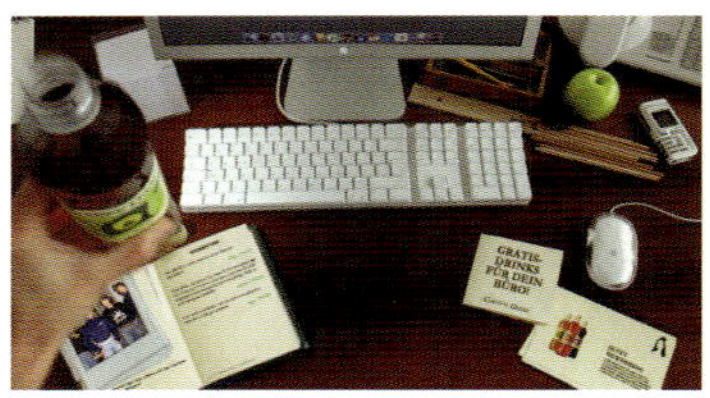

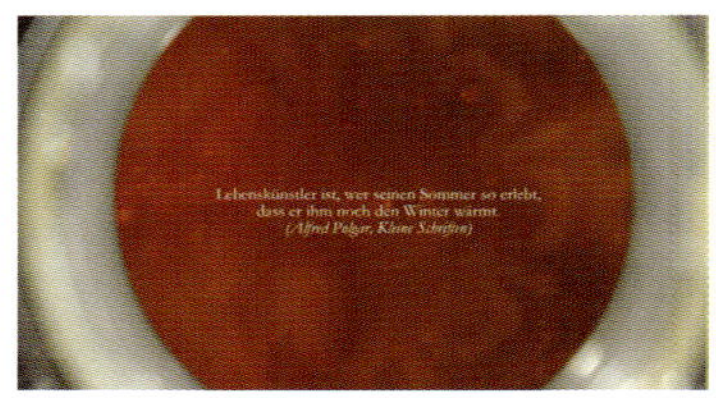

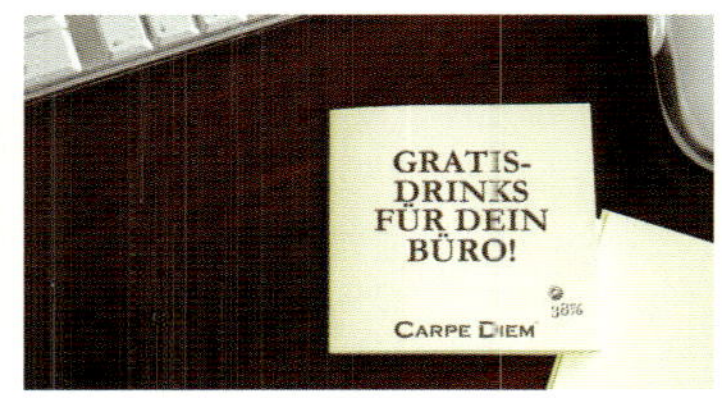

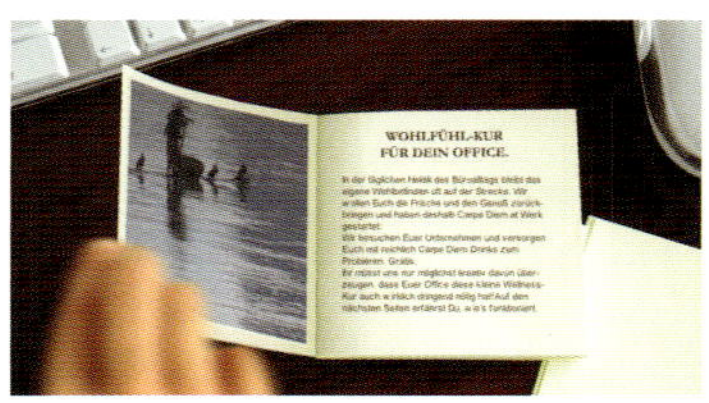

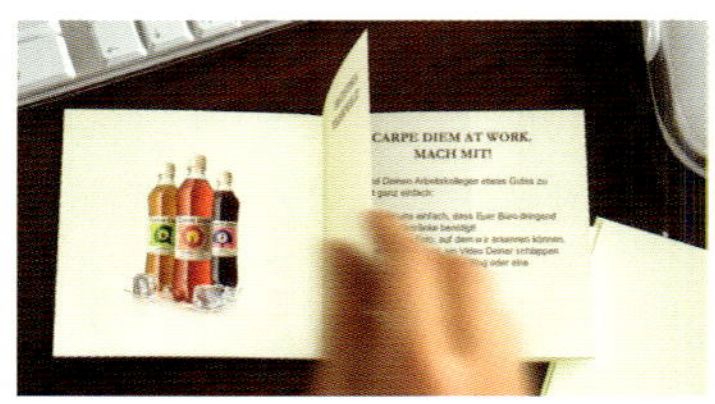

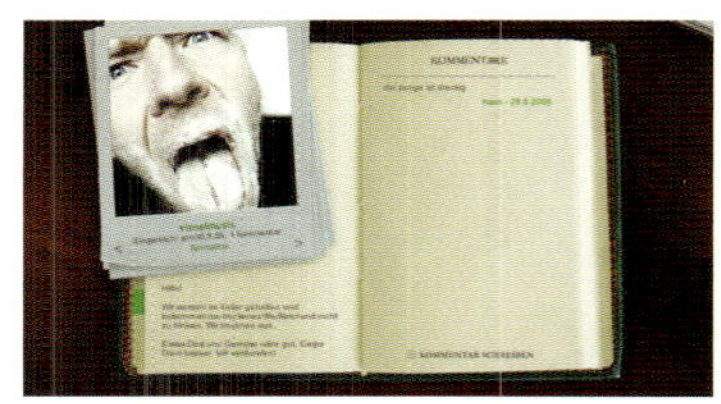

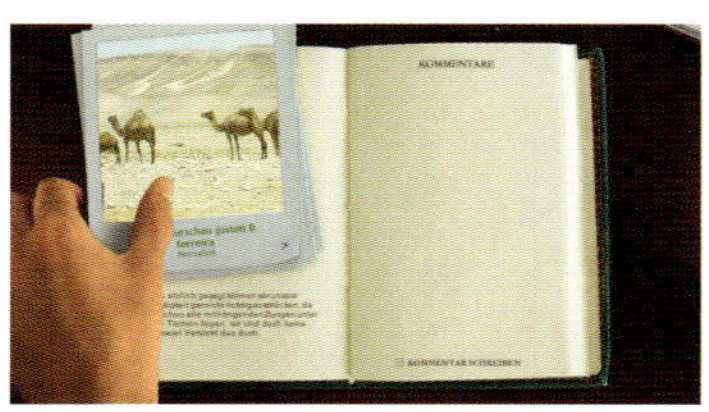

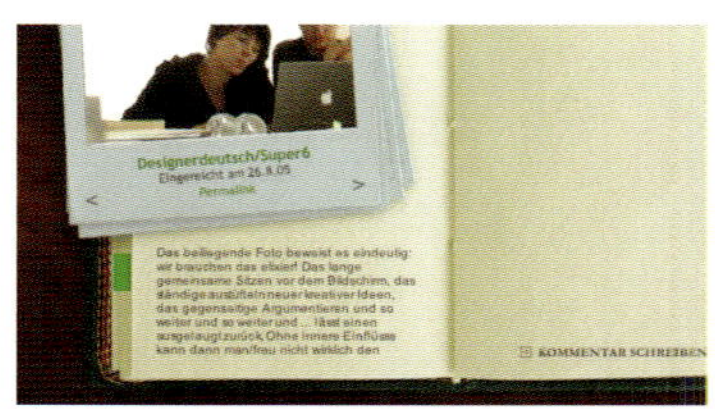

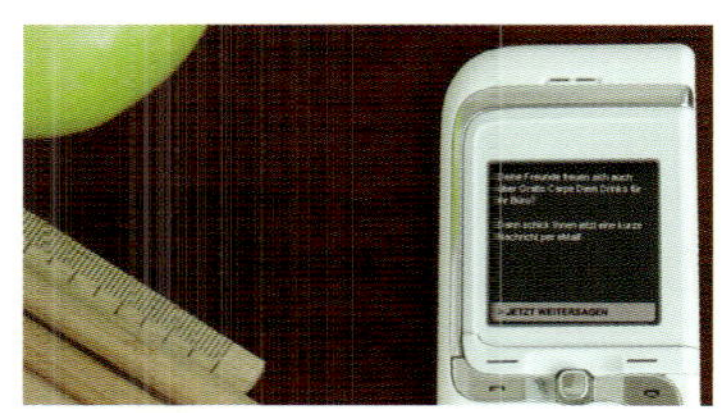

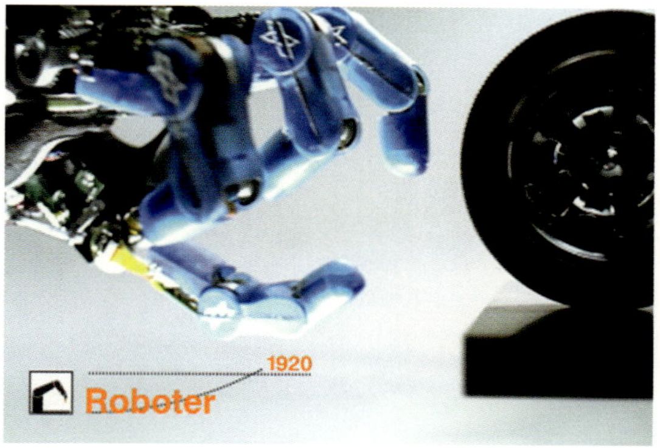

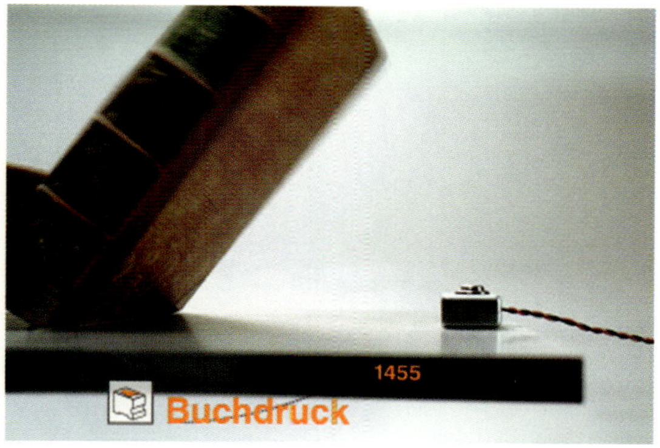

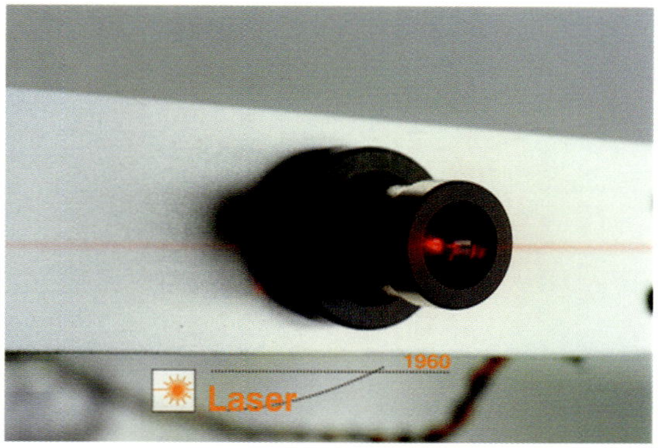

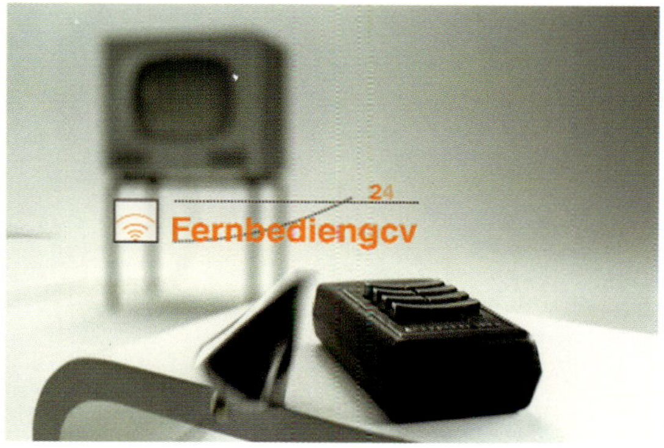

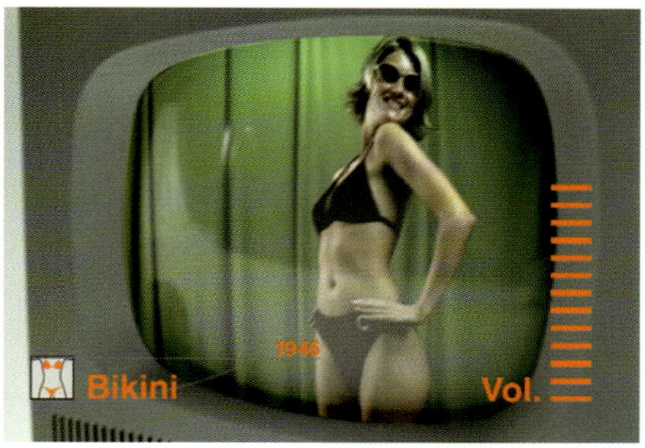

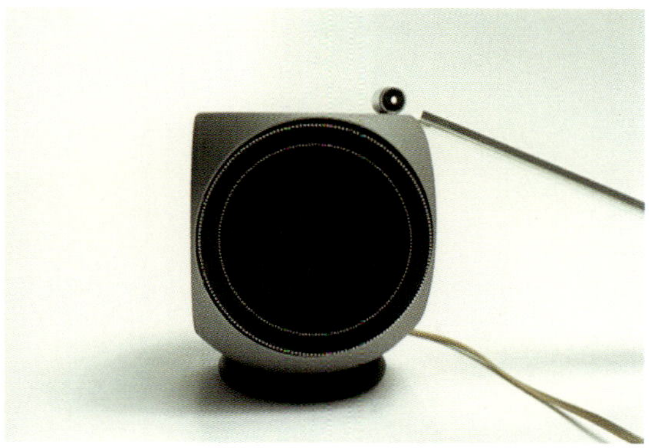

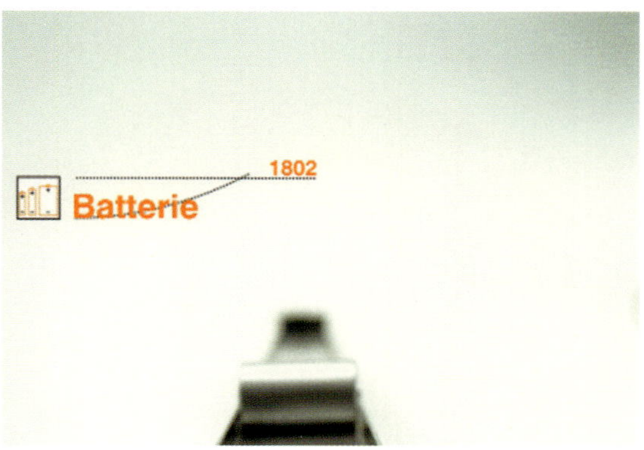

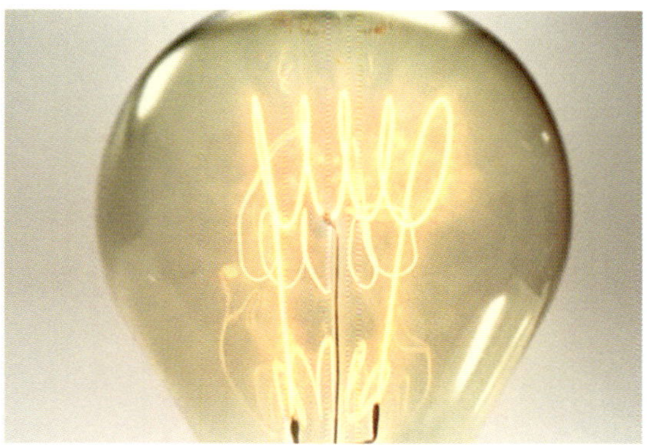

Auftraggeber Client
ZDF
Mainz, Germany

Projekt Project
ZDF – Unsere Besten „Die größten Erfindungen"
Vorspann
Opener

Design
Simon Mayer
Ralf Ott
Timo Fritsche
Heiko Leitsch
ACHT FRANKFURT GmbH & Co.KG
Frankfurt am Main, Germany
Frank Schneider
Filmstyler Pictures GmbH
Frankfurt am Main, Germany
Christiane Jähnel
ZDF – Corporate Design
Mainz, Germany

„Die größten Erfindungen" ist Teil einer Vorspannentwicklung der Sendereihe „ZDF – Unsere Besten" und wirft die Frage auf: Welche Erfindungen sind für die Menschheit am wichtigsten? Eine neue Erfindung funktioniert ohne das Wissen der vorherigen Erfindung nicht. Gedreht wurde vor weißem Hintergrund mit einer Motioncontrol-Kamera auf 35-mm-Film. Mit Hilfe einer spannend inszenierten Kettenreaktion repräsentieren ausgewählte Designobjekte und animierte Grafiken die großen Erfindungen anhand einer Timeline. Die unterschiedlichen Errungenschaften sollten die Zuschauer neugierig machen und sie animieren „die größte Erfindung" zu nominieren.

"The Greatest Inventions" is part of a title sequence developed for the broadcast-series "ZDF – Our Bests" and raises the question: Which inventions are most important to mankind? A new invention cannot function without the acquired knowledge of former inventions. The concept was realised with a motion-control-camera and shot on 35 mm film against a white background. In an excitingly staged chain reaction the greatest inventions are subjected like a timeline, represented by assorted design objects and motion graphics. The differences in achievement are supposed to animate the viewer to nominate "the greatest invention".

Auftraggeber Client
BMW Group
München, Germany

Projekt Project
BMW Group „Wie ein Fahrzeug entsteht"
www.bmwgroup.com/3Dprocess
3-D-Prozessdarstellung
3D process display

Design
Oliver Kratzer (Designkonzeption, Flash, 3D)
Andrea Sondermeier (Art Direction)
Silke Noetzel (Fachliche Konzeption, PM)
Ray Sono AG
München, Germany

Ziel war es, den Produktionsprozess in der Automobilfertigung der BMW AG vom Blechteil bis zum fertigen Fahrzeug für den Besucher der BMW Group Website erlebbar zu machen. Die Anwendung konzentriert sich auf die reine Darstellung des Prozesses; so realitätsnah wie möglich und so abstrakt wie nötig. Der Prozess steht im Fokus, nicht ein einzelnes Modell oder Werk. Die Modellierung des Automobils erfolgte daher auf Basis realer Konstruktionsdaten, die nachbearbeitet und auf die Optik technischer Zeichnungen reduziert wurden. Raum, Roboter und Werkzeuge wurden auf Basis von Bild- und Videomaterial nachmodelliert.

The goal of this application is to let visitors to the BMW Group website experience BMW's automobile production process – from sheet metal blanks to the completed vehicle. The application concentrates solely on representing the process – as close to reality as possible but as abstract as necessary to be generally valid. It is the process that is highlighted – not an individual model or plant. The automobile modeling reflects real construction data, reprocessed and reduced to resemble technical drawings. Space, robots and tools were modeled based on photo and video material.

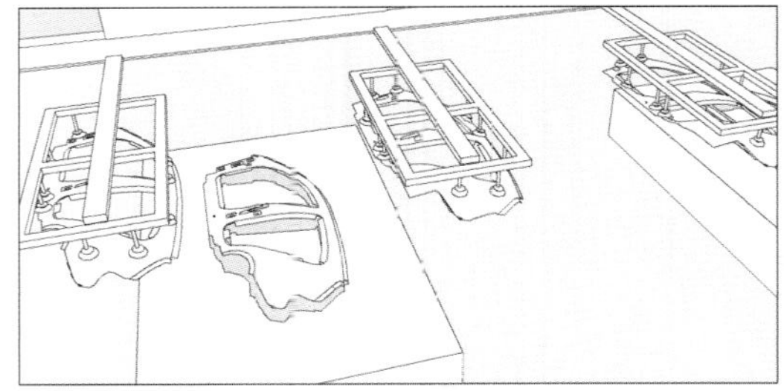

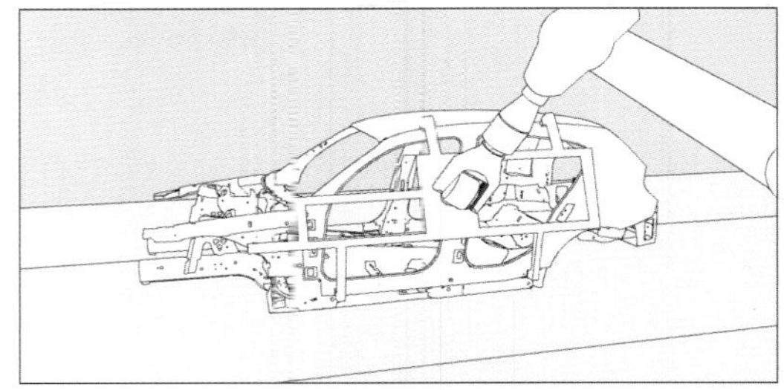

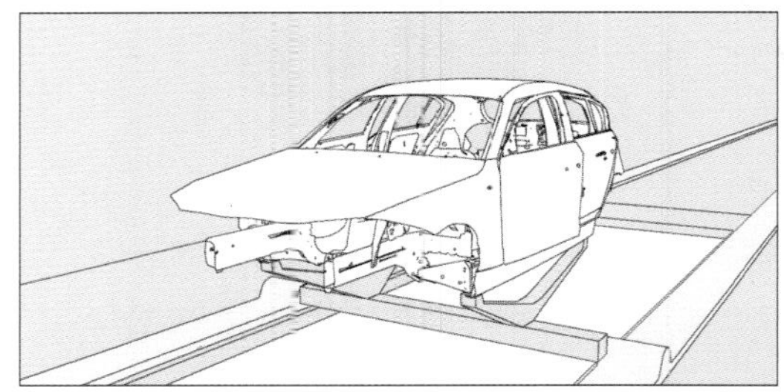

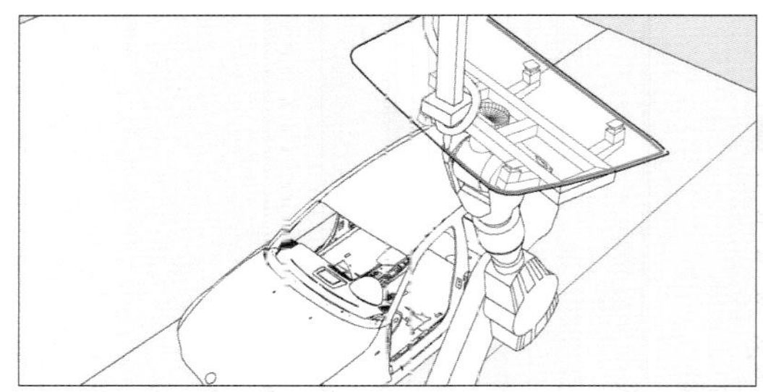

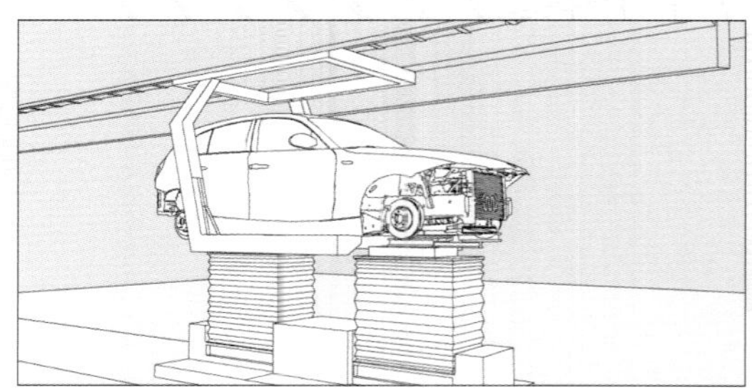

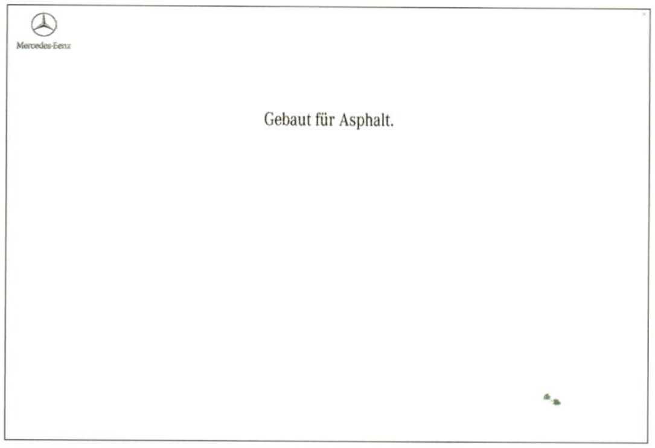

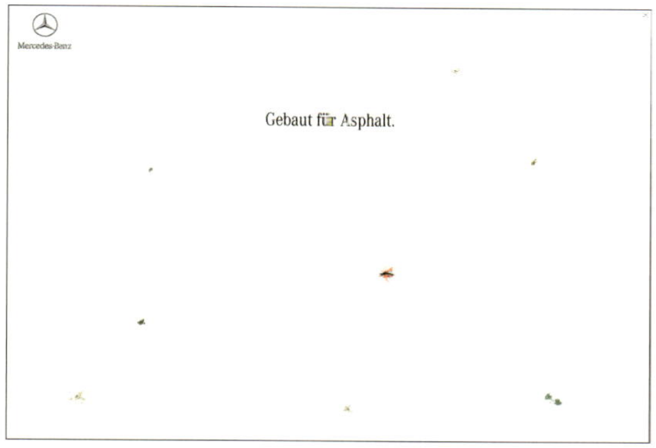

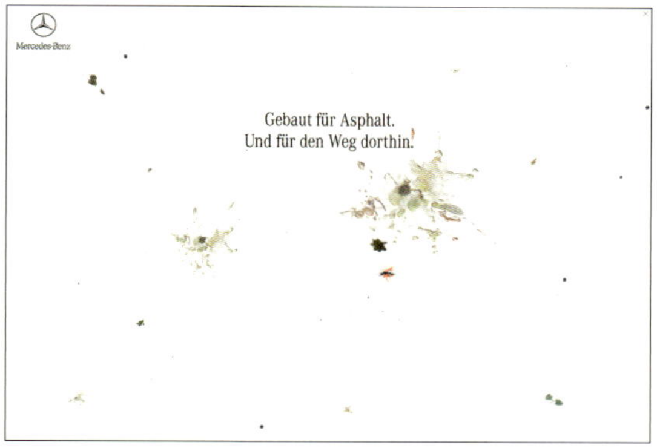

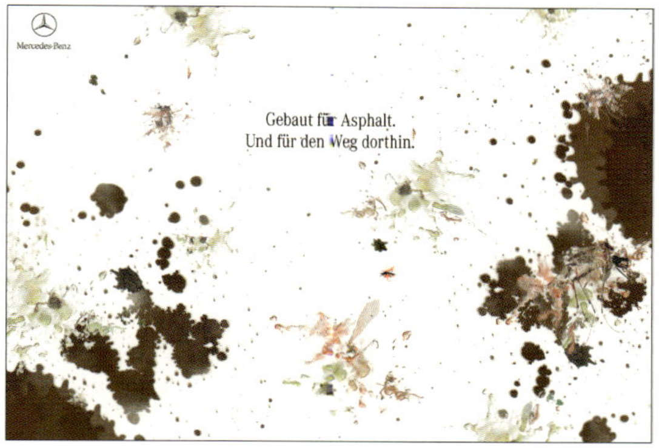

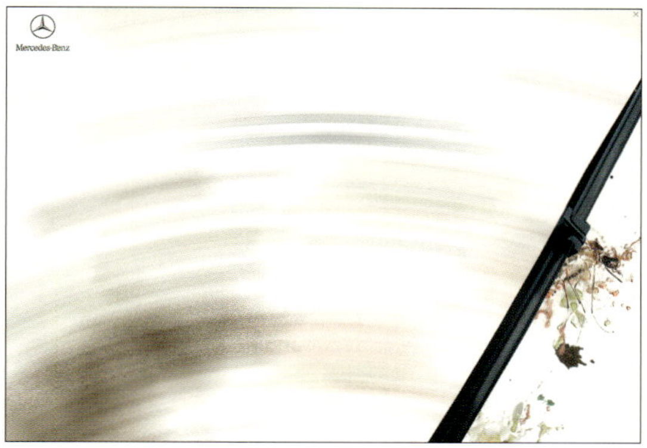

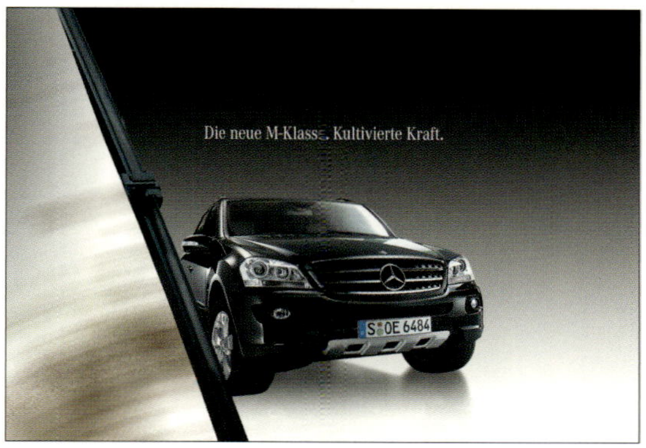

Auftraggeber Client
DaimlerChrysler Vertriebsorganisation Deutschland
Berlin, Germany

Projekt Project
Mercedes-Benz M-Klasse „Splatter"
Bildschirmschoner
Screensaver

Design
Dirk Kedrowitsch (GF)
Dirk Ollmann,
Daniel Richau (CD)
Kai Becker (AD)
Till Hinrichs (Grafik)
Benjamin Bruno (Text)
Rouven Laurien (Flash)
Matthias Mühlenhoff,
Claus Jacobsen (Account)
Klaus Burghauser,
Söhnke Wulff (Client)
Elephant Seven AG
Hamburg, Germany

Die Kampagne zur Markteinführung der neuen M-Klasse steht unter dem Motto: „Gebaut für Asphalt. Und für den Weg dorthin." Fernab typischer Fahrzeugpräsentation vermittelt dieser Screensaver die Kernaussagen dieser Devise – und weckt plakativ und lautstark Interesse, ohne das Fahrzeug sofort zu zeigen. Ursprünglich handelte es sich bei diesem Motiv um einen Banner mit erfolgreicher Klickrate. Grund genug, es den Anwendern als interaktives Geschenk unter www.mercedes-benz.de zur Verfügung zu stellen. Einerseits wurde die Homepage so um einen attraktiven Inhalt bereichert, andererseits eine Möglichkeit geschaffen, die M-Klasse auf heimischen Computern zu etablieren und die Vorfreude auf die Einführung zu steigern.

The campaign for the market launch of the new M-Class bears the strapline "Built for the road. And for getting that far." Taking a radically different approach to car advertising, the screensaver communicates the core elements of the strapline from a unique angle – generating interest in an immediate, bold way, without showing the vehicle instantly. This was originally a banner motif that generated lots of hits. And that was reason enough to make it available to users as an interactive gift at www.mercedes-benz.de. Firstly, this constituted an attractive addition to the homepage. And secondly, this made it possible to bring the M-Class to people's PCs and whip up enthusiasm ahead of the market launch.

Auftraggeber Client
Deutsche Energie-Agentur GmbH (dena)
Berlin, Germany

Projekt Project
Join the PowerScout
Internetwettbewerb
Online competition

Design
unit-berlin
Berlin, Germany

„Join the PowerScout" ist ein Internetwettbewerb zum Thema Energieverlust durch den Stand-by-Betrieb. Den Rahmen für den Wettbewerb bildet ein virtuelles Zimmer, das sich die Schüler der Klassen 6–10 bzw. Jugendliche im Alter von 10–16 Jahren selber einrichten und gestalten können. Dieses Zimmer wird von den Teilnehmern mit technischen Geräten bestückt, die entsprechend ihrer Funktion die verschiedenen Aufgaben des Wettbewerbs beinhalten. Werden die Aufgaben gelöst und als fertige Beiträge hochgeladen, entsteht ein jeweils einzigartiges Jugendzimmer, das spielerisch attraktiv die Wettbewerbsbeiträge in sich trägt.

"Join the PowerScout" is an online competition connected to the issue of wasted electricity and the value of stand-by devices. At its heart there is a virtual room that children between the ages of 10 and 16 can design and furnish in their own way. Participants fit the room with technical appliances that, according to their function, comprise the various tasks of the game. After completing the tasks and having uploaded their finished entries, participants are presented with a customized teenager's room – a light-hearted and appealing illustration of the competition entry.

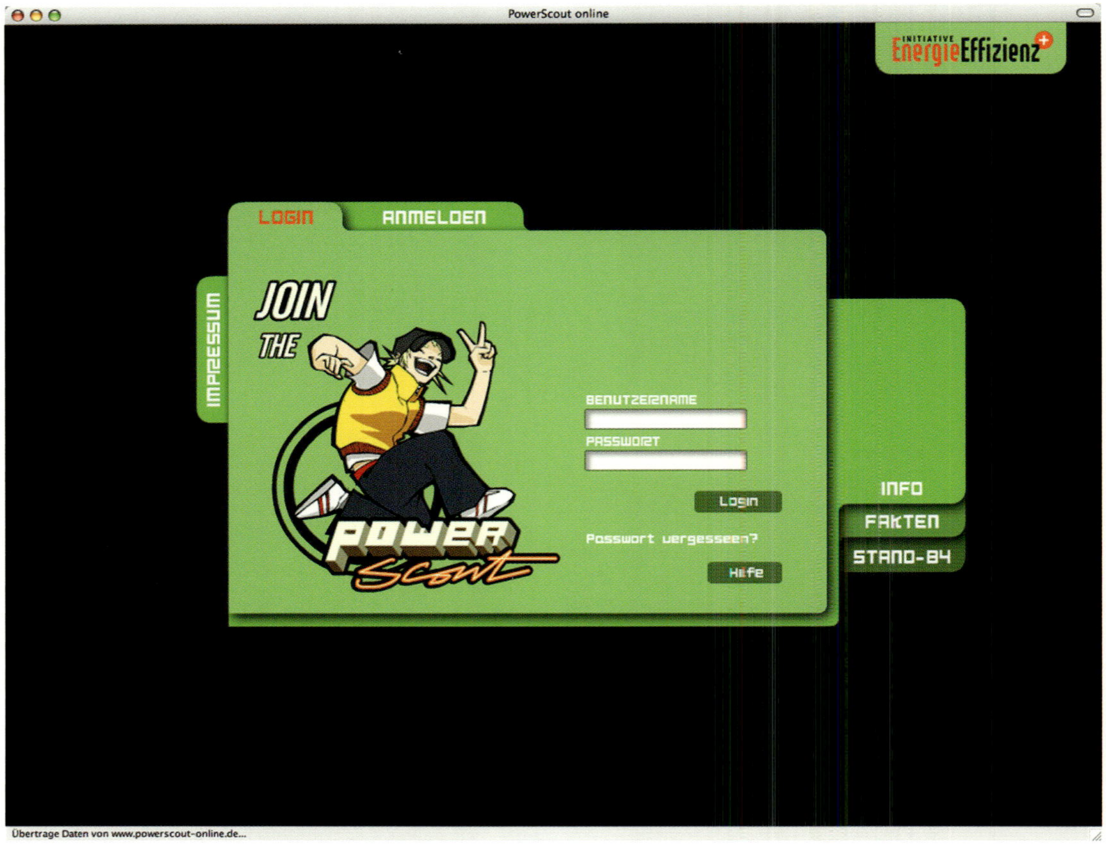

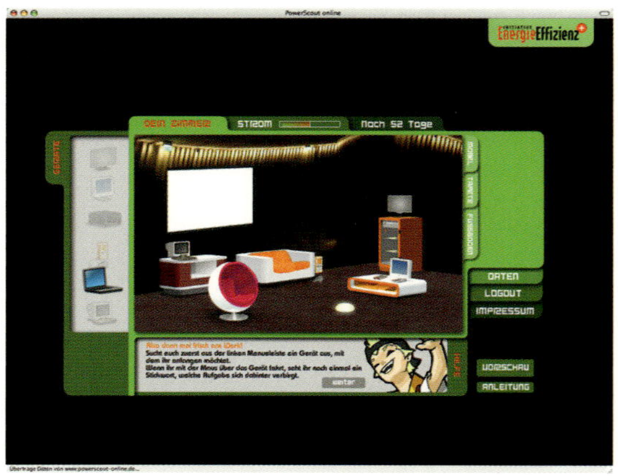

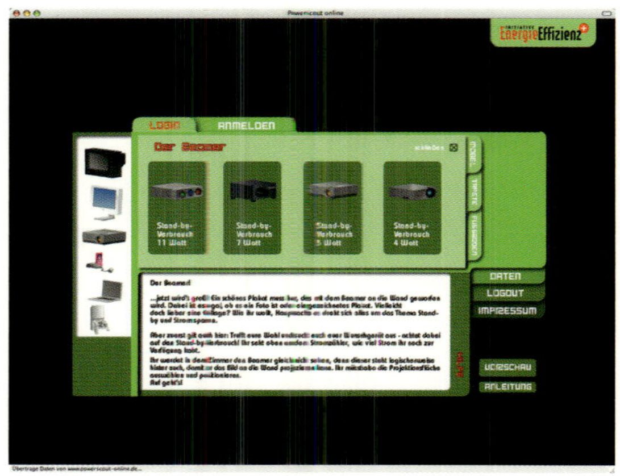

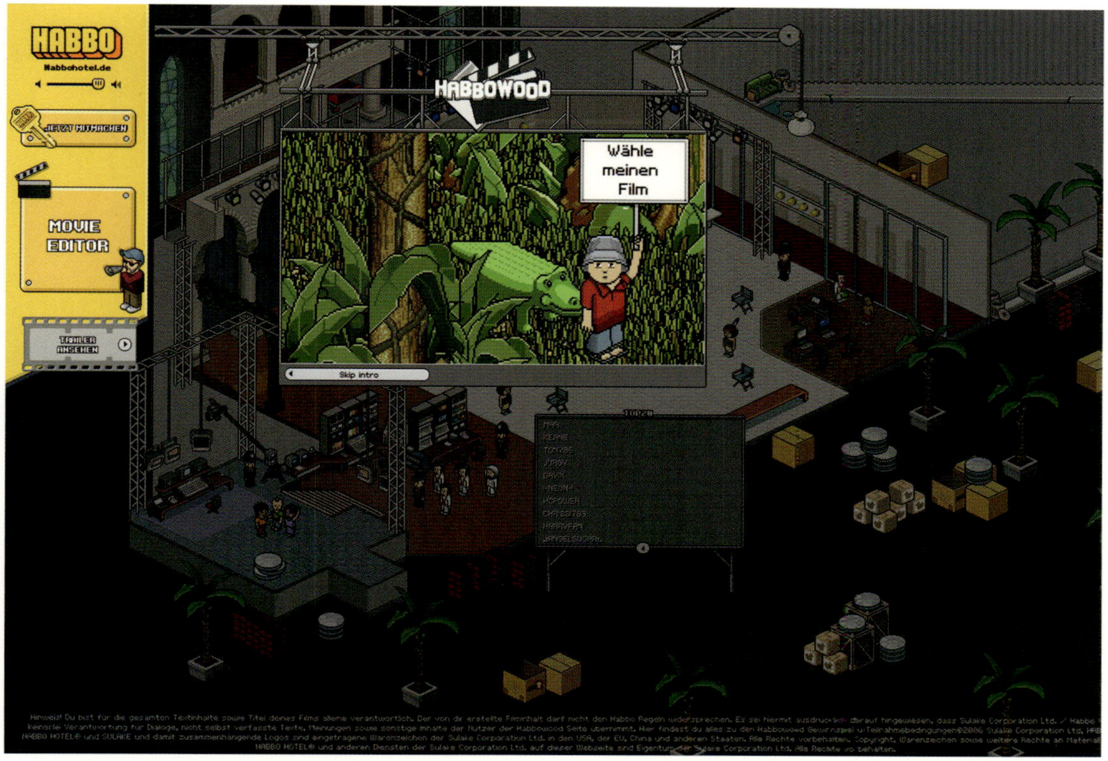

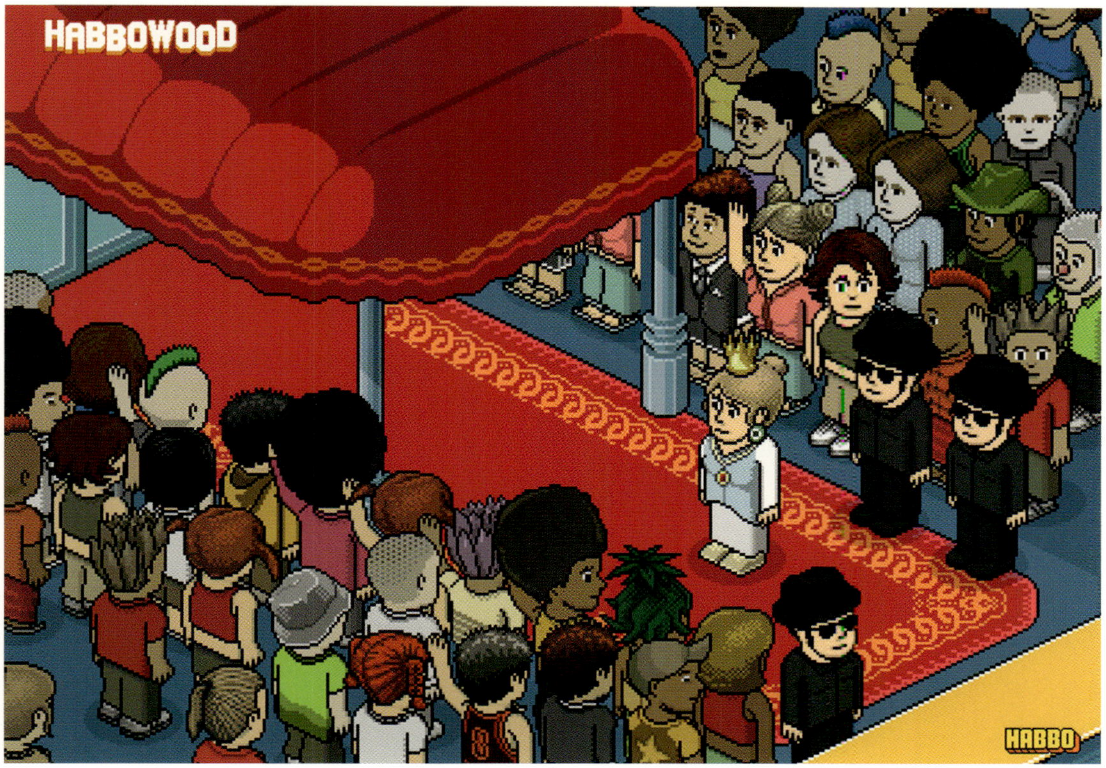

Auftraggeber Client
Sulake Deutschland GmbH
Frankfurt am Main, Germany

Projekt Project
Habbowood Campaign
Online-Kampagne
Online campaign

Design
Sami Lanu
Sulake Corporation Ltd
Helsinki, Finland

Das Habbo Hotel, eine der weltweit größten und beliebtesten Online-Erlebniswelten für Jugendliche, stellt den ultimativen Film- und Promotingcontest 2006 vor: „Habbowood"! Die Grundidee von „Habbowood" 2006 besteht in dem Vorhaben, dass die Teilnehmer sowohl die Rolle als Regisseur als auch die des Promoters übernehmen, fast wie im wahren Hollywood. Die Aufgaben sind einfach: die Habbo-Mitglieder entwerfen ihr eigenes Drehbuch, vergeben die Filmrollen und gestalten einen individuellen Film, indem sie die dazugehörigen Habbo Hotel-Filmhilfsmittel nutzen. Sie werben für ihren eigenen Film im Habbo Hotel. Wer die meisten Fans hat, gewinnt.

The basic idea behind the Habbowood Film Challenge was to offer Habbo users a new possibility to use their creativity. The participants were given the chance of being real film directors with the opportunity to name their movie and characters, write the screenplay and shape the genre by choosing a suitable soundtrack. Additionally, the users had to promote their movie. The short films included different setups and moods ranging from creepy horror to romance. Each user could create as many films as he or she liked. The films could also be sent to friends. The main goal was to get as many votes in favour of one's film as possible in order to win the Habbowood Film Challenge.

Auftraggeber Client
adidas AG
Herzogenaurach, Germany

Projekt Project
adidas Originals
Website

Design
Bejadin Selimi
NEUE DIGITALE
Frankfurt am Main, Germany

Tanzende Models führen in einem Breitband-Webspecial die Frühjahrs/Sommer-Kollektion 2005 von adidas Originals vor. Die Styles präsentieren sich im großformatigen Fullscreen-Video, dessen Auflösung höher ist als im Fernsehen.
Der User kann dabei aktiv werden und die Kleidungsstücke mit 3-D-Tiefeneffekt oder im Detail betrachten. Die interaktiven Videosequenzen setzen im Qualitäts-Ladezeit-Verhältnis neue Maßstäbe.

Dancing models present the spring/summer collection 2005 from adidas Originals on the Internet – in broadband full-screen video running at an even higher resolution than on television. The user can interact with the website and is able to view the styles in 3D or in detail. The interactive video sequences set new benchmarks in the online world, especially in regard to the ratio of high quality level to low loading time.

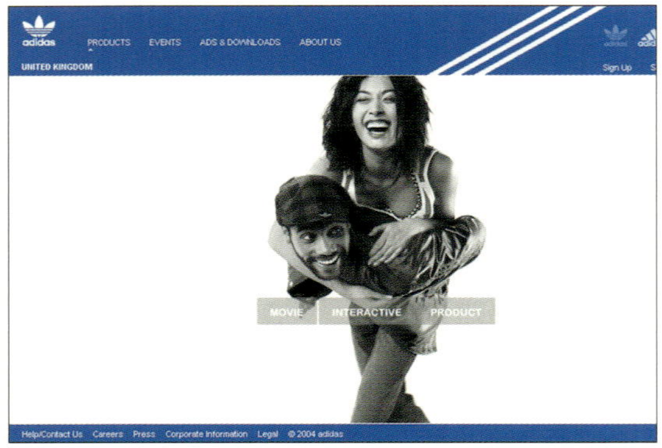

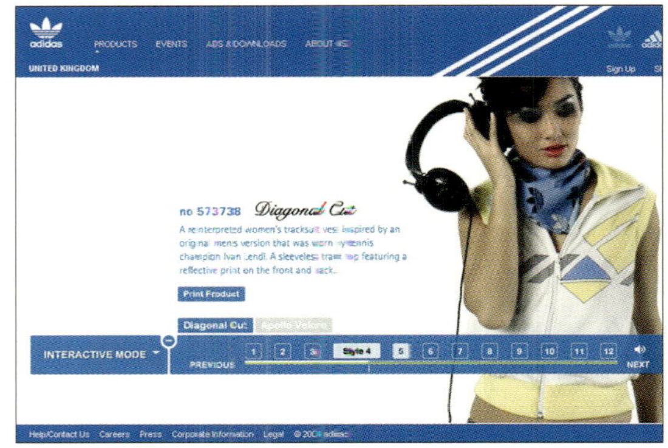

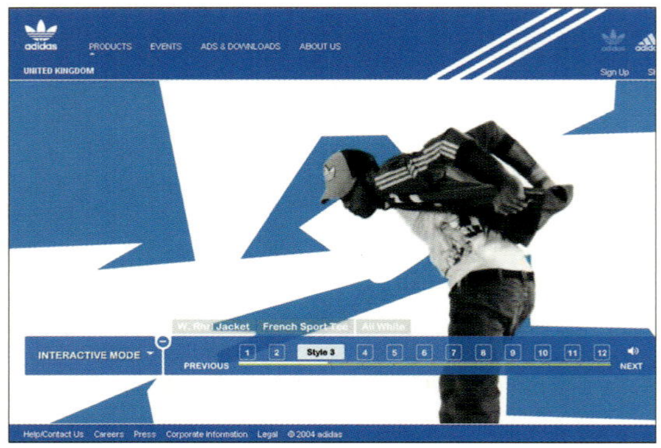

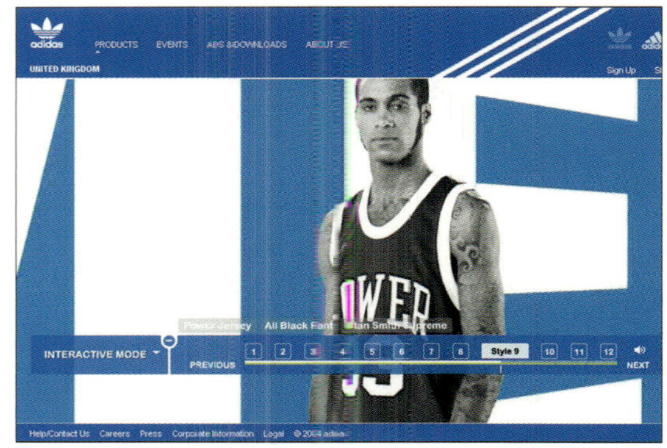

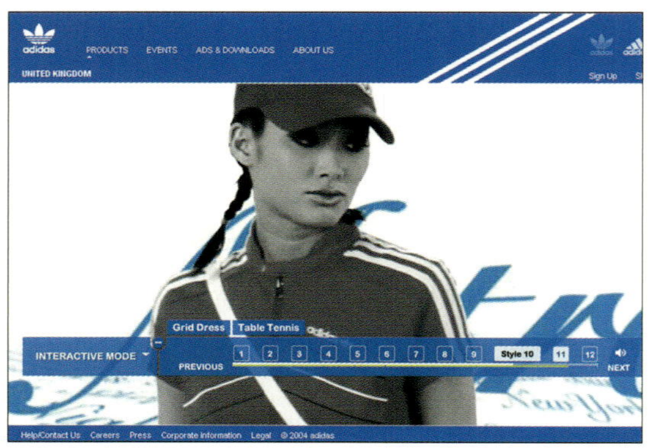

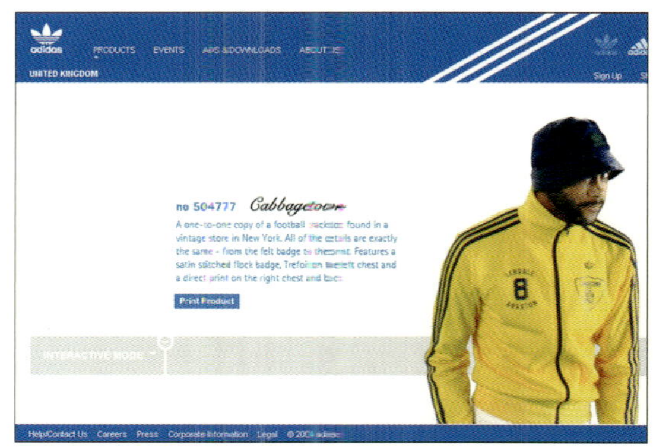

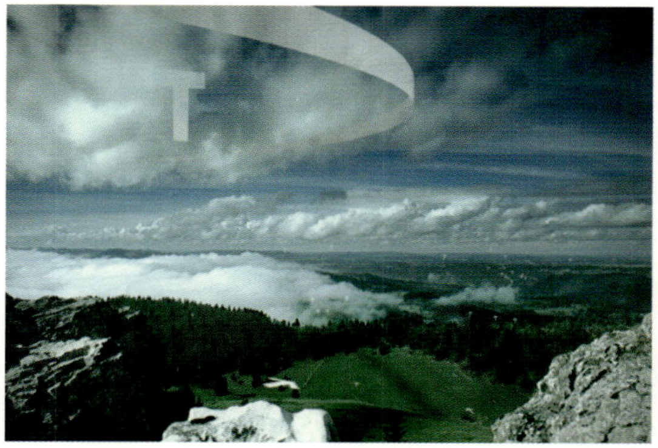

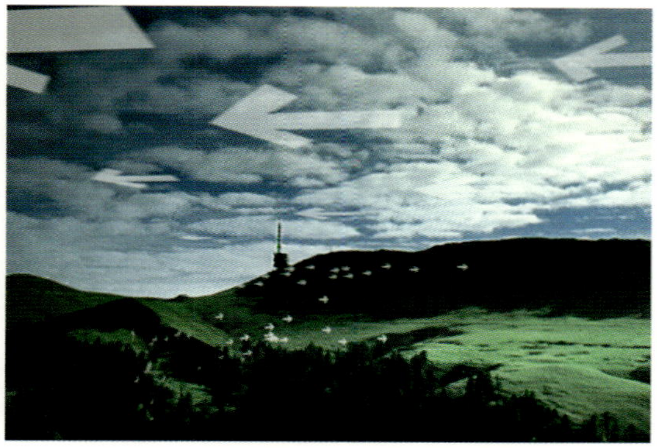

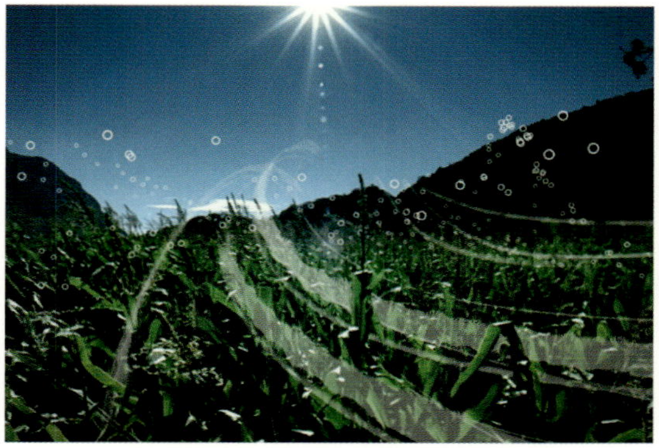

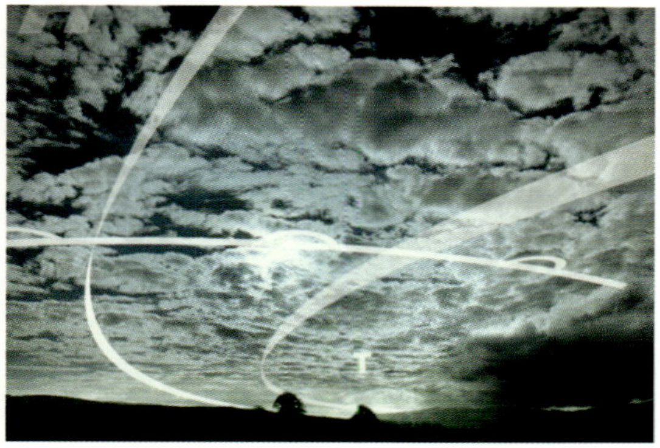

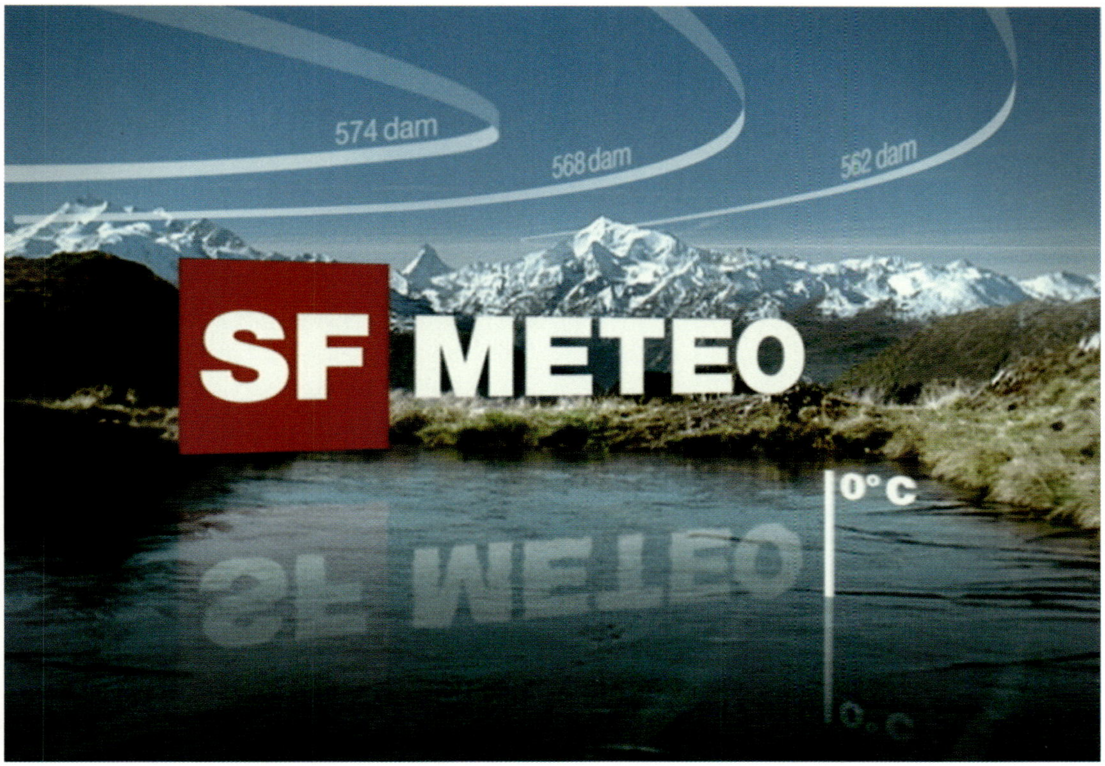

Auftraggeber Client
Schweizer Fernsehen
Zurich, Switzerland

Projekt Project
SF METEO
Vorspann
Opener

Design
Simon Mayer
Heiko Leitsch
Timo Fritsche
ACHT FRANKFURT GmbH & Co.KG
Frankfurt am Main, Germany
Frank Schneider
Filmstyler Pictures GmbH
Frankfurt am Main, Germany
Kevin Blanc
Schweizer Fernsehen
Zurich, Switzerland

„SF METEO" ist die dreiminütige Wettersendung des Schweizer Fernsehen und mit durchschnittlich 64 % MA eine der erfolgreichsten Sendungen von SF in der Primetime. Die Grundidee der Signation besteht darin, die Analytik und Interpretation von Wetterdaten auf der einen Seite mit den ausdrucksstarken Panoramen und Szenarien der Schweiz auf der anderen Seite in Verbindung zu setzen. So wurden landschaftlich aussagekräftige und reizvolle Szenen mit Hilfe der SF Meteorologen dahingehend überzeichnet, um für den Zuschauer unsichtbare Effekte wie Hoch- oder Tiefdruck, Temperaturen, Windströmungen etc. grafisch sichtbar zu machen.

"SF METEO" is Swiss Television's (SF) 3-minute weather broadcast and with a market share of 64 % one of SF's most successful shows during prime time. The basic idea behind the title sequence is the combination of the factual analysis and interpretation of weather-based data with the expressive panoramic views and scenarios of the Swiss landscape. With help from SF's meteorologists, images of attractive and significant landscapes were graphically complemented to visualize weather effects such as high or low pressure, temperature, air currents, etc.

multiple winner: digital media animation, digital media interface

Auftraggeber Client
Samsung Electronics
Suwon-city, South Korea

Projekt Project
FLOU
Home Gateway Interface

Design
Yi-sak Park
Si-Hyung Lee
Ji-Hoon Kim
Jung-Won Kang
So-Young Shin
Samsung Electronics
Seoul, South Korea

multiple winner

FLOU fungiert als ein Home Gateway, indem es die Funktionen für Sprache (Telefonie), Video (TV) und Daten (Internetdienste) umfassend implementiert. Die Benutzeroberfläche ermöglicht es dem Anwender, in ein bestimmtes aktives Menü zu springen. Hierdurch wird die Bedienung vereinfacht, da der Benutzer nicht mehr nach dem entsprechenden Menü suchen muss. Weil das Layout des Bildschirms frei kombiniert werden kann, können gleichzeitig zwei weitere Funktionen ausgeführt werden.

FLOU serves as a home gateway by implementing the functions of voice (telephony), video (TV), and data (Internet service) comprehensively. As regards the interface, it enables the user to jump to the specific menu activated according to user requirements. This way the inconvenience of operation is reduced and the operation as such is simplified, since the user no longer needs to search around for the menu. Additionally, two or more functions can be executed simultaneously on one screen because the layout of the screen can be combined freely.

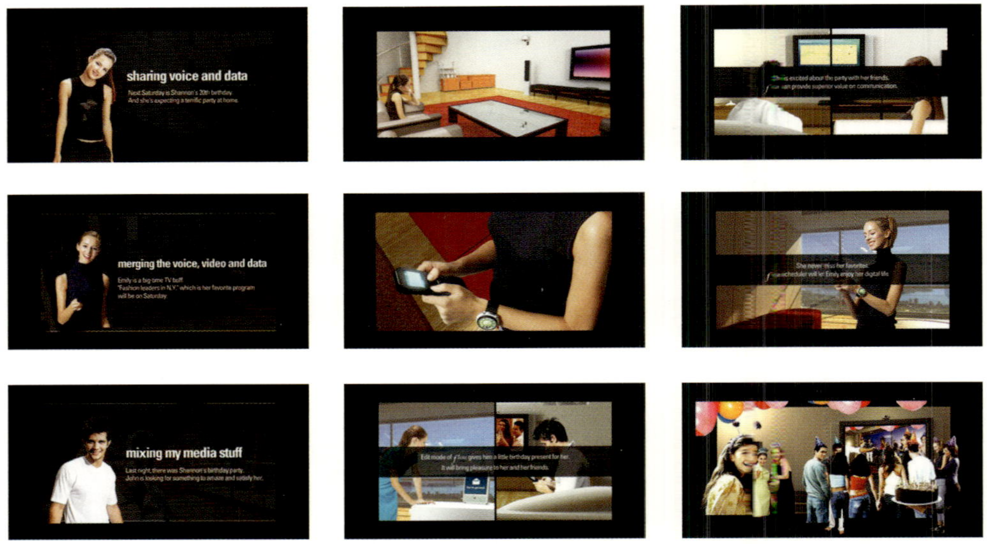

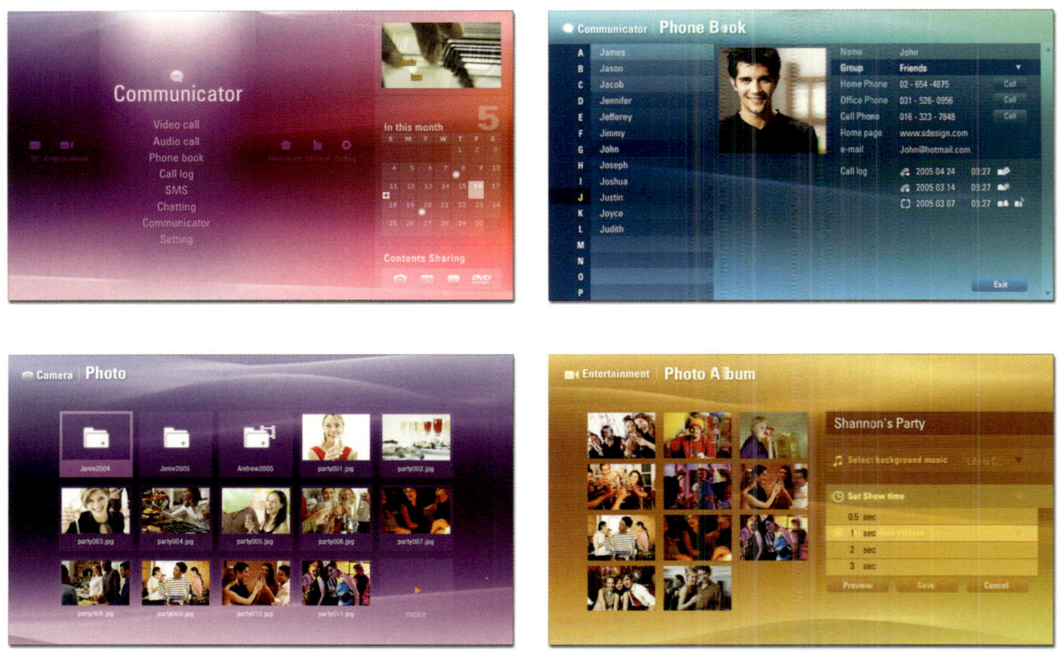

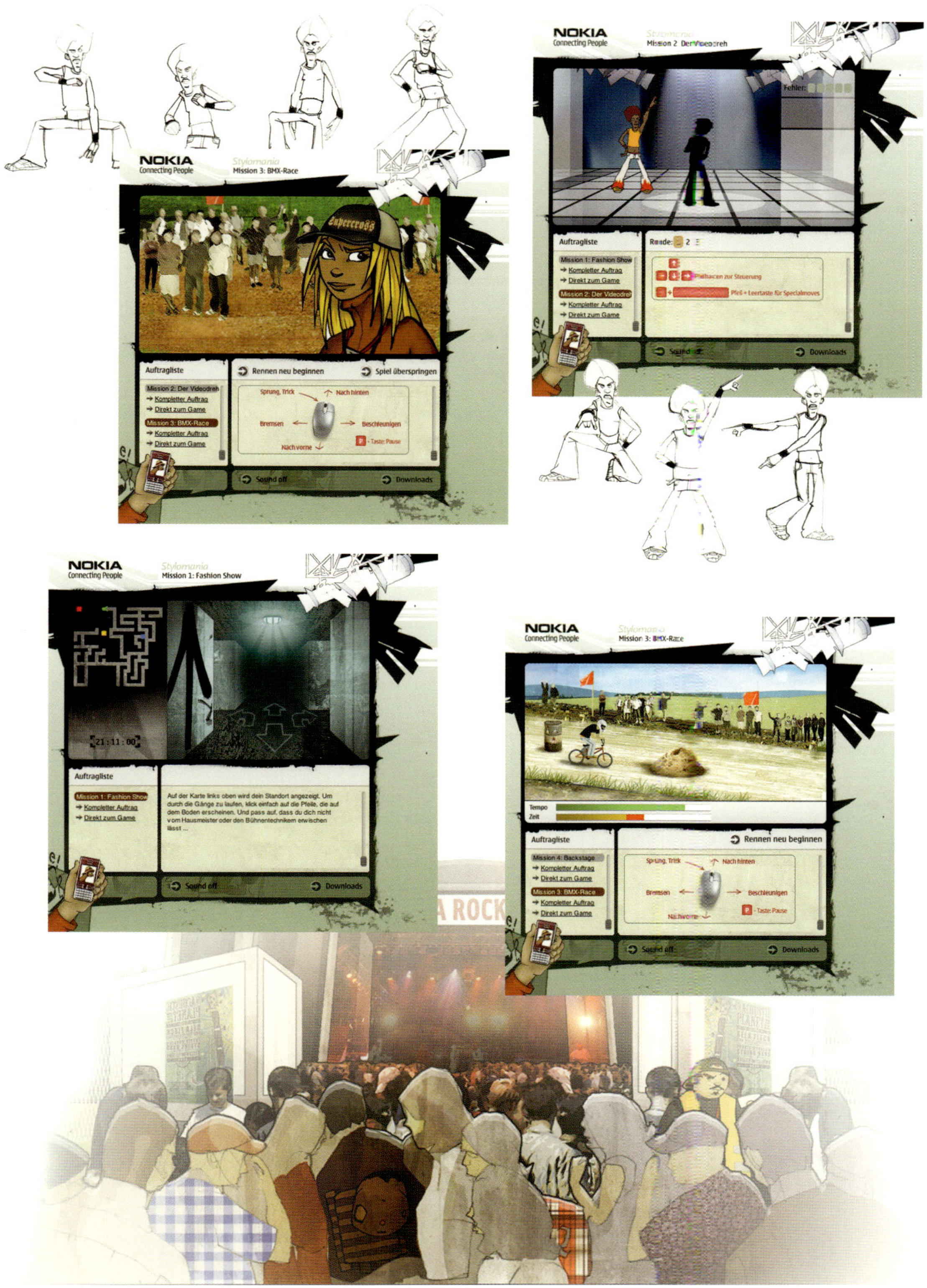

Auftraggeber Client
Nokia GmbH
Düsseldorf, Germany

Projekt Project
Stylomania
Microsite

Design
denkwerk
Köln, Germany

multiple winner

„It's a long hot summer!" Stylomania ist ein interaktives Spiel, das denkwerk zur Trafficsteigerung im Sommer entwickelt hat. Dabei steht das Nokia 3230 im Vordergrund. Die spielerische Darstellung der Produktvorteile verknüpft die Interessen der Zielgruppe mit den Funktionen und dem Image des Mobiltelefons. Trendbewusst und innovativ spielt sich der Nutzer in mehreren Episoden durch vier Städte. Um zu gewinnen, müssen einige Aufgaben gelöst werden – mit Unterstützung des Nokia 3230. Der Kommunikationsansatz konnte marken- und zielgruppenadäquat transportiert werden. Seit Launch verzeichnet die Stylomania Microsite eine hohe Zugriffrate.

"It's a long hot summer!" Stylomania is an interactive game, which was produced by denkwerk to increase the traffic during the less frequented summer season. The focus lies on the Nokia 3230. The playful description combines the interests of the target group with the function and image of the mobile phone. The user has to play several episodes while pacing through four European cities. In order to win the user has to solve several tasks – the Nokia 3230 assists. The communication approach was adequately transported to the brand and target group by using the online media. Since its launch, the microsite registers high access rates.

Auftraggeber Client
GM Europe / Adam Opel GmbH
Rüsselsheim, Germany

Projekt Project
Opel Antara GTC Reveal Film "The Fantasy Ride"
Multimediales Event-Design
Multimedia event design

Design
Norbert Fischbach
Simone Heckmann
pmg profile marketing group
Wiesbaden, Germany
Aleksej Schön
Mars Agardtha
Ben Schröteler
Roger Tilch
Regina Teichs
CON circus of now
Berlin, Germany
Philipp E. Kümpel
Andreas Moisa
Film Scoring
Leipzig, Germany

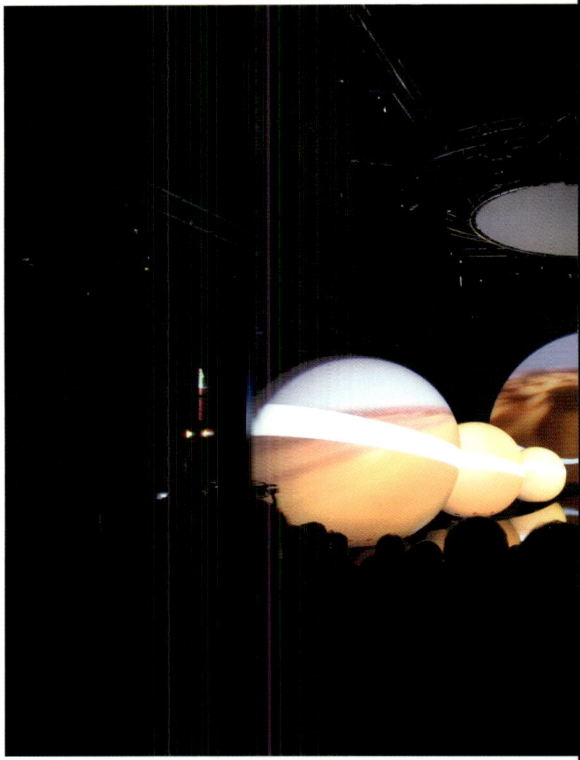

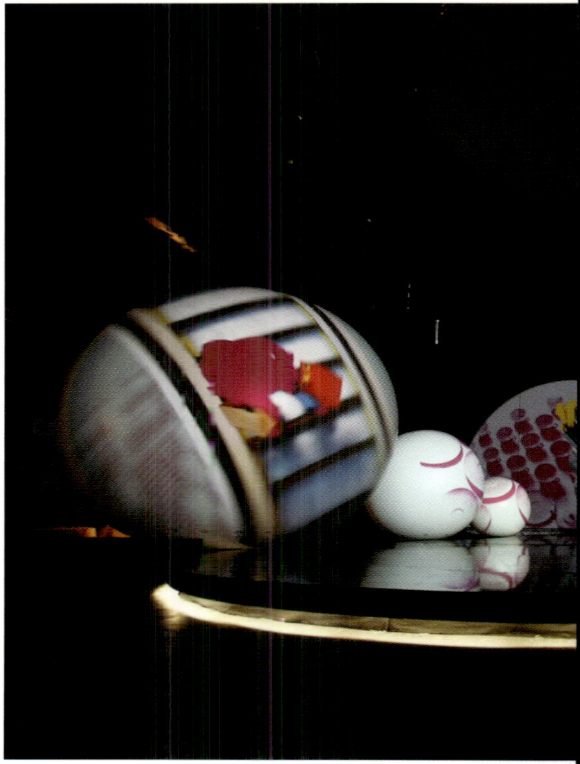

Zur IAA präsentierte General Motors (GM) internationalen Medienvertretern eine SUV Crossover Coupé Studie. Die Aufgabe: die Weltpremiere Opel Antara GTC passend zur neuen Formensprache inszenieren. Markanter, dynamischer, emotionaler. Den Designanspruch erlebbar und den Markenwert sichtbar werden lassen.
Ein dreidimensionales Broadcast-Design transportierte die Botschaft überzeugend. Rechts und links der halbrunden Bühnenfront ruhten zwei imposante Kugeln mit je fünf Metern Durchmesser. Auf einer Tiefe von 20 Metern wurden perspektivisch versetzt vier weitere Kugeln positioniert. Den Abschluss bildete eine gebogene, acht Meter hohe Rückleinwand. Diese ausdrucksstarken Elemente schufen ungewöhnliche Projektionsflächen für eine dynamische Show: Die räumlich angeordneten Projektionskörper und der Spiegeleffekt des Bühnenbodens ermöglichten die Darstellung eines raumgreifenden Videodesigns aus 360-Grad-Bildwelten und „virtueller Realitäten im Raum" – mit beeindruckender Tiefe und Qualität. Rasante Sportszenen, Eintauchen in die Welt des Power-Shoppings, meditative Wellness-Impressionen und urbane Lifestyle-Projektionen stimmten den Betrachter auf das zentrale Thema ein: den Opel Antara GTC in seiner Welt.

General Motors (GM) unveiled a SUV crossover coupé concept to international media representatives at the IAA motor show. The aim was to stage the world premiere of the Opel Antara GTC in keeping with the new design language – more distinctive, more dynamic and more emotional – and visually convey the design claim and brand value in a palpable way.
A three-dimensional broadcast design proved a compelling way to convey the message. There were two impressive-looking, five-meter wide balloons to the right and left of the semicircular stage. Four more balloons were positioned about 20 meters back and staggered at different eye levels. An arched, 8-meter-tall projection screen commanded the back of the stage.
These highly expressive elements created some unusual projection areas for a dynamic show. The staggered balloons used as projection surfaces and the mirror effect of the stage floor produced a 360-degree, all-encompassing video presentation and created "3D virtual realities" of impressive depth and quality. Action-packed sports scenes, forays into the world of power shopping, meditative wellness sequences and urban lifestyle projections all set the scene for the central theme of the show: the Opel Antara GTC in its natural environment.

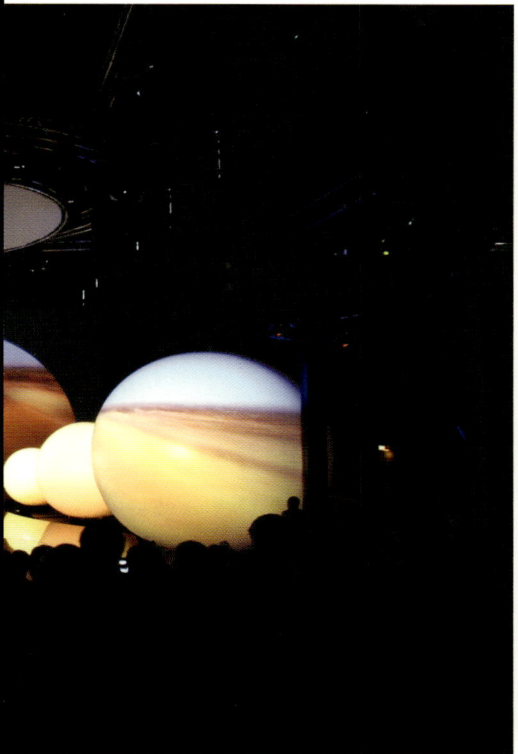

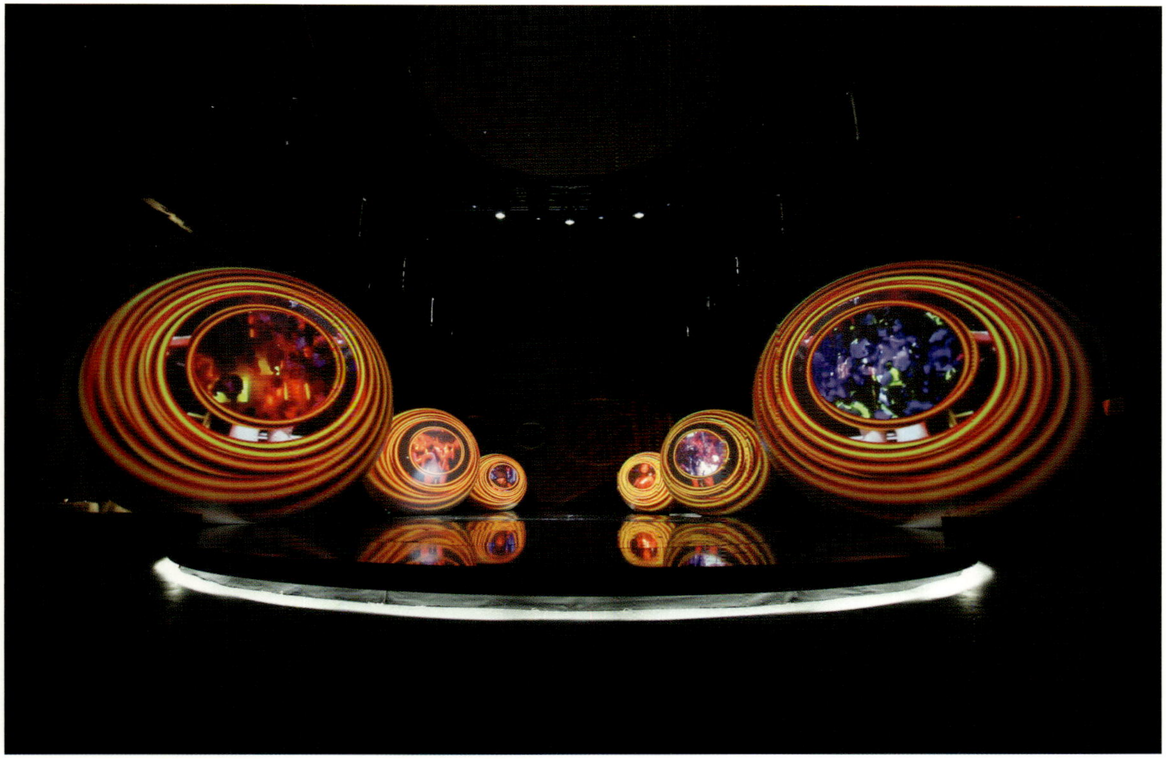

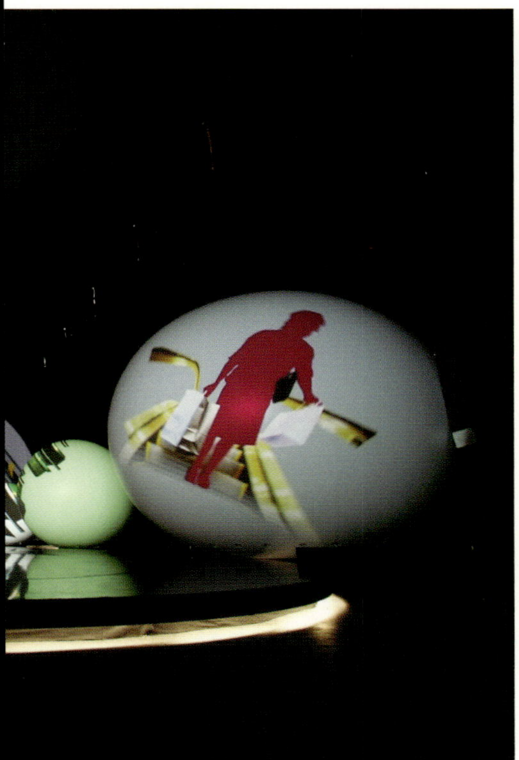

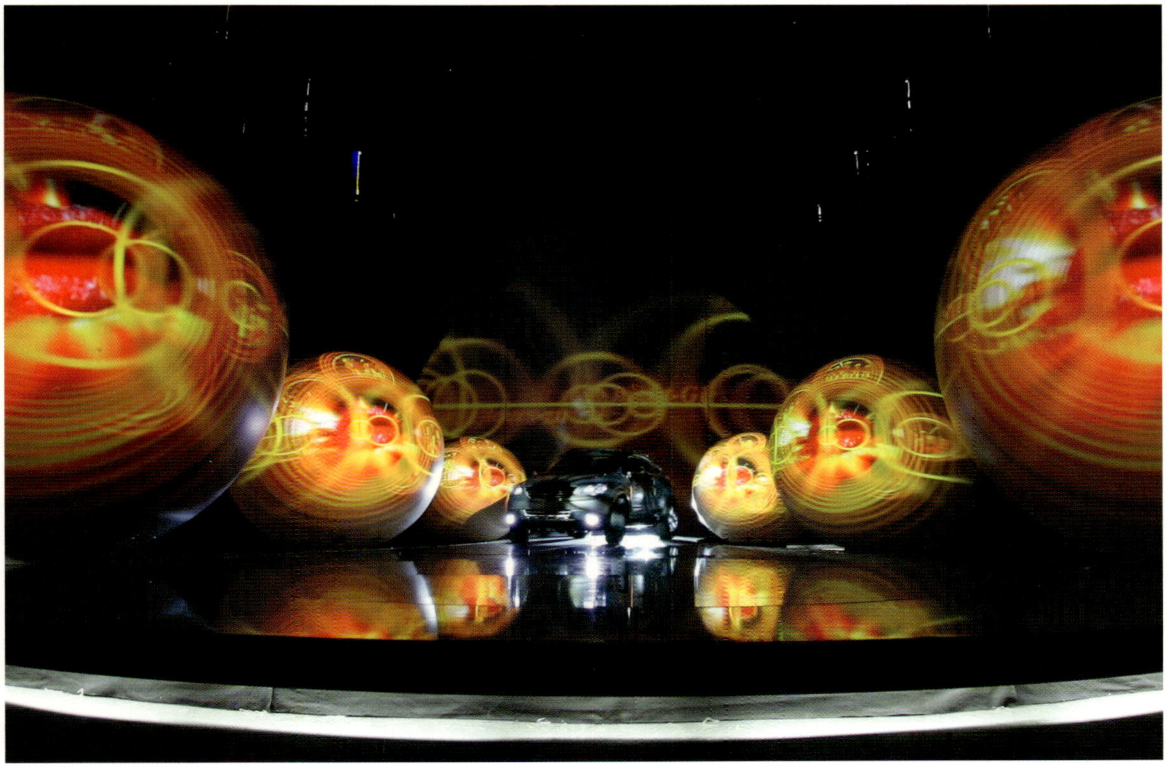

Auftraggeber Client
Wolff Kommunikation
Frankfurt am Main, Germany

Projekt Project
www.wolff-kommunikation.de
Website

Design
Carsten Wolff
Christiane Wolff
Thomas Rott
Wim Schafspelz
Wolff Kommunikation
Frankfurt am Main, Germany

Die Website soll einen repräsentativen und gut strukturierten Überblick über die realisierten Projekte unseres Designbüros geben und durch die Atmosphäre einen Eindruck von der Qualität unserer Arbeit vermitteln.

The website should provide a representative and well structured overview of the realised projects of our design office and give an impression of the atmosphere and quality of our work.

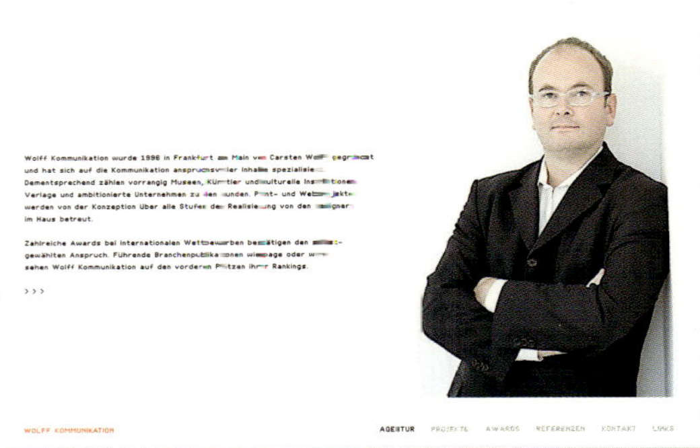

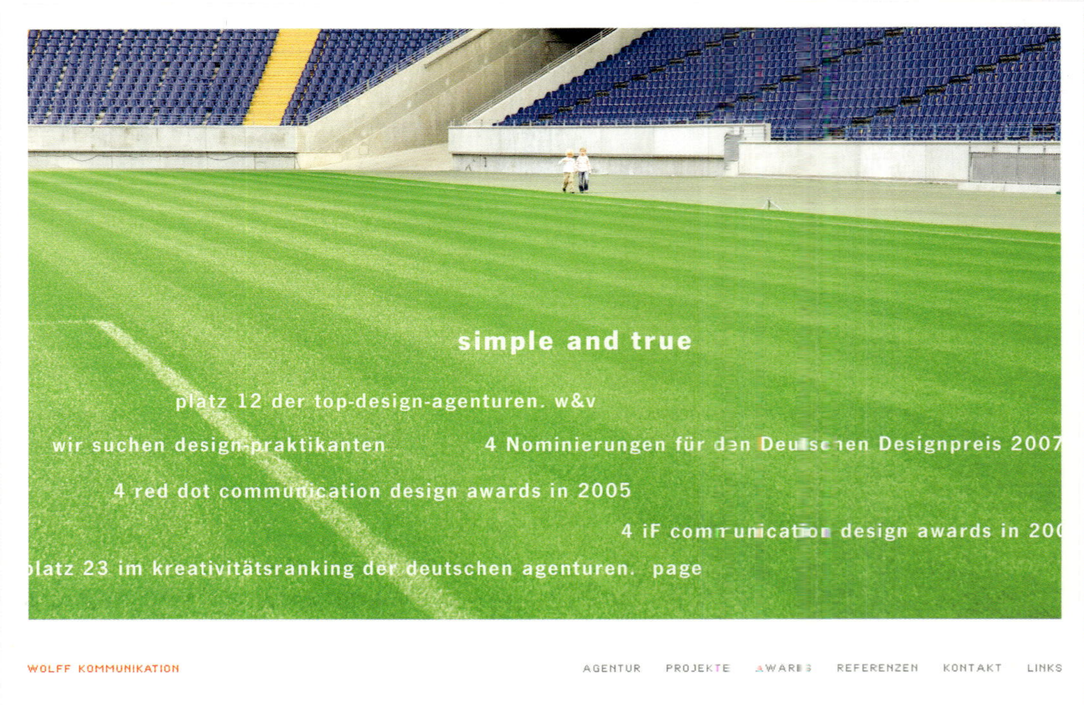

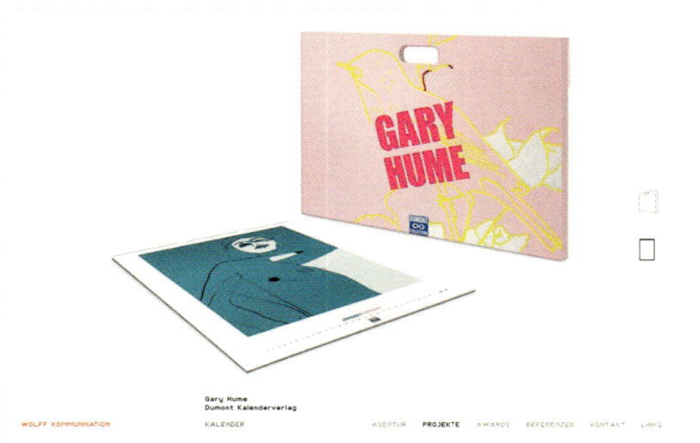

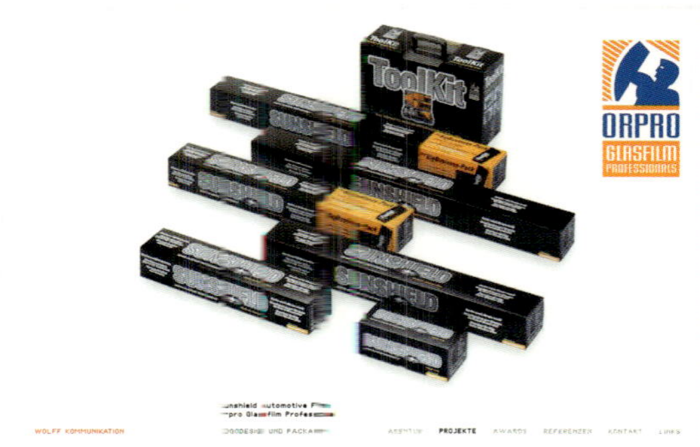

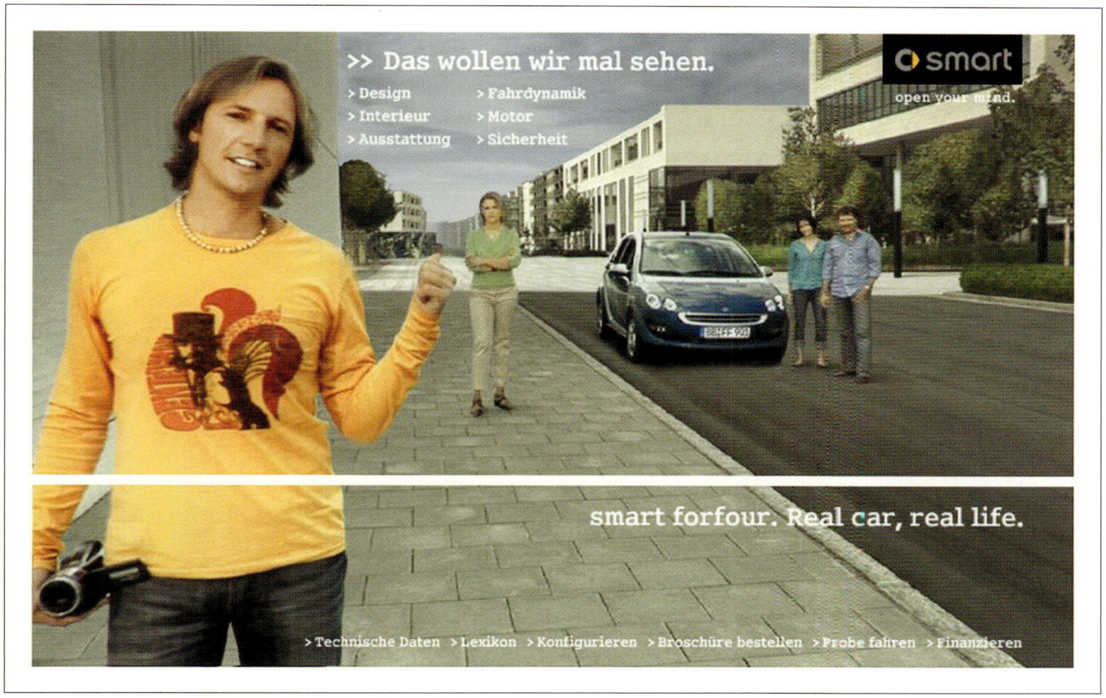

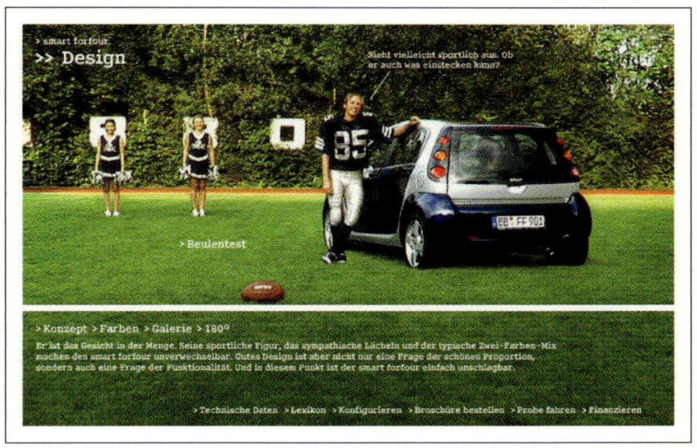

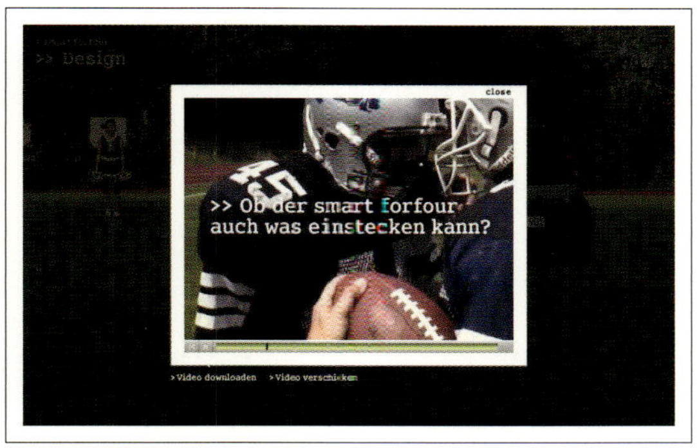

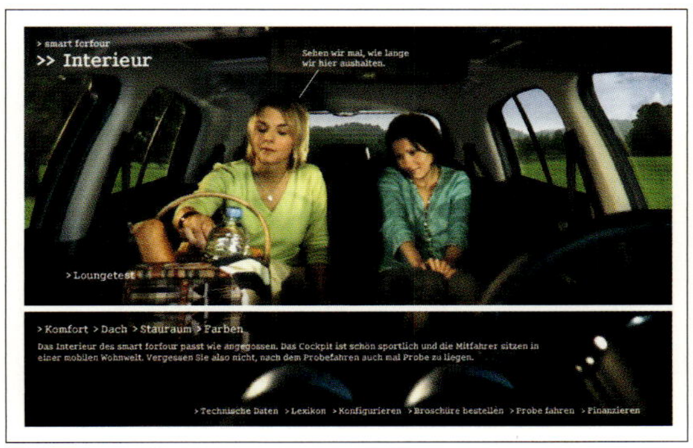

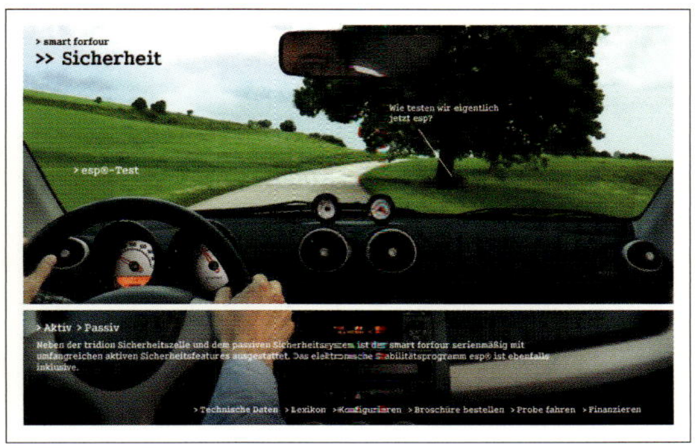

Auftraggeber Client
smart GmbH
Böblingen, Germany

Projekt Project
smart forfour
Website

Design
Anita Ragette (Producer)
Andrea zur Strassen (Production Manager, Creativity in Action, London)
Emil Graffmann (Director)
Oliver Kochs (Camera, OKAY Film, Munich)
Markus Nikowitsch (Sounddesigner)
Stefan Stabenow (Script)
Florian Eder (Storyboard)
Oliver Hering (Creative Director)
Georg Streibl (Art Director)
Richard Gabler, Mathias Temmen (Designer)
Christoph Kipp (Technical Director)
Robert Schlittenbauer (Account Supervisor)
Kriz Eideloth (Programmer)
H2OMEDIA AG
München, Germany

Die Webpage www.smart.com/forfour ist die Online-Fortsetzung der smart Kampagne „real car, real life". Die Internetkampagne startete aber nicht mit der Inszenierung des Fahrzeugs, sondern mit einem Casting. Denn die Darsteller sollten nicht nur authentisch wirken, sondern wirklich echt sein. In einem vorherigen Casting wurden also Sandra, Olivia, Jean-Pierre und Peter gefunden. Die vier Auserwählten sollten vor laufender Kamera den smart forfour auf Herz und Nieren testen und den Besucher durch die ganze Webpage begleiten. Folgerichtig wurde die Seite als eine einzige Kamerafahrt konzipiert – zum interaktiven Vor- und Zurückspulen.

The web page www.smart.com/forfour constitutes the online extension of the smart campaign "real car, real life". However, the Internet campaign did not initially start with focussing on the vehicle itself, it rather began with a casting. The site's presenters should be authentic and real. In the casting, Sandra, Olivia, Jean-Pierre, and Peter were found. They should put the smart forfour to the acid test in front of rolling cameras, and guide users through the entire web page. Consequentially, the page was designed as one single camera shot which can be forwarded and rewound interactively, giving its navigation a cinematic touch.

Auftraggeber Client
Hessischer Rundfunk
Frankfurt am Main, Germany

Projekt Project
Das Ende des Schweigens
Webspecial

Design
Dr. Anja Rau (Account)
Layla Keramat (CD)
Nicole Holzenkamp (AD)
Karsten Gerigk (Multimedia)
Tobias Kirchhofer (Producer)
BlueMars – Gesellschaft für digitale Kommunikation mbH
Frankfurt am Main, Germany

Im Frankfurter Auschwitz-Prozess von 1963 bis 1965 erfuhr die Weltöffentlichkeit erstmals Einzelheiten der Massenmorde und Verbrechen im Vernichtungslager Auschwitz. Das multimediale Webspecial „Das Ende des Schweigens" dokumentiert die Äußerungen von Überlebenden des Holocaust und die Versuche der Angeklagten, sich zu verteidigen. 67 interaktive Text-Ton-Bild-Collagen lassen scheinbar Unfassbares erlebbar werden. In einer Website, die durch die ungeschminkte Konfrontation mit dem Material ein intensives emotionales Erlebnis entstehen lässt.

It was not until the Frankfurt Auschwitz trials, which took place from 1963 to 1965, that the world first learned all the details involving the mass murders and crimes committed at the Auschwitz extermination camp. "The End of Silence" is a multimedia web special that documents statements made by Holocaust survivors and the attempts by the accused to defend themselves. Sixty-seven interactive collages of texts, sounds and images allow the seemingly incomprehensible to be experienced in a website whose unvarnished confrontation with the material produces an intense and emotional experience.

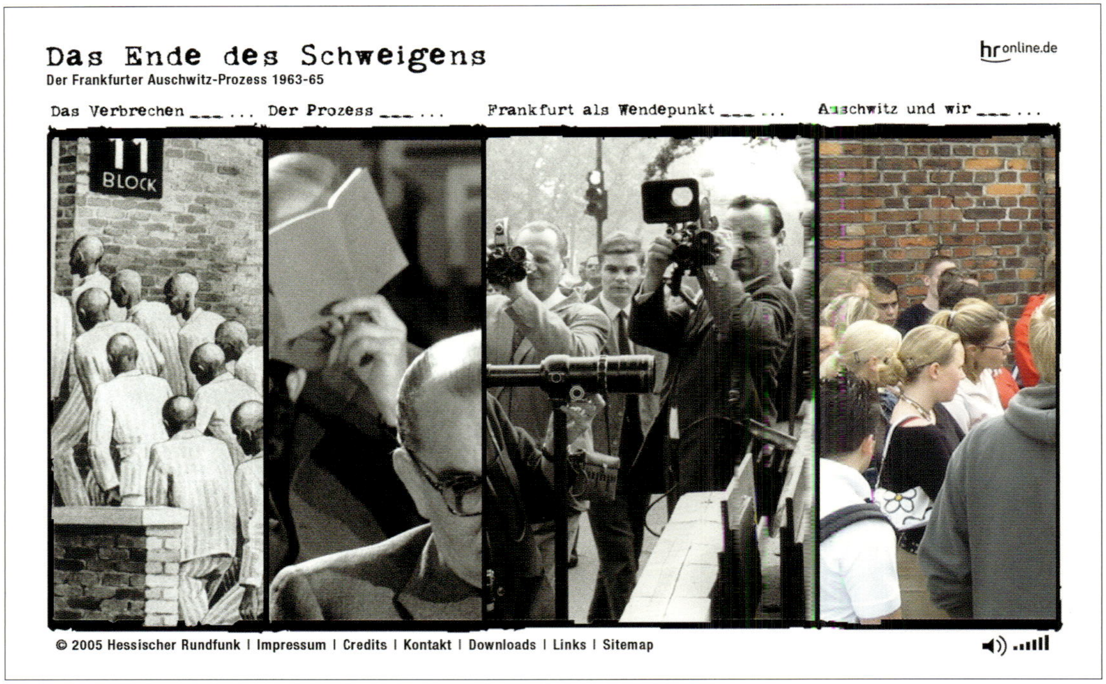

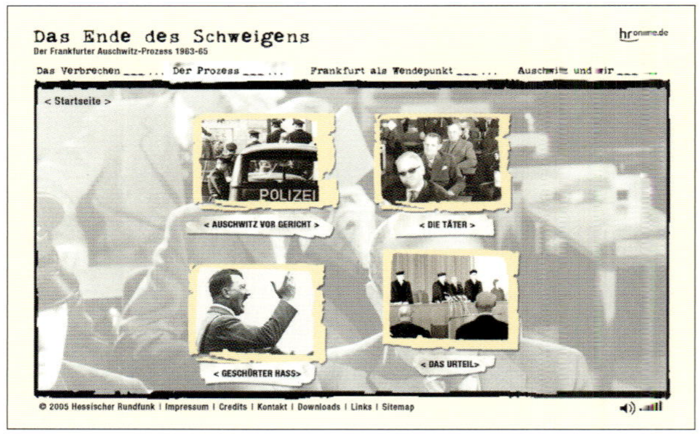

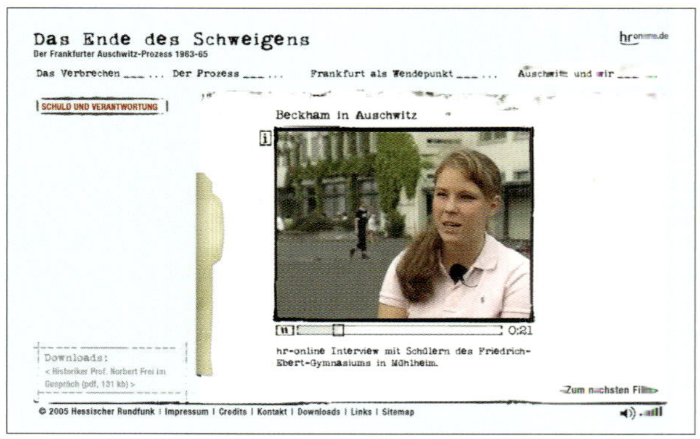

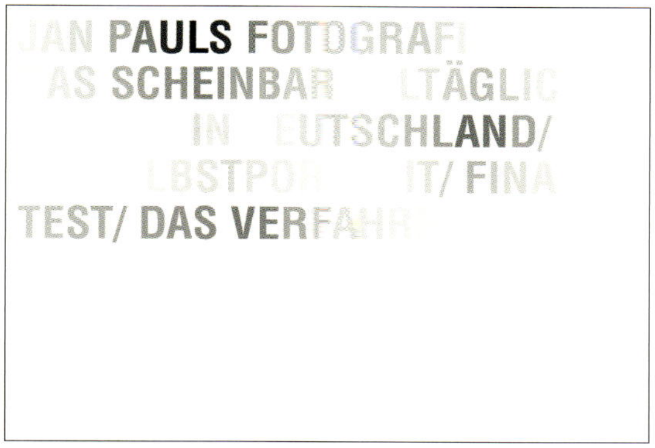

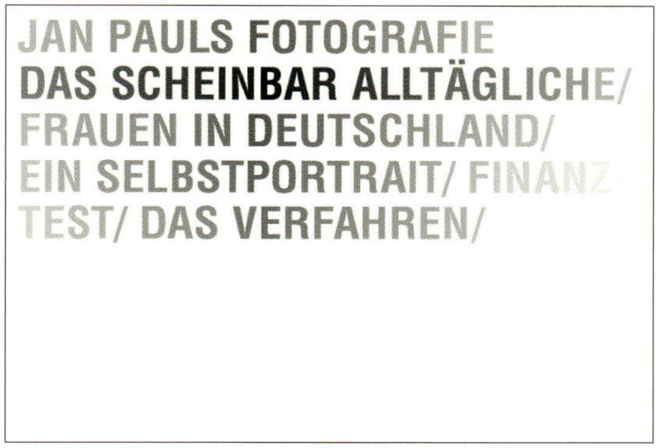

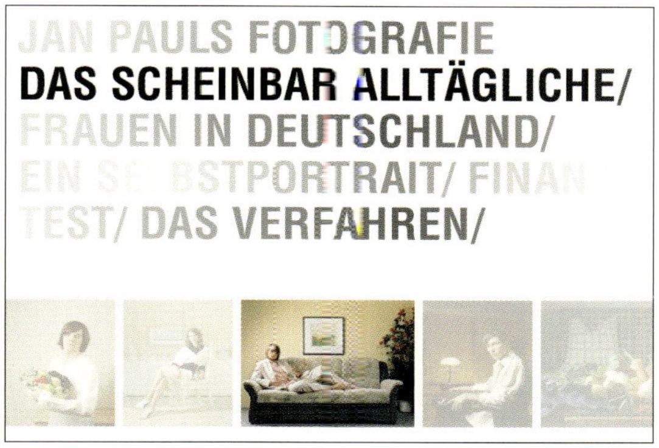

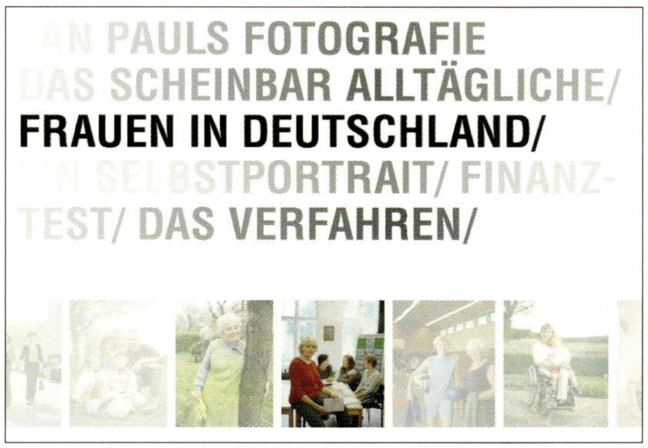

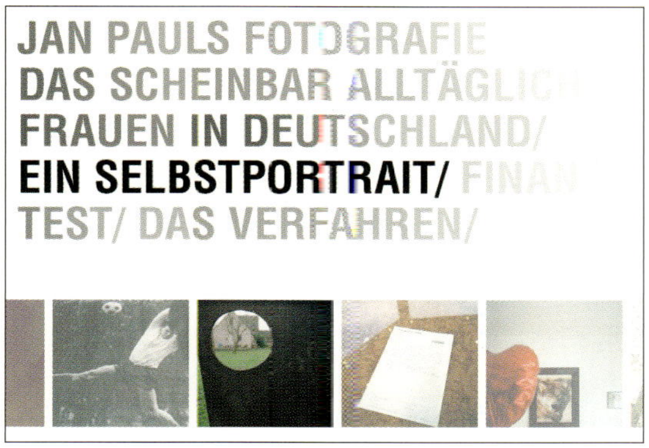

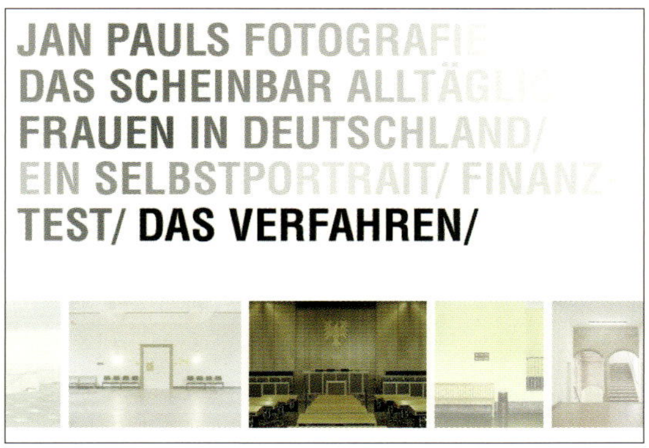

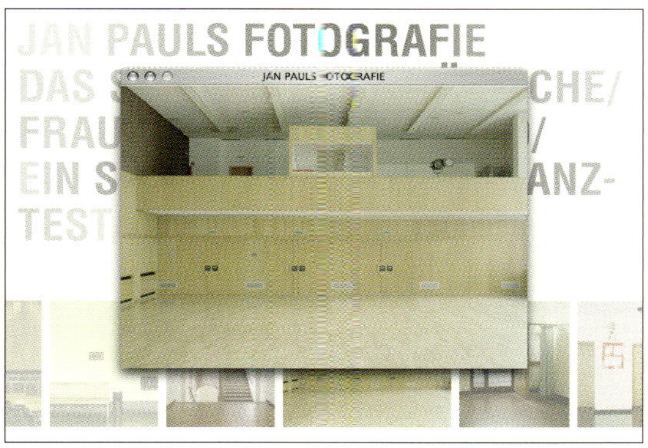

Auftraggeber Client
Jan Pauls
Berlin, Germany

Projekt Project
Jan Pauls Fotografie
Website

Design
Christoph Bebermeier
David Krause
BÜRO WEISS
Berlin, Germany

Internetauftritt in Anlehnung an die Ursprünge der Fotografie. Auf der Website des Berliner Fotografen Jan Pauls werden mit Hilfe der Maus die Text- und Bildinformationen graduell zum Vorschein gebracht. Das Erlebnis „Dunkelkammer", das Sichtbar-Werden des fotografischen Bildes, wird so auf den Bildschirm transportiert.

The Berlin-based photographer's website recalls the origins of photography. With the help of the mouse, words and pictures appear gradually, thus symbolizing the darkroom where the photographic picture is also unveiled step by step. That way the darkroom experience is brought to the computer screen.

Auftraggeber Client
DekaBank
Frankfurt am Main, Germany

Projekt Project
Skyper
Website

Design
Anne Wichmann
Jörg Waldschütz
NEUE DIGITALE
Frankfurt am Main, Germany

Dem Skyper-Claim „Up to the future" folgend, setzt die Seite für die Immobilie Skyper auf eine Navigation, die sich an der Gebäudehöhe von 153 Metern orientiert. Die Navigation über die Website vermittelt dem User das Gefühl, sich in einem Skyper-Fahrstuhl zu befinden und diesen selbst interaktiv steuern zu können. Während er sich über USPs des Gebäudes informiert, „erfährt" der User parallel die emotionale Komponente der Skyper-Büroräume. Ein virtuelles Fenster simuliert von jedem Stockwerk des Gebäudes aus einen einmaligen Ausblick über die Stadt Frankfurt.

The Skyper claim "Up to the future" is supported by an innovative form of navigation on the site for the real estate Skyper, based on the 153 meter height of the building. Through the navigation the user gets the feeling of moving up and down in the Skyper elevator. During the "elevator ride" the user receives information about the building's USPs, but at the same time "experiences" the emotional components of the Skyper office space. On every floor of the building a unique view over Frankfurt is simulated through a virtual window.

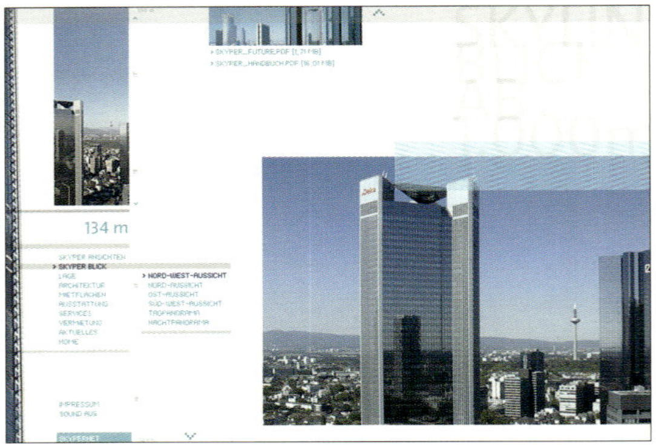

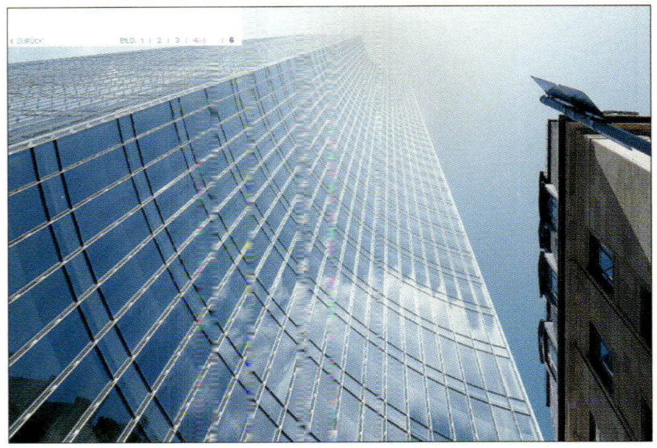

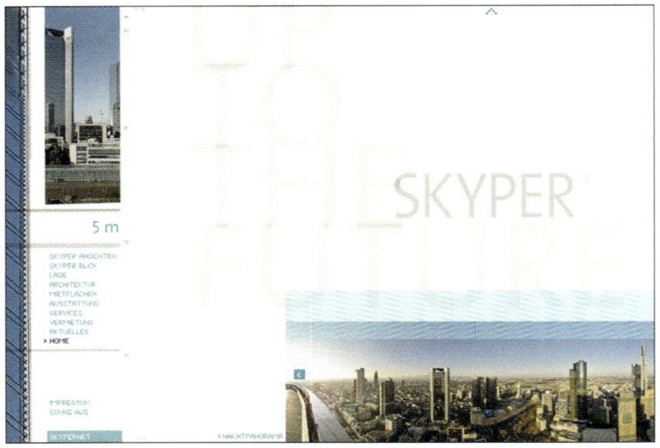

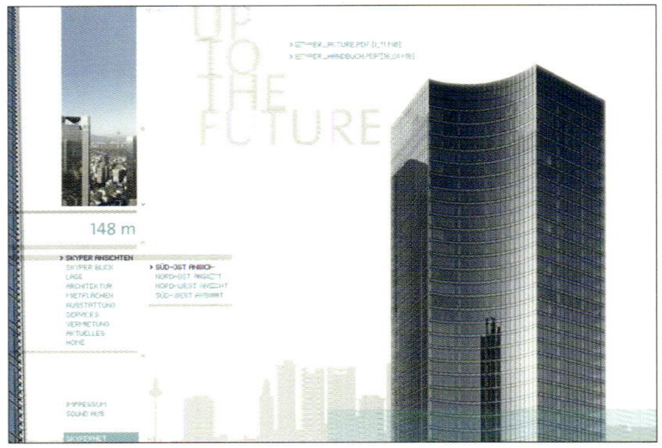

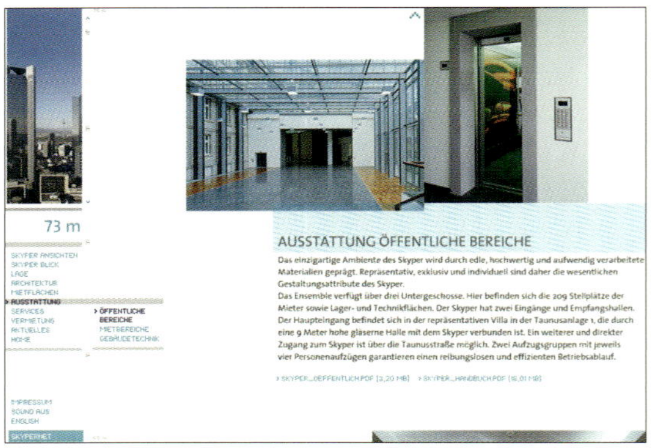

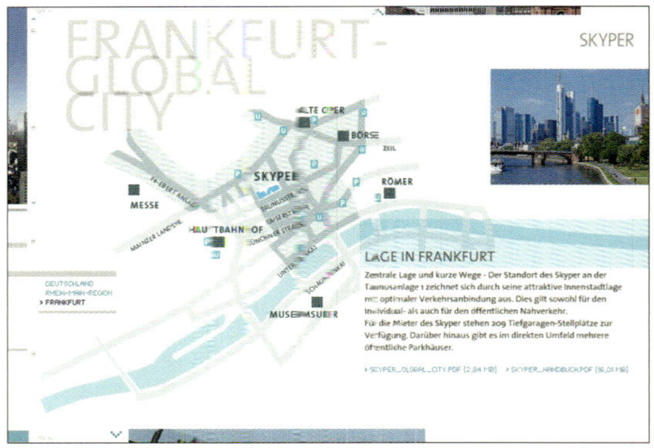

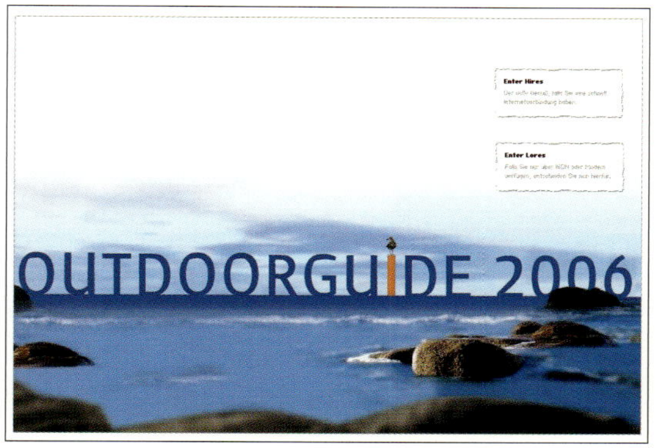

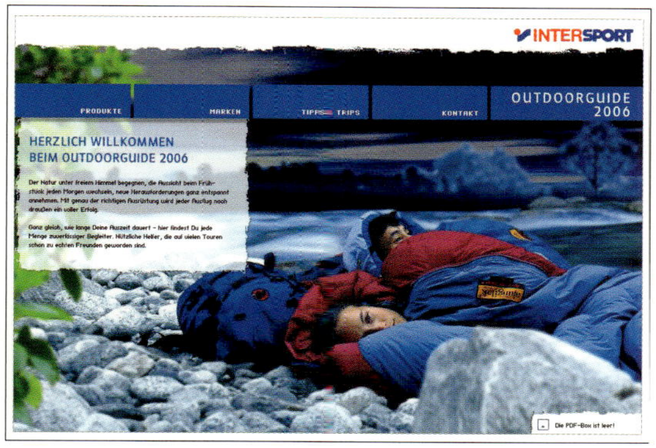

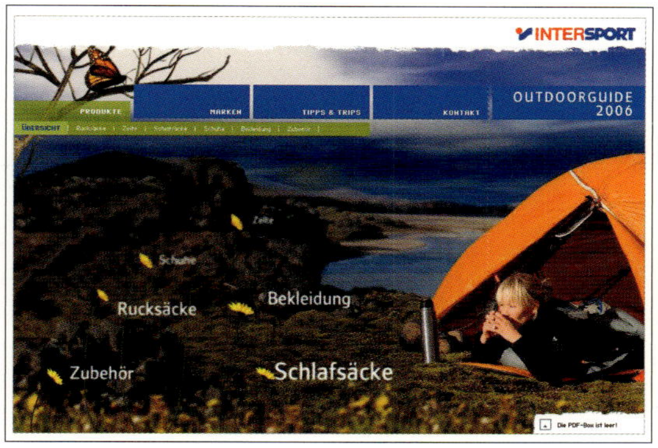

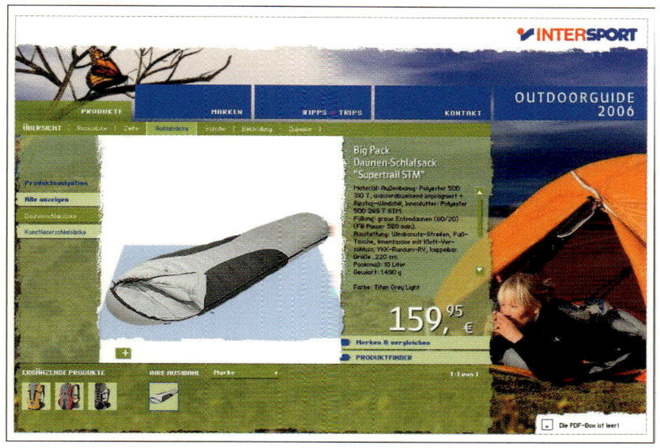

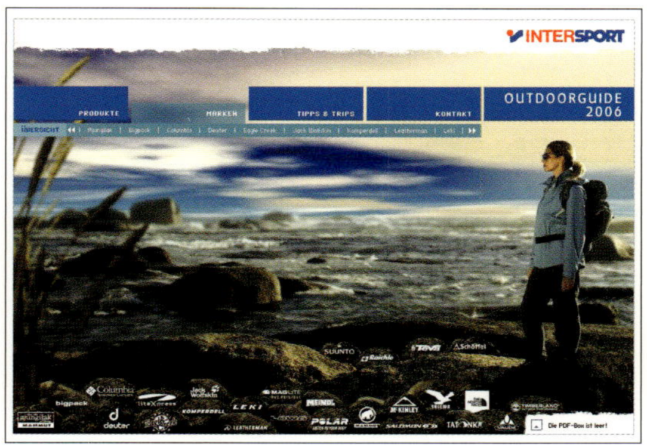

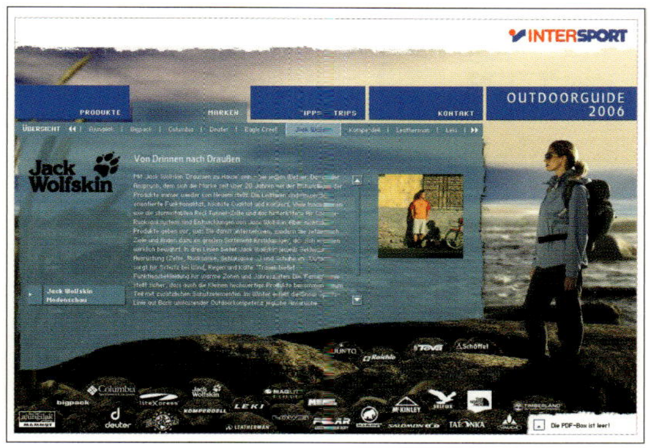

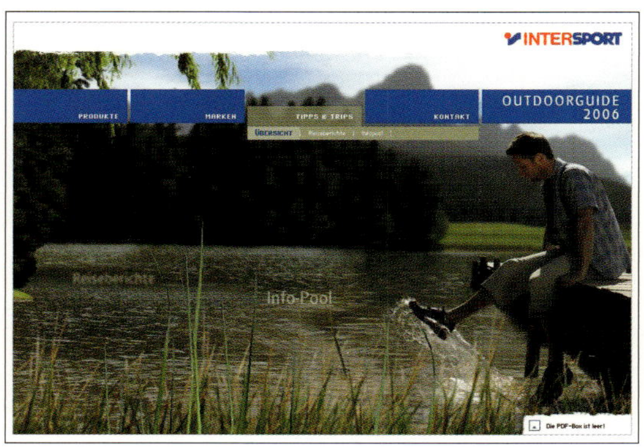

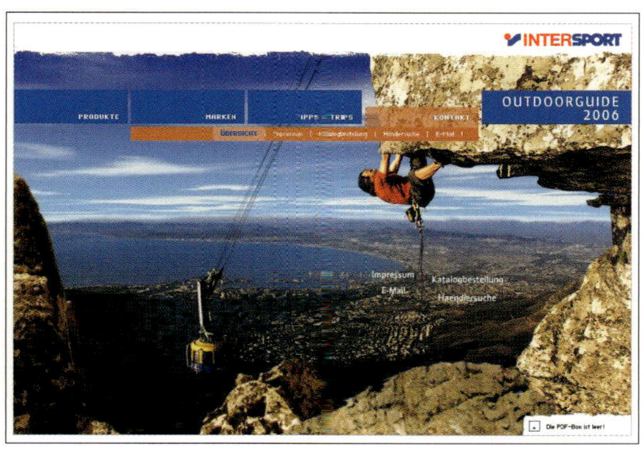

Auftraggeber Client
INTERSPORT Deutschland eG
Heilbronn, Germany

Projekt Project
Outdoorguide 06
Website

Design
Christoph Rüppel (CD)
Pia Hildebrand
Kim Oliver Hensler,
Florian Hucker (Design)
Michael Rehm (Programmierung)
work.id Werbeagentur
Villingen-Schwenningen, Germany

In animiertem Outdoor-Ambiente präsentiert der INTERSPORT Outdoorguide 06 ein breites Spektrum an Produkt- und Hintergrundinformationen zum Outdoorsport. Eine durchdachte Navigation ermöglicht die Produktsuche nach unterschiedlichen Suchkriterien. Komfortable Auswahl- und Ansichtsfunktionen erlauben es dem User, die einzelnen Produkte sehr direkt und sinnlich in Augenschein zu nehmen. Mithilfe der PDF-Box kann sich der User seine persönliche Produktwahl individuell zusammenstellen und über die interaktive Händlersuche direkt den Kontakt zum örtlichen Fachhändler aufnehmen.

An animated outdoor-landscape serves as a backdrop for the INTERSPORT Outdoorguide: The online-catalogue presents a wide spectrum of product and background information in the guise of a day in the outdoors. Well-devised site navigation offers effective product searches as well as customized search options. Intuitive selection and viewing devices give users a chance to experience the presented products in a highly unmediated and interactive way. A PDF-box allows users to individually select their outdoor-gear. An interactive dealer-search with full contact-options closes the gap between online presentation and PCS.

Auftraggeber Client
bulthaup Haus
Frankfurt am Main, Germany

Projekt Project
bulthaup – 5 Sinne
Website

Design
Peter Hartung
Nicole Hassert
Lisa Ferber
Georgios Margaris
QuinteX GmbH
Mannheim, Germany

bulthaup ist einer der führenden Hersteller von hochwertigen Designküchen. Die Internetseite für das neu eröffnete und exklusive bulthaup Küchenstudio in Frankfurt soll das besondere Konzept des Hauses vermitteln. Das heißt Verzicht auf die sonst üblichen Produktpräsentationen und Unternehmensdarstellungen. Stattdessen Vermittlung des sinnlichen Anspruchs der bulthaup Küchenwelt und Schaffung einer eigenen Welt. Umgesetzt mit einer aufwändigen multimedialen Präsentation, die charmant den Spieltrieb anregt.

bulthaup is a leading manufacturer of high-end designer kitchens. To accompany the opening of an exclusive new kitchen studio in Frankfurt, a website needed to be designed which would convey the company's special concept. This involved rejecting the conventional approach of merely presenting the company and its products; instead, the focus was on communicating the sensory qualities of bulthaup kitchens and creating a unique new universe. The mission was implemented in the form of an elaborate multimedia presentation which charmingly stimulates the user's playful impulses.

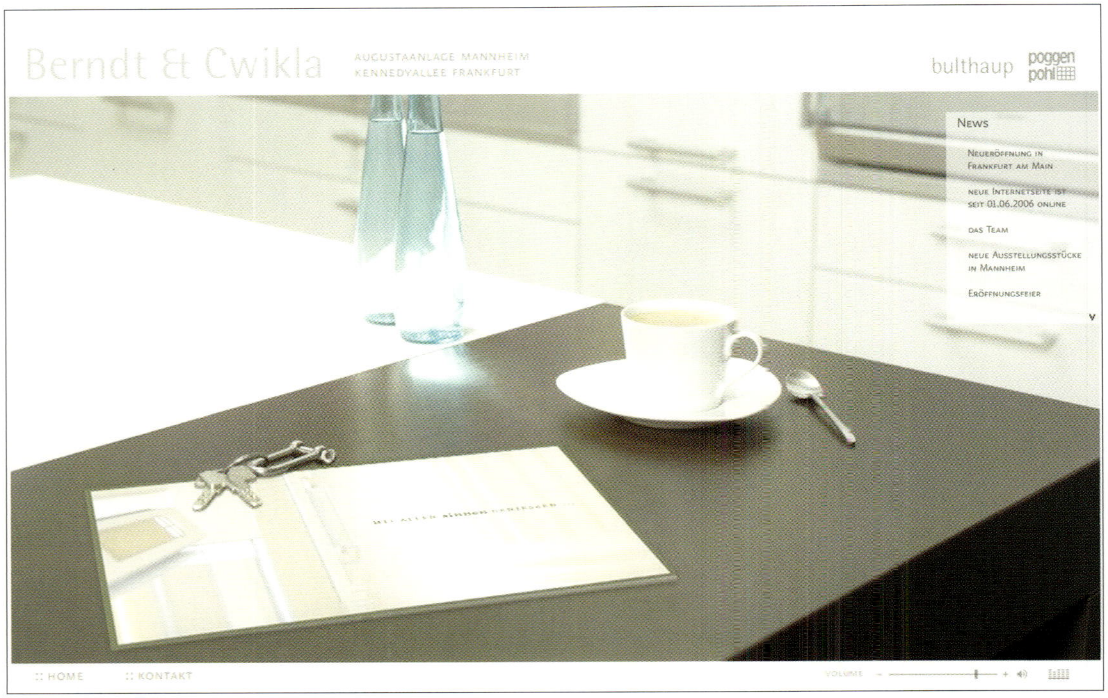

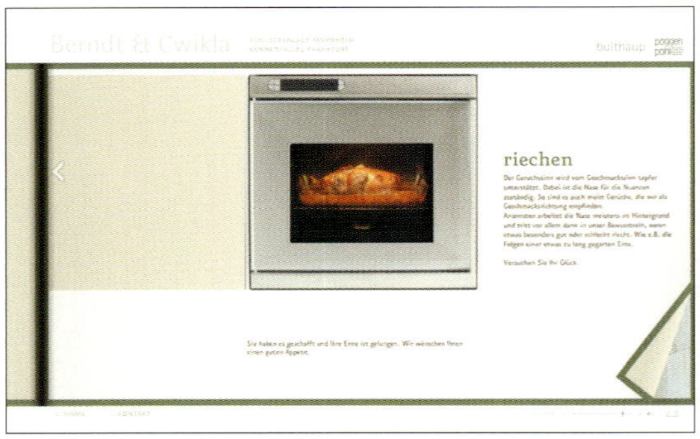

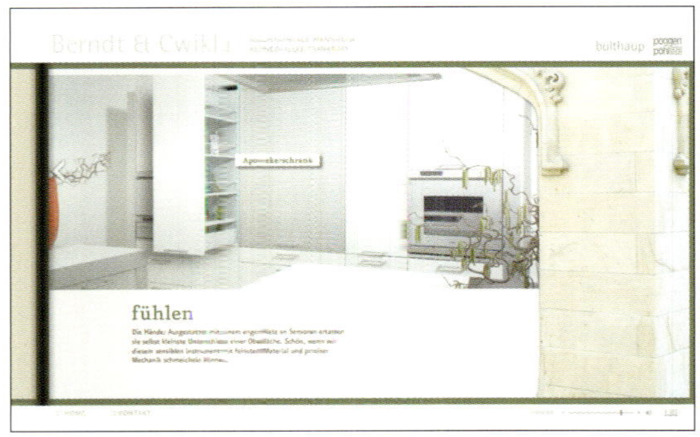

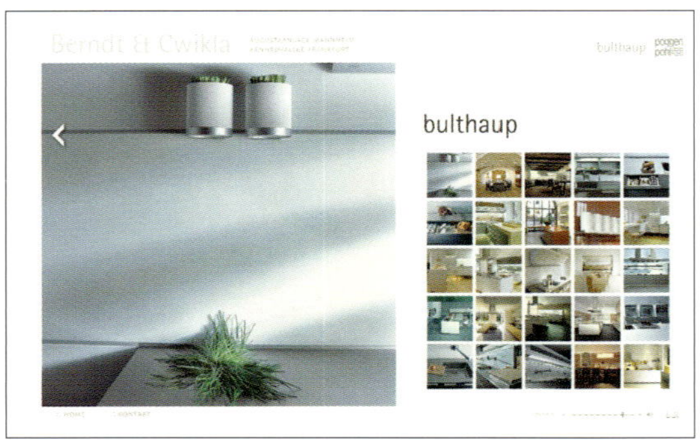

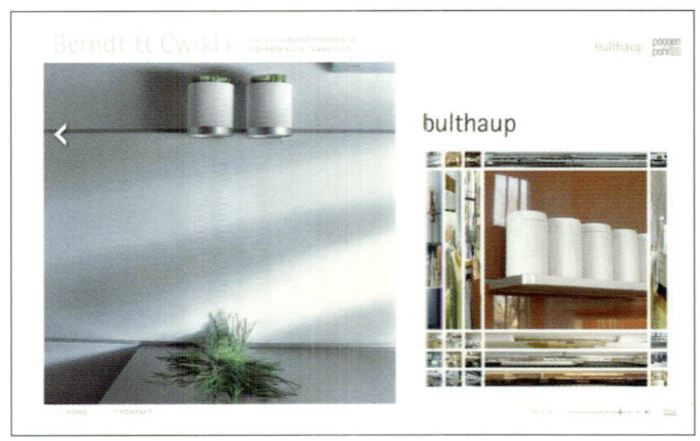

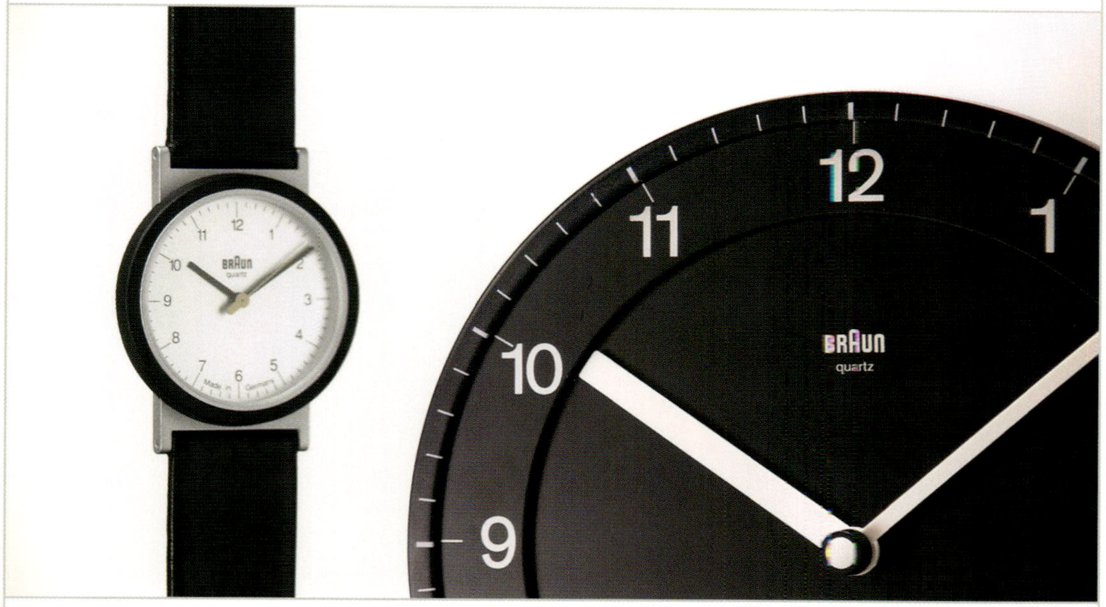

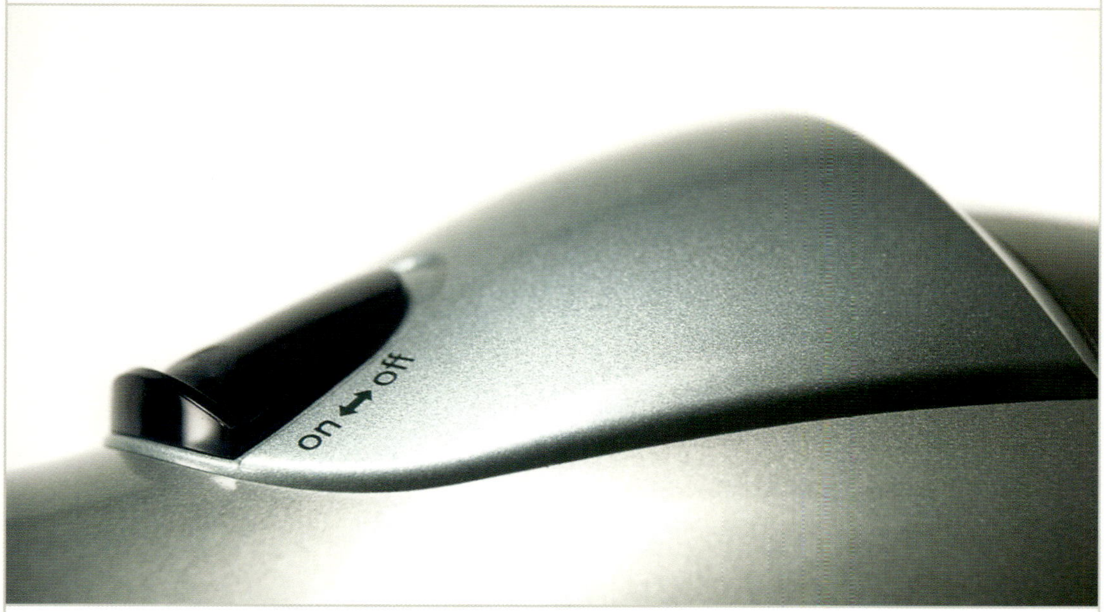

Auftraggeber Client
Braun GmbH
Kronberg/Taunus, Germany

Projekt Project
Braun Aesthetics
DVD

Design
Tom Fischer
ARTUS digitale Kommunikation GmbH
Frankfurt am Main, Germany

Braun, Hersteller von Elektrokleingeräten und Pionier des funktionalen Designs, feierte 2005 den 50. Geburtstag des Braun Designs. Aus diesem Anlass wurde ein Video in HDTV realisiert, das 50 Jahre Braun Designgeschichte in ästhetischen Nahaufnahmen ausgewählter Schlüsselprodukte darstellt. Es wendet sich an Besucher der Ausstellungen in Deutschland, London, Boston, Tokio, an Journalisten und alle Designinteressierten. Die High key-Bilder wurden auf eine speziell arrangierte Musik ohne zusätzliche Voice-Over montiert.

Braun, manufacturer of small electrical household appliances and pioneer of functional design celebrated its 50th design anniversary in 2005. For this occasion a HDTV video was created. 50 years of design history were depicted in exciting aesthetic close-up shots of key products. The video addresses exhibition visitors (Germany, London, Boston, Tokyo), journalists and everybody interested in design. The high-key images were edited to a specially remixed music without voice-over.

digital media interface

Auftraggeber Client
Clarion Co., Ltd.
Styling Department
Tokyo, Japan

Projekt Project
Clarion NAX963HD
Interface für Auto-Navigationsgerät
Interface for car navigation system

Design
Katsuyuki Kondo
Naoya Koga
Shuichi Obara
Tomohiro Harada
Takao Yahiro
Naoki Mori
Manabu Suwa
Clarion Co., Ltd.
Styling Department
Tokyo, Japan

Das Clarion Car Navigation System NAX963HD hat ein neu überarbeitetes Interface erhalten. Ziel ist es, die Fahrsicherheit zu erhöhen, indem die Verbindung zwischen Mensch und Maschine enger gestaltet wird. Als Grundlage dazu dienen Abläufe der Informationsverarbeitung im Gehirn in bestimmten Verkehrssituationen.
Das zentrale Element für die Übermittlung der Systeminformationen an den Fahrer ist das neue 3-D-Menü, dessen Angaben perspektivisch im virtuellen Raum angeordnet sind, ebenso wie Gegenstände, die die reale Welt bevölkern. Diese Art der Darstellung erlaubt es dem Fahrer auf sehr natürliche Weise, seine Konzentration vom Führen des Fahrzeugs auf die Wegweisung auf dem Display und umgekehrt zu lenken.

For enhanced safety and the closest possible relationship between car and driver, Clarion has redesigned its NAX963HD Car Navigation System Interface based upon mental models of driving situations centering on the way information is processed in the brain.
A central element involves the 3D menu, with the help of which the driver can comprehend information arranged in virtual space like an object existing in the real world, enabling him or her to shift attention from the actual driving to the motions on the screen and vice versa, all quite naturally.

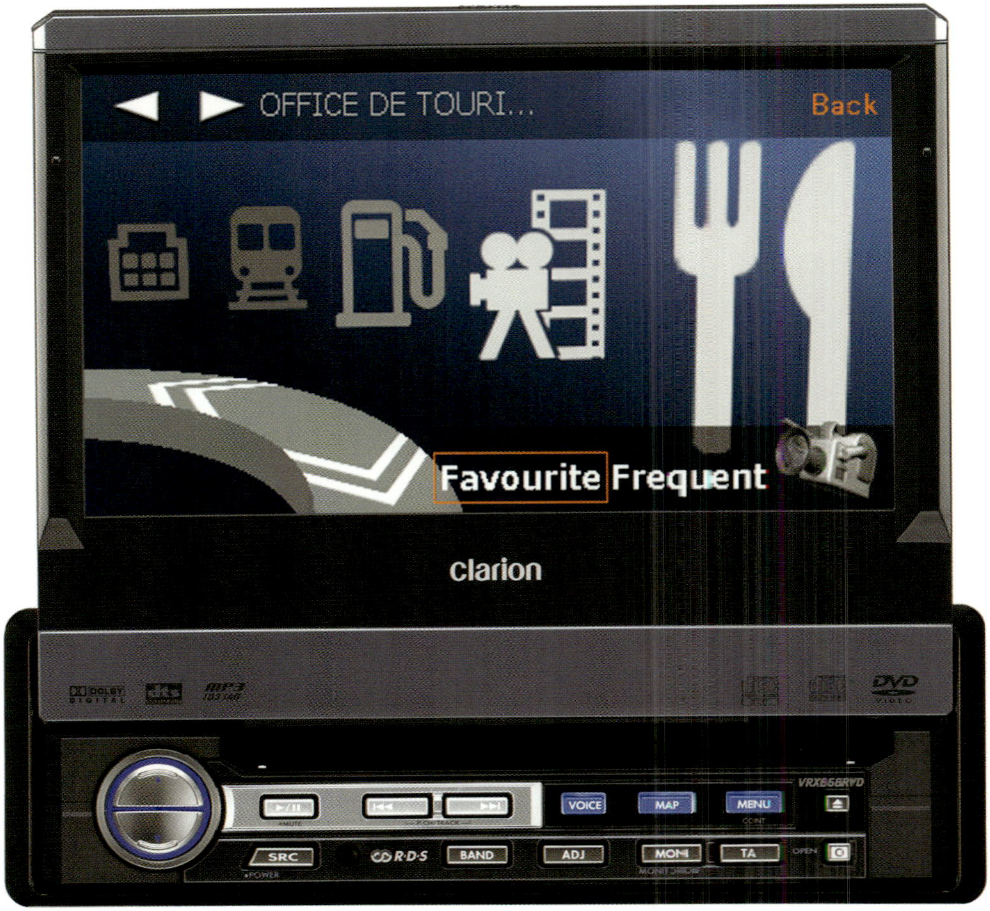

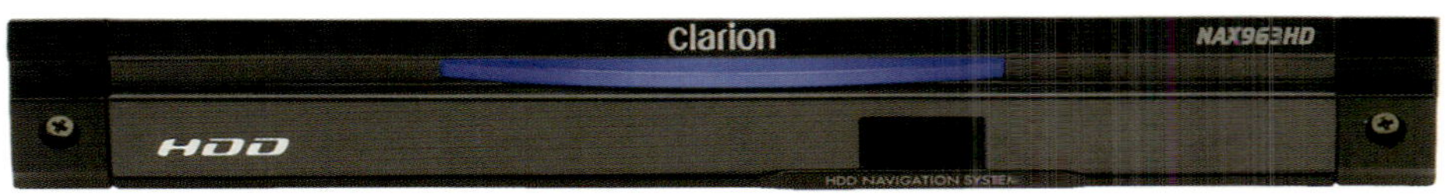

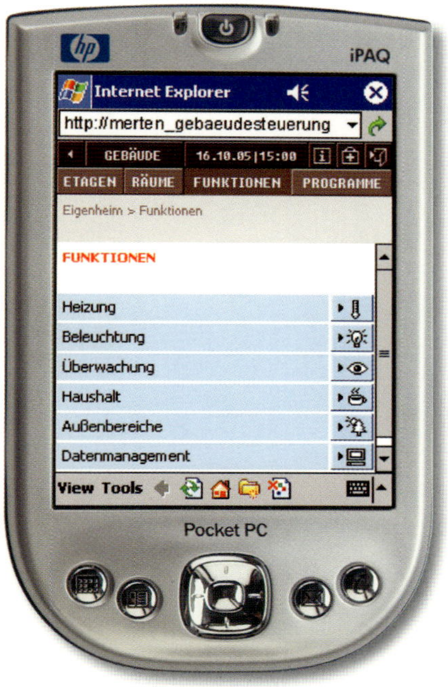

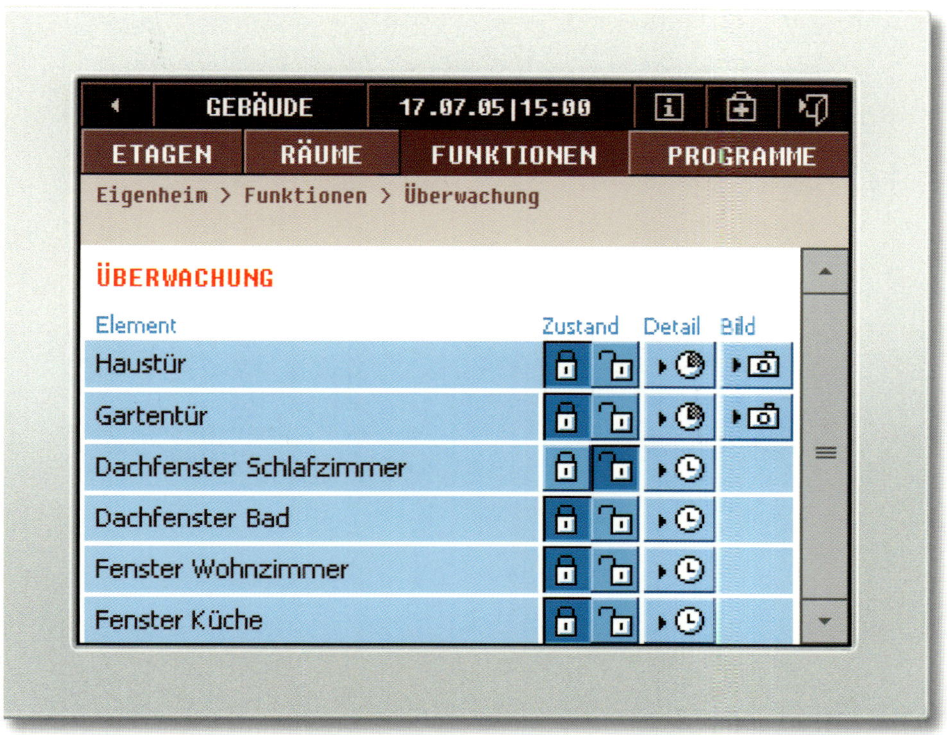

Auftraggeber Client
Merten GmbH & Co. KG
Wiehl, Germany

Projekt Project
merten@home
Bedienoberfläche zur Steuerung von Gebäudefunktionen
User interface for building control

Design
Christoph Kasmeridis (Konzept, Projektmanagement)
Anett Eickhoff (Design)
Jonas Gimatis (Konzept, Implementierung)
Meta Design AG
Berlin, Germany

Wäre es nicht schön, nach einer Geschäftsreise bereits am W-LAN Hotspot des Flughafens via PDA die Fußbodenheizung auf Wohlfühltemperatur hochzufahren? Nichts leichter als das und nur eine der vielen Möglichkeiten von merten@home. Mit der neuen Bedienoberfläche lassen sich sämtliche Gebäudefunktionen von Eigenheimen, Ferienhäusern oder Gewerbeobjekten komfortabel steuern. Ob Heizung, Beleuchtung, Jalousien, Überwachungssystem oder Haushaltsgeräte.
Ob vor Ort oder von unterwegs. Mit merten@home ist man immer und überall zuhause.

Wouldn't it be great if you could set your underfloor heating to a cosy temperature via PDA from the airport's W-LAN hotspot at the end of a business trip? Nothing could be simpler – and this is just one of the many options offered by merten@home. The new user interface enables users to conveniently control all building functions for private homes, holiday cottages or commercial properties. Whether it be heating, lighting, blinds, your surveillance system or your domestic appliances.
Whether locally or remotely. With merten@home, you are always at home, wherever you are.

Auftraggeber Client
Océ Technologies
Venlo, The Netherlands

Projekt Project
Océ Ocean look & feel
Corporate GUI Design

Design
Océ Design Team
Océ Technologies
Venlo, The Netherlands

Océ ist ein führender Hersteller und Dienstleister im Bereich Digitaldruck und Dokumentenmanagement. Moderne Produkte sind selten Einzellösungen. Unsere Anwendungen und GUI-Systeme vereinigen sich zu einer einheitlichen Océ Lösung. Um die Benutzerfreundlichkeit seiner Systempalette zu verbessern, entwickelte Océ sein neues „Océ Ocean look & feel"-Konzept. Das frische Erscheinungsbild basiert auf bewährten Standards und Arbeitsabläufen – trotz Anpassung an die unterschiedlichen GUI-Plattformen (Windows, Web und Touchscreens) tritt es doch als Produktfamilie auf. Das professionelle und freundliche Erscheinungsbild sowie die intuitive Navigation machen die Arbeit mit dem System zu einem angenehmen Erlebnis.

Océ is one of the world's leading suppliers of systems and services for digital printing and document management. Today products are rarely standalones. Applications and GUI engines join to create a single experience of an Océ solution. To enhance the usability of its total systems, Océ developed its new "Océ Ocean" look & feel. This fresh, user-friendly style deploys established standards and working methods. It is tuned to the various GUI platforms (windows, web and touch screens), but designed as a family. A professional, friendly appearance featuring Océ quality and intuitive navigation makes working with Océ solutions a pleasure.

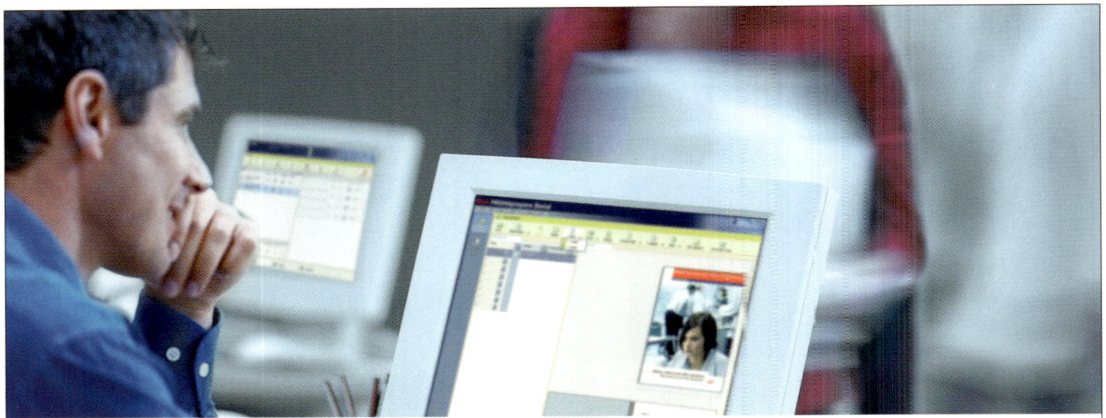

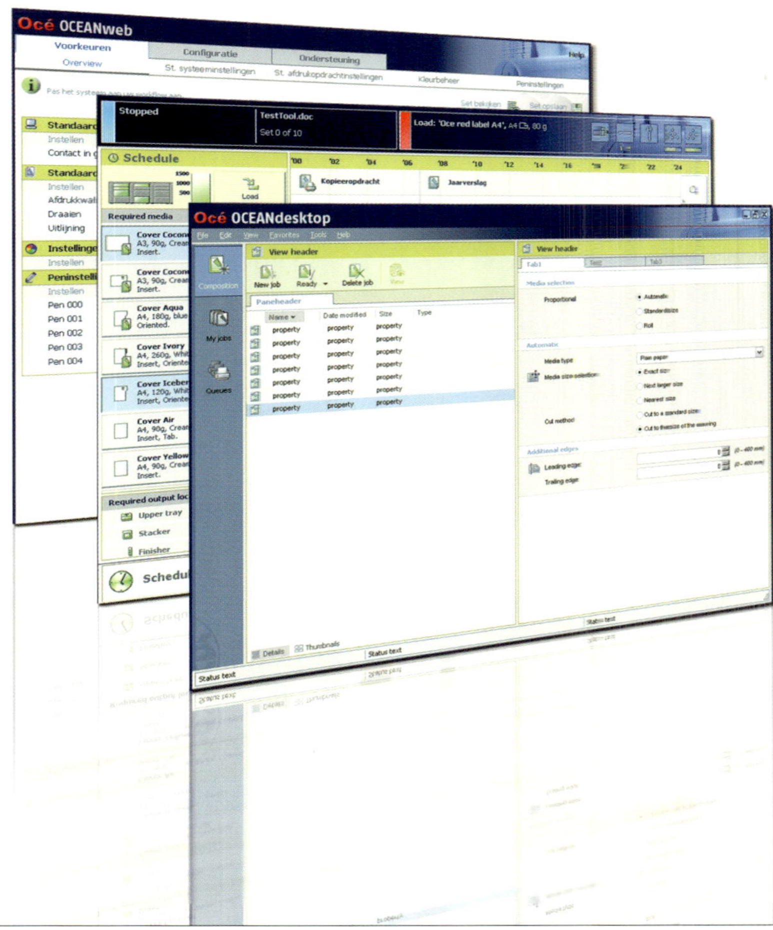

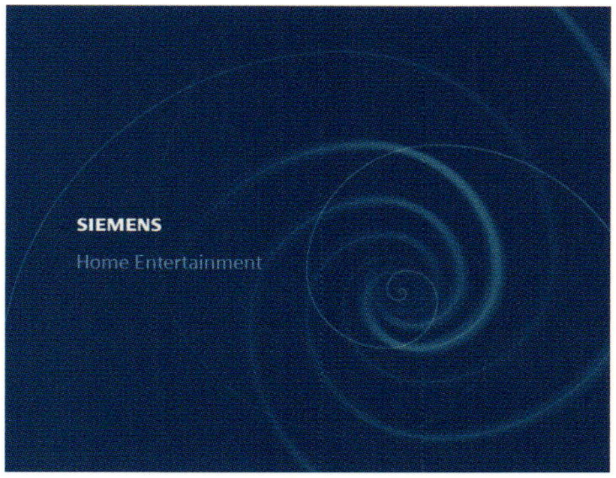

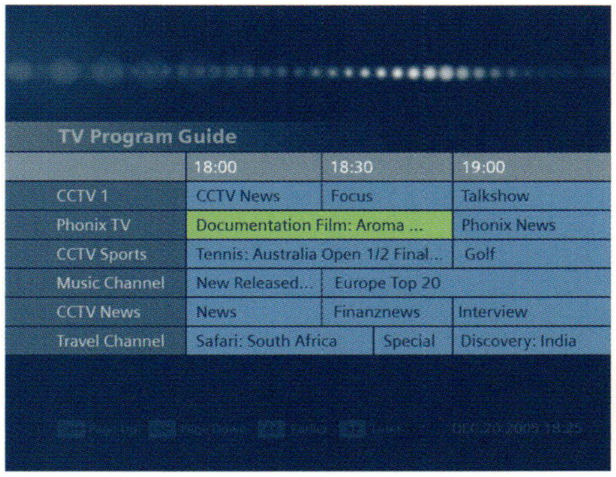

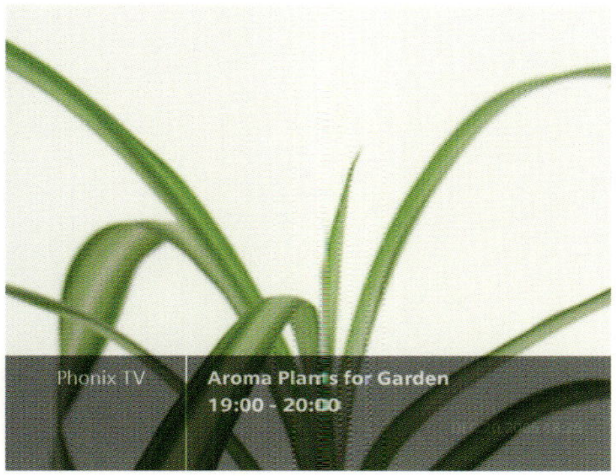

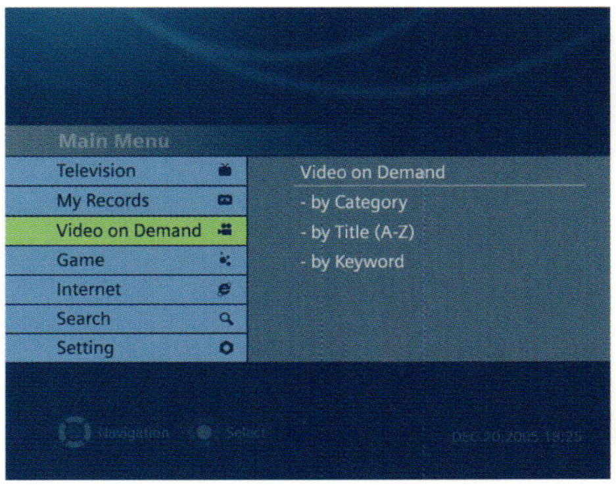

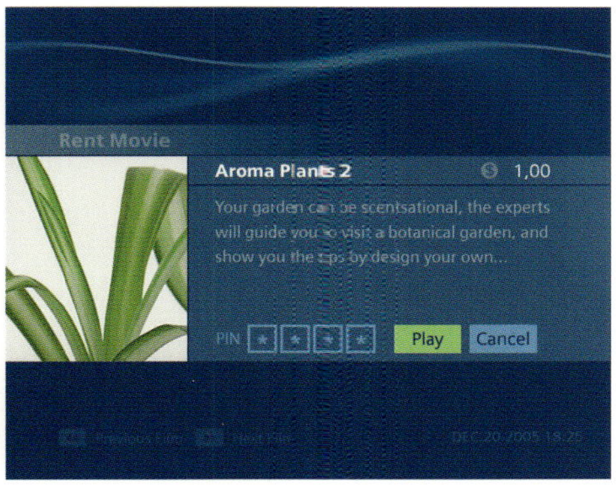

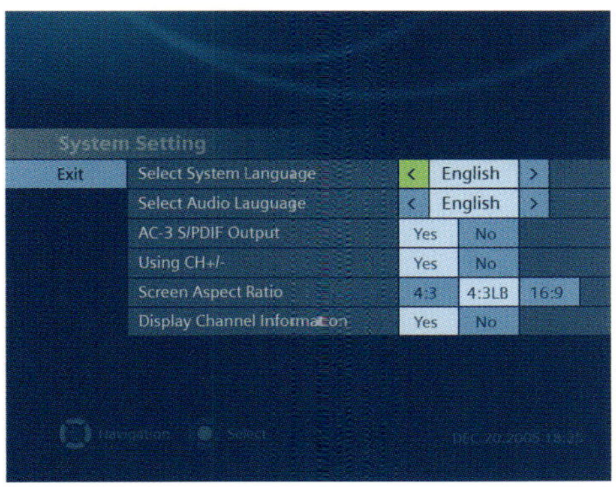

Auftraggeber Client
Siemens Communication Networks Ltd., Beijing (SCN-B)
Beijing, China

Projekt Project
Siemens Home Entertainment System Interface

Design
Zhou Bailiang
Siemens Communication Networks Ltd., Beijing (SCN-B)
Beijing, China

Die gesamte Bedienung der Siemens Home Entertainment Dienste erfolgt durch Menüs und Bedienelemente auf dem Fernsehschirm, die mit Hilfe einer Fernbedienung gesteuert werden. Diese eingeschränkte Interaktionsweise erfordert es, bei der Gestaltung der Screenlayouts in hohem Maße auf Benutzerfreundlichkeit und Bedienkomfort zu achten. Typografie und Farbdesign wurden hinsichtlich Lesbarkeit auf dem Fernsehschirm optimiert. Die systematische Anwendung von Farbkombinationen und -kontrasten unterstützt eine klare Trennung zwischen den Bedien- und Anzeigeelementen und den vielschichtigen Informationen.

The whole operation of the Siemens Home Entertainment services is done using menus and other interface elements on the TV screen, which are navigated by the remote control. This limited way of interaction requires paying strong attention to the user friendliness and ease of use when designing the screen layouts.
Typography and color concept are optimized with respect to readability on the TV screen. The systematic use of color combinations and contrasts supports a clear differentiation between operational and display elements and the multilayer information.

Auftraggeber Client
Eastman Kodak Company
Rochester, NY, USA

Projekt Project
Kodak Picture Kiosk
Bildterminal
Picture terminal

Design
Jo Gray
Tim Wood
Hye-Jin Nae
Eastman Kodak Company
Rochester, NY, USA

Für die Entwicklung der neuen Kodak Bildsprache stand bei allen Designentscheidungen die Bedienungsfreundlichkeit an erster Stelle. Konsequent wiederholen sich einfache geometrische Designkomponenten und Bildformen. Es wird eine reduzierte Farbpalette eingesetzt, wobei die Buttons durch stärkere Kontrastgebung besonders hervorgehoben werden. Zur Stabilisierung des Bildes von einer Darstellung zur nächsten kommt ein neuer Rasteraufbau zum Einsatz.

Ease of Use was at the forefront of design decisions while mirroring the new Kodak design language. Simple geometry of screen components and repeated forms are used throughout. A simplified color palette is used with more contrast on buttons to make them "pop". A new grid structure is employed to keep imagery stable from screen-to-screen.

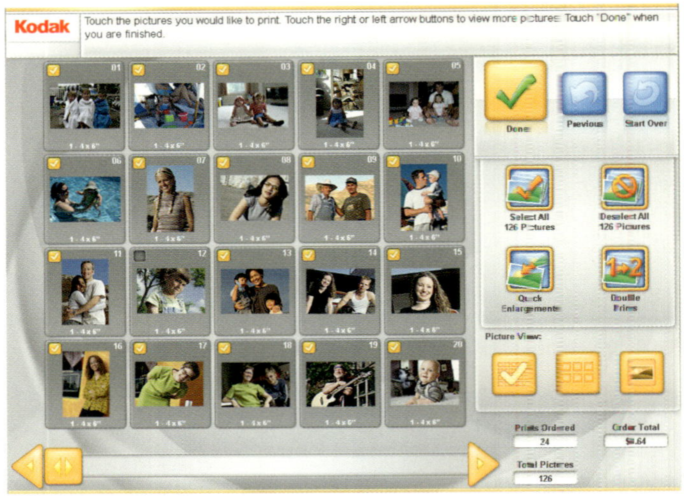

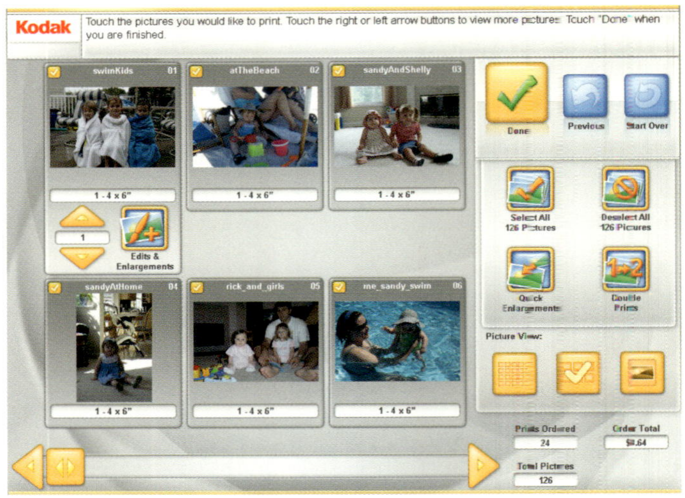

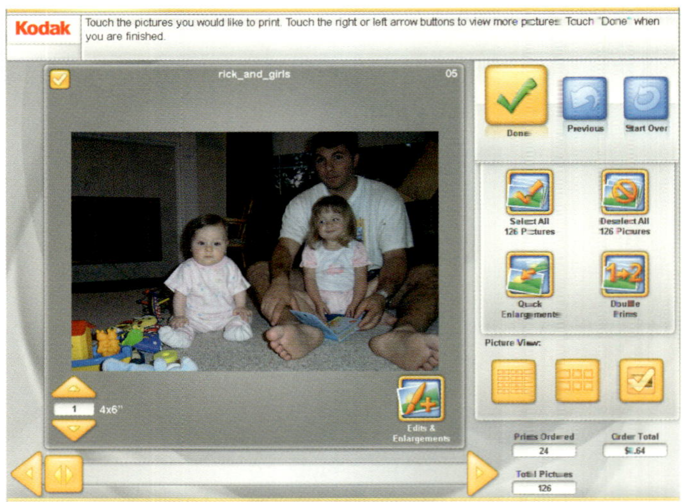

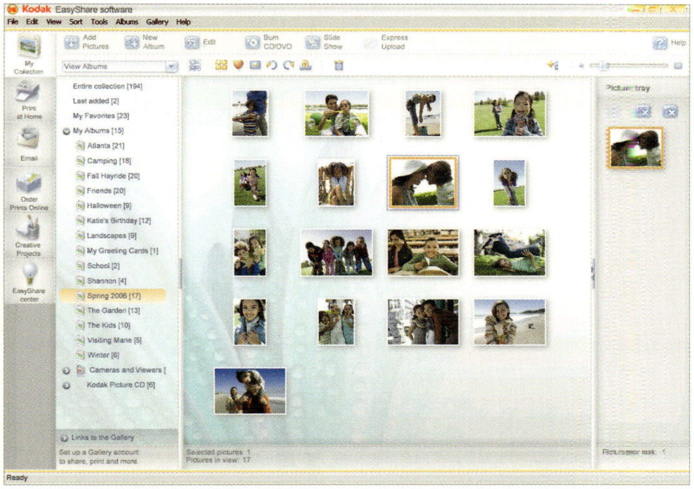

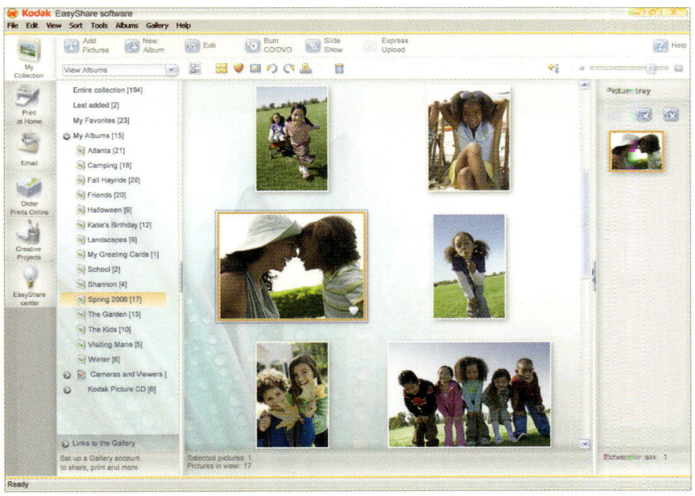

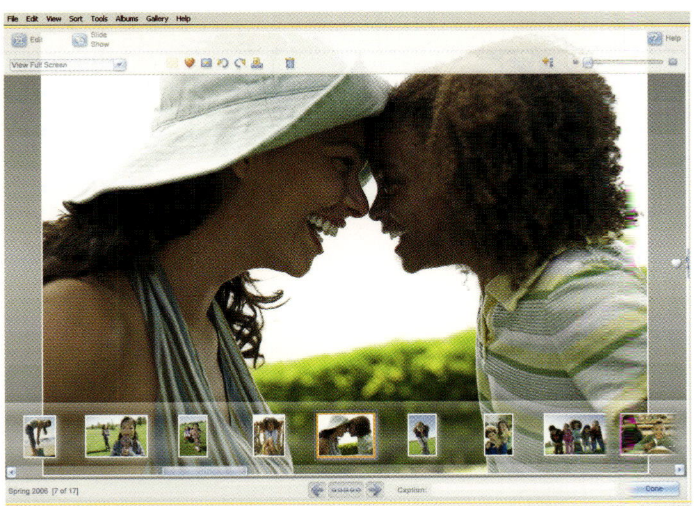

Auftraggeber Client
Eastman Kodak Company
Rochester, NY, USA

Projekt Project
Kodak EasyShare software V6.0
Software

Design
Tim Wood
Frank Marino
Hye-Jin Nae
Eastman Kodak Company
Rochester, NY, USA

Die Kodak EasyShare Software 6.0 ist eine von Grund auf neu entwickelte Version der weithin bekannten Kodak Photo Software. Ihre einfache Erfassbarkeit ist das Resultat einer harmonischen Kombination von Schlichtheit und Eleganz. Die Gegenüberstellung von Natur und Technik bereichert die unterkühlte Effizienz eines rein technischen Ansatzes um eine natürliche Note. Das erfrischend neue Design ist augenfreundlich und unterstreicht die Geradlinigkeit der Marke. Vollbildansicht mit Schnellnavigationsleiste, personalisierte Hintergründe, benutzerdefinierte Hintergrundbilder, übersichtliche Miniaturansicht, dynamische Symbole und integrierte Videoanzeige sind neben weiteren Features die Kernpunkte dieser neuen 6.0 Version.

Kodak EasyShare software version 6.0 is an all-new redesign of Kodak's popular photo software. The design embodies simplicity, elegance and harmony, all critical to the perception of "ease". The contrast and combination of nature and technology, tempers the cool efficiency of the sleek technological approach with a natural overtone. The fresh new design is easy on the eyes while retaining its trademark simplicity: full view with quick navigation bar, personalized backgrounds, customized background imagery, streamlined thumbnail view, dynamic icons, and embedded video viewing are many of the new features of version 6.0.

Auftraggeber Client
Samsung Electronics
Seoul, South Korea

Projekt Project
Samsung ICES 2006
Microsite

Design
Kangwook Choi
Jungkweon Yoon
graf
Seoul, South Korea

Das Ziel der Microsite war es, die drei Bereiche „Spielen", „Leben" und „Arbeit" in einem einzigen Konzept zu vereinigen, das die Bezeichnung „Imagine" erhielt. Der Stand von Samsung auf der ICES 2006 präsentierte sich als Schnittstelle zwischen Produkt und Lifestyle.

To unify the three dimensions of "play", "live", and "work" to one concept, which is "imagine". The ICES 2006 Samsung booth provides itself as the contact point between product and lifestyle.

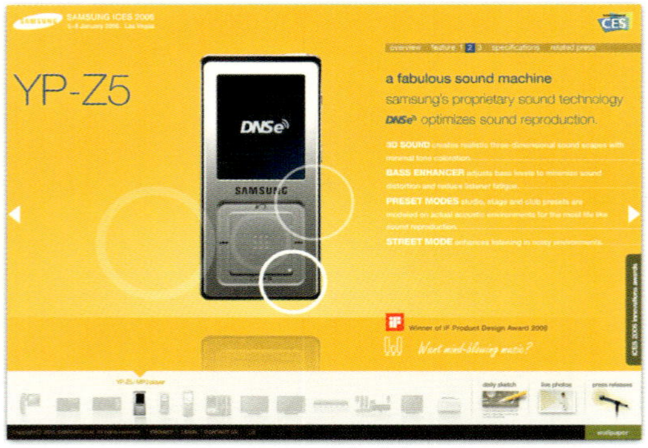

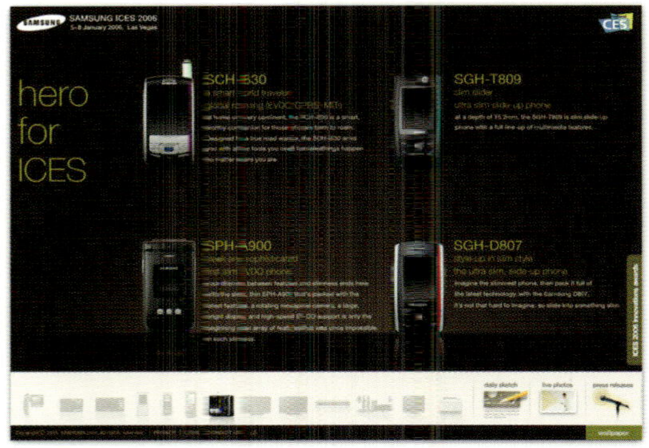

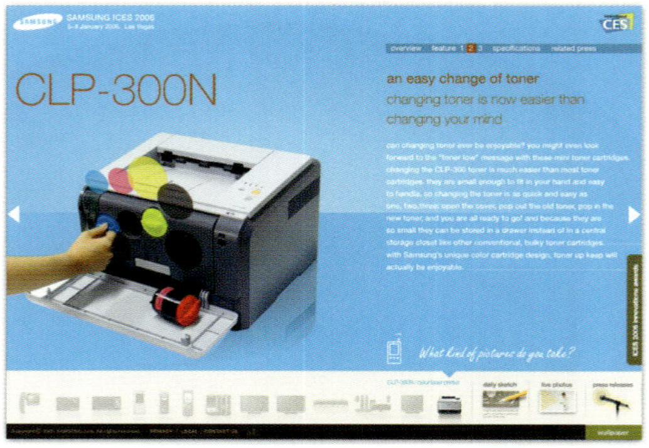

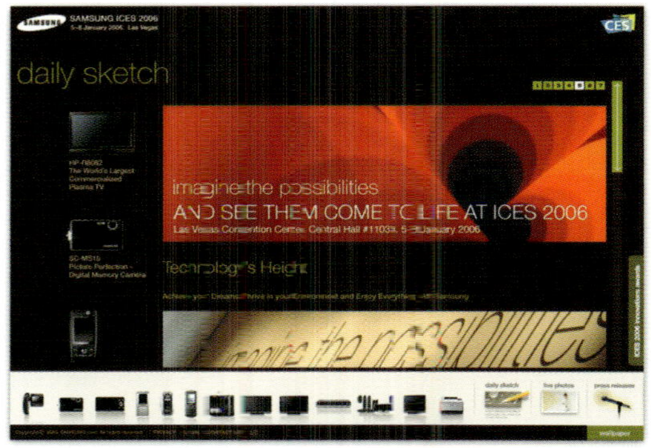

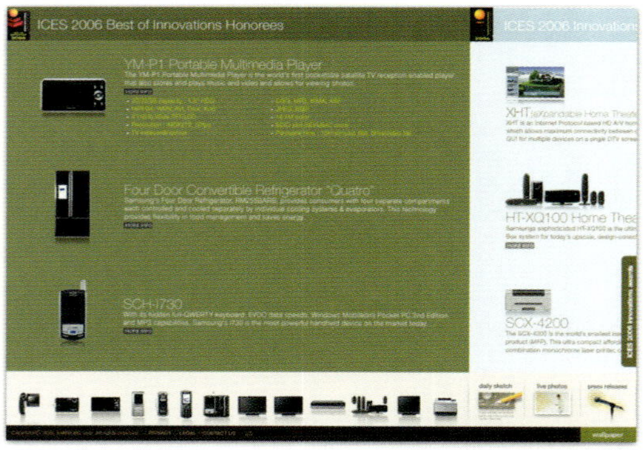

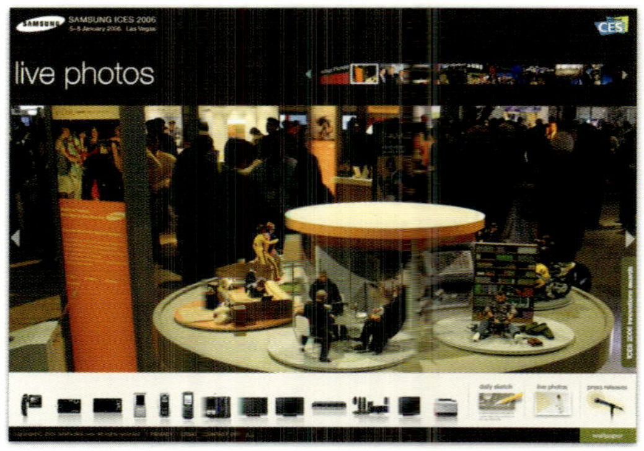

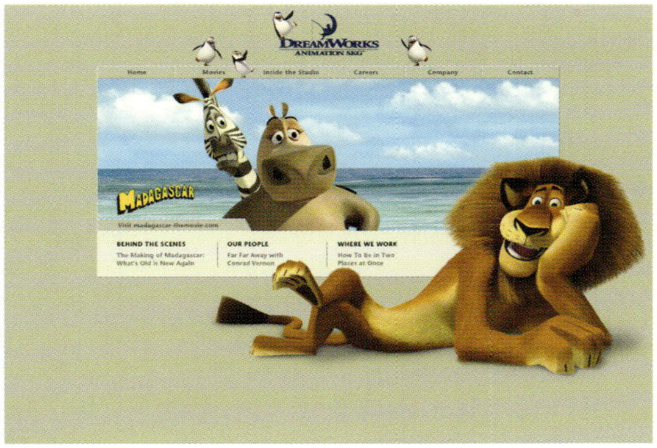

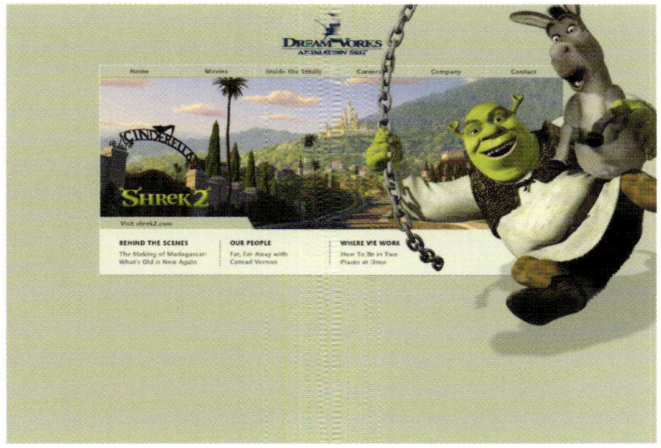

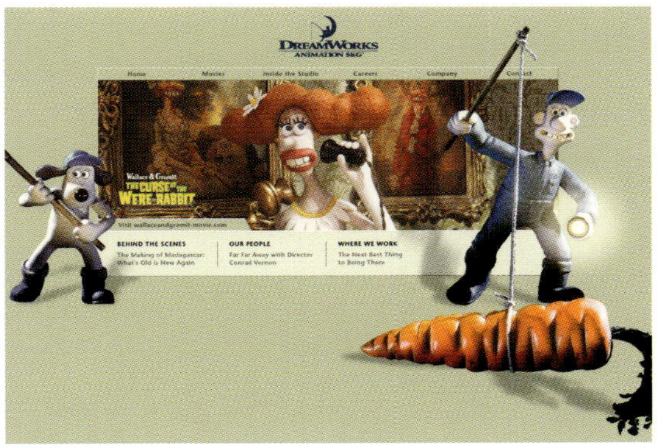

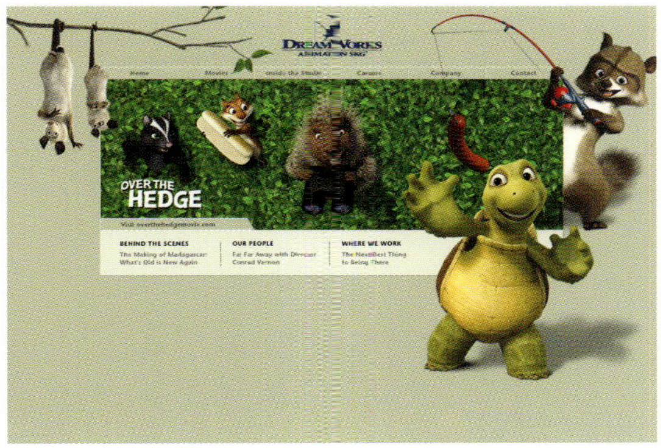

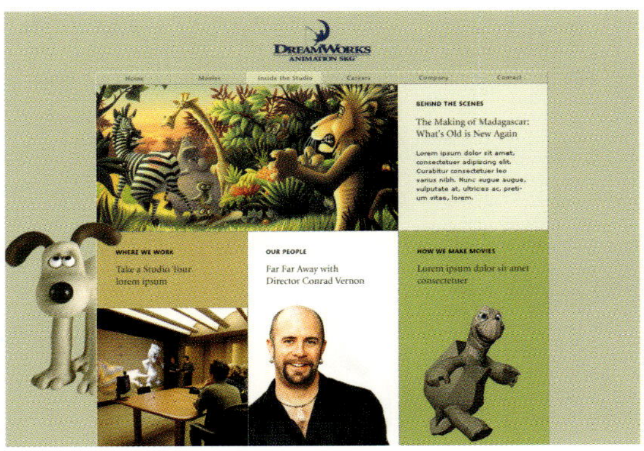

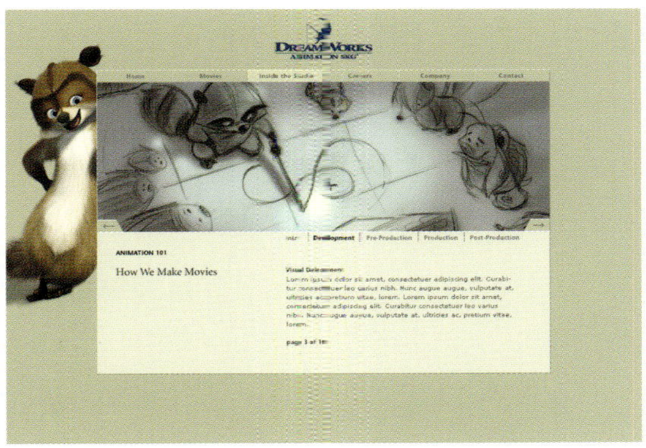

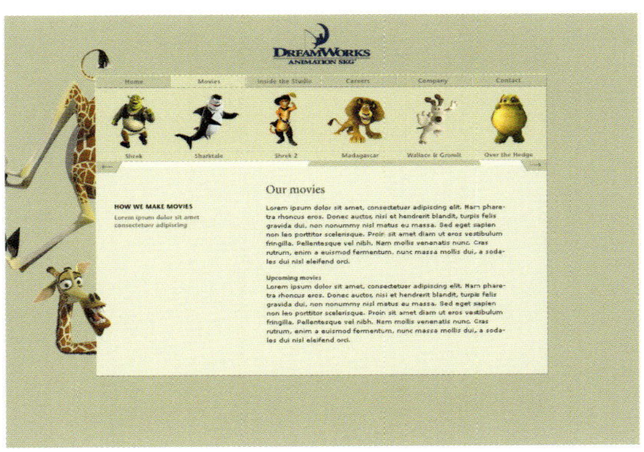

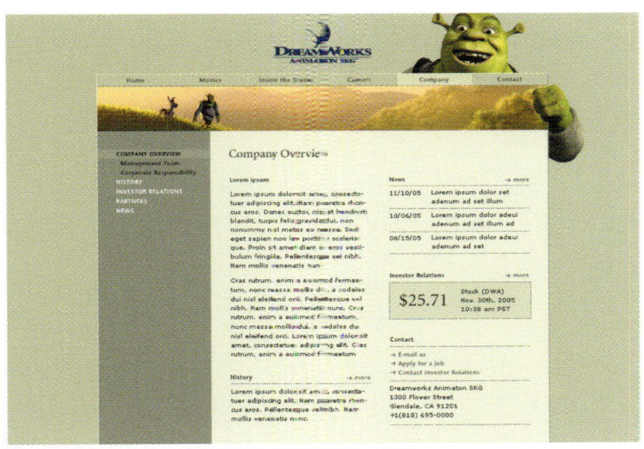

Auftraggeber Client
DreamWorks Animation SKG
Glendale, CA, USA

Projekt Project
DreamWorks Animation Website
Website

Design
Grøtting + Sauter
Hamburg, Germany

Ein internationales Team arbeitet von Hamburg aus an der Neugestaltung der Studio-Site für DreamWorks Animation.
„Die Filmstudios müssen an ihrem Profil arbeiten", erklärt Kreativdirektor John Grøtting. „In den 30er Jahren hatten Studios eine bestimmte Art, Filme zu machen. Die Verbindung des Films zum Studio half, den jeweils nächsten Film zu verkaufen."
Auf der neuen Seite finden sich nicht nur Informationen, Trailer und Previews aller Filme. Fans können sich auch hinter den Kulissen des Animationsbetriebs umsehen. Die Seite dient darüber hinaus der „Jagd nach Talenten" in diesem Wachstumsmarkt und soll Investoren informieren.

An international team from Hamburg worked on the new design of the DreamWorks Animation studio website.
"The film studios need to work on the brand", explains Creative Director John Grøtting. "In the 30s studios each had a unique style. The connection to the studios helped sell their next film."
On the new website you find not only information, trailers and previews. Fans can go behind the scenes and see the talent that worked on the films. Further, the site is a strong tool for recruiting and to educate potential investors.

Auftraggeber Client
Thomas Cook AG
Oberursel, Germany

Projekt Project
„Mix&Travel"-Komponente
Website-Navigationssystem
Website navigation system

Design
Petra Wöhrmann (Design)
Doris Weigl (Konzept)
Ray Sono AG
München, Germany

Mit der neuen „Mix&Travel"-Buchungskomponente will das Online-Reisebüro www.thomascook.de Kunden animieren, sich aus einer großen Anzahl von Flügen, Hotels und Mietwagen individuelle Reisen zusammenzustellen.
Thomas Cook beauftragte Ray Sono mit der Neukonzeption und Überarbeitung der Benutzersteuerung sowie der Logik des Buchungsprozesses der „Mix&Travel"-Komponente. Um den Nutzen und die Funktionalität der Komponente bessern zu erkennen und um den Prozess so benutzerfreundlich wie möglich zu gestalten, wurden neben einer Designoptimierung des Front-Ends die Logik weiterentwickelt und der Zugang zu den einzelnen Bausteinen neu konzipiert.

With its new "Mix&Travel" booking components, the online travel agency www.thomascook.de gives customers a wide assortment of flights, hotels and rental cars to choose from as they put together their own individual vacations.
Thomas Cook asked Ray Sono to develop a new concept for the site's navigation system as well as the "Mix&Travel" component logic, and to revise them. To increase component benefit and functionality and design the process as user-friendly as possible, the front-end design was optimized, site logic developed further and access to individual components redesigned.

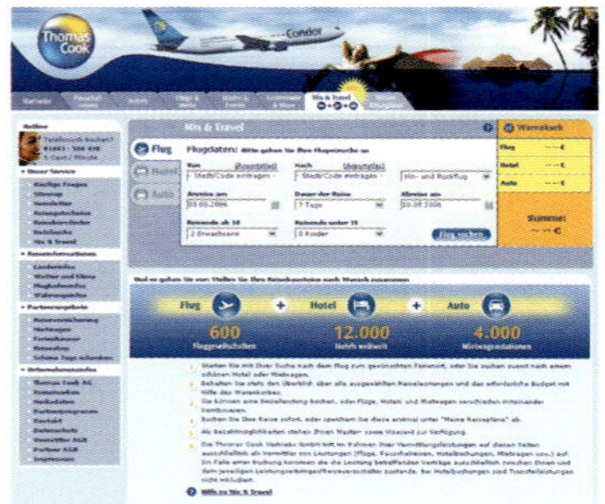

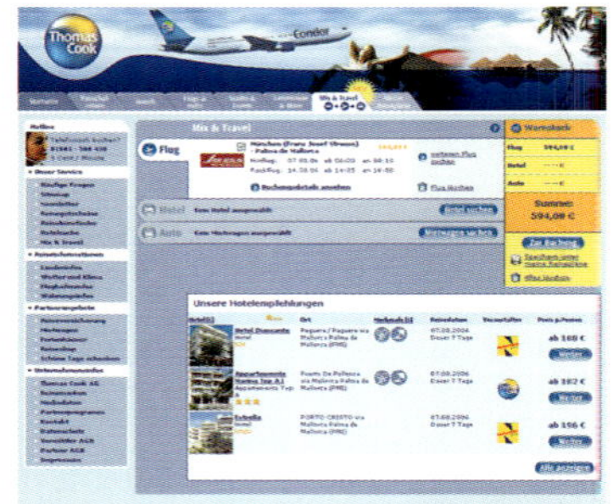

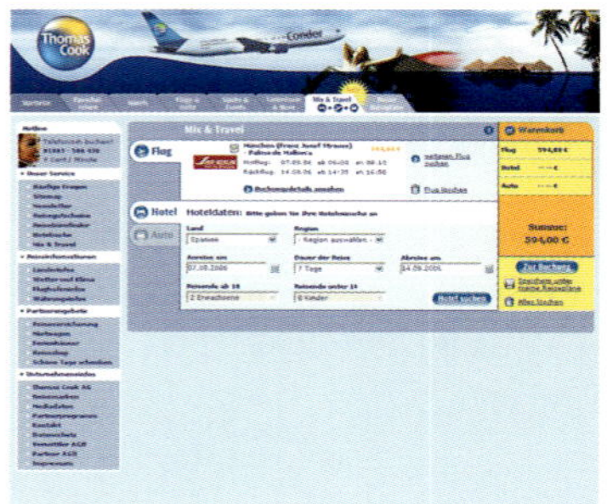

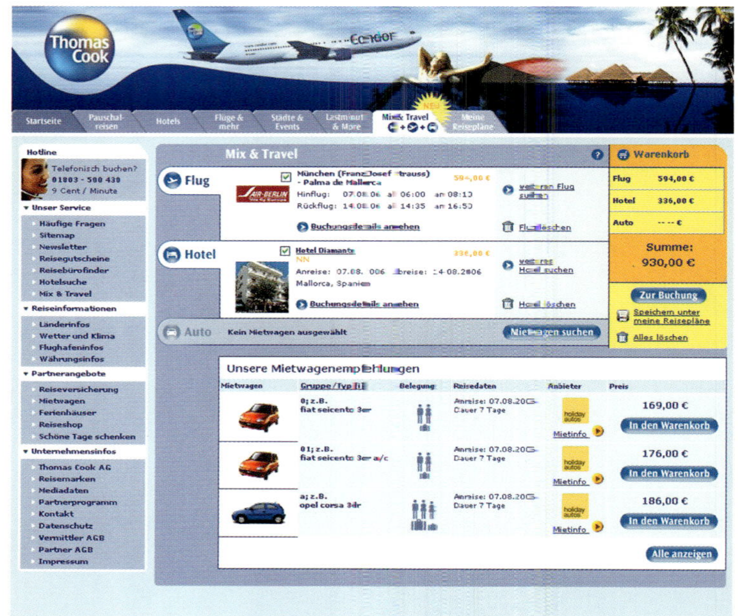

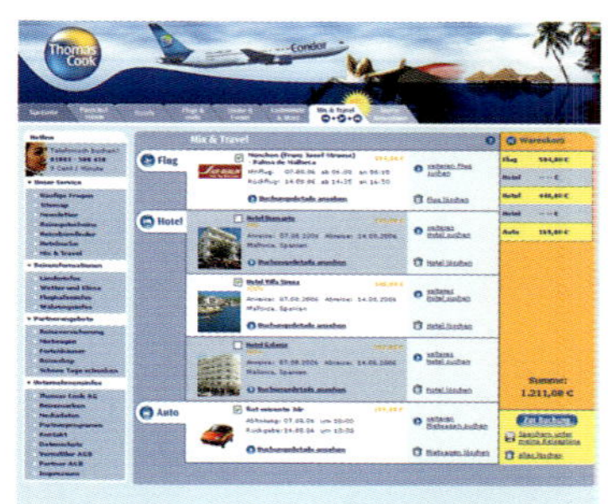

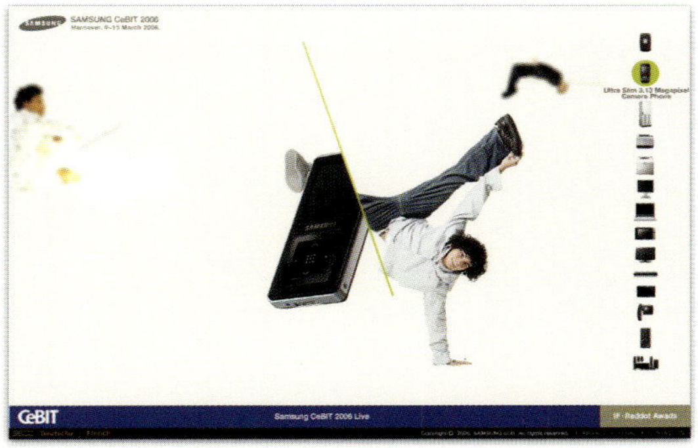

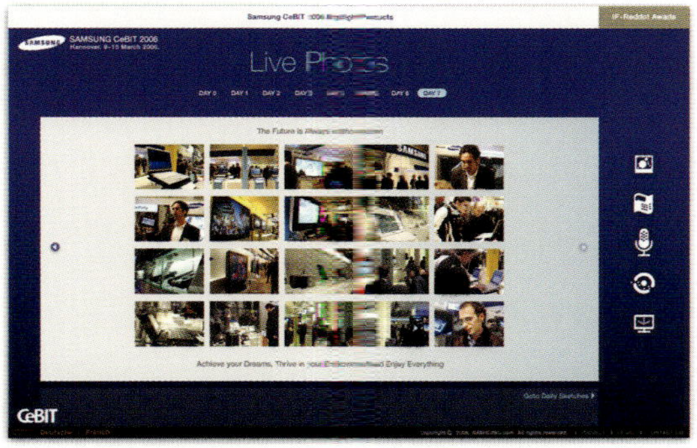

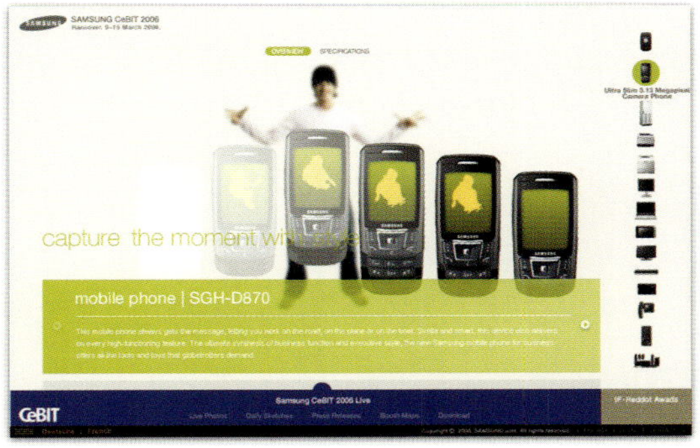

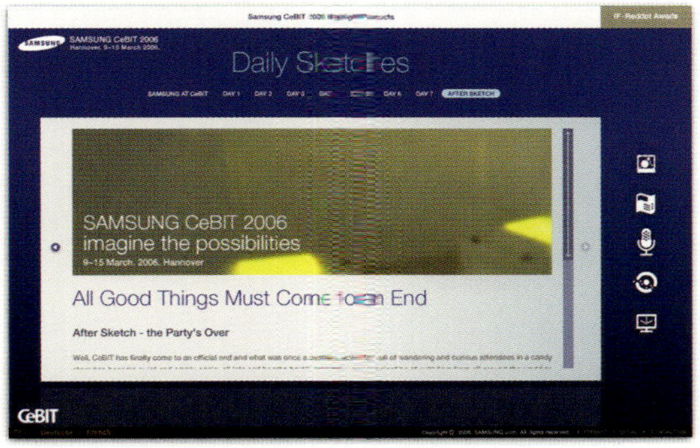

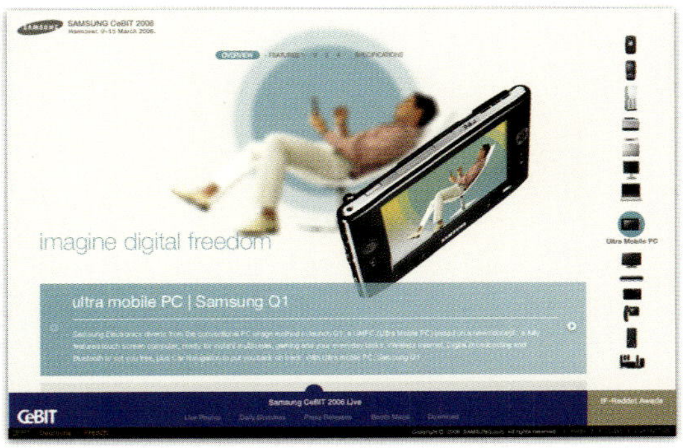

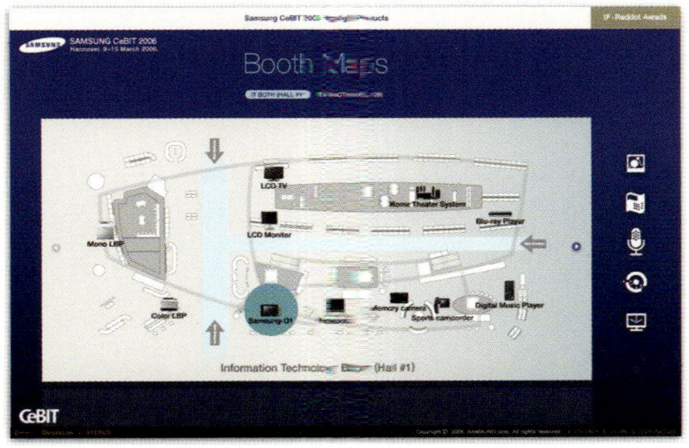

Auftraggeber Client
Samsung Electronics
Seoul, South Korea

Projekt Project
Samsung CeBIT 2006
Microsite

Design
Kangwook Choi
Jungkweon Yoon
graf
Seoul, South Korea

Um ein Produkt mit vielfältigen Anwendungsmöglichkeiten umfassend zu präsentieren, sollte es möglichst facettenreich dargestellt werden. Dazu werden diese Produkte bei „Imagine“ durch eine kontinuierliche Bildfolge unterschiedlicher Ansichten und Anwendungen illustriert. Die Microsite bietet eine Visualisierung des Konzepts, das von Samsung auf der CeBIT vorgestellt wurde.

A product's image needs to be multifaceted to reflect the product's many different aspects. "Imagine" illustrates these aspects, visualizing them by continually changing their dimensions, which was also projected at the Samsung booth during the CeBIT show. The microsite represents a visualization of the concept that was presented at the Samsung booth.

Auftraggeber Client
Open Business Club GmbH
Hamburg, Germany

Projekt Project
openBC.com: Get together – wherever you are
Online-Networking-Plattform
Online networking platform

Design
Christiane Brink
CB Kommunikation+Design
Hamburg, Germany
Nils Martens
Martens CM
Hamburg, Germany

openBC ist die weltweit größte offene Business-Networking-Community im Internet. Mehr als eine Million Netzwerker nutzen openBC in 16 Sprachen, um Geschäftsbeziehungen branchenübergreifend zu vertiefen und das Netzwerk zu erweitern. openBC unterstützt beim Entdecken von interessanten Kontakten, beim Wahrnehmen von geschäftlichen Möglichkeiten und bei der Pflege des Netzwerks. openBC ist ein modulares System und wird ständig weiterentwickelt. Durch die klare Struktur und kontextsensitive Optionen ist der Umgang mit der Applikation intuitiv. Das openBC-Design ist zurückhaltend und unaufdringlich, denn Mitglieder und Inhalte stehen im Fokus.

openBC is the world's largest open online business networking community. More than a million networkers around the globe use openBC in 16 languages to deepen their professional relationships across all industries and expand their network. openBC enables its active members to discover interesting contacts and new business opportunities, as well as maintaining their professional contacts. openBC is a modular system and is constantly being developed. The clear structure and context sensitive options make the handling of the application intuitive. The design of openBC is low key and functional because the focus is on members and content.

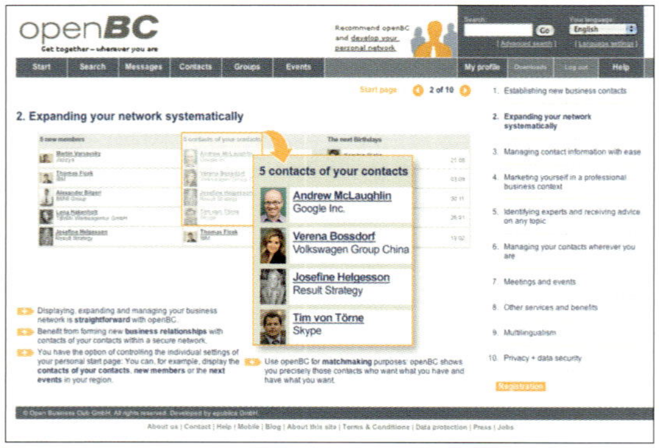

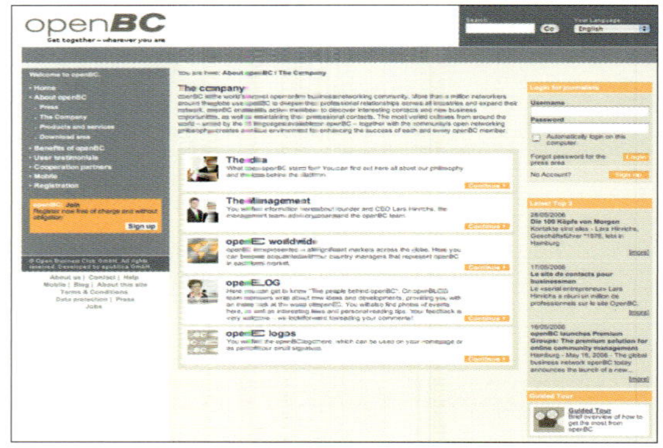

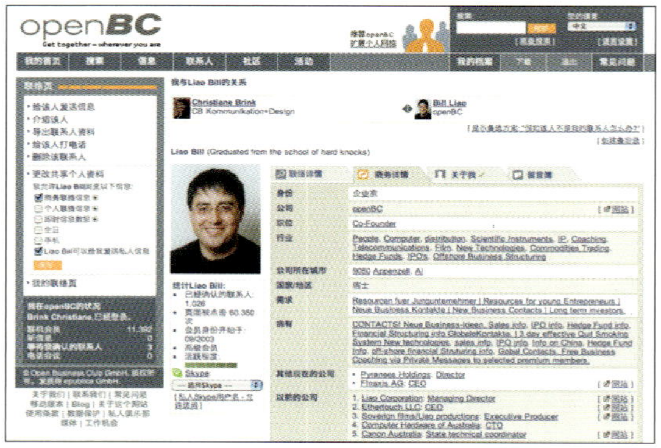

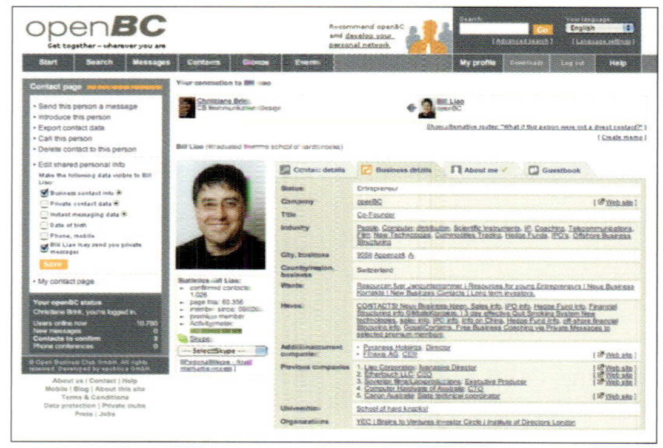

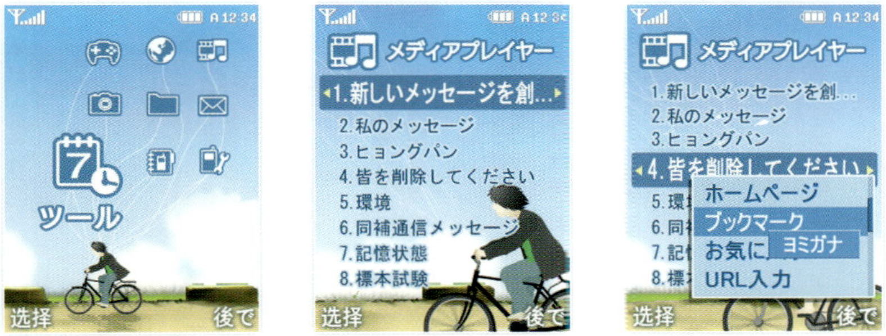

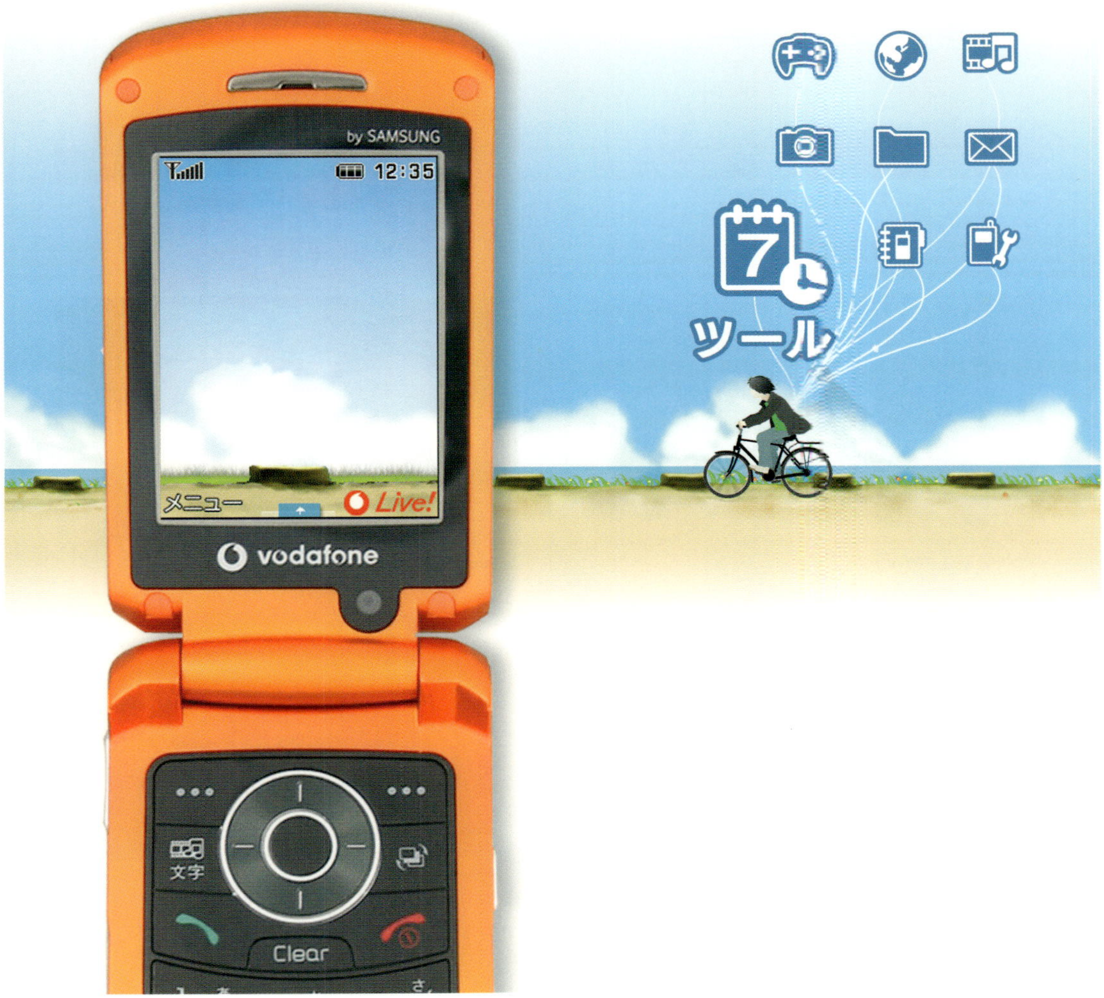

Auftraggeber Client
Samsung Electronics
Seoul, South Korea

Projekt Project
JOY
Graphical User Interface

Design
In-Won Jong
Kyungsik Yoon
Hyemi Hong
Minsuk Kim
Youjin Im
Heekyung Jeon
Heewoong Choi
Samsung Electronics
Seoul, South Korea

Mit JOY wird der funktionelle Zustand des Handys durch bewegliche Bilder angezeigt. Wenn der Nutzer neue Informationen eingibt, geraten die Bilddetails in unregelmäßige Bewegungen, was den Spaßfaktor bei der Benutzung des Programms erhöht.
Durch die Verwendung von Flash-Light 1.1 wurde eine interaktive Benutzeroberfläche geschaffen, wie sie im Bitmap-Format nicht möglich war.

JOY converts the existing static interface which displays the functional state of a mobile phone into an interface with objects changing and moving: The objects on the phone's screen react and move irregularly according to the key input or phone setting, thus offering users interactive fun and visible enjoyment. By applying the flash light 1.1 technology JOY provides an interactive GUI environment which can hardly be implemented in the existing bit map.

Auftraggeber Client
Samsung Electronics
Suwon-city, South Korea

Projekt Project
XHT (Extendable Home Theater)
Schnittstelle für Home A/V Network Solution
Interface for home A/V network solution

Design
Youngjae Ryu
Hyunwoo Kim
Hyunchung Kim
Samsung Electronics
Seoul, South Korea

Mit dem Extendable Home Theater System (XHT) von Samsung Electronics können selbst Neueinsteiger über eine einzige Fernbedienung ihre komplette Unterhaltungselektronik steuern. Bedienungsfreundlichkeit ist durch Einsatz von Standardsystemen gewährleistet, wobei die Skalierbarkeit von ergonomischen Gebrauchsflusssystemen zum Tragen kommt. Mit einem Tastendruck können Programme von diversen Medien wie HTV, STB und HDD abgespielt werden. In verschiedenen Räumen des Hauses können gleichzeitig die selben oder unterschiedliche Inhalte abgerufen werden. Eine spezielle Funktion erlaubt es, den anderen Bewohnern des Hauses Inhalte zu empfehlen, wodurch die Kommunikation untereinander und somit die Lebensqualität gefördert wird.

XHT allows you to play content via a single click, and is designed to provide novices with easy control of HTV, STB and HDD via an integrated remote controller. Standardized systems have been applied to enhance the ease of use, and the scalability of ergonomic, usage-flow design systems has also been taken into account.
The user can simultaneously enjoy the same or different content at multiple locations in the home, and view content in different rooms using a single device. The content recommendation function between family members encourages communication among them and enhances their quality of life.

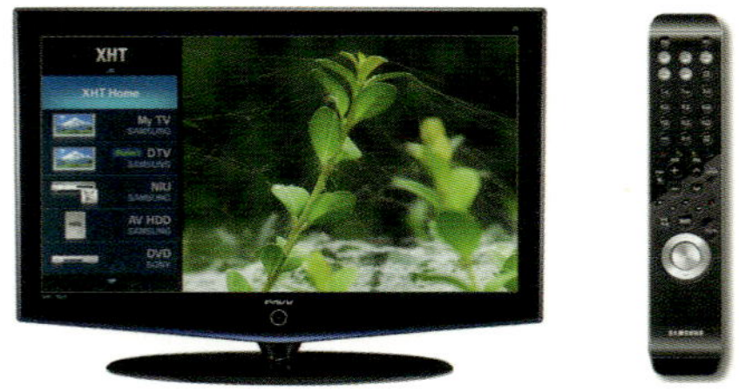

XHT Home

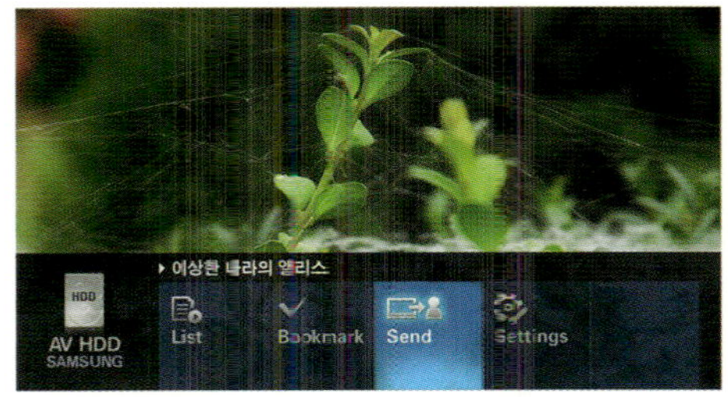

Device Menu (Function Set)

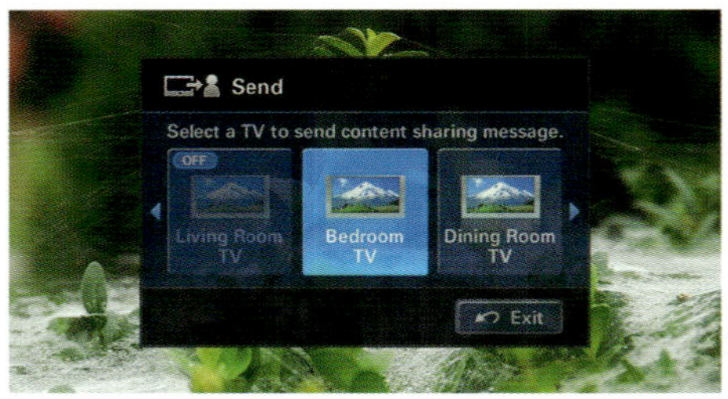

Send Menu (TV to TV sharing)

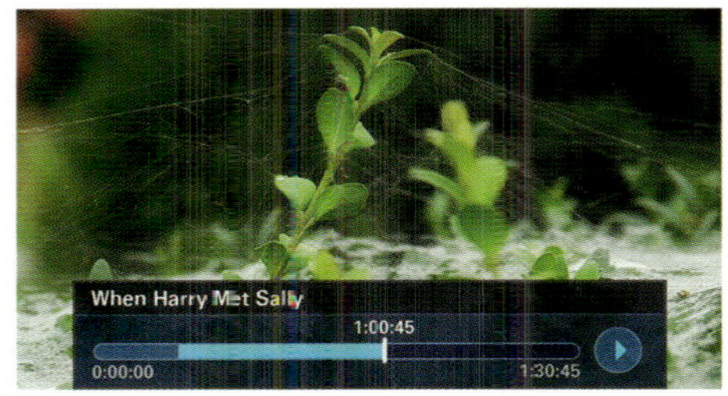

AV Control Feedback

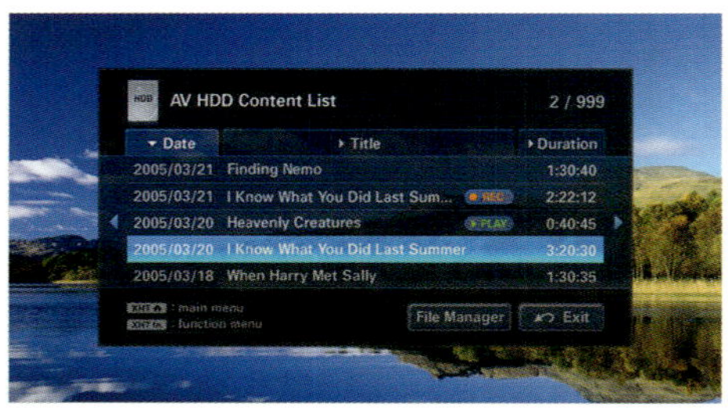

Content List (Movie)

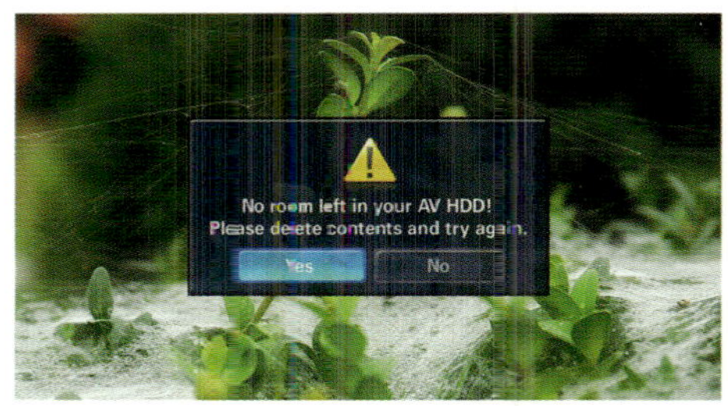

Warning Window

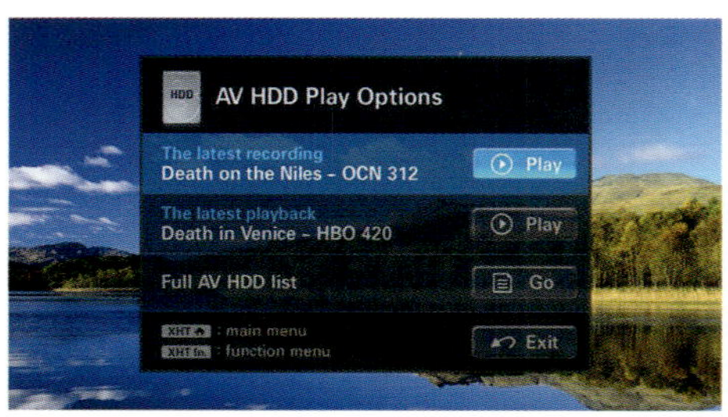

Play Option Menu

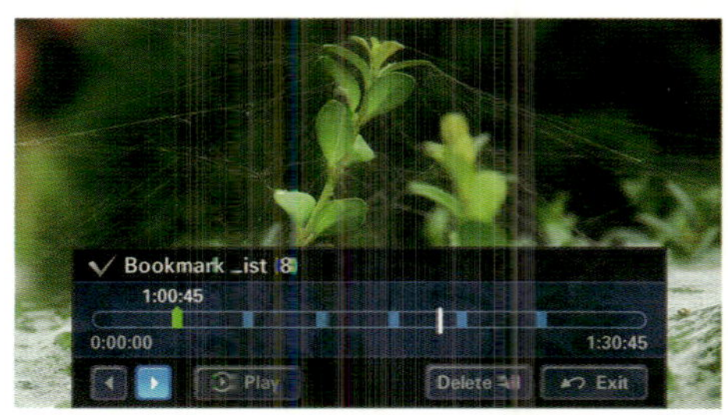

Bookmark List

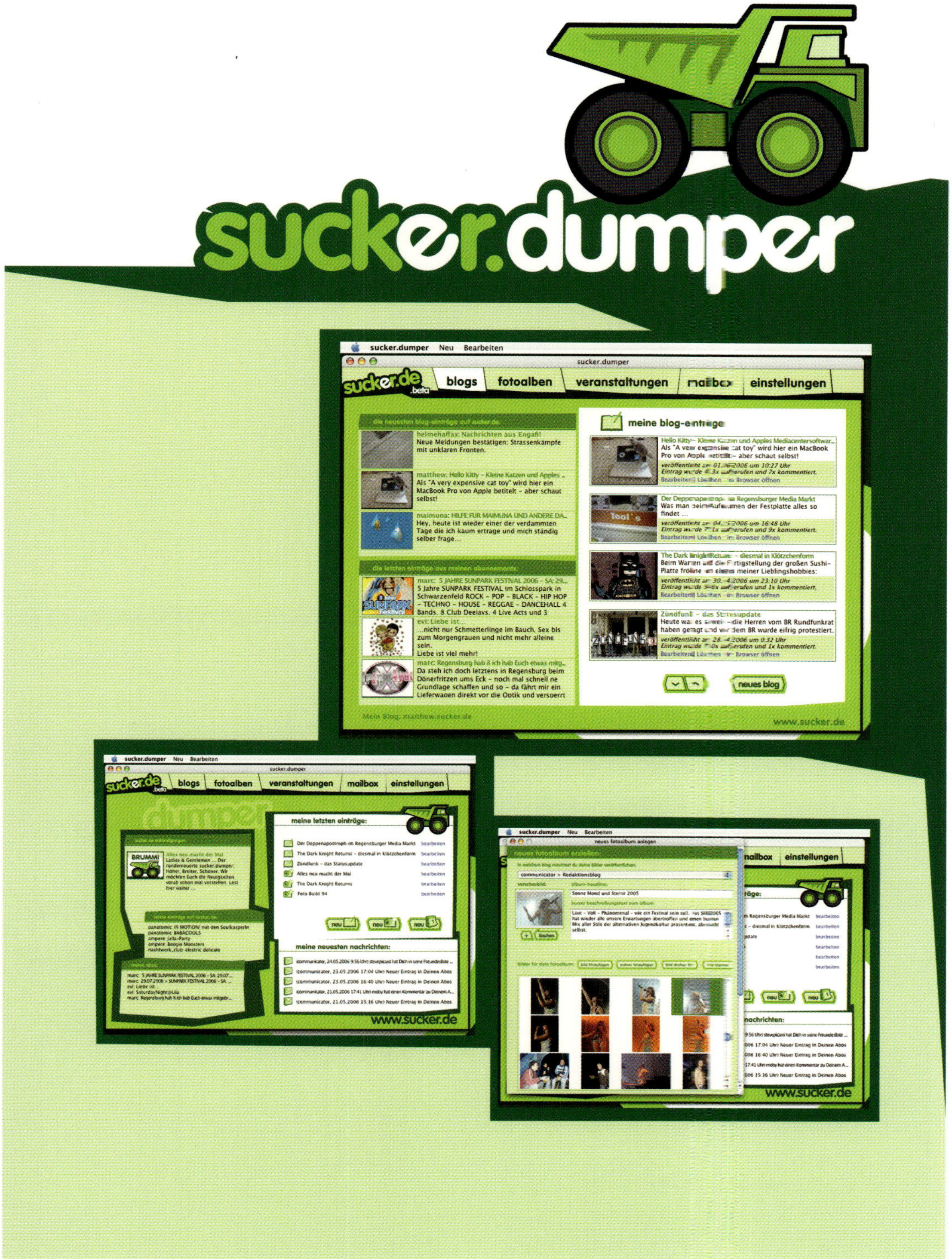

Auftraggeber Client
Kupferwerk GmbH human interface engineering
Regensburg, Germany

Projekt Project
sucker.dumper
Software zur Veröffentlichung von Blog-Beiträgen
Software for publishing blog entries

Design
Matthew Ulbrich
Andreas Gerauer
Gerald Trummer
Kupferwerk GmbH human interface engineering
Regensburg, Germany

Das Veröffentlichen von umfangreichen Mediendaten über einen Web-Browser ist ein sehr langwieriger und häufig benutzerunfreundlicher Prozess. Mediendaten werden in voller Auflösung auf den Server übertragen und anschließend von diesem auf das Ausgabeformat gerechnet.
Die Software sucker.dumper wurde entwickelt, um das Veröffentlichen von Blog-Einträgen auf der Jugend-Community sucker.de zu erleichtern. In einem leicht zu bedienenden und zielgruppengerechten Design ermöglicht die Anwendung das Erstellen von z.B. Fotoalben. Die Daten werden lokal auf das Ausgabeformat gerechnet und bandbreitenschonend auf den Server geladen und veröffentlicht.

Publishing of large media data via web browsers is a lengthy and in many cases user-unfriendly process. Data will be transmitted in full resolution onto the server and afterwards scaled down to the correct format.
The software sucker.dumper has been developed in order to simplify the publishing of blog entries at the youth community www.sucker.de. Featuring an easy-to-use and target-group-relevant design, the application enables the easy creation of photo albums. Data will be scaled down to the right format at clientside and transmitted onto the server in a bandwidth-friendly way.

Auftraggeber Client
A. Monforts Textilmaschinen GmbH & Co. KG
Mönchengladbach, Germany

Projekt Project
safe control
Anlagen- und Prozesssteuerung für Textilmaschinen
User-centred operations interface for textile finishing machines

Design
Prof. Detlef Rahe
Michael Wilde
Melanie J. Köhler
Marion Fröhlich
:i/i/d Institut für Integriertes Design
Bremen, Germany

Monforts ist ein Hersteller von hochkomplexen Anlagen zur Textilveredelung. Um die Bedienung der weltweit eingesetzten Anlagen schneller und sicherer zu machen, wurde ein einheitliches Interaktionsprinzip entwickelt, das die bisherige Vielzahl von Steuerterminals ablöst. Als Hauptnavigationswerkzeug dient eine horizontal verschiebbare virtuelle Lupe, die es dem Benutzer bei der Anlagensteuerung erlaubt, zwischen Gesamtansicht und einzelnen Bereichen zu wechseln. Das Interface nutzt eine klare und leicht verständliche Symbolik unter Verzicht auf Schriftzeichen und ermöglicht so eine schnelle Einarbeitung der Nutzer.

Monforts manufactures complex textile finishing machines. For faster and more secure operation of the worldwide used system, a new and innovative interface was developed and designed that replaced the previous system with several interfaces. The main navigation instrument is a virtual magnifier (comparable to a zoom). By scrolling horizontally, the operator can switch from complete overviews to detailed informations. The interface uses a clear and coherent codification, almost without text. It enables quick orientation and interaction. The operating becomes safer, faster and more attractive.

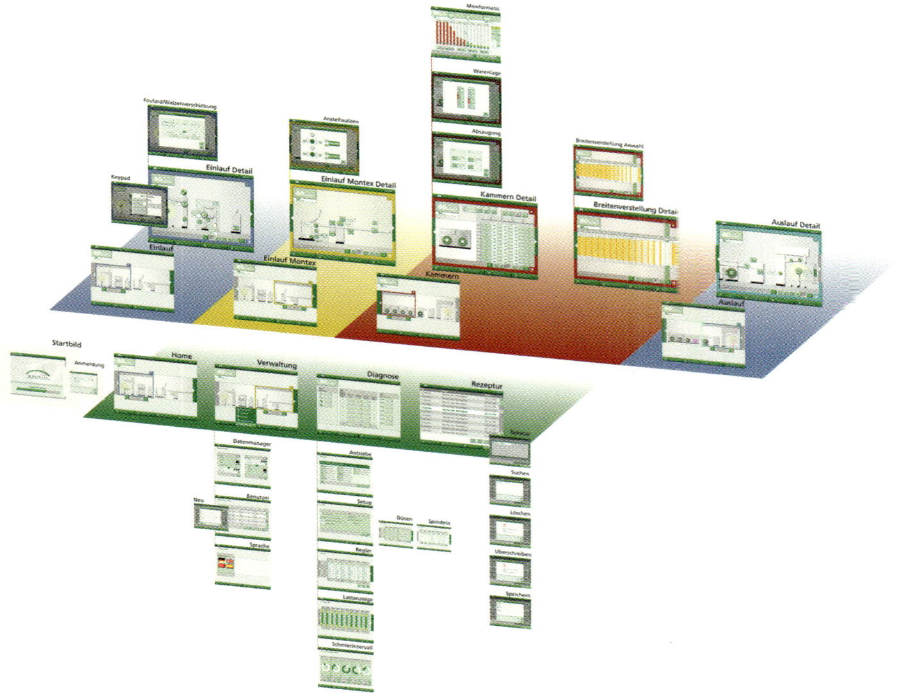

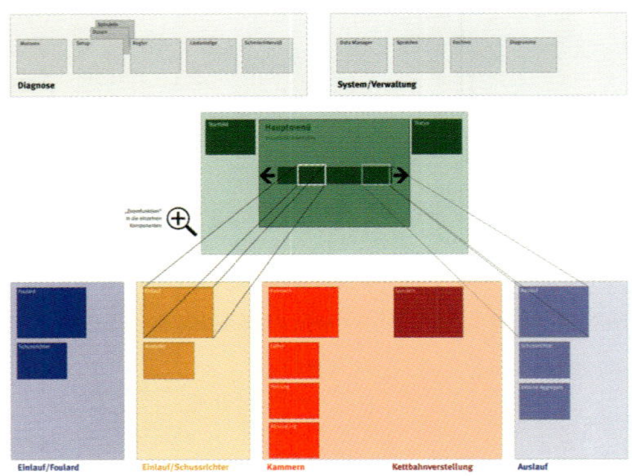

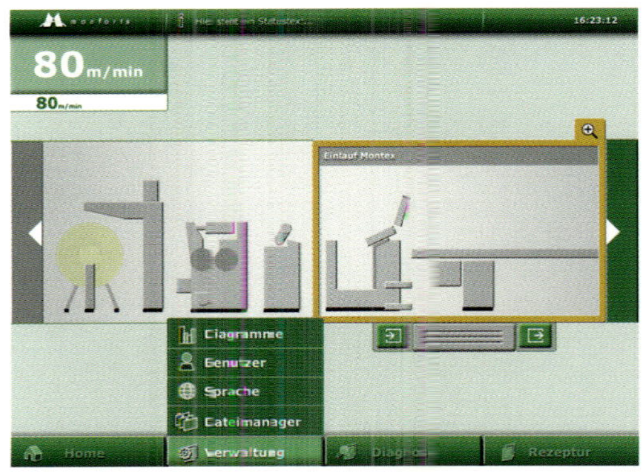

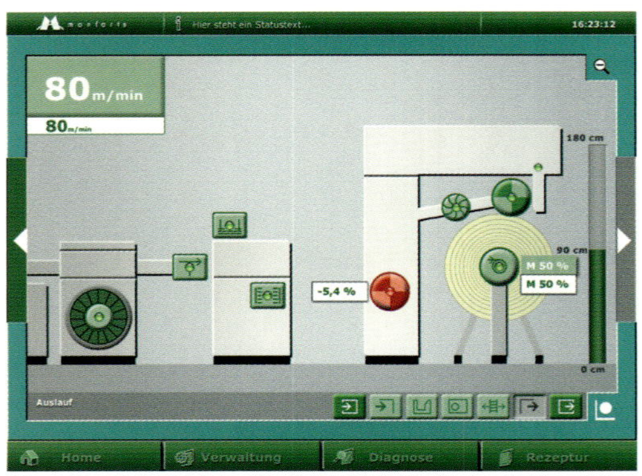

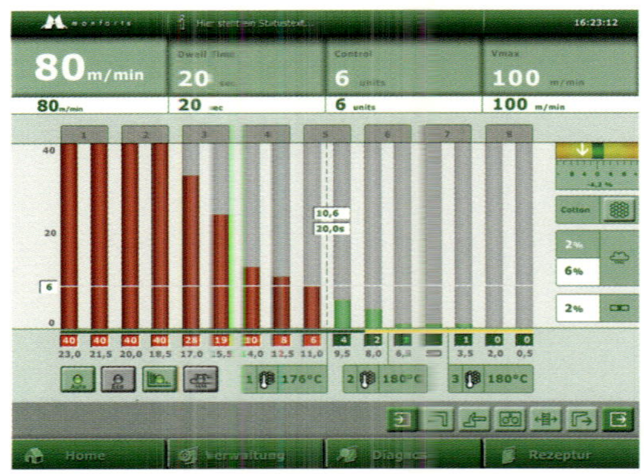

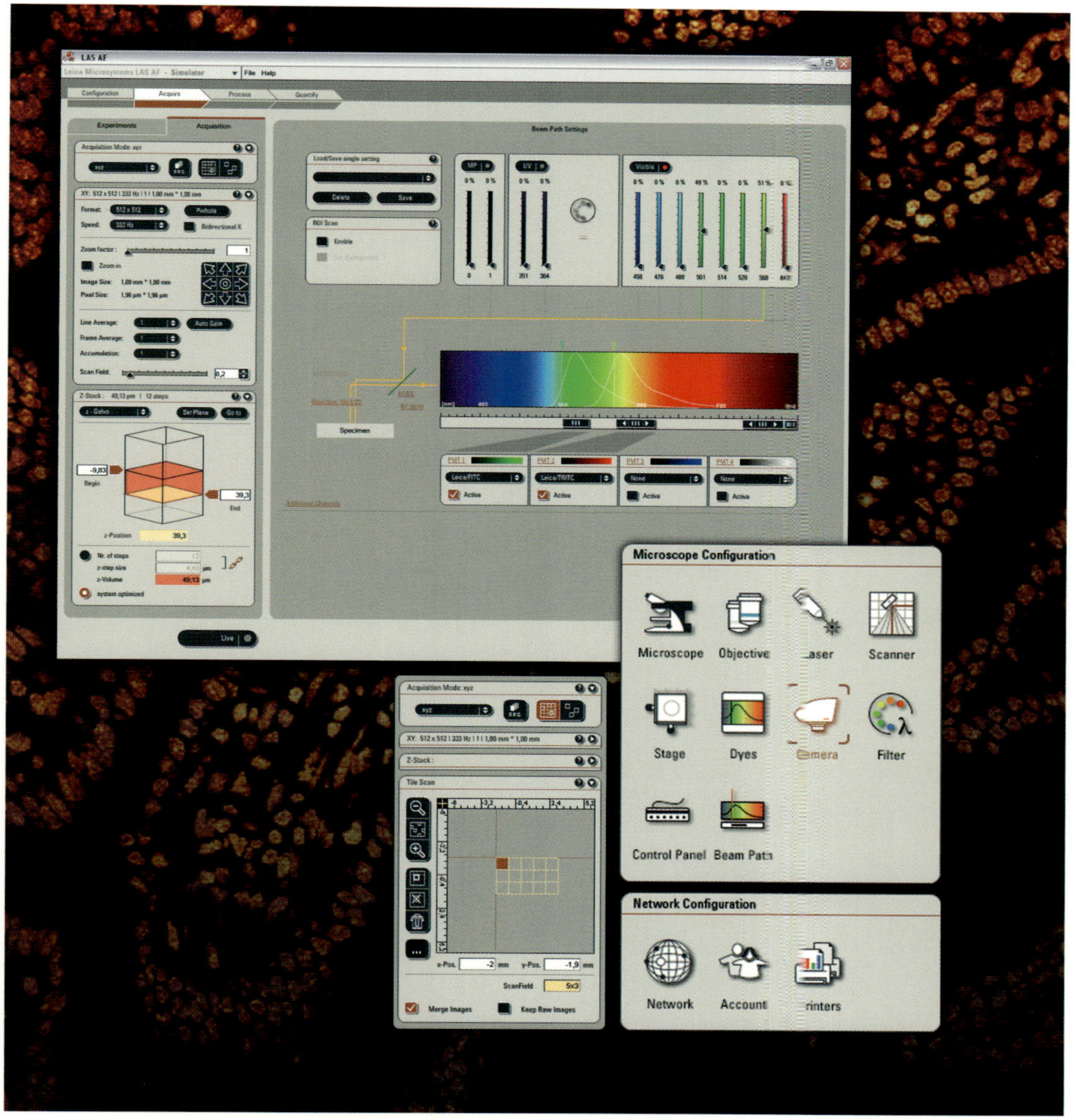

Auftraggeber Client
Leica Microsystems CMS GmbH
Mannheim, Germany

Projekt Project
Leica Application Suite AF
Software für konfokale Mikroskopie
Software for confocal microscopy

Design
ergon.design
Braunschweig, Germany

Die grafische Bedienoberfläche wurde auf der Grundlage von Usability-Untersuchungen einer umfassenden Neugestaltung unterzogen. Eine wesentliche Zielvorgabe war die effiziente Nutzbarkeit des komplexen Funktionsangebots durch unterschiedlich qualifizierte Anwender mit stark variierenden Anwendungszielen. Dazu wurden sämtliche Funktionen in Hinblick auf die gebräuchlichsten Anwendungsprozeduren neu strukturiert und Workflow-optimiert. Weiteres Gestaltungsziel war die Realisierung einer der Technik adäquaten, hochwertigen Qualitätsanmutung sowie der Aufbau eines Corporate Design für Softwareprodukte der Leica Microsystems CMS GmbH.

The graphics for this user interface underwent a comprehensive redesign based on usability research findings. One of the main aims was to make the complex functions easy and efficient for users who have varying degrees of expertise and widely differing objectives in using them. In addition, all the functions were restructured based on the most widely deployed application procedures, and optimized in terms of workflow. Another design objective was to create a high-quality look that matched the technology, as well as to develop a corporate design for software products from Leica Microsystems CMS GmbH.

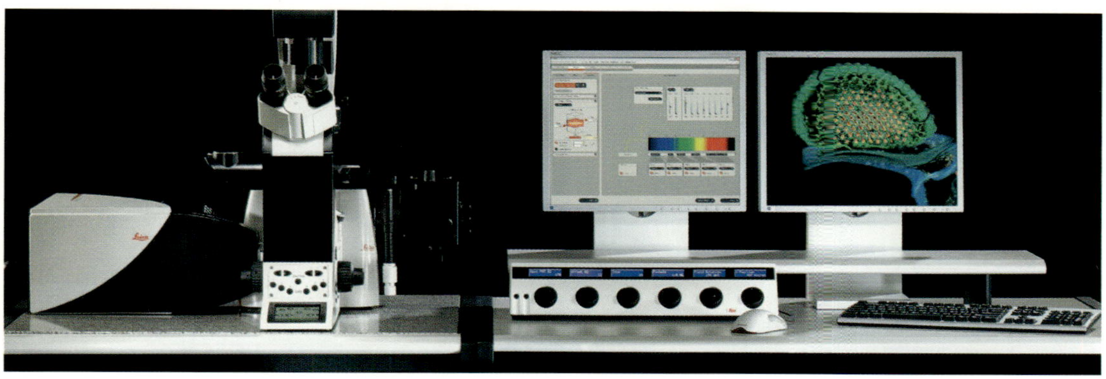

multiple winner: digital media interface, digital media screendesign

Auftraggeber Client
biozoom Services GmbH
Düsseldorf, Germany

Projekt Project
biozoom
Interaktives Produkt
Interactive product

Design
Claus Görner
Marcel Nero
User Interface Design GmbH
Ludwigsburg, Germany

multiple winner

biozoom ist ein biometrischer Scanner, der in Kombination mit einem Handy die Bestimmung von Wasserhaushalt, Körperfett sowie körperlicher und geistiger Fitness ermöglicht. Der Benutzer wird schrittweise durch die Messung geführt: Von der Profilerstellung bis zum Ergebnis erhält er laufend umfangreiche Hilfestellungen. Eindeutige Anweisungen und Benennungen erleichtern dem Benutzer den Umgang mit biozoom. Durch die klare grafische Darstellung kann der Benutzer die Messergebnisse, die auf dem Handy-Display angezeigt werden, schnell erfassen. Für eine leichte, intuitive Bedienung werden die Navigations- und Eingabetasten des Handys verwendet.

biozoom is a mobile biometric scanner which, in combination with a mobile phone, allows the user to determine body water content and body fat as well as physical and mental fitness. Supported by step-by-step instructions, the user is guided through the measurement process from the generation of a profile to the end result. Clear instructions and terminology make it easy for the user to operate biozoom. The measurement result is displayed on the mobile phone and can be quickly understood by means of a clear graphical design. The navigation buttons of the mobile phone ensure easy and intuitive operation.

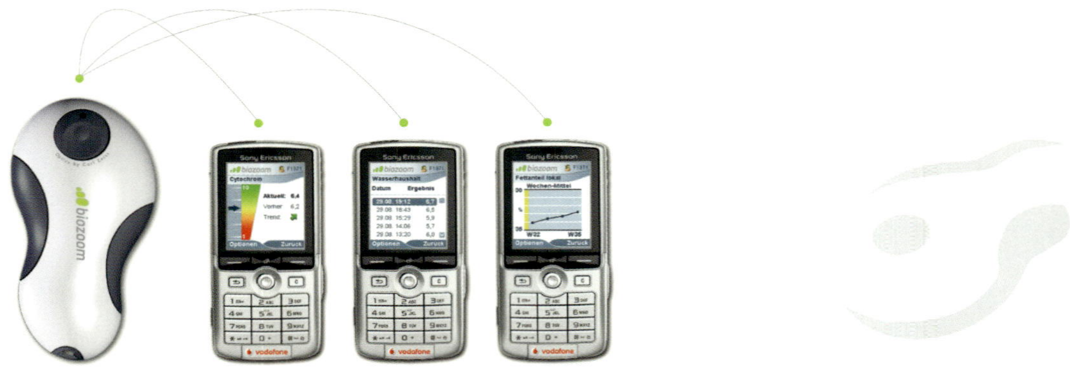

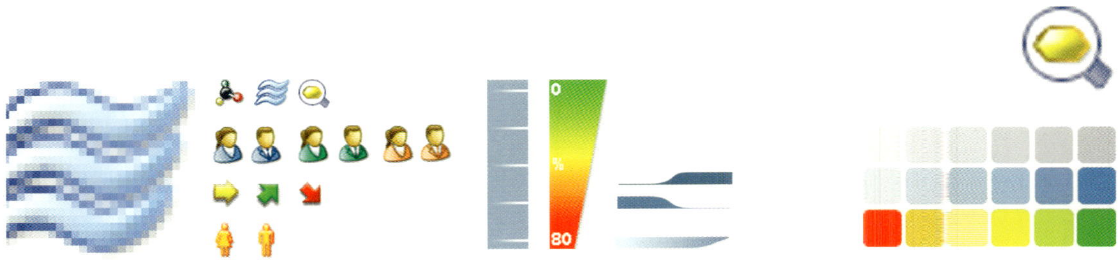

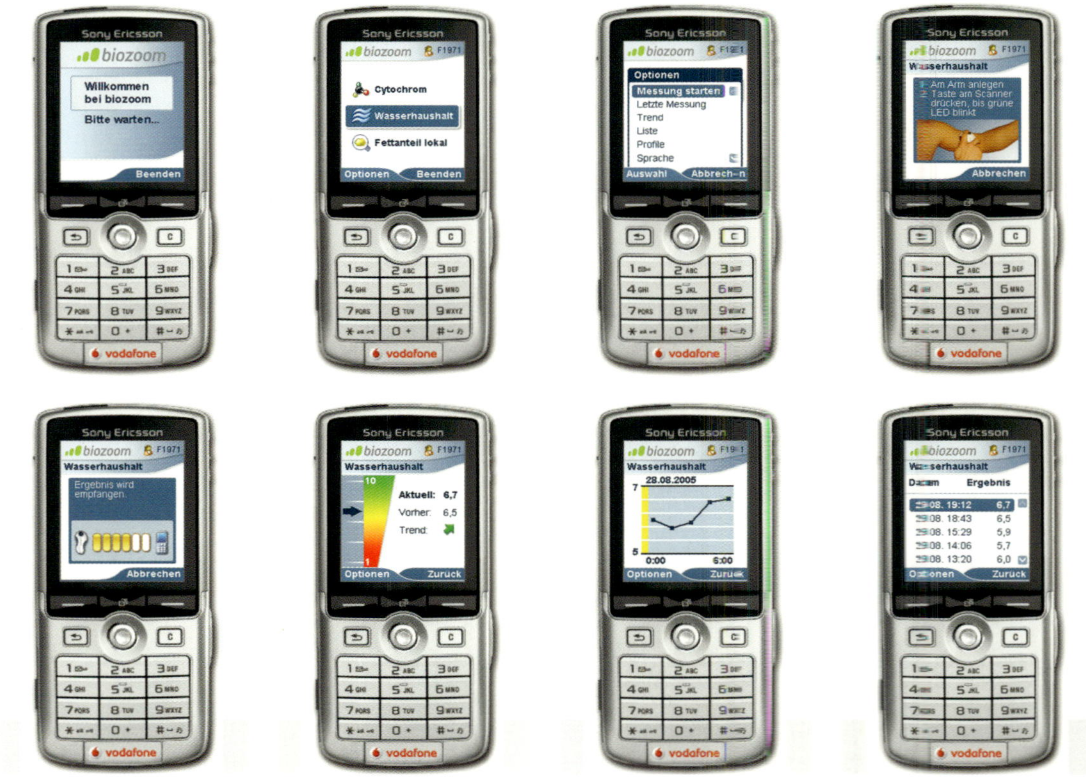

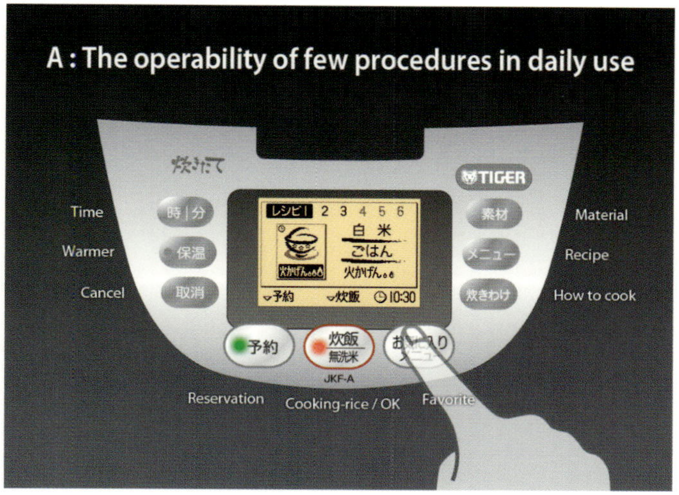

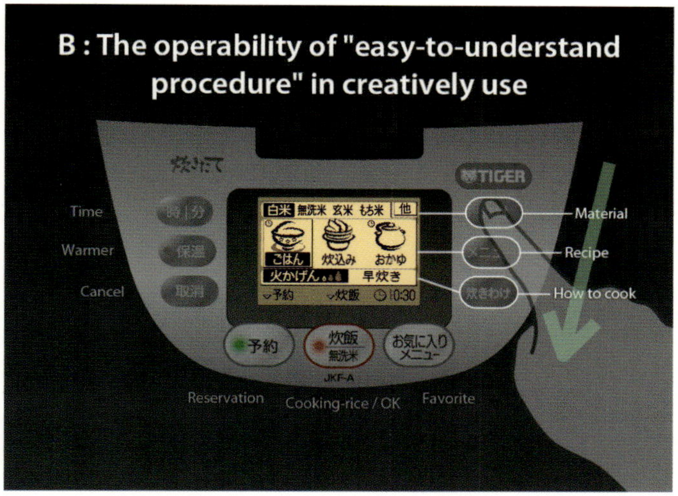

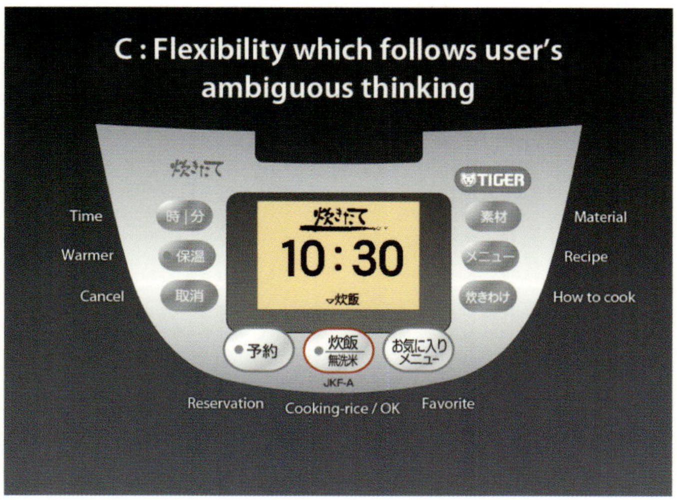

Auftraggeber Client
Tiger Corporation
Kadoma city, Osaka, Japan

Projekt Project
Electronic Rice Cooker with Warmer
Schnittstellen-Design
Interface design

Design
Fumiaki Kondo
Akiko Kitao
Takashi Ishihara
Emi Omoto
dmc.Co.,Ltd.
Yokohama city, Japan

multiple winner

Die Gestaltung wird dem Anspruch eines hochwertigen Produkts gerecht. Dank der Schönheit und der klaren Lesbarkeit der binären LCD-Anzeige, die eine Art „Sumi-e“ Ikonen verwendet, kommt sowohl die Einzigartigkeit des Geräts als auch seine Verbundenheit mit der feinen Küche zum Ausdruck.
Zur Festlegung der Ikonen wurde ein intelligent vernetztes, flexibles Modell von Denk- bzw. Entscheidungsprozessen herangezogen. Für den alltäglichen Gebrauch präsentiert die Anzeige eine leicht überschaubare Anzahl von Routinefunktionen. Dem kreativen Koch bietet die Anzeige leicht verständliche Anweisungen für komplexere Garprozesse. Dieser Ansatz ist ein Abbild der vielschichtigen Denkstrukturen der Benutzer.

This design is appropriate to absolute top-of-the-line products. Through its beauty and clear visibility on the binary LCD screen, the "sumi-e" style icon expresses both the product's originality and commitment to fine cooking.
A coherent and intelligent flexible mental model was deployed, resulting in a small number of routine procedures (important for normal usage) and easy-to-understand operation for more complex procedures (important for creative usage). This approach mirrors the user's own pluralistic thought patterns.

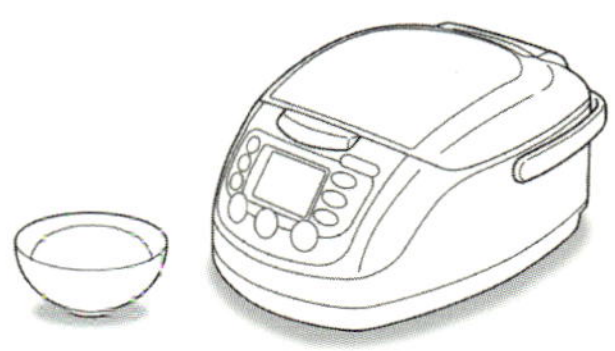

Auftraggeber Client
Montblanc International GmbH
Hamburg, Germany

Projekt Project
Montblanc 100 Jahre
Webspecial

Design
Dirk Kedrowitsch (GF)
Yvonne Grass,
Sven Mentel (CD)
Ulrich Scheele (Text)
Sylvia Videaux,
Christoph Klingebiel (Grafik)
Alexander Funke (Webdevelopment)
Claus Jacobsen,
Jost Thedens (Account)
Sebastian Hahn (Client)
Elephant Seven AG
Hamburg, Germany

Montblanc wird in diesem Jahr 100. Klar, dass eine der Luxusmarken der Welt dieses Jubiläum entsprechend begeht. Zum Beispiel mit einem eigenen, über Jahre entwickelten Diamanten: dem Montblanc Diamond. Dieser steht auch im Zentrum dieses Webspecials, denn er ist nicht nur das Symbol des Jubiläums, sondern dient auch dem Special als Opener, als Key Visual und als emotionale Klammer. Der Montblanc Diamant hat diverse Facetten – wie die Marke Montblanc selber auch. Diese Facetten prägen als optisches Element die Website – sie strukturieren die Seite, trennen Inhalte und fügen sie wieder zusammen. Und: Sie geben der Site ihren ganz speziellen, unverwechselbaren Look.

This year, Montblanc turns 100. And it goes without saying that one of the top luxury brands in the world will be celebrating its anniversary in fitting style. With its own diamond that took years to develop, for example – the Montblanc Diamond. And last but not least, with its own Web special to mark the big 100. This Web special is all about the Montblanc Diamond. This not only represents the brand's anniversary, but also serves as an opener, a key visual and an emotive framework for the special. The Montblanc Diamond is, just like the brand Montblanc itself, multifaceted. Its various facets serve as visual elements of the website – they give it structure, dividing the content and bringing it back together again.

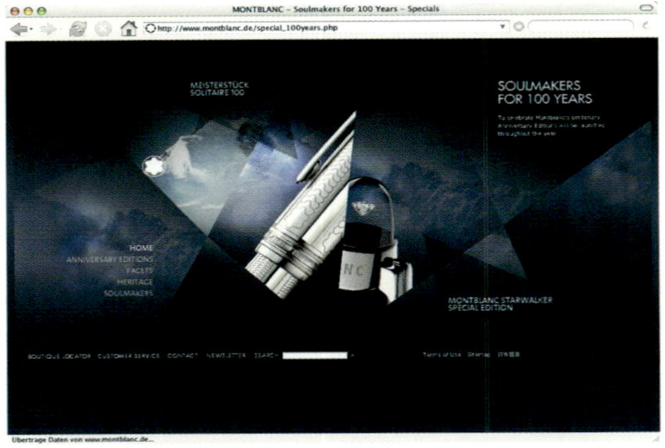

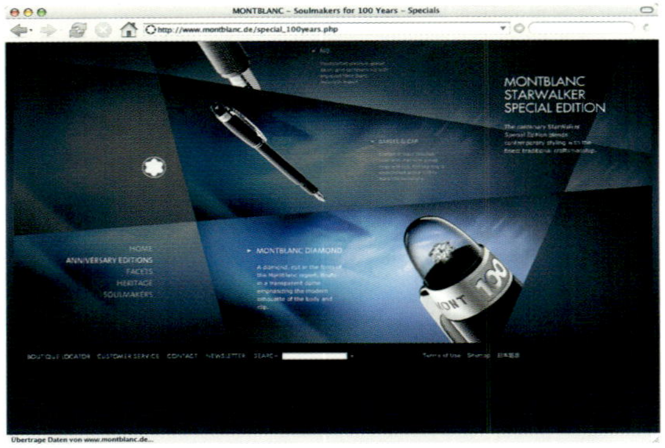

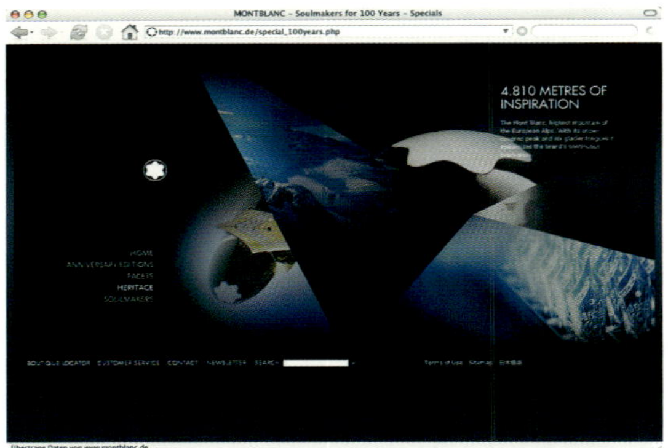

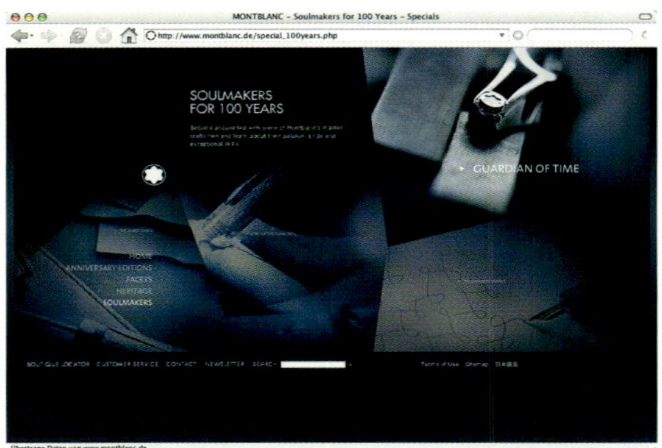

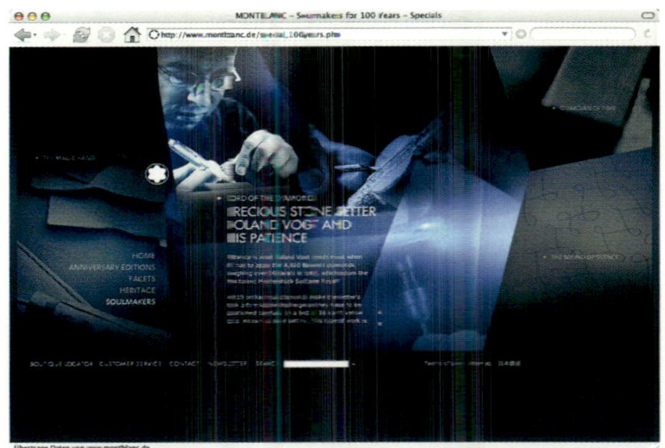

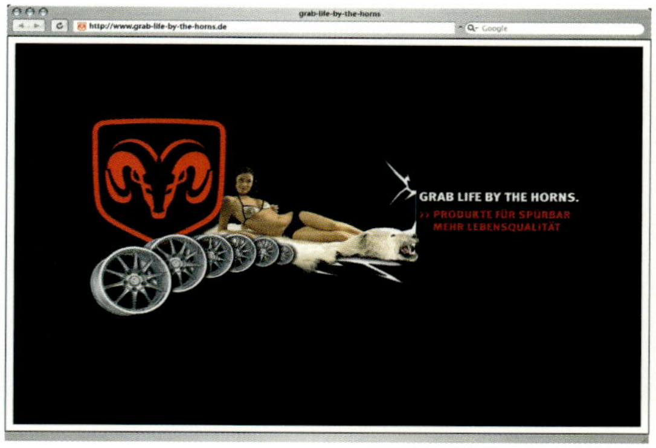

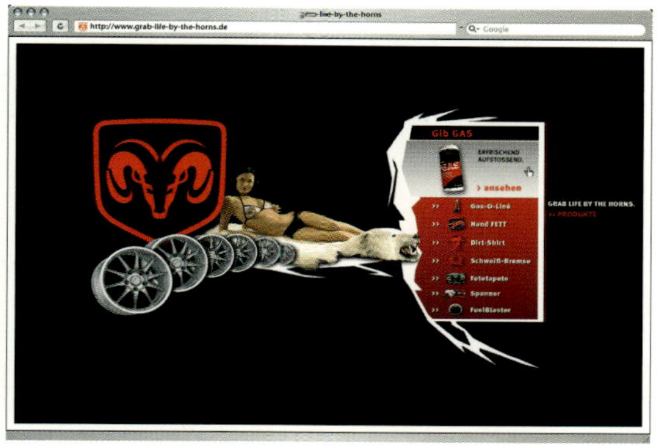

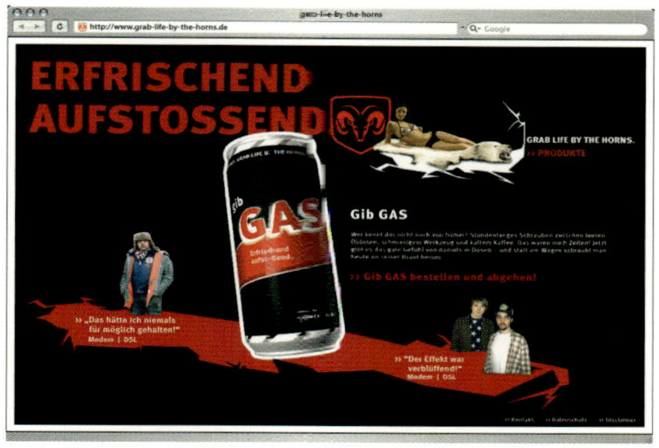

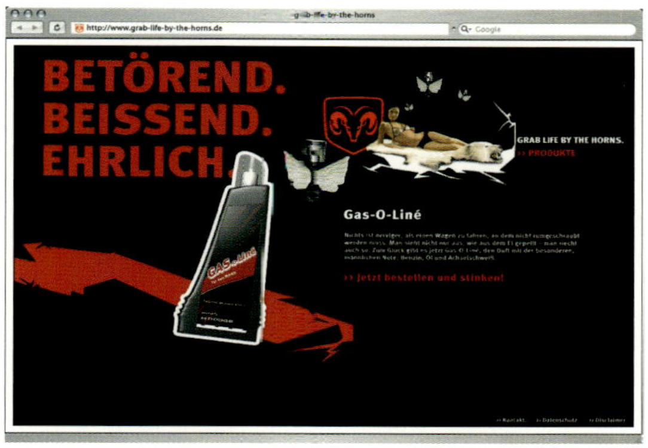

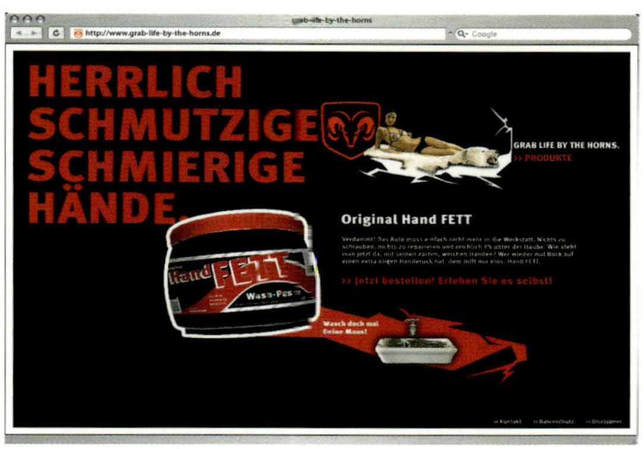

Auftraggeber Client
Chrysler Deutschland GmbH
Berlin, Germany

Projekt Project
Dodge www.grab-life-by-the-horns.de
Webspecial

Design
Dirk Kedrowitsch (GF)
Dirk Ollmann,
Daniel Richau (CD)
Sven Giese, Kai Becker (AD)
Till Hinrichs,
Markus Schmidt (Grafik)
Benjamin Bruno (Text)
Arne Otto (Flash)
Claus Jacobsen,
Stefanie Rohde,
Matthias Mühlenhoff (Account)
Robert Scheffler,
Matthias Möhler (Client)
Elephant Seven AG
Hamburg, Germany

Die Kampagne zur Einführung der Marke Dodge in Deutschland steht unter dem Motto: „Für den Mann". Plattform für alle kommunikativen Maßnahmen ist das Webspecial. Um den Nutzen von Dodge-Automobilen zu kommunizieren, werden hier in der ersten Phase nur skurrile Produkte vertrieben, die MANN braucht, wenn – Dodge sei Dank – Tuning und Schrauben nicht mehr nötig sind: Handfett für schmutzige Hände oder Gas-O-Liné für den Schraubergeruch werden über eine außergewöhnliche Navigation von Miss Tschechien kommentiert und begleitet. In der zweiten Phase wird die Marke über verschiedene Tools und Gimmicks spielerisch näher gebracht. In der dritten Phase wird der Dodge Caliber als Fahrzeug präsentiert und kommuniziert.

The market launch campaign for Dodge in Germany is clearly aimed at men. The platform for all communicative measures is the Web special. To communicate the advantage of Dodge cars clearly, in the first phase, the site only sells novel products that men might need when they no longer need to fiddle around with tuning and screws thanks to Dodge. These include hand grease for dirty hands or Gas-O-Liné for that mechanic smell, all of which are presented via an unusual navigator by Miss Czech Republic. In the second phase, users are familiarised with the brand in a fun way via various tools and gimmicks on the site. In the third phase, the actual Dodge Caliber is presented and advertised.

Auftraggeber Client
Audi AG
Ingolstadt, Germany

Projekt Project
Audi Roadjet Concept
Webspecial

Design
Dirk Kedrowitsch (GF)
Sven Mentel,
Oliver Viets (CD)
Markus Weiß,
Matthias Wagener (Account)
Michael Finke,
Sandy Zeng (Client)
Elephant Seven AG
Hamburg, Germany

Ihre Premiere feierte die Designstudie „Audi Roadjet Concept" auf der Detroit Motor Show Anfang 2006. Begleitend wurde die auf Komfort und Funktionalität ausgelegte Designstudie mit einem sechssprachig realisierten Webspecial vorgestellt. Das völlig neue Raumkonzept des Audi Roadjet spiegelt sich in der räumlich umgesetzten Menüführung wider. Gegliedert in fünf Hauptbereiche erhält der Anwender mit jedem Klick Aufschluss über das Fahrzeugkonzept mit seinem variablen, reise- und wellnessorientierten Innenraum und über Innovationen in Antriebstechnik und Car-to-Car-Kommunikation. Statischem Bildmaterial wurde mit einem durch den Anwender interaktiv steuerbaren Screen- und Sounddesign Dynamik verliehen.

The Audi Roadjet Concept celebrated its premiere at the Detroit motor show at the beginning of 2006. The launch of the study, which focuses on comfort and functionality, was accompanied by a Web special in six languages. The Audi Roadjet's all new space concept – its main characteristic – is reflected in the spaciously designed menus. The site is divided into five main sections, with each click offering the user an insight into the concept with its adaptable interior designed to maximise well-being during travel, and innovations in terms of the engine and car-to-car communication. Stills are given a sense of dynamism by the interactive screen and sound design.

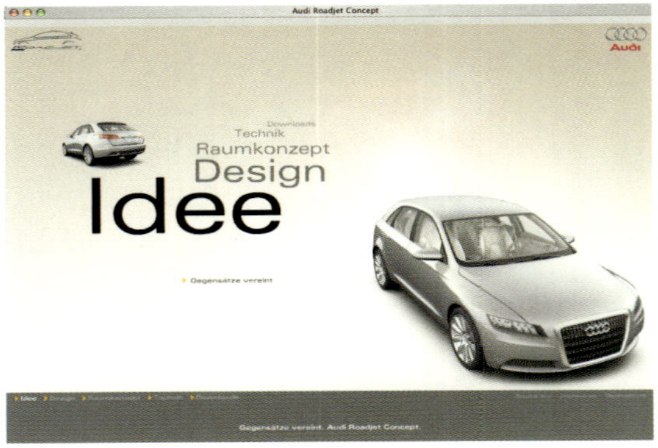

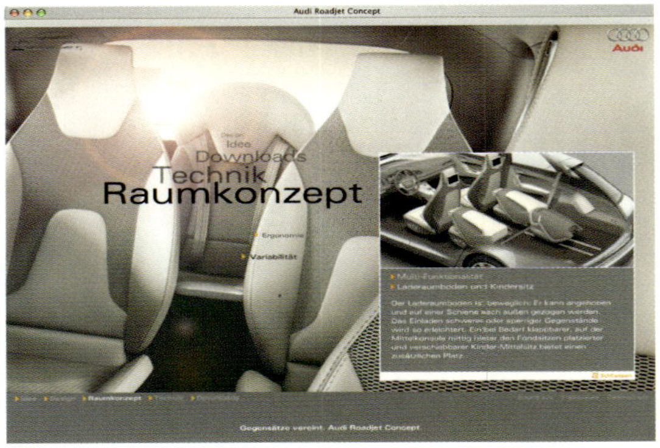

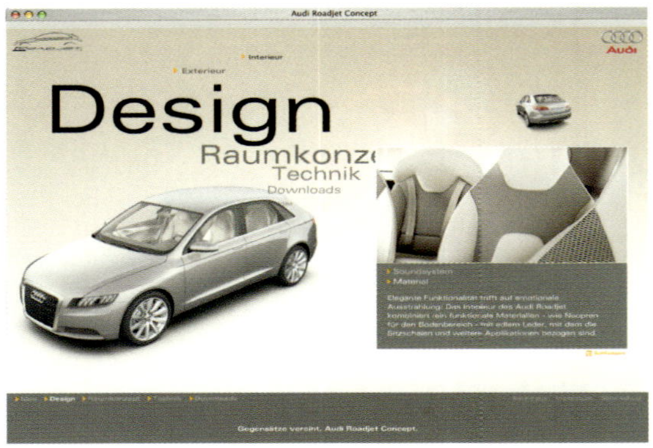

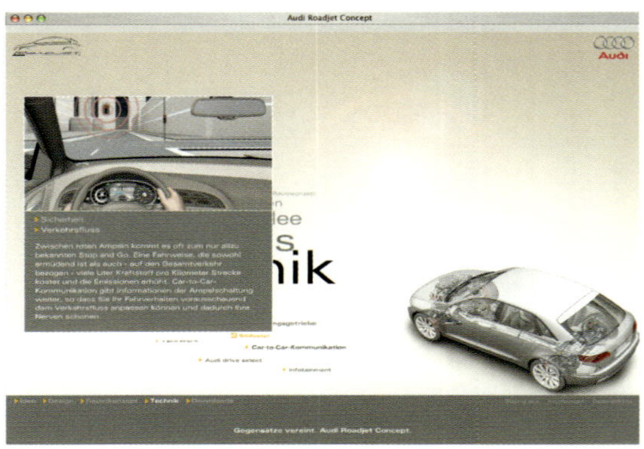

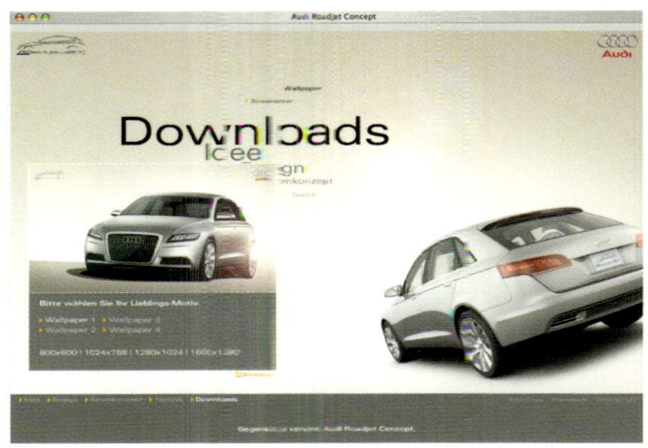

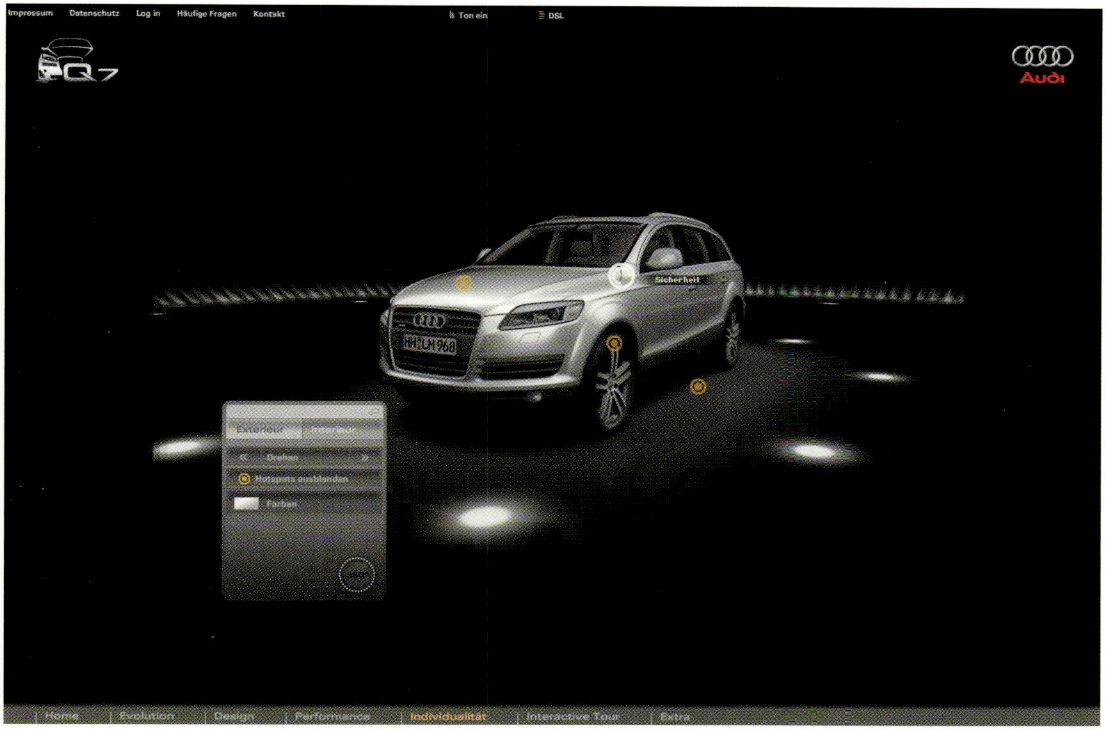

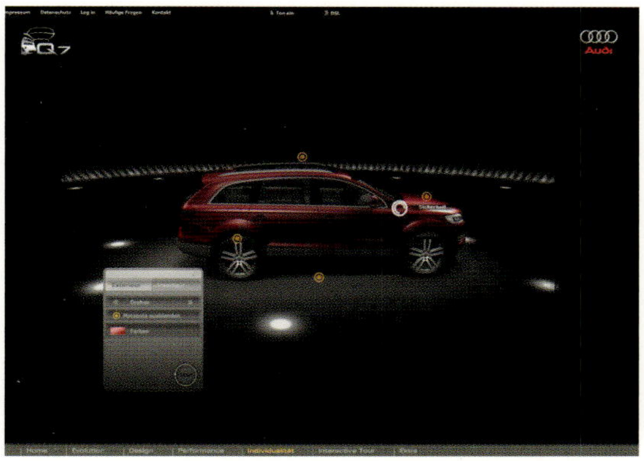

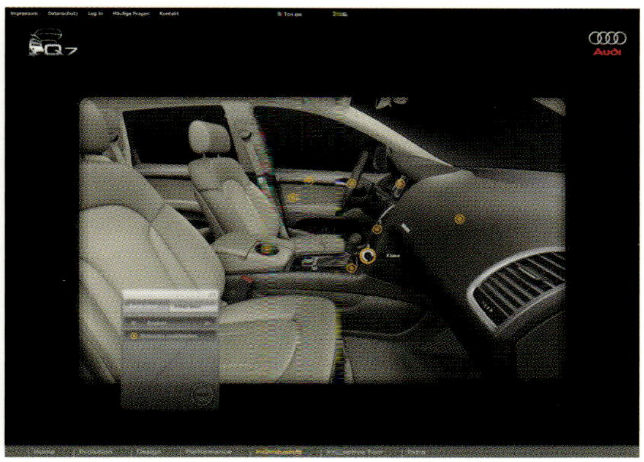

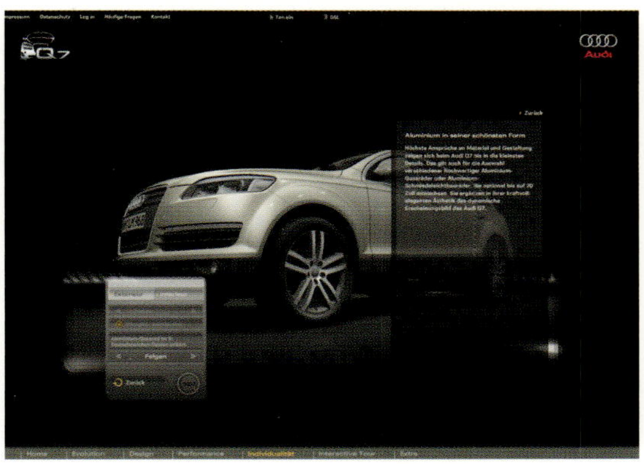

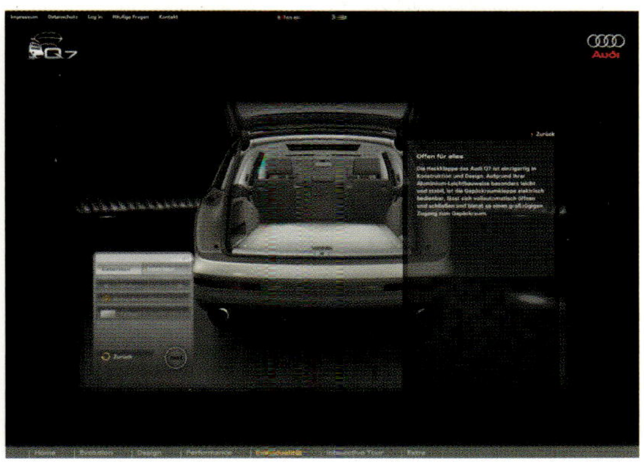

Auftraggeber Client
Audi AG
Ingolstadt, Germany

Projekt Project
Audi Q7-Globe Showroom
Microsite

Design
Sven Küster (CD)
Oliver Hinrichs (SAD)
argonauten360 GmbH
Berlin, Germany

Der „Q7-Globe Showroom" bietet dem User ein Informationserlebnis, das in seiner Intensität durchaus mit der Besichtigung des realen Audi Q7 bei einem Händler vor Ort vergleichbar ist. Im „Q7-Globe Showroom" kann der User den Audi Q7 sowohl von außen als auch von innen in durchgehenden 360°-Ansichten erleben. Die eingesetzten Renderings sind dabei von einer so annähernd fotorealistischen Qualität, dass man vermeint, die dargestellten Materialien wie Leder und Holz förmlich riechen zu können.

The "Q7-Globe Showroom" provides the user with an informative experience that is comparable in its intensity to viewing a real Audi Q7 in a dealer's showroom. In the "Q7-Globe Showroom" the user can experience both the inside and outside of the Audi Q7 via comprehensive 360° views. The renderings deployed in these are of such a high photorealistic quality that you could almost think you can smell the materials on display: the leather and wood of the interior

Auftraggeber Client
Procter & Gamble GmbH
Schwalbach am Taunus, Germany

Projekt Project
www.herbalessences.de
Website

Design
Sven Schlotfeldt (CD)
Mathias Huber (Design)
Martin Kirmaier (Technical Manager)
Christian Neuhäuser (Programmer)
Marcel Martin (Flash)
Thomas Kiemer (Text)
coma ag
creativ orientierte münchner agentur
München, Germany

Für die neue Kampagne von Herbal Essences wurde die Website www.herbalessences.de komplett neu gestaltet. Zentrale Aufgabe war, ein Design zu entwickeln, das den Markencharakter der Haarpflegeserie stärker betont und erlebbar macht. Das Thema Beauty steht jetzt noch mehr im Vordergrund. Die hochwertige Optik spricht das erlebnisorientierte Publikum auf emotionaler Ebene an und lädt zum spielerischen Kennenlernen der Produkte ein. Einzelne Bereiche wurden mit Flash animiert. Die Produkte präsentieren sich auf prominente und übersichtliche Weise. Der Produkt-Finder hilft bei der Suche nach dem passenden Shampoo für den eigenen Haartyp.

The website www.herbalessences.de has been completely redesigned for the new Herbal Essences campaign. The core task was to develop a design that placed greater emphasis on the haircare series' brand character and turned it into something that can be experienced. The theme of beauty is now much more prominent. The high-quality look appeals to an experience-seeking audience at the emotional level and invites it to get to know the products in a fun way. Individual areas have been Flash-animated. The products are presented clearly and prominently, and the product finder helps users search for the right shampoo for their own specific hair type.

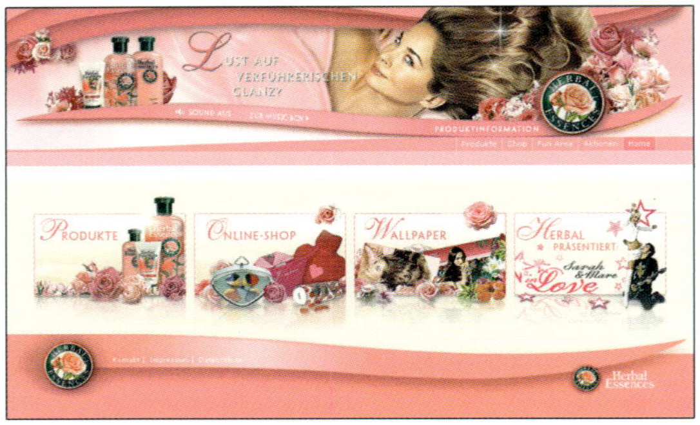

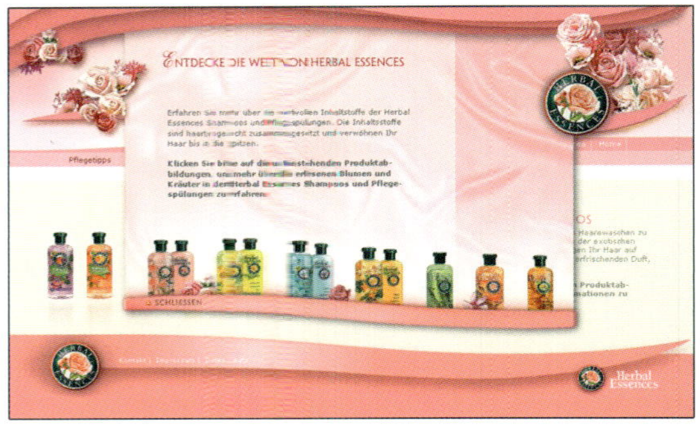

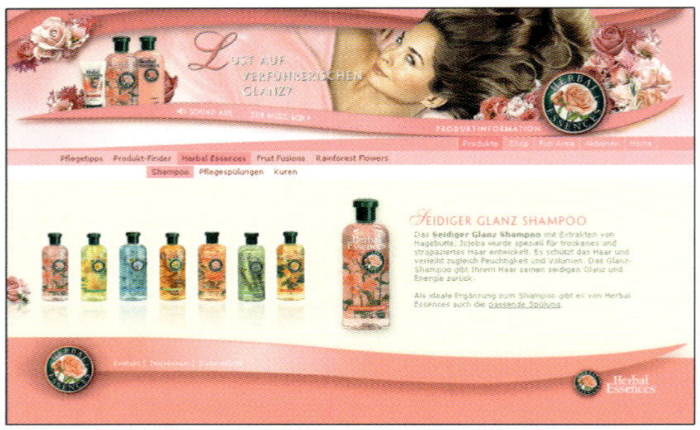

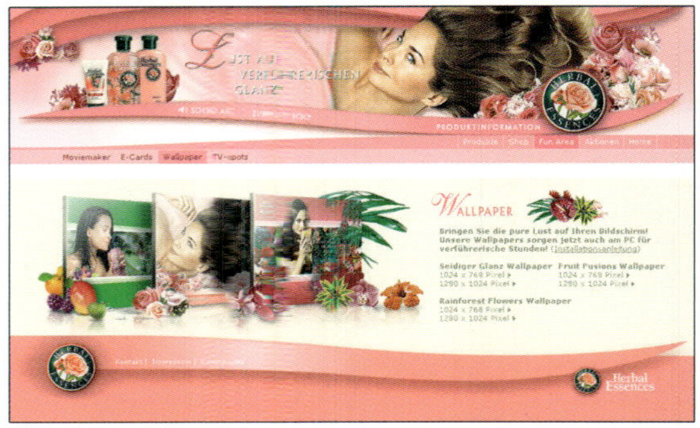

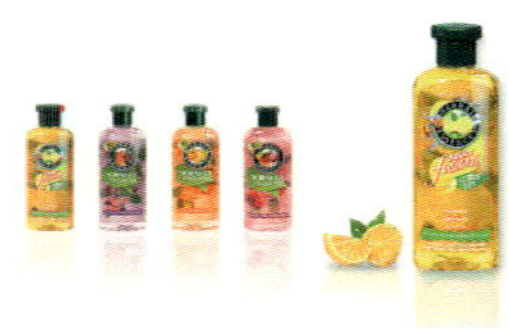

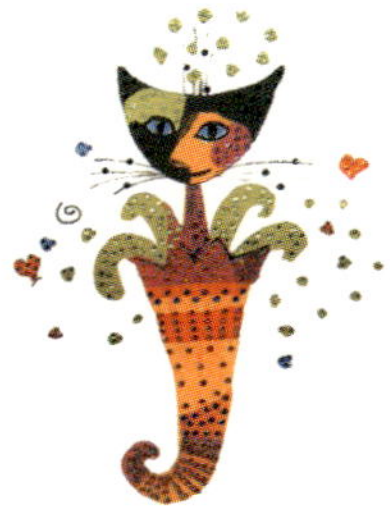

Auftraggeber/Client
Becker & Museum Licensing
München, Germany

Projekt/Project
www.rosina-wachtmeister.com
Website

Design
Sven Schlotfeldt, Jörg Herz (CD)
Franziska Bucher (Design)
Martin Kirmaier (Technical Manager)
Marcel Martin (Flash)
Thomas Kiemer (Text)
coma ag
creativ orientierte münchner agentur
München, Germany

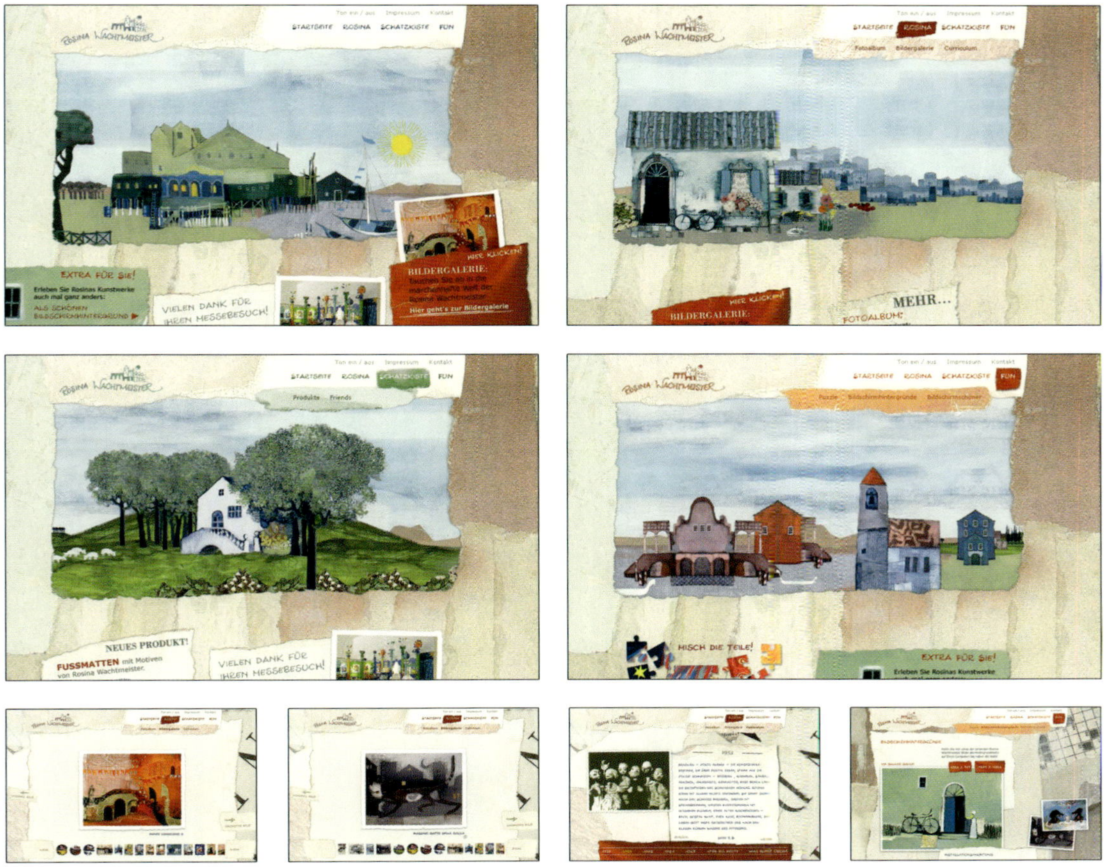

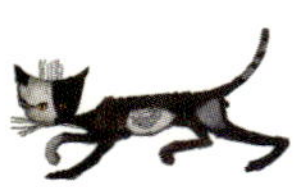

Der Internetauftritt von Rosina Wachtmeister www.rosina-wachtmeister.com wurde optisch und inhaltlich vollständig neu entwickelt. Weg vom Geraden und hin zum Spielerischen, so präsentiert sich die Website mit ihrem collagenartigen Design und spiegelt damit ganz deutlich den Stil der Künstlerin und den Charme ihrer Arbeit wider.
Ausschnitte aus ihren Lieblingsbildern wurden zu einer virtuellen Landschaft animiert, die den Besucher zu einer Entdeckungsreise durch die Bilderwelten der Künstlerin einlädt.
Das Spielerische kommt auch im Fun-Bereich zum Ausdruck. Hier kann der User unter anderem sein Glück beim Rosina-Wachtmeister-Puzzle versuchen.

The look and content of Rosina Wachtmeister's website www.rosina-wachtmeister.com has been completely redesigned. It has moved away from a direct approach towards a more playful one, with the website deploying a collage-type design, clearly reflecting the artist's style and the charm of her work.
Details from her favorite pictures have been animated into a virtual landscape, inviting the visitor to embark on a journey of discovery through the artist's picture universe.
The playfulness continues in a "fun" area where, amongst other things, the visitor can try his or her luck at a Rosina Wachtmeister jigsaw puzzle.

Auftraggeber Client
Reisenthel Accessoires
Puchheim, Germany

Projekt Project
www.reisenthel.de
Website

Design
Daniel Rothaug
Matthias Persch
Werner Goldbach
Alexander Dees
Kai Neugebauer
Sebastian Lenz
ZUM KUCKUCK
Würzburg, Germany

Relaunch des Online-Auftritts www.reisenthel.de. Zentrales Element der Website ist der umfangreiche Produktbrowser. Auf einen Blick können ca. 300 farbenfrohe Produkte erkannt und nach verschiedenen „weichen" Faktoren (z. B. Farbfiltern, Metakategorien) gefiltert werden. Die Echtzeit-Suche ermöglicht zusätzlich einen effektiven Zugriff auf bestimmte Produkte.
Die auf das Wesentliche reduzierte Gestaltung und Typografie der Website lässt das Design der Produkte maximal zur Geltung kommen. Das Format passt sich stets optimal in das Browser-Fenster ein. Die Hintergrundfarbe wechselt automatisch nach zwei Minuten durch die Farbpalette der Produkte.

Relaunch of the online presence of www.reisenthel.de. The crucial element of this website is its extensive product browser. An approximate of 300 products can be made out at a glance and filtered out by choosing various "soft" criteria (colour, meta-tags). On top of that, the realtime searching function allows the visitor an effective access to certain products.
Both the design as well as the typography of the website have been kept very basic and the format fits itself perfectly into the browser window. The background colour changes automatically and hardly noticeably after two minutes, triggered off by the colour palettes of the products.

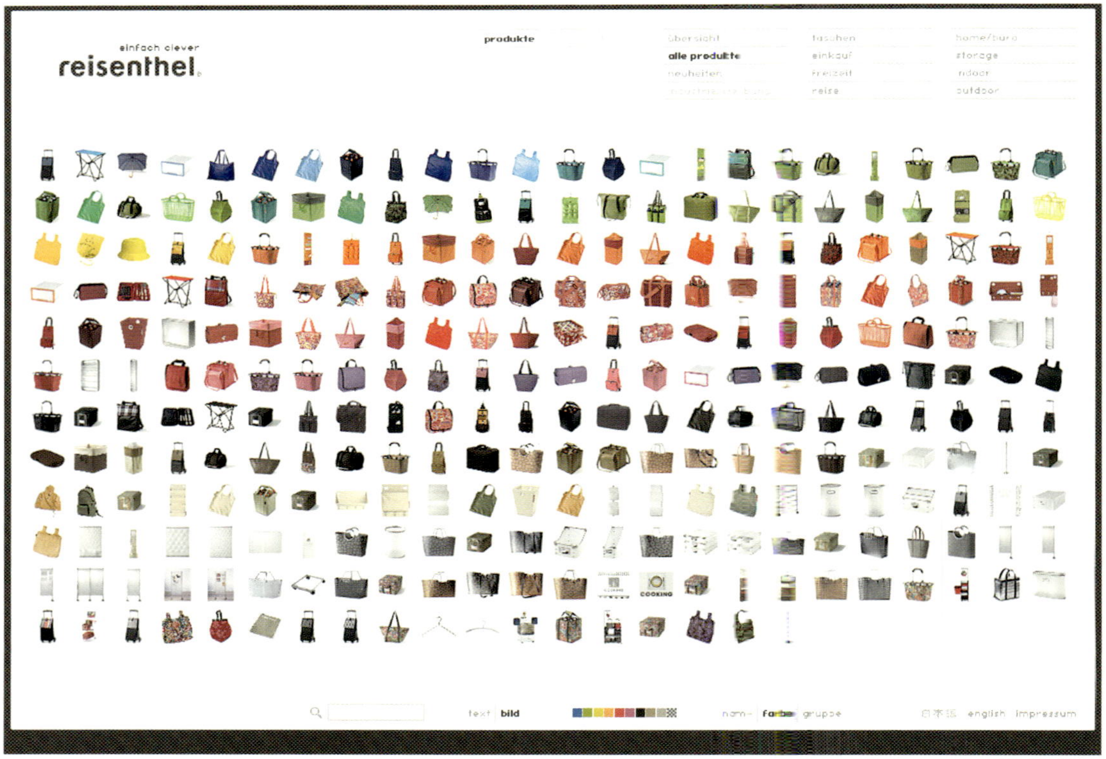

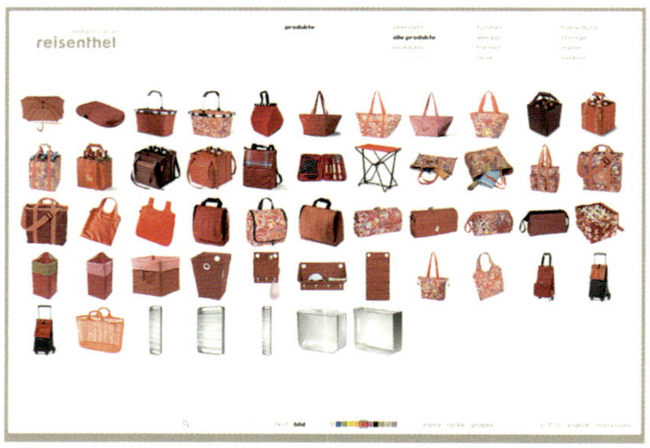

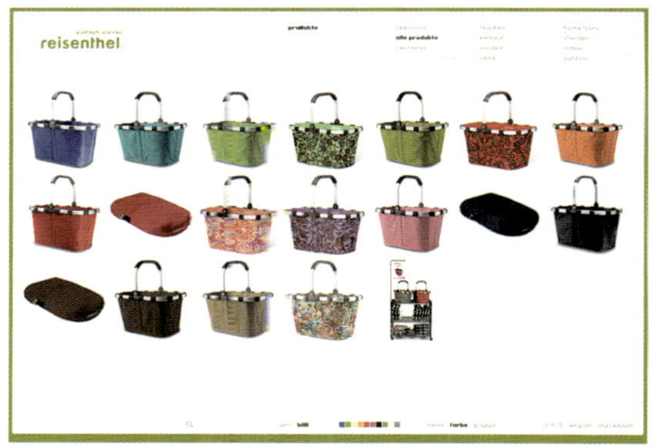

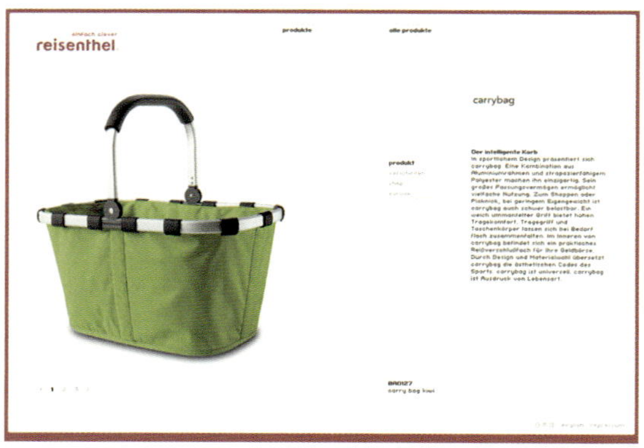

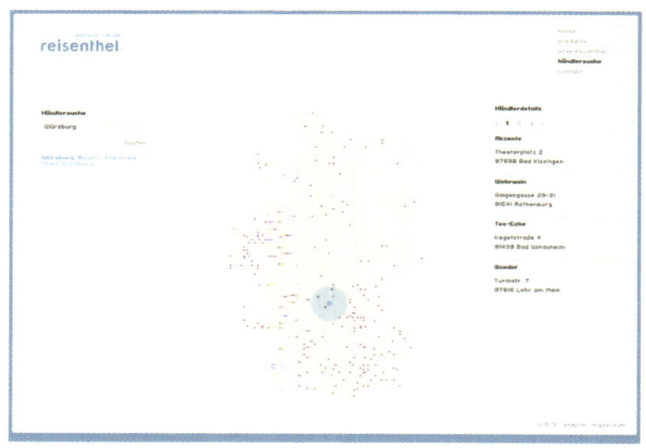

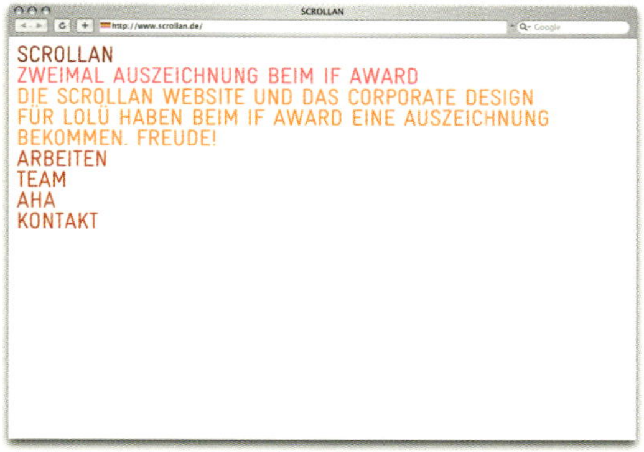

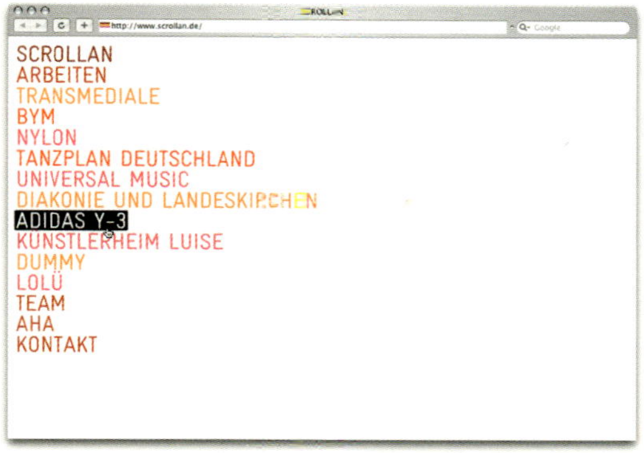

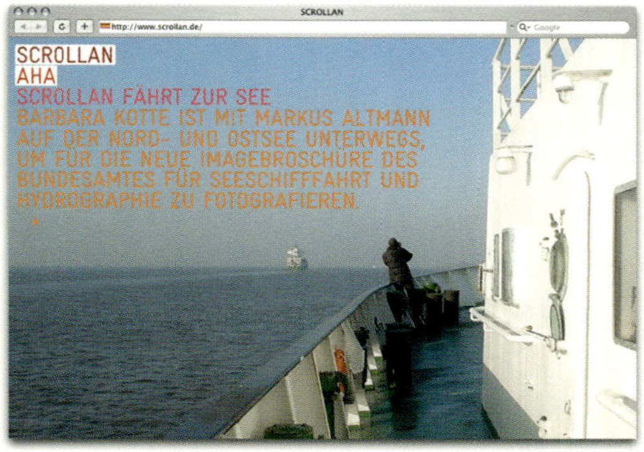

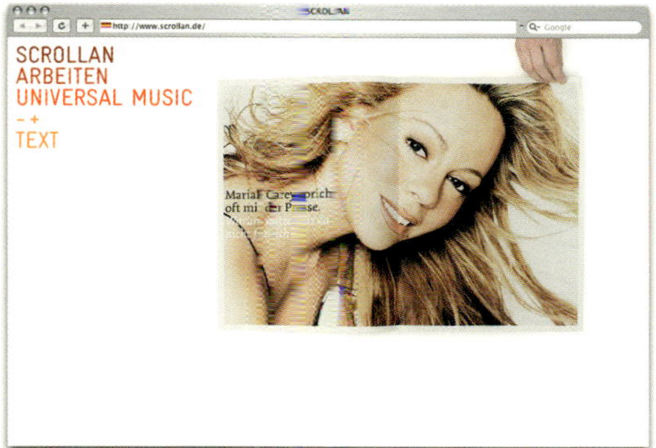

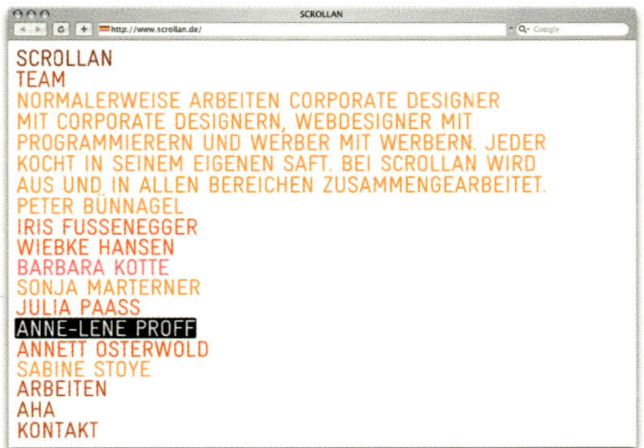

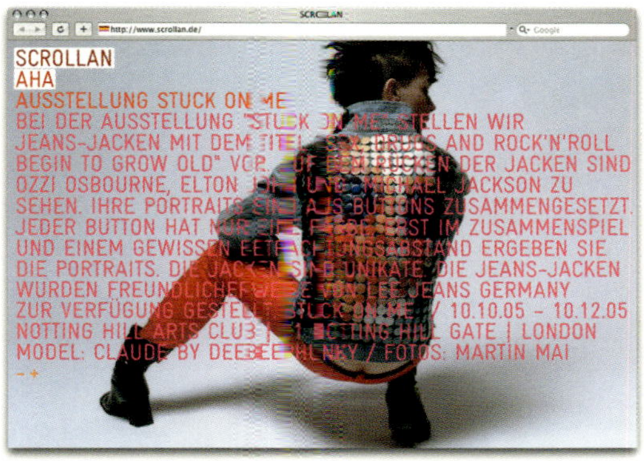

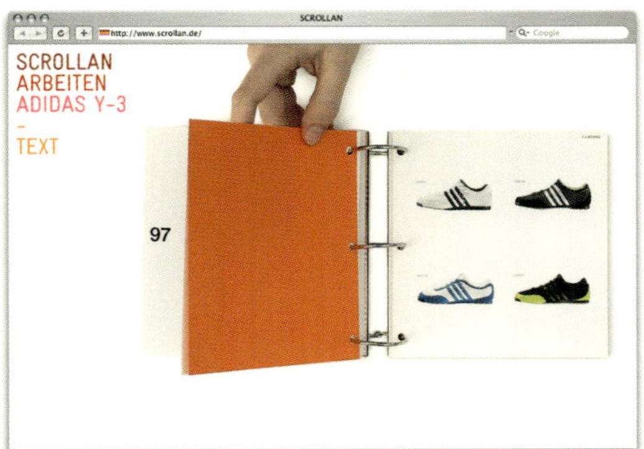

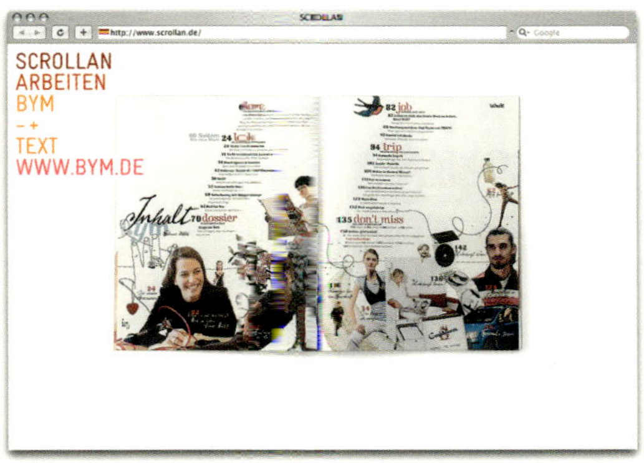

Auftraggeber Client
scrollan
Berlin, Germany

Projekt Project
scrollan
Website

Design
Peter Bünnagel
Iris Fussenegger
Barbara Kotte
Christian Kutschan
Anne-Lene Proff
Michael van Schaik
scrollan
Berlin, Germany

scrollan ist ein Designbüro, das nicht nur formal, sondern auch inhaltlich arbeitet. Der visuelle Eindruck der Website und der plakative Umgang mit Typografie sind angelehnt an die Geschäftsausstattung. Die Website anzuschauen soll Spaß machen – aber nicht durch pompöse Animationen, sondern eher durch schöne Details. In den Mouse-Overs sind kleine, freundliche Überraschungen versteckt. Die Arbeiten werden auf Weiß präsentiert und zum Teil von einer Hand gehalten. So können Größenverhältnisse leicht erfasst werden.

scrollan is a design team that not only works with formal design elements, but with inner content as well. The visual impression created by the website and its striking use of typography are based on the agency's stationery. Visiting the website is intended as a fun experience, not through the use of ostentatious animation, but by providing enjoyable details, including small, friendly surprises hidden in the mouseovers. The agency's projects are presented against a white background, and some are held in a hand to give you an idea of the respective size.

Auftraggeber Client
MTV Networks GmbH & Co. OHG
On Air
Berlin, Germany

Projekt Project
Euro Top 100
Programm-Package
Programme package

Design
Doris Fürst
Malte Haust
Andreas Jung
Bionic Systems
Düsseldorf, Germany

Die Vielfalt der europäischen Musikszenerie deklariert sich in der Ober- und Unterwelt eines abstrakten Makrokosmos.

The diversity of the European music scene is defined in surface and underworld of an abstract macrocosm.

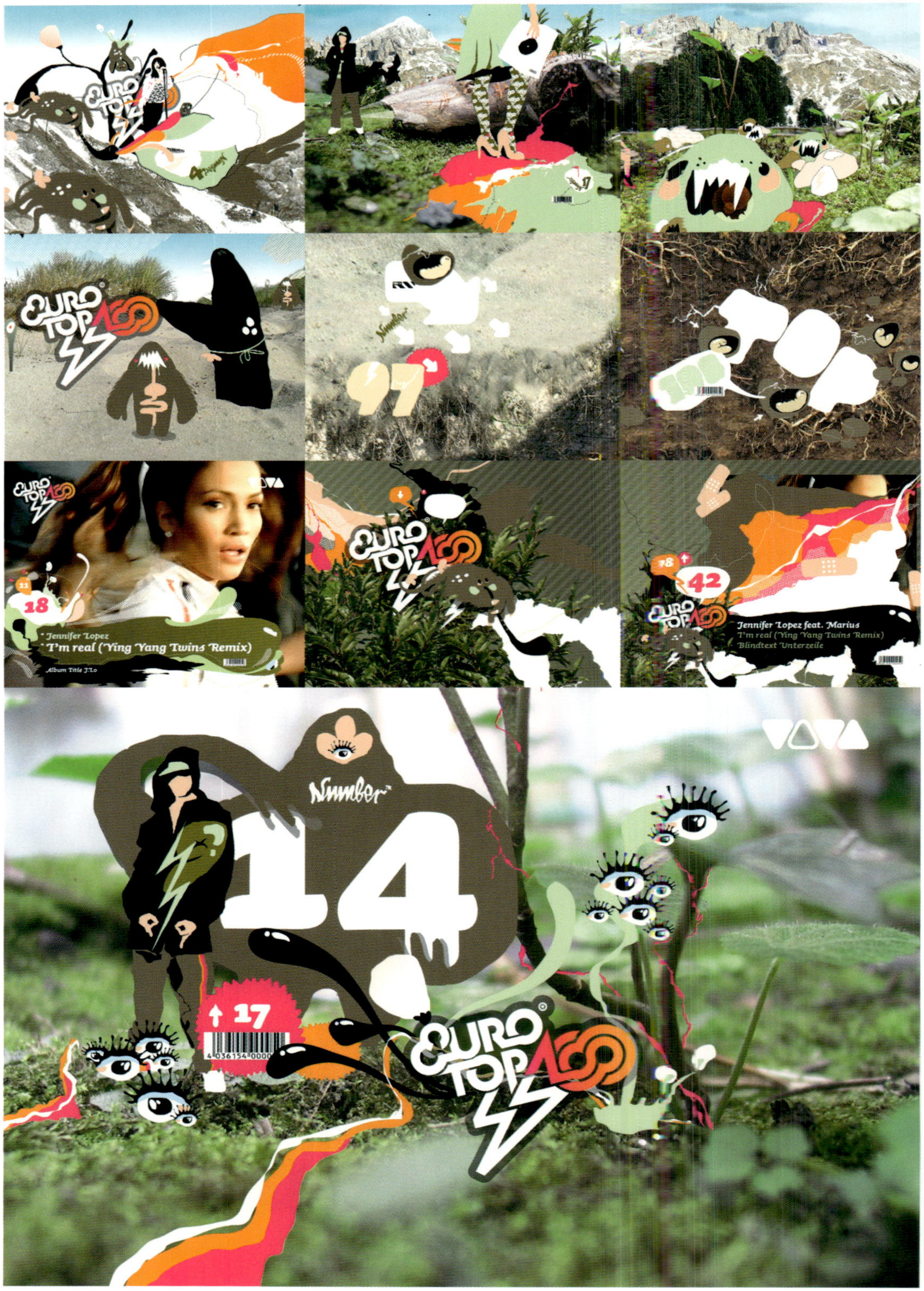

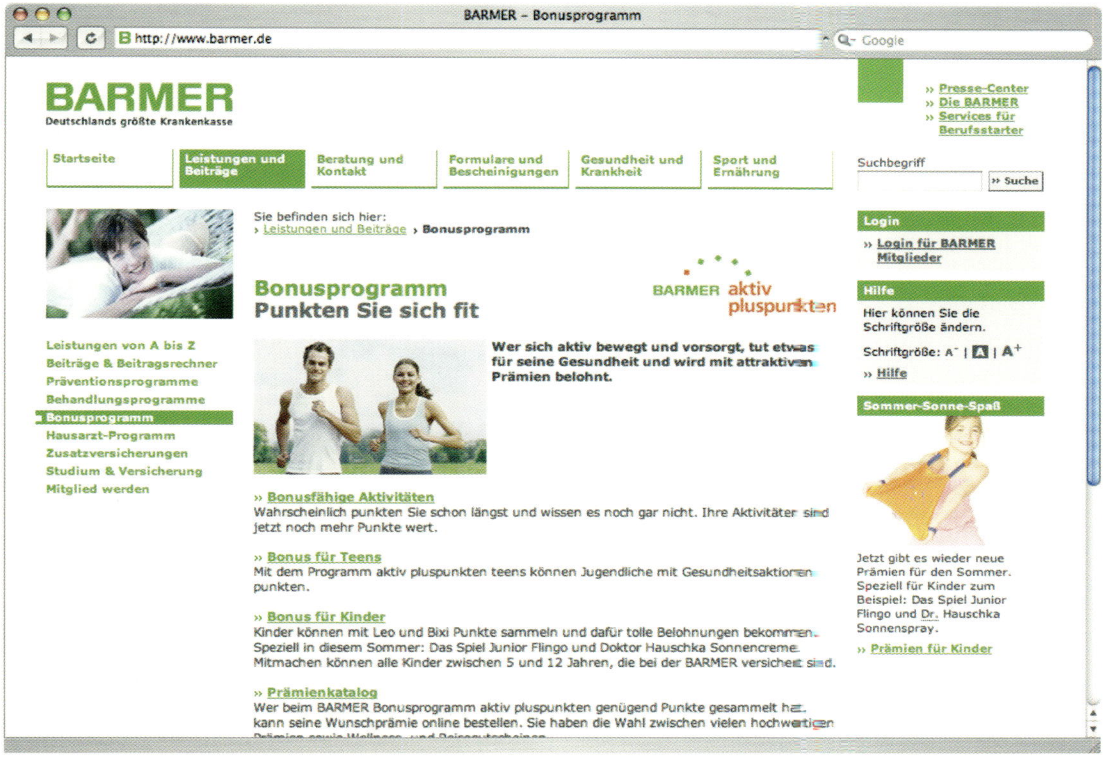

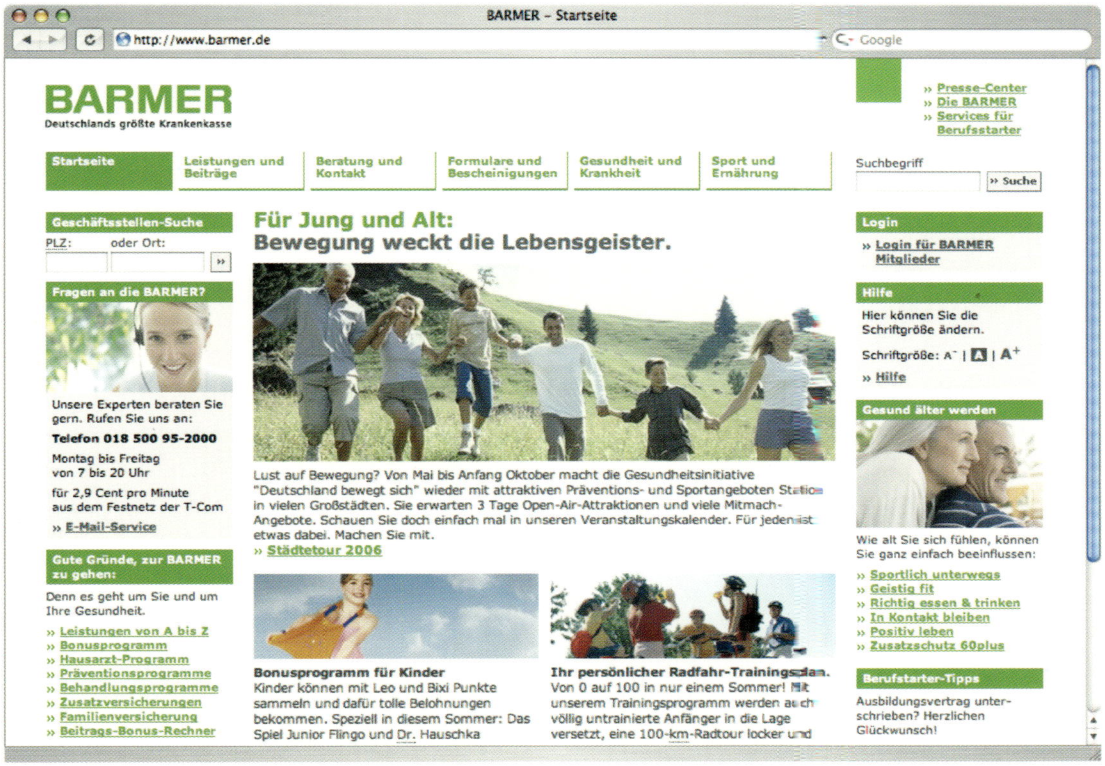

Auftraggeber Client
BARMER Ersatzkasse
Wuppertal, Germany

Projekt Project
www.barmer.de
Website

Design
Björn Hansen (Design)
Marion Böing (Konzept)
Przemyslaw Rychlewski (PM)
T-Systems Multimedia Solutions GmbH
Berlin, Germany

Die Internet-Geschäftsstelle der BARMER bietet Versicherten und Interessenten zahlreiche bedienfreundliche Online-Services und Informationen rund um den Versicherungsschutz. Darüber hinaus dient der Internetauftritt als allgemeine Informationsplattform zu den Themen Gesundheit und Lebensqualität. Die Konzeption und Gestaltung erfolgte in enger Abstimmung mit den Bereichen Marketing und IT der BARMER und berücksichtigt die Richtlinien für Barrierefreiheit von Internetanwendungen. Das Online-Design ist komplett in die Kommunikationsstrategie der BARMER eingebunden und greift bekannte Designelemente aus den Print-Publikationen auf.

BARMER's online branch offers policyholders and other interested parties a number of user-friendly online services and information involving all aspects of insurance coverage. In addition, the website serves as a general information platform on health and quality-of-life issues. The conception and design came about in liaison with BARMER's Marketing and IT departments, taking into account relevant guidelines on the "barrier-free" accessibility of Internet applications. The online design is fully integrated into BARMER's communications strategy and carries over familiar design elements from its printed publications.

Auftraggeber Client
Montblanc International GmbH
Hamburg, Germany

Projekt Project
Montblanc Relaunch
Website

Design
Dirk Kedrowitsch (GF)
Yvonne Grass,
Sven Mentel (CD)
Elke Maasdorff (AD)
Manfred Heider (Text)
Enrique V. Muros,
Christoph Klingebiel,
Roman Pelz (Screendesign)
Kim Christiansen,
Arne Otto (Flash)
Christian Koop,
Stefan Saasen,
Alexander Funke,
Ulf Teege,
Dan Bloch (Programmierung)
Atissa Tadjadod,
Sebastian Hahn (Client)
Elephant Seven AG
Hamburg, Germany

Der neue Internetauftritt von Montblanc ist an die Markenpositionierung und das aktuelle Erscheinungsbild von Montblanc angepasst. Das sich durch die gesamte Anwendung durchtragende Teaser-Konzept ermöglicht bereits auf der Homepage die Abbildung des gesamten Spektrums unterschiedlicher Themen. Das Herz des Auftritts bildet der Produktkatalog mit über 1.400 Artikeln in sieben verschiedenen Produktkategorien, der durch eine eigens entwickelte Datenbank gespeist wird. Hier kann man sich jeden einzelnen Artikel anschauen und anhand einer Toolbox Infos und Artikelvarianten aufrufen sowie Zusatzfunktionen wie Zoomen oder 3-D-Ansichten nutzen.

Montblanc's new website is designed to reflect the new brand positioning and Montblanc's current image. The teaser concept runs through the whole application and allows the whole spectrum of different topics to be incorporated into the homepage. The heart of the website is the extensive product catalogue with more than 1,400 articles in seven different product categories which draws on a specially developed database. This allows interested customers to look at any Montblanc item and call up information and different product versions using a toolbox. They can also use extra tools such as zooming in or 3-D views.

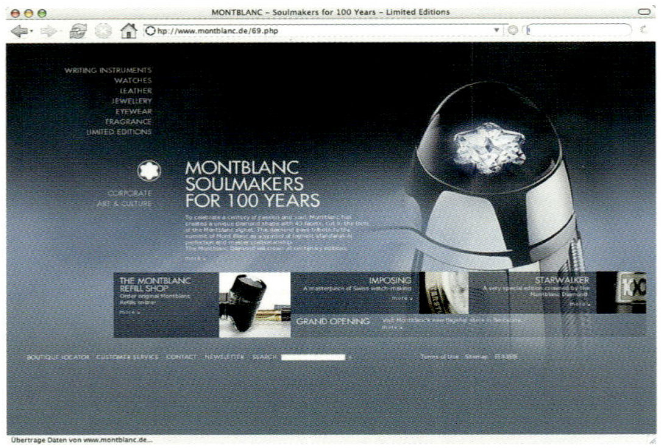

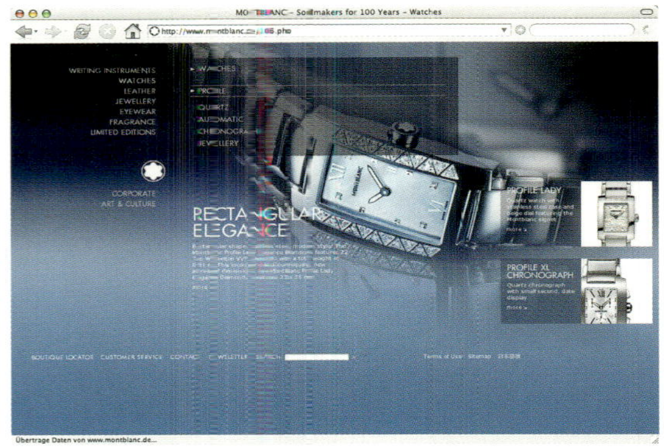

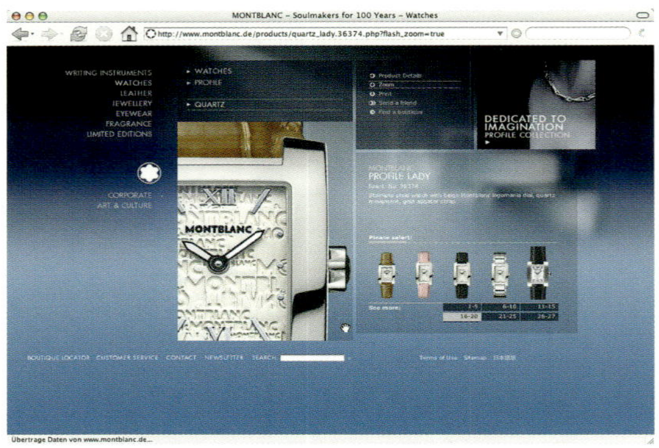

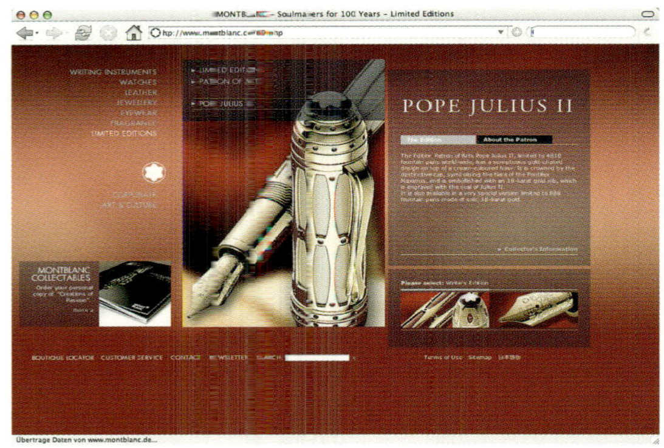

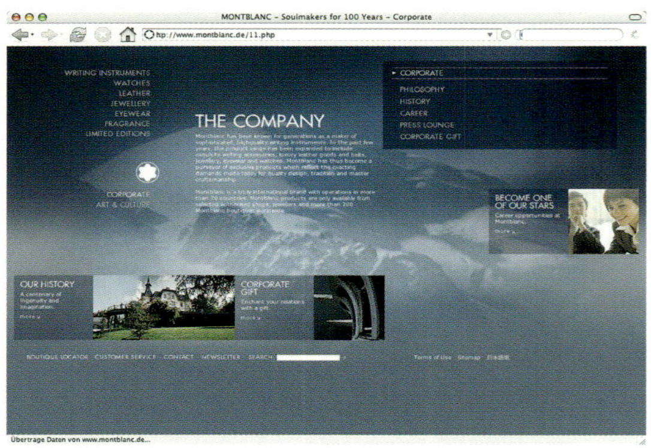

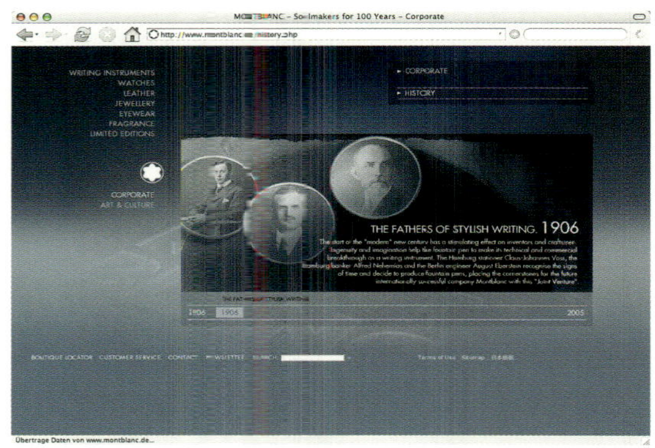

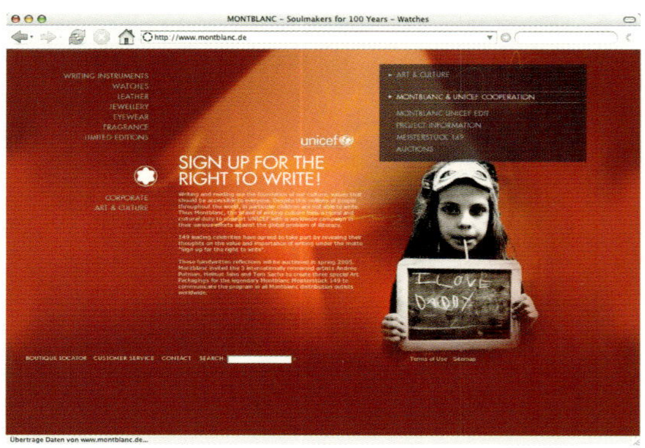

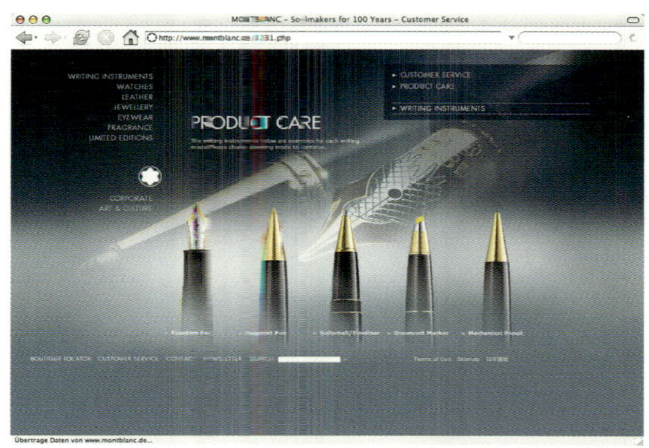

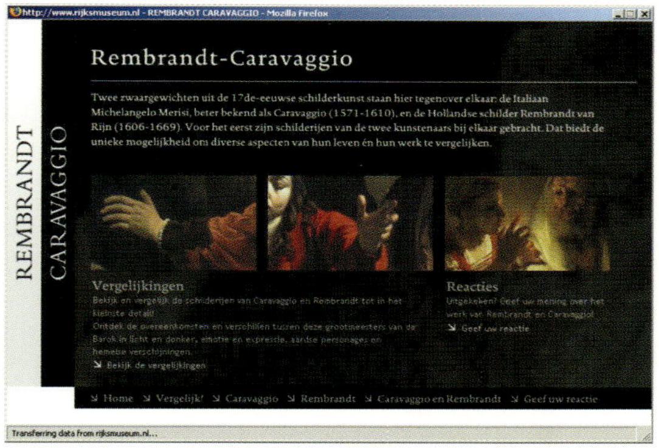
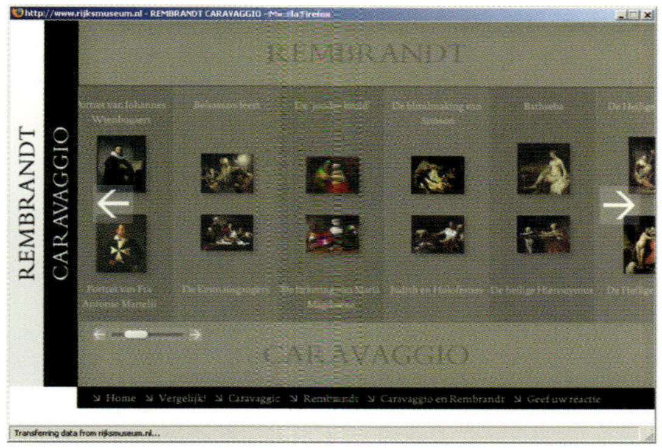
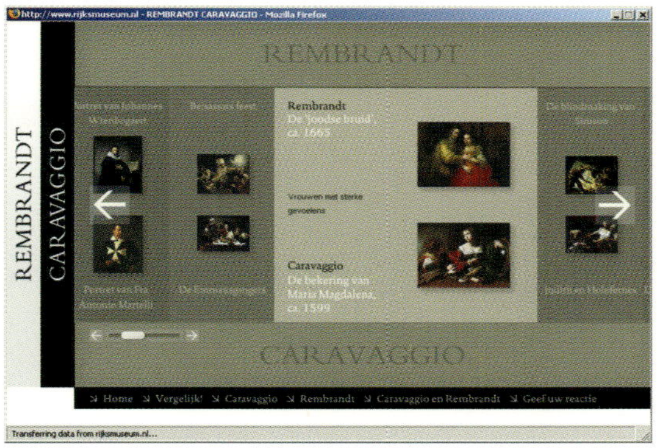
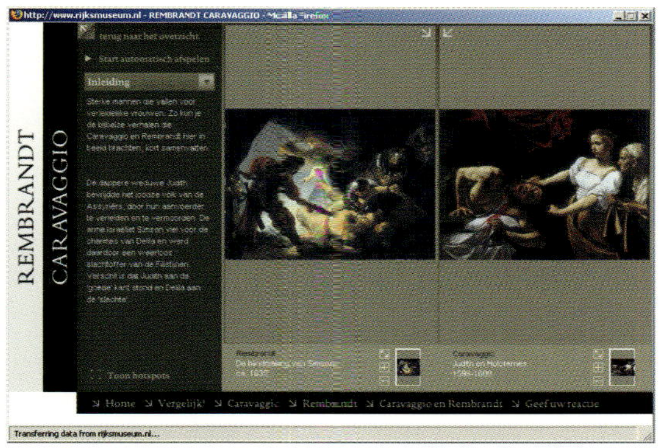
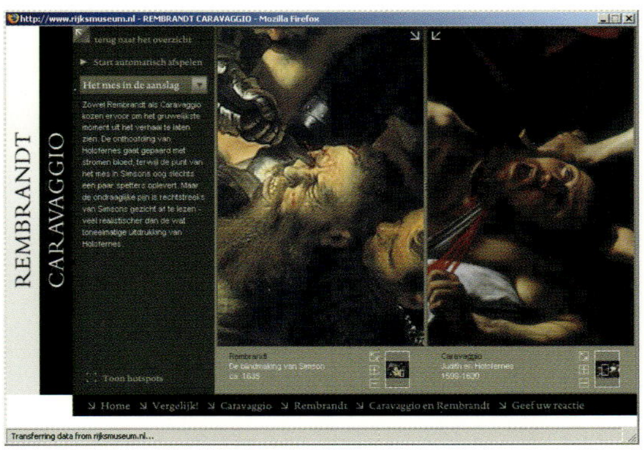
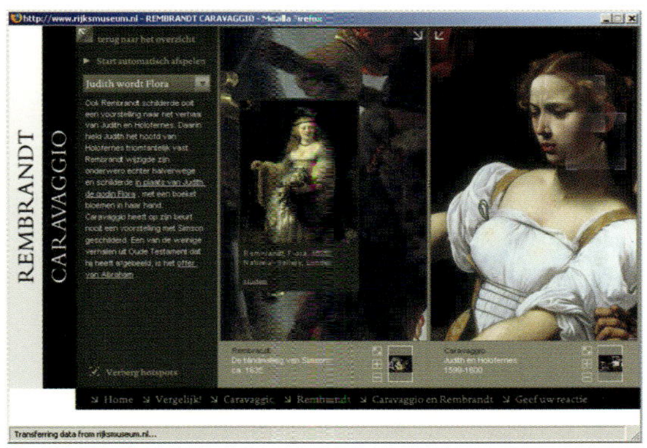
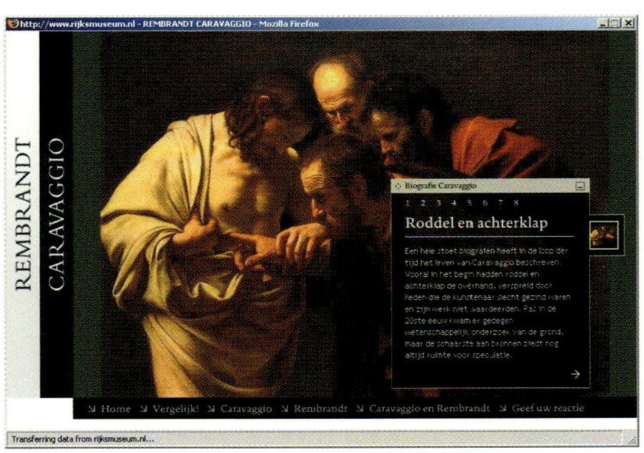
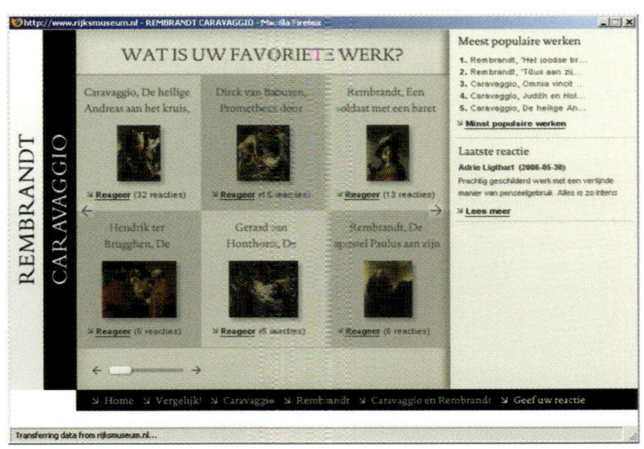

Auftraggeber Client
Rijksmuseum
Amsterdam, The Netherlands

Projekt Project
www.rembrandt-caravaggio.nl
Website

Design
Matthijs Klinkert
Jeroen van Erp
Louise Roose
Simone van Rijn
Daan Jobsis
Marcel de Jong
Tjeerd in 't Veer
Nils Adcink
Fabrique [design, communications and new media]
Delft, The Netherlands

Im Museum und auf der Website werden die Arbeiten der beiden Großmeister auf einmalige Weise verglichen. Vor allem die Betrachtung und der Vergleich der Gemälde auf Detailniveau ist online perfekt möglich. Besucher der Website können sich für eine automatische virtuelle Führung entscheiden, jedoch auch selbst Details und so genannte „Hotspots" vergrößern. Die Gemeinsamkeiten und auch die Unterschiede zwischen dem Werk von Rembrandt und Caravaggio werden anhand einiger Zusammenstellungen von Gemälden („Sets") veranschaulicht.

The works of the two great masters are compared to one another in a unique way in the museum and on the site. In particular, viewing and comparing the paintings right down to the detailed level can be carried out perfectly online. Visitors to the site can choose an automatic virtual guided tour, but can also zoom in themselves on details and "hotspots". The similarities and differences between the works of Rembrandt and Caravaggio are illuminated by means of a number of "sets" of paintings. It's also possible to register your response to the paintings themselves in this digital exhibition.

Auftraggeber Client
DaimlerChrysler AG
Vertriebsorganisation Deutschland
Berlin, Germany

Projekt Project
Mercedes-Benz Gebrauchtfahrzeug-Direktvertrieb
Website

Design
Claudia Steinhoff
Pixelpark AG
Berlin, Germany

Der Online-Shop für qualitätsgeprüfte, unfallfreie junge Gebrauchte von Mercedes-Benz überzeugt durch seinen kompletten Service: Alle Prozesse von der Fahrzeugauswahl über den Kauf bis zur Inzahlungnahme und Finanzierung werden online angeboten und offline durch telefonische Beratung, Lieferung nach Hause, Zulassung und Abholung des „Alten" optimal ergänzt. DaimlerChrysler bietet damit als erster Automobilhersteller an, den kompletten Kaufvorgang eines Gebrauchtwagens via Internet auszuführen. Mit dem Relaunch 2005 wurde die Plattform an den neuen Styleguide angepasst und alle Prozesse übersichtlicher und nutzerfreundlicher gestaltet.

The Mercedes-Benz online shop for quality-checked, accident-free, fairly recent used cars attracts customers by offering a full range of services: all the processes, ranging from vehicle selection, financing and purchase to trading in your own used car are offered online and supplemented perfectly with offline telephone advice, home delivery, registration and pickup of your "old car". This makes DaimlerChrysler the first carmaker to support the entire process involved in buying a second-hand car over the Internet. For its 2005 relaunch, the platform was adapted to the new style guide, and all processes were redesigned for enhanced transparency and user-friendliness.

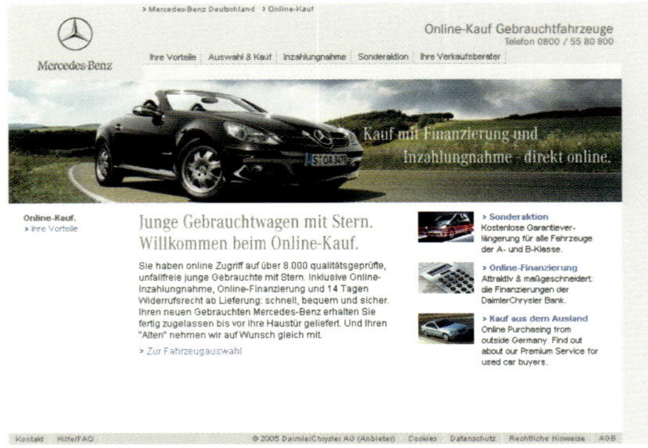

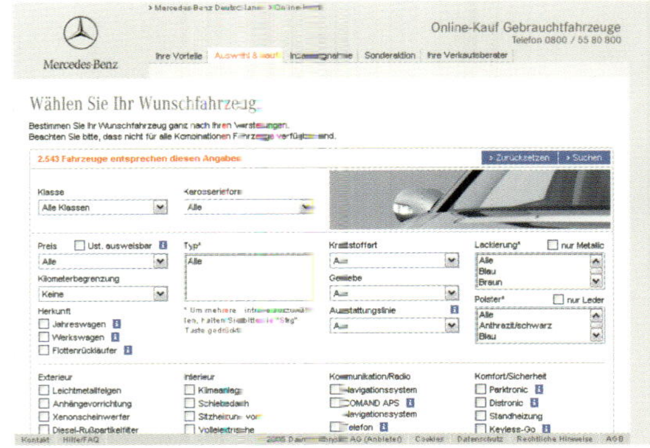

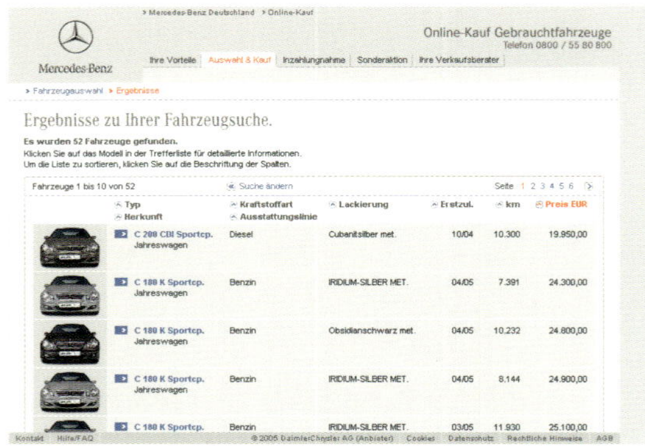

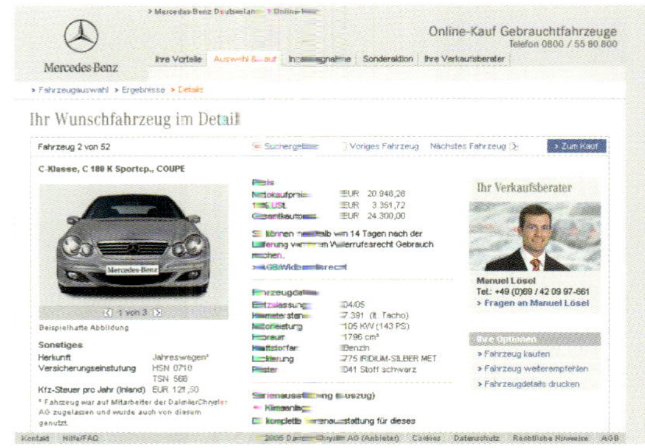

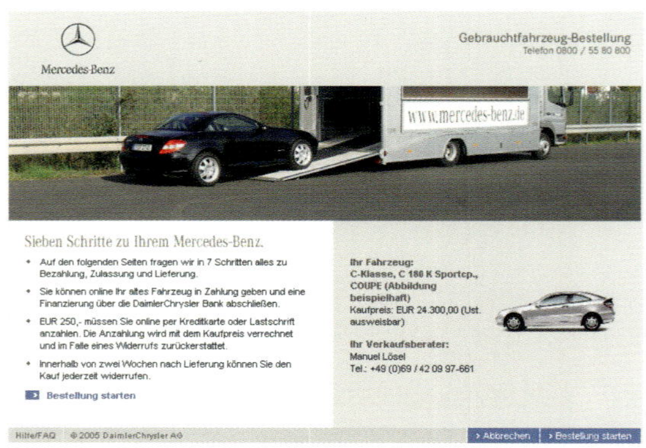

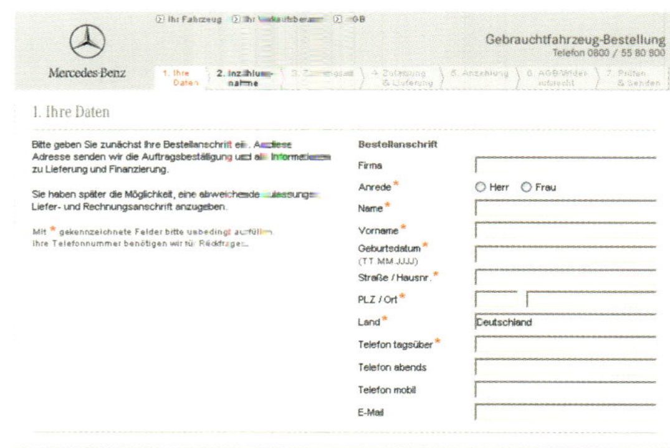

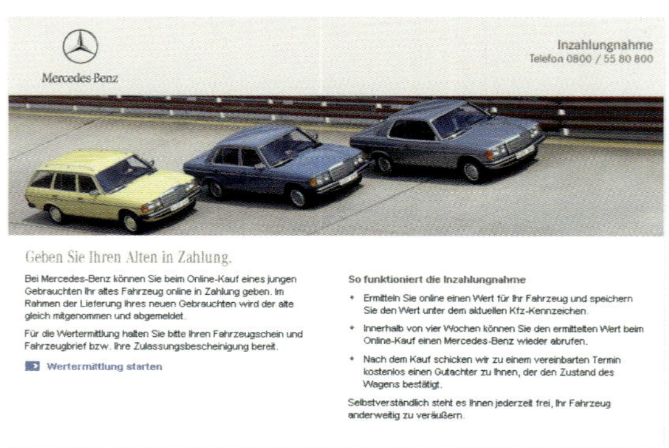

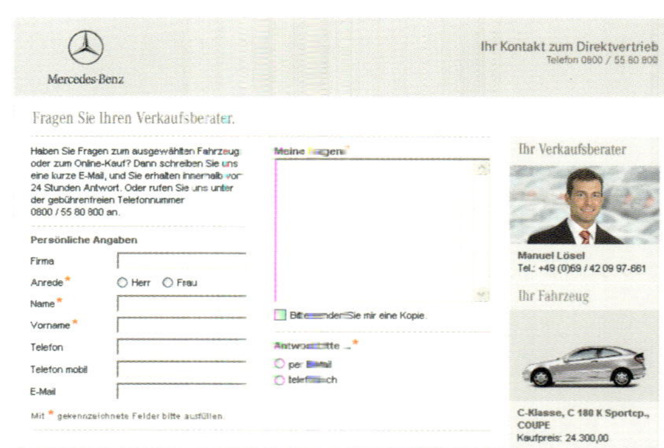

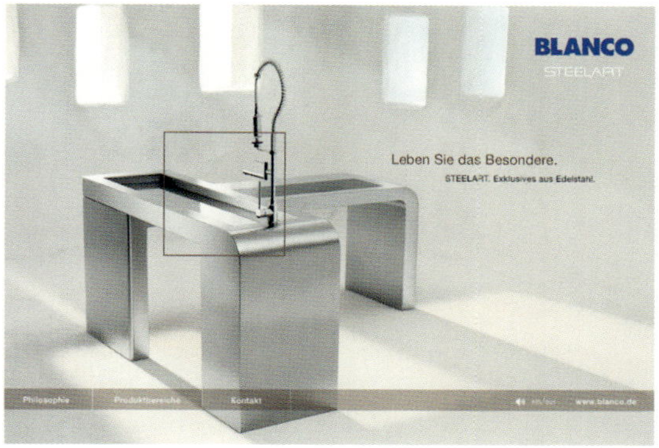

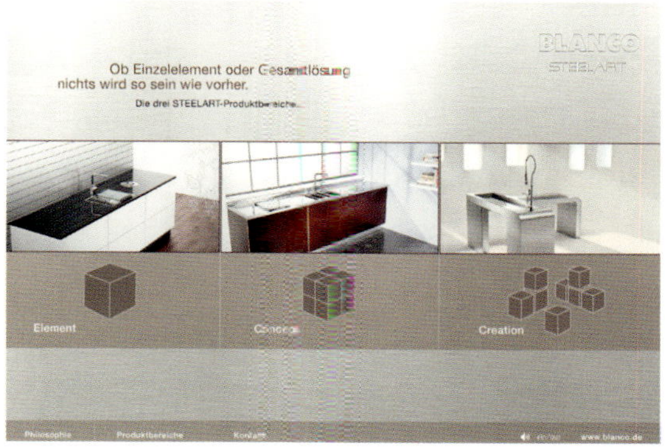

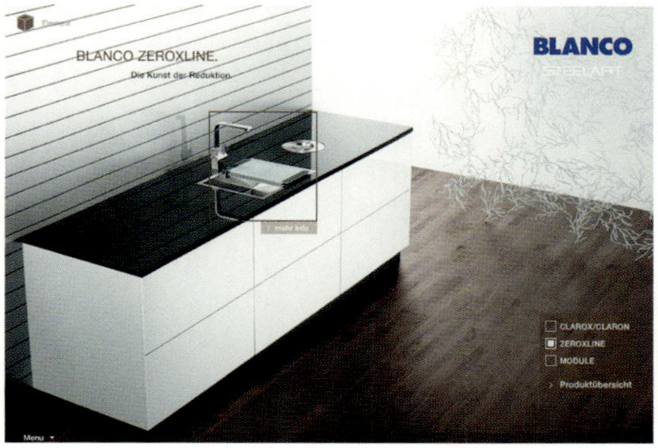

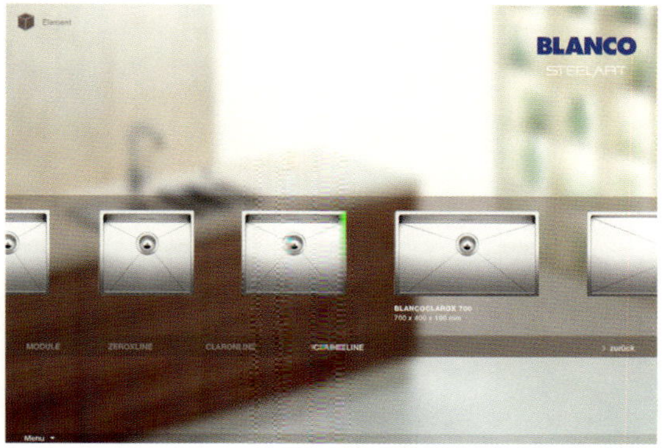

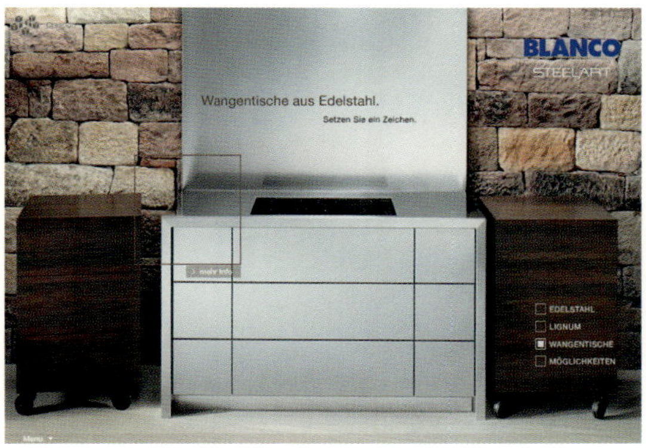

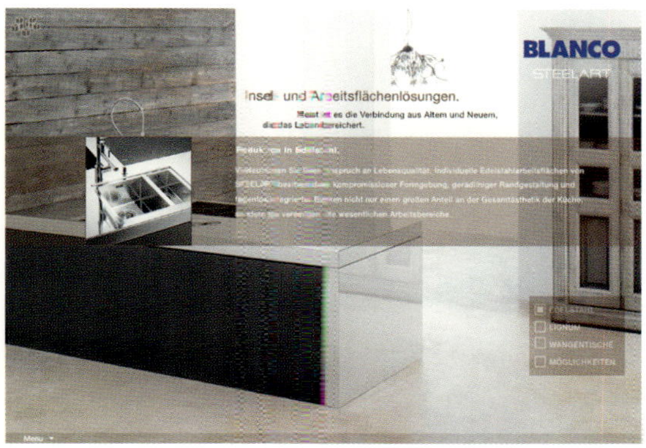

Auftraggeber Client
BLANCO GmbH + Co KG
Oberderdingen, Germany

Projekt Project
www.blanco-steelart.de
Microsite

Design
Christine Elsässer, Sabine Schmidt,
Sonja Schubert (Art)
Philipp Heimsch, Ila Stöckle (Text)
Stefan Horr, Harry Lépine (Technik)
Andreas Burz (Fotografie)
Beaufort 8 GmbH
Stuttgart, Germany

Bei der Microsite der hochwertigen Produktlinie STEELART von BLANCO steht abgesehen von der ansprechenden Darstellung der ästhetischen Edelstahlkonzepte, die im Grunde unbegrenzte Gestaltungsvielfalt der Produkte im Vordergrund. In ihrer natürlichen Umgebung arrangiert und in Szene gesetzt, mit Fokussierung besonderer Details zeigt sich die sehenswerte Produktpalette in voller Vielfalt und Ästhetik: übersichtlich strukturiert und beeindruckend fassbar. Die Flash-animierte Microsite ist ein virtueller Rundgang durch die Welt von STEELART, die mit dem Spiel von Schärfe und Unschärfe Visionen in der Küchengestaltung klar werden lässt.

Besides the attractive presentation of aesthetical stainless steel concepts the main purpose of the STEELART microsite is to demonstrate the basically unlimited possibilities of arrangement, which this high-class product line by BLANCO offers. Shown in their natural environment and with focus on special details the whole product range is presented in its full diversity and aesthetics: clearly structured and impressively palpable. The Flash-animated microsite is a virtual tour through the world of STEELART: playing with blurring effects it gets visionary kitchen-arrangements become real.

Auftraggeber Client
Schirn Kunsthalle Frankfurt
Frankfurt am Main, Germany

Projekt Project
Summer of Love
Website

Design
Bejadin Selimi
NEUE DIGITALE
Frankfurt am Main, Germany

Die Website zur „Summer of Love"-Ausstellung stellt nicht nur ein begleitendes Medium dar, sondern bietet dem Nutzer darüber hinaus neue individuelle Möglichkeiten in den „Summer of Love" einzutauchen. Ein besonderes Highlight der Seite ist der LSD-Modus: Der User klickt auf einen so genannten „Blotter" und aktiviert dadurch eine visuelle Interpretation eines LSD-Rausches – in diesem Modus verändert sich durch Lichteffekte, Farbspiele und Verzerrungen die komplette Website. So kann der User live erleben, inwiefern Psychedelica die Entstehung von Kunst einer ganzen Ära beeinflusst haben.

The website doesn't just offer an accompanying medium to the exhibition "Summer of Love", but rather offers the user the opportunity to experience new highlights in the "Summer of Love": A particular highlight of the site is the LSD mode. The user clicks on the so-called "Blotter" and activates a visual interpretation of an LSD trip. In this mode the webpage changes to display light effects, colour tricks and distortion. In this way the user can experience live how psychedelic substances influenced the art of an entire era.

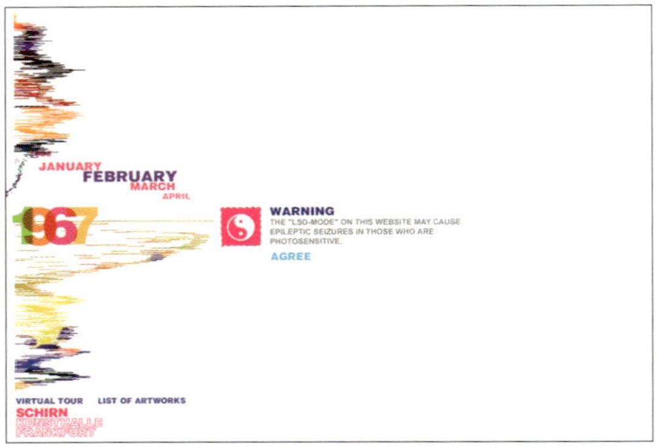

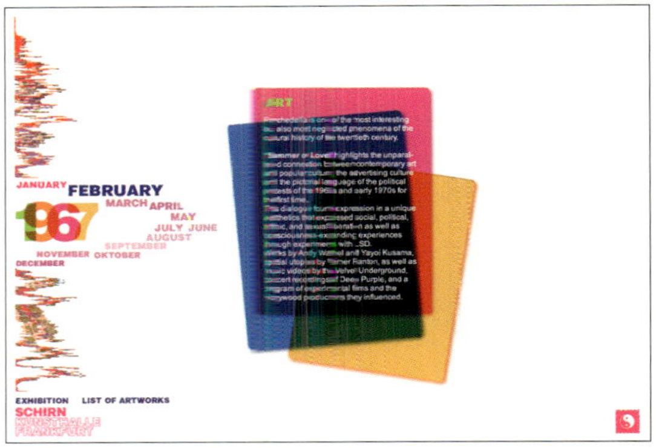

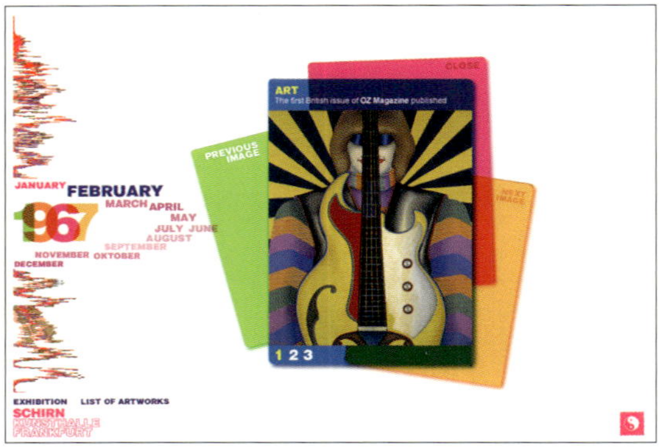

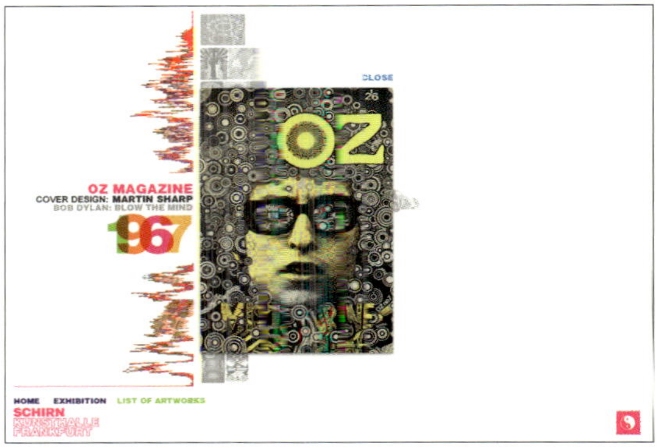

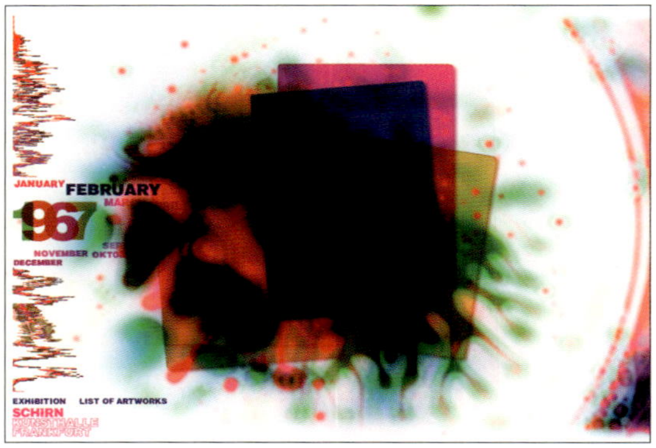

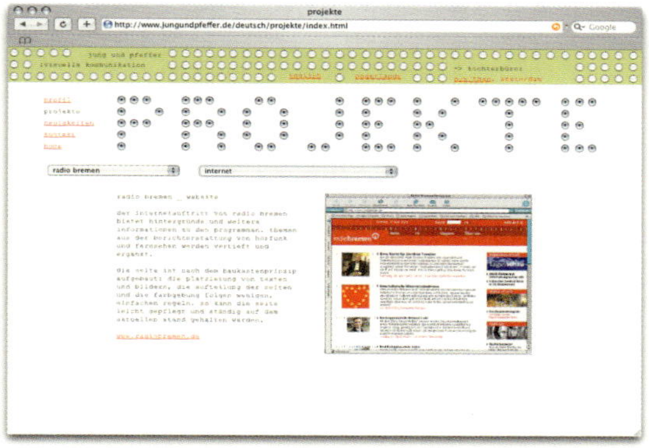

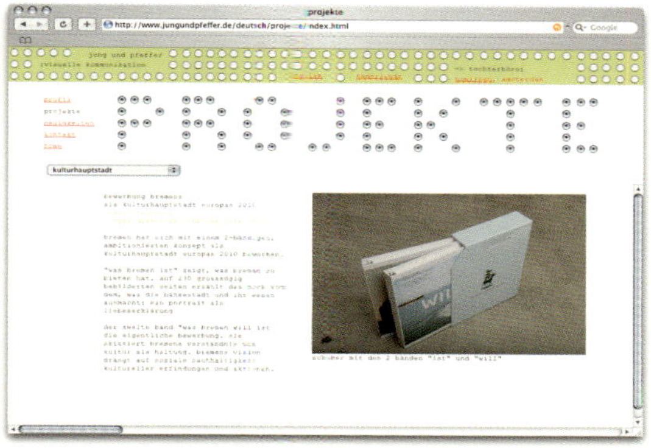

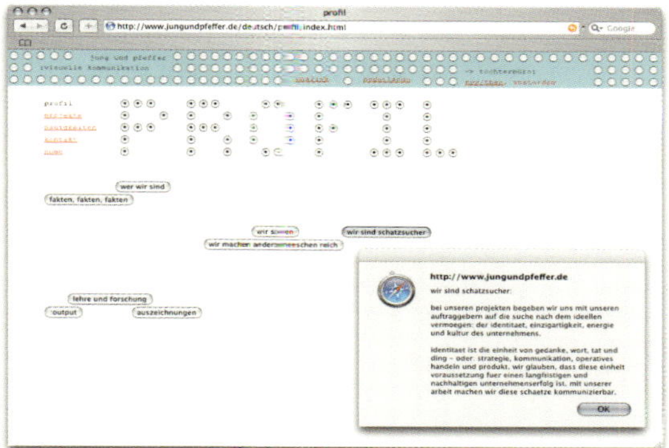

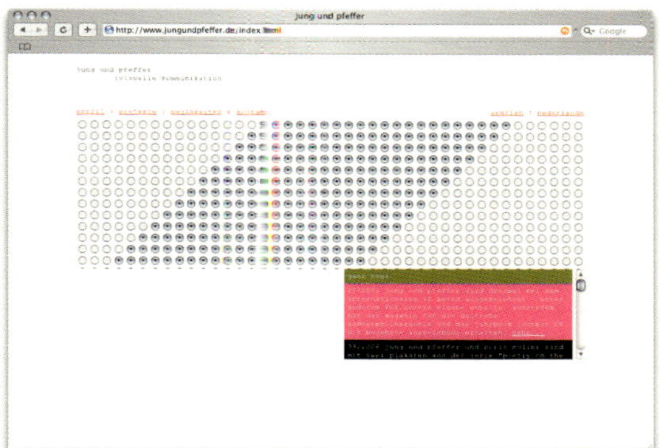

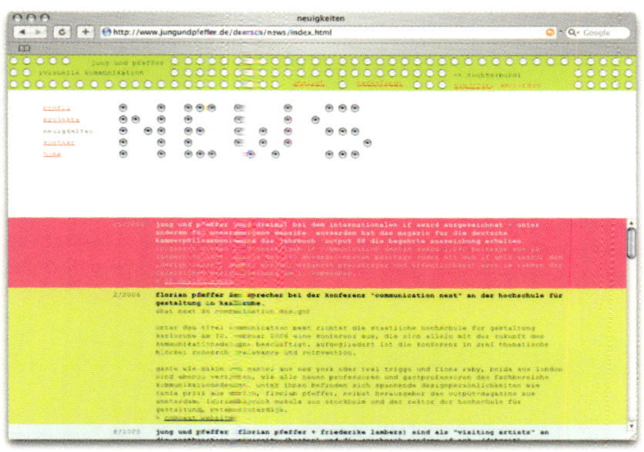

Auftraggeber Client
jung und pfeffer
visuelle kommunikation
Bremen, Germany

Projekt Project
www.jungundpfeffer.de
Website

Design
jung und pfeffer
visuelle kommunikation
Bremen, Germany

Die Website von jung und pfeffer verzichtet auf Bewegung, Flash, komplexe Menüstrukturen und ist bewusst reduziert und spröde gestaltet.
Ausgangspunkt war die selbstauferlegte Beschränkung, nur HTML-Elemente für die Gestaltung zu verwenden. Die Möglichkeiten von HTML sollten allerdings voll ausgeschöpft und neu interpretiert werden. Aus Radio-Buttons werden so Bilder bzw. Schriften und aus Alert-Boxen werden Pop-up-Fenster. Die Seite spielt bewusst mit User-Gewohnheiten und erzeugt eine Ästhetik, die aus dem Medium selbst kommt.

jung und pfeffer's website does not use motion, Flash or complex menu structures. It has an intentionally reductionist design.
The starting point for the design was the self-imposed constraint that only HTML elements could be used in the design. However the options supported by HTML were to be exploited to the hilt and reinterpreted. So, radio buttons turn into pictures and words, and alert boxes become pop-up windows. The website consciously plays with the user's habits, creating an esthetic drawn entirely from the medium itself.

Auftraggeber Client
Willy Bogner GmbH & Co. KG
München, Germany

Projekt Project
Bogner Online-Auftritt
Website

Design
Tobias Ulmer
Matthias Erb
WERBEWELT Interactive
Ludwigsburg, Germany

Der Name Bogner steht nicht nur für ein erfolgreiches Modeunternehmen, sondern mit Sônia und Willy Bogner für einen besonderen Lifestyle: Die Welt von Bogner. Die Website spiegelt auf der einen Seite die Historie des Unternehmens und der Person Willy Bogner als Unternehmer und Filmemacher wider und auf der anderen Seite die Produktwelt, die exklusive Qualität und Luxus verkörpert. Der Kollektionsbereich präsentiert sich als nutzerfreundlicher, moderner Katalog. Die Models erscheinen in einer Slideshow vor ruhigem Hintergrund. Ein Vorschaumodus erlaubt Eingriffe in die Bildfolge. Darüber hinaus ist die Seite direkt mit dem Bogner Online-Shop verknüpft.

The brand with Sônia and Willy Bogner stands for an exclusive lifestyle: the world of Bogner. This bilingual website in a glossy magazine look presents more than fashion. Extensive, atmospheric backgrounds correspond to Bogner's aesthetics. The home page is generated randomly from the topics of the bogner-homeshopping-magazine, the user receives a full screen slideshow. The menu levels start short full screen video sequences. An archive shows Bogner's work as a film maker. The collections create a user-friendly catalog. A slideshow shows the models in front of a static background. The user can purchase any article by double-clicking on it.

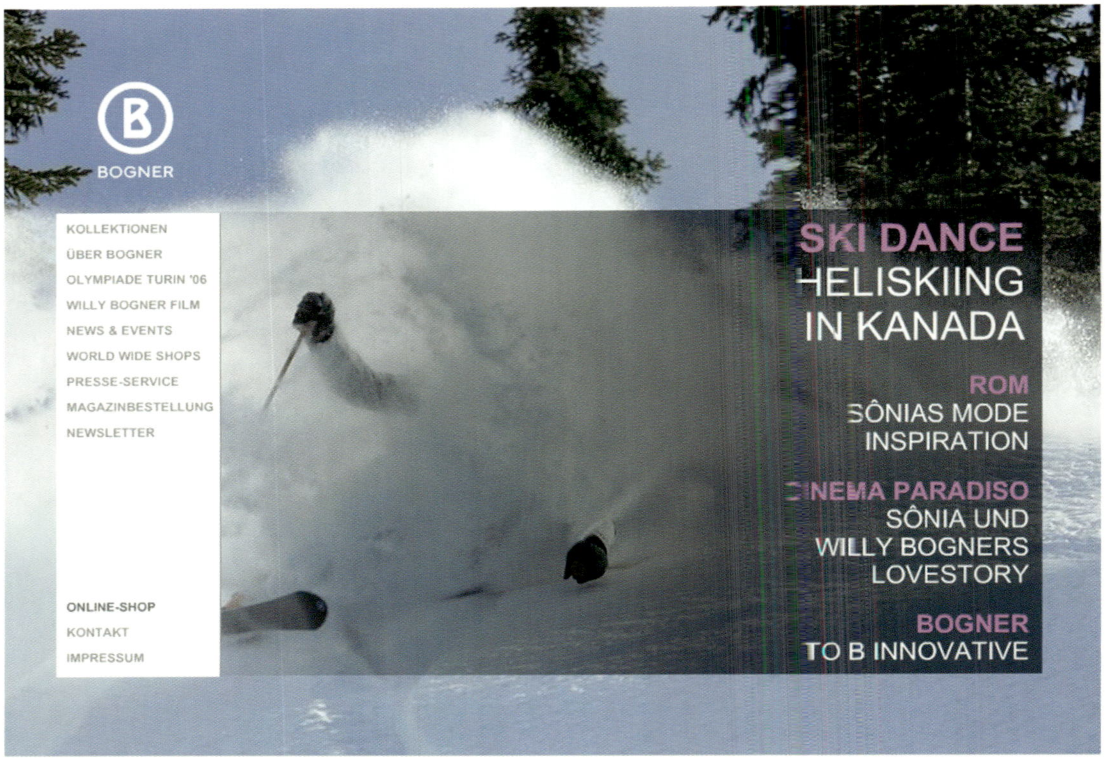

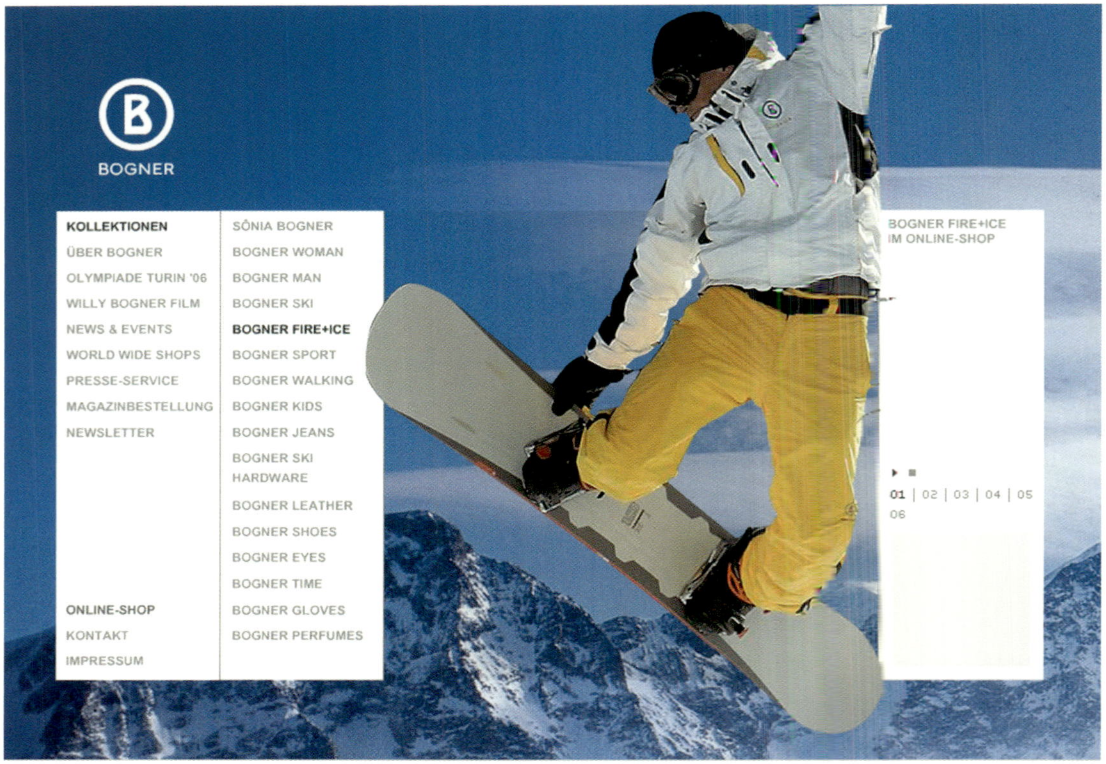

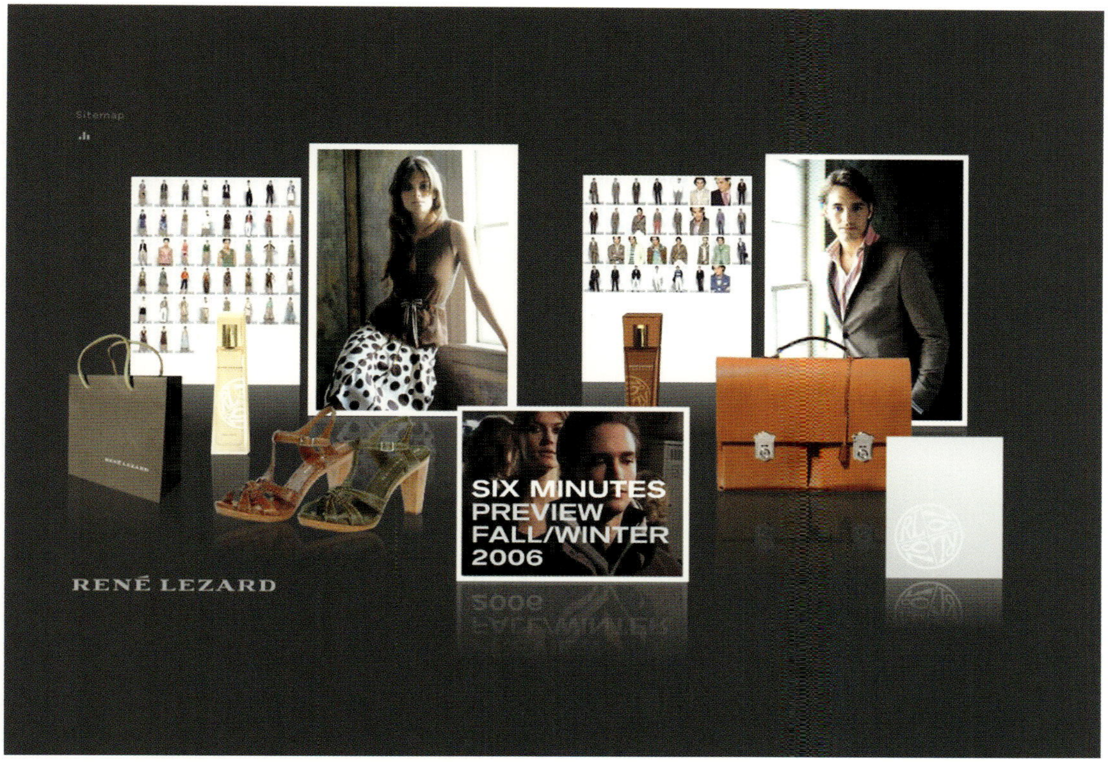

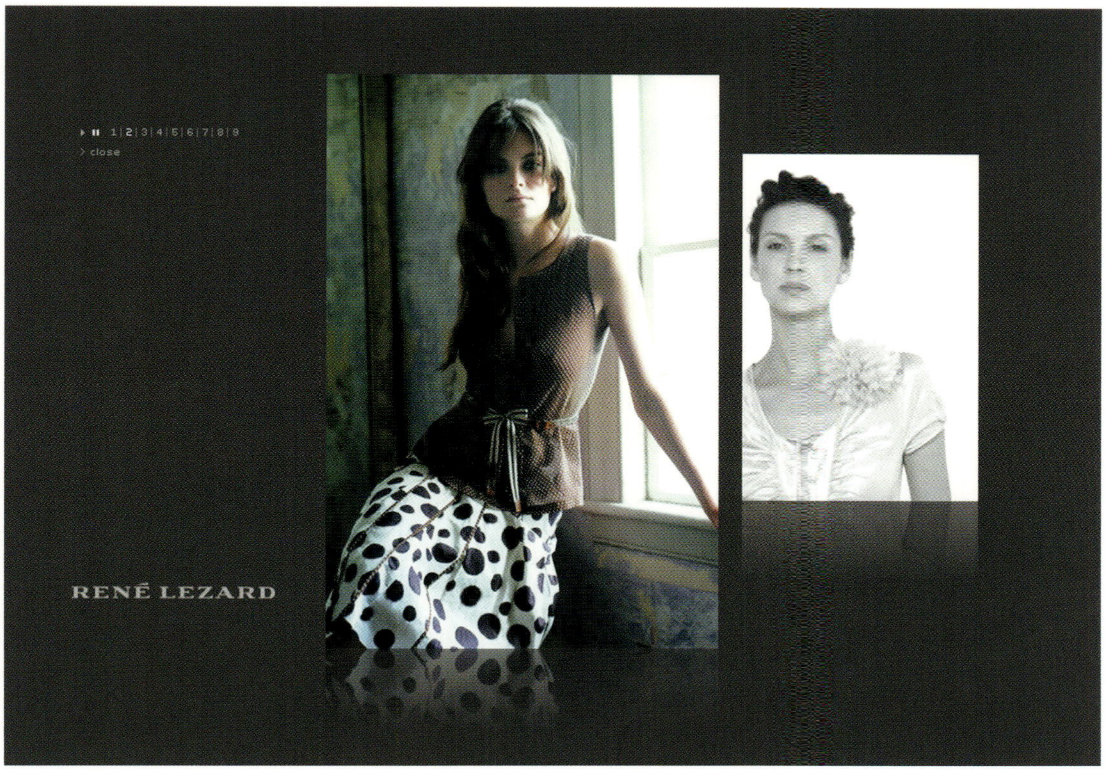

Auftraggeber Client
RENÉ LEZARD Mode GmbH
Schwarzach, Germany

Projekt Project
RENÉ LEZARD Online Relaunch
Website

Design
Tobias Ulmer
Matthias Erb
WERBEWELT Interactive
Ludwigsburg, Germany

Die Website stellt einen neuen Teil in der integrierten Kommunikation von RENÉ LEZARD dar. Sie nimmt Kampagneninhalte auf und führt diese medienspezifisch innerhalb der Website weiter. Die Markenwerte „Gelassenheit mit Stil“ und „Authentizität“ werden über Backstage-Videos und Online-Specials vertieft. Die Lizenzprodukte werden unter dem Markendach kommuniziert und RENÉ LEZARD als Lifestylemarke positioniert. Um die ganze Lifestylewelt von RENÉ LEZARD abzubilden, wurde ein räumliches Szenario entwickelt, in dem die ganze Vielfalt der Produktpalette und Informationen dargestellt wird.

The website presents a new part of the integrated communication of RENÉ LEZARD incorporating current campaigns and communicating them within the online medium. The brand values “Relaxed with Style” and “Authenticity” are being intensified through the backstage videos and online specials. All licensed products are communicated through the brand itself whereby RENÉ LEZARD is taking its position as lifestyle brand. To display the whole world of RENÉ LEZARD a three-dimensional scenario was developed to show the whole variety of products and information.

Auftraggeber Client
Holzmedia GmbH
Burgstetten, Germany

Projekt Project
Holzmedia – www.holzmedia.de
Website

Design
Danijela Djokic
Martin Grothmaak
Jürgen Späth
Projekttriangle
Stuttgart, Germany

Die Holzmedia GmbH verbindet zwei Komponenten, die zusammen in der Synergie ganz neue Lösungen ergeben: Möbel und Medientechnik. Holzmedia ist nicht nur Anbieter von hochwertigen Medienmöbeln, sondern auch Planer, Gestalter und Umsetzer von kompletten Besprechungs- und Konferenzräumen. Die Website fungiert als dialogfähiges Informationstool. Das bestehende Corporate Design wurde durch Bewegung und Interaktivität auf das digitale Medium erweitert. Mit dem Broschürenkonfigurator kann der User das Produktportfolio individuell konfigurieren und als PDF-Broschüre downloaden. Animierten Einblick in die Welt von Holzmedia bietet der Virtual Showroom.

Holzmedia combines fine furniture and multimedia technology. Two elements, whose synthesis leads to complete new solutions. Holzmedia offers top-of-the-line multimedia furniture, and also assists in the planning, design and installation of complete meeting and conference rooms.
The website acts as an interactive information tool. The existing corporate design has been upgraded to the digital medium by interactivity and animation. With the brochure configurator the user can put together customized product catalogs which can be downloaded as a PDF-file. Animated insight into the world of Holzmedia is provided by the Virtual Showroom.

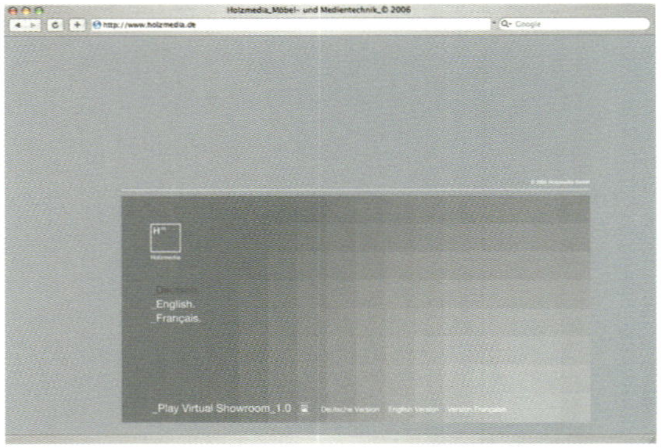
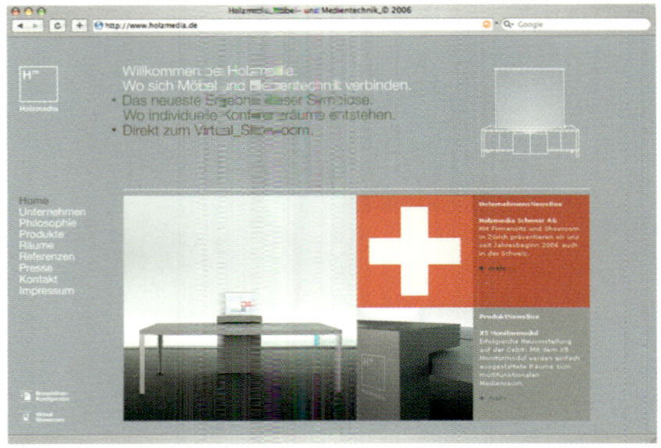
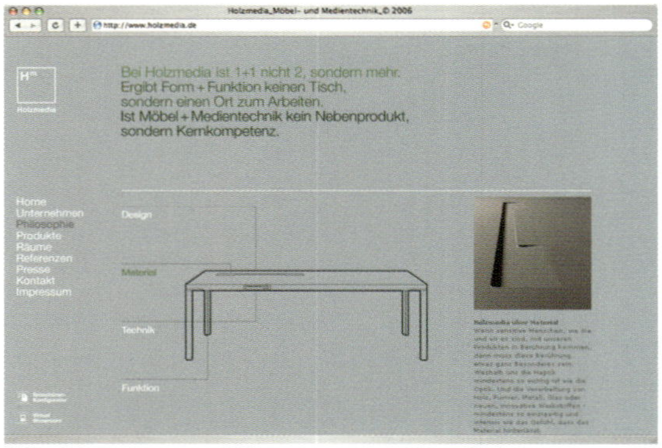
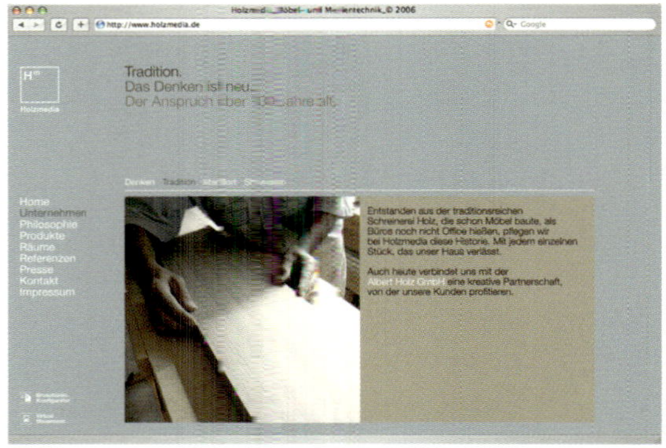
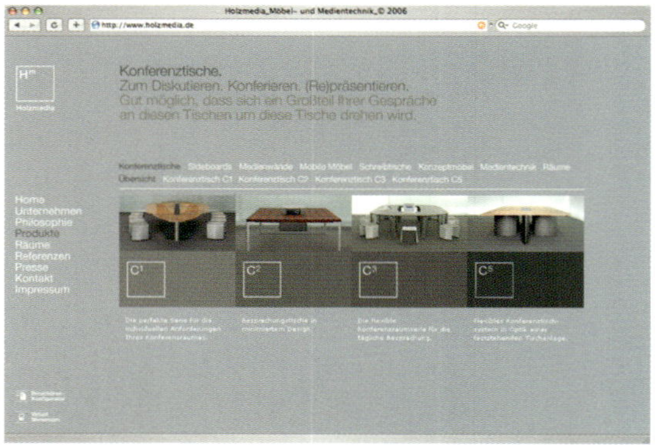
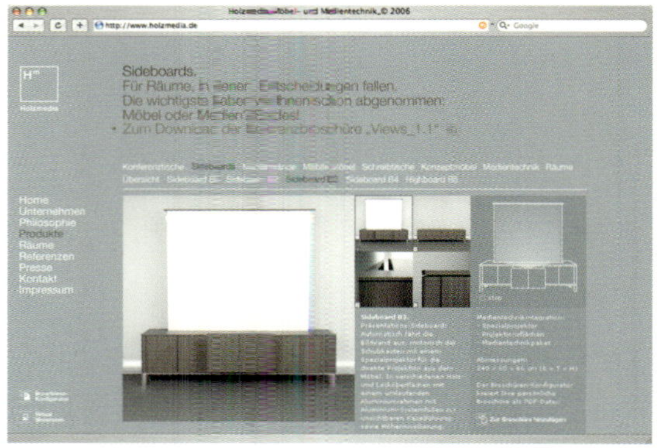
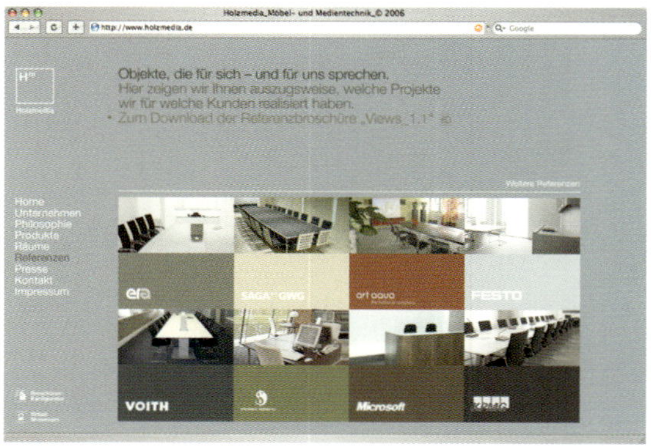
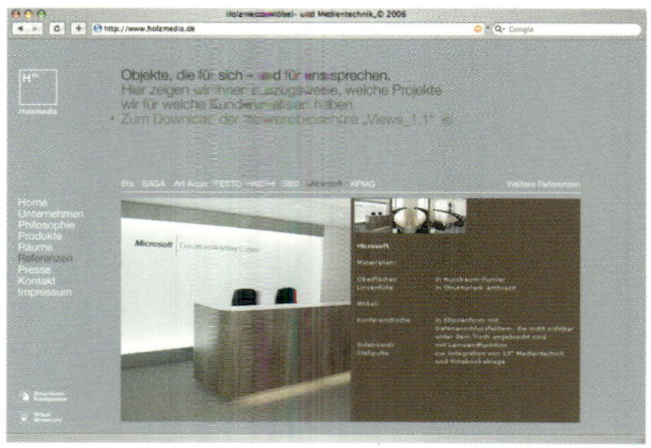

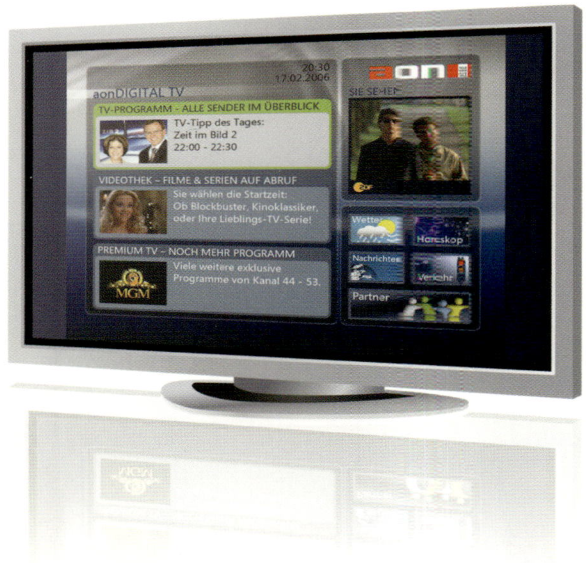

Auftraggeber Client
Telekom Austria AG
Vienna, Austria

Projekt Project
aonDigital TV
User Interface für digital TV –
Portal, EPG, VoD, Zusatzdienste
User Interface for digital TV –
Portal, EPG, VoD, Services

Design
Simone Keller-Kurz
Franz Koller
Joachim Machate
Marcel Nerz
Petra Prestel
User Interface Design GmbH
Ludwigsburg, Germany
Günter Strasser, Michael Pocher
Telekom Austria AG
Vienna, Austria

Die Bereiche von aonDigital TV sind durch eine eindeutige Anordnung klar voneinander getrennt. Unterstützt wird die übersichtliche Struktur durch den gezielten Einsatz von Umrahmungen und Farbe sowie aussagekräftiger Headlines. Die intuitive Bedienung, die sich dadurch ergibt, garantiert somit eine komfortable Benutzerführung. Die ausgewählten Elemente werden im TV-Bildschirm durch ein kräftiges Grün hervorgehoben. Diese Farbe ist nicht nur Primärfarbe der Telekom Austria. Sie wirkt frisch und freundlich und hilft dem Benutzer zugleich, sich optimal zurechtzufinden. Der visuelle Gesamteindruck des Screendesigns betont die innovative Technik.

A clear structure, colors and meaningful labeling ensure intuitive operation and convenient navigation of aonDigital TV. Important elements are emphasized in a strong green color – the primary color of Telekom Austria – generating a fresh and friendly association and ensuring optimal user navigation. The visual impression of the screen design intensifies the innovative technology. Users perceive the dark background colors and soft distribution on the TV as comfortable and not distracting at all. The rounded frames ensure a more dynamic and modern screen. Color, form and structure give aonDigital TV the appropriate look and feel.

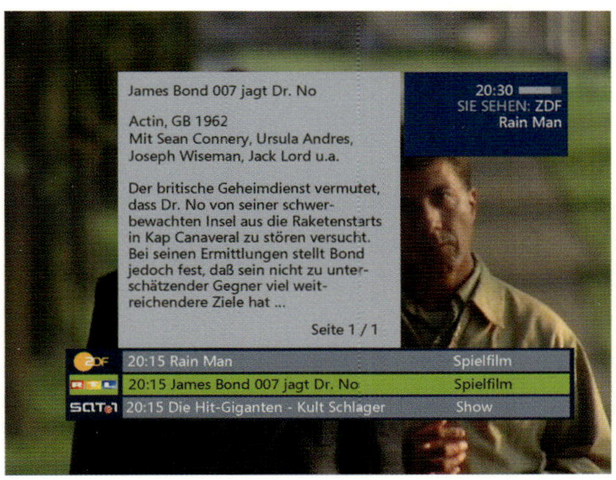

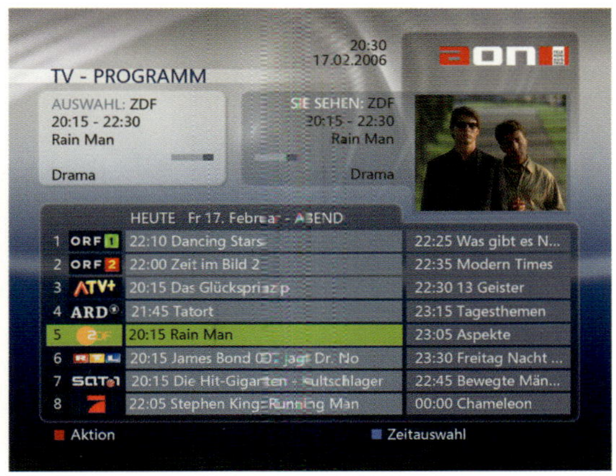

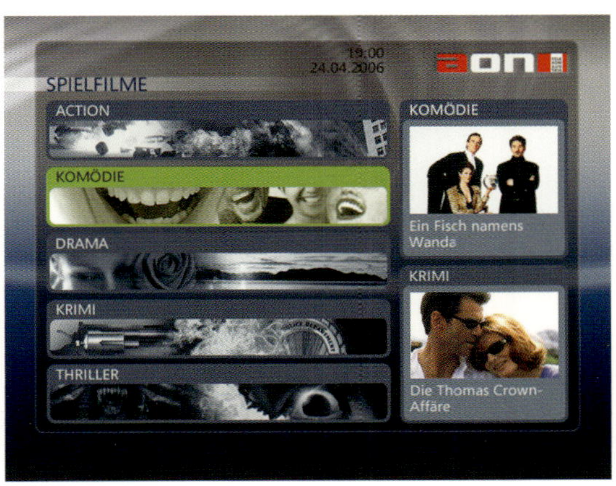

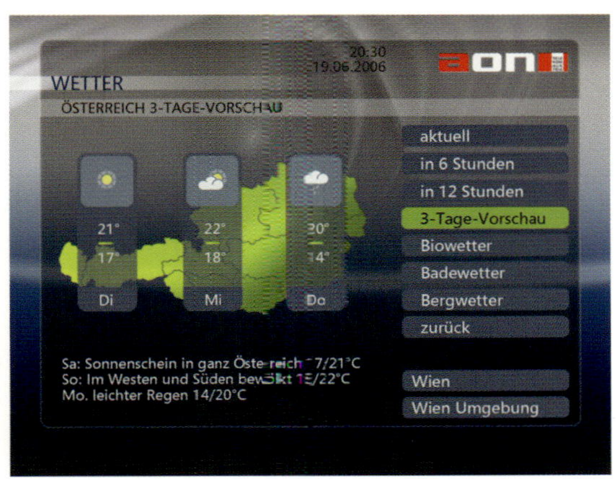

Auftraggeber Client
Samsung Electronics
Suwon-city, South Korea

Projekt Project
Sonic Motion Screensaver
Bildschirmschoner
Screensaver

Design
Soo-Hyeon Koh
Hee-Jung Kim
Samsung Electronics
Seoul, South Korea

Büromenschen, die in der eintönigen Office-Umgebung durch eine Vielzahl an Informationen überlastet werden, bietet der Sonic Motion Bildschirmschoner einen erfrischenden Ausflug in die Natur. Um naturechte Geräusche und Bewegungen zu erzeugen, werden „Visual", „Sound" und „Motion" in einer Technologie zusammengeführt, die zum strategischen Ziel „Sensory Branding" führt.
Mit diesem Design, das eigentlich der Bildschirmschonung dient, bieten wir den PC-Benutzern und -Benutzerinnen zugleich das wunderbare Erlebnis, sich zu fühlen, als wären sie in der freien Natur und würden frische Luft atmen.

The Sonic Motion Screensaver helps refresh PC users' weary bodies and minds as they cope with tons of information in a monotonous office environment. It uses the three languages of "Visual + Sound + Motion" to reproduce nature's vivid sounds and motions on the basis of a simple visual image. It is an effective expression of the strategic goal of Samsung-oriented Sensory Branding.
Sonic Motion Screensaver serves as a "window" enabling PC users to enjoy a breath of nature within the confines of their office, even as it works to protect their monitor.

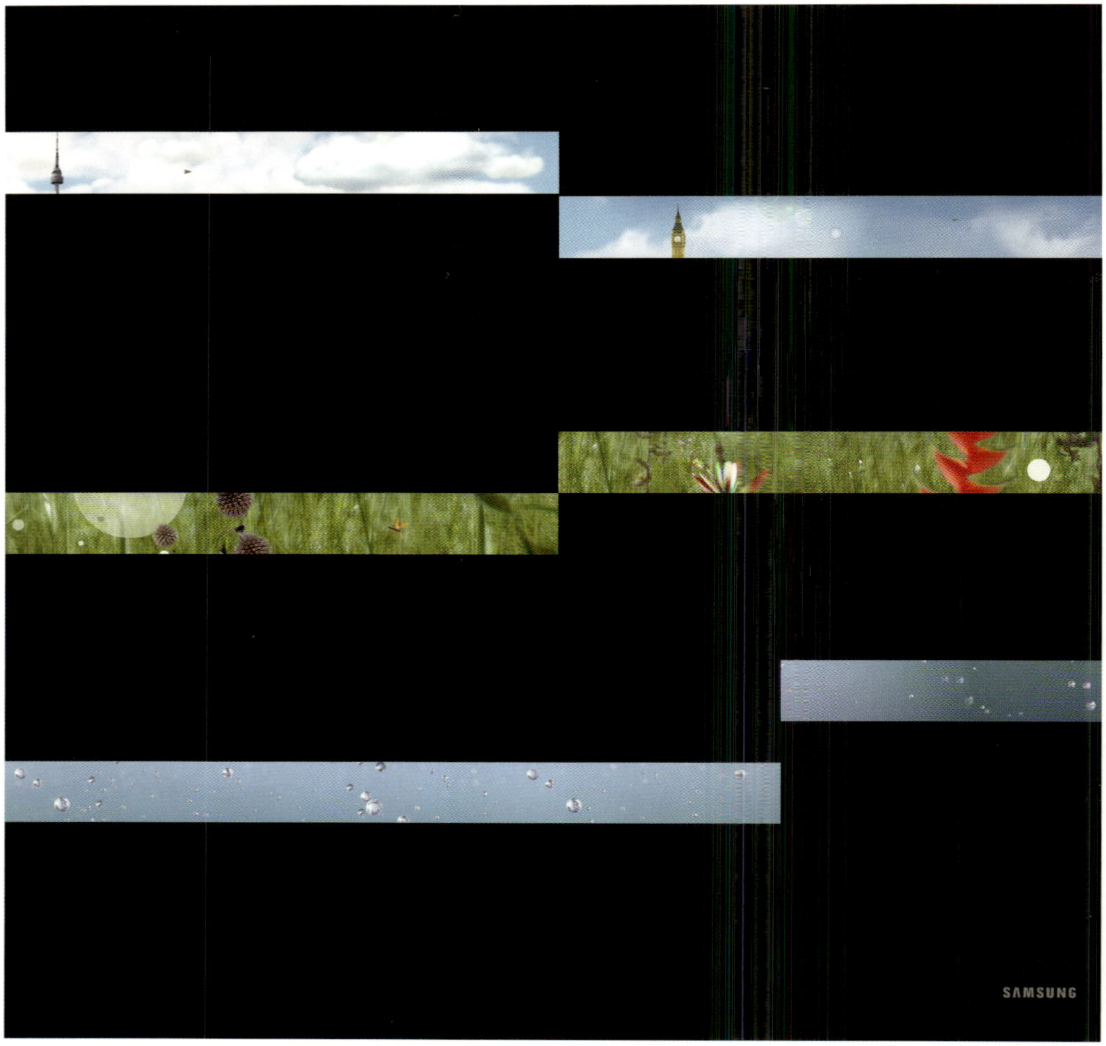

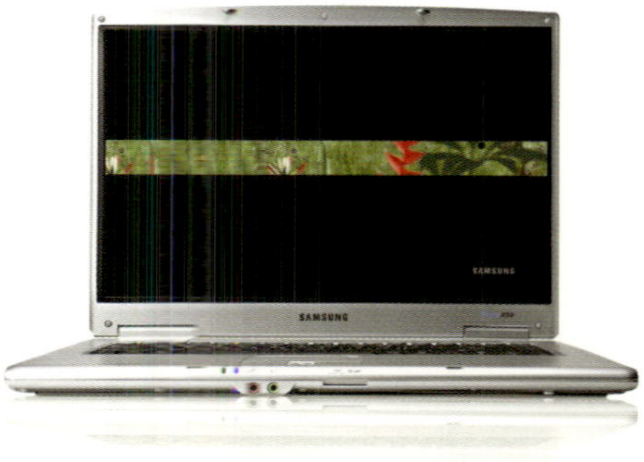

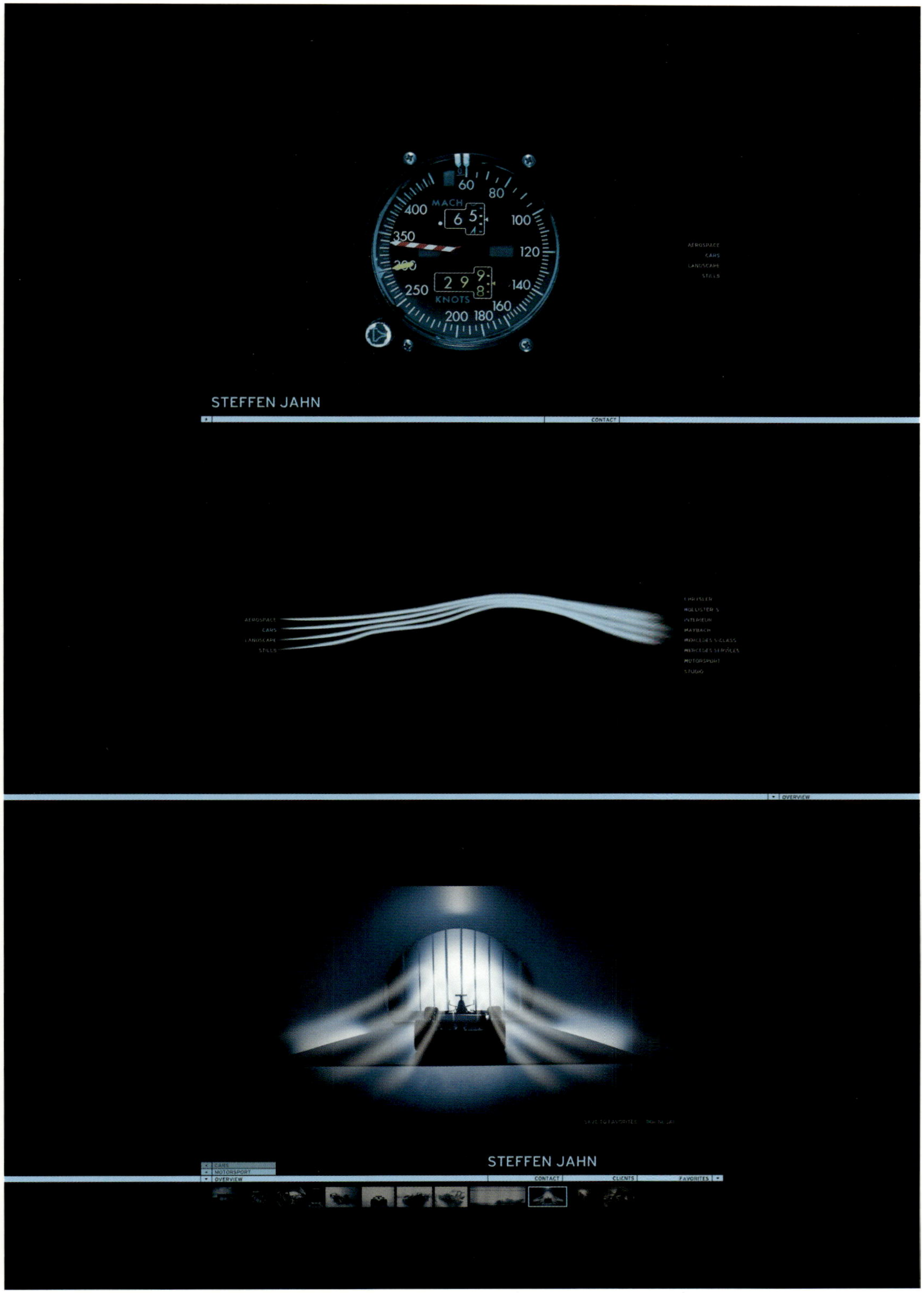

Auftraggeber Client
Steffen Jahn
Fotografie
Stuttgart, Germany

Projekt Project
Steffen Jahn Fotografie – Online Portfolio
Portfolio-Website und Photo-Management-System
Portfolio website and photo management system

Design
Wolfgang von Geramb
Robin zer
Superior Design Studio
Hamburg, Germany

Aufgabe: Entwickeln Sie eine Website mit CMS für den Auto-, Luftfahrt- und Landschaftsfotografen Steffen Jahn
Idee: Mit dem Ziel der optimalen Form testen Firmen ihre Transportmittel auf Luftwiderstand im Windkanal. Wir bedienten uns dieser Metapher und ließen Luft die Konturen bilden. Die Kategorien „Cars", „Aerospace", „Landscape" werden mithilfe einer Windkanal-Engine dynamisch generiert.
Lösung: Wir entwickelten den Photofresher, ein Photo-Management-System, mit dem Bilder, Texte, Kategorien flexibel geändert werden können. Eine Introfunktion präsentiert per Zufall neue Bilder. Alle Bilder-Untergrößen werden generiert.

Task: Online portfolio including CMS for the car-, aerospace- and landscape photographer Steffen Jahn.
Idea: With the goal of finding the optimal form, companies test their means of transport in the wind tunnel on air resistance. Using this metaphor we let air shape the outlines. Through a wind tunnel engine the categories "cars", "aerospace", "landscape" are generated dynamically.
Technique: We developed the Photofresher. It is a photo management system whereby pictures, text, categories can be changed flexibly. All other image sizes will be generated automatically. An intro function presents new pictures at random

Auftraggeber Client
Staatskanzlei des Landes Nordrhein-Westfalen
Referat für Internationale Angelegenheiten
Düsseldorf, Germany

Projekt Project
Der NRW-Kubus
Multimediapräsentation
Multimedia presentation

Design
Thomas Müller
Ralf Schroeder
Wenzel Spingler
Die PR-BERATER GmbH
Agentur für Kommunikation
Köln, Germany
Dietmar Gapp
bluelemon Interactive GmbH
Köln, Germany

Die deutsche und die japanische Regierung haben das „Deutschlandjahr in Japan" initiiert. Als Teil dieses Events will Nordrhein-Westfalen mit einer Multimediapräsentation, dem „NRW-Kubus", gezielt junge Japaner ansprechen und ihnen die innovative und lebenswerte Region vorstellen. Der im Raum „schwebende" Würfel lockt das Publikum mit technischer Vielfalt und klarem Design. An vier Navigationsstelen können Betrachter vier Welten (Life Style & Kultur/ Umwelt & Energie/Life Science/High Tech) interaktiv erleben – von Musik über Animationen, Fotos, Filme bis zu virtuellen Räumen. Durch den Würfel und seine Navigation erlebt der Benutzer NRW räumlich, thematisch, klanglich und visuell.

The German and the Japanese governments initiated the "German Year in Japan". As part of this event, NRW aims to directly approach young Japanese people with a multimedia presentation introducing them to the innovative and attractive area. The "floating" cube attracts the audience with technical variety and lucid design. At four navigation points observers can experience interactive themes from four worlds (Lifestyle & Culture/Environment & Energy/ Life Science/High Tech) – from music and animation, photos, film to virtual rooms. With the cube and its navigation the visitor experiences NRW spatially, thematically, acoustically and visually.

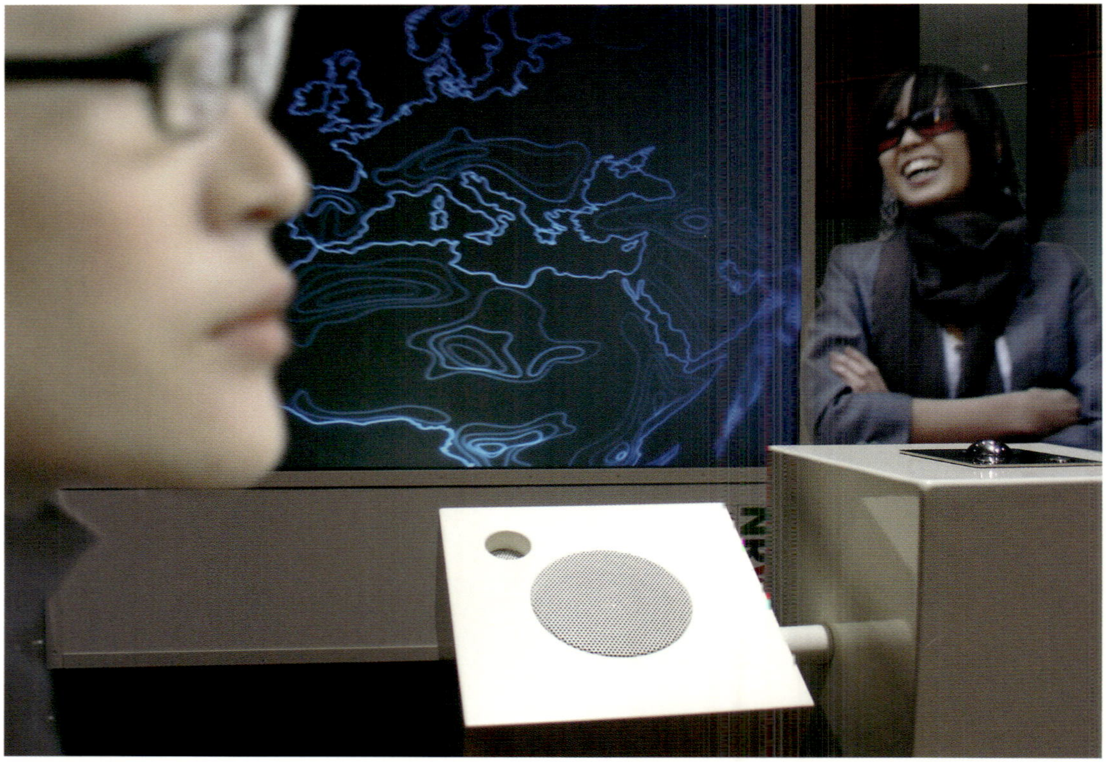

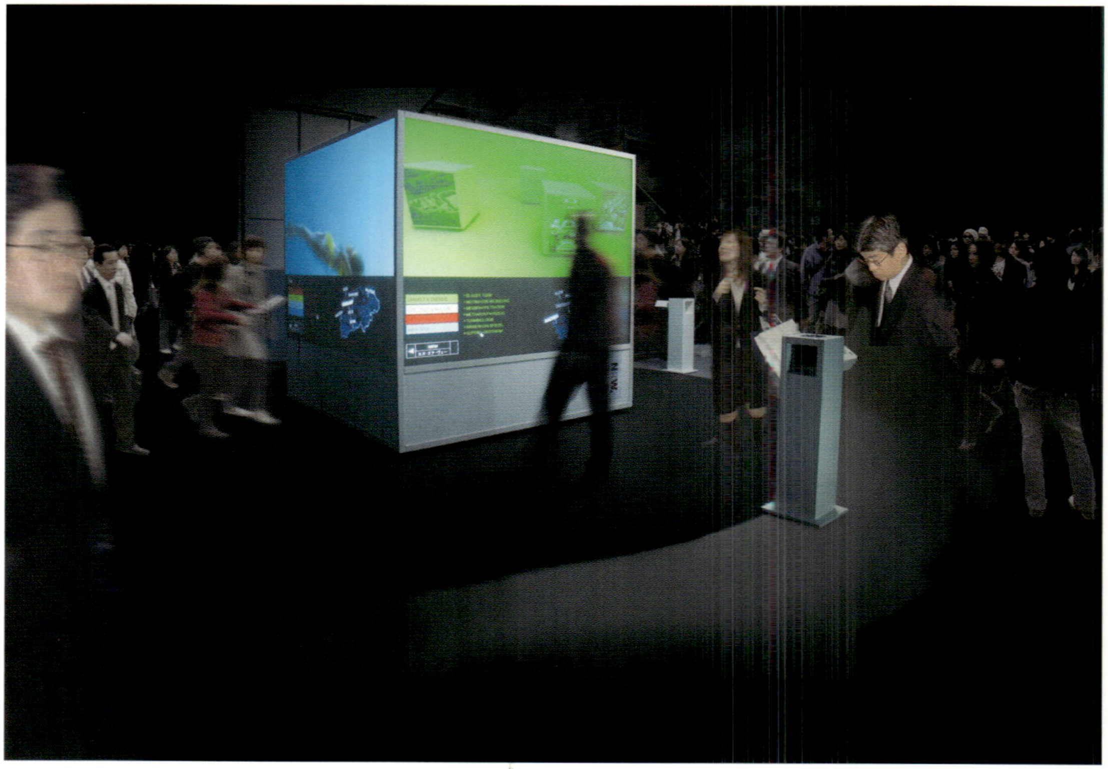

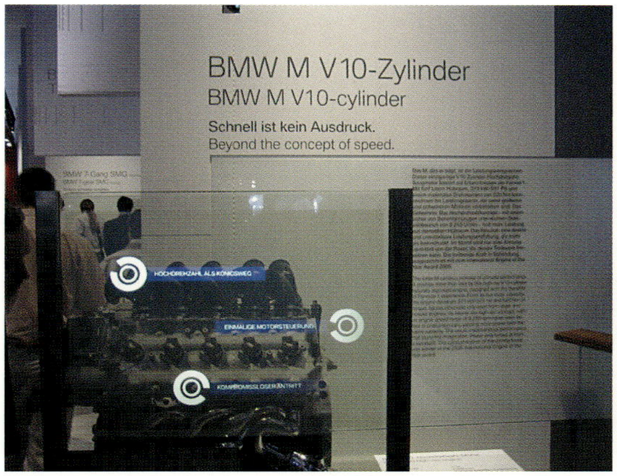

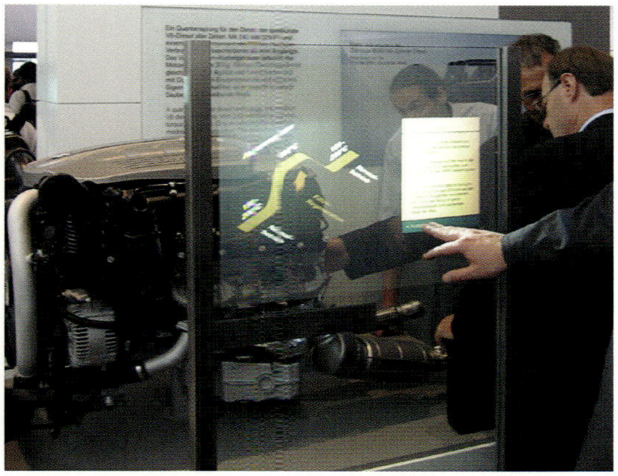

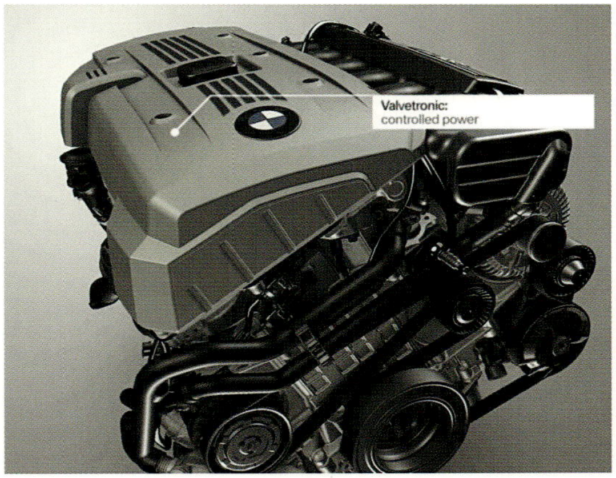

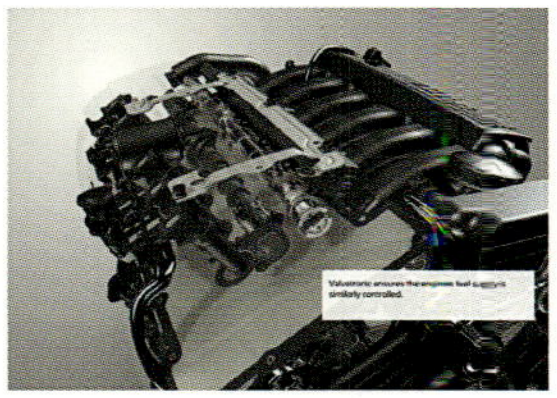

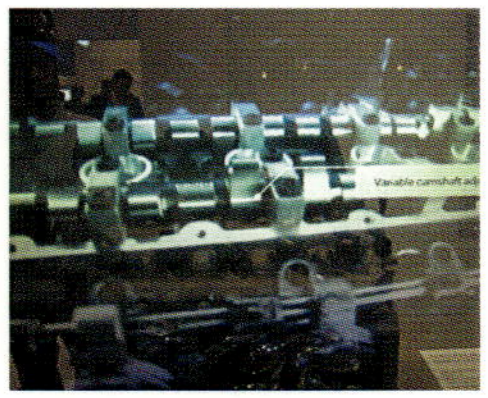

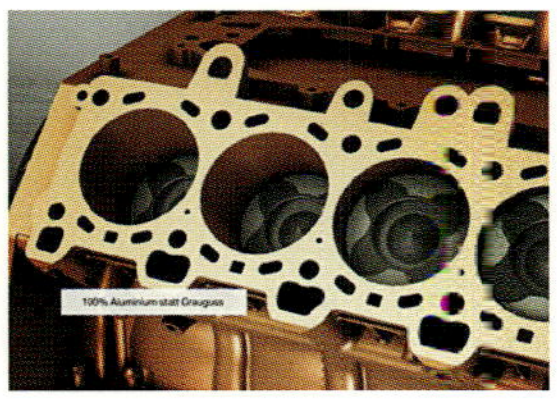

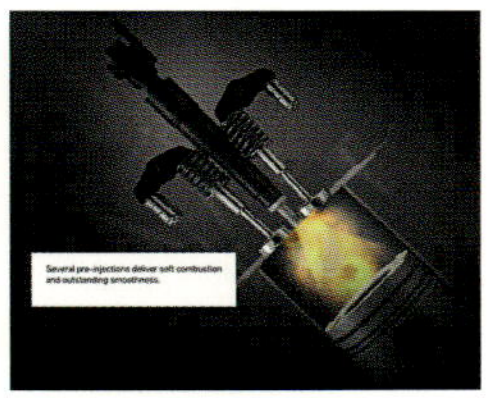

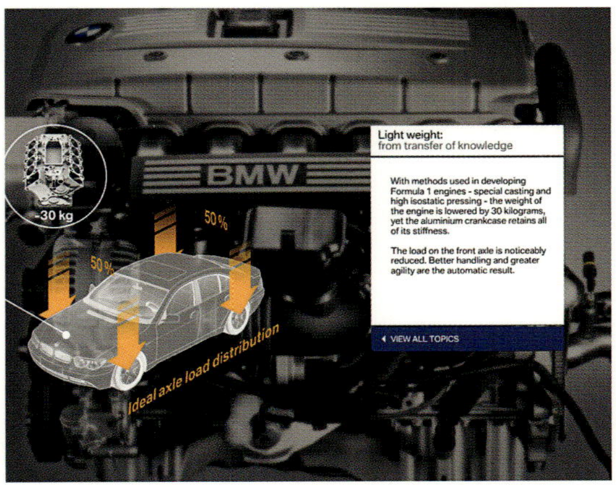

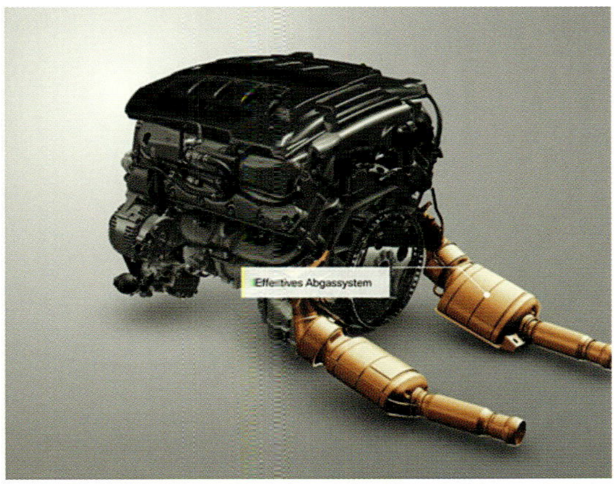

Auftraggeber Client
BMW AG
München, Germany

Projekt Project
BMW Mixed Reality Exponate
Interaktive Messeinstallation
Interactive exhibit

Design
Patrick Fonger
Lorena Beltrami
Jan Koepke
Jochen Röhling
John DuBois
Michael Plog
Oliver Schwenk
Interone Worldwide GmbH
Hamburg, Germany

Die BMW Mixed Reality Exponate sind interaktive und teiltransparente Touchscreen-Anwendungen, die auf der IAA 2005 in Frankfurt am Main eingesetzt wurden. Sie verdeutlichen und demonstrieren technologische Besonderheiten von verschiedenen BMW Motoren. Vor den ausgestellten Motoren werden 3-D-Animationen und Filme auf durchsichtige Projektionsflächen projiziert. Abwechslungen in der Durchsichtigkeit geben den Blick auf den realen Motor immer wieder frei und führen dabei zu einer Mischung aus virtuellem und realem Erleben. Dadurch kann die Funktion der Motoren direkt und interaktiv am ausgestellten Exponat erläutert werden.

The BMW Mixed Reality exhibits are interactive and partly transparent touchscreen applications that were displayed at the 2005 Frankfurt Motor Show. They demonstrate and provide insights into technical innovations in varying BMW engines. 3-D animations and films were projected onto transparent surfaces in front of the engines. By alternating the transparency, views of the real engine behind the screen were given at different points, creating a mix of the virtual with the real. Visitors to the exhibit were thus given a demonstrative explanation of how the engines function – directly and interactively.

Auftraggeber Client
DeTeImmobilien
Münster, Germany

Projekt Project
DeTeImmobilien Immobilienplattform
Gewerbeimmobilien-Marktplatz
Real estate marketplace

Design
Die Firma GmbH
Innovative Kommunikation
Wiesbaden, Germany

Das Gewerbeimmobilienportal der T-Com Tochter unterstützt die mediengerechte Vermarktung des jeweils aktuellen Immobilienbestands im Internet. Der Zugriff auf die umfangreichen Details der zahlreichen und unterschiedlichen Mietobjekte wird im Nutzungsprozess frühzeitig und interessensgerecht über die dynamische, flashbasierte Suchabfrage ermöglicht. Die Präsentation der komplexen Informationen und Verbundenheit einzelner Objekte wird über klar strukturierte, einfach erschließbare Suchergebnislisten- und Detailseitendesigns erreicht. Die geradlinige, reduzierte Gestaltung transportiert den technischen und informativen Charakter der Präsenz.

This T-Com subsidiary's commercial property website supports property marketing over the Internet in a way that is perfectly suited to this type of media. The website's dynamic, Flash-based search facility gives the user quick access to extensive details on a wide variety of rental properties. Complex information on individual properties and the connections between them is presented by means of clearly structured, easy-to-read hit lists and detail views. The reduced linear design of the website conveys its technical and informative character.

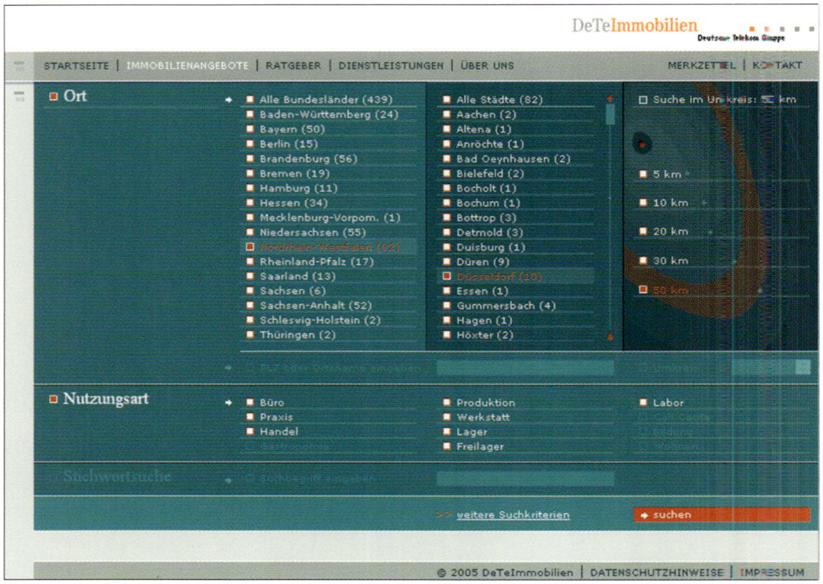

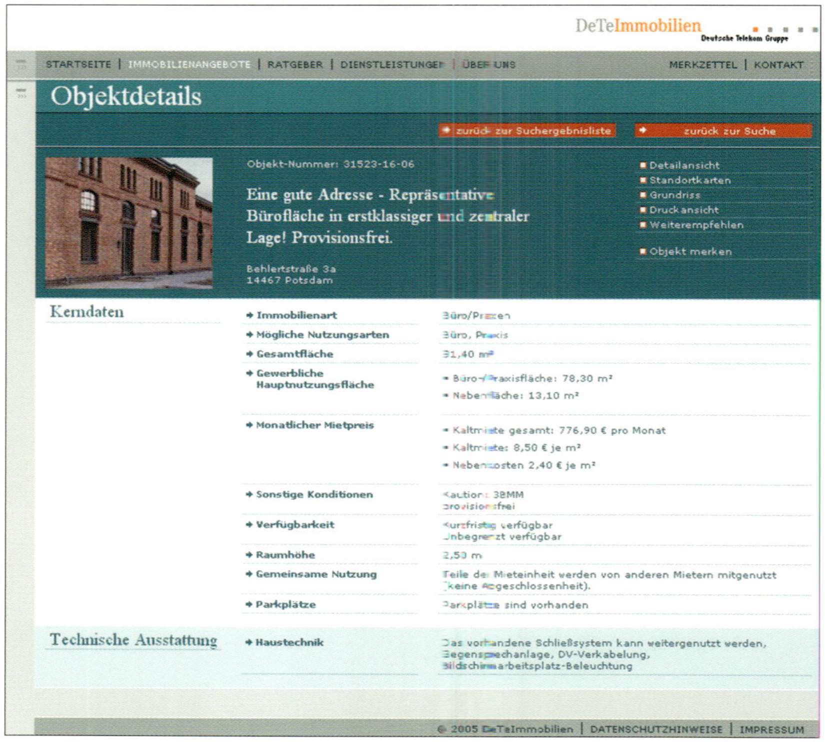

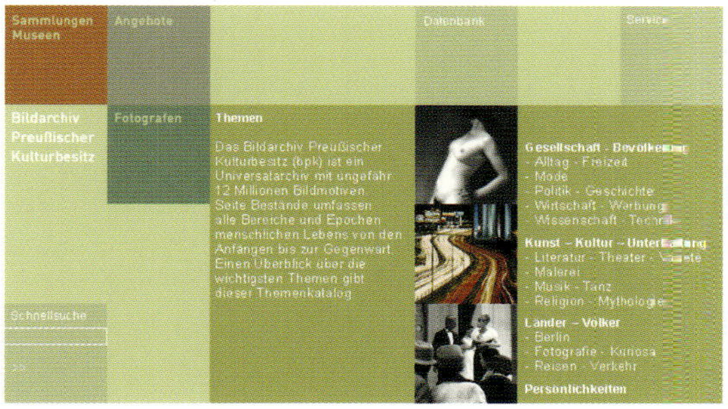

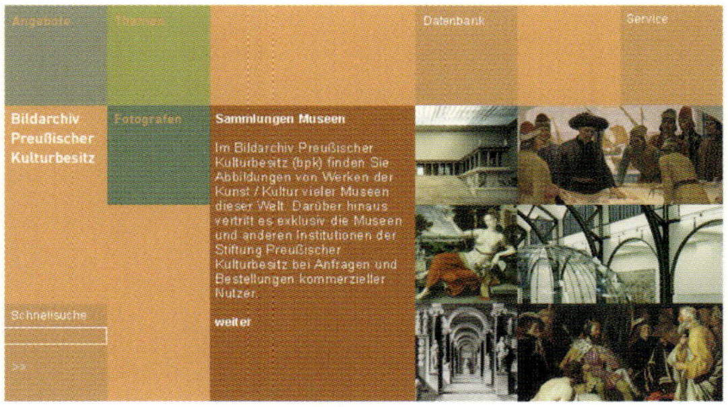

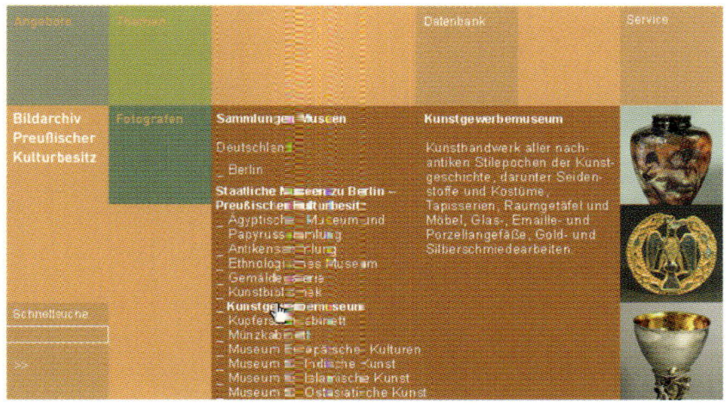

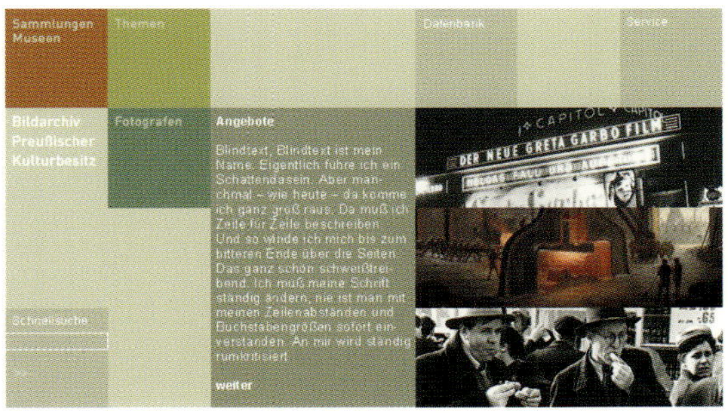

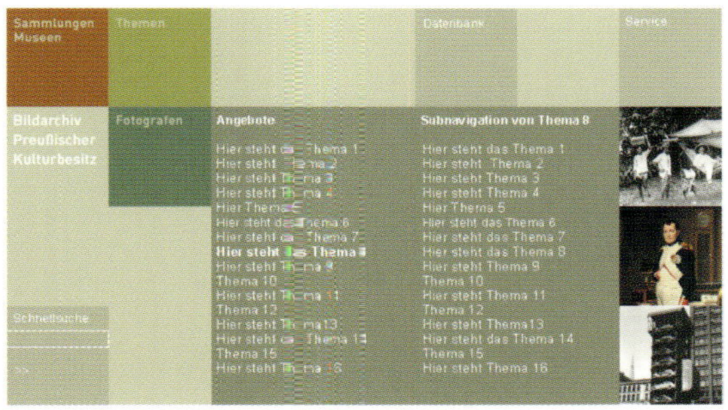

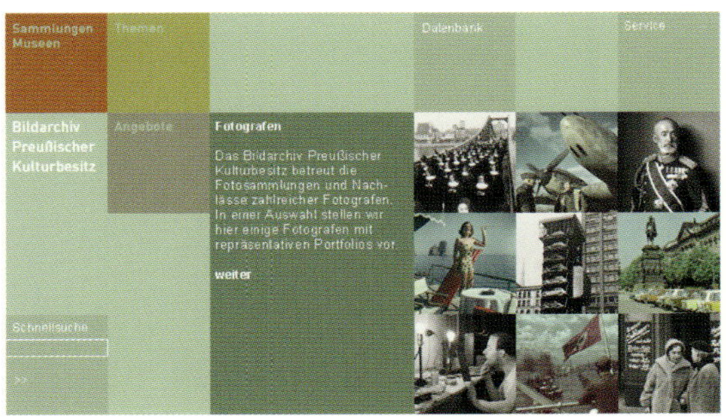

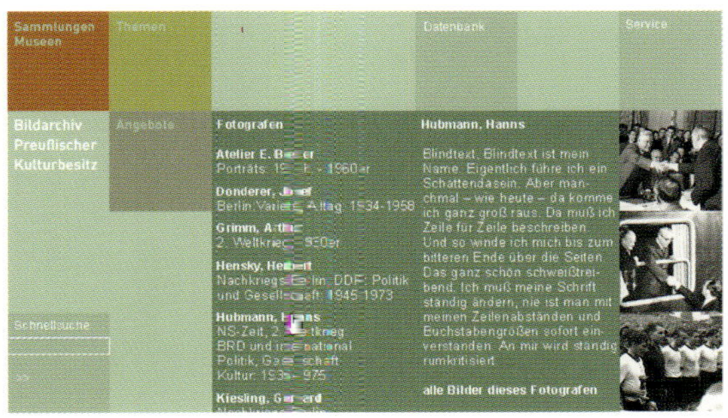

Auftraggeber Client
Bildarchiv Preußischer Kulturbesitz (bpk)
Berlin, Germany

Projekt Project
Bildarchiv Preußischer Kulturbesitz
Website

Design
Rabea Hashagen
langzeitwirkung™
Berlin, Germany

Das Bildarchiv Preußischer Kulturbesitz (bpk) hat zwei Aufgaben: Es ist Agentur am Bildermarkt sowie zeitgeschichtliches und kulturhistorisches Fotoarchiv. Das bpk vertritt Museen, Bibliotheken und Kulturinstitutionen aus aller Welt und bietet Abbildungen von deren Werken zur kommerziellen Nutzung an.
Das Spezialarchiv mit über 12 Millionen Bildmotiven bietet bereits 75 000 ausgewählte Motive unter www.bpk-images.de zur Onlinerecherche und zum Download. Der Webshop ist bei ca. 50 Bildagenturen eingesetzt und dadurch kommerziellen Nutzern bereits vertraut.

Bildarchiv Preussischer Kulturbesitz (bpk), the picture library of the Prussian Cultural Heritage Foundation in Berlin, performs two tasks: It is a photo agency serving the media and it is an archive preserving images as a cultural source. bpk represents museums, libraries and other cultural institutions from all over the world and offers photographic reproductions of their collections for publishing and other commercial use.
The specialized archive contains more than 12 million images and already offers 75,000 selected images for online research and download under www.bpk-images.de. As the Web shop technology is used by about 50 photo agencies, it is well-known by commercial users.

Auftraggeber Client
Siemens Audiologische Technik GmbH
Erlangen, Germany

Projekt Project
Visualization of Digital 5 Family Features
Flash-Animationen / CD-ROM
Flash-Animations / CD-ROM

Design
Michael Ebner
Tilo Schattschneider
portamundi GmbH & Co. KG
Nürnberg, Germany

Das jüngste High-End-Hörgerät der Siemens Audiologische Technik GmbH bietet eine nie da gewesene Hörqualität. Es setzt den neuesten Stand der Technik in sechs verschiedenen Funktionen, so genannten Features, um. Siemens Audiologische Technik beauftragte portamundi mit der Erstellung und Umsetzung eines Konzepts zur Visualisierung der Features. Visualisierungen sollen es Hörgeräteakustikern weltweit ermöglichen, dem Endkunden die Funktionen/Vorteile des Geräts anschaulich darzustellen. Die Animationen stellen auditive Features visuell dar und sind Bestandteil eines SW-Tools (CONNEXX), das Hörgeräteakustikern auf CD-ROM zur Verfügung gestellt wird.

The newest high-end hearing instruments from Siemens Audiological Engineering Group offer previously unknown hearing quality, setting new industry standards in six functional features. Siemens Audiological Engineering Group has commissioned portamundi to develop and implement a concept to visually explain these advanced features. The visualization should make it possible for hearing aid professionals worldwide to graphically display these functions and their advantages to end users. Animation to visually display audio features will be integrated in SW-Tools (CONNEXX) on CD-ROM.

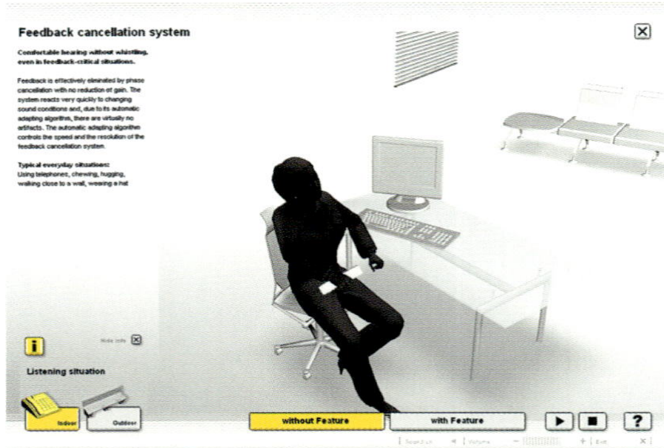

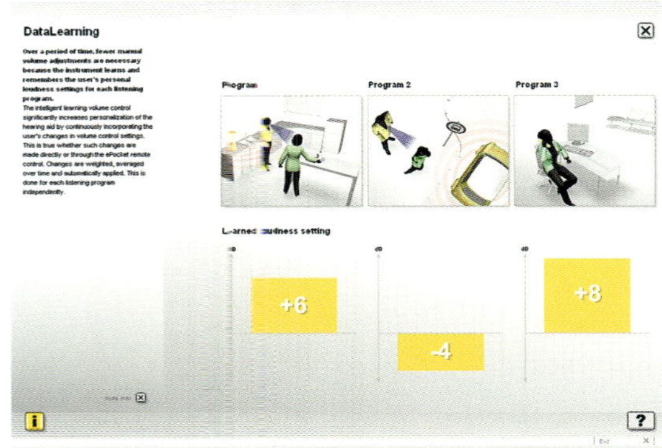

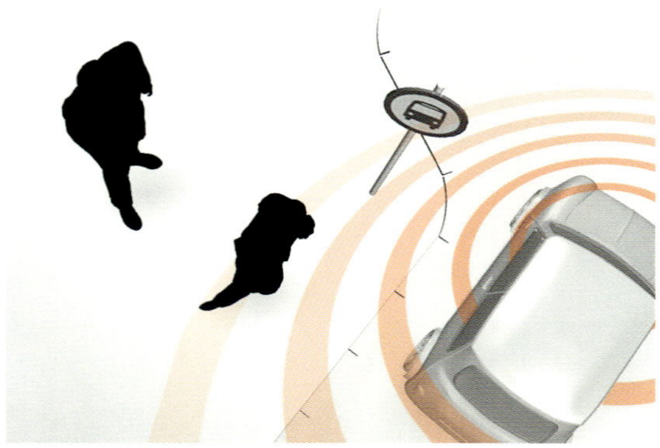

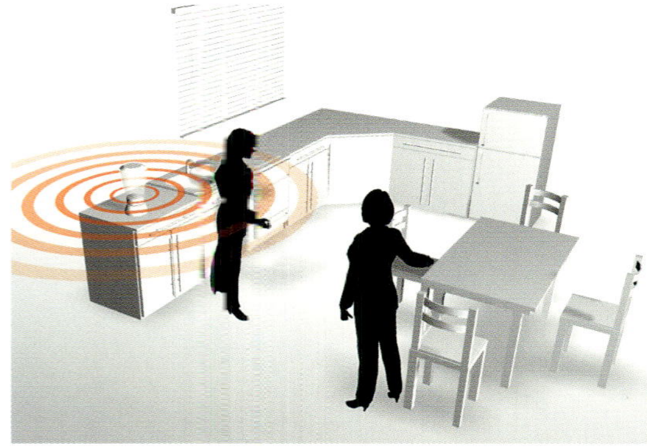

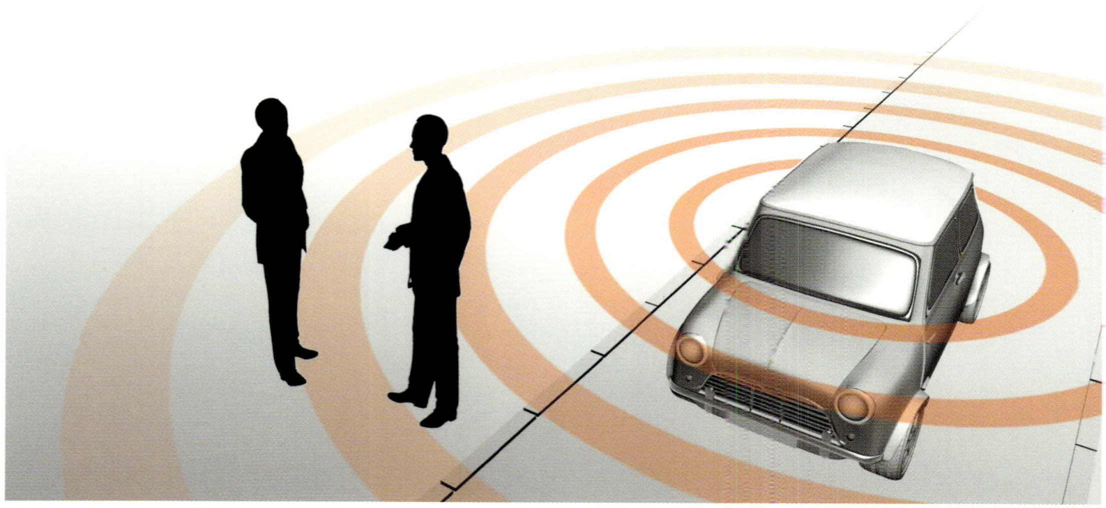

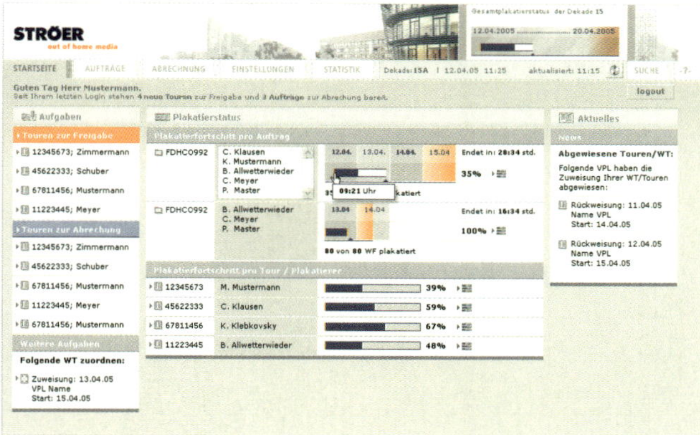

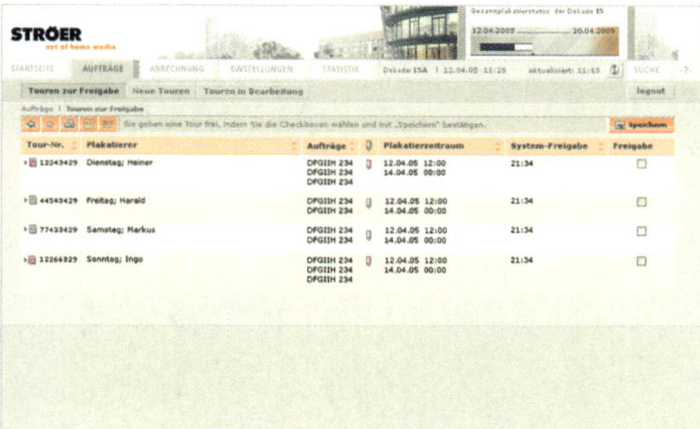

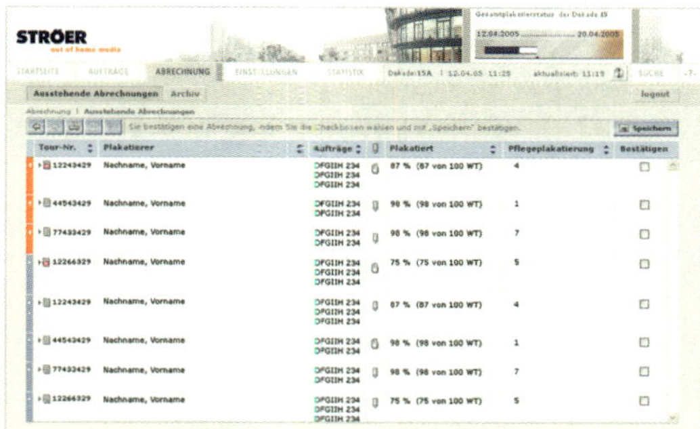

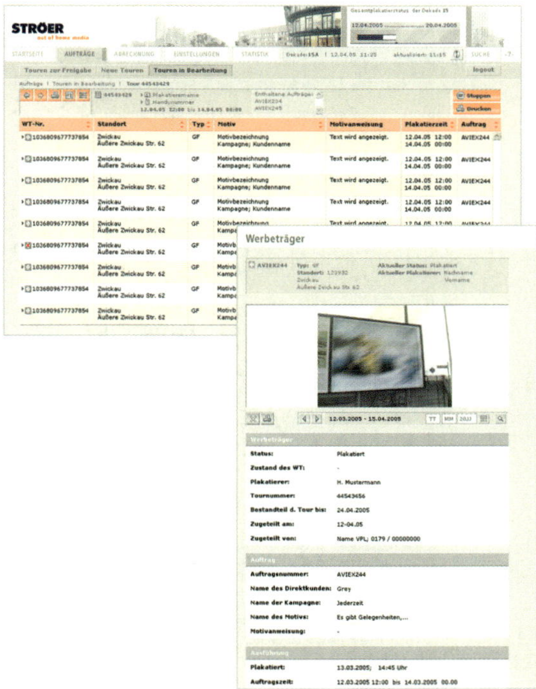

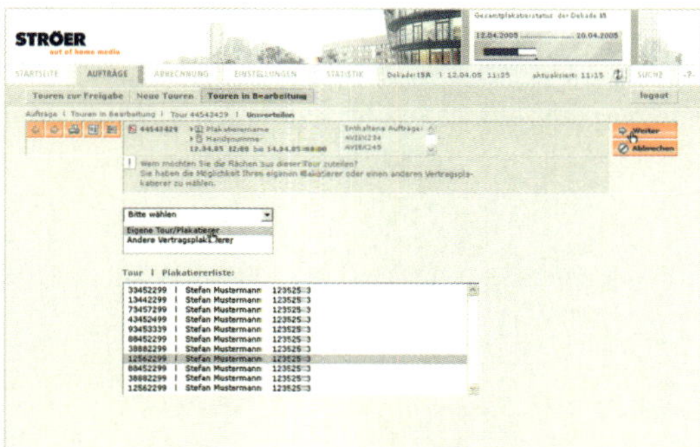

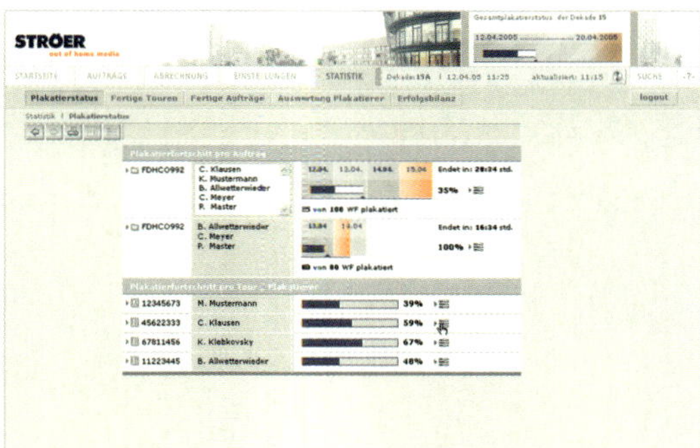

Auftraggeber Client:
Ströer Out-of-Home Media AG
Köln, Germany

Projekt Project
Ströer Wireless Posting Control
Website

Design
denkwerk
Köln, Germany

Ströer Wireless Posting Control ist eine eigens für Ströer entwickelte, integrierte Mobil-/Webapplikation zur Steuerung, Überwachung und Dokumentation der Plakatier-Prozesse in Deutschland. Mit Hilfe der Webanwendung lassen sich alle Plakatierungen vom Auftrag über die Zusammenstellung der Touren bis zur Abrechnung komplett digital organisieren und live überwachen. Vor Ort wird jede Plakatierung per Handyscan und Foto dokumentiert und an das System zurückgespielt. Damit stellt Ströer eine hohe Qualität der Bewirtschaftung sicher.

Ströer Wireless Posting Control is especially developed for Ströer, is an integrated mobile/web application made for controlling, monitoring and documentation of the posting process in Germany. With this support, all postings can be completely digital organized and observed live starting at the request for the tour configurations up to the invoicing. On location, every posting will be documented via mobile phone scan and photo and transmitted to the system. Thereby, Ströer assures the high quality of the posting process.

Projekt Project
Kaleidotype
Computer-Programm
Computer programme

Design
Michael Buttgereit
Wolfram Heidenreich
Martin Winter
Jörg Fassmann
Buttgereit und Heidenreich GmbH
Haltern am See, Germany

Die Idee zu Kaleidotype entstand während der Arbeit für verschiedene christliche Institutionen und Kirchen, die aufgrund des Briefings ein Kreuz in der Bildmarke führen sollten. Dafür entwickelten wir ein einfaches Computer-Programm, das es ermöglicht, Buchstaben zu Kreuzformen zusammenzuführen.
Aus den vielfältigen Möglichkeiten des Gestaltens mit „Buchstabenbildern" entstand das Kaleidotype-Computer-Programm, mit dem symmetrische Muster aus Schriftzeichen in einem interaktiven Prozess dynamisch erzeugt werden können. Die Muster können in unterschiedlichen Dateiformaten ausgegeben und in anderen Programmen weiterverarbeitet werden.

The idea for Kaleidotype was born during our continuous work for different Christian institutions and churches. Their briefing was to develop a figurative brand that shows a cross. First we produced a simple computer programme, which offered the possibility to create a cross composed of simple letters.
From the various options of creating images composed of letters we developed the Kaleidotype computer programme. It makes it possible to create symmetric patterns on the basis of letters in an interactive process. The patterns can be transferred into different file formats and worked with in different programmes.

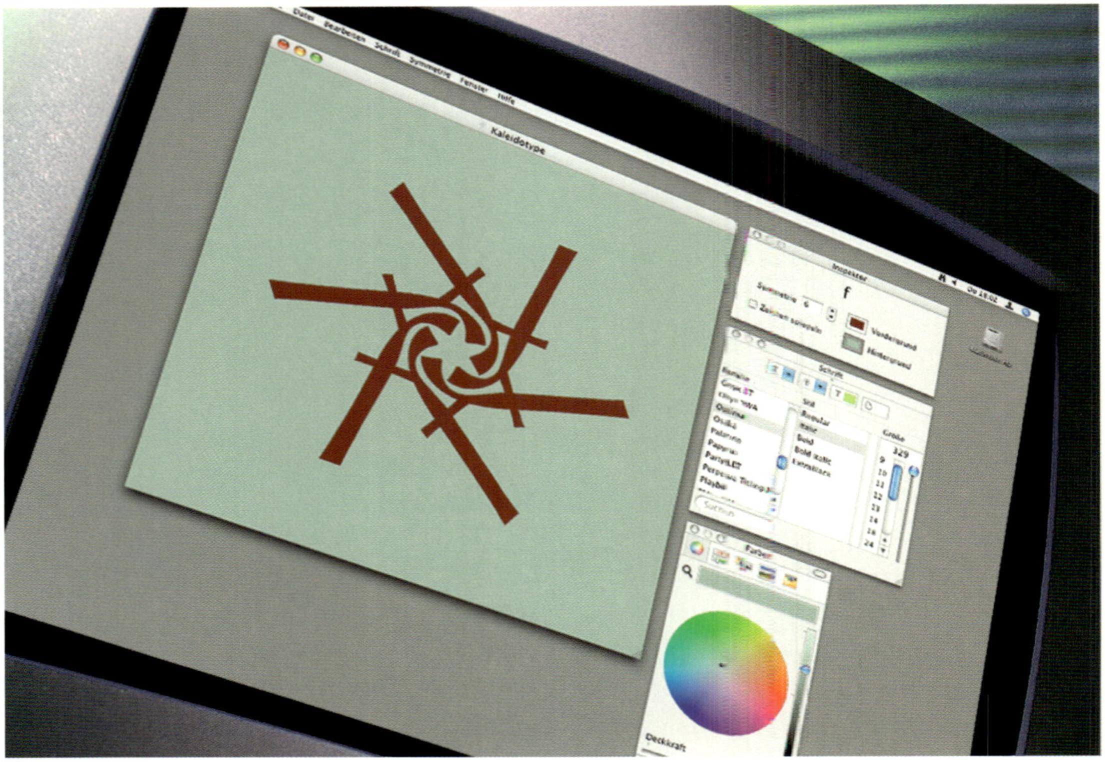

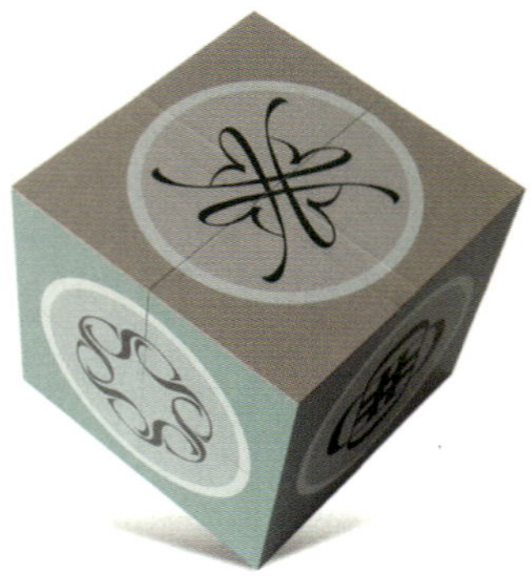

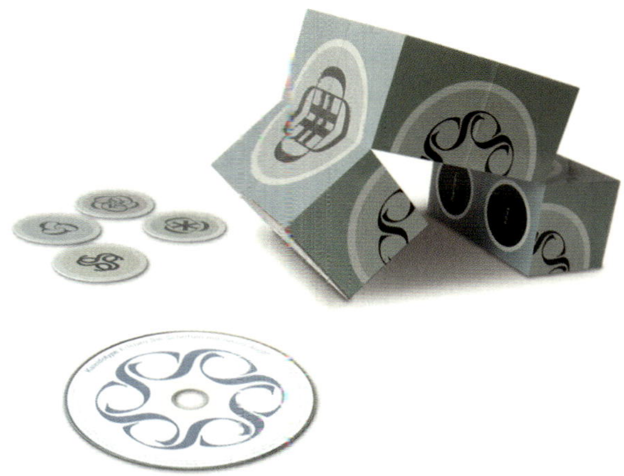

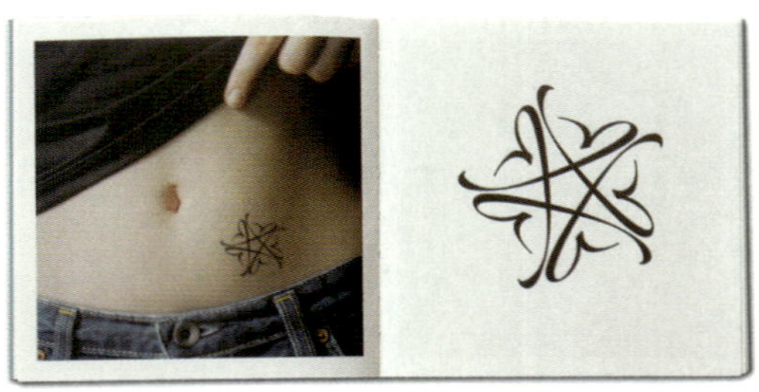

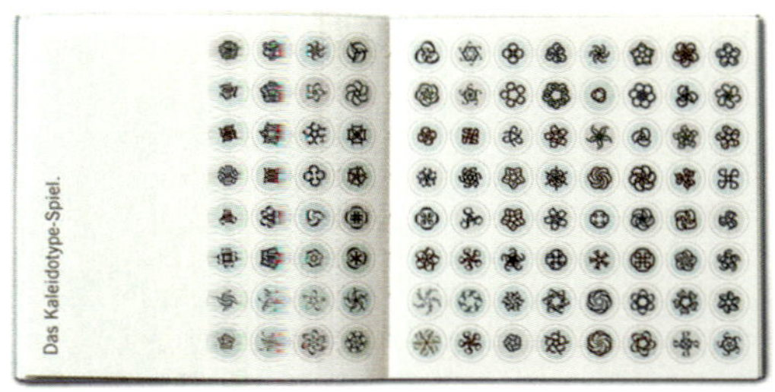

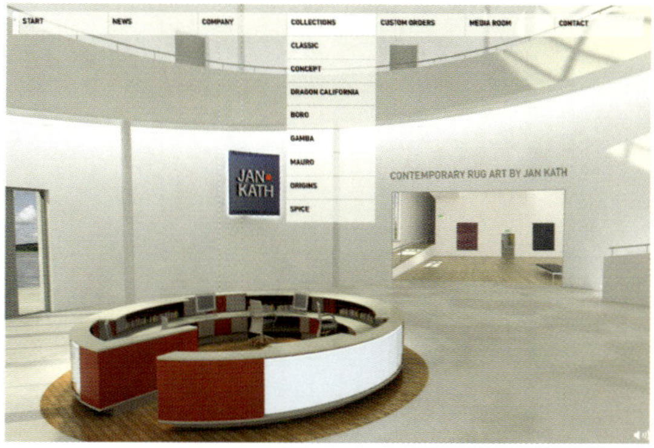

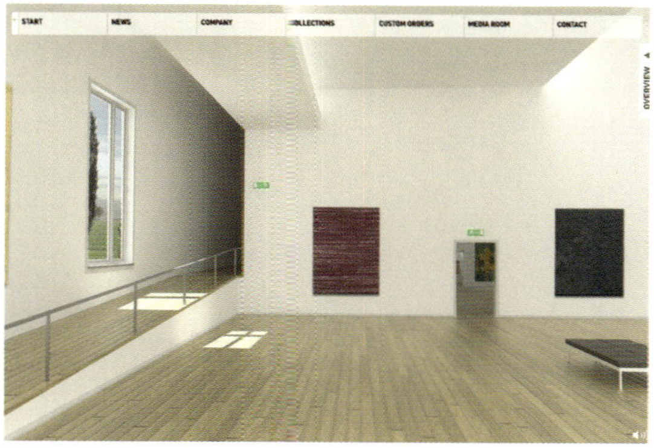

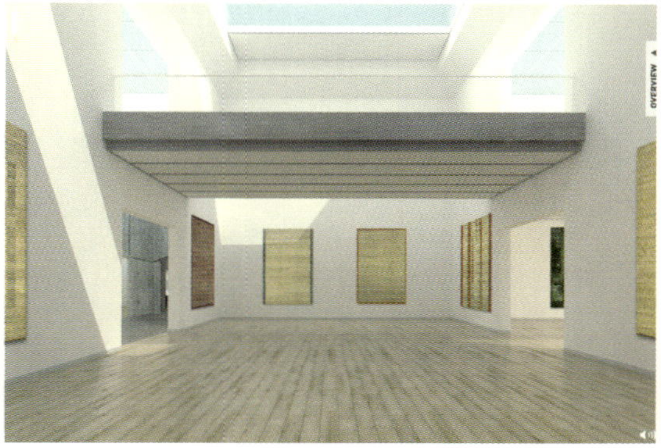

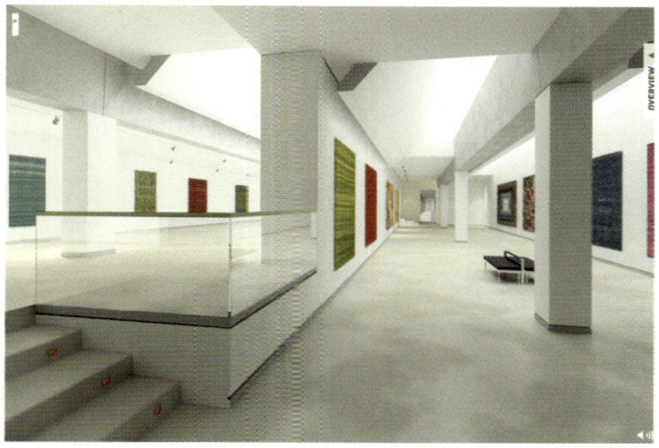

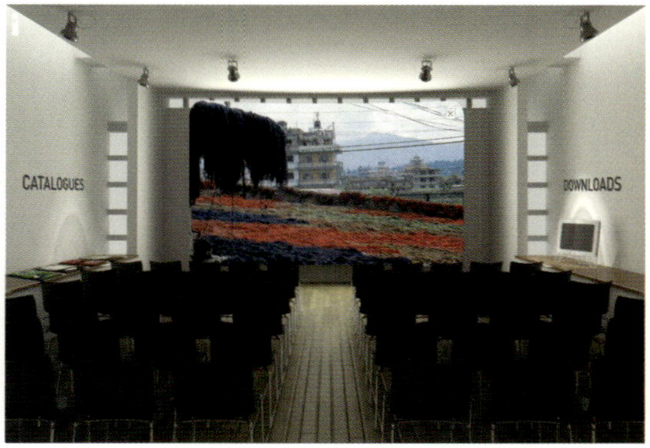

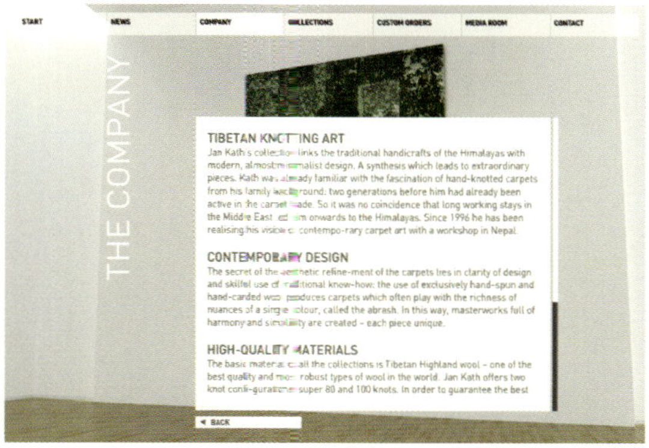

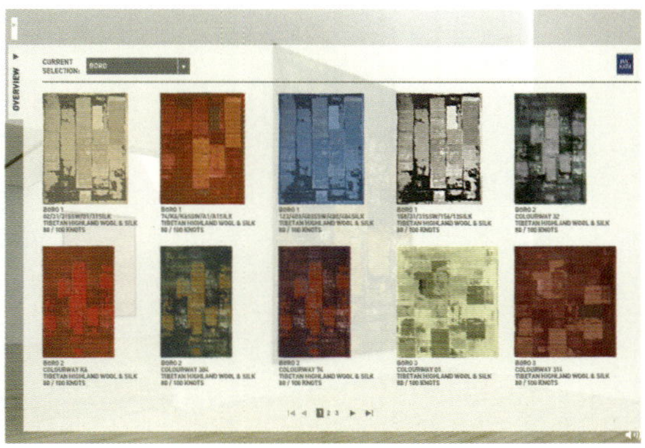

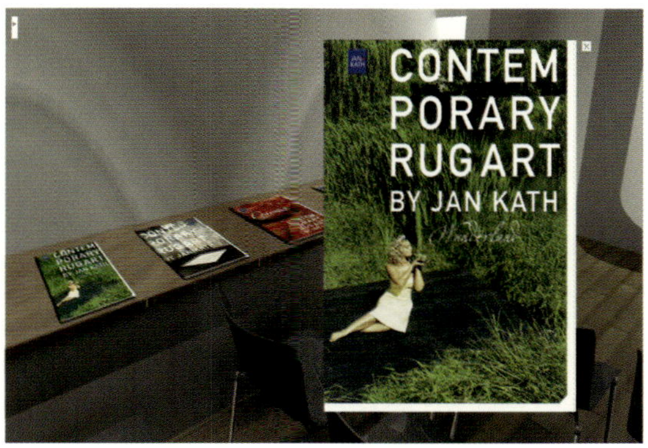

Auftraggeber Client
Jan Kath Design GmbH
Bochum, Germany

Projekt Project
www.jan-kath.de
Website

Design
René Wynands
Silke Löhmann
Tom Schülke
Andreas Hucks
Oktober Kommunikationsdesign GmbH
Bochum, Germany

Ein virtuelles, in 3D gebautes Museum für den Teppichdesigner Jan Kath. Der Besucher kann in dem Museum frei umhergehen und sich die Teppiche anschauen.
Für den Zugang wird ein Passwort benötigt: Name: oktober, Passwort: oktober.

A virtual museum, completely built in 3D, for the rug artist Jan Kath. In the 3D space the visitor can move around freely and look at the rug art.
To log in the following password is needed: name: oktober, password: oktober.

Auftraggeber Client
BMW Group AG
München, Germany

Projekt Project
BMW Welt Film – von der Vision zur Realität
Image- und Markenfilm
Image and brand film

Design
Steffen Armbruster
Hagen Thomas
Ingo Leuschner
Christian Liedtke
Sven Strassner
:: dan pearlman markenarchitektur
Berlin, Germany

Ein gigantisches Projekt wie die BMW Welt ist ohne veranschaulichende Mittel nicht vorstellbar. Durch den Film, der auf Grundlage eines interaktiven Walkthroughs erstellt wurde, ist das BMW Erlebnis- und Auslieferungszentrum schon vor der Bauabnahme in allen Dimensionen erfahrbar. Die Fahrzeugausstellung veranschaulicht durch bewegte Exponate, liebevoll gestaltete Fahrzeuge, reale Menschen, Filme im Film und interessante Kamerafahrten die komplexen Zusammenhänge zwischen Architektur und Markenerlebnis.
Trotz hoher Bildqualität gelang es, die Renderzeiten verträglich zu halten und einen authentischen „Look & Feel" der BMW Welt zu vermitteln.

It is hard to imagine a huge project like BMW Welt ("BMW World") without seeing images of what it will actually look like when it is finished. Thanks to this film, which has been created as an interactive walkthrough, the viewer can now experience every dimension of BMW's brand experience and distribution center before final completion. The vehicle display illustrates the complex interrelationship between architecture and brand experience using moving exhibits, lovingly designed vehicles, real people, films within a film and interesting tracking shots.
Despite the high quality of the images, the rendering times remain tolerable and the film manages to convey the genuine look and feel of BMW Welt.

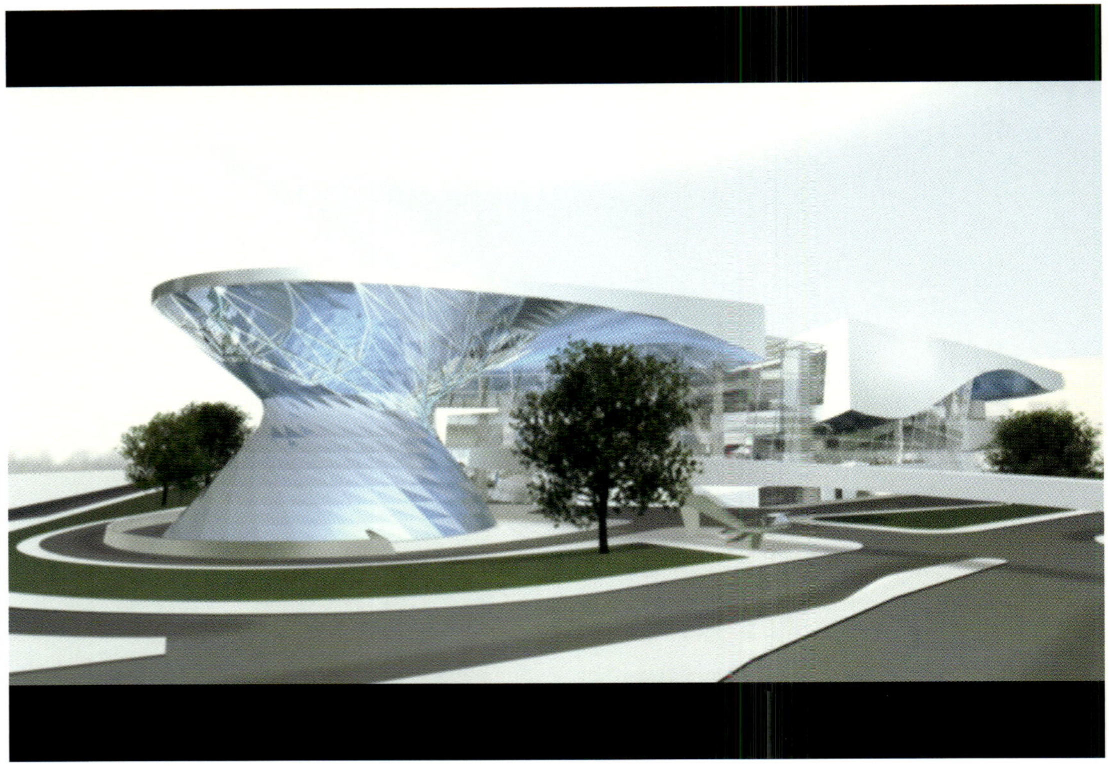

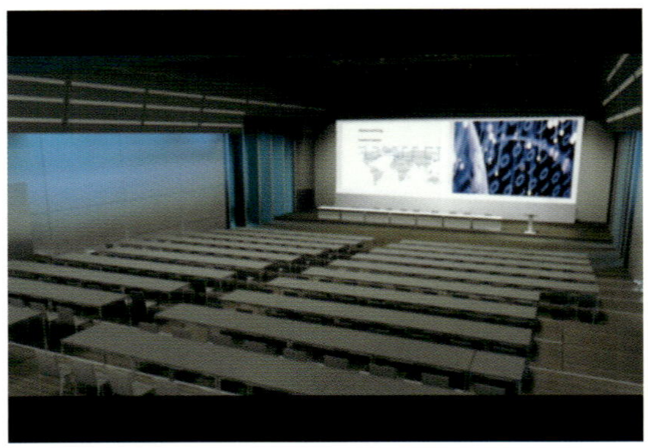

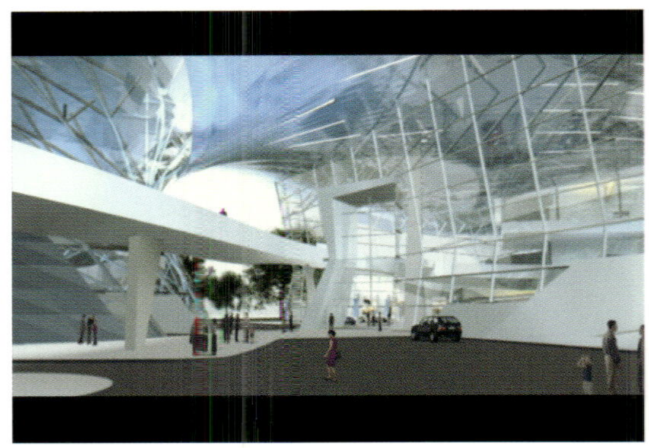

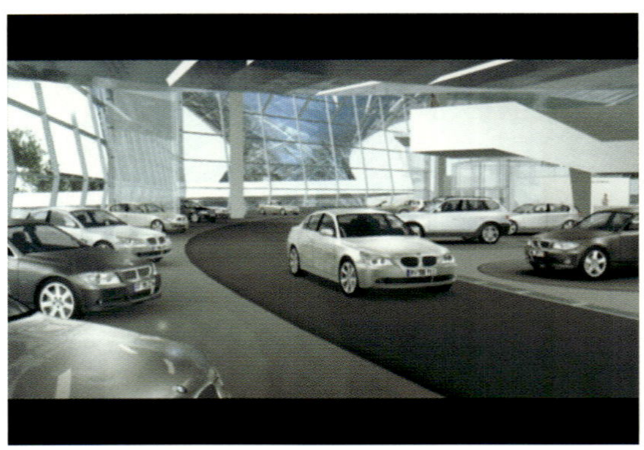

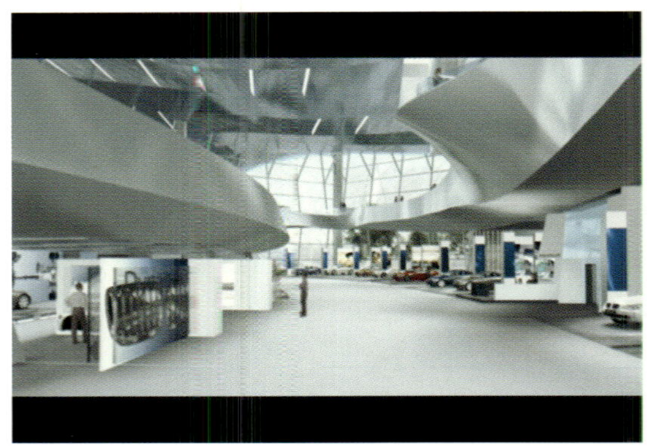

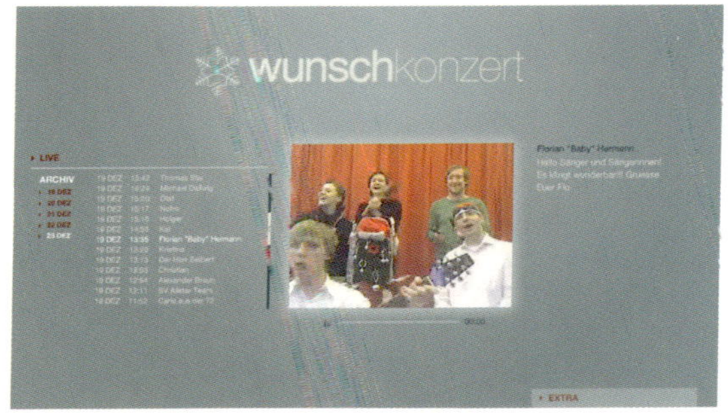

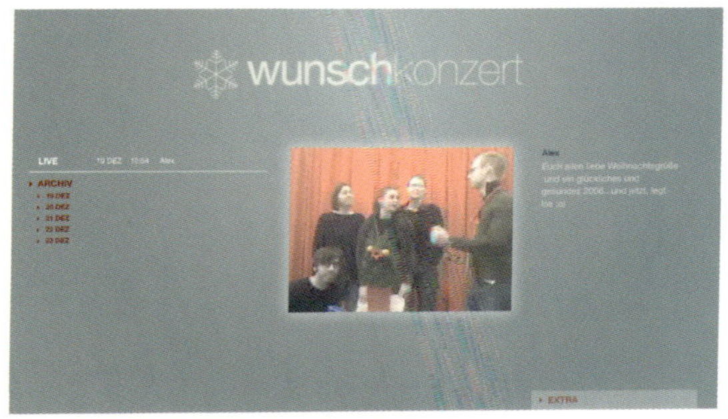

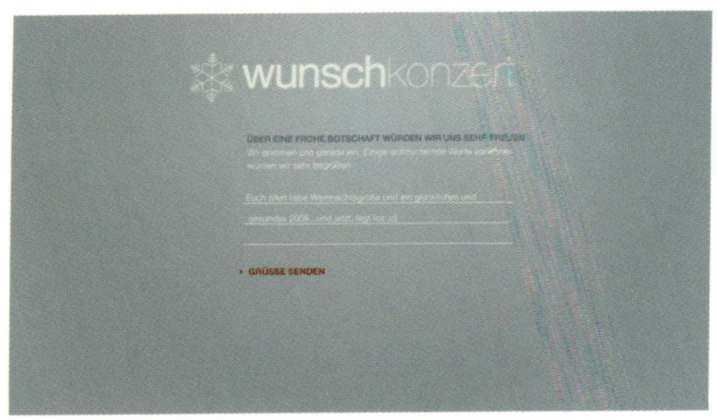

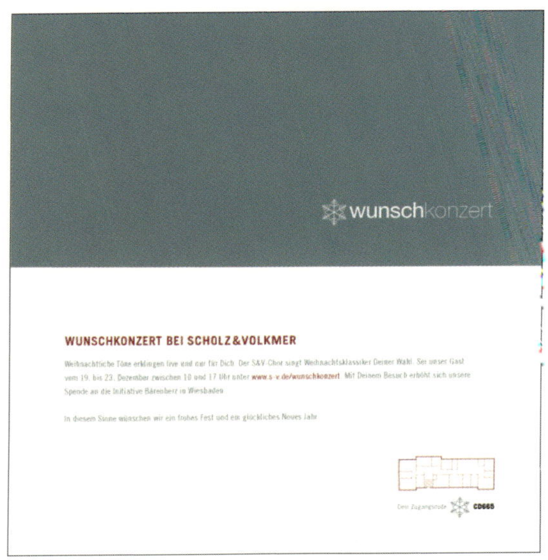

Auftraggeber Client
Scholz & Volkmer GmbH
Wiesbaden, Germany

Projekt Project
Scholz & Volkmer Wunschkonzert
Weihnachtsaktion mit Live-Event im Internet
Christmas mailing with accompanying real-time activity on the Internet

Design
Scholz & Volkmer GmbH
Wiesbaden, Germany

Weihnachtsaussendung mit begleitendem Live-Event im Internet. Kunden und Freunde wurden eingeladen, über eine Website einen von drei Weihnachtssongs auszusuchen. Live konnten sie miterleben, wie Mitarbeiter von Scholz & Volkmer für sie persönlich das Lied gesungen, musikalisch umrahmt und spontane Grüße übermittelt haben. Jeder Song bedeutete eine Spende an eine karitative Einrichtung für krebskranke Kinder.

Christmas mailing with accompanying real-time activity on the Internet. Clients and friends were invited to choose one of three Christmas songs via a website. Live they were able to watch employees from Scholz & Volkmer singing and playing the song for them and conveying spontaneous greetings. Every song represented a donation to a hospice for children with cancer. Implementation: The studio was presented online via a webcam. Users were able to watch Scholz & Volkmer employees personally singing for them – in real time. This action was saved as a CD.

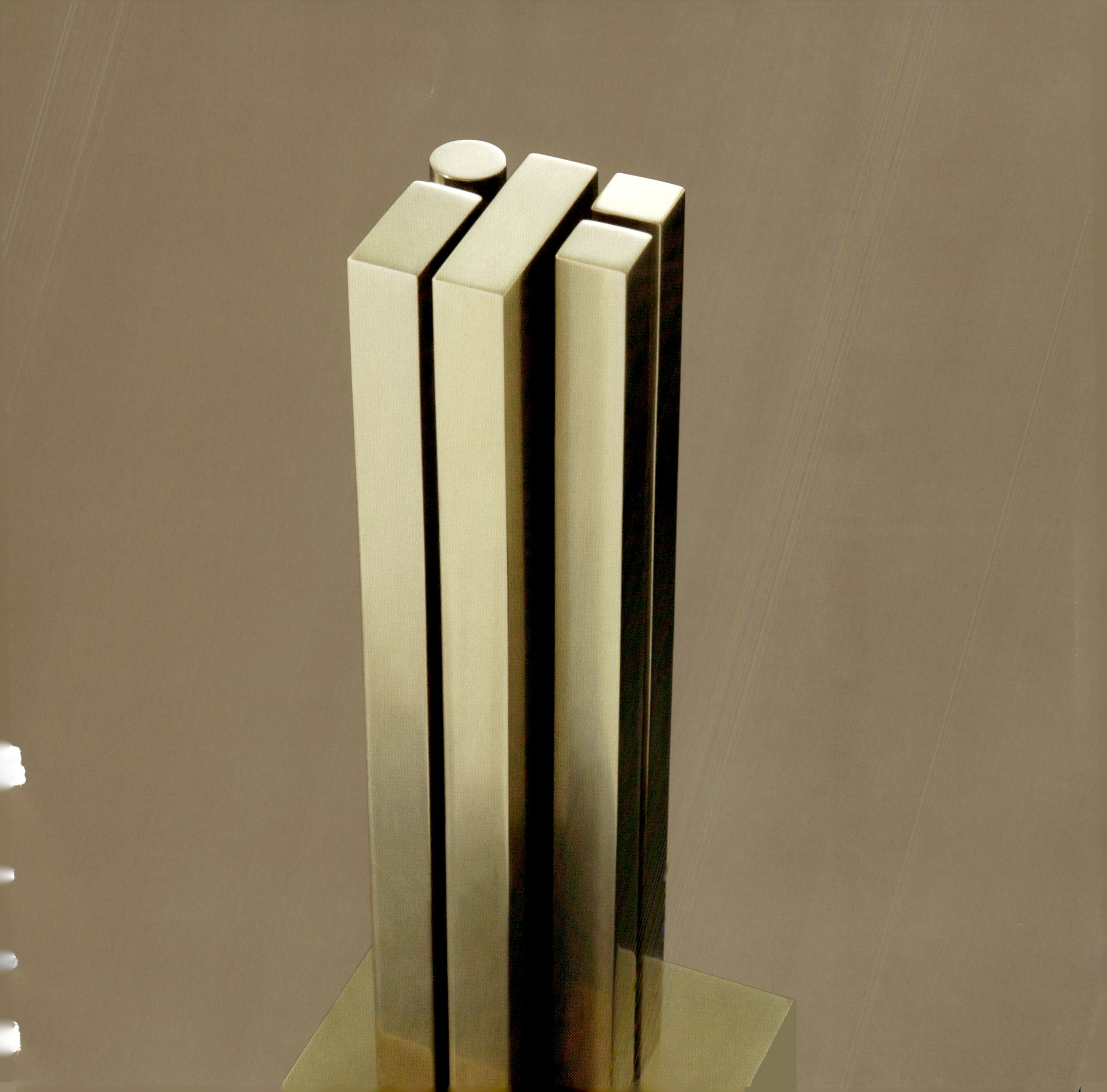

iF gold selection 2006

iF gold selection 2006

■ Eine Auszeichnung in einem der iF Wettbewerbe ist seit mehr als 50 Jahren ein Gütesiegel für herausragendes Design. Wer im iF product design award, im iF communication design award oder im iF material award von der jeweils fachkundigen, internationalen Jury das begehrte iF Label erhält, gehört zu den Besten seiner Branche. Zudem gibt es in jedem iF Wettbewerb die Chance, eine ganz besondere Trophäe nach Hause zu tragen: einen der iF gold awards. Diese eigens von Herbert H. Schultes entworfene iF Skulptur ist der „Design Oscar" – und wer ihn erhält, hat mit einem ganz besonderen Produkt, einer Materialinnovation oder einer Kommunikationsleistung die Jury überzeugt und sich so an die Spitze der Ausgezeichneten gesetzt. Alle Gewinner der iF gold awards bilden die iF gold selection des Jahres, die noch einmal in besonderer Form in der iF online exhibition dargestellt und auch als repräsentative Ausstellung zu sehen sein wird.

■ For over 50 years winning an iF award has served as a symbol of outstanding design achievement. Anyone who manages to convince our international jury of experts to issue an iF product design award, an iF communication design award or an iF material award belongs to the crème de la crème in their sector. And every iF competition also offers its contestants a chance to win a special trophy as well: the iF gold award. These iF statuettes are a special creation of Herbert H. Schultes, and are considered the "Design Oscar". Winning one of them means the recipient was particularly successful in convincing the jury of the exquisiteness of his or her product, material innovation or communication project, and therefore deserving of a special status among the other winners. All winners of an iF gold award automatically become members of the distinguished group known as the iF gold selection for the year, which is specially featured in the iF online exhibition and forms the basis of a high-class exhibition as well.

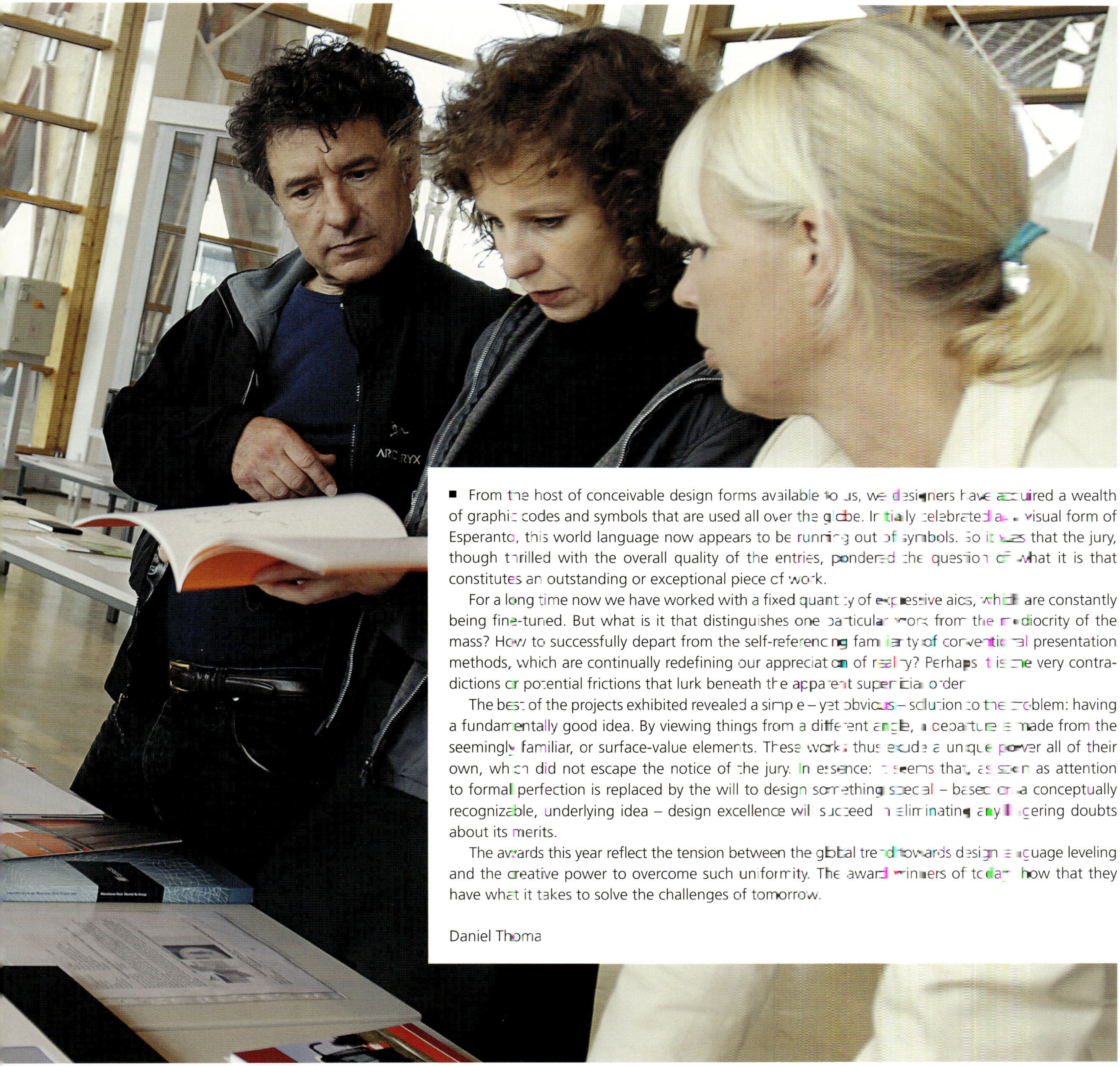

■ From the host of conceivable design forms available to us, we designers have acquired a wealth of graphic codes and symbols that are used all over the globe. Initially celebrated as a visual form of Esperanto, this world language now appears to be running out of symbols. So it was that the jury, though thrilled with the overall quality of the entries, pondered the question of what it is that constitutes an outstanding or exceptional piece of work.

For a long time now we have worked with a fixed quantity of expressive aids, which are constantly being fine-tuned. But what is it that distinguishes one particular work from the mediocrity of the mass? How to successfully depart from the self-referencing familiarity of conventional presentation methods, which are continually redefining our appreciation of reality? Perhaps it is the very contradictions or potential frictions that lurk beneath the apparent superficial order.

The best of the projects exhibited revealed a simple – yet obvious – solution to the problem: having a fundamentally good idea. By viewing things from a different angle, a departure is made from the seemingly familiar, or surface-value elements. These works thus exude a unique power all of their own, which did not escape the notice of the jury. In essence: it seems that, as soon as attention to formal perfection is replaced by the will to design something special – based on a conceptually recognizable, underlying idea – design excellence will succeed in eliminating any lingering doubts about its merits.

The awards this year reflect the tension between the global trend towards design language leveling and the creative power to overcome such uniformity. The award winners of today show that they have what it takes to solve the challenges of tomorrow.

Daniel Thoma

■ Daniel Thoma (*1969) absolvierte 1998 an der FH Darmstadt den Studiengang Kommunikationsdesign als Diplom Designer und war von 1996 bis 1999 als freier Mitarbeiter u. a. bei der Neue Digitale, Frankfurt am Main, und im Büro Grappa Dor, Berlin, tätig. Seit 1999 ist er Designer bei MetaDesign, Berlin, seit 2003 in der Funktion des Design Directors für die Konzeption, Gestaltung und Implementierung komplexer Unternehmensidentitäten und Erscheinungsbilder verantwortlich.

■ Daniel Thoma (*1969) majored in Communication Design at the Darmstadt University of Applied Sciences, graduating in 1998. From 1996 to 1999 he worked as a freelance designer for the Frankfurt-based Neue Digitale company and the Berlin-based Grappa Dor company. Since 1999 he has been a designer at MetaDesign, Berlin, serving since 2003 as Design Director for the conception, design and implementation of complex corporate identities and appearances.

■ Pippo Lionni (*1954) entstammt einer New Yorker Architektenfamilie. Bevor er sich Ende der 70er Jahre dem Designfach verschrieb, suchte er Zuflucht in der Philosophie und der Mathematik an der Portland State und der New York University und war Jazzmusiker in New York und Paris. Sein konzeptioneller Designansatz ist aus verschiedensten Einflüssen gespeist und kommt in komplexen, multidimensionalen Systemen zum Ausdruck, wie Schilder, Set Design, Corporate Identity Projekte, Umwelt Design u.v.m. Im Jahre 2001 erhob ihn der französische Kultusminister in den Stand eines „Chevalier de l'Ordre des Arts et des Lettres".

■ Pippo Lionni (*1954) grew up in a New York family of architects, escaped into Philosophy and Mathematics at Portland State University and New York University, played jazz in New York and Paris, and became a designer in the late 1970s. He brings to design a conceptual approach, of diversified influences as formalized in complex multidimensional systems: such as signage, set design, corporate identity programs, environmental design … He was awarded the distinction of "Chevalier de l'Ordre des Arts et des Lettres" by the French Ministry of Culture in 2001.

■ Manu Lange setzte ihre anfänglich in Basel begonnene Designausbildung im Jahre 1991 am Fachbereich Design der FH Köln fort. Während des Studiums arbeitete sie in der Modellwerkstatt des Architekten O.M. Ungers und war an der Setgestaltung eines Peter Greenaway Films beteiligt. Für ihre Abschlussarbeit erhielt sie 1996 den Sonderpreis der Jury des Lucky Strike Junior Designer Award. Seit 1998 betreibt sie in Köln mit ihrer Partnerin Elli Durach das „Büro für Gestaltung, Lange + Durach".

■ Manu Lange transferred 1991 from the design school in Basel to the Cologne Design School. During her studies she worked as an assistant in the model shop of the architect O.M. Ungers and for the settings of a film by Peter Greenaway. For her thesis, Manu Lange was awarded a special distinction from the jury of the Lucky Strike Junior Designer Award 1996. Since 1998 together with her partner Elli Durach she runs the graphic design studio "Lange + Durach" in Cologne.

Daniel Thoma

Pippo Lionni

Manu Lange

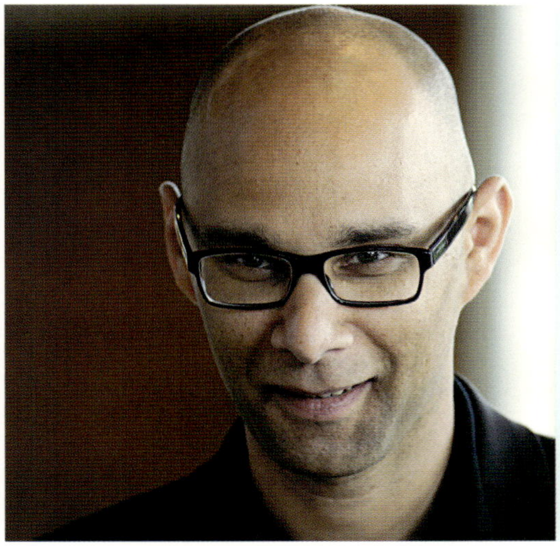

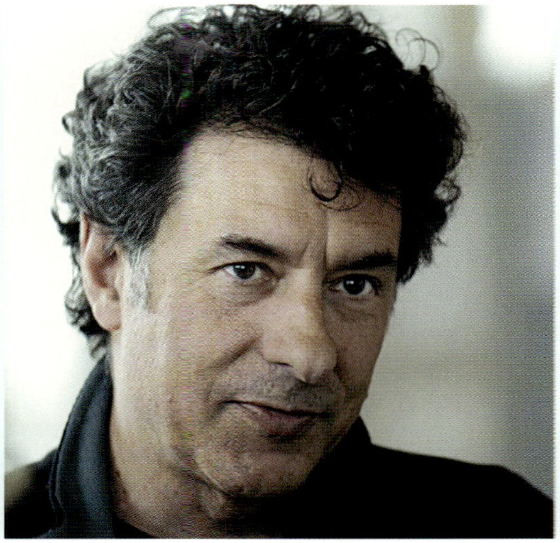

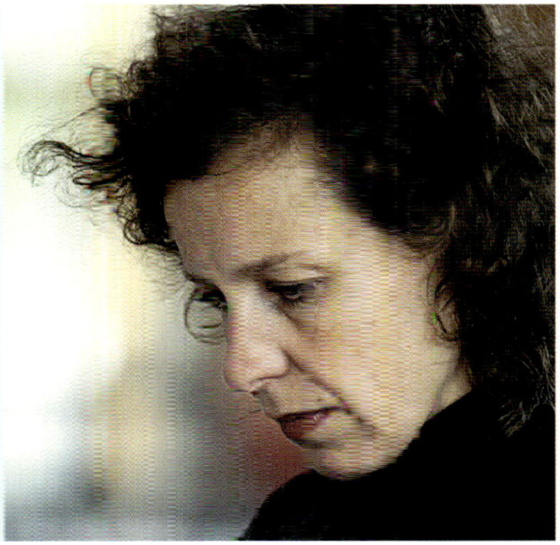

gold award: print media image

Auftraggeber Client
feldmann+schultchen design studios
Hamburg, Germany

Projekt Project
DEUTSCHBAND®
Klebeband
Adhesive tape

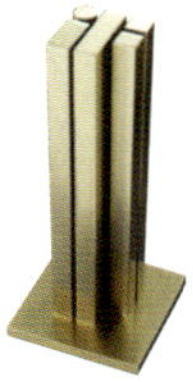

Design
feldmann+schultchen design studios
Hamburg, Germany

DEUTSCHBAND® steht für einen entspannten Umgang mit den deutschen Landesfarben. 66 Meter Schwarz-Rot-Gold von der Rolle kleben überall: auf Autos, Wänden und Kleidung.
Weitere Anwendungsbeispiele und Klebeideen gibt es unter www.deutschband.de. Die deutsche Trikolore wird zum Symbol für Freiheit, Demokratie und Kreativität. Damit erfreut das DEUTSCHBAND® nicht nur Patrioten.

DEUTSCHBAND® is a convenient way of making Germany's colours visible. 66 metres of black-red-gold adhesive tape on a reel that stick to everything – to cars, to walls and to clothes.
Further ideas how to use DEUTSCHBAND® can be found at www.deutschband.de. The German tricolour becomes a symbol of freedom, democracy and creativity – and DEUTSCHBAND® doesn't just cheer patriots.

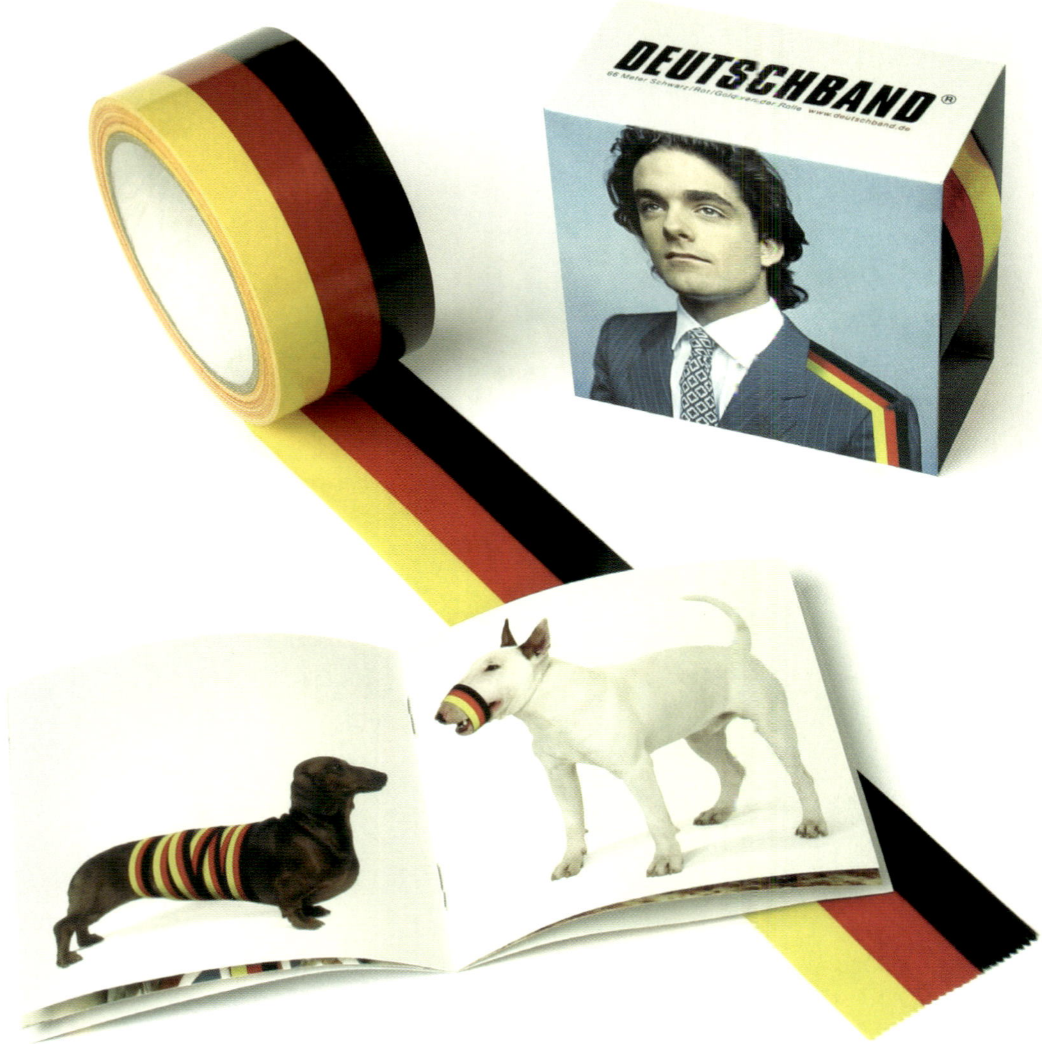

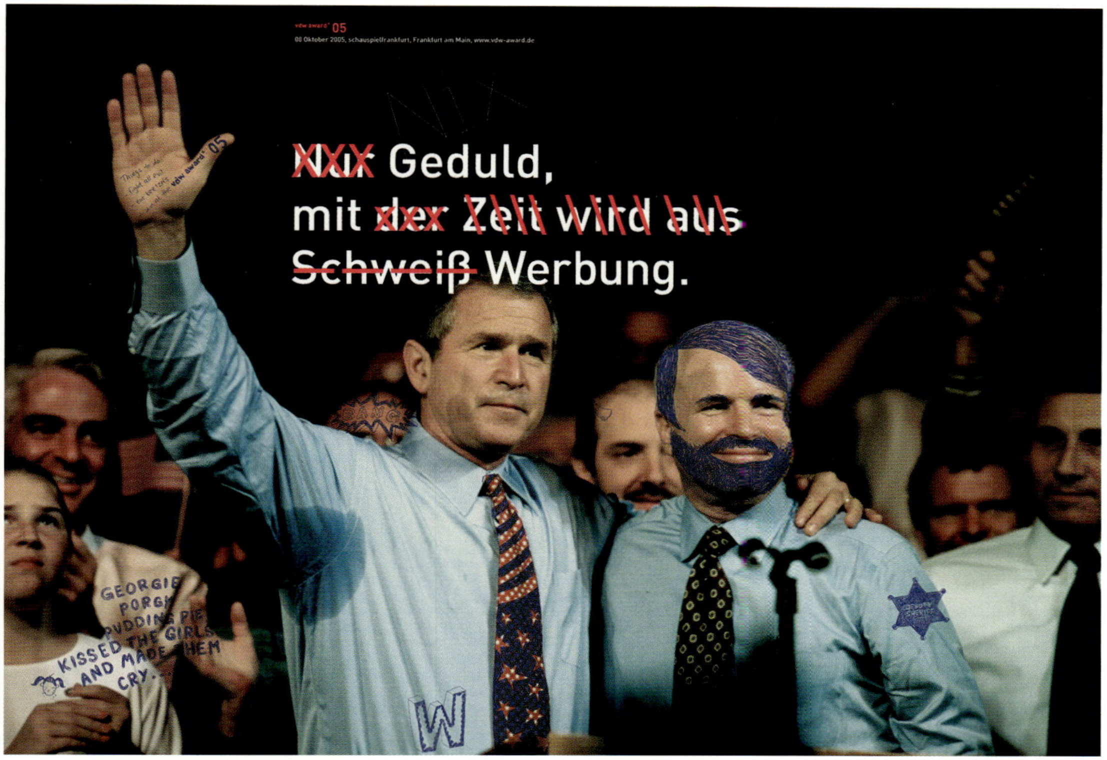

Auftraggeber Client
Verband Deutscher Werbefilmproduzenten e. V.
Frankfurt am Main, Germany

Projekt Project
vdw award 05 – nur geduld
Kampagne (Poster, DVD-Packaging, Postkarten, Tickets, Programm)
Campaign (poster, DVD packaging, postcards, tickets, programme)

Design
Kai Bergmann
dreizueins architektur designgrafik
Frankfurt am Main, Germany
Catrin Altenbrandt
Adrian Nießler
Oliver Hesse
Group.IE
Frankfurt am Main, Germany

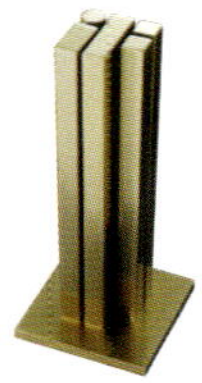

Wie die letzten drei Jahre davor stellte der VDW uns auch dieses Jahr die Aufgabe, für den vdw award ein ganzheitliches Kommunikationskonzept zu finden. Ein Konzept, das sowohl Sponsoren wie auch alle Kreativen anspricht, die mit dem Medium Werbefilm zu tun haben.
Unser kreativer Ansatz war hier, dass der VDW und die ganze Werbefilm-Branche – gerade in der heutigen Zeit – als Vorreiter und Motor für die gesamte Wirtschaft wirken. Oder anders gesagt: Wir brauchen keine Leute mit Geduld, sondern Macher. Also haben wir diese Kampagne gemacht. Mit Ungeduld und Spucke. Und siehe da, mit etwas Geduld wird aus einer Aufgabenstellung eine Kreativ-Auszeichnung.

As they had done in the previous three years, the VDW, the Association of German Advertising Film Producers, asked us to create an integrated communication concept for the vdw award 05. The concept was to appeal to sponsors as well as to creatives working in the fields of the advertising film.
Our creative approach was based on the fact that the VDW and the whole advertising branch with their innovative strength serve as a motor for the whole economy, especially nowadays. In other words: we don't need people with patience, we need doers. That's how we created our campaign. With impatience and verve – and lo and behold, with a little patience the objective turns into a creative concept that receives a design award.

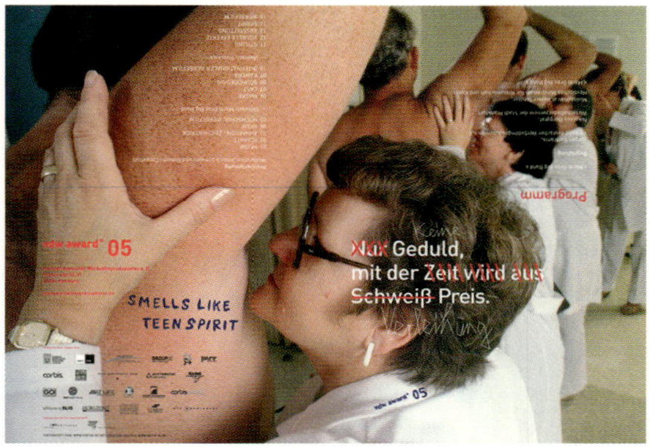

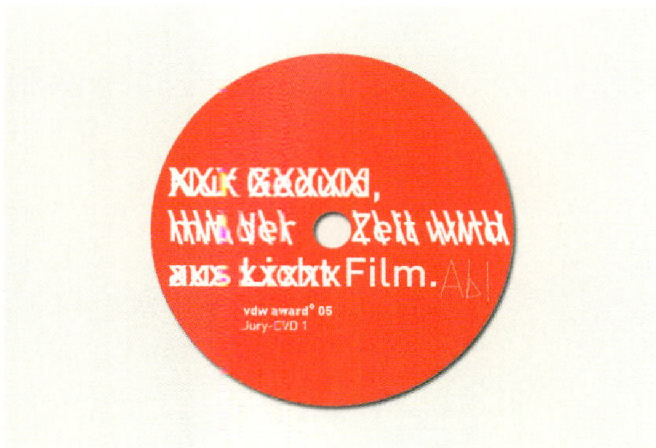

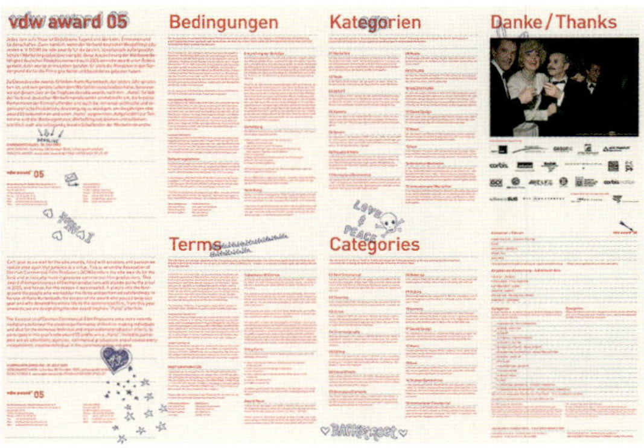

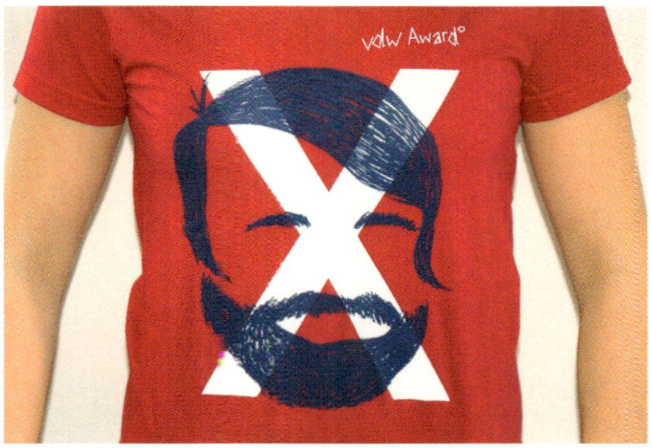

gold award: print media image

Auftraggeber Client
Suzuki International Europe GmbH
Bensheim, Germany

Projekt Project
Suzuki Handschuhfachbuch 2006
Buch
Book

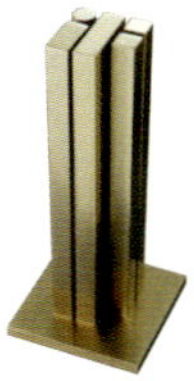

Design
Sven Voelker
Nils Voelker
Kay Michalak
Sven Voelker Studio
Berlin, Germany

Das Suzuki Handschuhfachbuch 2006. In jedem neuen Suzuki liegt eines davon. Das Buch begleitet die Marke in Deutschland nun schon in der zweiten Ausgabe auf ihrem neuen Weg. Mit einer Auflage von 25.000 Stück erreicht das jährlich erscheinende Buch jeden Suzuki Fahrer. Es erzählt von Menschen, die auf die eine oder andere Weise mit dem Unternehmen und den Fahrzeugen verbunden sind. Das Buch wurde von Sven und Nils Voelker in Berlin konzipiert, geschrieben und gestaltet. Die Fotografien stammen von Kay Michalak.

The Suzuki Glove Box Book 2006. Each new Suzuki car comes with one of them. This is the second edition of the book, presented by the brand in Germany. With 25,000 copies it reaches every Suzuki driver and builds up trust in the Japanese car manufacturer. It tells stories about people who are in most different ways related to Suzuki. The book is written and designed by Sven and Nils Voelker. The photographs are by Kay Michalak.

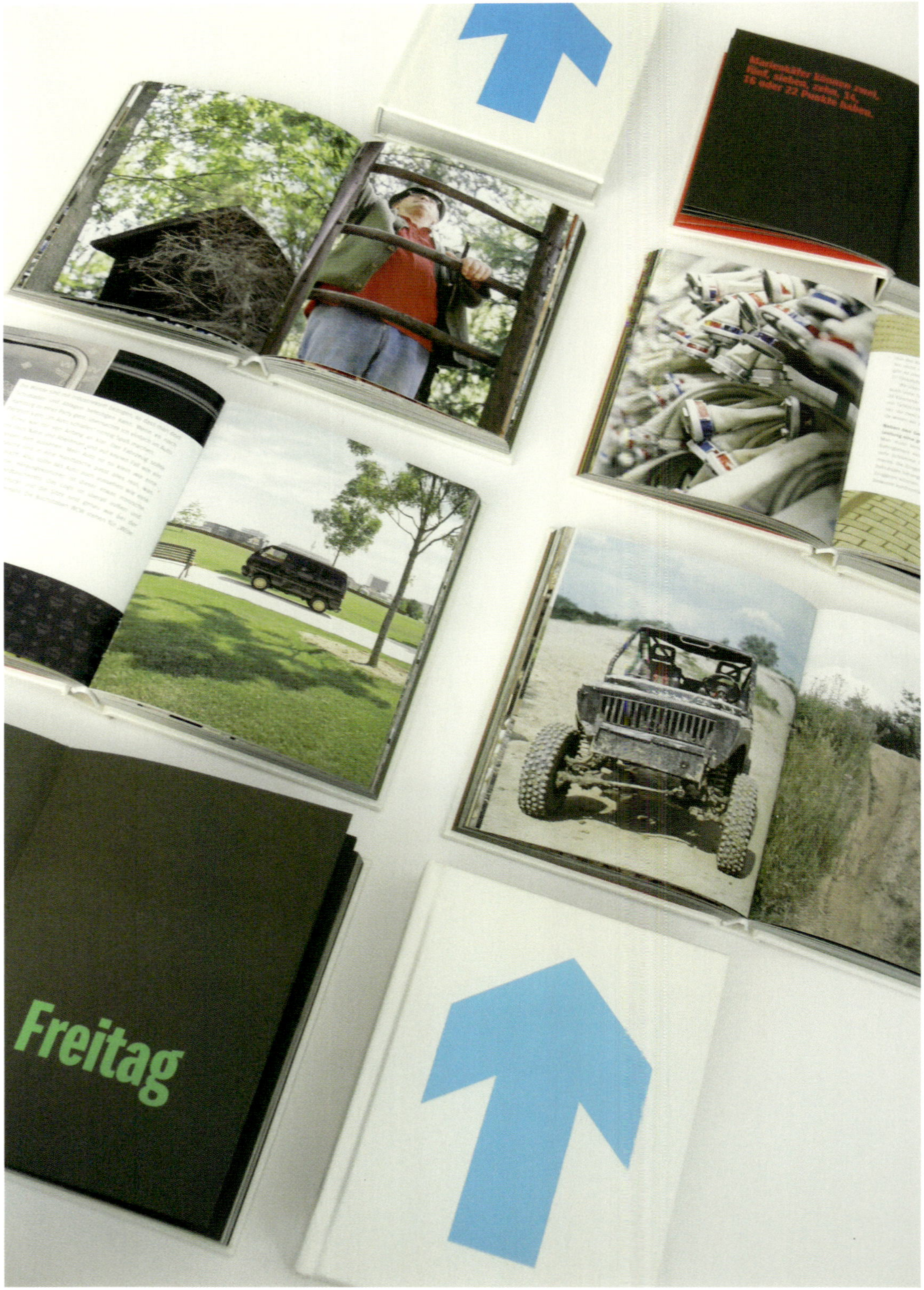

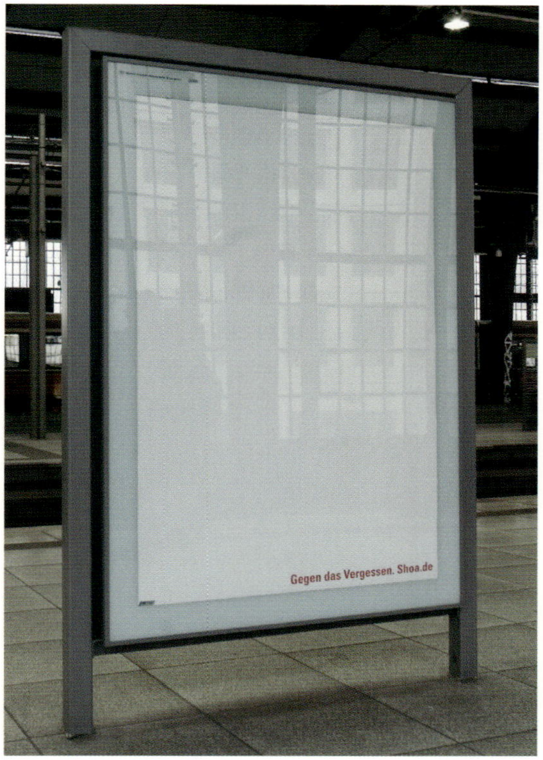

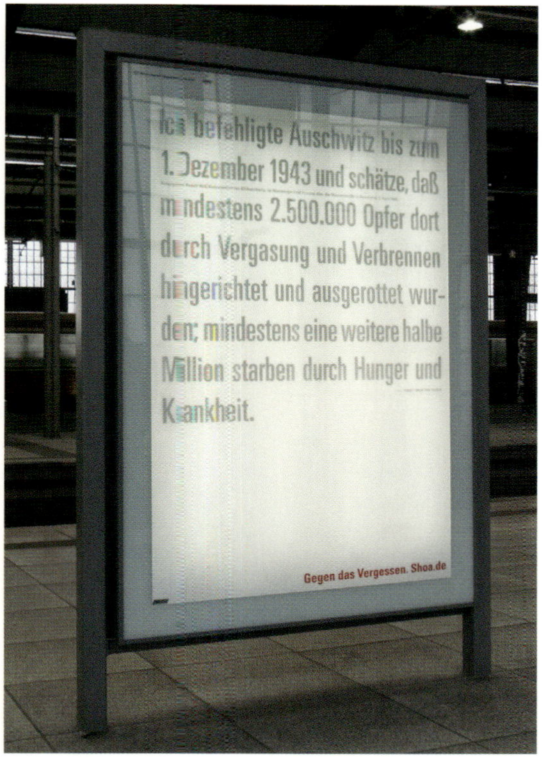

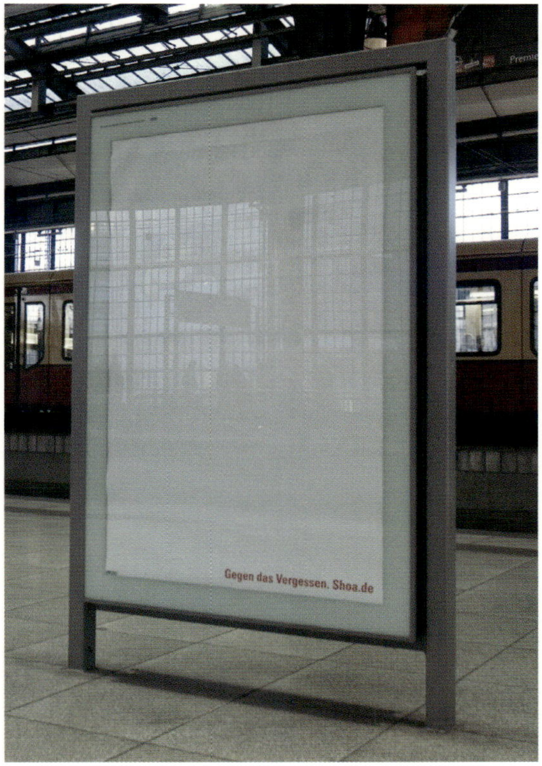

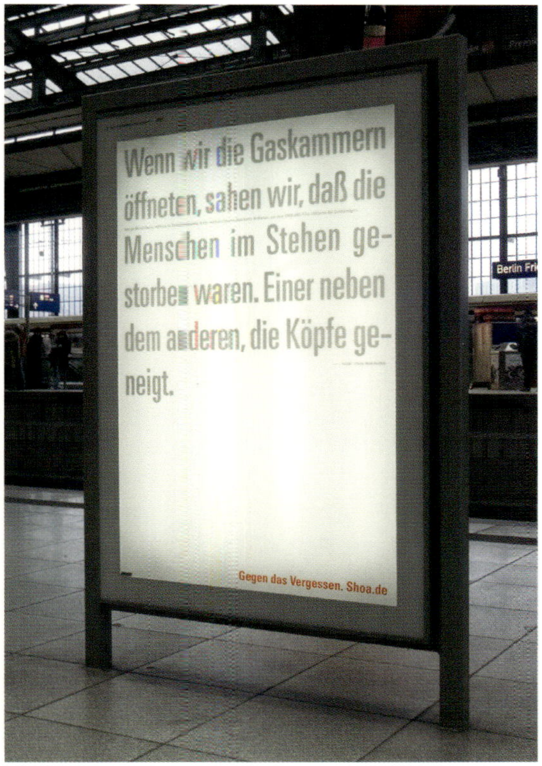

Auftraggeber Client
Arbeitskreis Shoa.de e.V.
Berlin, Germany

Projekt Project
Kampagne „Gegen das Vergessen"
Plakatkampagne
Poster campaign

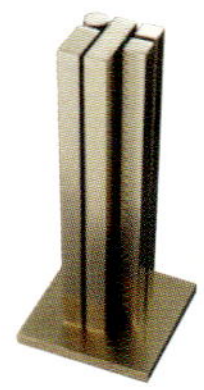

Design
Christoph Bebermeier
BÜRO WEISS
Berlin, Germany

Die Kampagne „Gegen das Vergessen" richtet sich gegen das Vergessen und Verdrängen der deutschen Vergangenheit zur Zeit des Nationalsozialismus, sie steht für Information und Auseinandersetzung mit diesem Teil der deutschen Geschichte. Der Gegensatz von An- und Abwesenheit historischer Information ist das kennzeichnende Merkmal der Kampagne. Auf scheinbar unbedruckten Citylight-Postern werden erst durch die Durchleuchtung Zitate aus der Zeit des Nationalsozialismus sichtbar: Shoa.de kommt seiner aufklärenden Aufgabe nach und arbeitet immer wieder „Gegen das Vergessen".

The "Against Forgetting" campaign was directed against displacing and forgetting the Holocaust. A main characteristic of the campaign is working with the contrast of presence and absence of historical information. In daylight the Citylight-Posters displayed the slogan "Against Forgetting. Shoa.de" at a relatively small size on a mostly empty surface. This way, the posters stressed the danger of forgetting the Holocaust. In the evening, as soon as the Citylight-Posters lit up (back-lit), quotations from the national socialist era became visible. Thus, Shoa.de fulfilled its call "Against Forgetting" again and again.

gold award: print media sales promotion

Auftraggeber Client
adidas AG
Herzogenaurach, Germany

Projekt Project
Materials of the World
Order Booklet

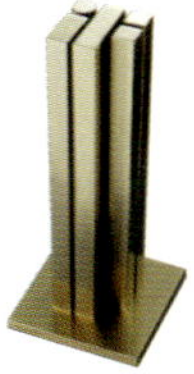

Design
designliga
München, Germany

In der exklusiven „Materials of the World" Kollektion von adidas finden sich formal sowie in der Auswahl der Materialien die Einflüsse verschiedener Regionen aus aller Welt wieder.
Das Order Booklet schafft den passenden Rahmen für die nach Ländern geordneten Kollektionsteile, die wir in einer Art Reisetagebuch in Szene setzten. Im Work-in-Progress-Stil scheint das Buch die Dokumentation einer für diese Kollektion vollzogenen Weltreise zu sein. Pflichtinhalte wie Texte, Artikelbeschreibungen oder Stoffmuster wurden händisch eingeklebt oder geheftet.

Formally and fabric-wise the exclusive "Materials of the World" collection by adidas shows influences from various regions all over the world.
With the order booklet we created the adequate framework for the fashion pieces that are categorised according to their origin and are staged like in a traveller's diary. The work-in-progress style makes the booklet seem like the documentation of a journey around the globe. Required contents such as text, product descriptions, or cloth patterns were bonded and tacked by hand.

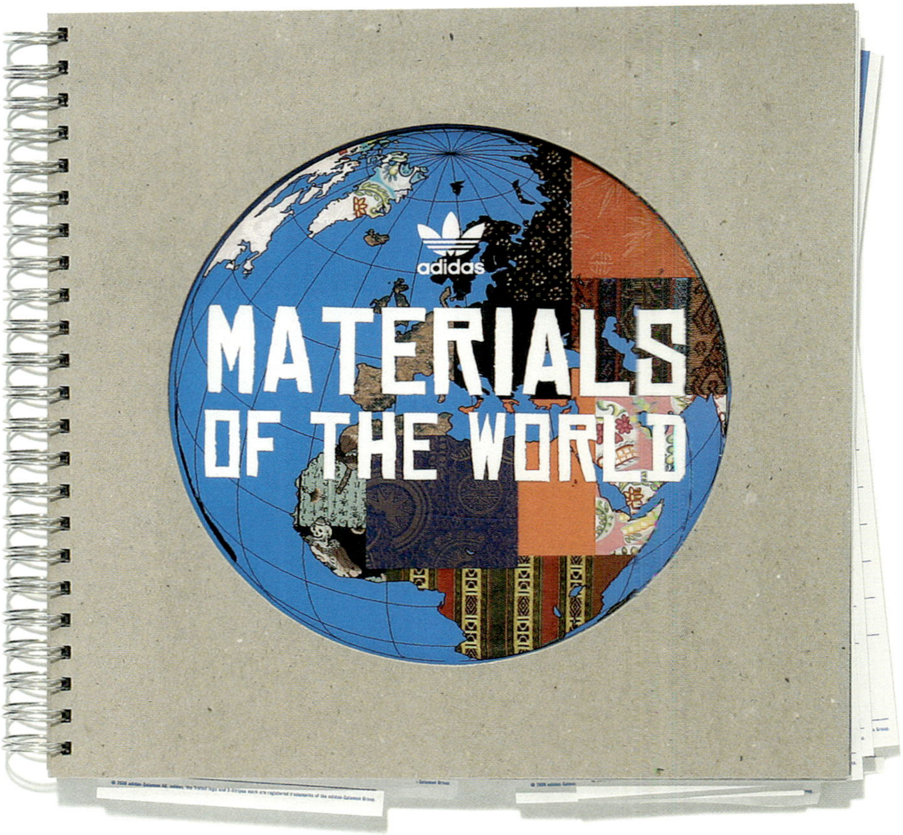

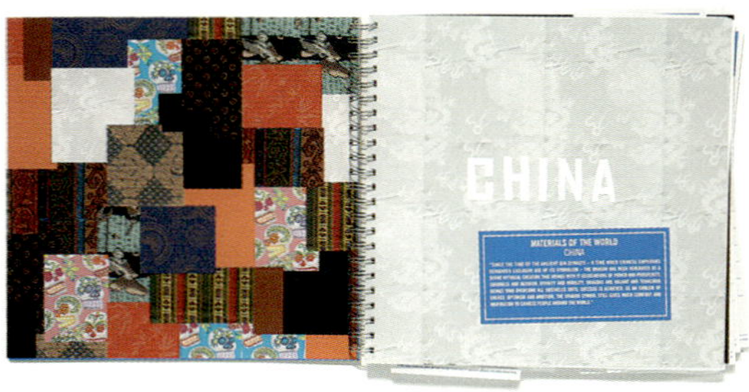

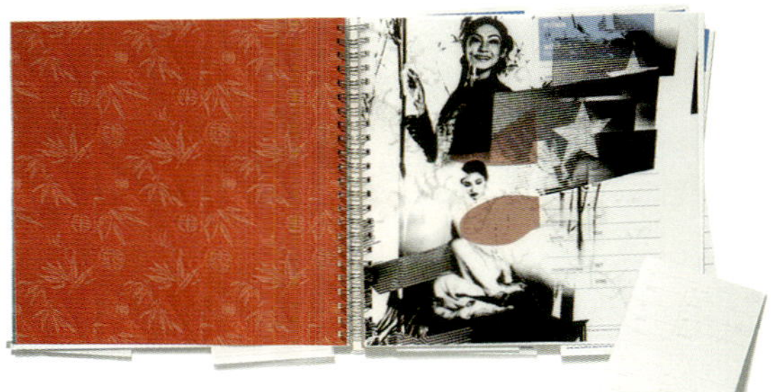

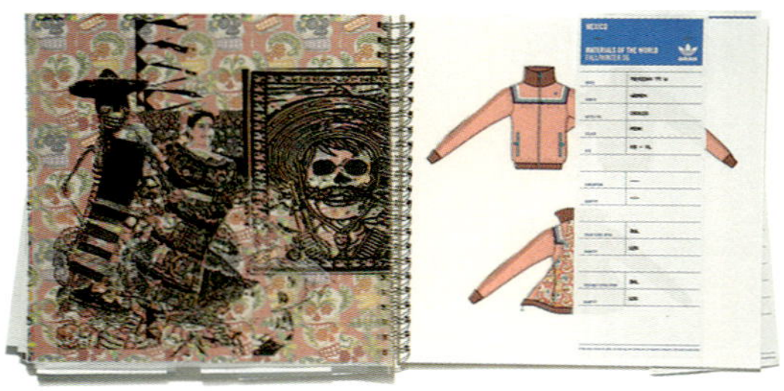

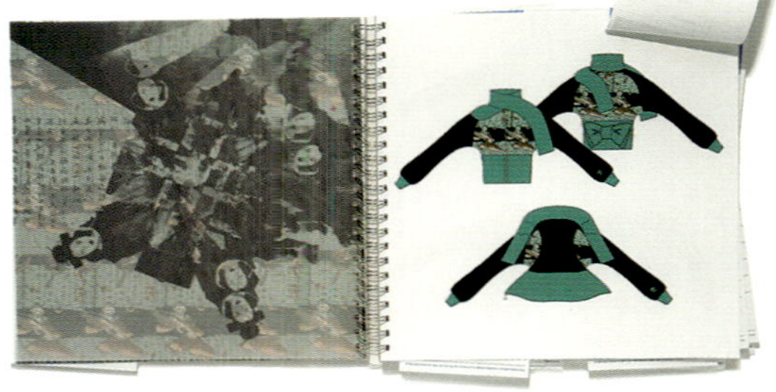

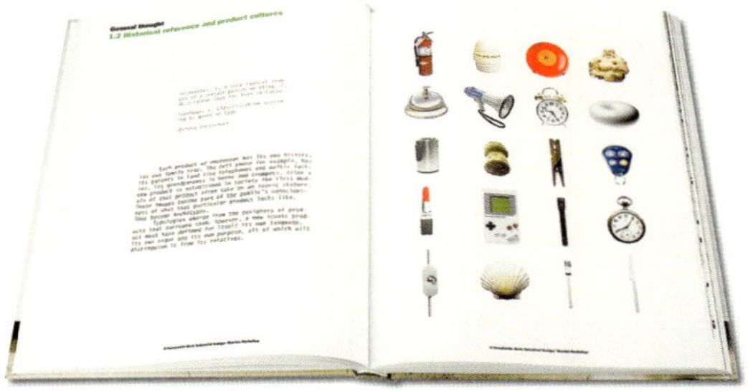

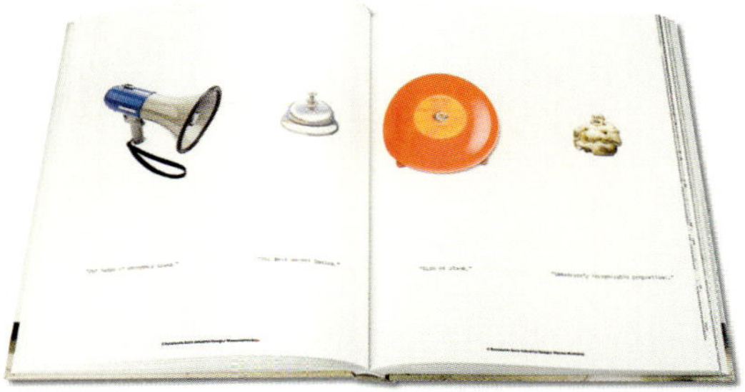

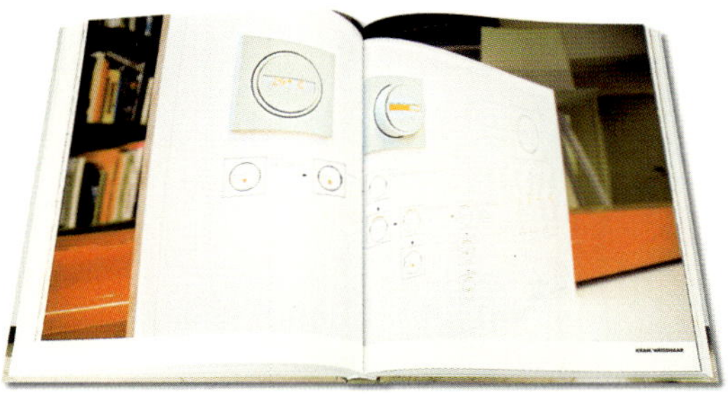

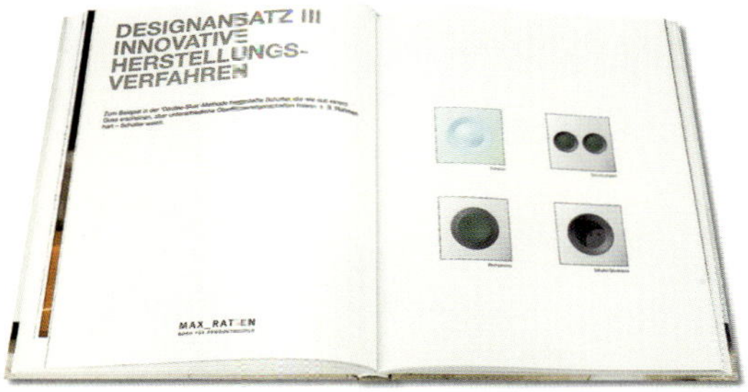

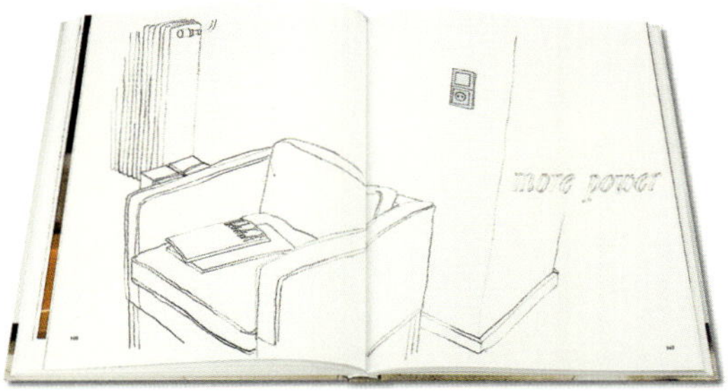

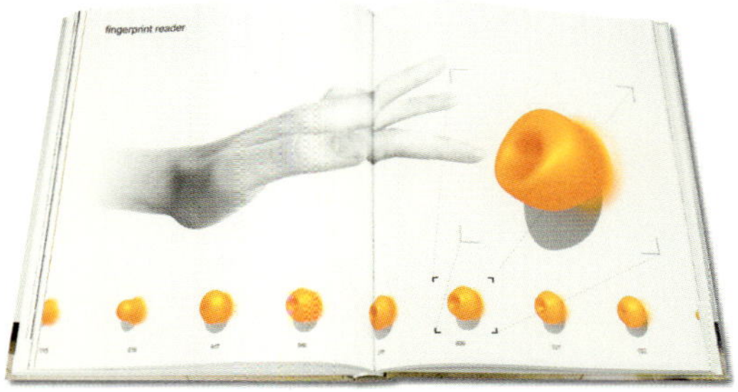

Auftraggeber Client
Merten GmbH & Co. KG
Wiehl, Germany

Projekt Project
Switch – Von der Kultur des Schaltens /
Switch – The Culture of Switching
Buch
Book

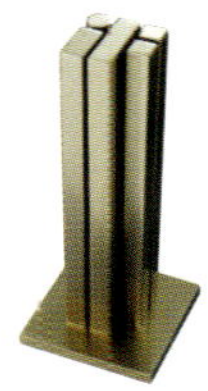

Design
SIGN Kommunikation GmbH
Frankfurt am Main, Germany

Wie oft kommt der Mensch pro Tag in Berührung mit einem Schalter? Müssen Schalter immer wie Schalter aussehen? Welchen sinnlichen Gewinn bietet der Akt des Schaltens? Und hat sich das Schalten im Laufe der Jahre verändert? Das Thema Schalten wirft viele Fragen auf. Schalten und der Mythos des „Licht-Werdens" hat Kulturwissenschaftler, Psychologen und Philosophen beschäftigt, ist und bleibt aber auch ein Thema für Design und Architektur. Mit dieser Themenstellung beschäftigten sich auf Einladung von Merten namhafte Gestalter dieser Zeit. Die sehr unterschiedlichen Ergebnisse des Workshops sind in der Publikation dargestellt.

How often do we come into contact with a switch? Do switches always have to look like switches? Which sensation causes the act of switching? And did switching in the course of the years change by technical innovation? The topic of switching brings up many questions. Switching and the myth of "letting there be light" has occupied culture scientists, psychologists and philosophers, but also and remains a central consideration of design and architecture. At the invitation of Merten, renowned designers of our time addressed this topic. The very varied results of the workshop are shown in the publication.

gold award: print media publishing media

Auftraggeber Client
Philips Design
Eindhoven, The Netherlands

Projekt Project
past tense, future sense –
Competing through creativity
Buch
Book

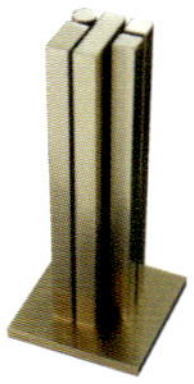

Design
Stefano Marzano & Team
Philips Design
Eindhoven, The Netherlands

„past tense, future sense" ist entstanden, um 80 Jahre Design bei Philips zu feiern. Das Buch zeichnet die reiche Designgeschichte von Philips nach, ist aber dennoch mehr als nur der feierliche Akt eines einzelnen Unternehmens. Designer, Experten, Manager, Bildungsinstitute, Mitarbeiter von Philips, Medienvertreter und viele andere werden es als ein Werk schätzen, das acht herausragende Jahrzehnte schildert, als ein Nachschlagewerk und vor allem als spannende und einnehmende Lektüre. Die Erstellung des Buchs war ein umfangreiches Unterfangen: Es enthält mehr als 80 Einzelgeschichten und rund 1.300 Bilder auf 808 Seiten.

"past tense, future sense" has been created in honor of 80 years of design at Philips. The book chronicles the rich history of design at Philips, yet is more than just an act of celebration by one particular company. Designers, design professionals, managers, education institutions, Philips employees, media personnel and many others will undoubtedly appreciate its value as a reflection on eight remarkable decades, a reference source and above all, an interesting and engaging read. Producing the book was quite an undertaking. The book consists of more than 80 separate stories, approximately 1,300 images and a total of 808 pages.

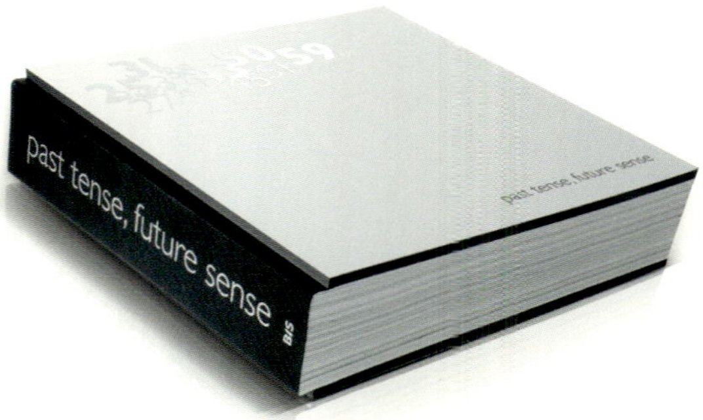

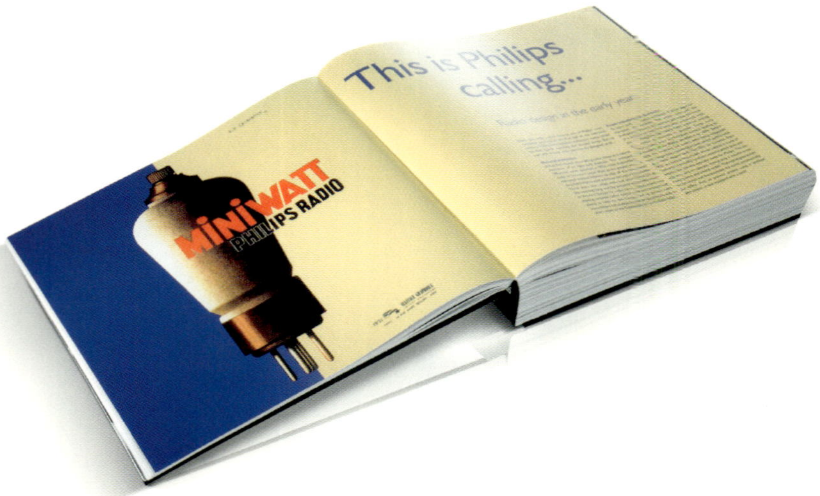

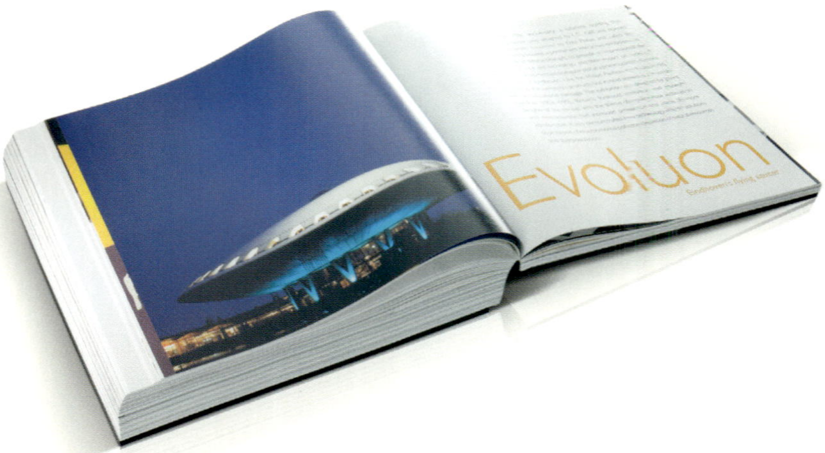

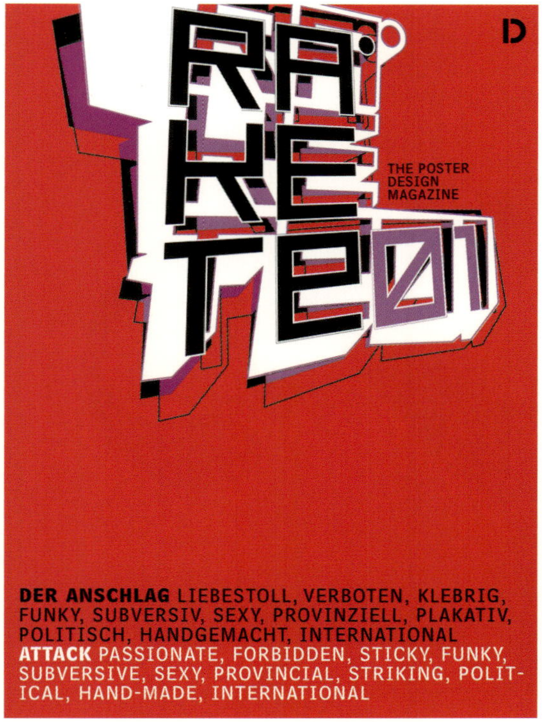

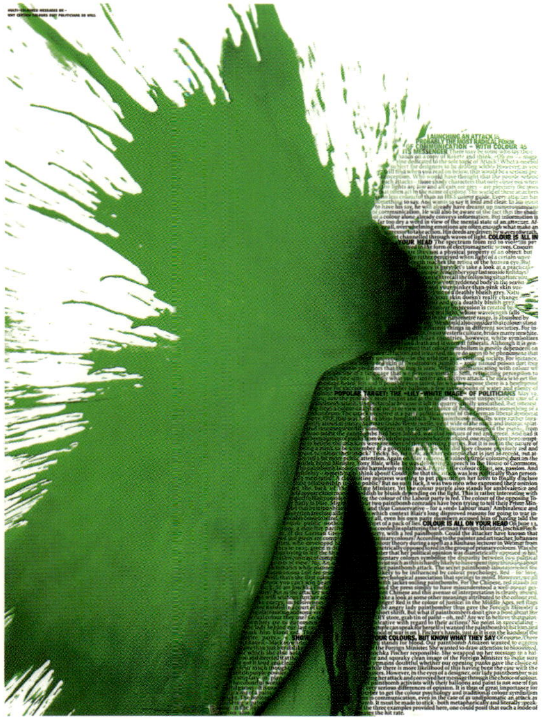

Auftraggeber Client
Fachhochschule Düsseldorf
Fachbereich Design
Düsseldorf, Germany

Projekt Project
Rakete – The Poster Design Magazine
Magazin
Magazine

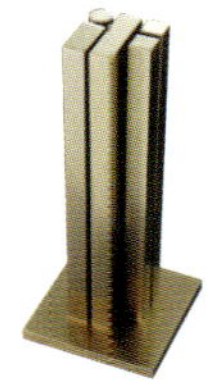

Design
Katrin Spohr
Anja Stehning
Redaktion Rakete FHD
Düsseldorf, Germany

Rakete ist das erste Posterdesignmagazin! Eine so ernsthafte wie amüsante Zeitschrift, die ein kommunikatives Thema mit einem experimentellen Designansatz aus verschiedenen Perspektiven beleuchtet. Dabei werden Bezüge zu Produktdesign, Architektur, Mode, Literatur oder Musik hergestellt. Rakete zeigt keineswegs nur Projekte und Themen der Düsseldorfer Hochschule, sondern wagt sich in fremde Gebiete, macht Interviews und Reportagen. Denn Raketen fliegen weit, um Bekanntes aus der Ferne zu betrachten und Unbekanntes zu entdecken. Mit Rakete haben wir ein eigenes Format geschaffen, um Standpunkte neuartig darzustellen. Alles in allem brachten 40 Studenten in 15 „Splitterredaktionen" Rakete01 an den Start.

Enter Rakete (German for "rocket"), the first design magazine in poster format! This light-hearted as well as serious periodical explores a communicative topic from a variety of perspectives. This is necessarily an interdisciplinary process, touching on such fields as product design, new media, architecture, fashion, literature or music. And Rakete goes beyond the university realm. It dares to chart unknown terrain (Remember, rockets travel wide to view the familiar from afar and rub shoulders with the unknown!). With Rakete, we have created our own novel and profound format for presenting different points of view. No less than 40 students worked on this project! And a total of 15 editorial splinter groups set about launching Rakete01.

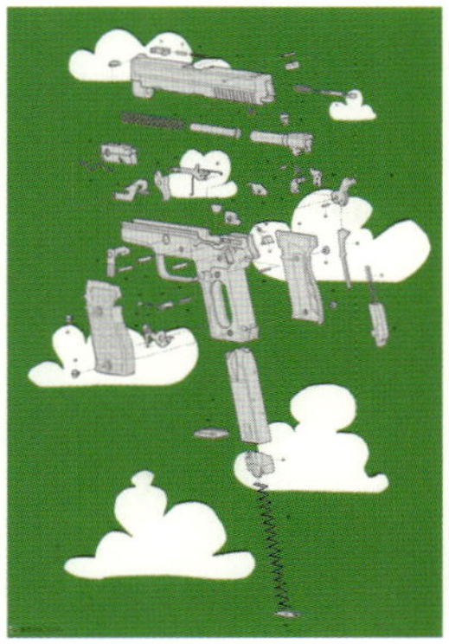

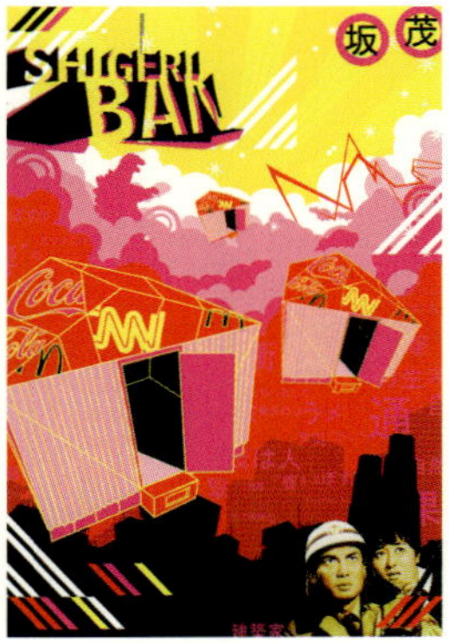

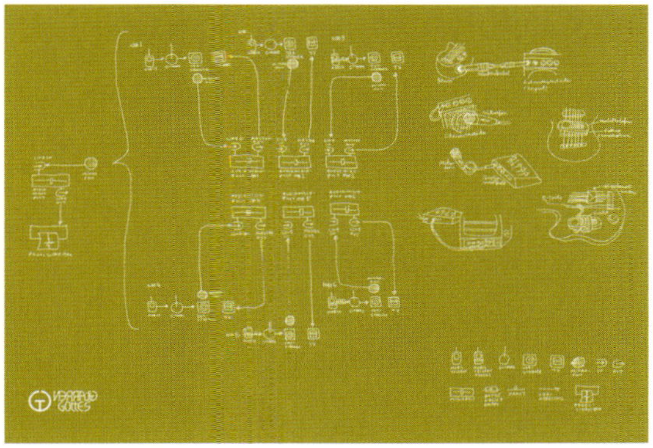

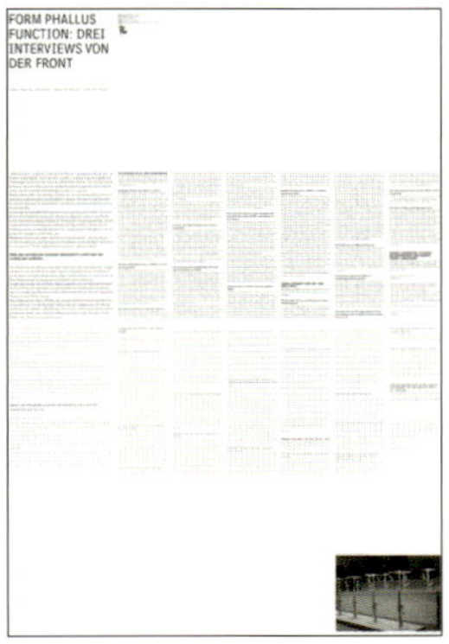

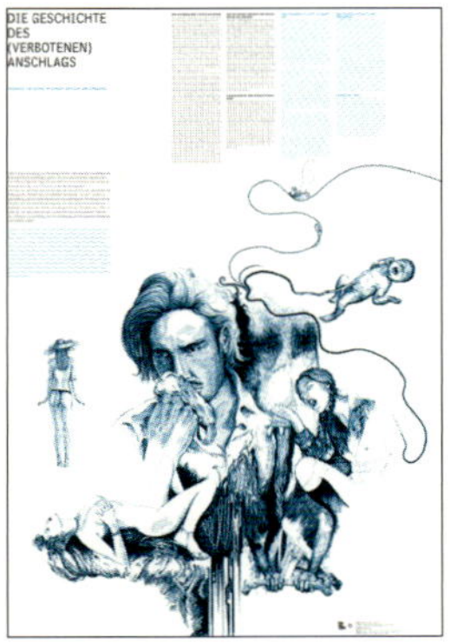

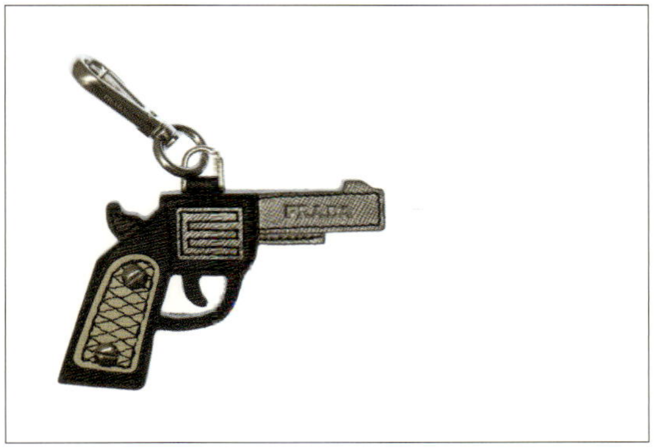

gold award: print media publishing media

Auftraggeber Client
Design2context
Institut für Designforschung der Hochschule für Gestaltung und Kunst Zürich (hgkz)
Zurich, Switzerland
Hochschule für Grafik und Buchkunst Leipzig/ Academy of Visual Arts (HGB)
Leipzig, Germany
Institut für Interdisziplinäres Design Leipzig (2id)
Leipzig, Germany

Projekt Project
Das Gesetz und seine visuellen Folgen
Buch
Book

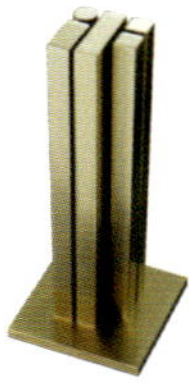

Design
Severin Wucher
Design Severin Wucher
Berlin, Germany

Das Buch dokumentiert ein interdisziplinäres Forschungsprojekt (Design2context/hgkz, HGB Leipzig, 2id Leipzig) unter Leitung von Prof. Ruedi Baur. Es untersucht in Bildessays und Textbeiträgen die Wechselwirkungen zwischen Normierung und Gestaltung. Im Bildteil werden die Beziehungen zwischen Gesetzen und ihren visuellen Ausdrucksformen dokumentiert, der Textteil beleuchtet das Thema aus der Perspektive der Rechtswissenschaft, Philosophie, Ethnologie und anderen. Severin Wucher übernahm die grafische Gestaltung des Buchs (Lars Müller Publishers), das seine typografischen Gestaltungsregeln offen legt und somit auch ein Buch über das Büchermachen ist.

This book documents an interdisciplinary research project (Design2context/hgkz, HGB Leipzig, 2id Leipzig) under the direction of Prof. Ruedi Baur. It examines the reciprocal impact of standardization and design in a series of photo essays and text contributions. The photo essays document the relationship between laws and their visual expression, while in the text, the overall theme is examined from the standpoint of law, philosophy, ethnology and other branches of science. Severin Wucher was responsible for the graphic design of the book (Lars Müller Publishers), which reveals its own typographical design rules, thus simultaneously representing a book on the art of making books.

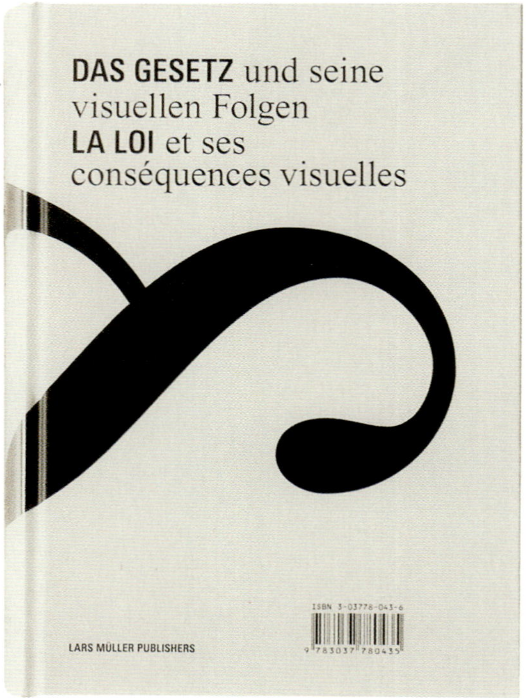

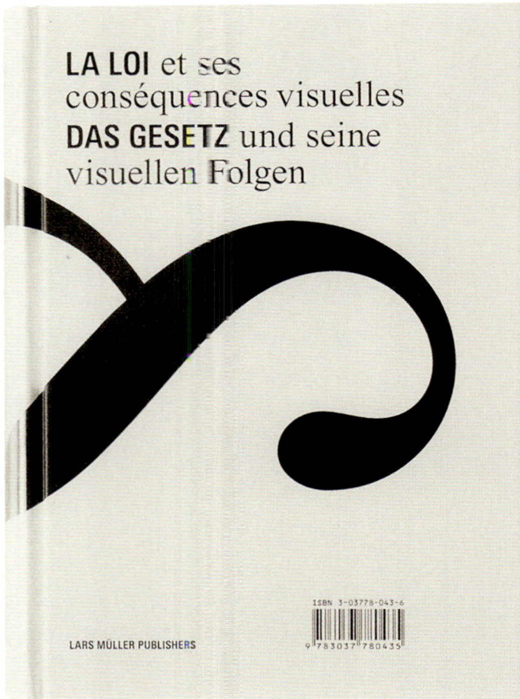

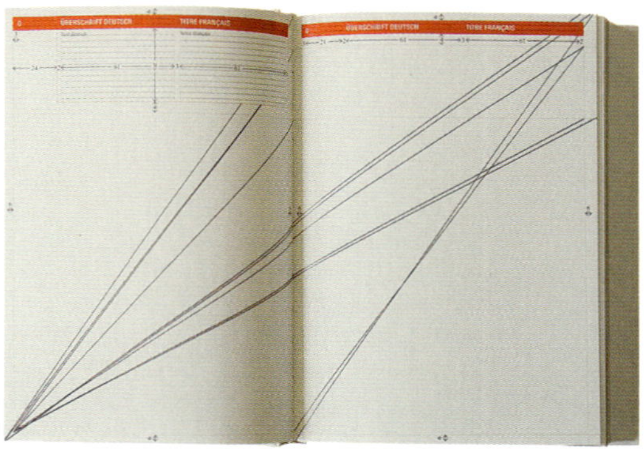

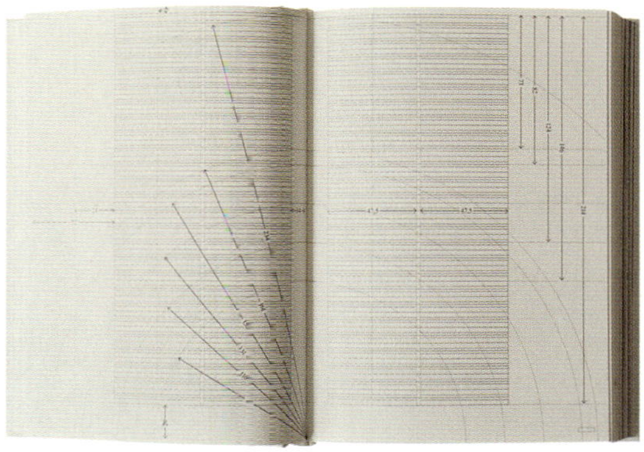

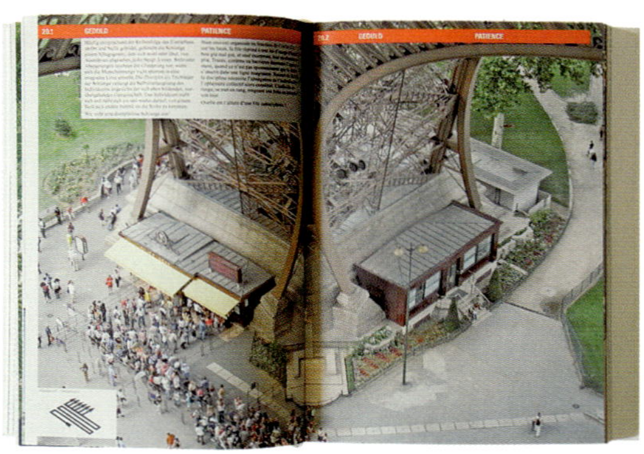

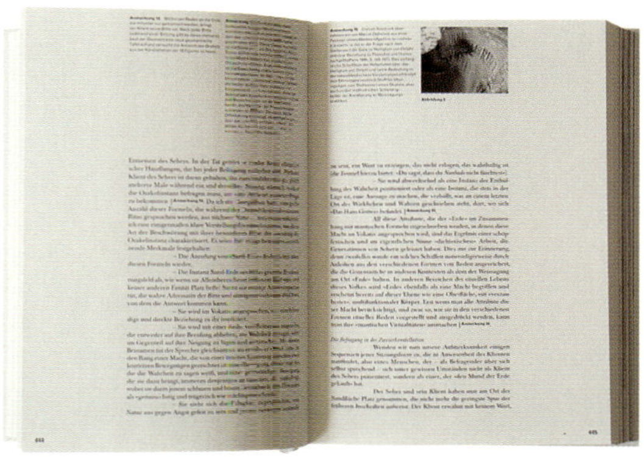

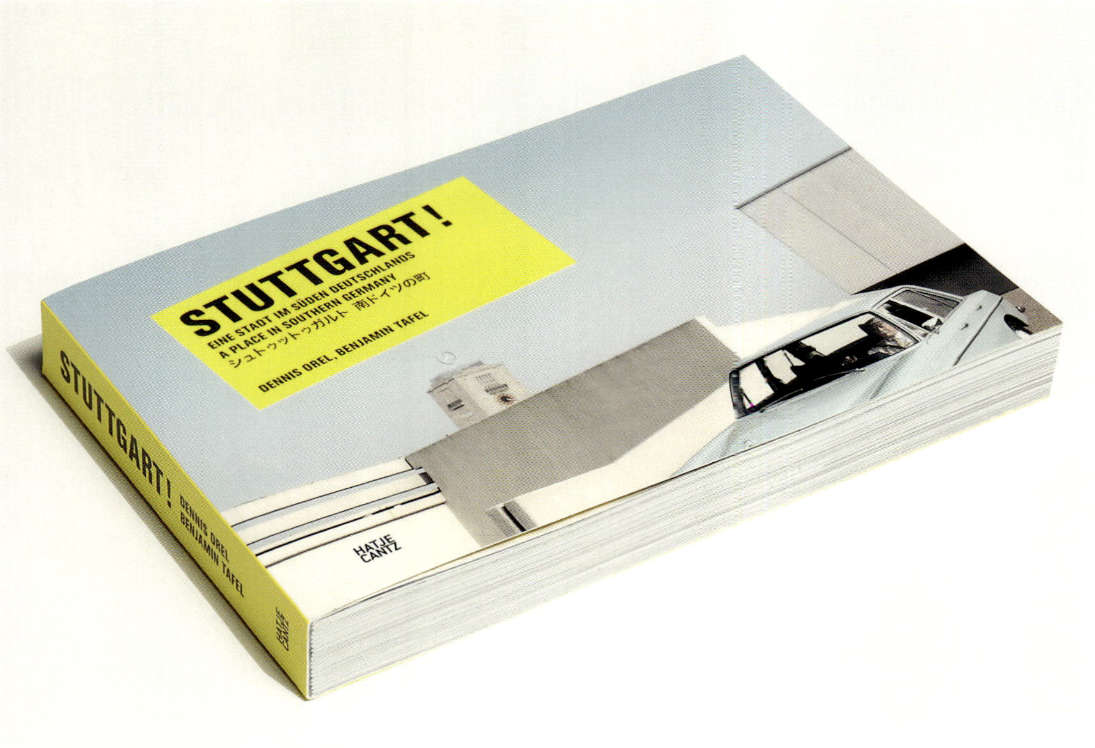

Auftraggeber Client
Hatje Cantz Verlag
Ostfildern, Germany

Projekt Project
STUTTGART! A Place in Southern Germany
Buch
Book

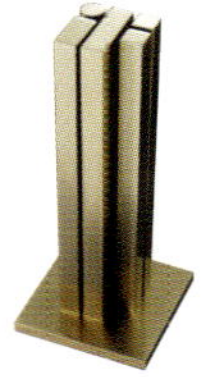

Design
Dennis Orel
Benjamin Tafel
orel:tafel
Fotografie und Kommunikationsdesign
Stuttgart, Germany

Ein fotografischer Streifzug durch eine deutsche Stadt, die Kamera gerichtet auf das Schauspiel des Alltags. Bilder von türkischen Schachmeistern beim Boulespiel, einheimischen Keglern, Daimlerfahrern im Glanz der Discokugel, bayerischen Trachten auf schwäbischem Boden, innerstädtischen Bundesstraßen, Schweizern auf Eis, chinesischen Touristen im Blitzlichtgewitter und Deutschlands ältestem Hundesalon. Stuttgart! A Place in Southern Germany.

A photographic expedition through a German city with the camera focused on the spectacles of everyday life. Pictures of Turkish chess champions playing boules, local bowlers, a disco-glazed Mercedes, Bavarian costumes on Swabian territory, highways cutting through the inner city, Swiss people on ice, Chinese tourists in a storm of popping flashbulbs, and Germany's oldest dog salon. Stuttgart! A Place in Southern Germany.

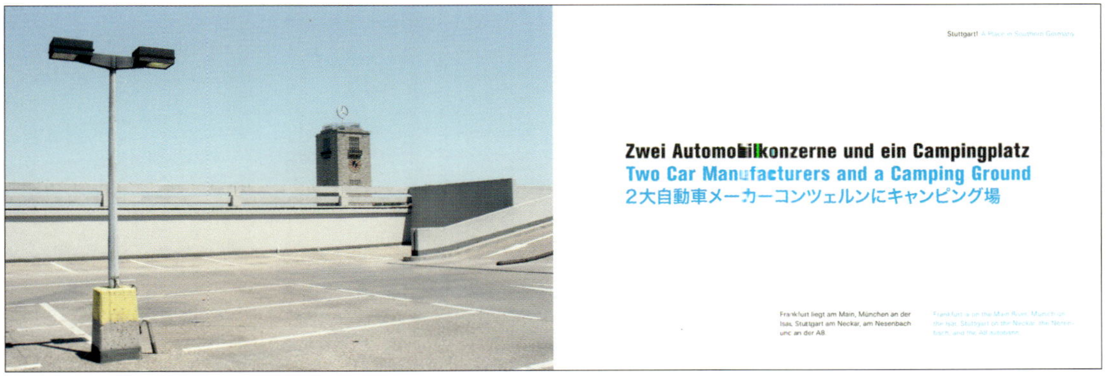

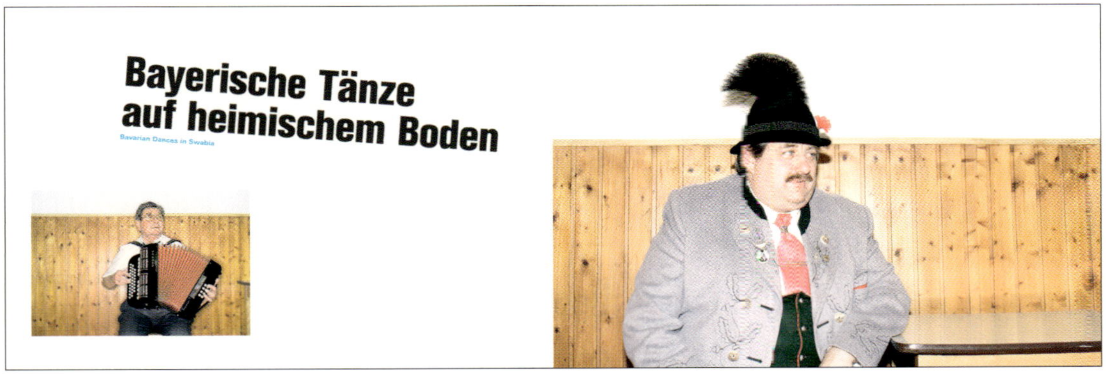

Auftraggeber Client
Theater Bielefeld
Bielefeld, Germany

Projekt Project
Theater Bielefeld
Abendprogramme
Evening programmes

Design
Hans Dirk Schellnack
Nodesign
Essen, Germany

Auszug aus den bereits veröffentlichten Abendprogrammen des Theaters Bielefeld. – Indem wir die Rückseite des achtseitigen Leporellos zugleich als Stückplakat nutzen, konnten wir anstelle der vorher – aufgrund des knappen Kulturetats – üblichen schwarz-weißen und werbelastigen Programme nach dem Relaunch vierfarbige Programmhefte ohne Werbung produzieren. Aufgrund der späten Kostümproben gibt es im Theater Bielefeld keine der sonst bei Abendprogrammen üblichen Bühnenfotos – aus dieser Not (und aus dem immensen Zeitdruck bei der Produktion der Hefte, die teilweise binnen weniger Stunden entstehen) ergaben sich Abendprogramme, die häufig mit assoziativen Bildern und Texten arbeiten und den Theaterbesucher so nicht nur primär informieren sollen, sondern als Baustein der Inszenierung dazu animieren, einen individuellen Zugang zum Stück zu entdecken. In enger Zusammenarbeit mit Dramaturgen und Fotografen entsteht so eine Bandbreite verschiedener Programmhefte, die jedes für sich den Charakter der jeweiligen Stücke betonen, spielerisch sind, selbst zum Teil des Theatererlebnisses werden, gemeinsam aber durch verbindende Elemente ein homogenes Bild ergeben.

Excerpt from the already published evening programmes of the Bielefeld Theatre. – By using the entire rear of the eight-page fold-out leaflet simultaneously as a poster for the play we were able to produce, after the relaunch, a four-colour programme leaflet without any advertising. Before the relaunch it was usually a black and white programme filled with advertising (due to the tight culture budget). The late costume rehearsals at the Bielefeld Theatre mean that customary stage photos usually included in evening programmes are not available; this circumstance (and the tremendous time pressure in producing the leaflets which are partly created within hours) helped to create evening programmes that frequently rely on associative images and texts. They do not only aim at providing information to the theatre-goer, but – as a component of the production – help to find an individual access to the play. In close cooperation with the literary and artistic directors as well as photographers this leads to a wide variety of programme leaflets. Each leaflet emphasises the character of the respective play, is playful itself and thus becomes part of the theatre experience, while all the leaflets together produce a homogeneous image thanks to the bonding elements.

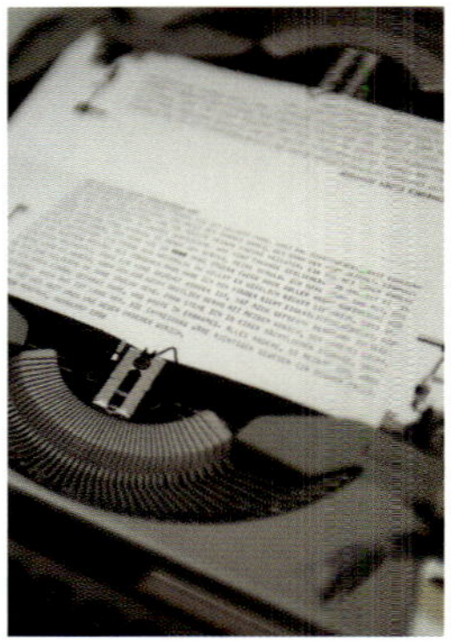

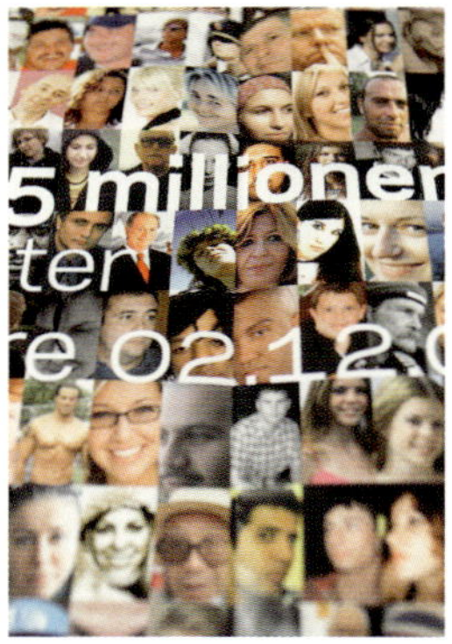

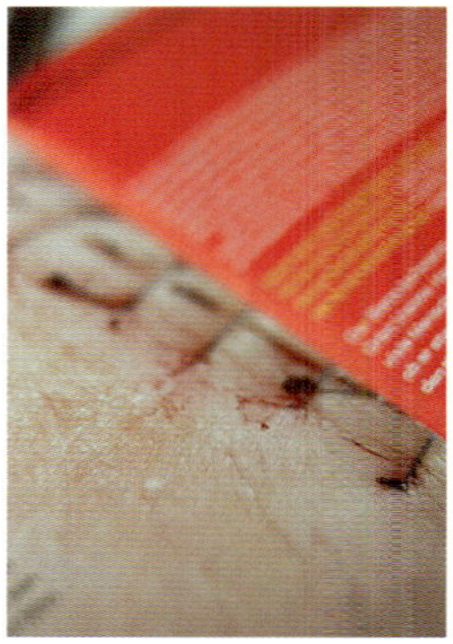

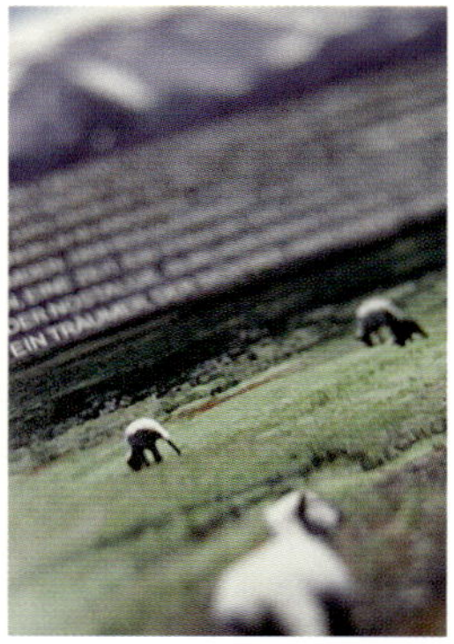

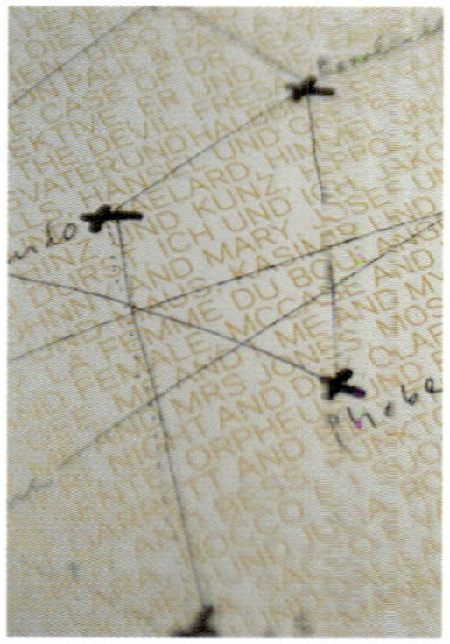

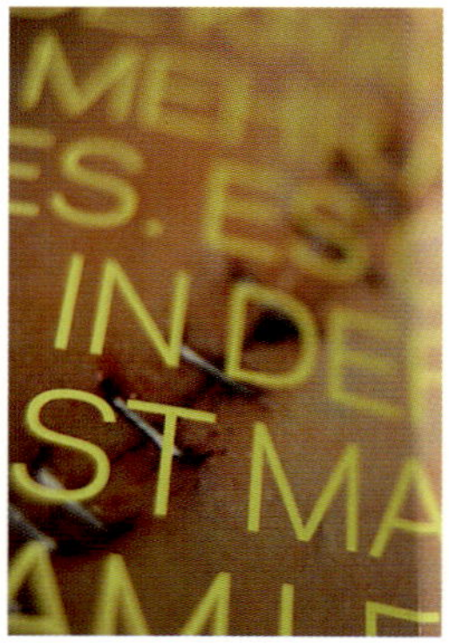

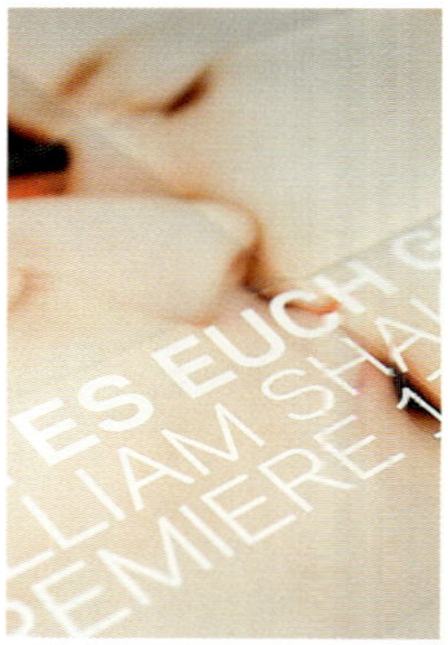

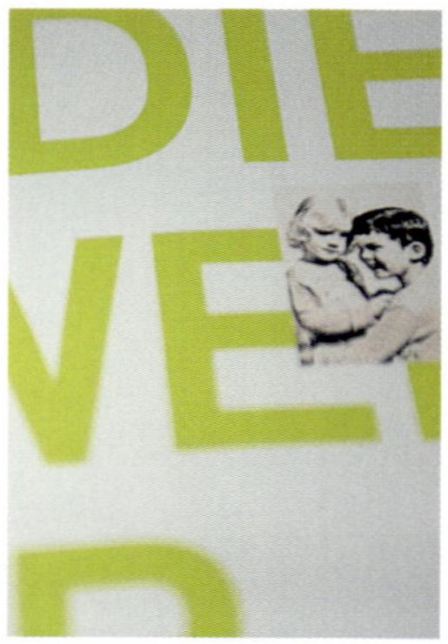

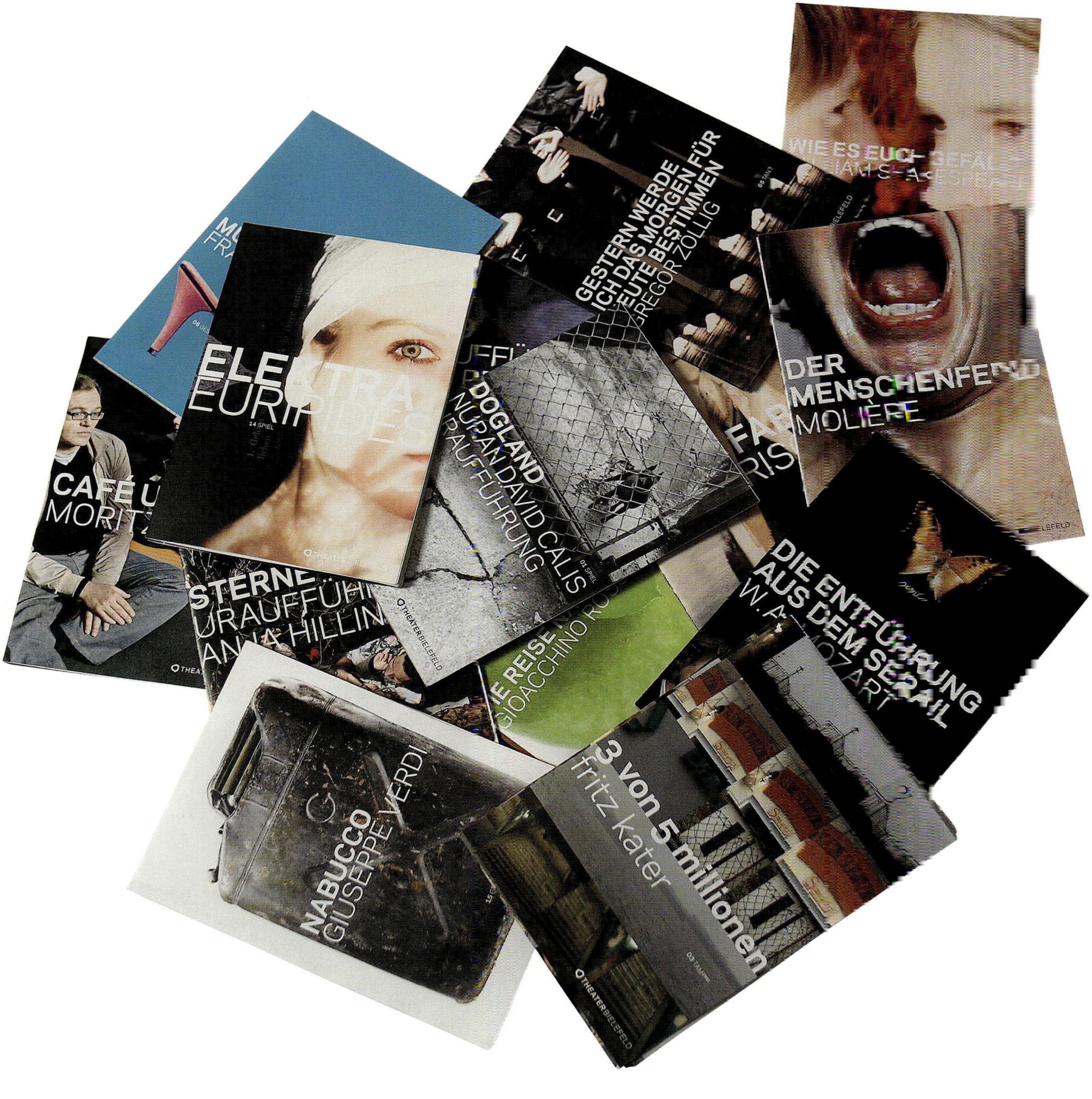
ELEKTRA
EURIPIDES
14 SPIEL
DOGLAND
NURAN DAVID CALIS
URAUFFÜHRUNG
01 SPIEL
GESTERN WERDE
ICH DAS MORGEN FÜR
DER
MOLIÈRE
DIE ENTFÜHRUNG
AUS DEM SERAIL
GIOACCHINO
NABUCCO
GIUSEPPE VERDI
3 von 5 millionen
fritz kater
THEATER BIELEFELD
CAFÉ
STERNE

Auftraggeber Client
gutenberg beuys gmbh
feindruckerei
Hannover, Germany

Projekt Project
raster. – magazin für feines zeug
Kundenmagazin
Customer magazine

Design
Rainer Gärtner
G73 Werbeagentur
Hannover, Germany

raster. ist das halbjährlich erscheinende Magazin der gutenberg beuys feindruckerei. Es richtet sich vornehmlich an Werbeagenturen und sieht sich als Gestaltungs-, Unterhaltungs- und Bildungsmagazin, gewürzt mit eigenwilligem Humor und Nonsens. Durch die inhaltliche und gestalterische Vielfalt wird es oft zur Hand genommen und beeindruckt dabei ganz nebenbei durch die drucktechnisch formale Qualität. Das Image der Druckerei als unkonventioneller, aber qualitativ hochwertig arbeitender Dienstleister findet seine stetige Festigung. Zusätzlicher Effekt der ausgezeichneten Ausgabe: die Verpackung in einer Kunstfelltasche (Modell Fred Feuerstein).

raster. is a semiannual magazine put out by gutenberg beuys specialist printers. It is aimed primarily at advertising agencies and views itself as a design, entertainment and educational magazine, spiced up with original humor and nonsense. The diversity and variety of its content and design means that it is often picked up and read, and readers are incidentally impressed by the technical quality of the printing, reinforcing the printer's image as an unconventional, yet high-quality service provider. An additional feature of this outstanding publication is the Fred Flintstone-type artificial fur case it comes in.

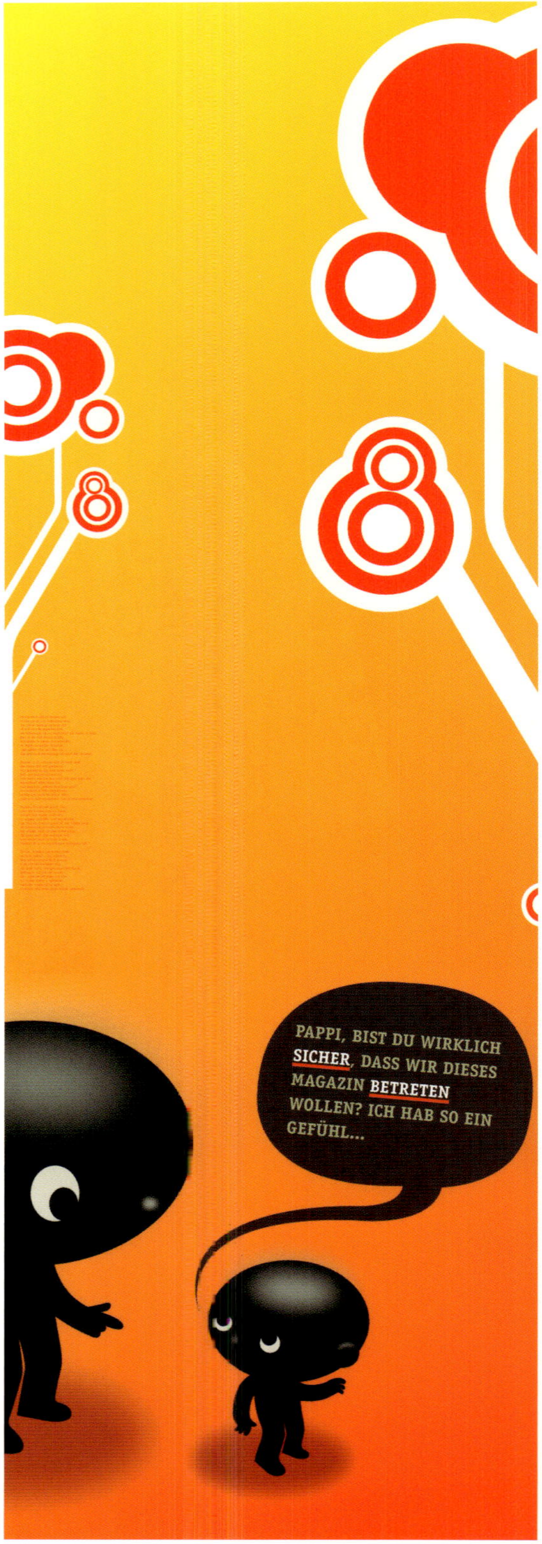

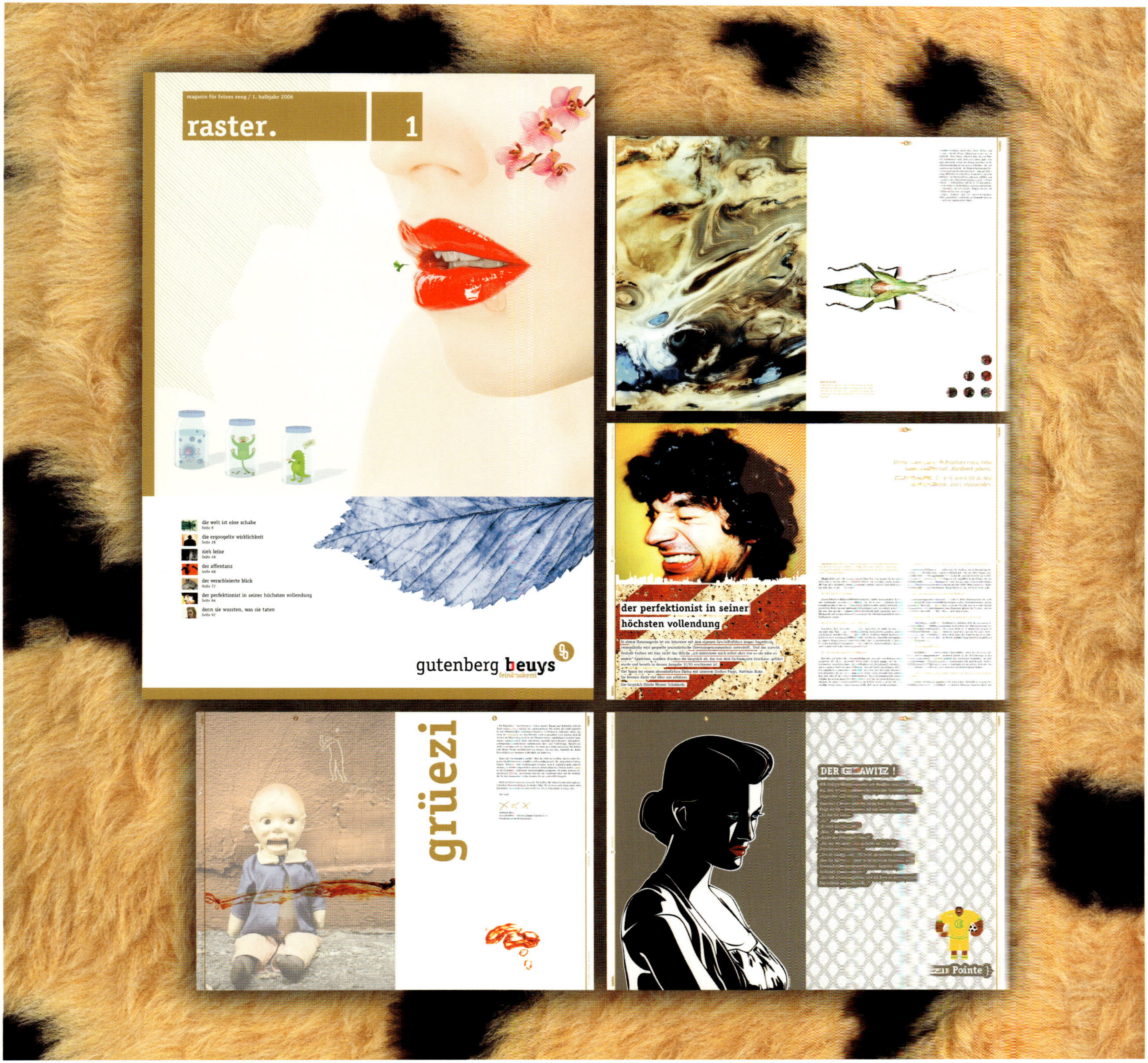
magazin für feines zeug / 1. halbjahr 2006
raster.
1
die welt ist eine schabe
Seite 8
die ergoogelte wirklichkeit
Seite 28
zieh leine
Seite 48
der affentanz
Seite 68
der verschleierte blick
Seite 72
der perfektionist in seiner höchsten vollendung
Seite 84
denn sie wussten, was sie taten
Seite 92
gutenberg beuys
feindruckerei
der perfektionist in seiner
höchsten vollendung
grüezi
Pointe

print media image

Auftraggeber Client
ThyssenKrupp AG, ThyssenKrupp Steel
Düsseldorf, Germany

Projekt Project
Discover Steel #2 China, Begegnungen
Imagebuch
Image book

Design
Annette Häfelinger
Kerstin Weidemeyer
Kurt Steinebrunner
häfelinger+wagner design
München, Germany

Die Imagebuchreihe „Discover Steel" stellt die internationalen Standorte von ThyssenKrupp vor, möchte Denkanstöße geben und überraschen. Unter dem Titel „Begegnungen" besuchten ein östliches und ein westliches Fotografen- und Autorenteam Menschen in China, die mit dem Werkstoff Stahl in Verbindung stehen. Dabei beleuchtete jedes Team die andere Kultur. Das duale Thema „Begegnungen" wird durch die Gestaltung, Farbsystematik und Papierwahl interpretiert, indem sich deutsche (silber-graue) und chinesische (rote) Schriftzeichen auf halbtransparentem Papier überlagern. Silber-glänzende Prägungen reproduzieren die ästhetische Anmutung von Stahl.

The image book series "Discover Steel" throws light on ThyssenKrupp Steel international locations and is intended to be thought provoking and surprising. Under the title "Encounters", Eastern and Western teams of photographers and authors visited Chinese people connected with steel as a working material. Each team looked at the "other" culture. The dual subject "Encounters" is interpreted by the form, system of colours and choice of paper; German (silver-grey) and Chinese (red) characters are superimposed on semi-transparent paper. Shiny silver embossing effects reproduce the aesthetic impression of steel.

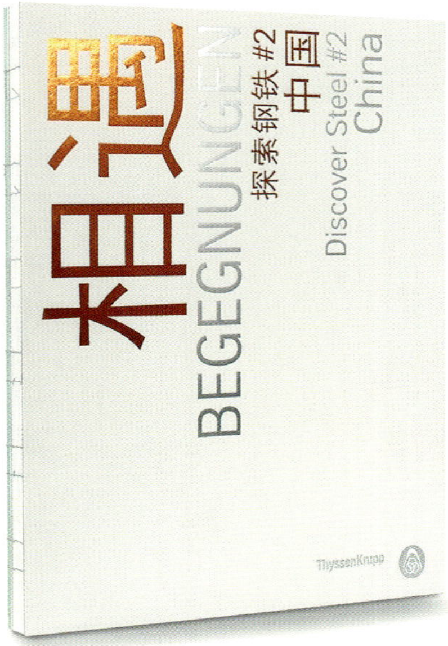

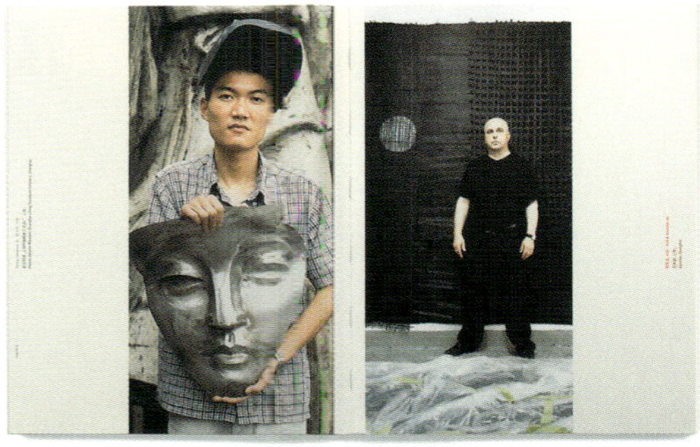

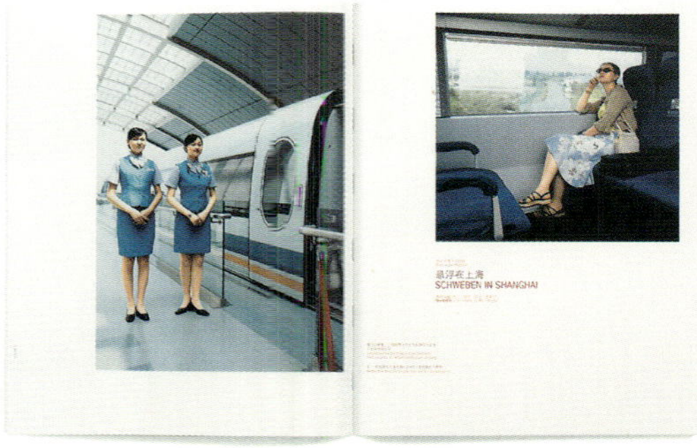

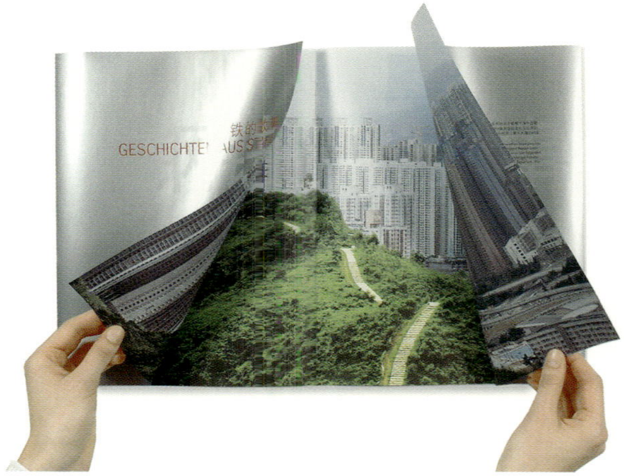

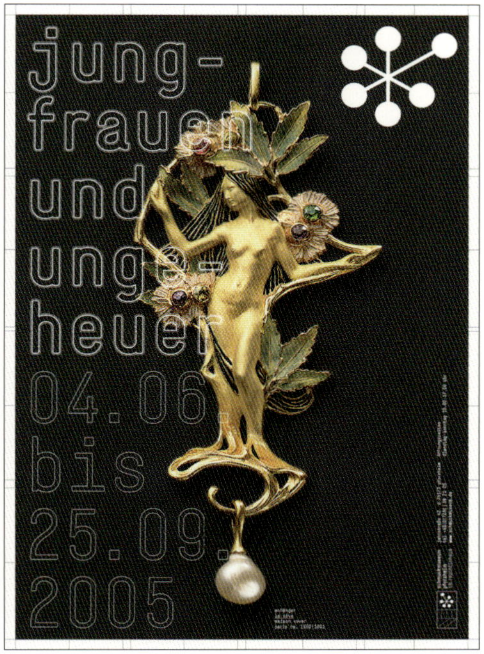

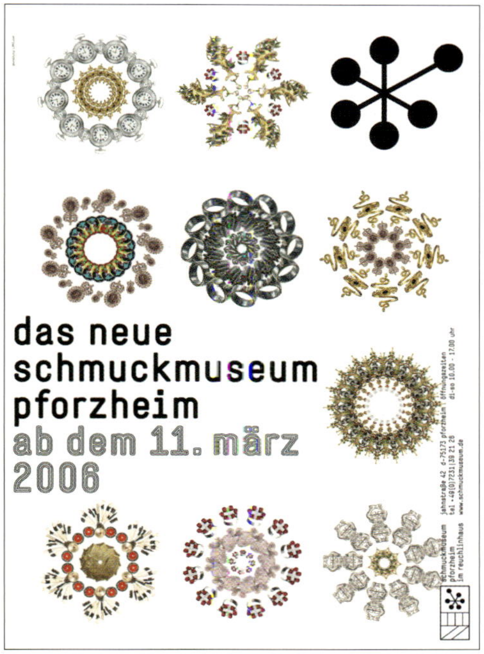

Auftraggeber Client
Schmuckmuseum Pforzheim
Pforzheim, Germany

Projekt Project
Printmedien für das Schmuckmuseum Pforzheim
Plakate und Flyer
Posters and flyer

Design
Sascha Lobe
Gitte Bohr
Jan Maier
Dirk Wachowiak
L2M3 Kommunikationsdesign GmbH
Stuttgart, Germany

Ausgangspunkt für die Neuentwicklung des gesamten Corporate Designs ist neben der weltweit einmaligen Sammlung das in den 60er Jahren gebaute, inzwischen denkmalgeschützte Haus. Aus ihm leiten sich die Modulhaftigkeit und die variierende Ornamentik der Oberflächen ab. Schraffuren, Punktraster und Outlineschriften schaffen einen barocken Modernismus.

The starting point for the new development of the entire corporate design is in addition to the worldwide unique collection the now landmarked house built in the 1960s. The modularity and the varying ornamentation of the surfaces are derived from it. Hatchings, dot screens and outline typefaces create a baroque modernism.

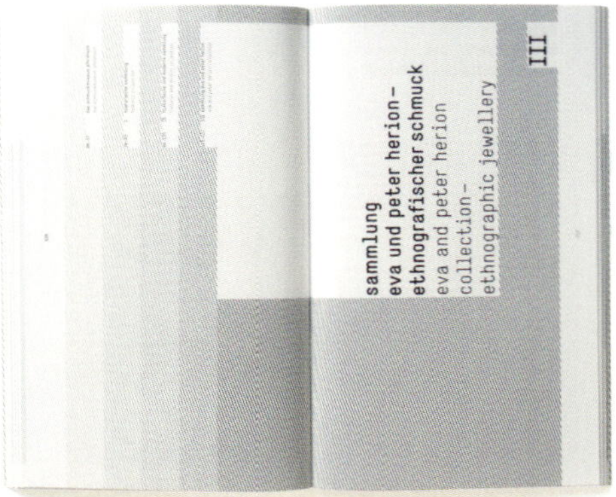

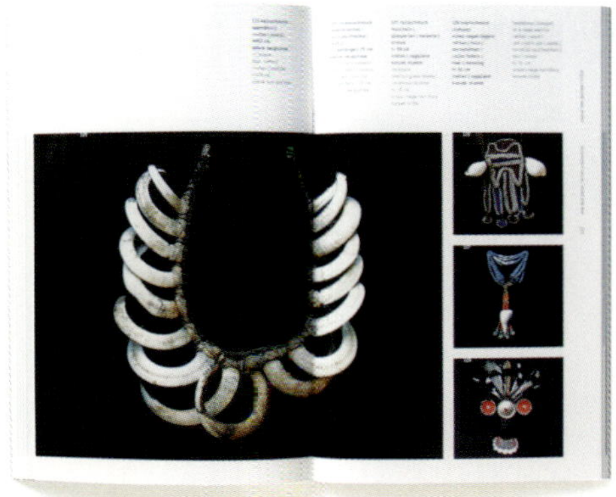

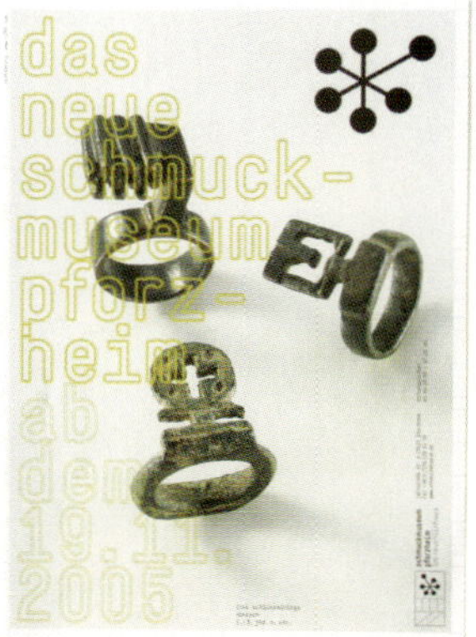

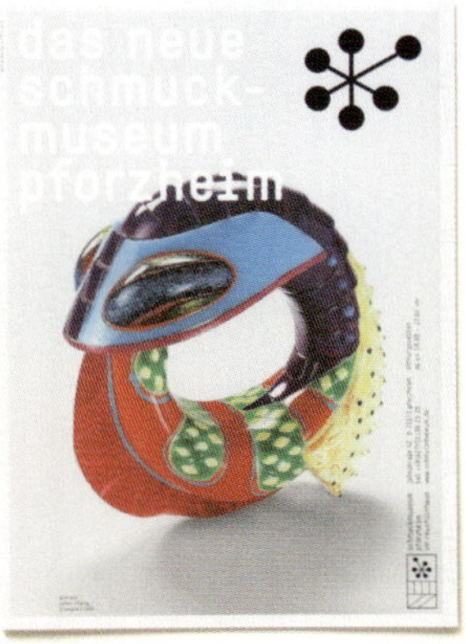

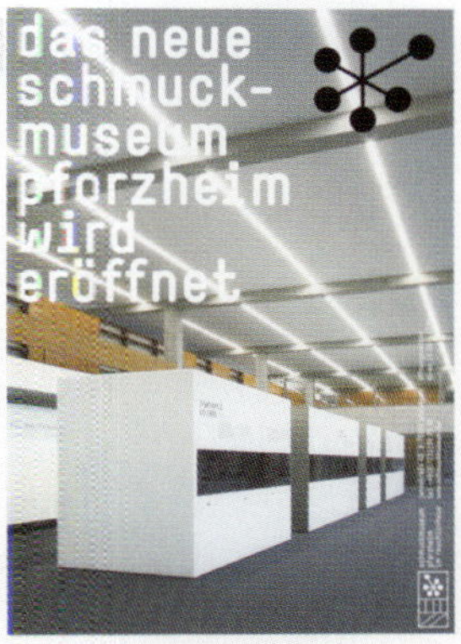

Auftraggeber Client
Heine/Lenz/Zizka
Frankfurt am Main, Germany

Projekt Project
iPray
Weihnachtsgruß
Holiday greetings

Design
Heine/Lenz/Zizka
Frankfurt am Main, Germany

iPray. Nie wieder Platzprobleme.
Jetzt kann man die Weihnachtskrippe unterwegs mitnehmen und hat immer noch Platz für alles andere. Es ist möglich, die Figuren auf der Tischplatte zu bewegen und neue Szenen auf Basis der emotionalen Disposition der eigenen Person zu erfinden: zum Beispiel „Hinstellen" oder „Vor- und Zurückrücken". Erleben Sie Ihre Lieblingsszenen der heiligen Geschichte. Erstellen Sie unterwegs ein Theater „out of the box". Starten Sie Ihr eigenes Weihnachten. Synchronisieren Sie sich mit anderen Gläubigen. Mit dem iPray sind alle Optionen sofort verfügbar.

iPray. Christmas on the go.
Now you can take the Nativity scene along – and still have enough space for everything else. Move around the little figures across the table. Deploy new scenes according to your personal emotional disposition: i.e. standing them up-right, pushing them back or forth. Make up your favorite scenes of the Holy Family. Create a theater out of the box en route. Start your own personal Christmas. Sync yourself with other believers. With iPray all options are instantly at hand.

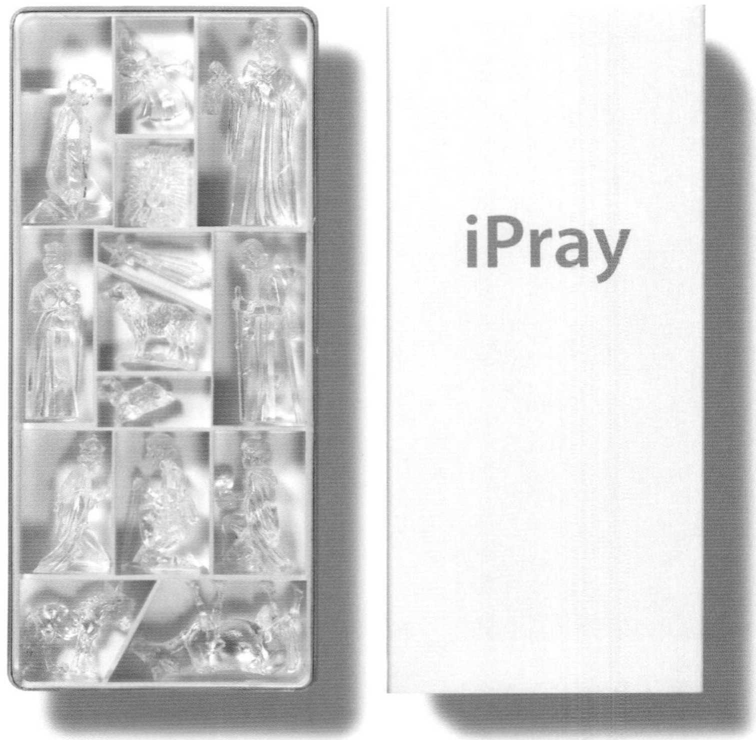

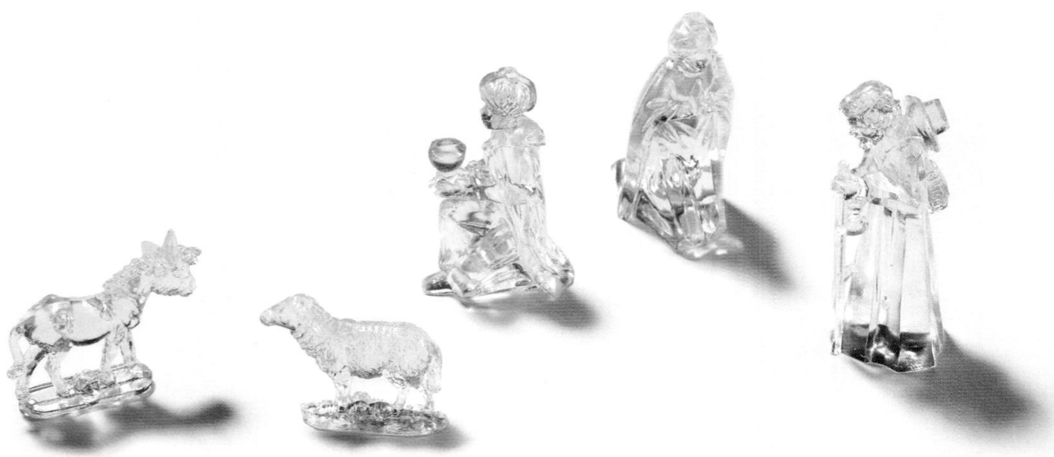

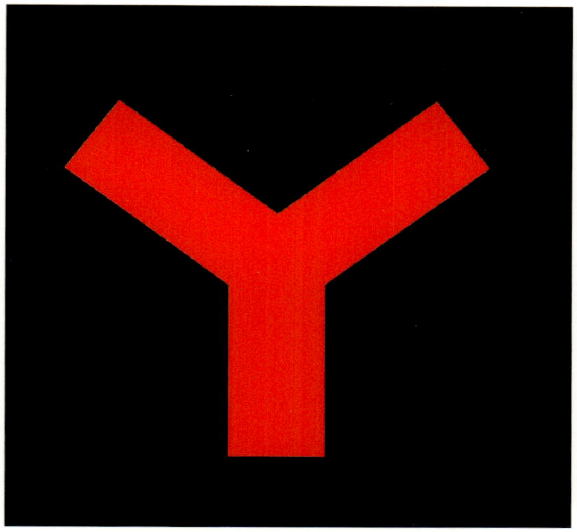

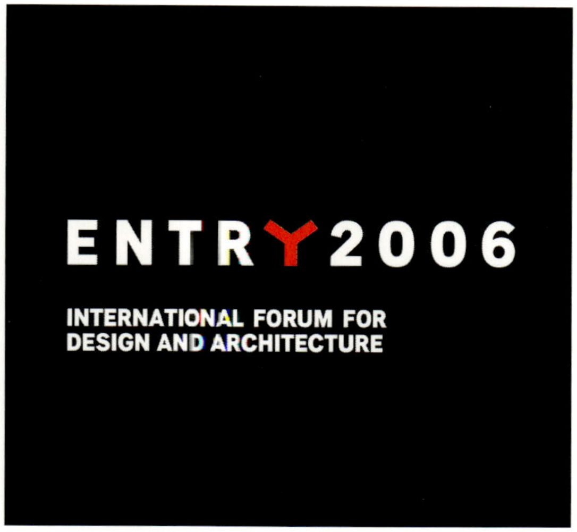

Auftraggeber Client
Ausstellungsgesellschaft Zollverein GmbH
Essen, Germany

Projekt Project
ENTRY2006
Corporate Design

Design
Knut Maierhofer
Patrick Märki
Bruno Marek
Peta Kobrow
Constanze Knoese
Christina Baur
Bernd Müller
KMS Team GmbH
München, Germany

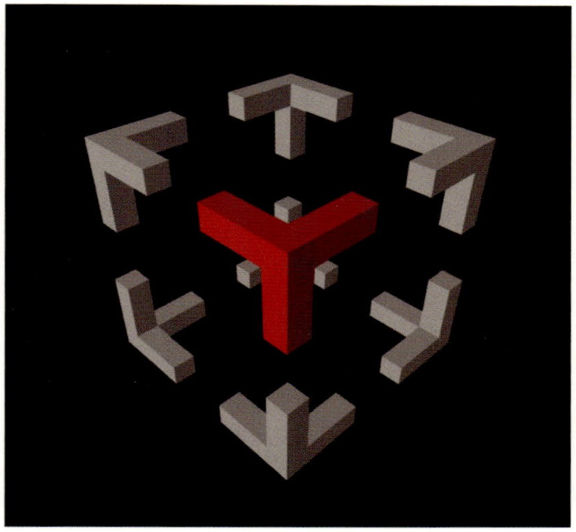

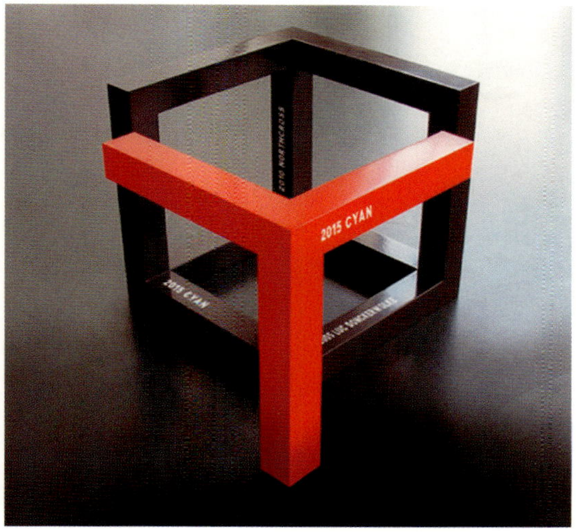

Im Sommer 2006 findet in Essen die erste Weltausstellung für Design und Architektur statt: ENTRY2006. Aktuelle und visionäre Projekte führender Gestalter und Architekten geben Einblick in die Entwicklungen der kommenden Jahrzehnte.
Ausgangspunkt unseres Konzepts ist das markante Bildzeichen: ein rotes Y (dem Endbuchstaben von ENTRY), das dreidimensional als Koordinatenkreuz realisiert wird. Es verbindet so die Vorstellungen des Grafischen und des Räumlichen, den jeweiligen Ur-Domänen von Design und Architektur.

In summer 2006 Essen hosts the first World Exposition for Design and Architecture ENTRY2006. Visionary projects by leading designers and architects offer insights into the developments of the coming decades. The starting point of our conception is the distinctive symbol of a red Y (the final letter in ENTRY), which, rendered in three dimensions, appears as a system of coordinates. It thus connects the ideas of the graphic (plane) with that of the spatial, the respective original domains of design and architecture.

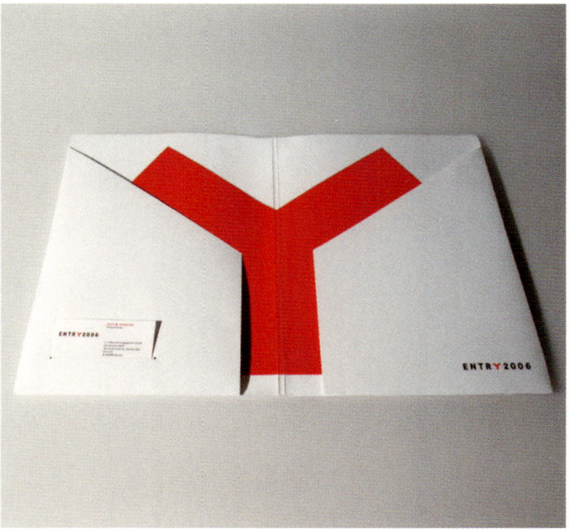

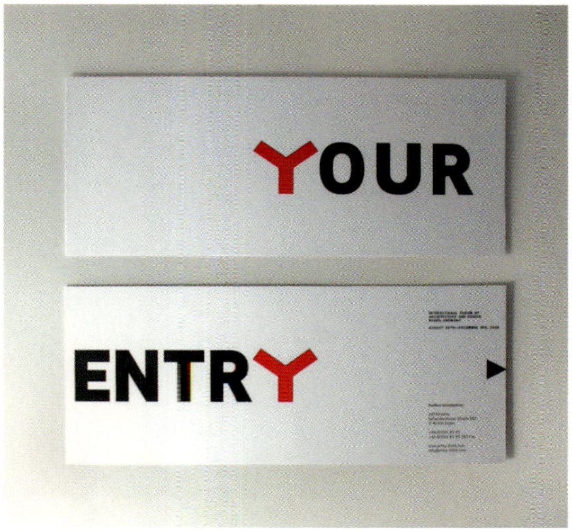

print media image

Auftraggeber Client
Deutscher Werkbund Bayern e.V.
München, Germany

Projekt Project
Werkbundsiedlung Wiesenfeld
Corporate Design

Design
Michael Keller
Julia Just
Carola Gerlach
Axel Sanjosé
Silke Streppelhoff
Christina Baur
Christian Ring
KMS Team GmbH
München, Germany

Zum 100-jährigen Bestehen des Deutschen Werkbunds entsteht in München die Werkbundsiedlung Wiesenfeld. Wir haben für die Werkbundsiedlung das Erscheinungsbild entworfen. Das Konzept basiert auf einem System von Rechtecken (deren Proportionen aus dem Werkbundlogo abgeleitet sind), die einen imaginären „Gedankenraum" bilden: die Welt der Grundsätze, Themen und Projekte des Werkbunds. Die „Ideenfelder" verändern den Blick auf die Realität, schaffen neue Perspektiven und bilden zugleich Referenzpunkte für neue Ansätze in Architektur und Gestaltung.

On the occasion of the centennial of the Deutscher Werkbund, 2007 will see the groundbreaking for the construction of the Werkbundsiedlung Wiesenfeld in Munich. For this project we designed the public image. The concept is based on a system of rectangles (in proportions derived from the Werkbund logo), which form an imaginary "conceptual space": the world of the principles, topics and projects of the Werkbund. The "fields of ideas" alter the outlook on reality, creating new perspectives and providing reference points for new approaches in architecture and design.

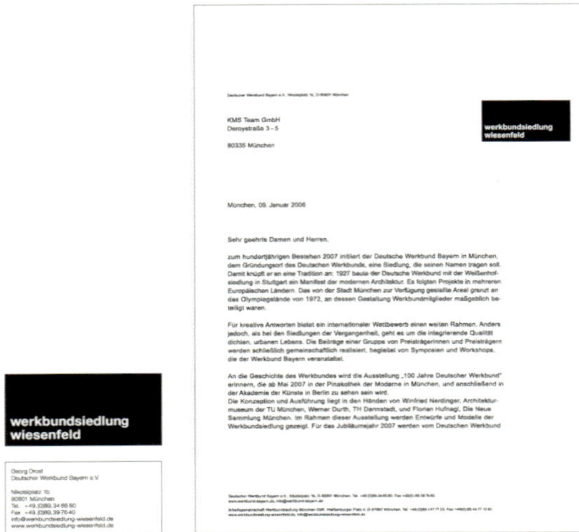

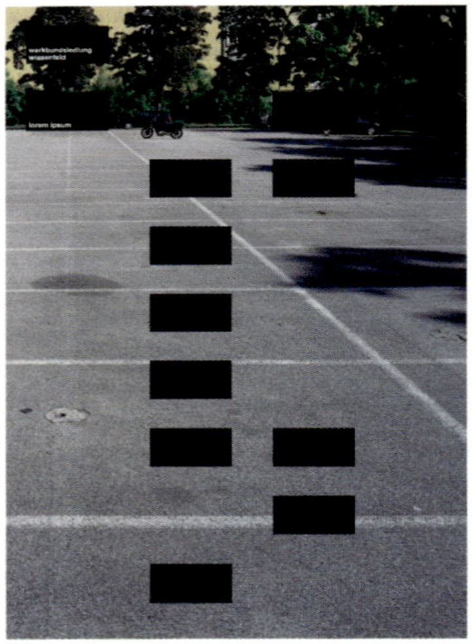

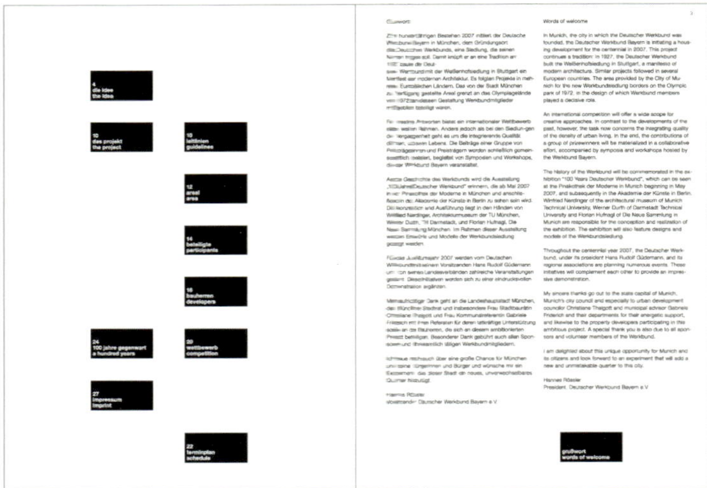

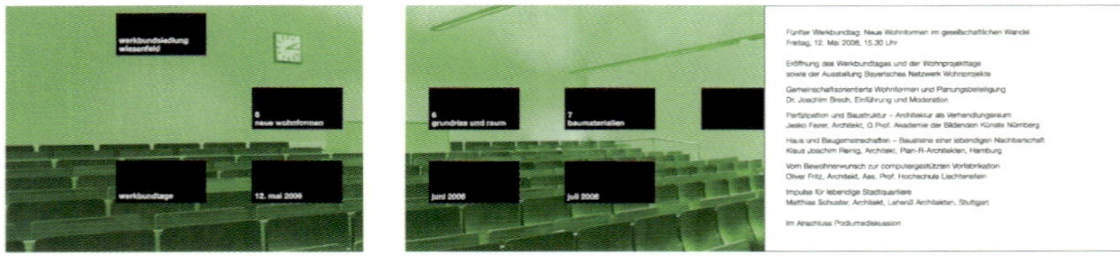

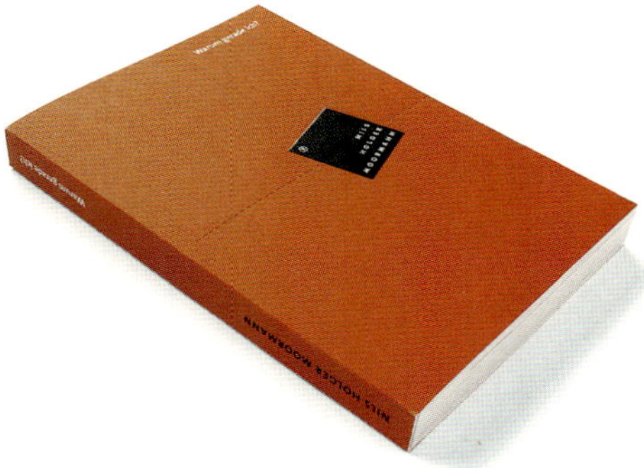

Auftraggeber Client
Nils Holger Moormann GmbH
Aschau im Chiemgau, Germany

Projekt Project
Warum gerade ich?/Why me?
Firmenbroschüre
Corporate brochure

Design
Jäger & Jäger
Designer AGD
Überlingen, Germany

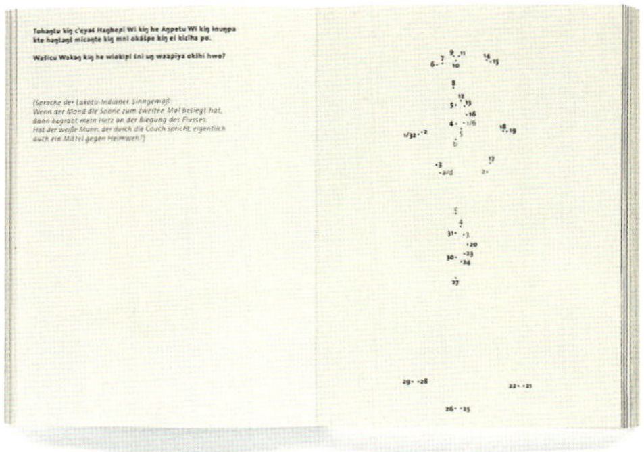

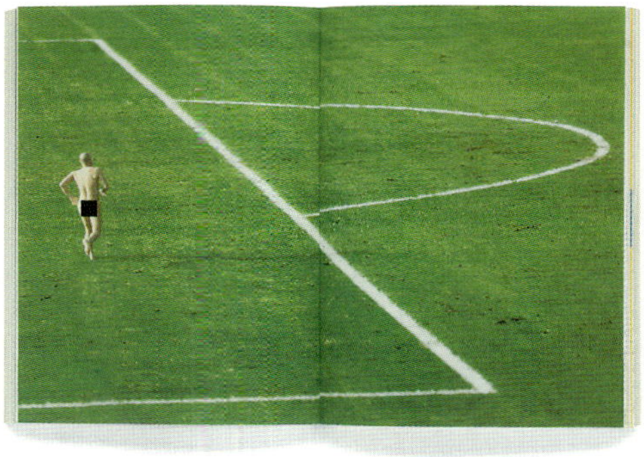

Die zweijährig erscheinende Broschüre „Mobro" fragt sich in diesem Jahr, warum ausgerechnet ihr so etwas passieren müsse: Um die Besonderheiten der einzelnen Möbel herauszuarbeiten, wird jedem Möbelstück ein Charakter – eine „Seele" – zugeordnet. Der Katalog übernimmt gezwungenermaßen die Rolle des Psychologen: Jedes Möbel darf ihm seine Probleme und Nöte, Sorgen und Freuden erzählen. So hat die Leiter „HochAcht" Höhenangst und Garderobe „Pin Coat" joggt gerne nackt. Es gibt auch Möbel wie „Taurus", die sich von Nils Holger Moormann leider völlig missverstanden und gänzlich fehl am Platze fühlen…

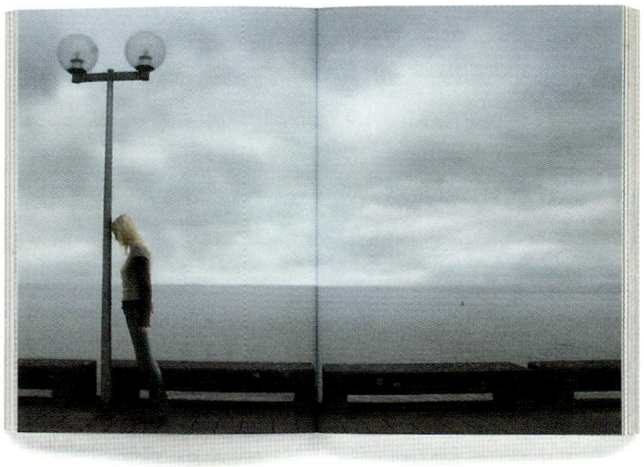

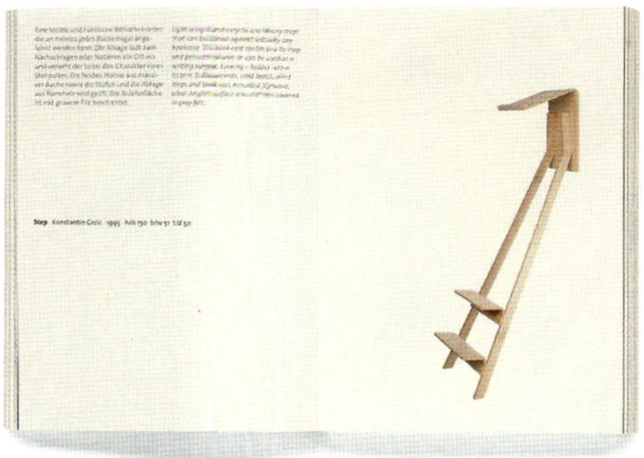

The "Mobro" brochure, which appears every two years, is this year asking why it has to be like this: In order to find out their true characters, every single piece of furniture is given a personality or a "soul". And so, in a way, the catalogue has to take on the role of psychologist. Each piece of furniture is allowed to confide its problems and worries, fears and joys. The "HochAcht" ladder, for example, suffers from vertigo and the "Pin Coat" wardrobe likes jogging naked in the woods. And there is also "Taurus" which unfortunately feels totally misunderstood by Nils Holger Moormann and therefore thinks it doesn't belong…

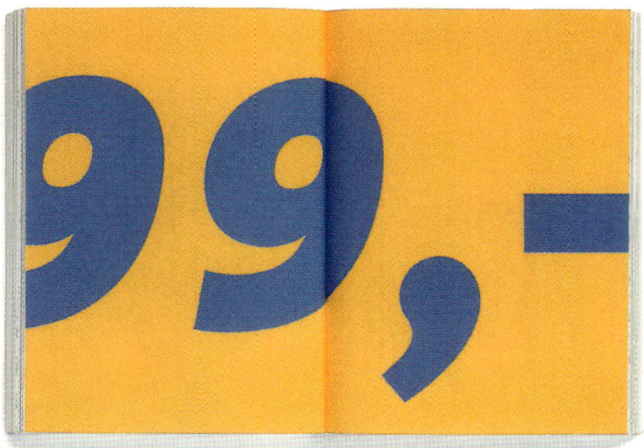

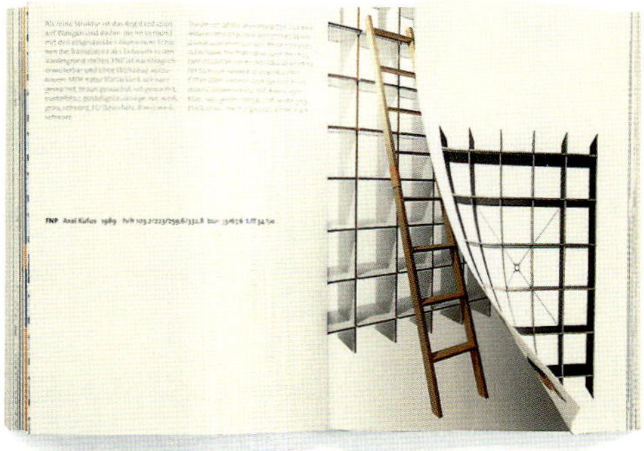

print media image

Auftraggeber Client
Typotron AG
St. Gallen, Switzerland

Projekt Project
12 Minuten
Broschüre
Brochure

Design
merkwürdig GmbH
Frankfurt am Main, Germany

Im März 2005 durften wir Albert Kriemler, dem Designer des Schweizer Modelabels AKRIS, über die Schulter schauen und das Defilee in Paris live miterleben. Entstanden ist eine Broschüre, die für den Betrachter die gleiche Spannung erzeugt, wie wir sie vor Ort empfunden haben. Wegen der speziellen Bindeart ist es zunächst nicht möglich, die Bilder zu betrachten. Nur durch die kleinen Löcher (angelehnt an das Erscheinungsbild von AKRIS) ist zu erahnen, dass sich mehr dahinter verbirgt. Die perforierten Seiten müssen getrennt werden, um die Bilder anzuschauen. Der Umschlag der Broschüre kann auch als Poster genutzt werden.

In March 2005 we were able to witness the Paris Défilé live and look over the shoulder of Albert Kriemler, the designer for Switzerland's AKRIS fashion label. The result was a brochure which creates the same tension for the reader as if he or she had actually been there. Due to the brochure's special binding, it is impossible to view the pictures at first. Only by looking through the small holes (which are a reference to the AKRIS look) can the reader tell there is more in back. The reader then has to tear the perforated pages to actually view the pictures. The cover of the brochure can also be used as a poster.

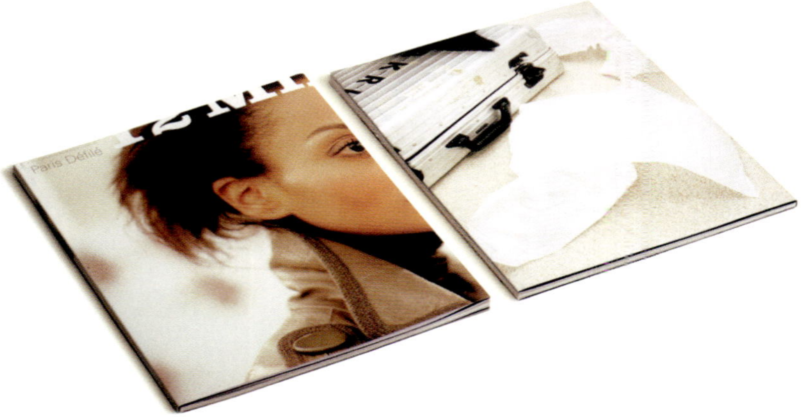

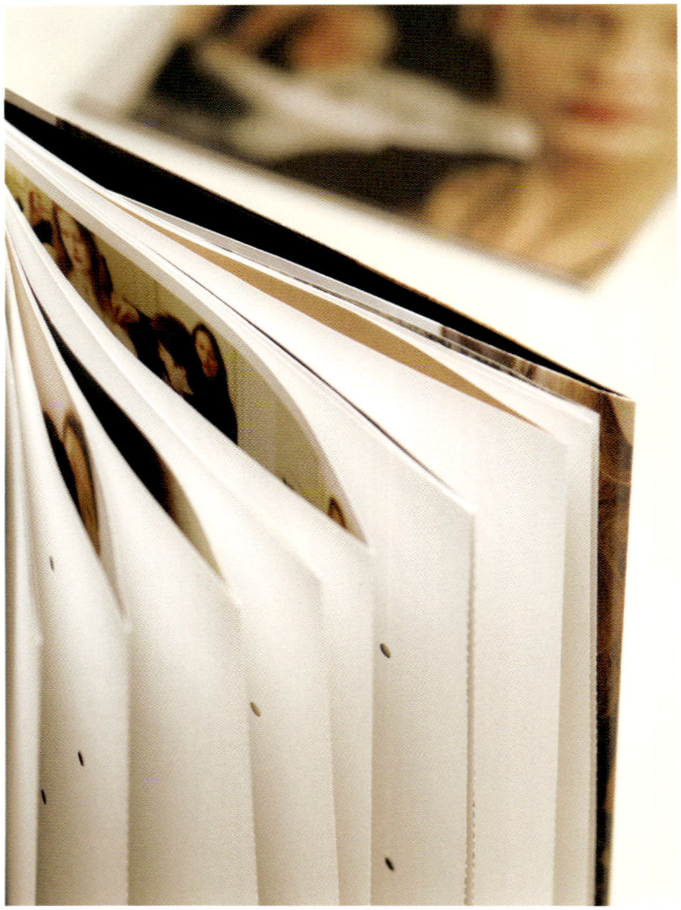

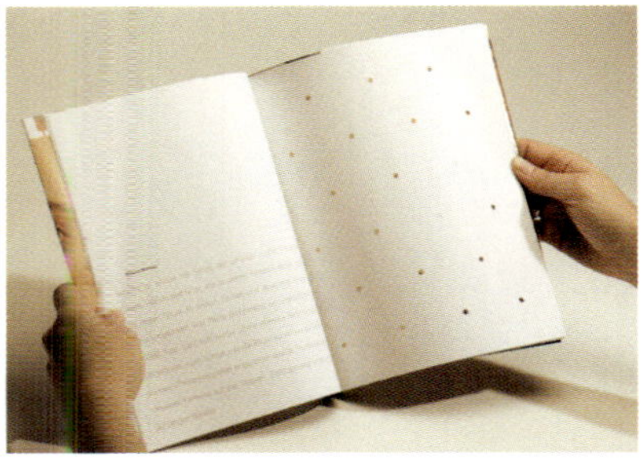

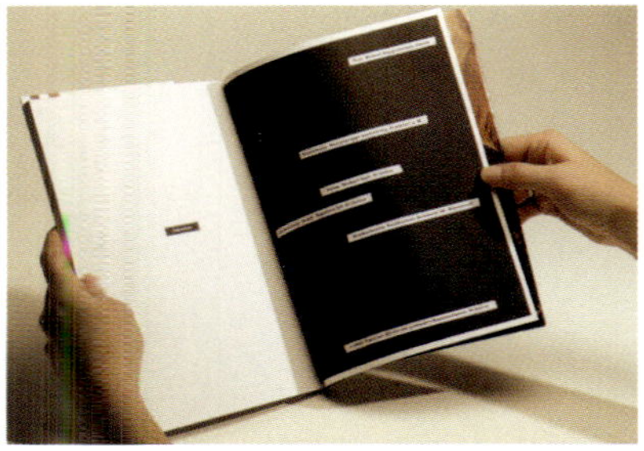

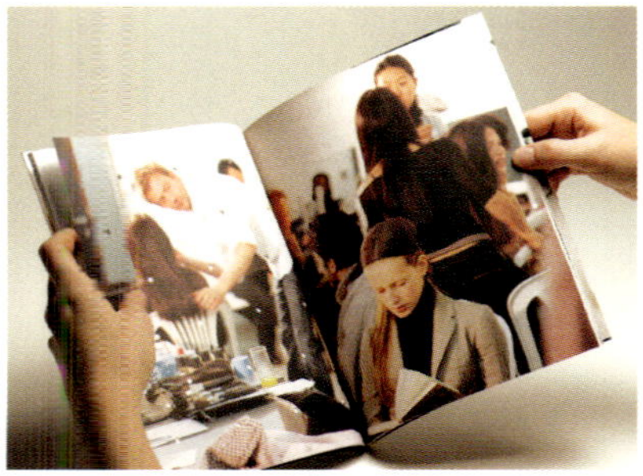

Auftraggeber Client
Typotron AG
St. Gallen, Switzerland

Projekt Project
Herzsprung
Broschüre
Brochure

Design
merkwürdig GmbH
Frankfurt am Main, Germany

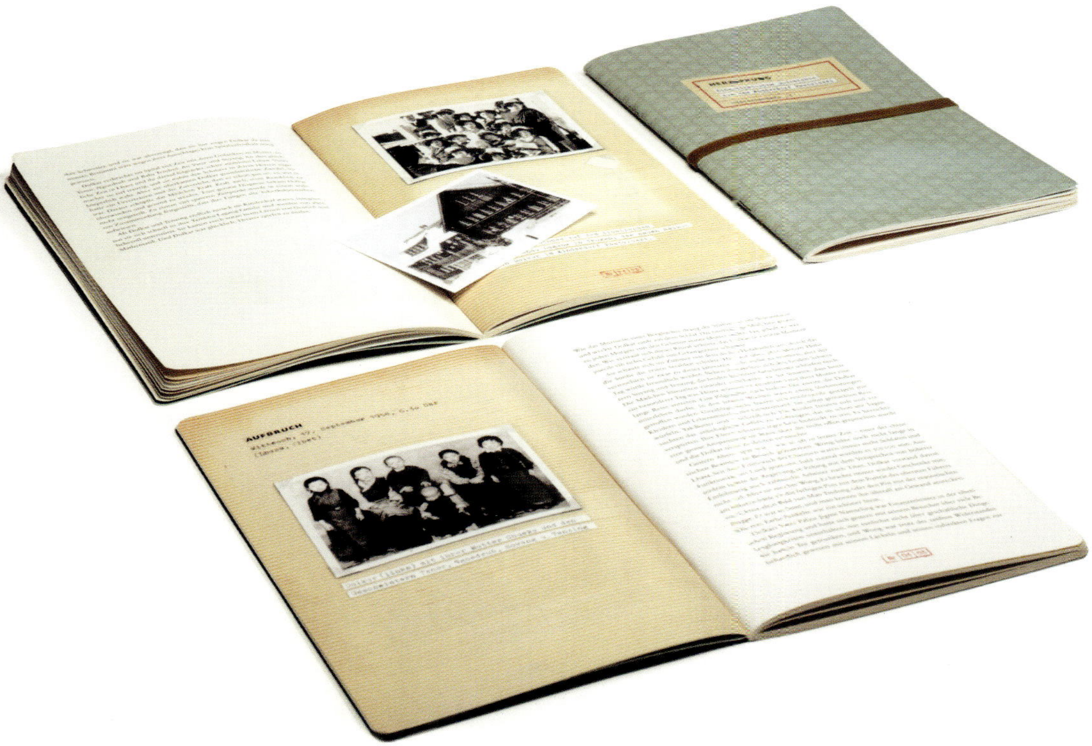

Diese Broschüre beschreibt und zeigt die Lebensgeschichte einer Tibeterin, die als Kind aus ihrer Heimat flüchten musste und im Kinderdorf Pestalozzi im appenzellischen Dorf Trogen (Schweiz) eine neue Heimat fand. Viele der Originaldokumente wurden in einer Art Tagebuch zusammengefasst. Mehrere Seiten erinnern an typische Fotoalben. Auf einer Seite „fehlt“ sogar ein Foto, das lose zwischen die Seiten gelegt wurde. Der Umschlag gewinnt durch seine „verschmutzte“ Optik an Authentizität. Das Etikett auf dem Umschlag wurde von Hand in eine dafür geprägte Form eingeklebt. Ein Gummiband hält das Heft zusammen, damit auch nichts verloren geht.

This illustrated brochure describes the life story of a Tibetan woman who had to flee her homeland as a child, finding a new home in the Pestalozzi Children's Village in Trogen, Switzerland. Many of the original documents were assembled into a kind of diary. Several pages resemble a typical photo album. On one page a photo is even “missing” and is placed loosely inside the pages of the book. The cover has a slightly smudged look, enhancing the impression of authenticity. The label on the cover is glued in by hand in a specially recessed space. The whole booklet is bound together with a rubber band so that nothing can get lost.

print media image

Auftraggeber Client
e15 Design und Distributions GmbH
Oberursel, Germany

Projekt Project
WIR FEIERN 10 JAHRE e15
Buch
Book

Design
Antonia Henschel
SIGN Kommunikation GmbH
Frankfurt am Main, Germany

Der Name ist Programm: Zum zehnjährigen Bestehen der Firma e15 entstand das Bilderbuch „WIR FEIERN 10 JAHRE e15". Fotos und Skizzen der letzten zehn Jahre werden von Essays von Markus Frenzl begleitet. Statt Orange wird im Jubiläumsjahr die Sonderfarbe Gold eingesetzt.

It's all in a name: The book WIR FEIERN 10 JAHRE e15 ("We're celebrating 10 years at e15") was produced for the tenth anniversary of the e15 company. Photos and sketches of the last 10 years are accompanied by essays by Markus Frenzl. In place of the traditional orange, the jubilee color of gold has been used.

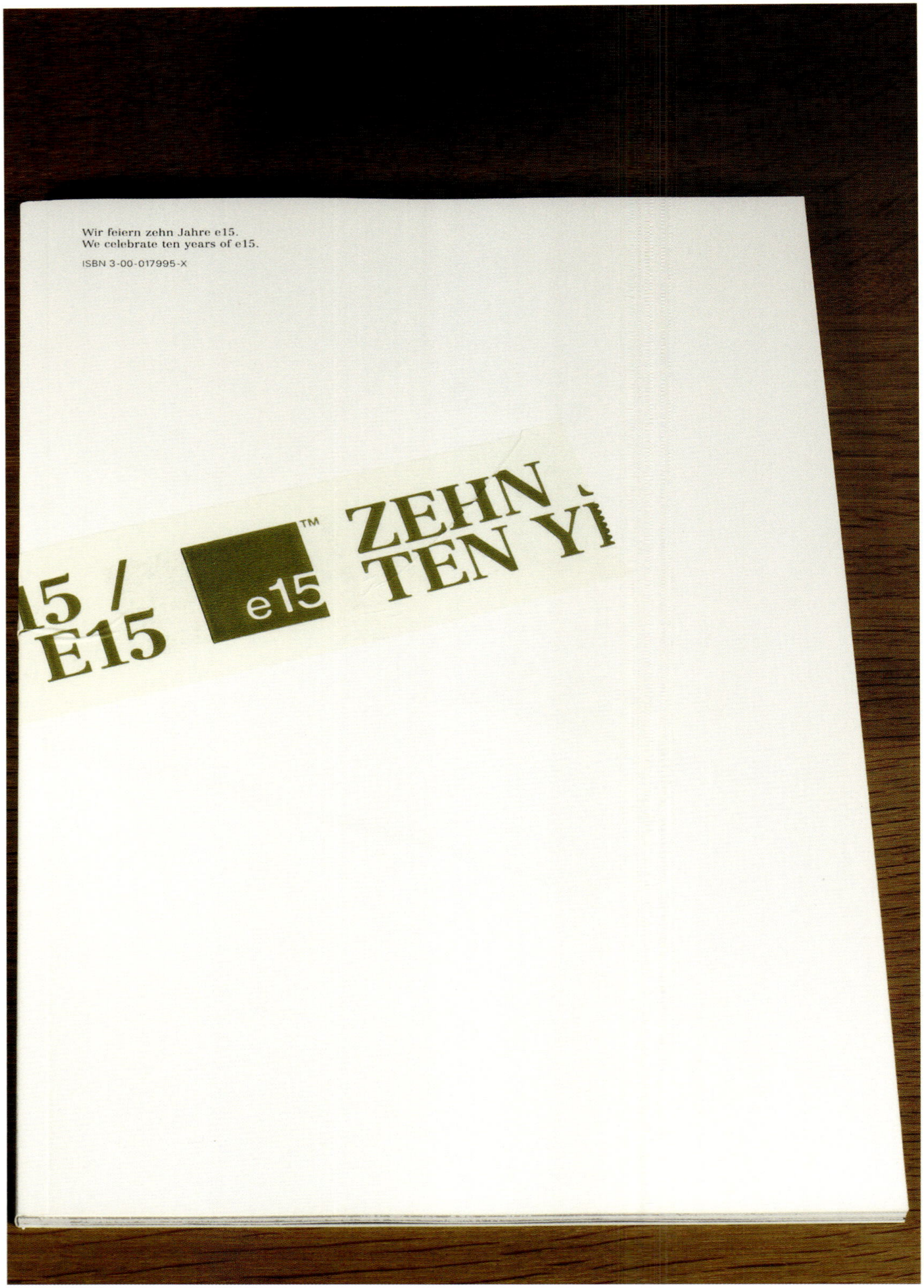

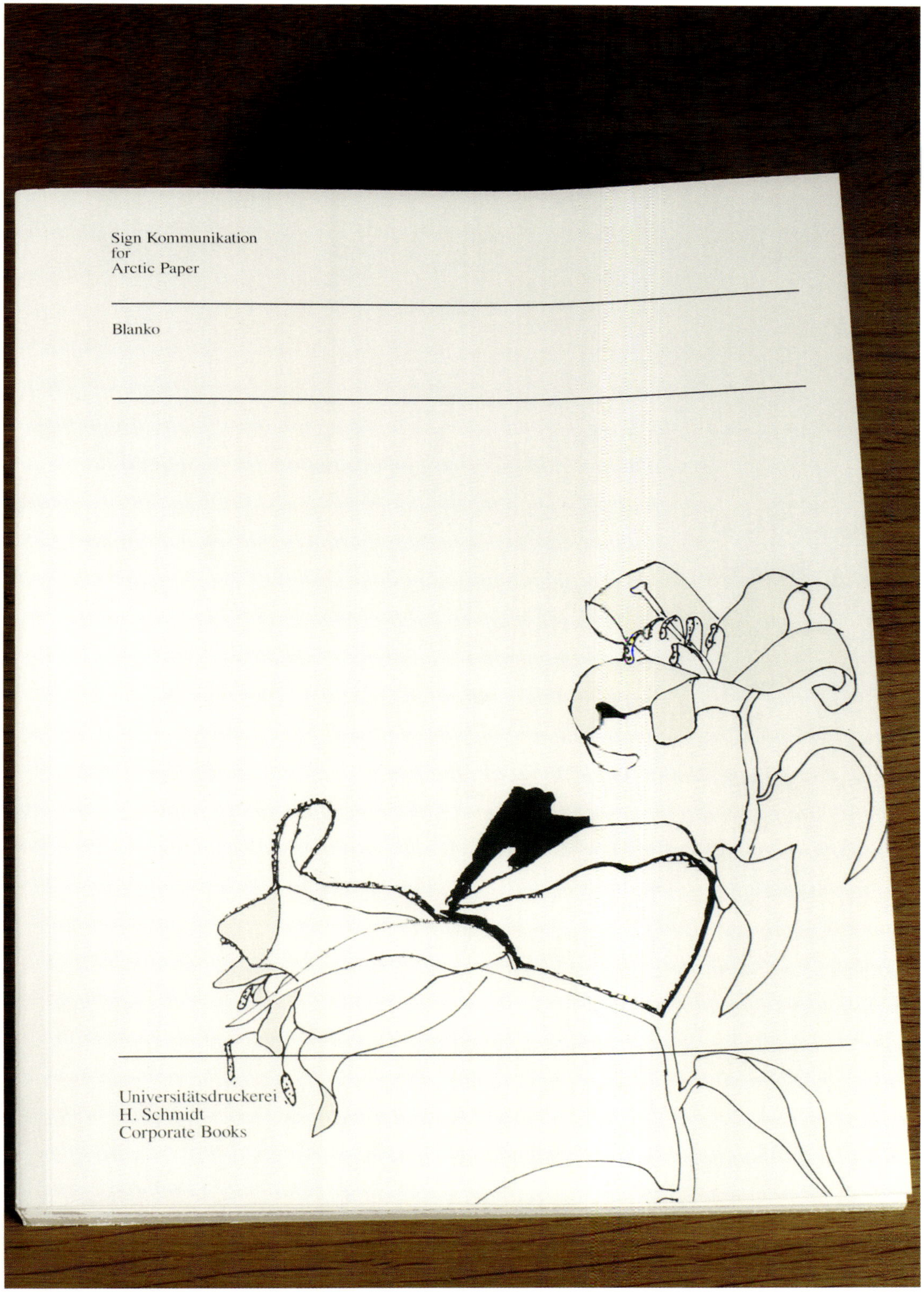

Auftraggeber Client
Arctic Paper Deutschland GmbH
Hamburg, Germany

Projekt Project
BLANKO
Buch
Book

Design
Antonia Henschel
Karl W. Henschel
Jörn Hofmann
Martin Kaufmann
Kiril Golovchenko
SIGN Kommunikation GmbH
Frankfurt am Main, Germany

„BLANKO" entstand im Auftrag der Papierfirma Arctic Paper Deutschland. Das Buch ist Projektionsfläche und Freiraum für eigene Ideen. „BLANKO" steht für den Raum, den es gedanklich zu füllen gilt, für eine Gestaltung, die den Nutzer einbezieht, weil sie ihm Anregungen gibt, ihm aber nichts vorschreibt. Eine Gestaltung, die nicht bis ins Letzte zu Ende gedacht und damit vorbestimmt und eindimensional ist.

"BLANKO" was commissioned by the paper company Arctic Paper Deutschland. The book serves as a surface and a space onto which you can project your own ideas. "BLANKO" stands for the space you need to fill with your own thoughts, for a design that draws in the user because it provides him or her with a stimulus, yet without prescribing anything. A design that is not thought out down to the last detail. A design that is therefore neither predefined nor one-dimensional

print media image

Auftraggeber Client
Slade School of Fine Art
University College London
London, United Kingdom

Projekt Project
Slade School of Fine Art
Studienbuch, Broschüre, Plakat
Prospectus, brochure, poster

Design
Nanni Goebel
Christopher Wiehl
paarpiloten
London, United Kingdom

Die Slade School of Fine Art ist eine der renommiertesten Kunstschulen Englands, die BA- und MA-Studiengänge anbietet. Das neu entwickelte Erscheinungsbild spiegelt den Austausch zwischen BA und MA mit zwei parallel laufenden Farbbalken wider: Cyan für MA, Yellow für BA, die Mischfarbe Grün wird für die Slade selbst verwandt, so auch für Informationen, die die Slade als Ganzes betreffen. Die „Slade Summer School" hat einen hellen Grünton.

The Slade School of Fine Art is one of the best-known art schools in England, offering BA and MA degrees. Its redesigned look reflects the give-and-take between the BA and MA spheres, with two color bars running in parallel: cyan for the MA, yellow for the BA. A blended color of green is used for the Slade itself, as well as for information concerning the Slade as a whole. The "Slade Summer School" is portrayed in light green.

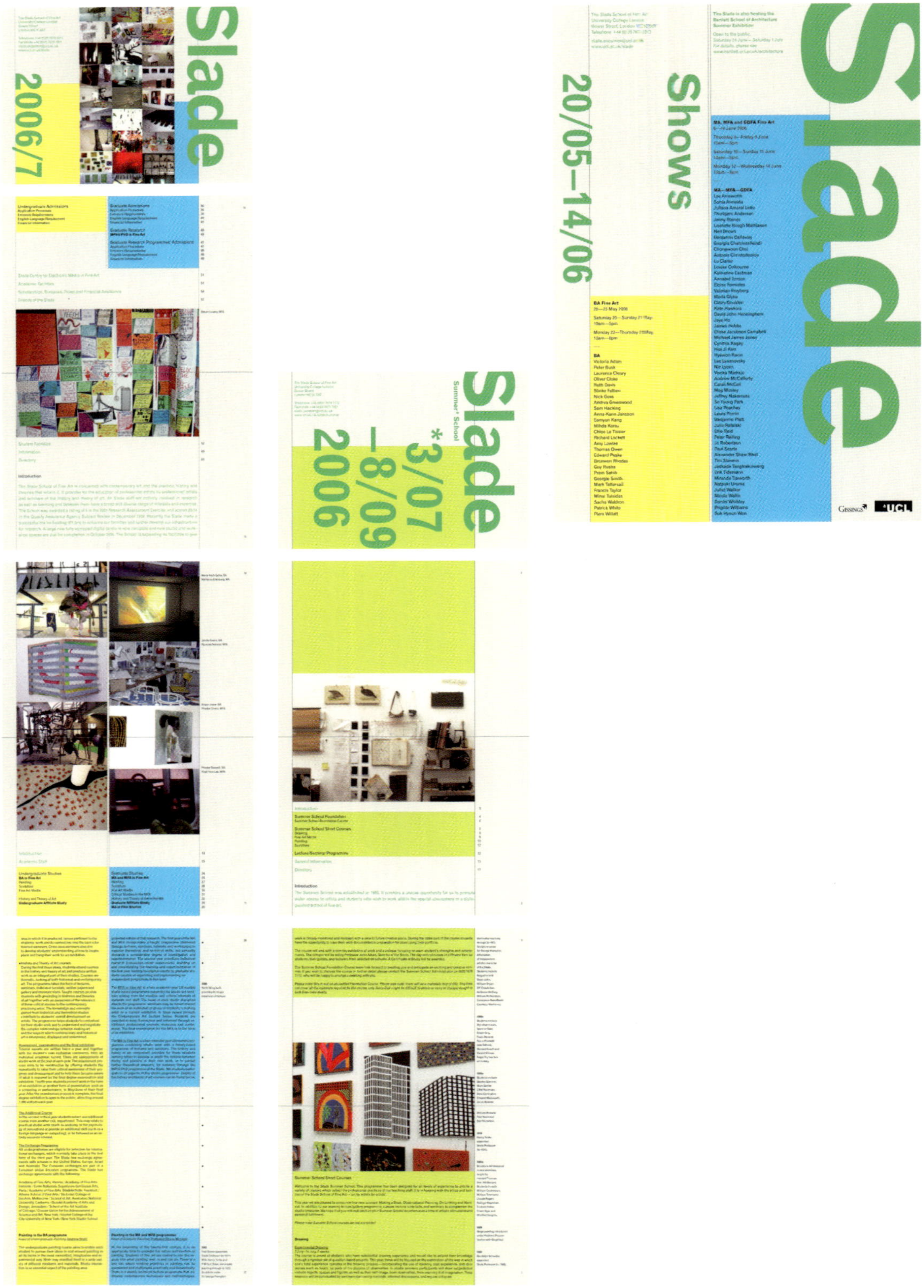

Auftraggeber Client
Hessen Agentur GmbH
Wiesbaden, Germany

Projekt Project
„Hessen inspiriert" Teaser-Package
Verpackung und Folder
Packaging and folder

Design
Frank Rocholl
Sven Kils
Jesse Kearney
Andreas Hambuch
Rocholl Selected Designs
Frankfurt am Main, Germany

Als erste Teasermaßnahme für die Einführung des Hessen Produktes 01 („Hessen inspiriert"-Duft) wurde an Partner der Hessischen Landesregierung das Set verschickt. Idee war hierbei der Transport eines jüngeren Hessenbildes in Kombination mit einem irritierenden Medium für eine Landesregierung (Parfüm), aber auch die Einbeziehung der Partner in den Entwicklungsprozess des Duftes.

As the first step in a teaser campaign for the "Hessen inspires" fragrance (Hessen Product 01), this set was mailed to partners of the hessische state government. The idea was to convey a youthful impression of Hessen in combination with an unusual medium for a state government (perfume), as well as to involve the partners in the development process for the fragrance.

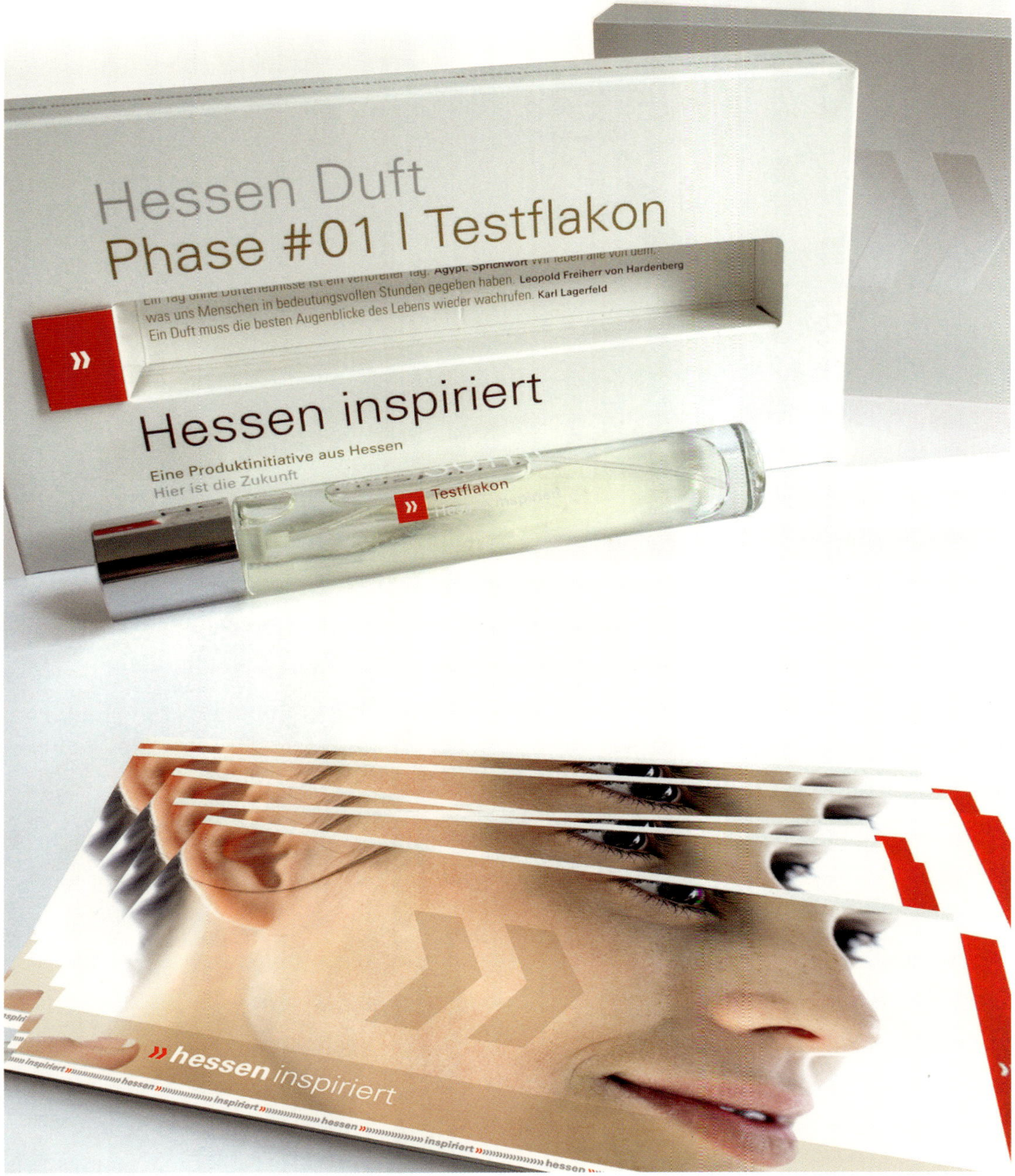

Auftraggeber Client
Kulturgemeinschaft Stuttgart
Stuttgart, Germany

Projekt Project
Kultur
Zeitung
Magazine

Design
Kurt Ranger
Peter Fajt
Jürgen Schöller
Ranger Design
Stuttgart, Germany

Zeitung „Kultur" – Seit 1991 gibt die Kulturgemeinschaft Stuttgart zehnmal pro Jahr „Blätter für Kenner und Neugierige" heraus. Oder kurz gesagt: die „Kultur". Eine Zeitung mit Kultcharakter. Renommierte Journalisten, Feuilletonisten und Querdenker geben hier Ansichten zum Besten und sagen, was aktuell (Kultur-)Sache ist. Von der ersten Ausgabe an begleitet Ranger Design die Geschicke dieser Zeitung, ohne die Stuttgarts Kulturlandschaft viele, viele Seiten ärmer wäre. Die „Kultur" wurde über all die Jahre hinweg bereits mehrfach für ihre Designqualität ausgezeichnet.

Magazine "Kultur" – Ten times a year since 1991 the Kulturgemeinschaft Stuttgart has published "Pages for Connoisseurs and the Curious". Or to put it more succinctly: the magazine "Kultur". A magazine with cult status. Renowned journalists, belletrists and unconventional thinkers give their views and tell us what is currently culturally in vogue. From the very first issue Ranger Design has accompanied the fortune of this magazine, without which the cultural scenery of Stuttgart would not only be lacking many pages on culture literally, but also many disparate viewpoints figuratively. Over the years the "Kultur" has received several awards for its quality of design.

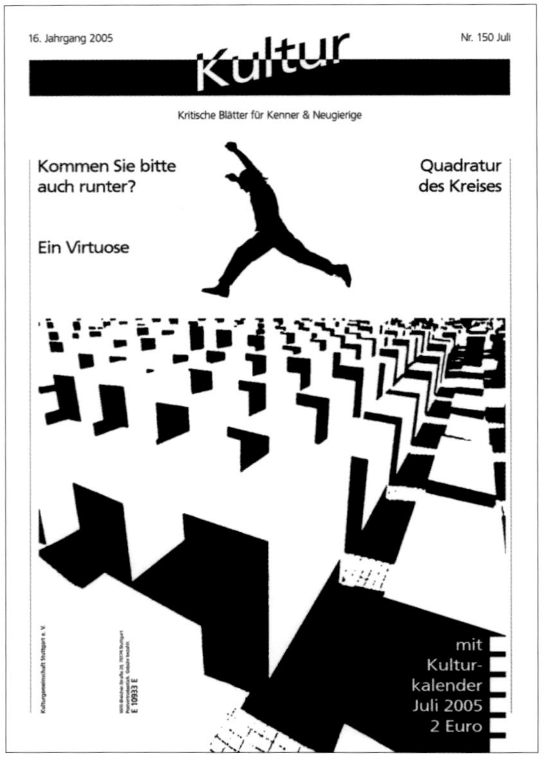

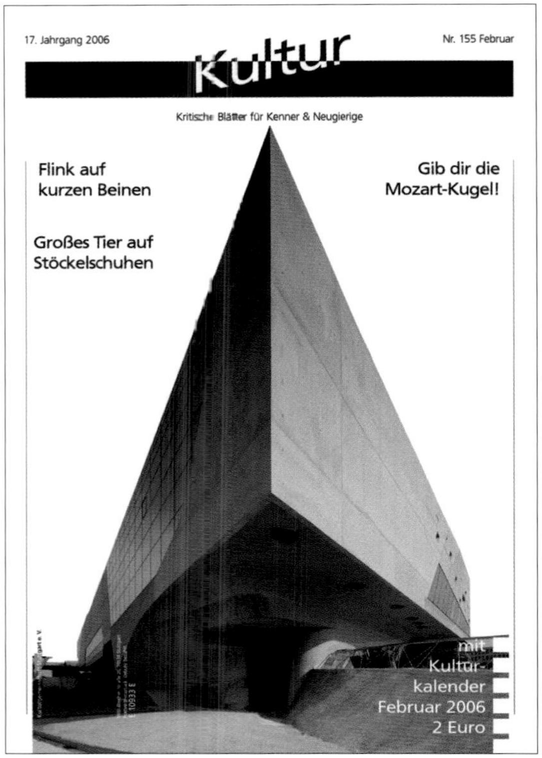

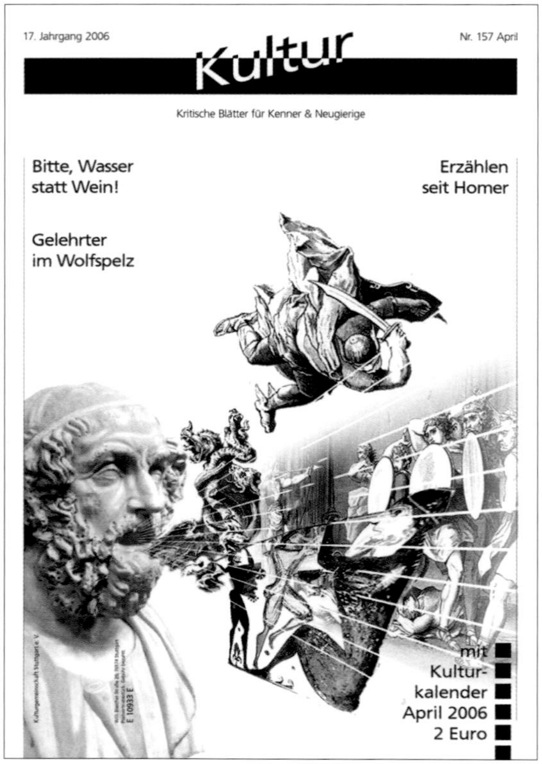

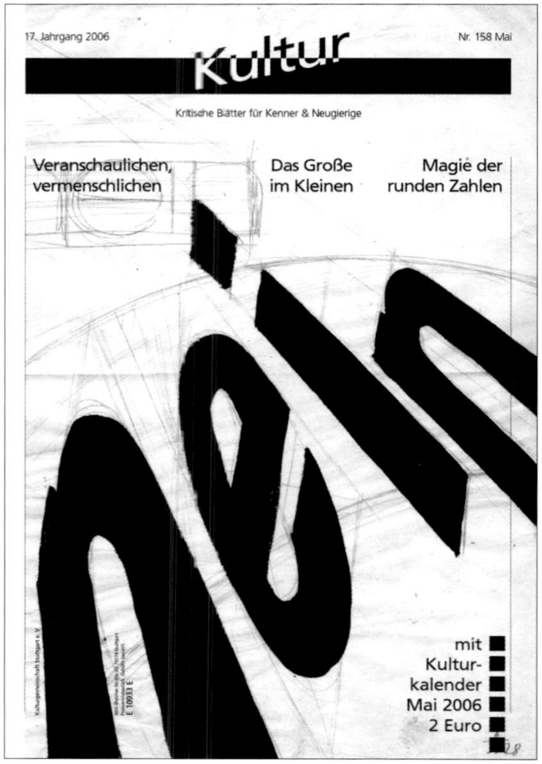

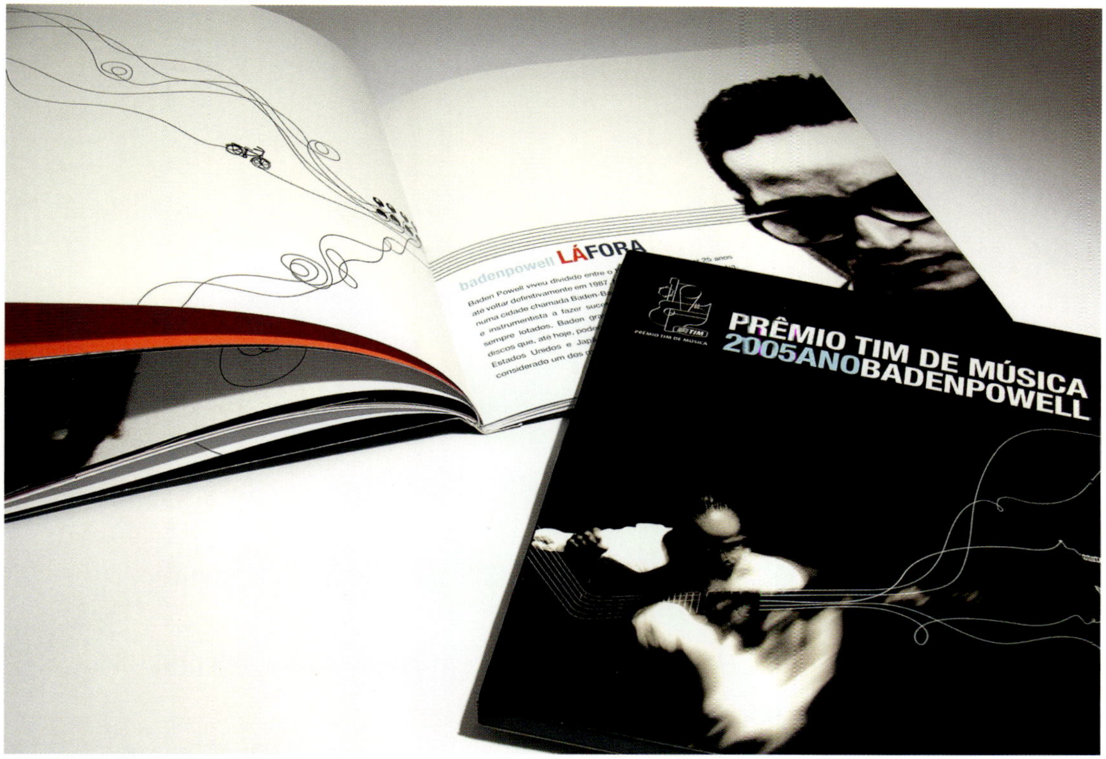

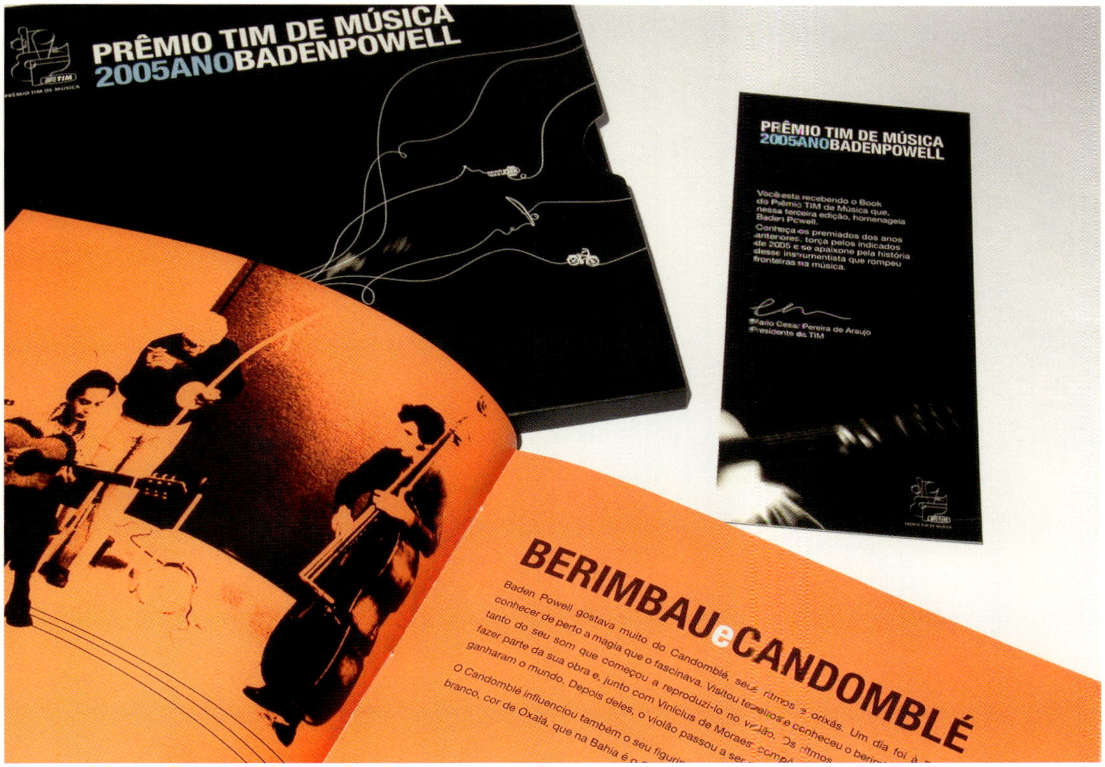

Auftraggeber Client
TIM Brasil
Rio de Janeiro, Brazil

Projekt Project
Prêmio TIM de Música 2005 – Baden Powell
Imagedesign für den TIM Musikpreis 2005
Image design for the 2005 TIM Music Awards

Design
Fred Gelli
Alessandro Ávila
Paulo Caetano
Tania Savaget
Paulo Henrique
Barbara Fornagic
Tátil Design
Rio de Janeiro, Brazil

Der TIM Musikpreis ist ein Projekt der Telefongesellschaft TIM in Brasilien. Tátil Design arbeitet seit 2003 an seiner grafischen Gestaltung. Unsere Herausforderung für das Jahr 2005 war es, das Ansehen des Preises noch weiter zu festigen. Dazu entwickelten wir eine grafische Sprache, die einem der berühmtesten Gitarristen der Welt gerecht wird: Baden Powell. Die geschwungenen Linien stehen für die Laufbahn des brasilianischen Musikers, während die geraden Linien an die Saiten seiner Gitarre erinnern. Die freien, durch Wind geformten Linien symbolisieren den kreativen Prozess des Komponisten, der glaubte, dass „Musik in der Luft liegt".

TIM Brasil initiated the TIM Music Awards. Tátil Design has been working on the Awards' graphic conception since 2003. Our challenge in 2005 was to further consolidate the prestige of the Awards by creating a graphic language that would measure up to one of the world's greatest guitar players: Baden Powell. Sinuous lines illustrate the career of the Brazilian musician, who broke frontiers through his art. Straight lines allude to the strings of Baden Powell's guitar. Free lines – as if being shaped by the wind – reveal the creative process of the composer who believed that "music is in the air". The lines are simple but sophisticated, just like Baden's work.

print media image

Auftraggeber Client
Messe Frankfurt Exhibition GmbH
Frankfurt am Main, Germany

Projekt Project
Outlook – Building Perspectives 04
Magazin
Magazine

Design
Fuenfwerken Design AG
Wiesbaden, Germany

„Outlook – Building Perspectives" ist eine Kommunikationskampagne der Messe Frankfurt für Architekten, Designer und Ingenieure – ihr zentrales Medium ein Magazin. Das zweisprachige Periodikum erscheint zweimal im Jahr mit einer Auflage von 14.000 Stück und wird international an Fachleute versendet. Neben einem aktuellen Schwerpunktthema, Kurzinformationen zu Seminaren und Ausstellungen werden in Porträts und Interviews Architekten, Designer und ihre Projekte vorgestellt. Ziel ist es, die Messe als Kompetenzzentrum für Architektur und Design zu kommunizieren.

"Outlook – Building Perspectives" is a communications campaign staged by the Messe Frankfurt exhibition company and aimed at architects, designers and engineers, featuring a magazine as its central medium. This bilingual periodical appears twice a year with a print run of 14,000 and is distributed to specialists worldwide. Apart from focusing on a current topic and providing short updates on seminars and exhibitions, the magazine features architects, designers and their projects in a series of profiles and interviews. The aim is to convey the exhibition company as a center of competence for architecture and design.

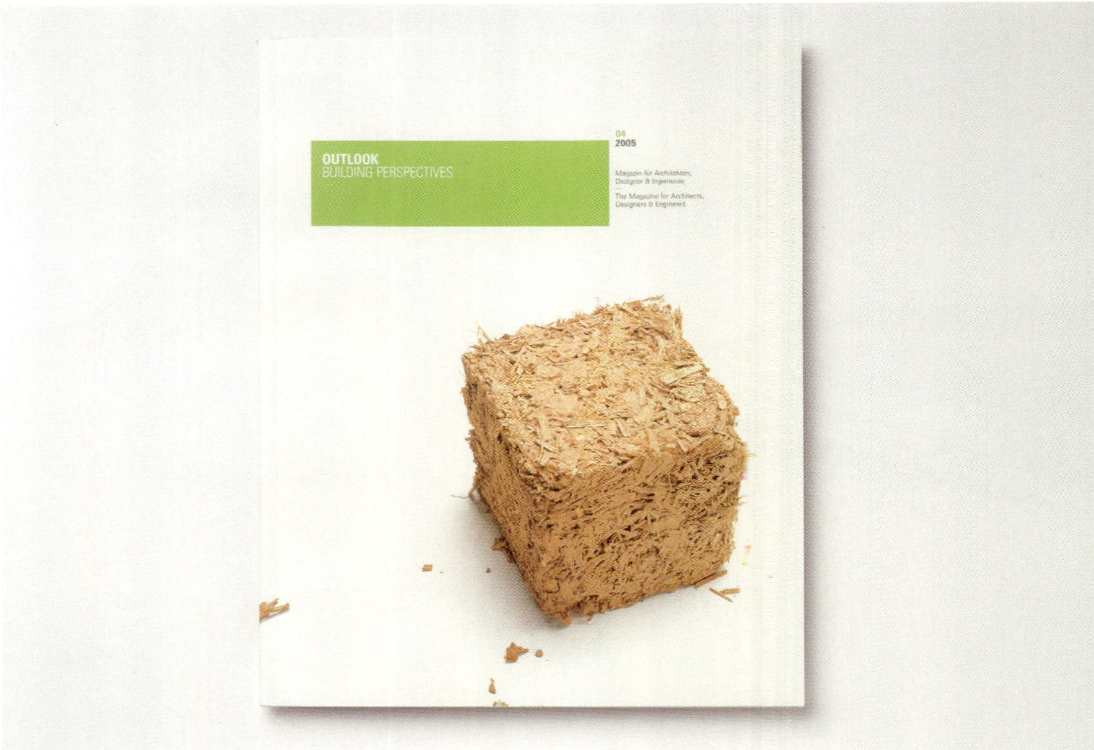

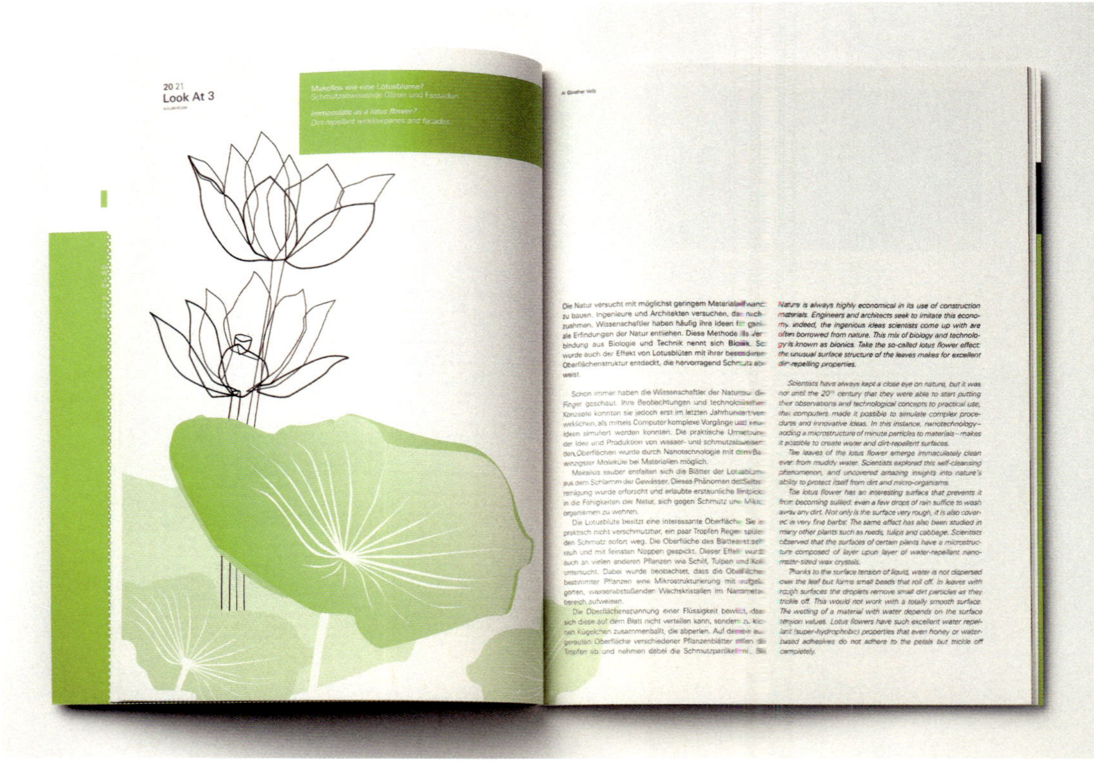

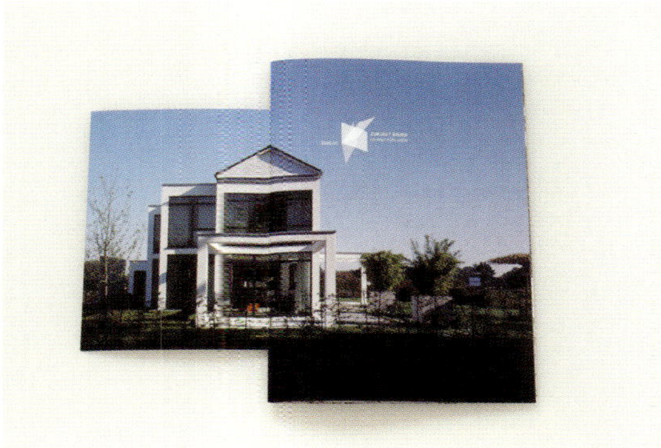

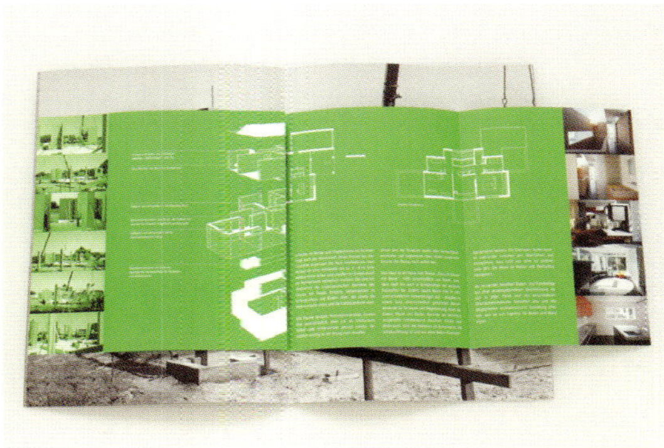

Auftraggeber Client
Betonmarketing Deutschland GmbH
Erkrath, Germany

Projekt Project
ZUKUNFT BAUEN 1 – Ein Haus fürs Leben
Imagebroschüre
Image brochure

Design
Ulrich Nether
Gloria Cutura
netherblu gestaltcrossover
Düsseldorf, Germany

Mit neuer Technik kann man mit Beton flexibel, innovativ, schnell und kostengünstig bauen. Für die Imagekampagne ZUKUNFT BAUEN hat Betonmarketing ein erstes Projekt entwickelt, das zunächst der Bauherren-Zielgruppe dargestellt wurde. Für die ästhetisch verwöhnte Zielgruppe Architekten sollte eine eigene Kommunikation entstehen. Dafür wurde ein Medienkonzept entwickelt, die Broschüre ist so das erste Produkt von ZUKUNFT BAUEN mit eigener Identität innerhalb des Markenbildes BETON: Innovationsprojekt Bauen mit Beton, veränderbar, dynamisch, modular, intelligent und in hoher Qualität und Ästhetik. Das setzen Logo und Broschüre greifbar um.

New technologies for concrete construction make it possible to build in a flexible, innovative, fast and cost-efficient way. With the campaign "FUTURE OF CONSTRUCTION" Concrete Marketing developed a project, which was initially geared towards building owners. For the aesthetically fastidious target group of architects an extra communication campaign should be created. This brochure is the first in this campaign with its own identity within the trademark "CONCRETE": the innovation project building with concrete is modifiable, dynamic, modular and intelligent – with high quality and aesthetic. Logo and brochure tangibly realize these claims.

print media image

Auftraggeber Client
Deekeling Arndt Advisors in Communications GmbH
Düsseldorf, Germany

Projekt Project
Deekeling Arndt Advisors
Corporate Design

Design
Knut Maierhofer
Sarah Graf
Patrick Märki
Bernd Müller
Sandra Ehm
Melanie Sauer
KMS Team GmbH
München, Germany

Deekeling Arndt Advisors ist eines der führenden Büros in der strategischen Kommunikationsberatung. Den Ausgangspunkt unserer Gestaltung bildete die Frage, wie man Kommunikation visualisiert und symbolisiert. Aus dieser Fragestellung heraus verwendeten wir als zentrale grafische Elemente Satzzeichen, die in der geschriebenen Sprache zur Steuerung des Sprachverständnisses dienen. Durch den Zusatz eines Doppelpunkts entsteht aus dem Namen des Unternehmens das Schriftzeichen; die einzelnen Kommunikationsmittel wie Pressemappe, Unternehmensbroschüre oder Publikationen tragen stilisierte Anführungszeichen.

Deekeling Arndt Advisors is one of the leading firms for strategic communications consulting.
Our design took its starting point from the question of how to visualize and symbolize communication. With this in mind, we used punctuation marks, which in written language help to guide linguistic understanding, as our central graphical elements. The addition of a colon to the name of the firm produces the logotype, while the individual means of communication – such as press kit, the company brochure or publications – feature stylized quotation marks.

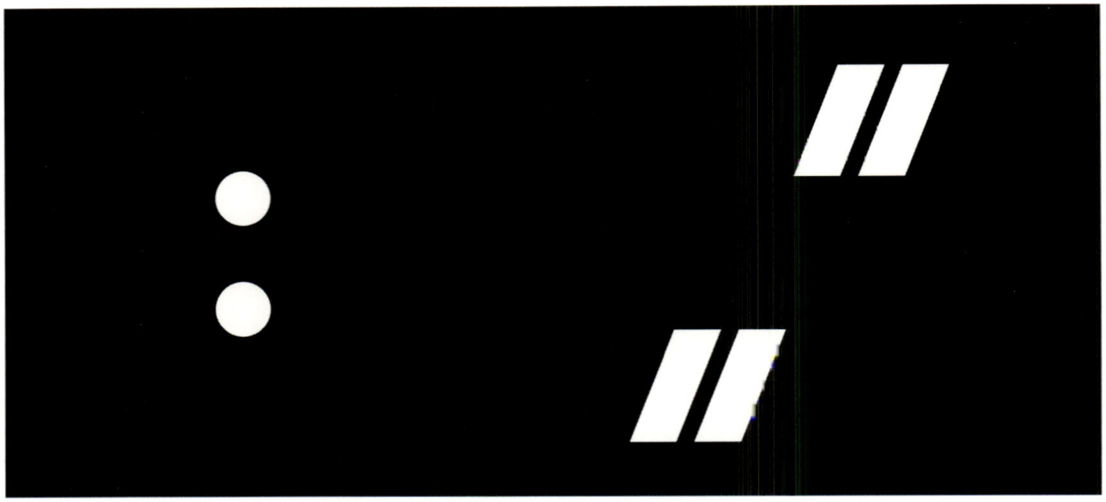

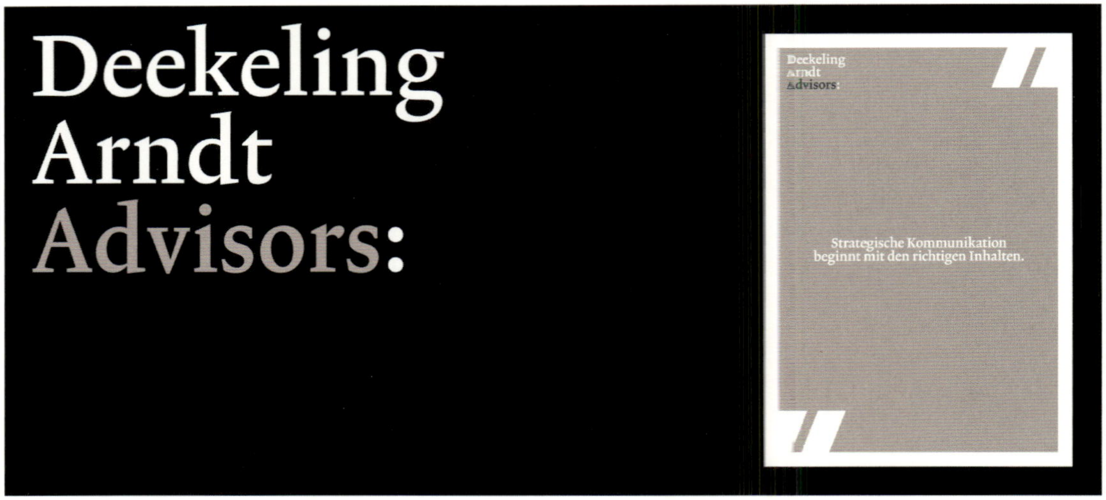

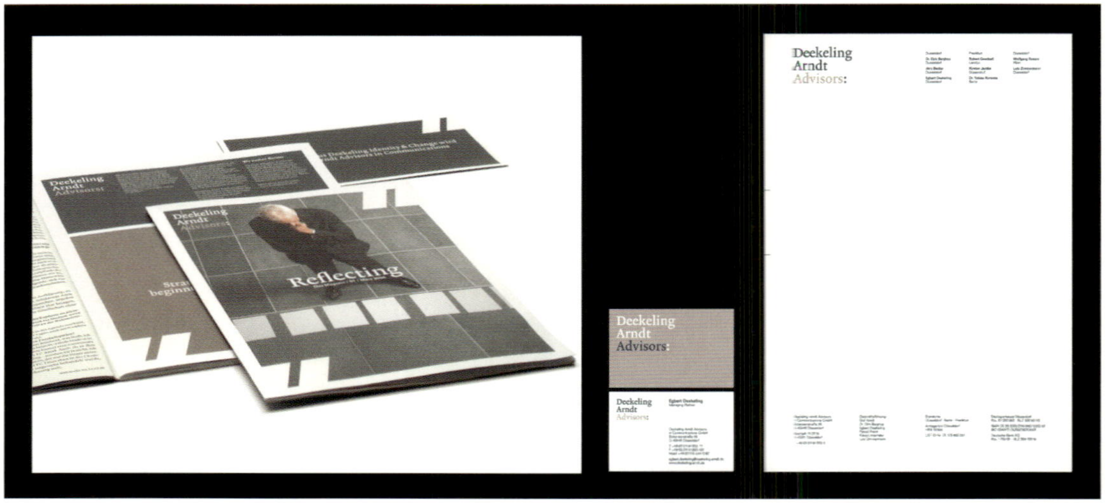

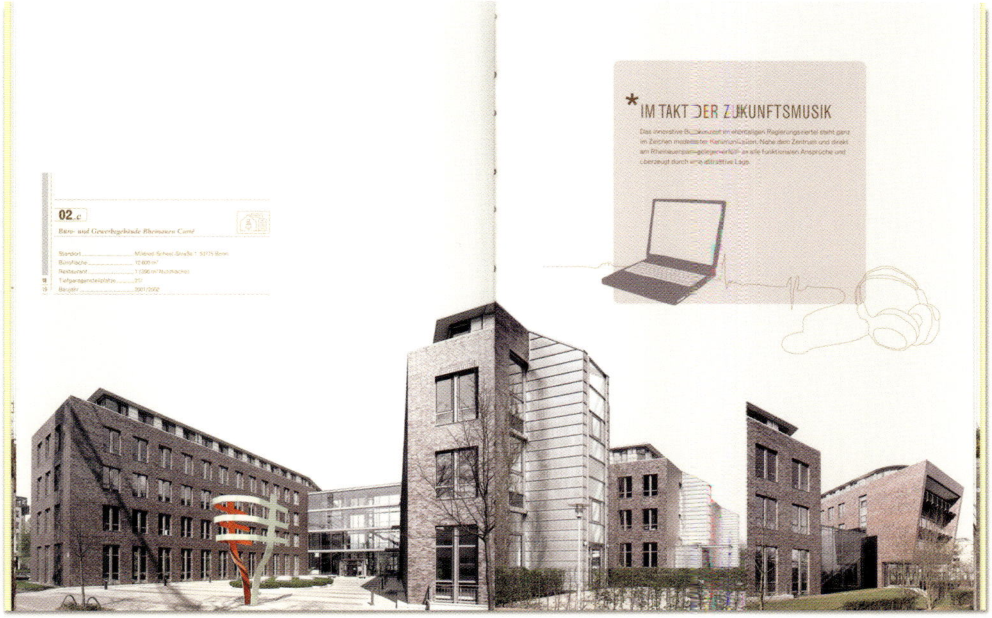

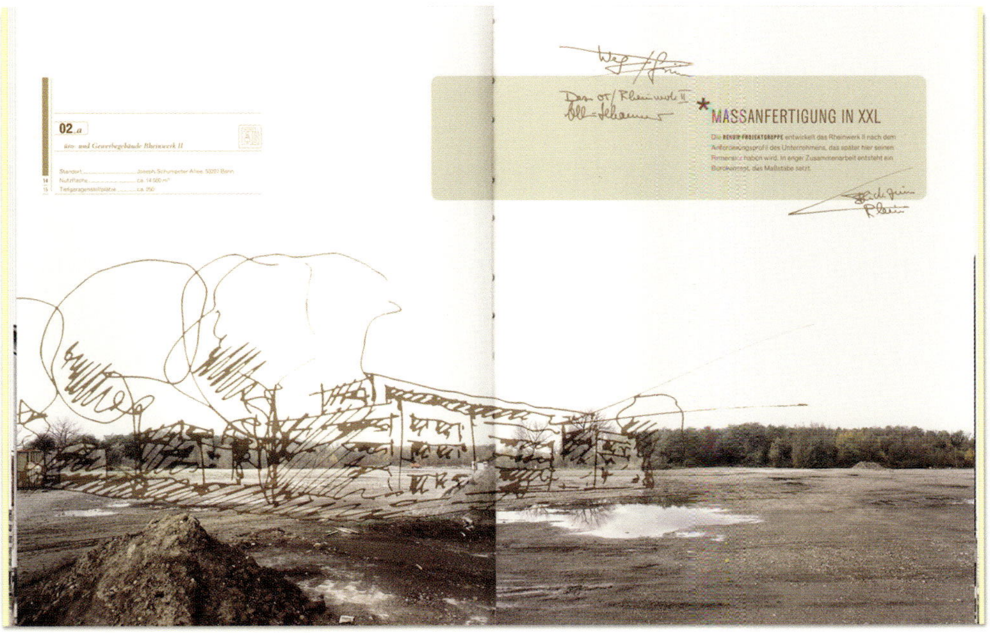

Auftraggeber Client
Renum Projektgruppe
Bonn, Germany

Projekt Project
Renum
Imagebroschüre
Image brochure

Design
Peter Specht
Irene Mohr
Petra Steuns
Oliver Dreyer (Text)
BRANDIT
Köln, Germany

Für seinen Kunden Renum entwickelte und realisierte BRANDIT Marketing & Kommunikation eine 32-seitige Imagebroschüre. Das Unternehmen hat sich auf die Entwicklung, Realisierung und Vermarktung von hochwertigen Immobilienlösungen spezialisiert. Um die Kompetenz von Renum zu unterstreichen, wurde der Fokus konsequent auf die Projekte der Firma gelegt. Die Gedanken und Wünsche der Menschen, die in diesen Immobilien leben und arbeiten, waren dabei der konzeptionelle Aufhänger. Illustrationen, die auf die individuellen Charakteristika der jeweiligen Immobilie eingehen, flankieren die Bildkomposition und laden die kreative Inszenierung emotional auf.

BRANDIT Marketing & Kommunikation, Cologne, created a 32-page image brochure for their client Renum. The Bonn-based company specialises in the development, construction and marketing of quality real estate. The brochure underlines Renum's expertise by focusing on the company's projects. The thoughts and wishes of people living and working in Renum buildings form the core concept. Illustrations emphasize the individual character of every building. They sit alongside the picture layout and add to the overall emotional appeal.

Auftraggeber Client
3Z Informationsgrafik und Architektur
Domat/Ems, Switzerland

Projekt Project
Grischata, Graubünden zwischen Wahn und Wirklichkeit
Buch
Book

Design
Christina Luzzi
3Z Informationsgrafik und Architektur
Domat/Ems, Switzerland

Stellen wir uns vor, die Masse Graubündens wird zu einem Berg aufgetürmt. Oder wir planieren alles. Der größte Schweizer Kanton eine Hochebene. Wie sieht der Bergkanton aus, wenn man aus ihm ein tiefes Tal meißelt und alles Wasser in dessen Sohle sammelt? Drei imaginäre, grafische Modelle, drei Möglichkeiten dem realen Graubünden zu begegnen, zu hinterfragen, darüber zu schmunzeln.
Christina Luzzi hat mit diesem Thema an der Hochschule für Gestaltung in Schwäbisch Gmünd diplomiert. Nun liegt eine erweiterte und überarbeitete Arbeit vor mit Grafiken sowie straffen, oft ironischen Texten, Anekdoten und Geschichten rund um das Thema Zahlen.

Imagine that the entirety of the Swiss canton of Graubünden was bulldozed together to form a giant mountain. Or that it was leveled off to turn Switzerland's biggest canton into a plateau. What would this mountainous canton look like if a deep valley were chiseled into it and all the water collected at the bottom? Three imaginary graphic models present three different options for encountering the real Graubünden, questioning it and smiling about it.
This was the subject of Christina Luzzi's thesis at the University of Design in Schwäbisch Gmünd. Now an extended, revised version of it has been produced with graphics as well as tight, often ironical texts, anecdotes and stories involving the topic of numbers.

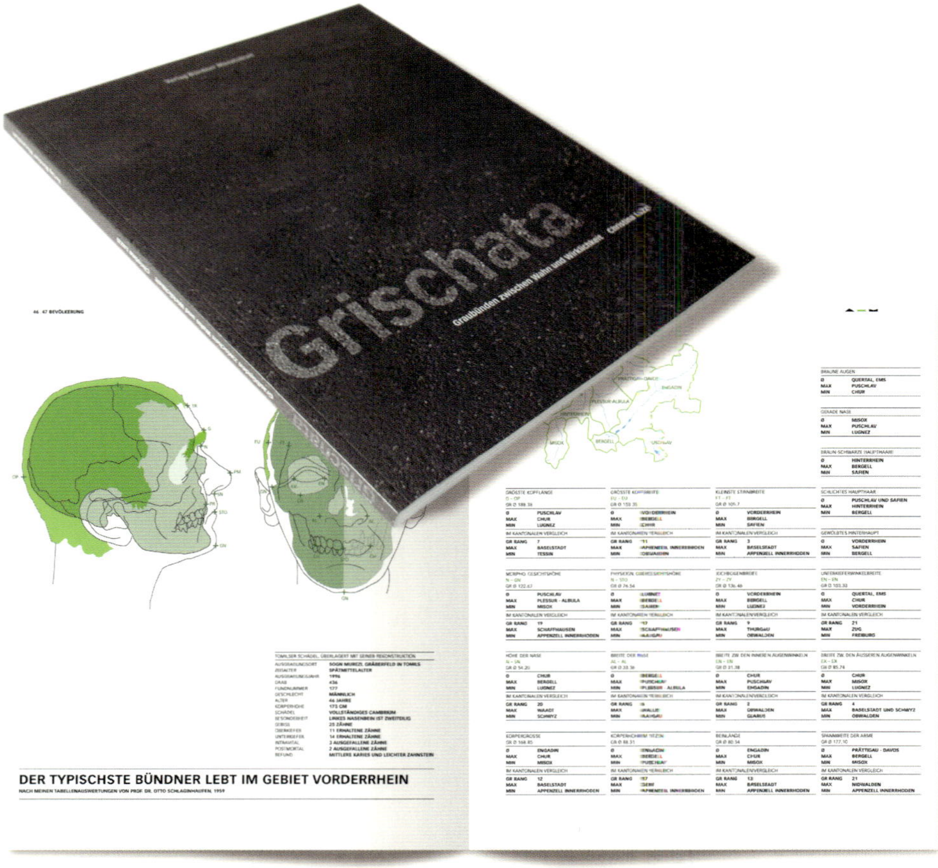

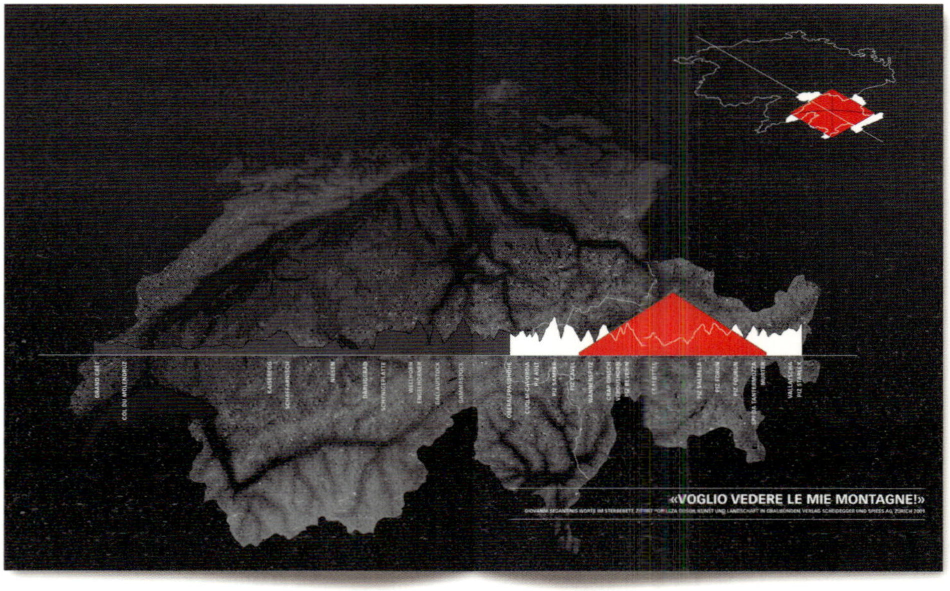

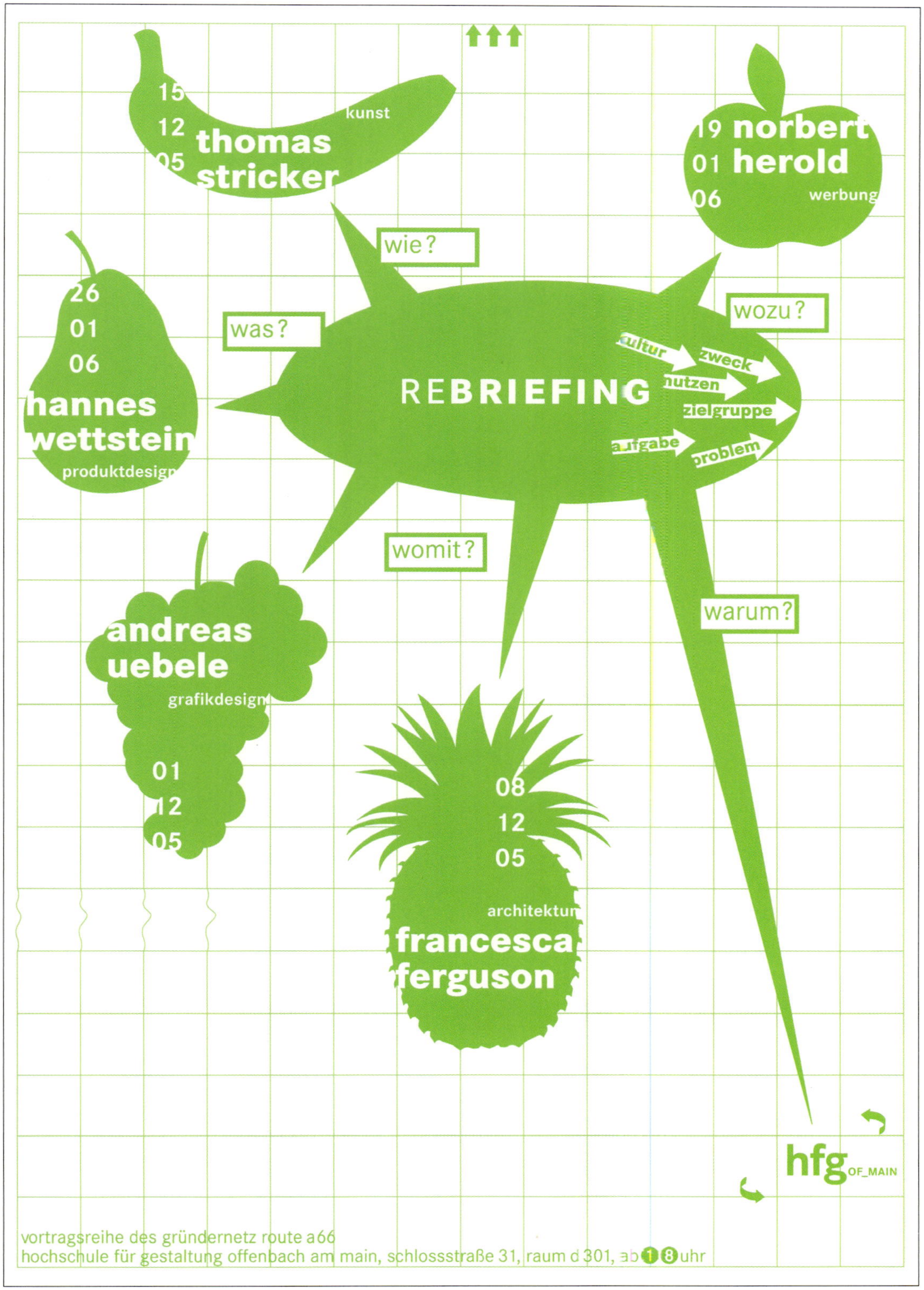

Auftraggeber Client
HfG Offenbach
Offenbach am Main, Germany

Projekt Project
Re-Briefing
Plakat
Poster

Design
Klaus Hesse
Hesse Design GmbH
Erkrath, Germany

Anliegen dieser Vortragsreihe ist es, aus den Bereichen Kunst, Architektur, Produktdesign, Grafikdesign und Werbung den Prozess von der ursprünglichen Aufgabenstellung bis hin zum realisierten Ergebnis mit allen Missverständnissen und Widersprüchen darzustellen.

This series of lectures seeks to portray the development process that takes place in the worlds of art, architecture, product design, graphic design and advertising between the initial job brief and the completion of the final result, complete with all the accompanying misunderstandings and contradictions.

Auftraggeber Client
design hoch drei GmbH & Co.KG
Stuttgart, Germany

Projekt Project
Maul-Taschen
Papiertragetaschen
Paper bags

Design
Rosa Pupillo
Wolfram Schäffer
Alexander Stertzig
Susanne Wacker
Markus Wichmann
Angelika Stehle (Text)
design hoch drei GmbH & Co.KG
Stuttgart, Germany

Sagen, was Sache ist. – Wie schafft man es, dass die Bürger sich einmischen in die Politik und Tacheles reden? Man braucht ein neues Medium zur Kommunikation. Die Maul-Tasche ist die Antwort auf Politikverdrossenheit. Insgesamt sieben Tüten mit Sprüchen zum Thema Politik sind seit Mai 2005 in Stuttgart im Umlauf. Neben dem Aktionslogo „Sag, was Sache ist" wurden die gedruckten Statements plakativ-illustrativ gestaltet. So verschwimmen bei „Wenn immer nur Politiker reden, verstehen wir bald gar nichts mehr" die Wörter zu einer wahren Buchstabensuppe. Fett gold glänzt das Wort „Speck" auf der Tüte mit der Aufforderung „Ran an den Diätenspeck".

Make your point! – How do you manage to get citizens involved into politics again and talk turkey? One needs a new communication medium. The "Maul-Tasche" is the answer to the disenchantment with politics. Altogether seven paper bags with bold slogans on the issue politics have been circulating in Stuttgart since May 2005. In addition to the logo "Make your point!", the printed statements were designed with striking illustrative typography.

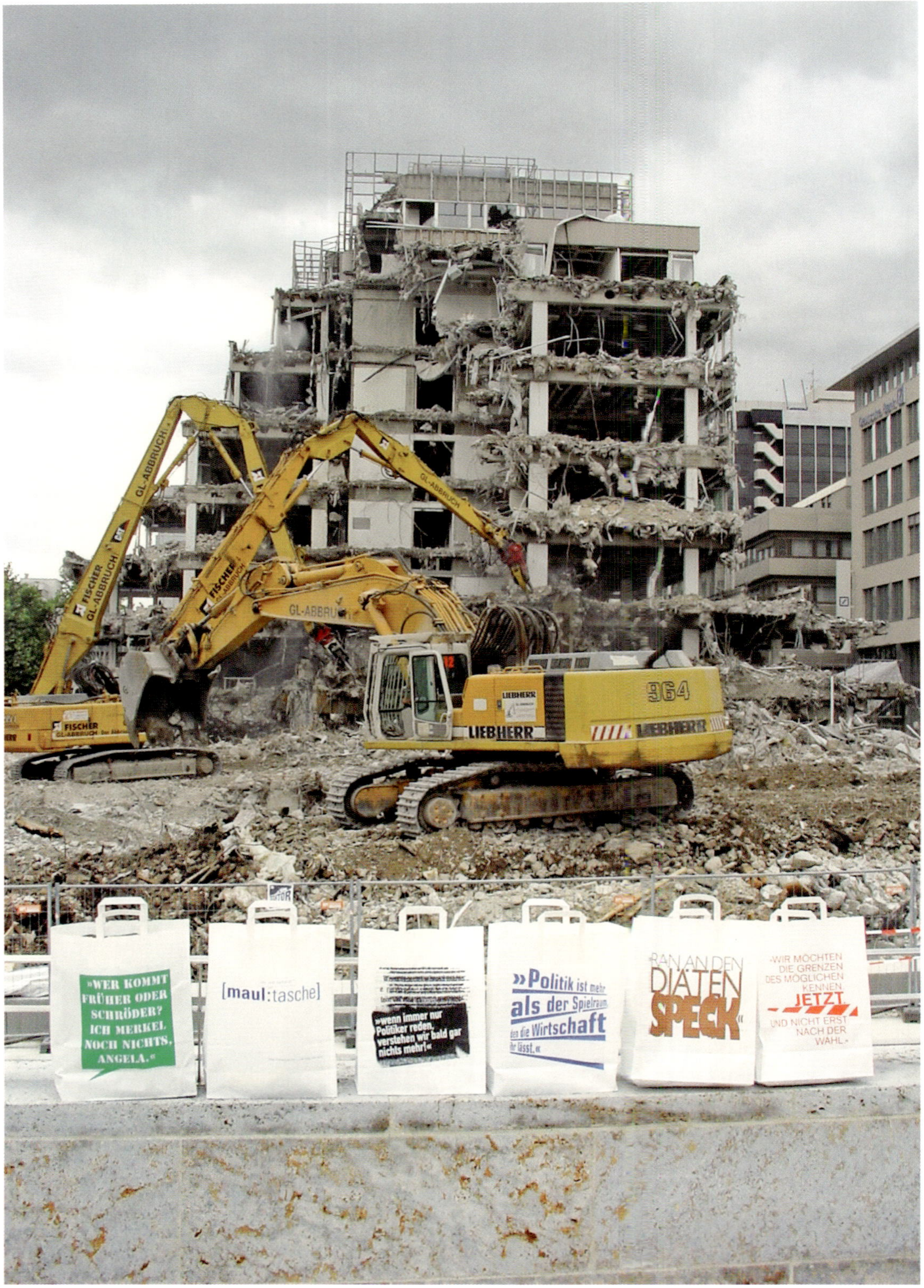

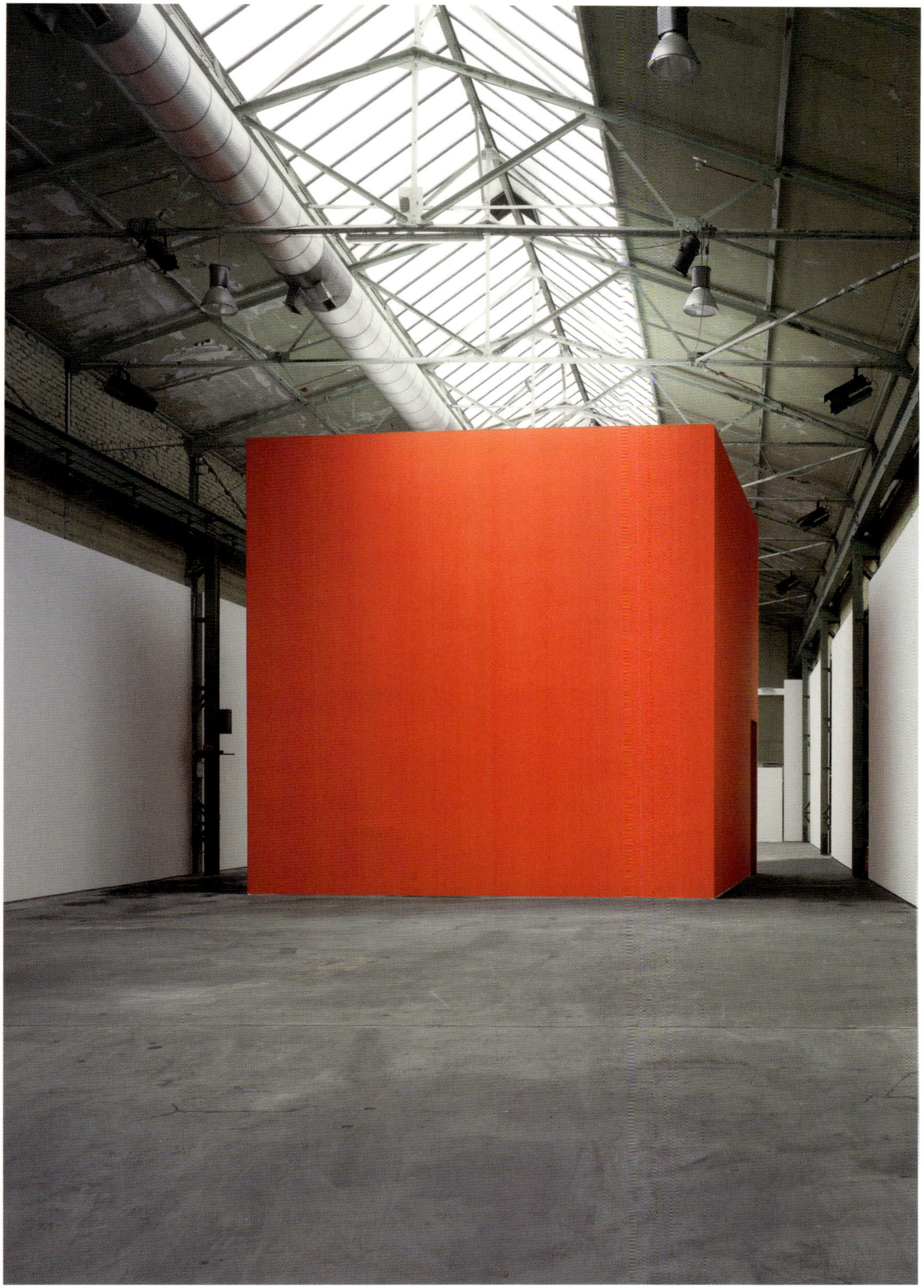

Auftraggeber Client
Prof. Otto Piene
Groton, MA, USA
Halle 6 – Galerie Christine Hölz
Düsseldorf, Germany

Projekt Project
Otto Piene Solo
Ausstellungskatalog
Exhibition catalogue

Design
Margarethe Saxler
Matthias Herzogenrath
herzogenrathsaxler communikationsdesign
Düsseldorf, Germany

„Otto Piene Solo“ ist Einladung zum Gebrauch, zum Sehen und Lesen. Entsprechend äußern sich Haptik und Format.
In einer kaleidoskopisch vielfältigen, farbigen Bildfolge wird der Entstehungsprozess einer neuen Inszenierung des Künstlers, beginnend bei der ersten Skizze, endend bei der fertigen Ausstellungssituation, dargestellt; der Leser ist gegenwärtiger Teilnehmer. Ein monochromer Bildteil markiert den Beginn des „Alten“, den textlichen Streifzug durch 50 Jahre künstlerische Arbeit.

“Otto Piene Solo” is an invitation to use, to see and to look, expressed in surface feel and format.
Immersed in kaleidoscopic diversity, the reader becomes part of the creative process of the staging, from the first sketch up to the complete exhibition situation. The “monochrome” marks the beginning of the “old”. Images from 50 years of artistic work, fused to formal unity by duplex transformation, accompany the art historical text.

Auftraggeber Client
Institut für Auslandsbeziehungen e.V. – IFA
Stuttgart, Germany

Projekt Project
anders als immer
Ausstellungskatalog
Exhibition catalogue

Design
Sibylle Schneider
Heine/Lenz/Zizka
Frankfurt am Main, Germany

„anders als immer" – Thema der weltweit tourenden Ausstellung, die Gewohnheiten in Frage stellt. Die eckige Brezel weist den Weg. Ausstellung anschauen, Bildband blättern, Textband lesen, Katalog individuell gestalten: Abbildungen mit Aufklebern markieren.

"Somewhat different" is the title of a unique design exhibition that questions everyday habitudes on its tour around the world. "Somewhat different" – a rectangular pretzel signposts the way. View the exhibition, flick through the catalogue, read the texts, design a unique catalogue: mark selected images with text excerpts.

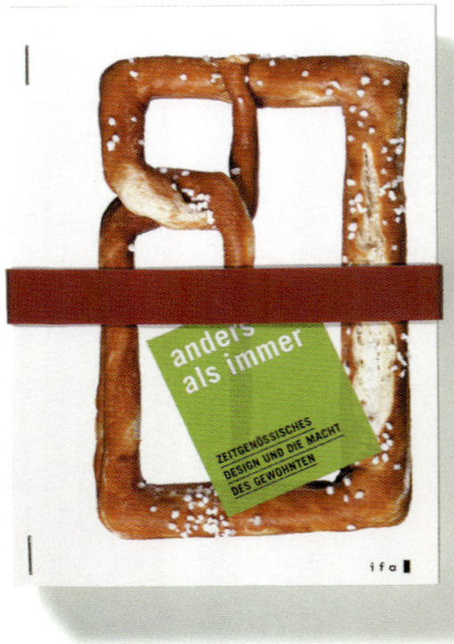

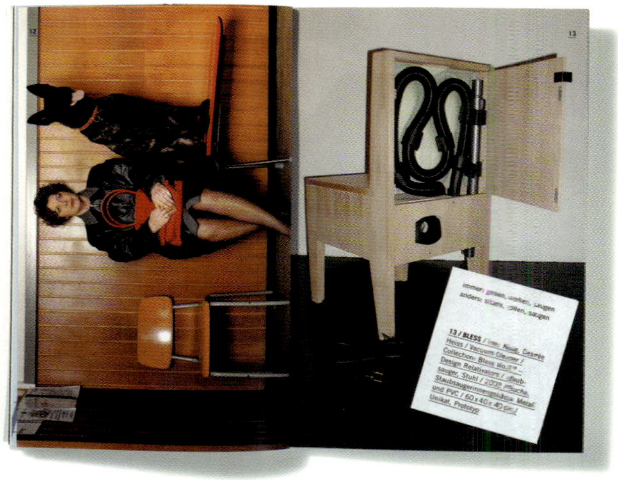

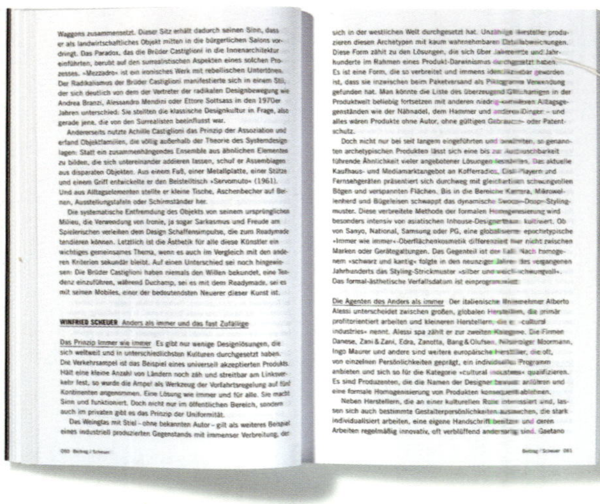

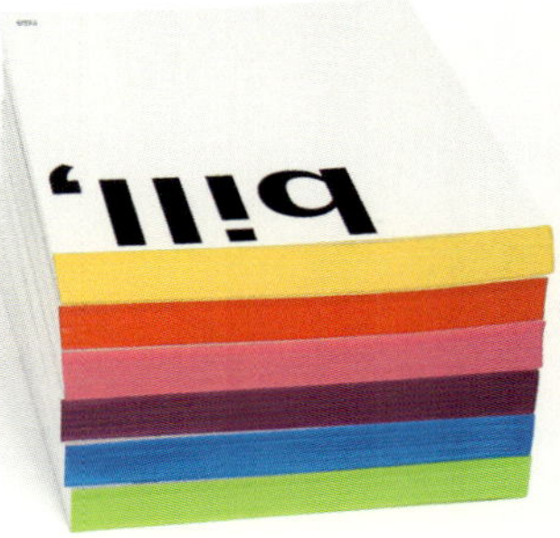

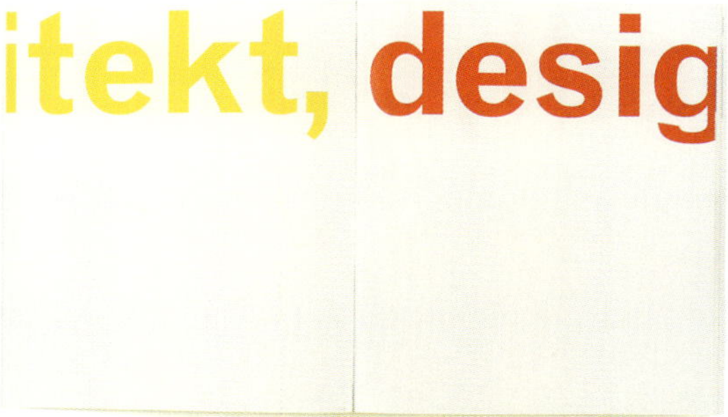

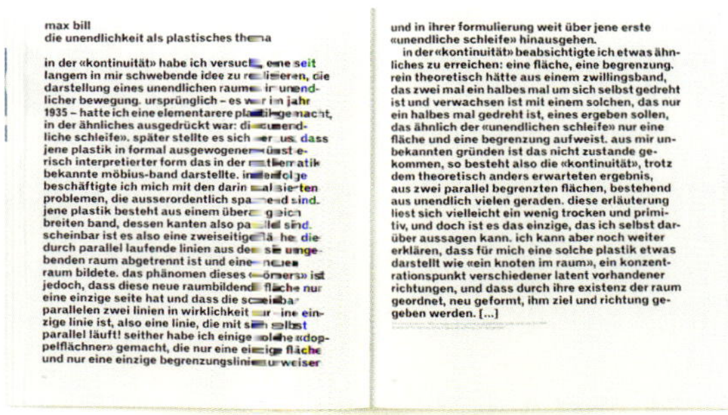

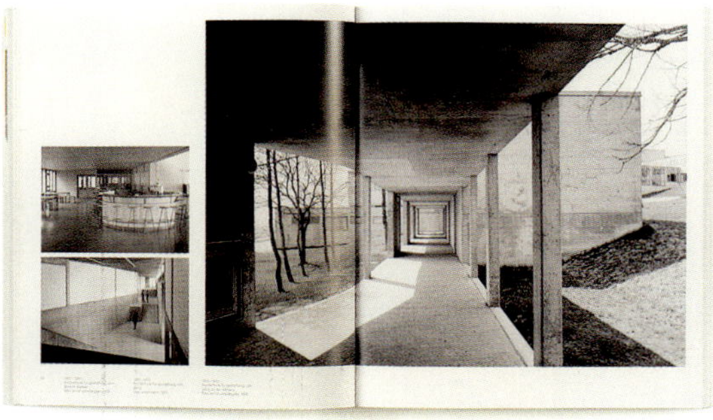

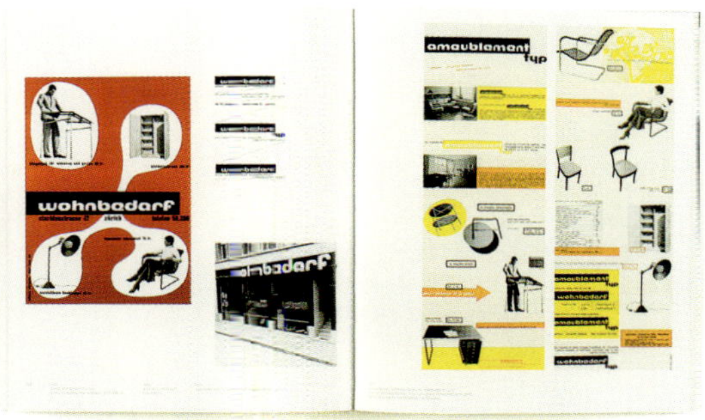

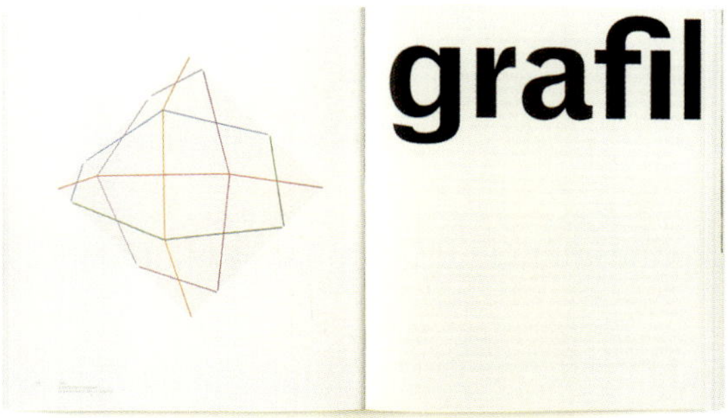

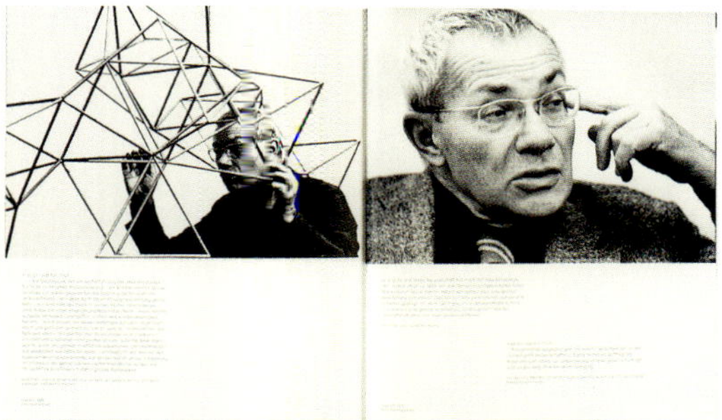

Auftraggeber Client
Kunstmuseum Stuttgart
Stuttgart, Germany

Projekt Project
Max Bill, Maler, Bildhauer, Architekt, Designer
Katalog
Catalogue

Design
Sascha Lobe
Ina Bauer
L2M3 Kommunikationsdesign GmbH
Stuttgart, Germany

70 Jahre nach seinen typografischen Hauptwerken einen Billkatalog zu gestalten, bedeutet für uns, Bills Gestaltungsauffassung nicht zu imitieren, sondern sie in eine zeitgemäße Form zu übersetzen.
Für die Unterscheidung von Text- und Bildseiten haben wir unterschiedliche Papiere gewählt und innerhalb des Layouts eine klare, stark strukturierte Typografie entwickelt. Die „Breite" von Bills Schaffen als Maler, Designer, Bildhauer und Architekt findet ihren Ausdruck in großen Schriftzeilen, die im Buchintro über das Seitenformat hinauslaufen und in den einzelnen Kapiteln als Ein- und Ausstieg zitiert werden.

For us, 70 years after his typographical main works, designing a Bill catalog doesn't mean imitating Bill's design concept, but translating it into a contemporary form.
For the differentiation between text pages and picture pages we chose different types of paper, and within the layout we developed a clear, strongly structured typography. The "breadth" of Bill's work as painter, designer, sculptor and architect is expressed in large-sized lines of writing that, in the book introduction, run off the page format and are cited in the individual chapters as introductions or closings.

Auftraggeber Client
HfG Offenbach
Offenbach am Main, Germany

Projekt Project
Lost Gender
Plakat
Poster

Design
Klaus Hesse
Benjamin Schulte
Hesse Design GmbH
Erkrath, Germany

Die Welt nach den Geschlechtern. Eine Vortragsreihe von Prof. Rotraut Pape. Es wurden Künstler und Künstlerinnen eingeladen, die mittels einer Mischung aus digitalen und analogen Medien die Vision einer Welt entwerfen, in der die klassischen Geschlechter neu definiert werden.

The world according to both sexes: a series of seminars by Prof. Rotraut Pape. Various artists of both sexes are invited to develop, using a mixture of digital and analog media, a vision of the world that redefines the classical outlook of man and woman.

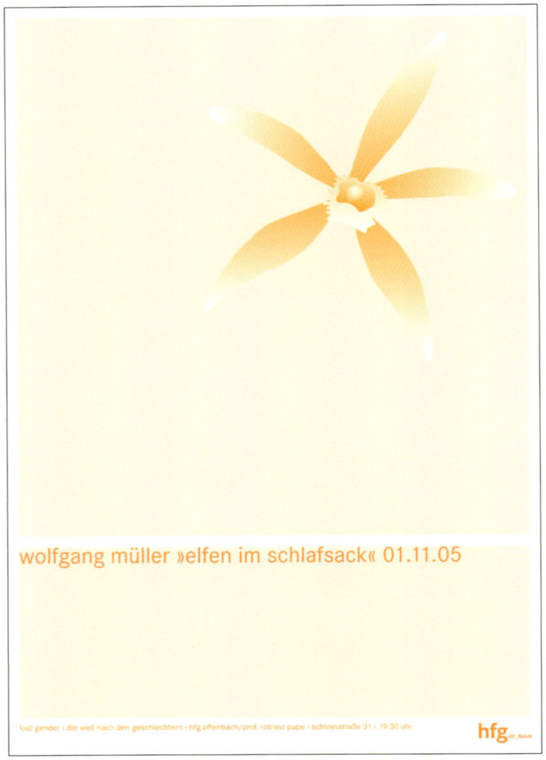

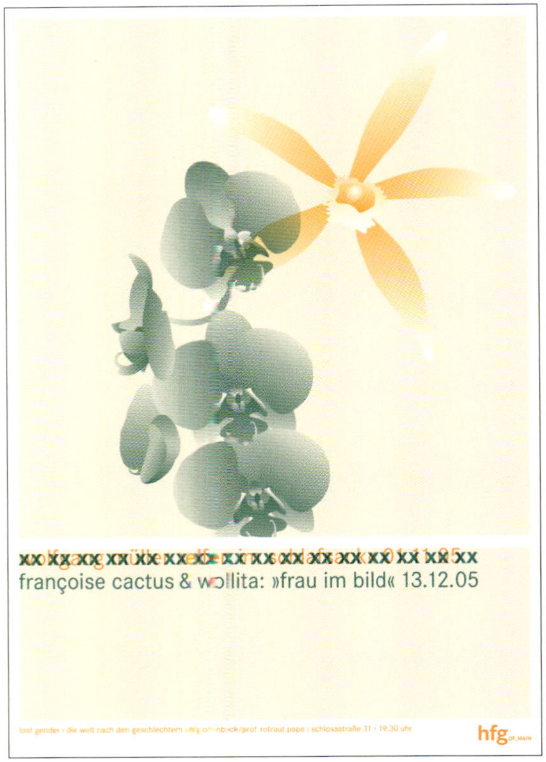

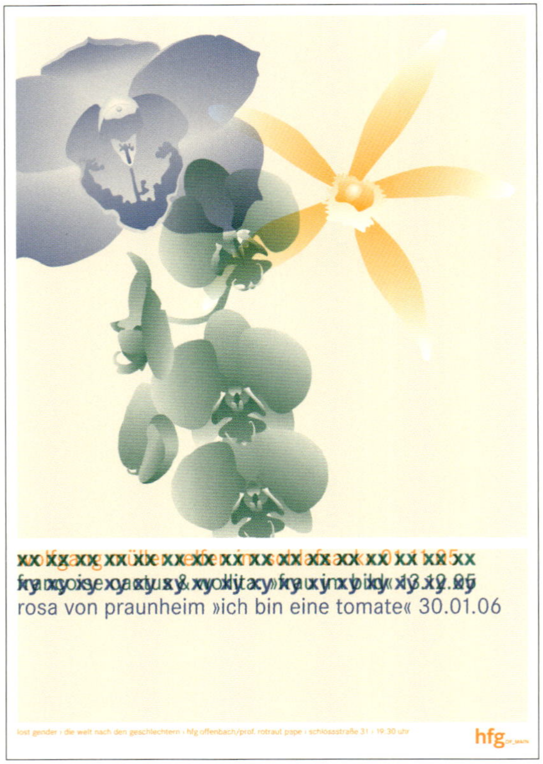

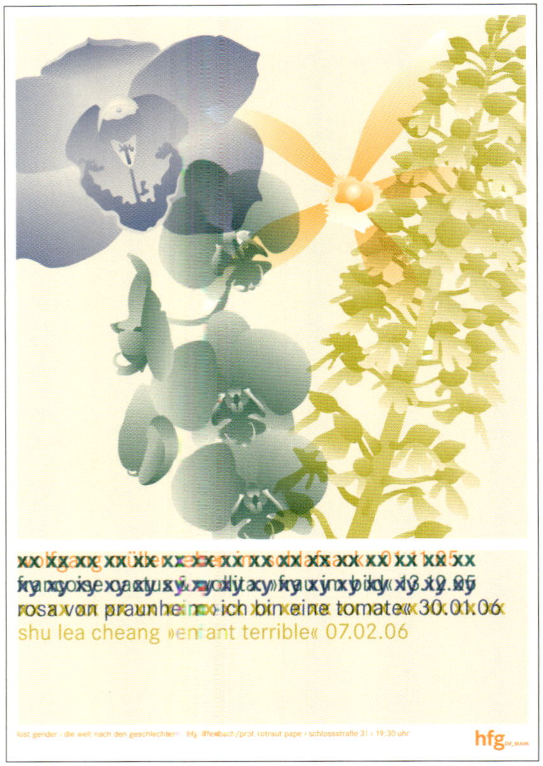

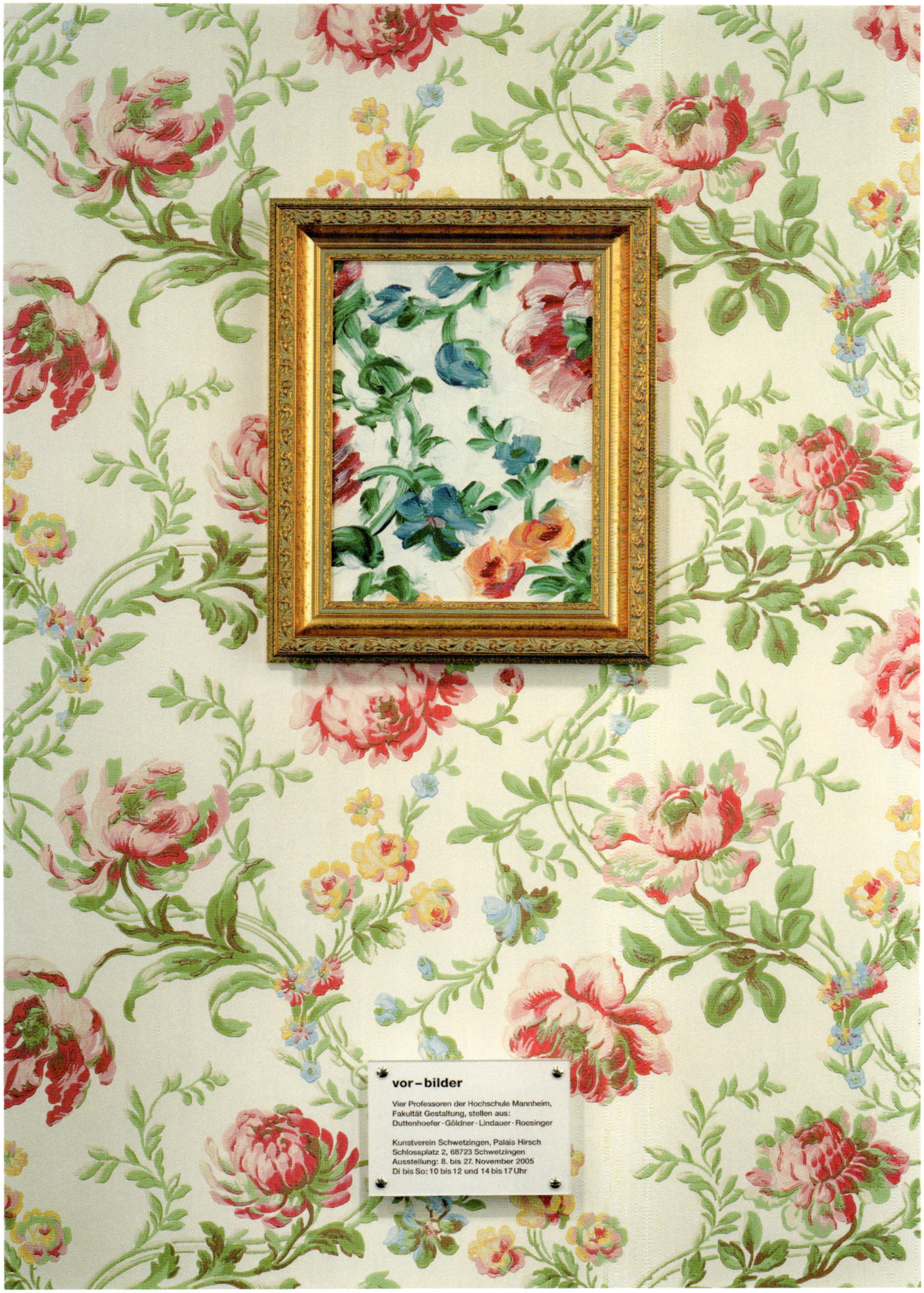

Auftraggeber Client
Kunstverein Schwetzingen
Palais Hirsch
Schwetzingen, Germany

Projekt Project
vor – bilder
Plakat
Poster

Design
Armin Lindauer
Mannheim, Germany

Plakat zur Ausstellung von vier Professoren der Hochschule Mannheim.

Poster for an exhibition by four professors of the Mannheim University of Applied Sciences.

Auftraggeber Client
Laboratorium
Zagreb, Croatia

Projekt Project
First-person singular
Ausstellungskatalog
Exhibition catalogue

Design
Ivana Vucic
Orsat Frankovic
Sasa Stubicar
Laboratorium
Zagreb, Croatia

„In erster Person (Ichform)" ist ein Ausstellungskatalog über die zeitgenössische kroatische Kunst von 1960 bis heute, der die Rede des Künstlers in der Ichform behandelt. Inspiriert wurde der Katalog durch die berühmten Moleskine-Tagebücher, die europäische Künstler und Intellektuelle nutzten. Der Buchumschlag ist aus reflektierendem, spiegelähnlichem Papier gemacht und wird dadurch ein interaktives Mittel, das jeden Betrachter dazu verleitet, sein Selbstporträt zu interpretieren.

"First-person singular" is a catalogue for an exhibition about the artist's first-person speech shown by examples of Croatian contemporary art from the 1960s to the present. The project was inspired by "Moleskine", the legendary notebook of European artists and intellectuals. For the cover we used reflective, mirror-like paper so that the reader of the catalogue is encouraged to interpret his or her self-portrait.

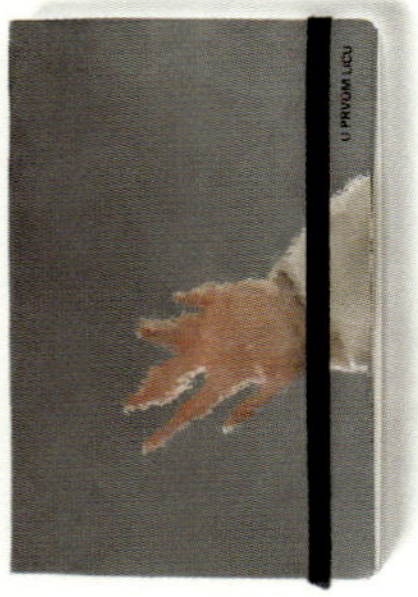

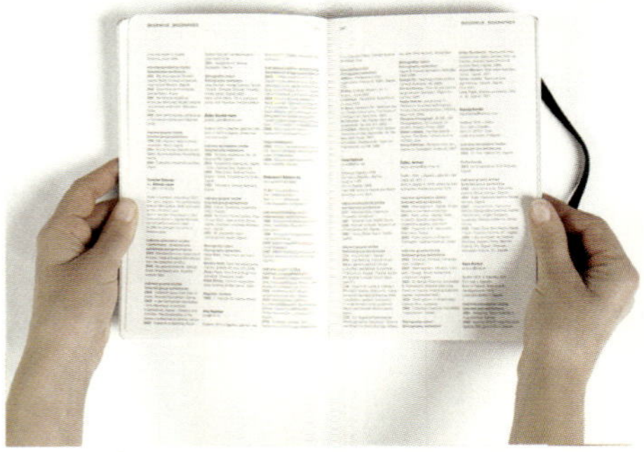

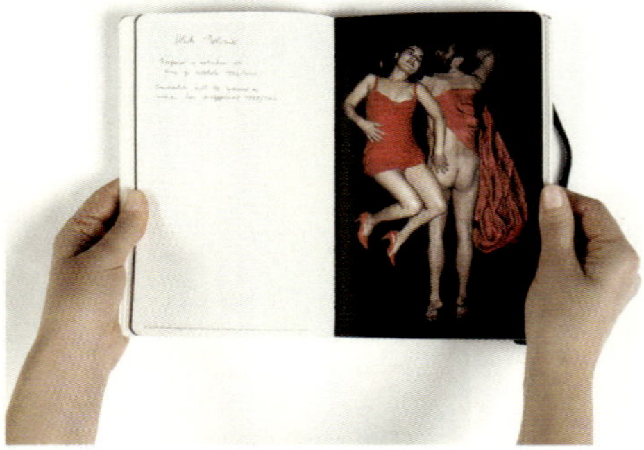

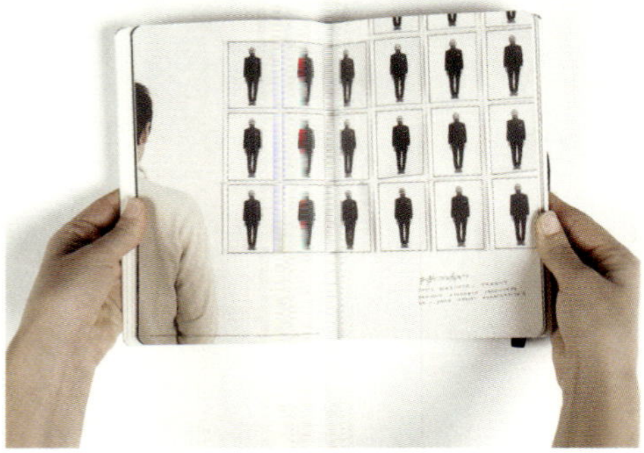

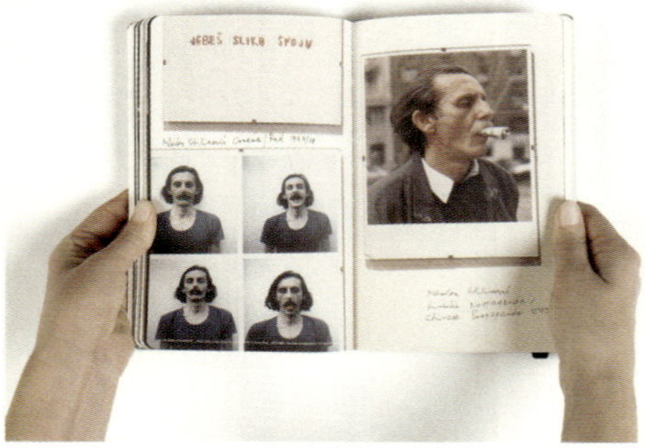

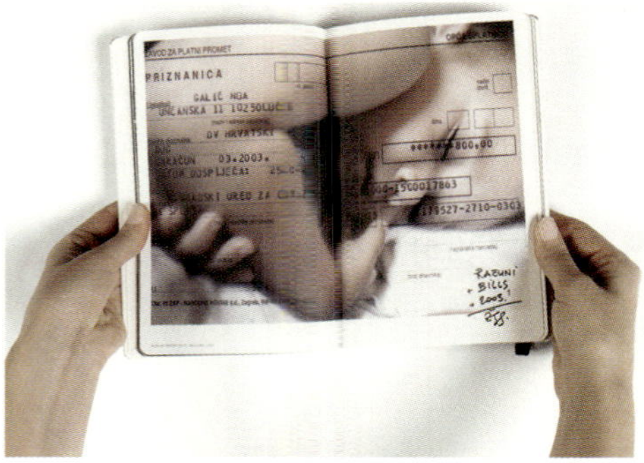

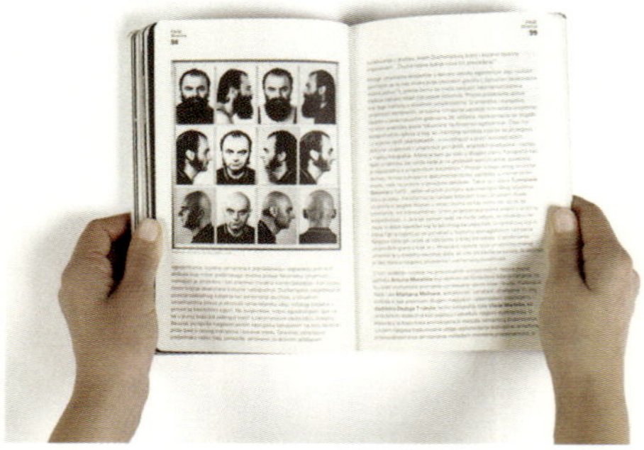

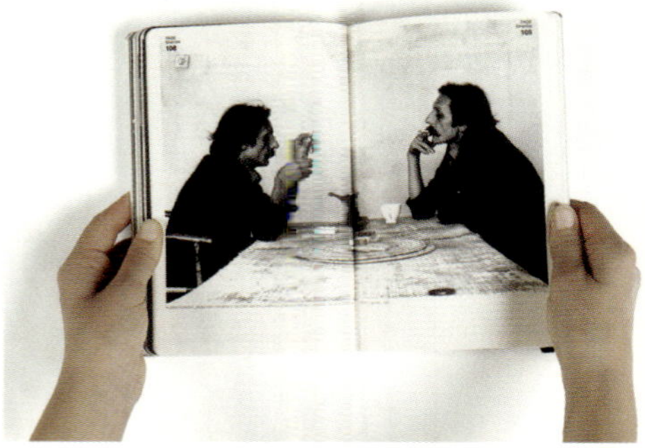

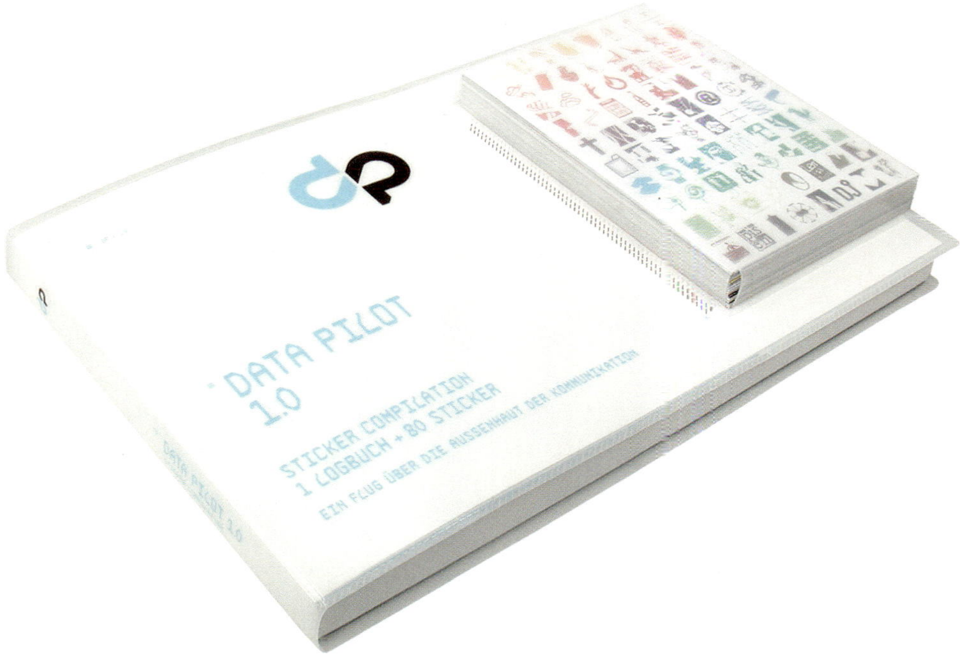

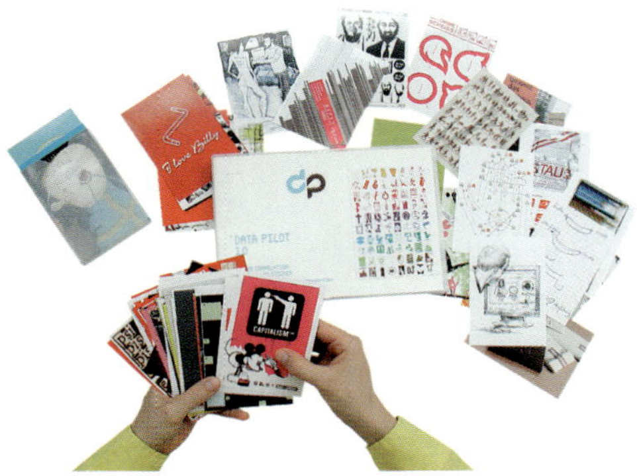

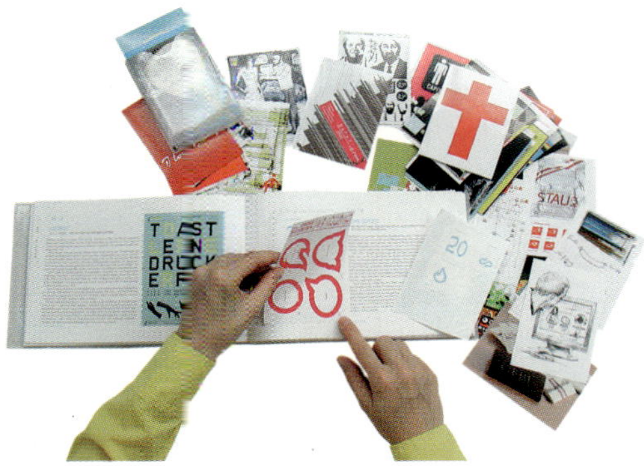

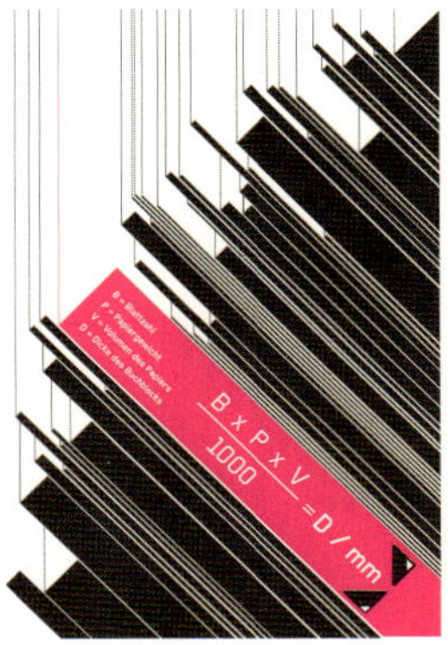

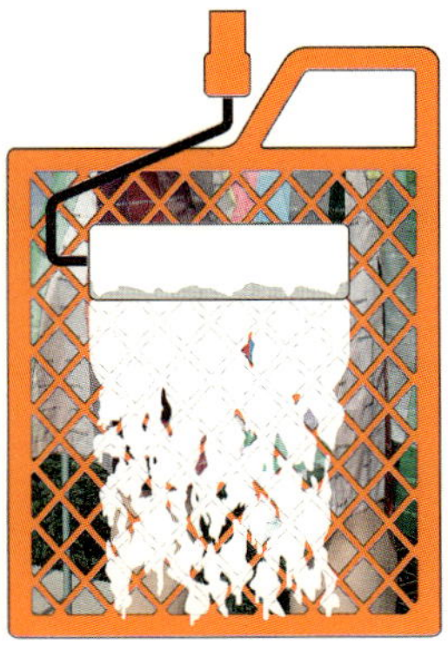

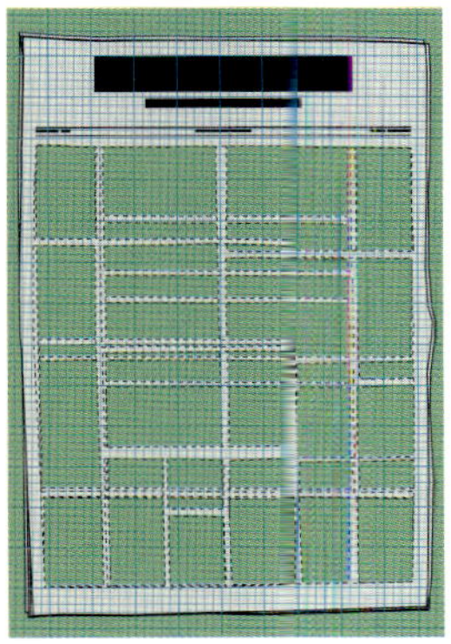

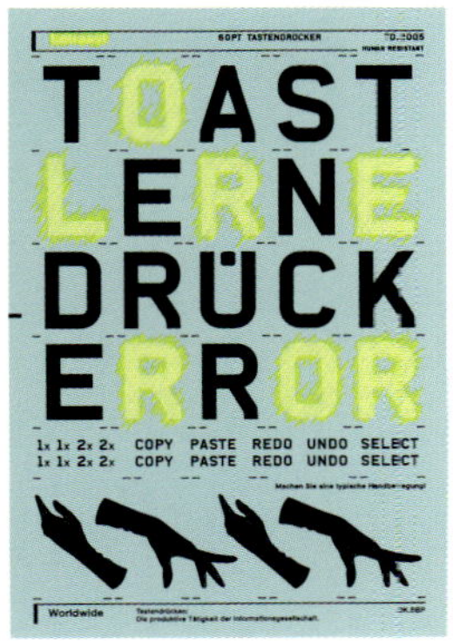

Auftraggeber/Client
DATA PILOT c/o Designklinik
Stuttgart, Germany

Projekt/Project
DATA PILOT 1.0 / STICKER COMPILATION
1 Logbuch und 80 Sticker
1 logbook and 80 stickers

Design
Designklinik
Büro für Design und Kommunikation
Stuttgart, Germany

Der DATA PILOT 1.0 ist die weltweit erste Grafikdesign-Sticker-Compilation. Mit 80 Grafik-Stickern, die 80 Texte zur Kommunikation wunderbar, überraschend, kühn, aber atemberaubend visualisieren. Sticker und Texte sind mit Liebe gemacht von 150 internationalen Designern, Gestaltern, Textern, Fotografen, Autoren aus der Schweiz, Deutschland, USA und Österreich. Darunter klangvolle Namen wie Kurt Weidemann, Fons Hickmann, Beat Gloor, Büro Destruct und viele andere.
Das Prinzip ist denkbar einfach. Jeder kennt es von der Grundschule her: Sticker-Päckchen rausnehmen. Passenden Text suchen und finden. Richtigen Sticker dazu kleben. Fertig.

A flight over the external skin of communications. – DATA PILOT 1.0 is the first ever compilation of graphical stickers. 80 stickers that visualize 80 texts in a glorious, surprising, bold and breathtaking manner. The stickers as well as the texts are crafted affectionately by 150 designers, writers, photographers and authors from Switzerland, Germany, the USA and Austria. Among them many illustrious names such as Kurt Weidemann, Fons Hickmann, Beat Gloor, Büro Destruct and many more.
Principle. Conceivably simple. Everybody knows it from primary school days. Take the stickers, find a matching text and paste the correct sticker to it. That's it.

Auftraggeber Client
Schöck Bauteile GmbH
Baden-Baden, Germany

Projekt Project
Nonverbale Einbauanleitungen
Illustrierte technische Anweisung
Illustrated technical instruction

Design
Cornelia Willand
Volkmar Hoppe
Britta Pfeifhofer
bluequest Visual Communication
Darmstadt, Germany

Schöck ist einer der führenden Hersteller von Bauelementen in Europa. Gesucht wurde eine Lösung, die ohne Sprache die komplexen Anweisungen zum Einbau von Bauelementen vermittelt: einfach verständlich und hochpräzise über alle Einbauschritte hinweg. Für die Gesamtkonzeption waren zudem produktionstechnische Anforderungen relevant. Kommunikationsunterstützend fungiert die Figur „Ebi".
Die Konzeption wurde für den ISOKORB entwickelt und wird sukzessive auf alle anderen Schöck-Produkte übertragen. Damit erhält jedes Produkt eine international verständliche Anleitung, die den Handwerker fachlich genau, aber auch leichtfüßig beim Einbau begleitet.

Schöck is a major manufacturer of components for the construction industry in Europe. The aim was a solution able to communicate complex instructions for the installation of components in a non-verbal form: easily understandable and highly precise at every step of the installation. The character "Ebi" acts as communication support.
The concept was originally developed for the ISOKORB and is going to be assigned gradually to all other Schöck products. As a result each product comes with its own internationally understandable instructions, which are not only technically correct, but also make installation work a breeze.

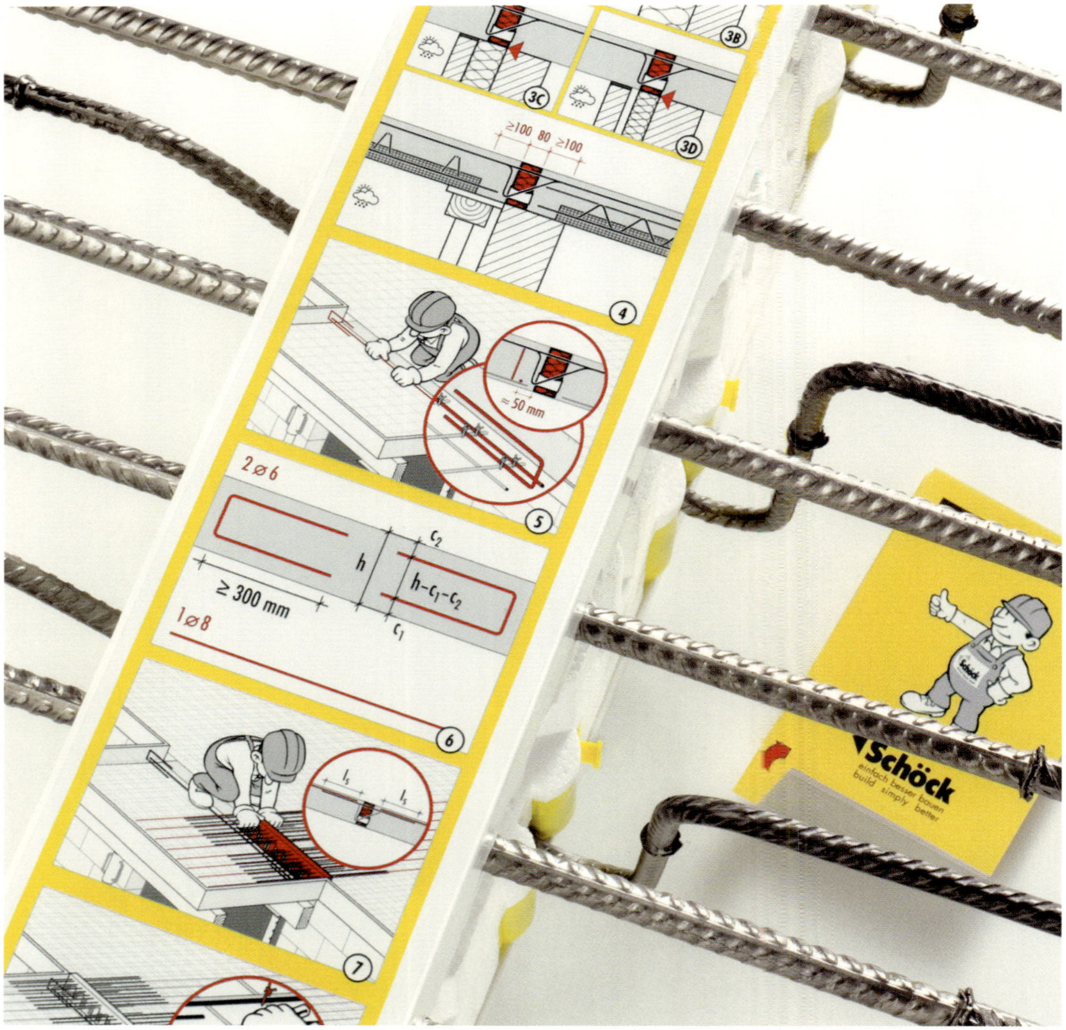

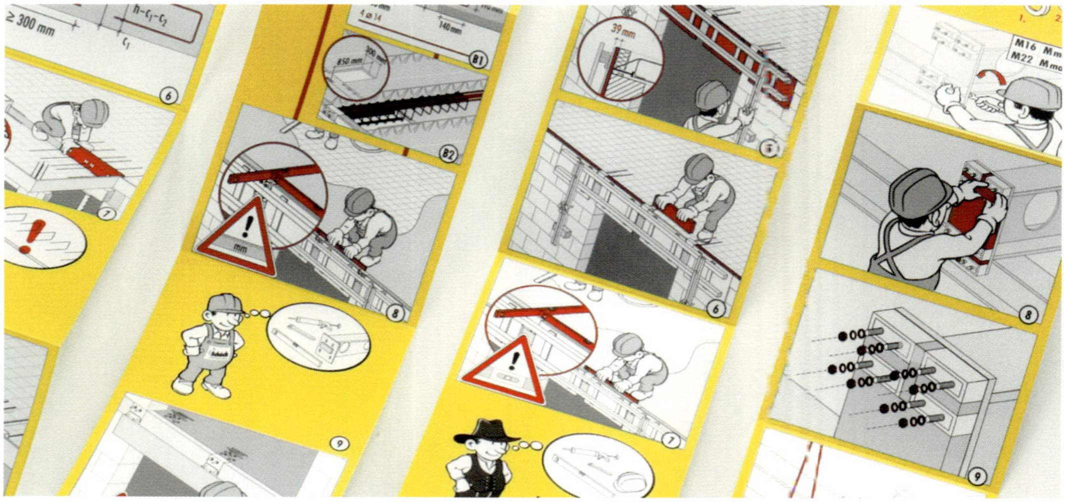

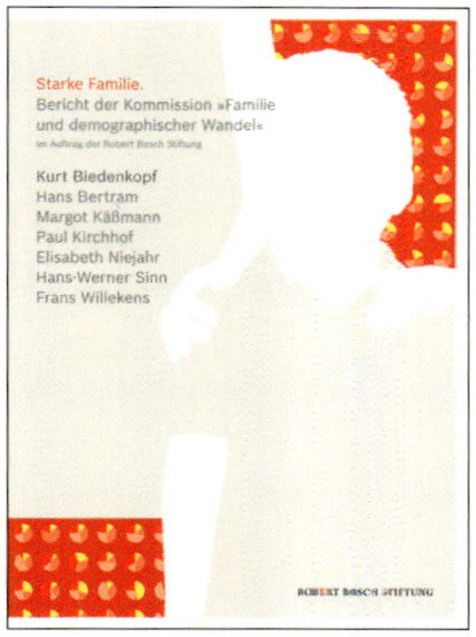

Auftraggeber Client
Robert Bosch Stiftung
Stuttgart, Germany

Projekt Project
Starke Familie
Broschüre
Brochure

Design
Christoph Zielke
Nils Jaedicke
Klaus Hesse
Hesse Design GmbH
Erkrath, Germany

Mit ihrem Bericht „Starke Familie“ hat die von der Robert Bosch Stiftung eingesetzte Kommission Familie und demographischer Wandel unter der Leitung von Professor Kurt Biedenkopf im Dezember 2005 ein von Politik, Medien und Gesellschaft viel beachtetes Werk vorgelegt und eine Reihe von Handlungsempfehlungen ausgesprochen. Die Robert Bosch Stiftung will mit dem Bericht nicht zuletzt dafür werben, im kinderentwöhnten Deutschland Vertrauen und Wertschätzung für Kinder und Familie wiederherzustellen.

With their “Strong Families” report, published in December 2005, the Commission on Families and Demographic Change brought into being by the Robert Bosch Foundation under the leadership of Professor Kurt Biedenkopf has produced a document which has earned widespread praise from politicians, the media and society at large and which includes a wealth of suggestions for action. One of the goals of the Robert Bosch Foundation is to help re-establish trust in and esteem for, children and families in a Germany where children are becoming ever less common.

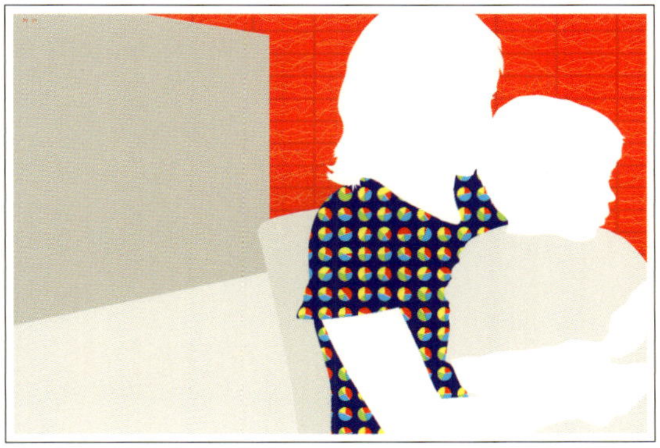

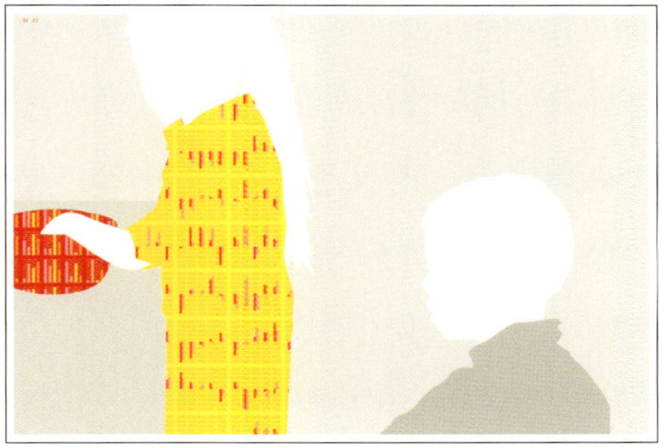

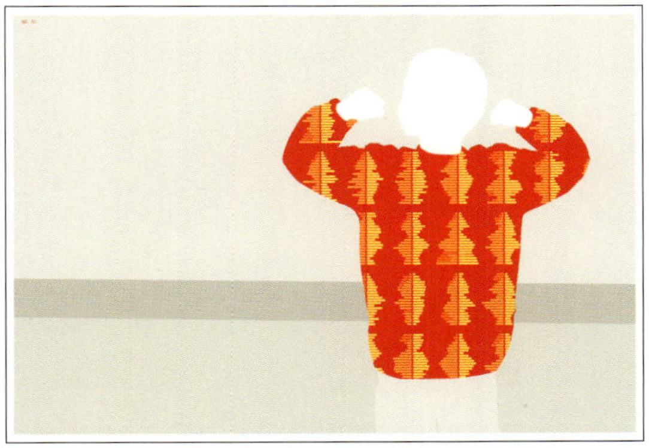

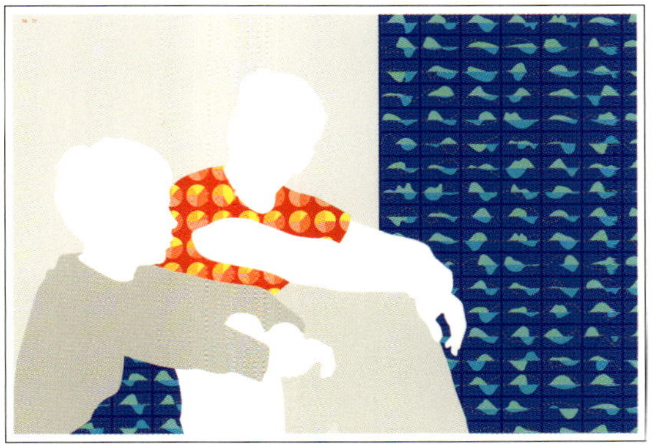

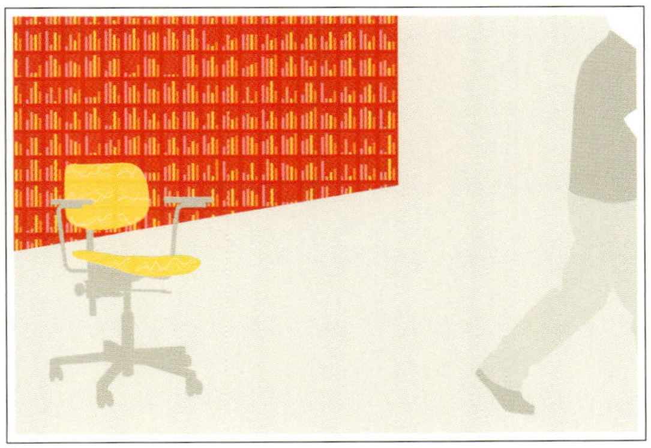

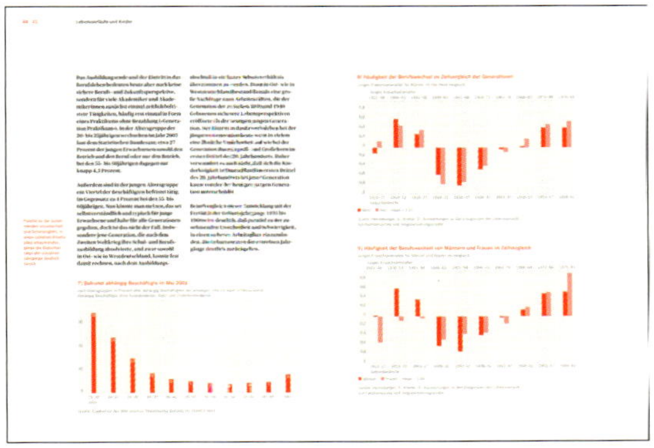

Auftraggeber Client
Düsseldorfer Schauspielhaus
Düsseldorf, Germany

Projekt Project
Düsseldorfer Schauspielhaus
Spielplan
Theatre programme

Design
Claudia Fischer-Appelt
Petra Matouschek
Martina Massong
Martin Schwatlo
Hana Sedelmayer
Ligalux
Hamburg, Germany

Im Zuge des Intendanzwechsels am Düsseldorfer Schauspielhaus entwickelte Ligalux einen neuen visuellen Auftritt. Die Corporate Identity macht den ganzheitlich kreativen Anspruch des Hauses auch über Design spürbar. Neben dem plakativ kraftvollen Logo ist das Spielplanheft mit über 160 Seiten gestalterischer Höhepunkt: Jeder Teilbereich trägt ein individuelles Design – und verspricht ein Feuerwerk an Theatererlebnissen.

In the wake of the change of theatre directors at the Düsseldorfer Schauspielhaus, Ligalux developed a new overall look. The corporate identity made the holistic-creative standard of the theatre also tangible in the design. In addition to the striking, powerful logo, the theatre programme, consisting of more than 160 pages, is a design highlight: each section has its individual design – promising a firework of theatre experiences.

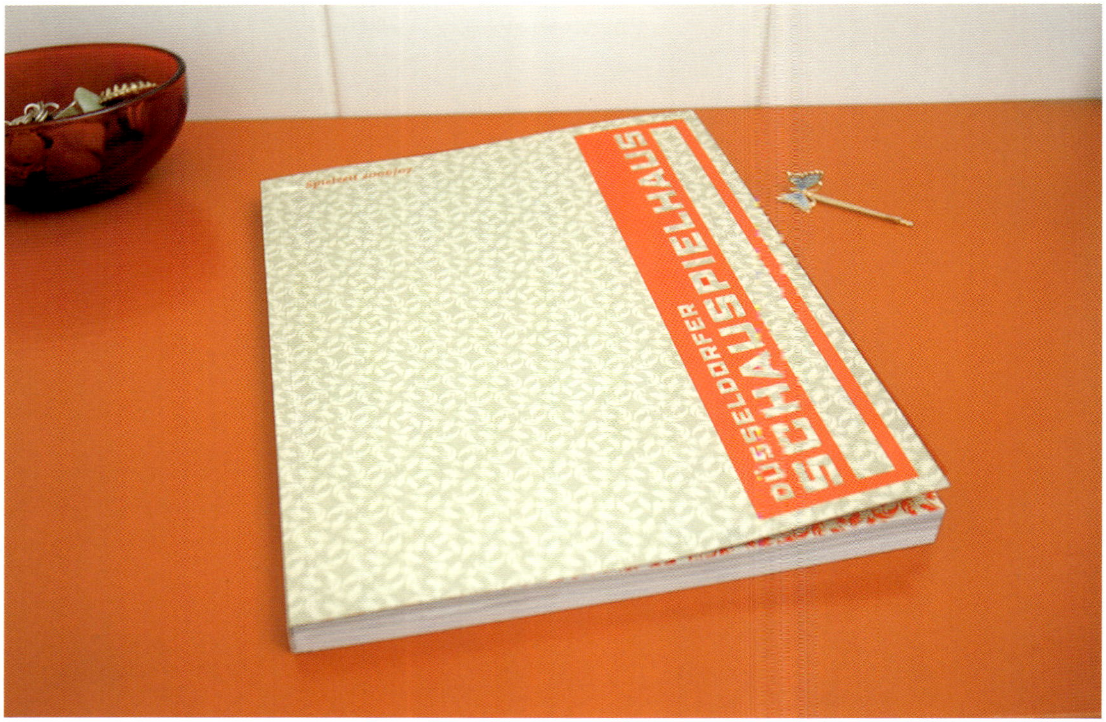

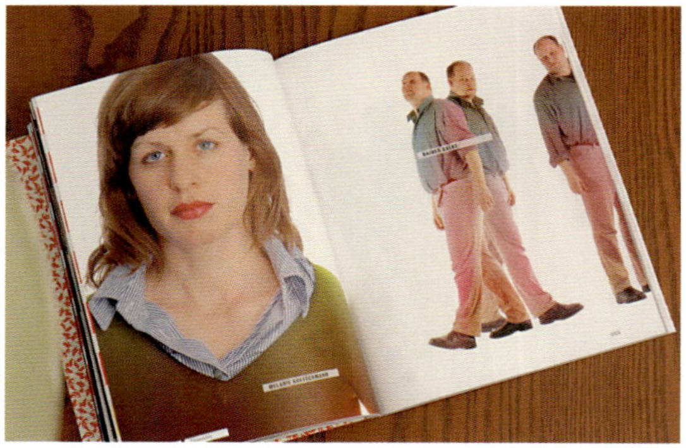

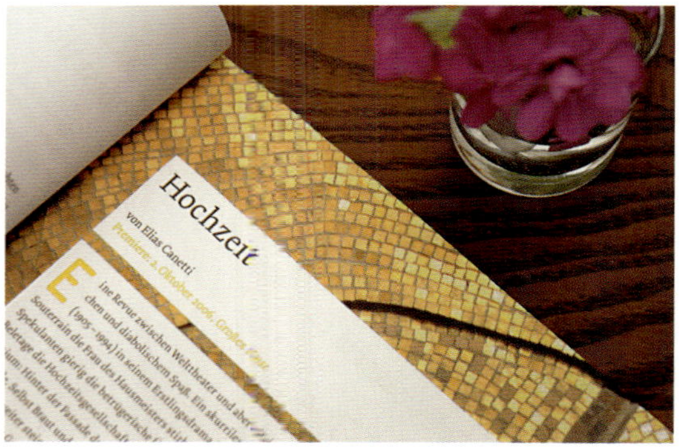

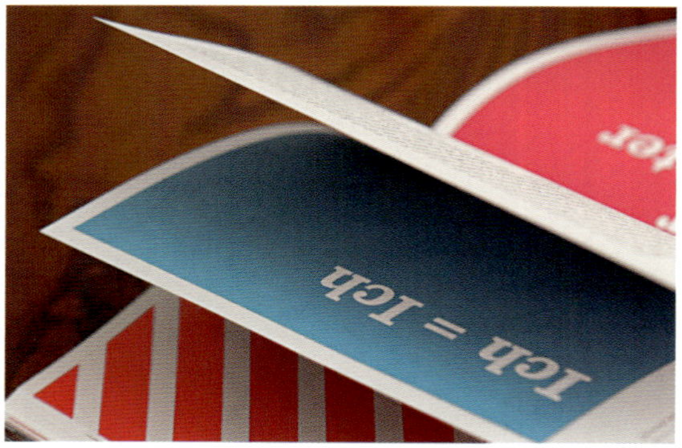

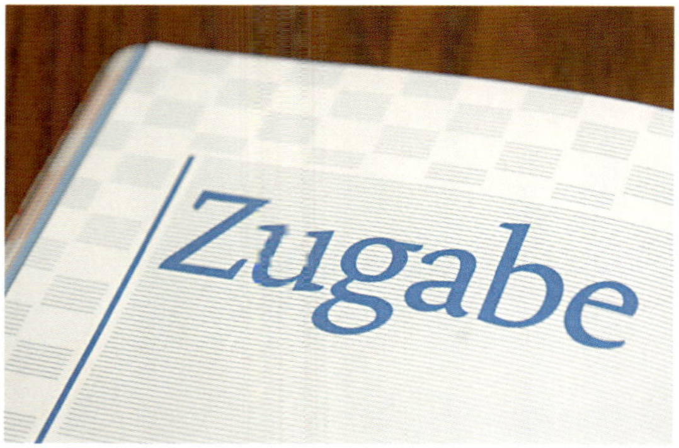

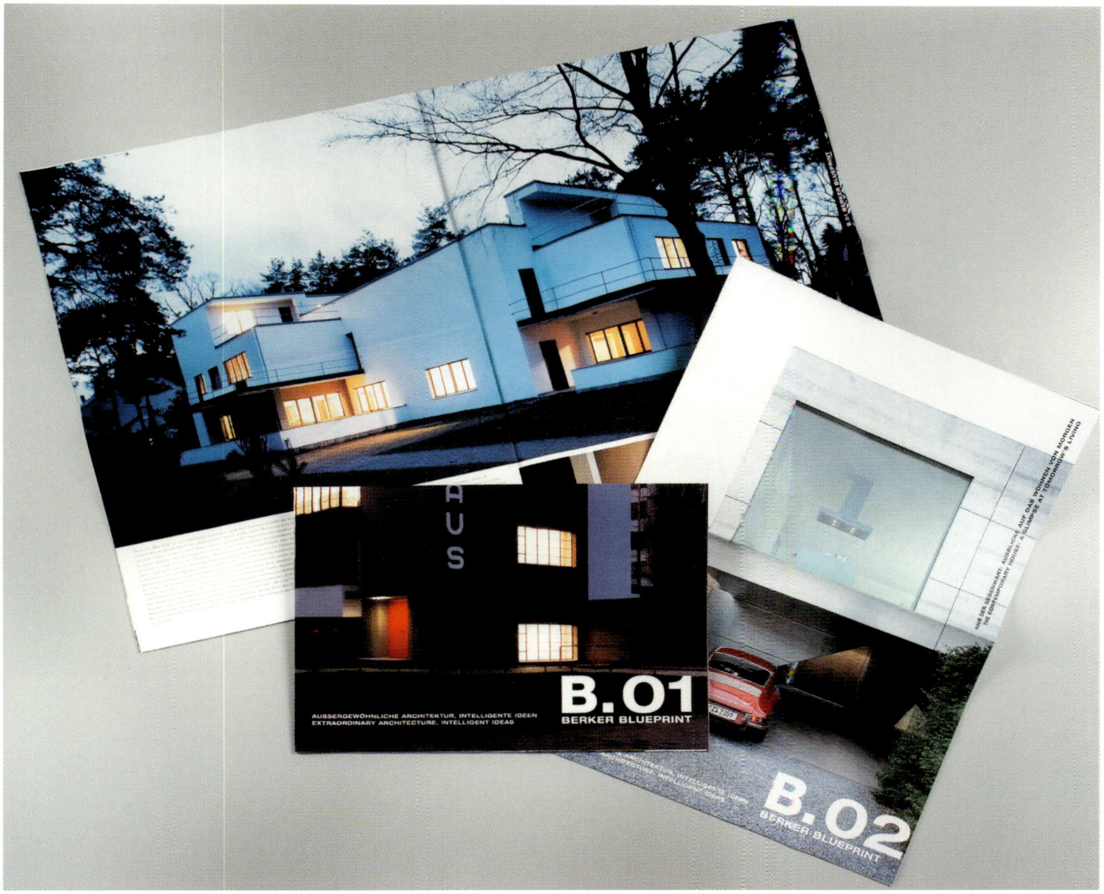

Auftraggeber Client
Berker GmbH & Co. KG
Schalksmühle, Germany

Projekt Project
berker blueprint
Architektenzeitung
Newspaper for architects

Design
Thomas Biswanger Design
Ingolstadt, Germany

Architekten interessieren sich für außergewöhnliche Architektur. Berker, Hersteller hochwertiger Schalter und Systeme, ist weltweit in vielen außergewöhnlichen Bauwerken vertreten. Was liegt näher, als diese den Architekten in außergewöhnlicher Form vorzustellen? „berker blueprint" präsentiert zwei Mal pro Jahr herausragende architektonische Objekte in hochwertiger Fotografie und auf einem unkonventionellen Oversize-Format. Die Architektur (Bauhaus Dessau, Haus der Gegenwart München und „langer Eugen", Bonn) ist dabei immer der Star; en passant werden die Berker-Schalter in den jeweiligen Objekten vorgestellt.

Architects are interested in extraordinary architecture. High-quality switches and systems made by Berker can be found all over the world in many extraordinary buildings. So what could be more natural than to present them to architects in an extraordinary form? Twice yearly, "berker blueprint" presents outstanding architecture in the form of high-quality photographs in an unconventional oversize format. The star is always the architecture (the Bauhaus in Dessau, the Haus der Gegenwart in Munich and the "Langer Eugen" in Bonn), while Berker switches found in the respective buildings are presented en passant.

Auftraggeber Client
Roset Möbel GmbH
ligne roset
Gundelfingen, Germany

Projekt Project
ligne roset Katalog 2006
Katalog
Catalogue

Design
Ulrike Beling
Nina Thoenen
Beling Thoenen Design
Zurich, Switzerland

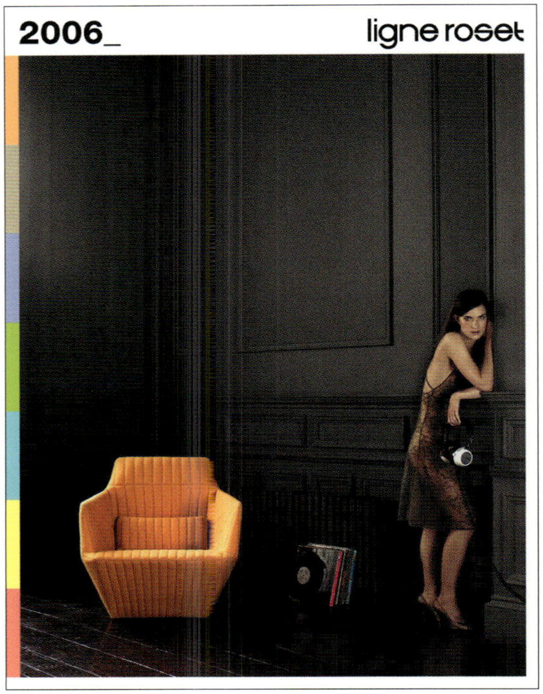

Für die französische Herstellermarke ligne roset ist der international distribuierte Katalog, der jährlich erscheint, eines der zentralen Kommunikationselemente. Entsprechend war die Aufgabenstellung an die neu beauftragte Agentur Beling Thoenen Design, Zürich, eine doppelte: Zum einen sollen Interessenten durch die Lektüre an die Marke herangeführt werden. Zum anderen ist der Katalog ein unterstützendes Instrument bei Verkaufsgesprächen im Handel.
Der klaren inhaltlichen Strukturierung sowie dem markanten Layout und einer starken Typografie kommen dabei größte Bedeutung zu. Darüber hinaus wird der Einstieg in den Katalog mit redaktionellen Elementen magazinartig angereichert.

A key communication element for the French manufacturing brand "ligne roset" involves its catalogue, which appears annually and is distributed worldwide. The mission for the newly appointed agency, Beling Thoenen Design in Zurich, was therefore two-fold: the catalogue needed to introduce its readers to the brand, and should also serve as a support tool for retail sales conversations.
This meant it was extremely important for the contents to be clearly structured, as well as having a striking layout and a strong typography. In addition to this, the catalogue is enhanced by journalistic elements in magazine style.

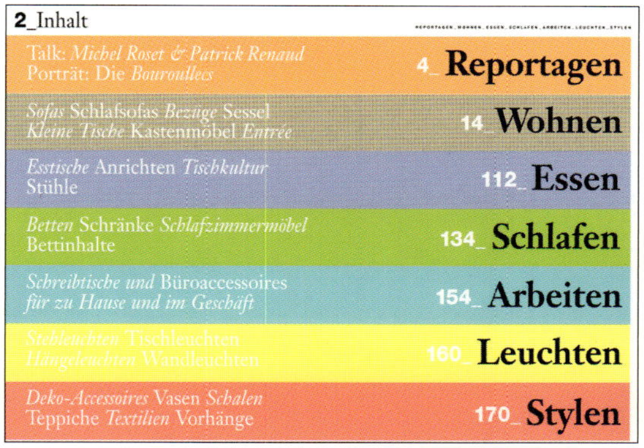

2_Inhalt

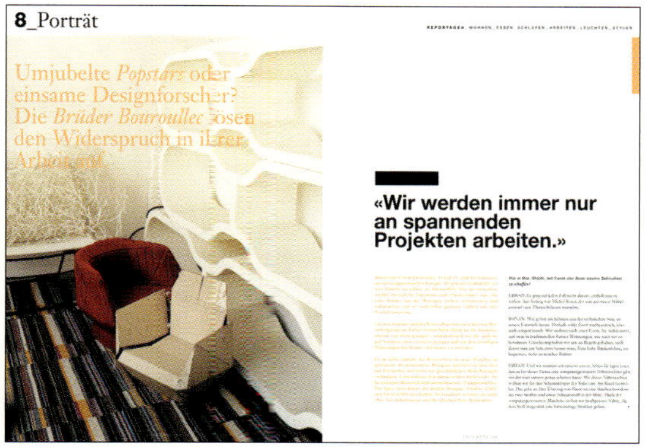

8_Porträt

Umjubelte *Popstars* oder einsame Designforscher? Die *Brüder Bouroullec* lösen den Widerspruch in ihrer Arbeit auf.

«Wir werden immer nur an spannenden Projekten arbeiten.»

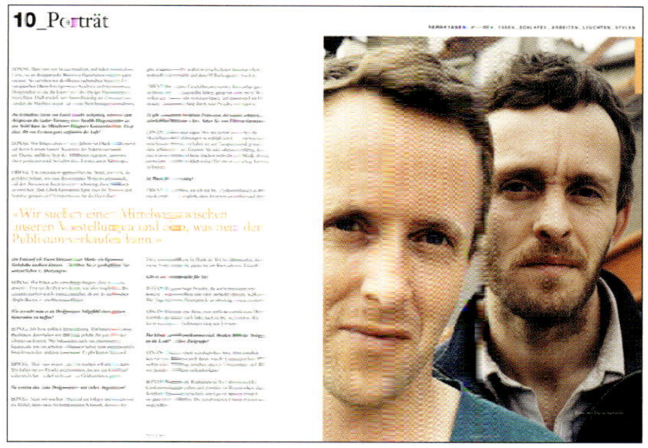

10_Porträt

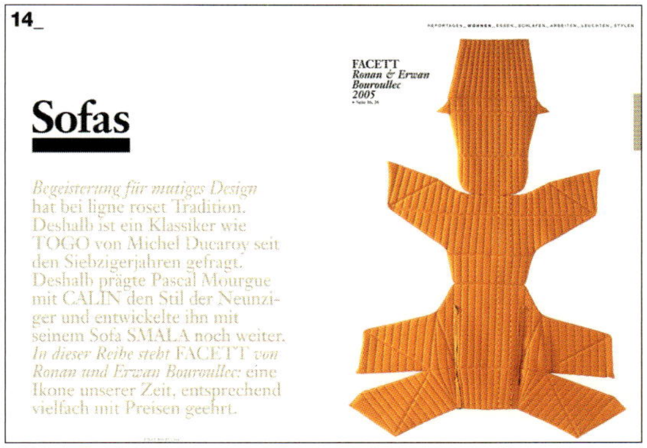

14_

Sofas

Begeisterung für mutiges Design hat bei ligne roset Tradition. Deshalb ist ein Klassiker wie TOGO von Michel Ducaroy seit den Siebzigerjahren gefragt. Deshalb prägte Pascal Mourgue mit CALIN den Stil der Neunziger und entwickelte ihn mit seinem Sofa SMALA noch weiter. *In dieser Reihe steht* FACETT *von Ronan und Erwan Bouroullec:* eine Ikone unserer Zeit, entsprechend vielfach mit Preisen geehrt.

FACETT *Ronan & Erwan Bouroullec 2005*

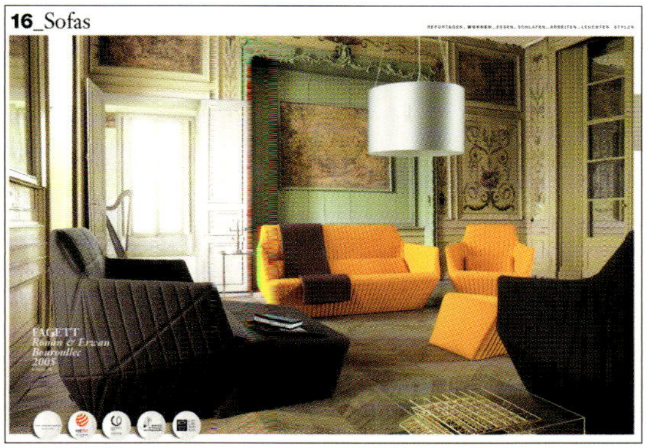

16_Sofas

FACETT *Ronan & Erwan Bouroullec 2005*

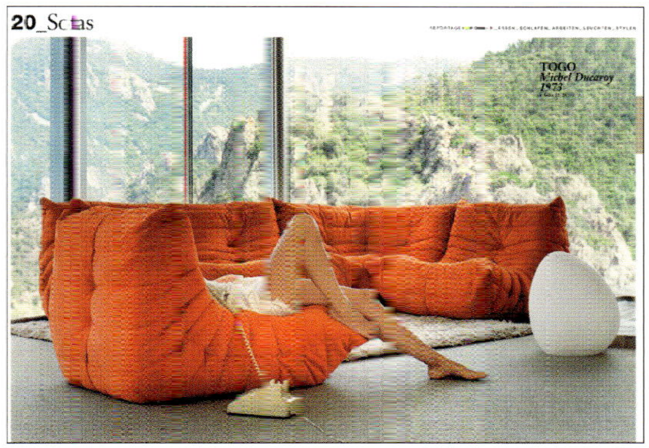

20_Sofas

TOGO *Michel Ducaroy 1973*

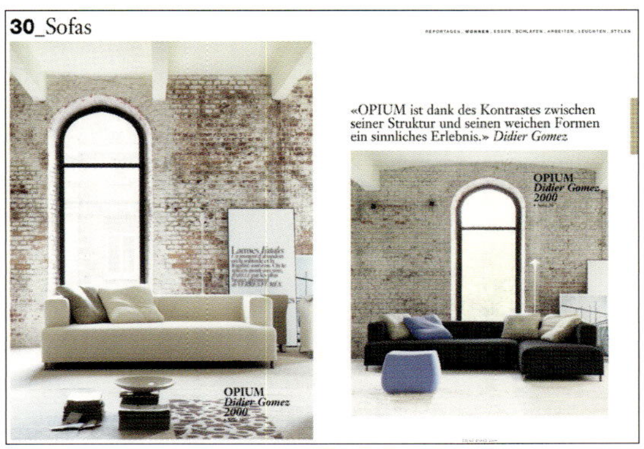

30_Sofas

«OPIUM ist dank des Kontrastes zwischen seiner Struktur und seinen weichen Formen ein sinnliches Erlebnis.» *Didier Gomez*

OPIUM *Didier Gomez 2000*

OPIUM *Didier Gomez 2000*

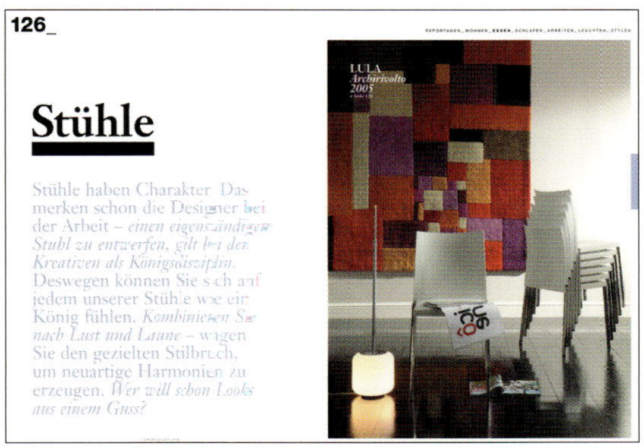

126_

Stühle

Stühle haben Charakter. Das merken schon die Designer bei der Arbeit – *einen eigenständigen Stuhl zu entwerfen, gilt bei den Kreativen als Königsdisziplin.* Deswegen können Sie sich auf jedem unserer Stühle wie ein König fühlen. *Kombinieren Sie nach Lust und Laune* – wagen Sie den gezielten Stilbruch, um neuartige Harmonien zu erzeugen. *Wer will schon Looks aus einem Guss?*

LULA *Archirivolto 2005*

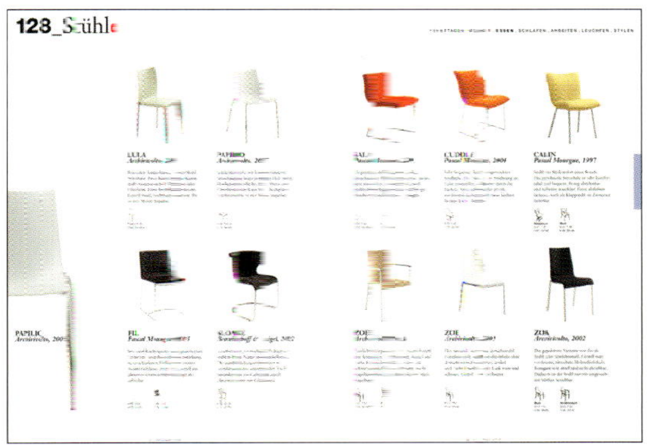

128_Stühle

Auftraggeber Client
Deutsche Post AG
Bonn, Germany

Projekt Project
Schöne Grüße! – 365 Postkarten aus dem Alltag
Postkartenkalender
Postcard calendar

Design
Sabrina Lyhs-Webelhaus
Jane Sauer
lyhssauer GbR
Düsseldorf, Germany

Der Postkartenkalender „Schöne Grüße!" begleitet mit 365 Motiven durch das Jahr und regt täglich zum Versenden eines kurzen Grußes an. Dabei wird das Medium Postkarte neu interpretiert: Banales, Kurioses, Fundstücke und Momentaufnahmen des Alltags sind das Thema dieser ungewöhnlichen Sammlung. Durch verschiedene Bild-Text-Verknüpfungen entstehen täglich neue Schreibimpulse. Der Kalender soll Anregungen liefern und gleichzeitig Spielraum für eigene Assoziationen und Interpretationen lassen.

The postcard calendar "Schöne Grüße!" accompanies the writer through the year with 365 motifs and provides a daily stimulus to send a short greeting. Here, the medium postcard is reinterpreted: every-day banal, odd items, and things discovered by chance and snapshots are the topic of this unusual collection. Every day, new stimuli to write are generated with various image-text links. The calendar is to provide ideas and, at the same time, scope for the writer's own associations and interpretation.

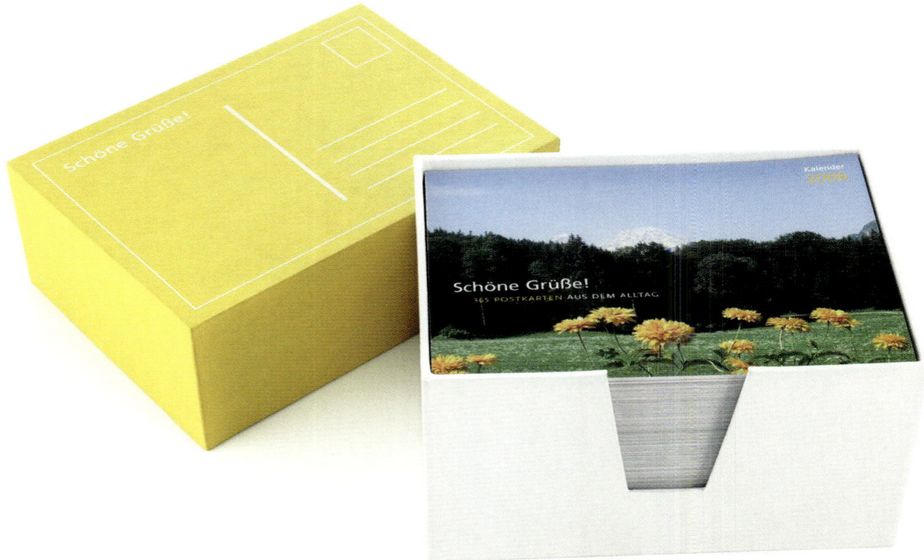

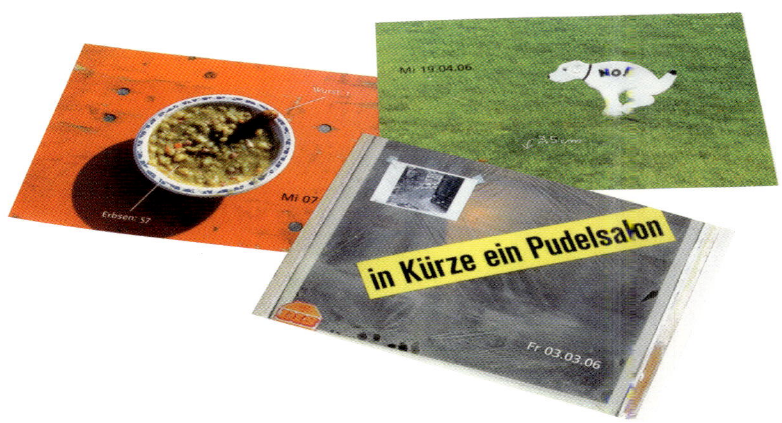

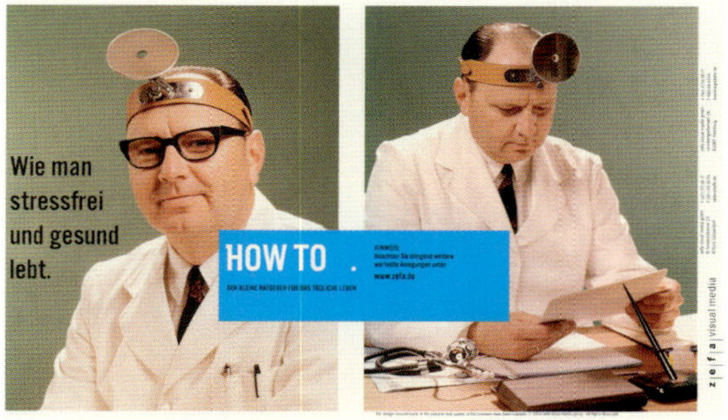

Auftraggeber Client
zefa visual media GmbH
Düsseldorf, Germany

Projekt Project
HOW TO
Leporello-Mailing
Accordion-folded mailing

Design
Christopher Wiehl
Marcus Berressem
paarpiloten
Düsseldorf, Germany

Die Bildagentur zefa kaufte die Rechte an einer amerikanischen RetroFile-Kollektion mit ca. 2.500 Bildern. Durch ein Mailing sollte die Kollektion bekannt gemacht werden. „HOW TO" ist ein humorvoller Ratgeber für das tägliche Leben. Bilder wurden zu Kapiteln wie „Bei der Arbeit" oder „Beim Sport" zusammengestellt, kommentiert und mit einem Rat versehen.

The zefa photo agency bought the rights to an American RetroFile collection containing around 2,500 pictures. A mailing was planned to popularize the collection. "HOW TO" is a humorous primer for daily living. The pictures were concatenated into individual chapters with titles like "At Work" or "Work out," including accompanying comments and advice.

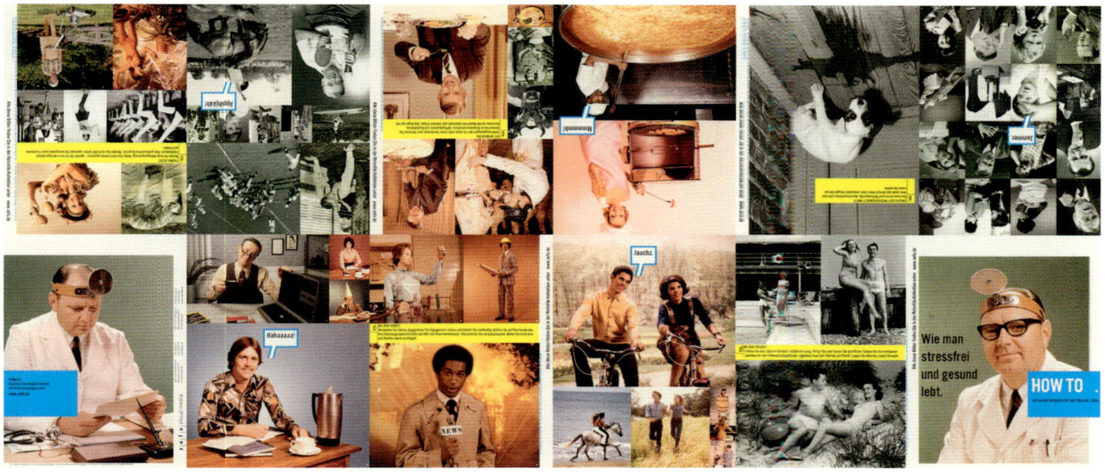

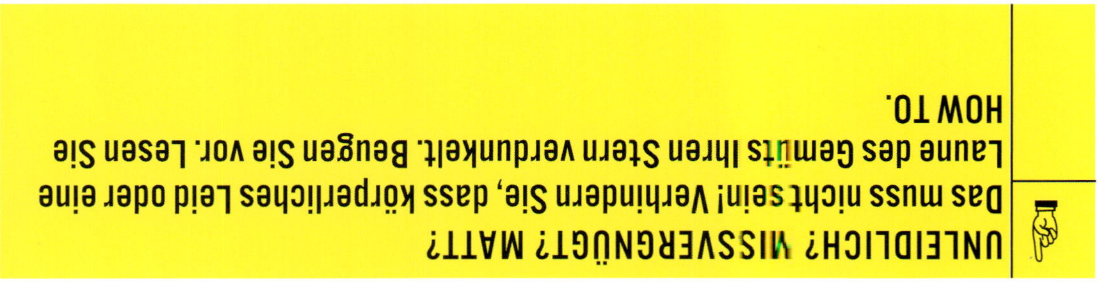

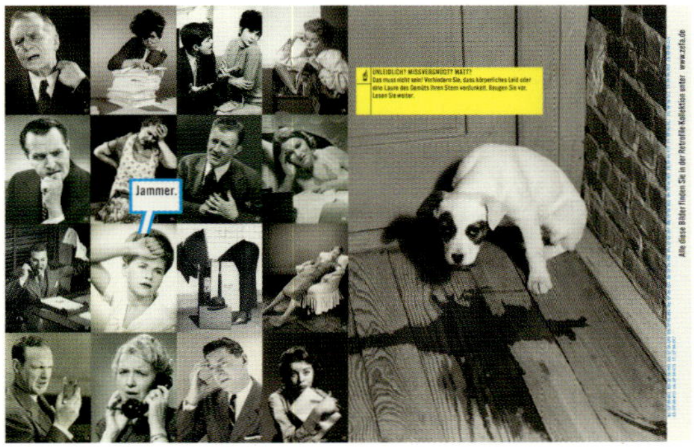

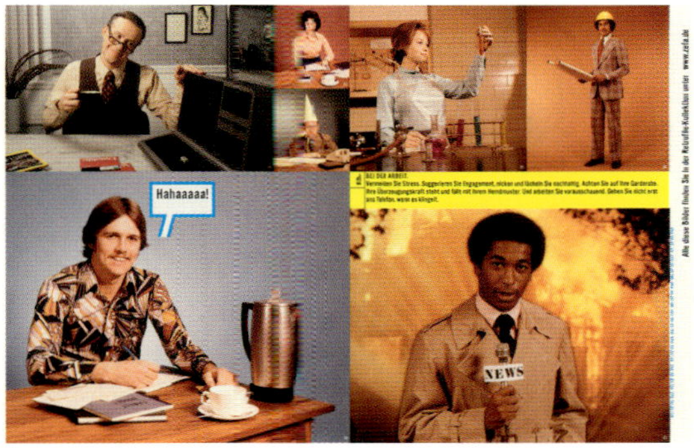

Auftraggeber Client
brauner.microphones
Hamminkeln, Germany

Projekt Project
The Art of Microphones
Folder in Metallschachtel
Folder in a metal case

Design
da®da designprojekt
Düsseldorf, Germany

Die Brauner Röhrengerätemanufaktur baut High-End-Röhrengeräte für den professionellen Anwender. Vom geheimen Kultstatus hat sich das Unternehmen in den letzten Jahren seinen Platz unter den weltweit führenden Herstellern für Kondensator-Röhrenmikrofone geschaffen. Handgemachte Präzision sollte erlebbar werden. Dazu werden die einzelnen, ansprechenden Produktfolder in einer lasergravierten Metallschachtel präsentiert.

brauner.microphones is one of the leading manufacturers for high end tube microphones. Great care is taken to ensure that all their products have their own unique qualities to achieve incredibly high levels of authenticity. Several product folders are presented in a laser engraved metal case to communicate that experience.

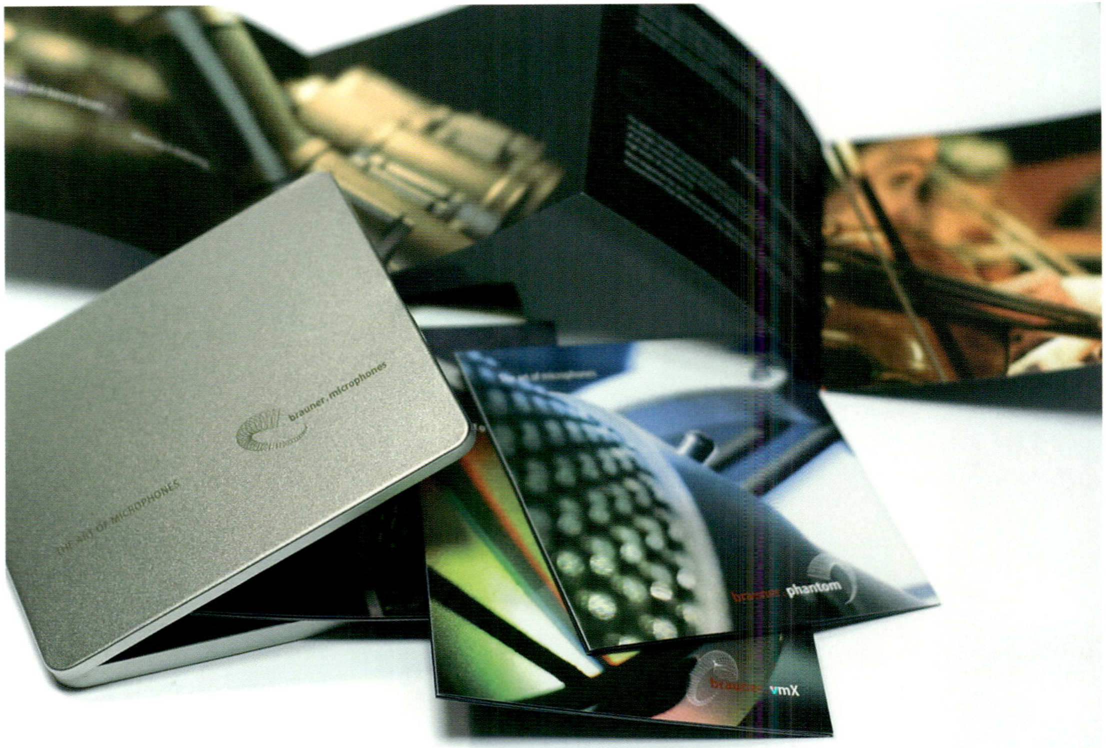

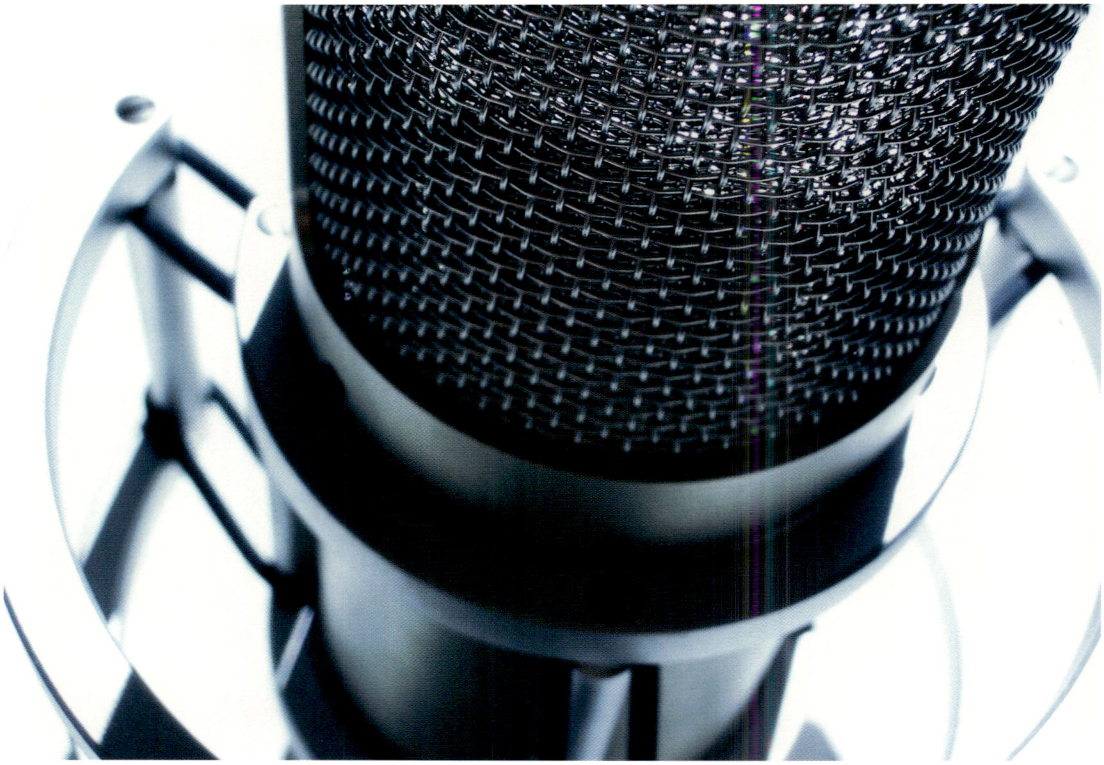

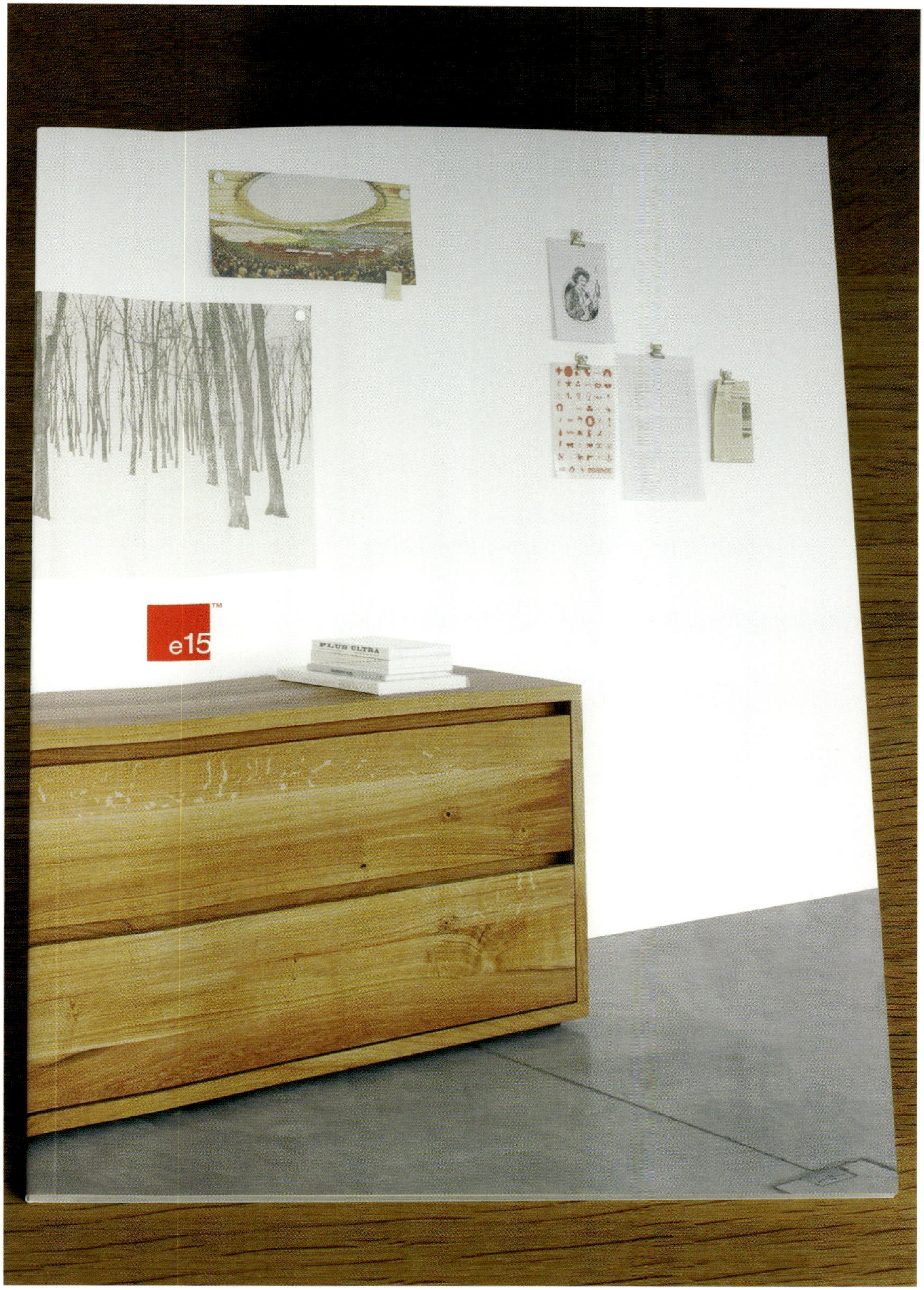

Auftraggeber Client
e15 Design und Distributions GmbH
Oberursel, Germany

Projekt Project
e15 Hauptkatalog 2006
Buch
Book

Design
Antonia Henschel
SIGN Kommunikation GmbH
Frankfurt am Main, Germany

Der neueste Katalog von e15 erscheint im Design der letzten Jahre, zeichnet sich aber durch mehr Kompaktheit aus. Die Fotografie von Ingmar Kurth und die Grafik von Antonia Henschel zeigen auf über 120 Seiten die Produkte und die Welt von e15. Die Möbel von e15 sind modern, die Herstellung ist traditionell. Entworfen um über kurzlebigen Trends zu stehen, überzeugen die Produkte durch Bestimmtheit, Detail und Schlichtheit und erlauben es den Materialien, sich in ihrer reinen Form und Schönheit zu entfalten.

e15's latest catalogue comes in the same design as in previous years, but now stands out for its greater compactness. Ingmar Kurth's photography and Antonia Henschel's graphics show the products and world of e15 over more than 120 pages. e15's furniture is modern, while the production methods are traditional. Designed to outlast passing trends, the products are clear-cut, refined and uncomplicated, allowing the natural beauty of the materials to be appreciated in its most elegant form.

Auftraggeber Client
Corbis
London, United Kingdom

Projekt Project
LOOP
Buch
Book

Design
Christopher Wiehl
Nanni Goebel
Kerstin Gerhards
Josephine Kreuz
paarpiloten
Düsseldorf, Germany

Für die Bildagentur Corbis sollte ein Buch zu einer neuen, extravaganten Bildkollektion gestaltet werden. Farbstreifen schaffen Verbindungen von Bild zu Bild und lassen das Auge über die Oberfläche gleiten. Vektorillustrationen sorgen als Kapiteltrenner für kleine Pausen. Icons geben assoziative Kommentare. Das Buch ist eine Endlosschleife der Bilder ohne Anfang und Ende.

The mission was to design a book describing an extravagant new picture collection from the Corbis photo agency. Colored stripes connect the pictures, allowing the eyes to glide over the surface. Vector illustrations serve as chapter dividers for small intermezzos. Icons deliver associative commentary. The book represents an infinite loop of pictures, without a beginning or end.

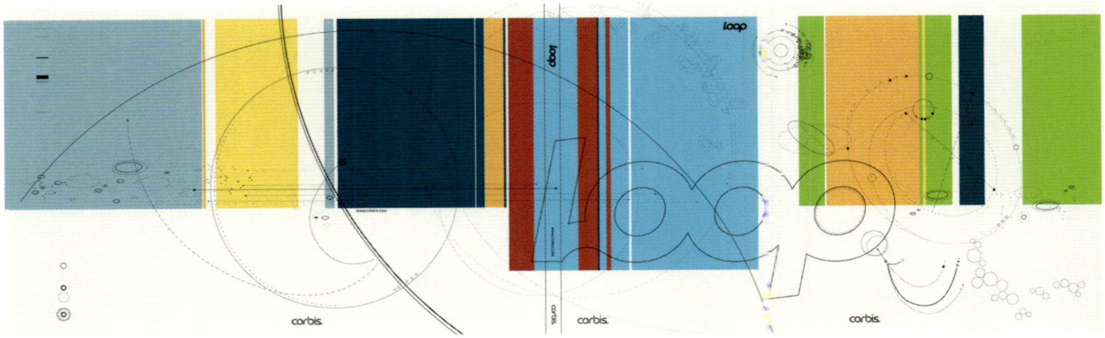

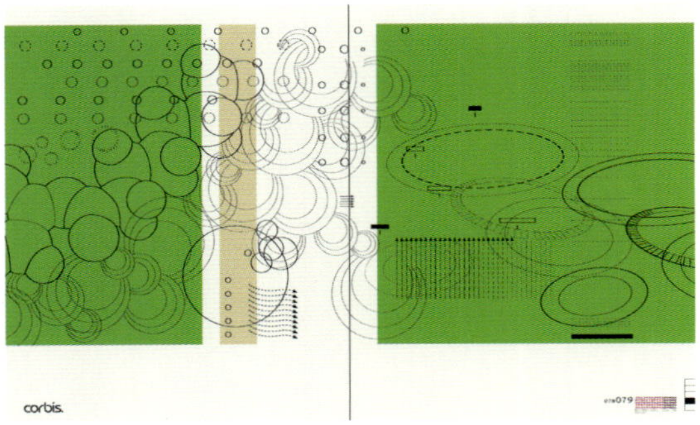

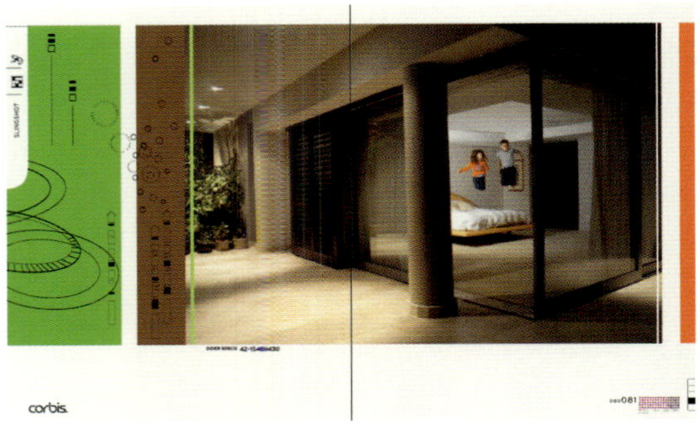

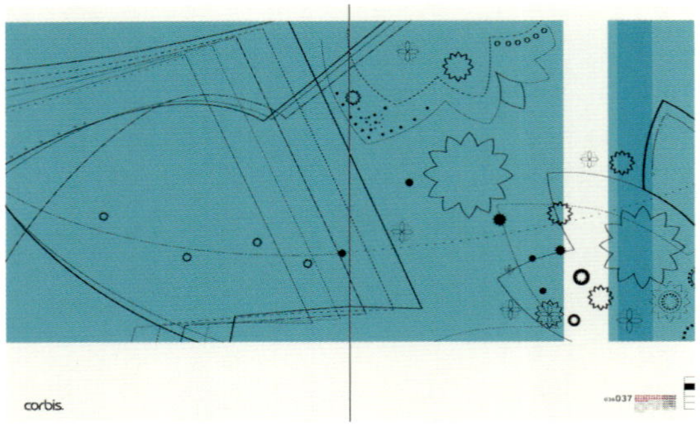

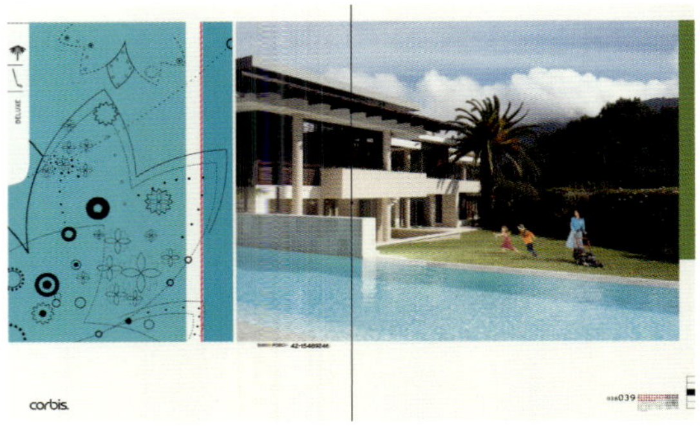

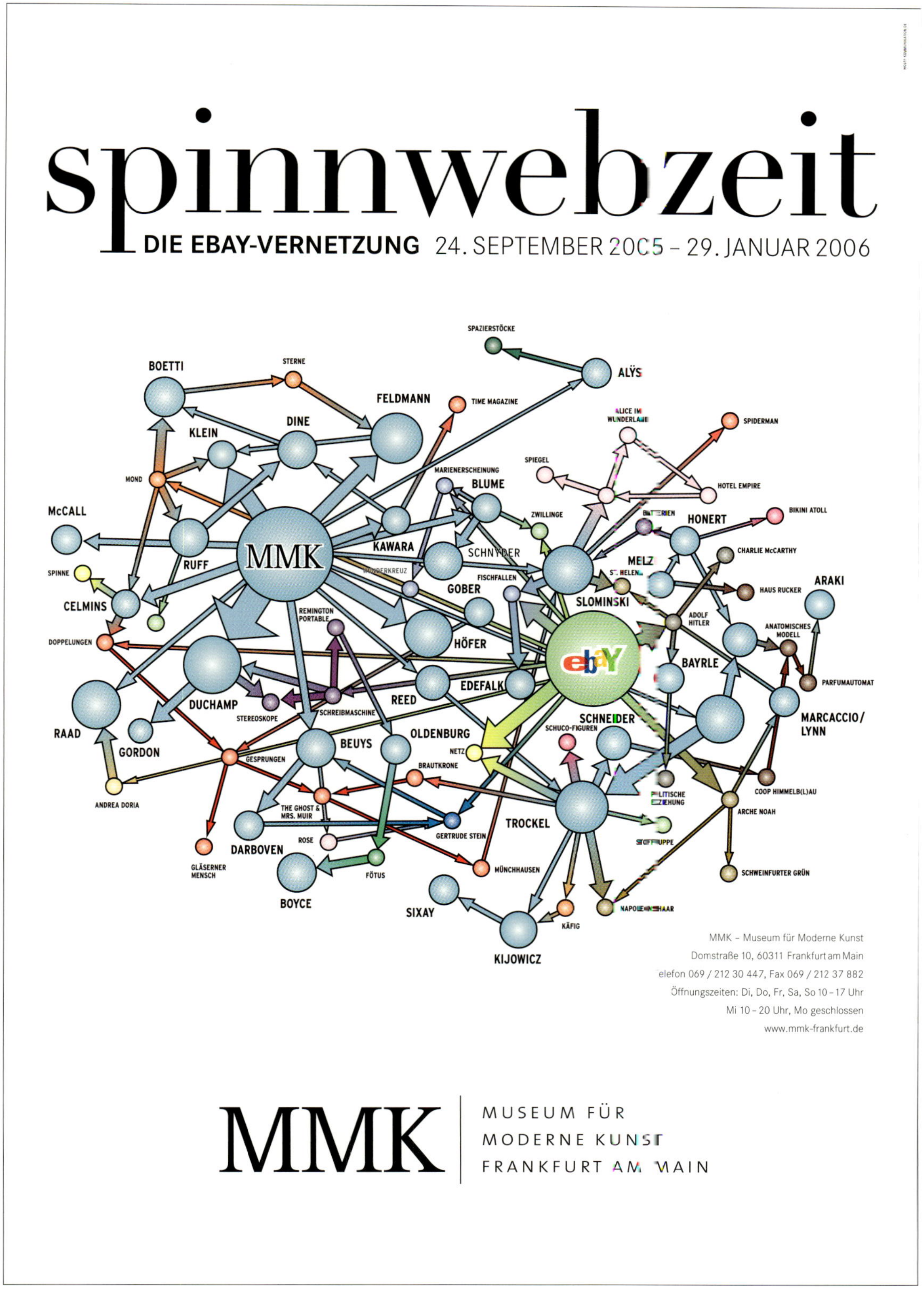

Auftraggeber Client
MMK Museum für Moderne Kunst Frankfurt am Main
Frankfurt am Main, Germany

Projekt Project
Spinnwebzeit. Die eBay-Vernetzung
Plakat und Key Visual
Poster and key visual

Design
Carsten Wolff
Christiane Wolff
Thomas Rott
Wim Schafspelz
Wolff Kommunikation
Frankfurt am Main, Germany

Die Kunst bildet heute ein selbstreferenzielles System, in dem viele Werke Bezug nehmen auf andere Werke. So entsteht ein Netzwerk, das nur für den Eingeweihten nachvollziehbar und lesbar ist.
Wie das Museum ein Sammlungsort für Kunstwerke ist, ist das Netzwerk eBay, das sich als Online-Marktplatz versteht, ein Sammlungsort für Dinge, die nur eines gemeinsam haben: Sie suchen einen Käufer.
In den Räumen des Museums für Moderne Kunst (MMK) finden diese beiden Netzwerke nun zusammen und werden – Raum für Raum – im Bedeutungszusammenhang kombiniert.

Art today forms an auto-referential system where many works relate to other works. Thus, a network evolves that is comprehensible and readable to insiders only.
In as much as a museum can be called a site where artworks are collected, the eBay network, which sees itself as an online marketplace, must be called a site where objects gather that have only but one thing in common: They desire to be bought.
On the premises of the Museum for Modern Art Frankfurt (MMK) those two networks entangle and are – room by room – combined in new fields of meaning.

print media sales promotion

Auftraggeber Client
BASF AG
Ludwigshafen, Germany

Projekt Project
heat
Buch
Book

Design
IDEO GmbH
design + innovation
München, Germany

Eine Materialentwicklung von BASF ist das Thermoplast Ultrason, dessen hochfeste Materialeigenschaften bei Haushaltsprodukten bislang nicht effektiv eingesetzt worden sind. BASF beauftragte daher IDEO mit der Entwicklung von Anwendungen für den häuslichen Bereich. IDEO erarbeitete fünf neuartige Konzepte, die wohl auf bekannten und bewährten Produkten beruhen, für die sich jedoch dank Ultrason unkonventionelle Möglichkeiten eröffnen. Dieses Buch dokumentiert, wie ein unverstellter Blick zu tieferen Einsichten und zündenden Ideen für neue Produktlösungen führen kann.

A material developed by BASF: Ultrason, which has powerful properties not currently being effectively utilised for domestic products. BASF came to IDEO with the challenge of bringing the material properties of Ultrason to life in the domestic world. IDEO designed five concepts – products we know and understand but redesigned with fresh eyes and an innovative approach afforded by Ultrason®. This book is a response to that challenge, and documents how insights can spark ideas for product solutions.

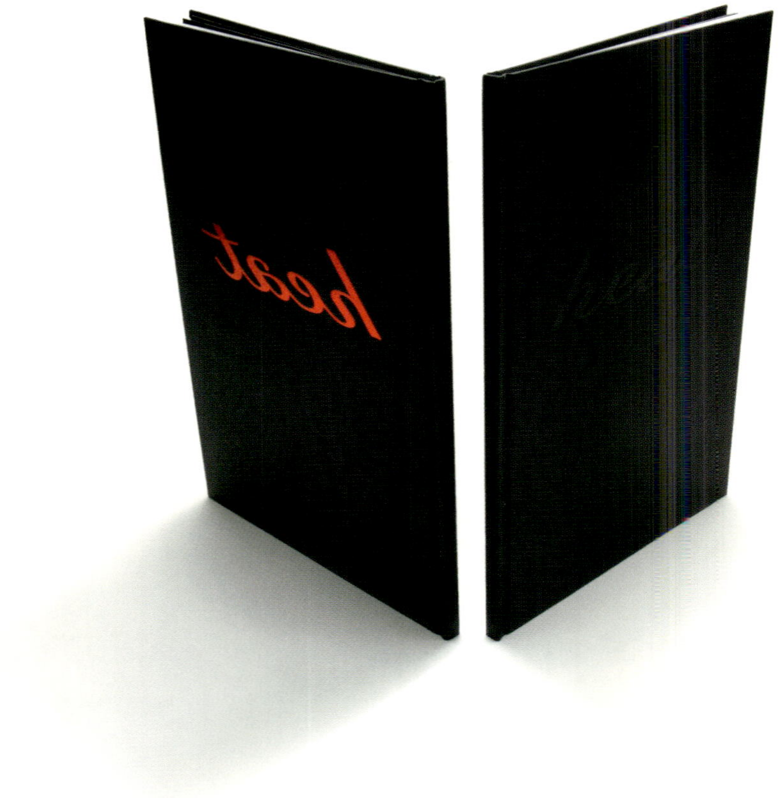

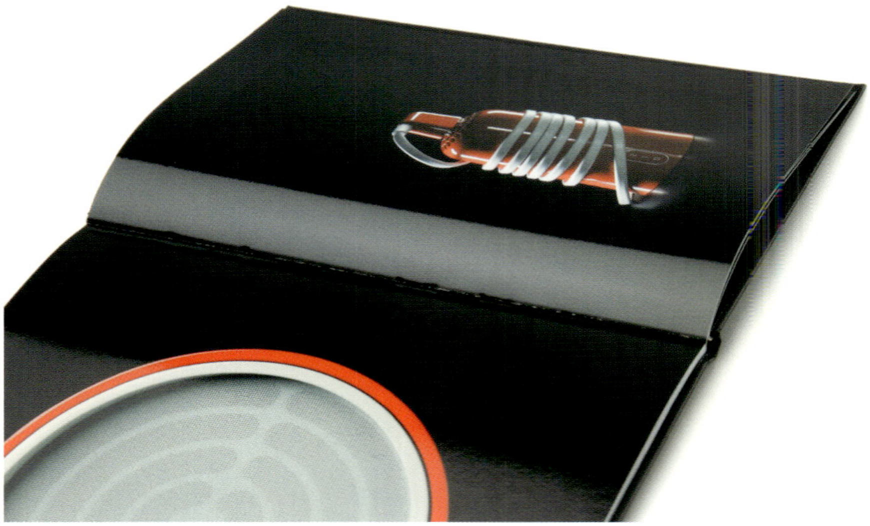

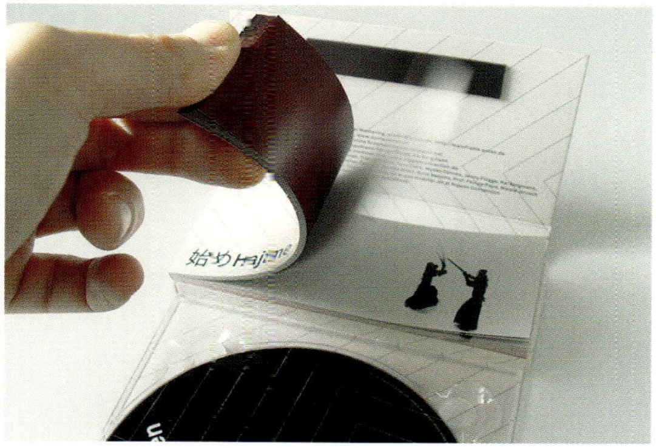

Auftraggeber Client
das modular music and publishing
Hattersheim, Germany

Projekt Project
Nippon Connection – Exchanging Tracks
CD-Packaging
CD packaging

Design
Kai Bergmann (CD/AD)
Jörg Baumann (Foto)
dreizueins architekturdesigngrafik
Frankfurt am Main, Germany
Christina Grasmann
Daniela Mladenic
Fachhochschule Mainz
Mainz, Germany

multiple winner

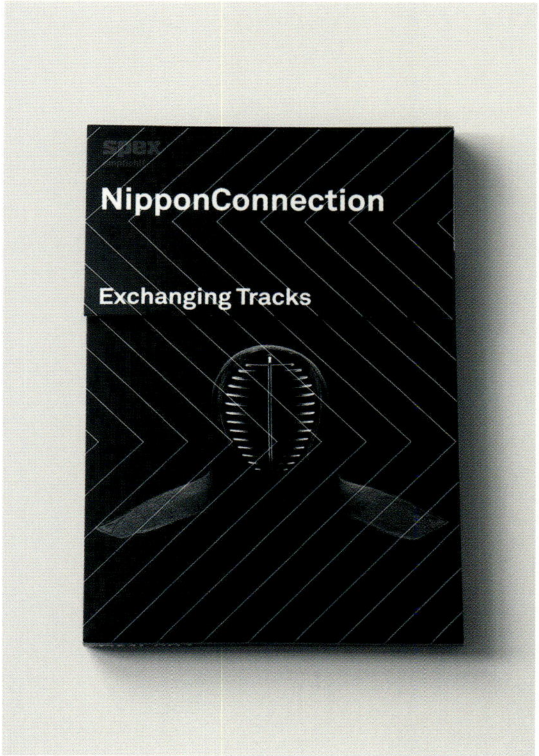

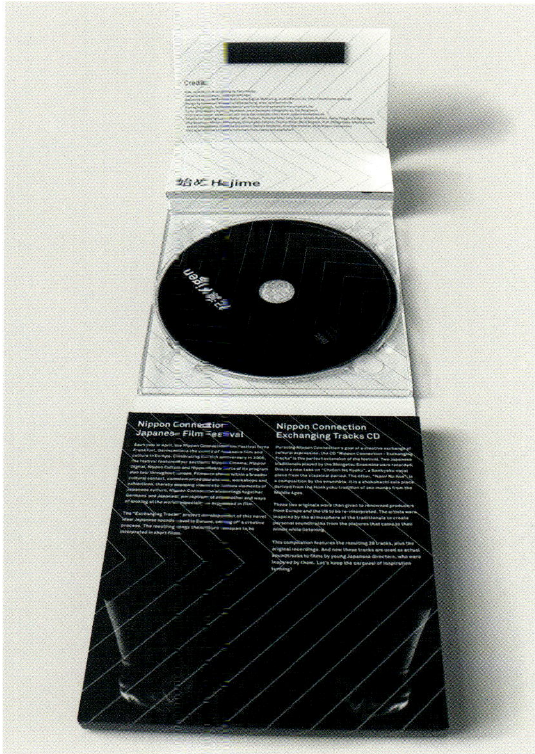

Die 3er CD-Box mit elektronischen Re-Interpretationen traditioneller japanischer Musik ist das Debüt des Labels. Da sie in Zusammenarbeit mit dem Filmfestival Nippon Connection entstanden ist, enthält das aufwändige Packaging – angelehnt an das Medium Film – ein Daumenkino, das die Sequenz eines Kendo-Kampfes wiedergibt. Der japanische Schwertkämpfer steht als Parabel für die drei musikalischen Ebenen: Kigen (Tradition), Moku So (Meditation) und Shikake (Angriff). Sich treffende Diagonalen werden zu richtunggebenden Zeichen, die den interkulturellen Austausch symbolisieren. Die visuelle Ästhetik kommuniziert den musikalischen Anspruch.

Nippon Connection – Exchanging Tracks, a three-CD compilation presenting re-interpretations of traditional Japanese music, is the debut of the music label. It is the result of a cooperation between the label and Frankfurt's Japanese film festival Nippon Connection. Coming with an exceptional packaging inspired by movies it contains a flip-book with a sequence of a kendo fight. The Japanese swordfighter is a parable for the three musical levels of the project: Kigen (tradition), moku so (meditation) and shikake (attack). Crossing diagonals symbolize the intercultural exchange, the all-over visual concept communicates the musical aspiration of the project.

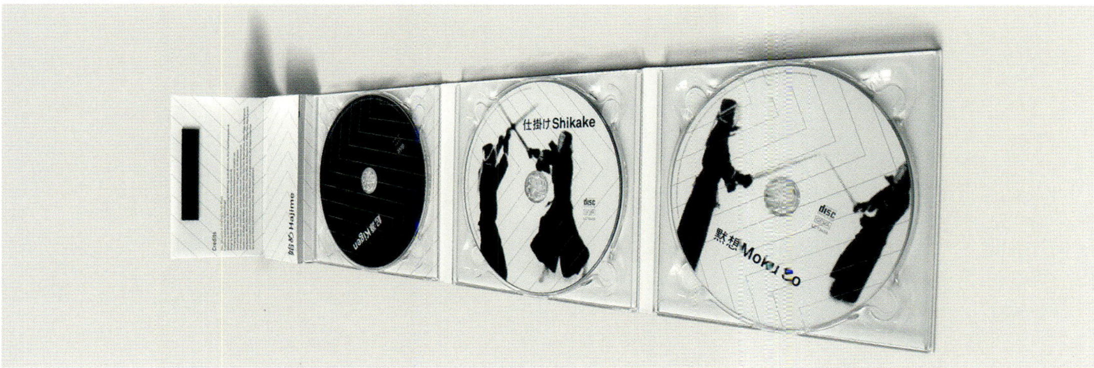

Auftraggeber Client
Kunstmuseum Stuttgart
Stuttgart, Germany

Projekt Project
Kunstmuseum Kommunikation
Einladungen, Quartalsprogramme, Ausstellungsreihe „Frischzelle“
Invitations, calendars of events, exhibition series "Life cell"

Design
Sascha Lobe
Kathrin Löser
L2M3 Kommunikationsdesign GmbH
Stuttgart, Germany

Ein Vierteljahr vor Eröffnung des Hauses wird das Erscheinungsbild eingeführt: Die Fotografie eines verpackten Kunstwerks dient in der Weihnachtsaussendung als Vorankündigung. Hier werden auch die Grundzüge für das weitere CD definiert: Abstrakte Titelblätter, die durch Faltungen entstehen, und eine selbstbewusste, moderne Typografie bringen auch unterschiedlichste Bildmotive und Kunstwerke in einem Erscheinungsbild zusammen und zeigen so die vielfältigen Aktivitäten des Kunstmuseum Stuttgart.

A quarter of a year before the grand opening the corporate design will be introduced: the photograph of a packaged work of art serves as a preview in the Christmas mailing. This is where the main features for the up-and-coming corporate design will be defined: Abstract title pages, created through folding, and a self-confident, modern typography get the most diverse picture themes and works of art to form one impression and thus show the diversity of activities of the Stuttgart Art Museum.

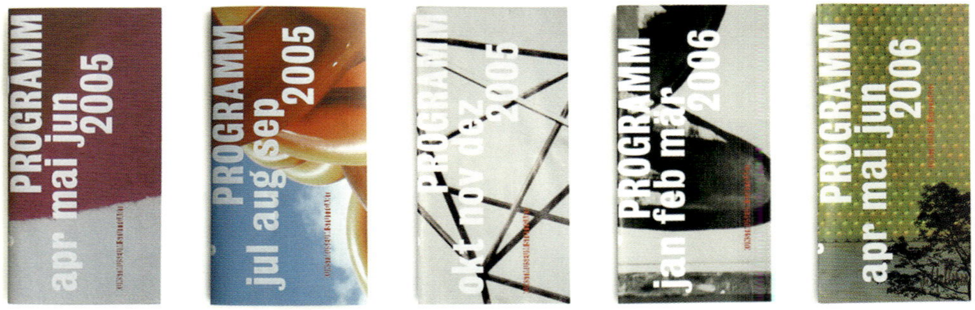

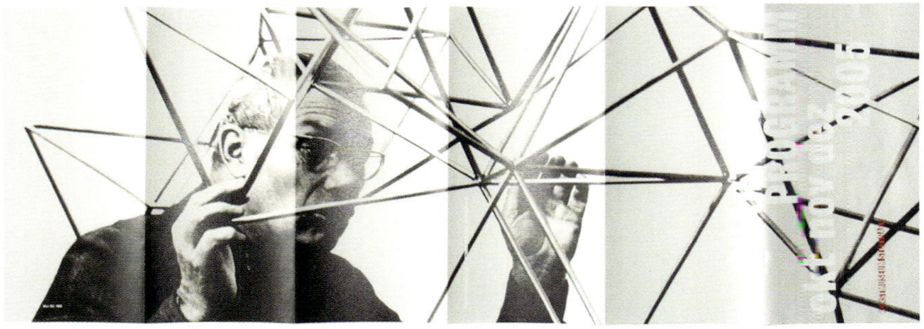

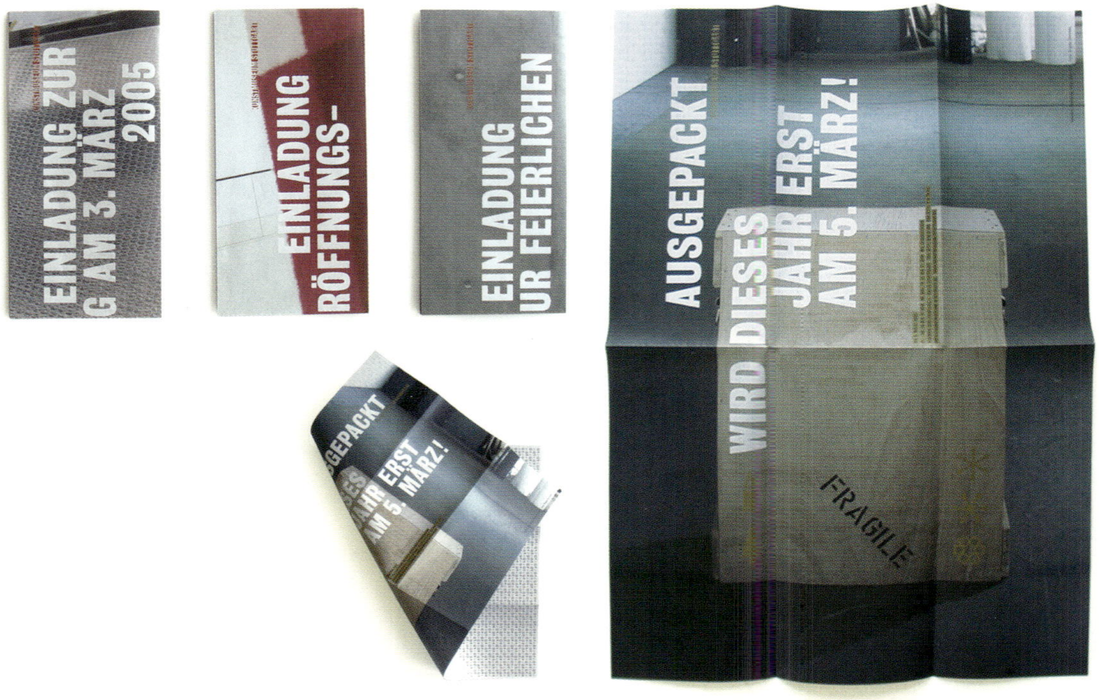

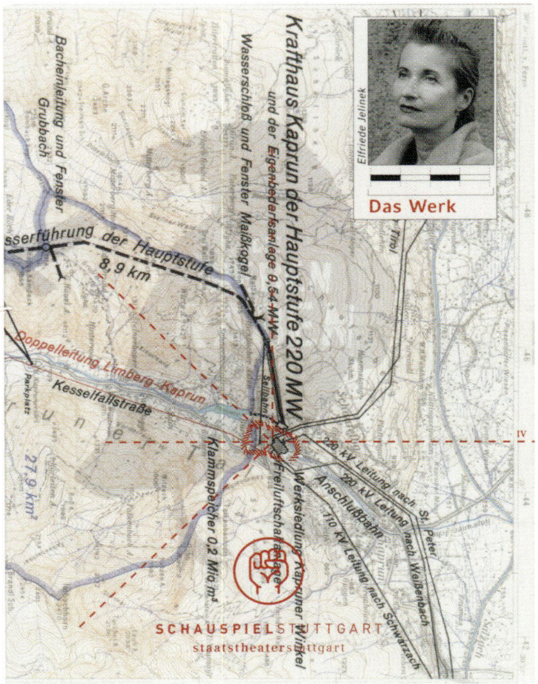

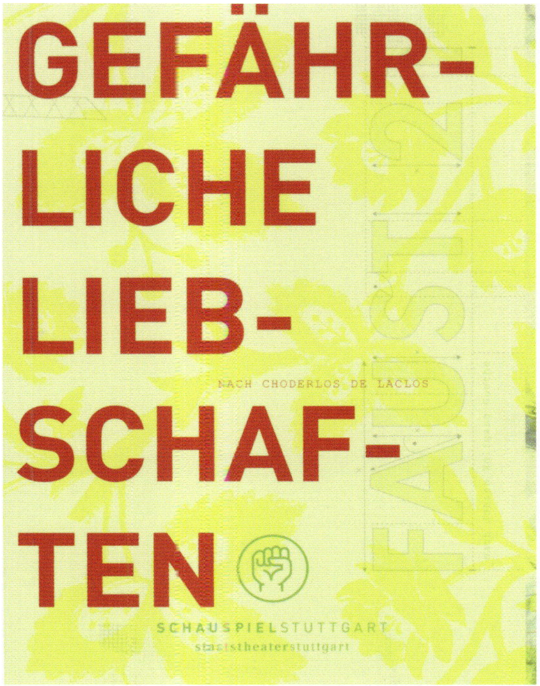

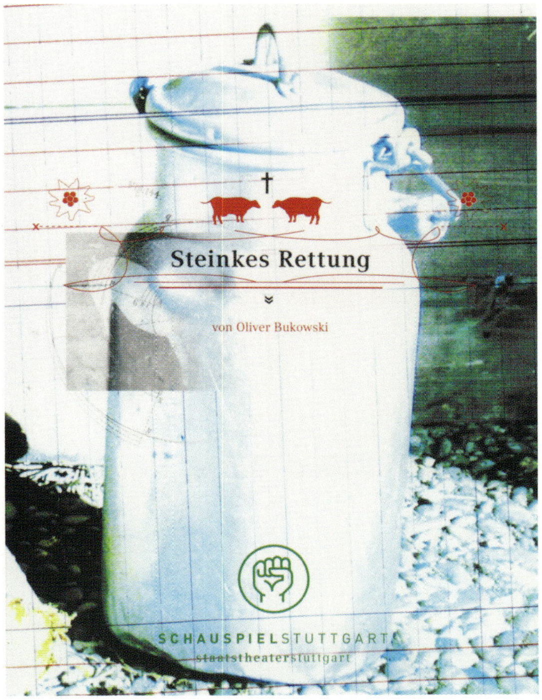

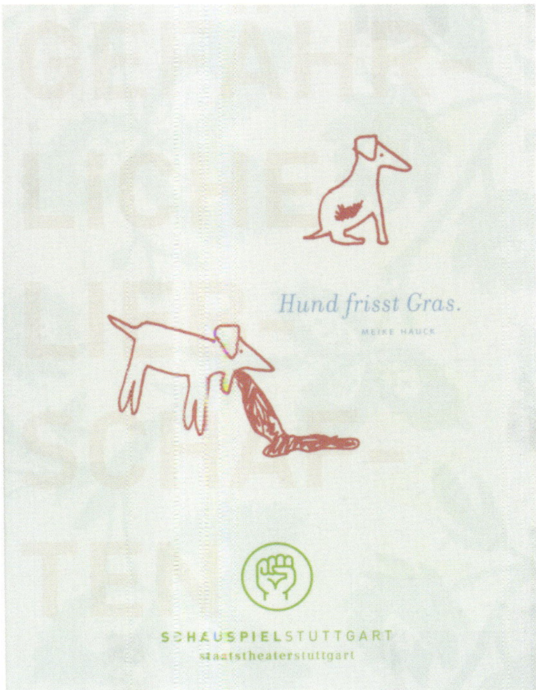

Auftraggeber Client
Staatstheater Stuttgart Schauspiel Stuttgart
Stuttgart Germany

Projekt Project
Schauspiel Stuttgart
Programmhefte
Programmes

Design
Anders Bergesen
Kirsten Dietz
Anika Marquardsen
Jochen Rädeker
Strichpunkt
Stuttgart Germany

Programmhefte für das Staatstheater Stuttgart Schauspiel Stuttgart

Programmes for the Staatstheater Stuttgart Schauspiel Stuttgart

Auftraggeber Client
lollekundbollek
Düsseldorf, Germany

Projekt Project
lollekundbollek
Geschäftspapierausstattung
Business stationery

Design
Christopher Wiehl
paarpiloten
Düsseldorf, Germany
Lars Loick
lollekundbollek
Düsseldorf, Germany

lollekundbollek ist ein Gestaltungskollektiv mit dem Schwerpunkt „Multimedia". Die beiden Gestalter Lars Loick (Lollek) und Thomas Baumhoff (Bollek) arbeiten sowohl als Team als auch einzeln. Ihr Humor und ihr Hang zu einer gewissen Ost-Nostalgie sind in ihrer Arbeit stilprägend.

lollekundbollek is a design collective that focuses on multimedia. The two designers Lars Loick (Lollek) and Thomas Baumhoff (Bollek) work together as a team and individually. Their humor and propensity towards a certain Eastern Bloc nostalgia are defining characteristics of their work.

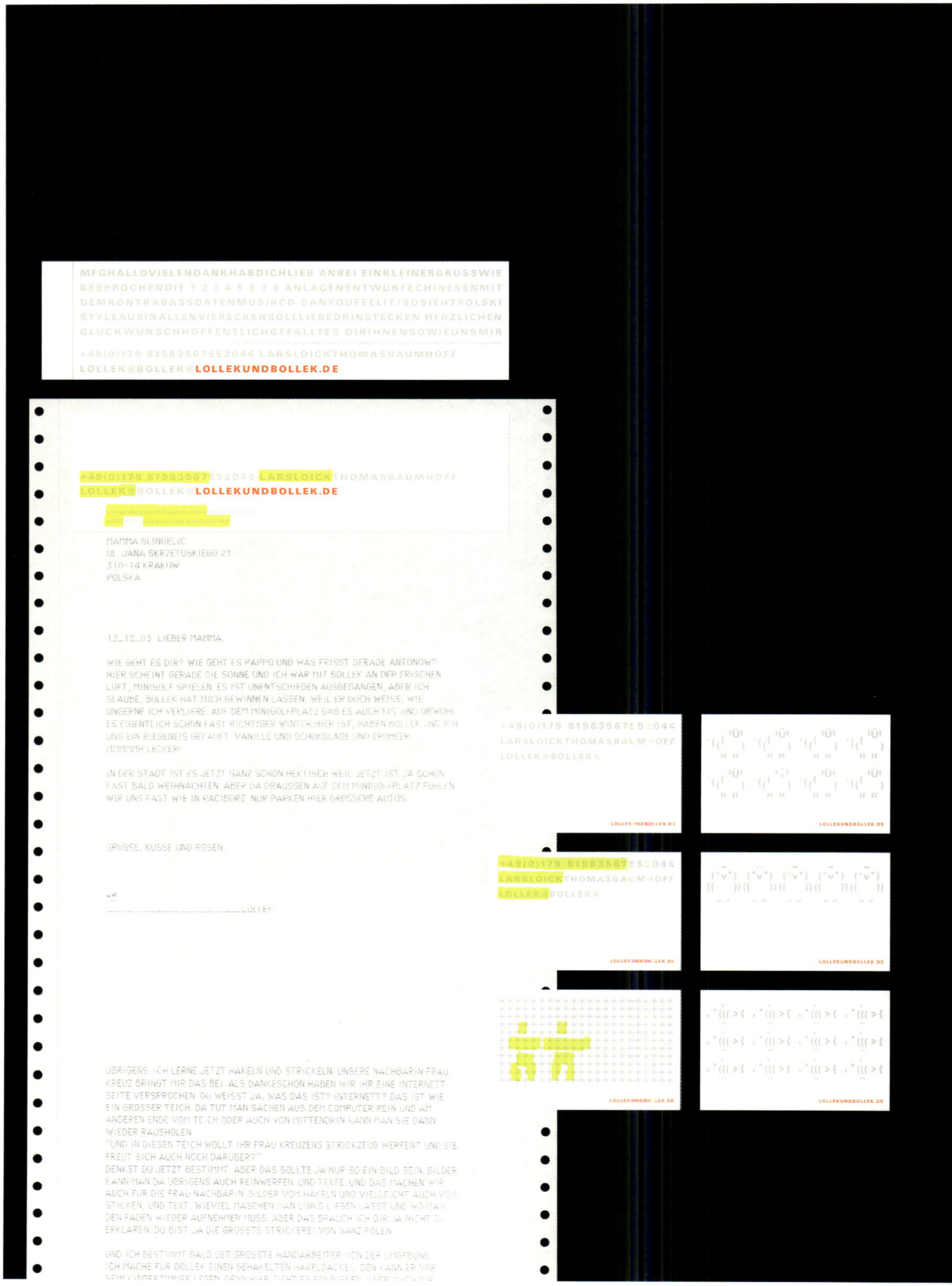

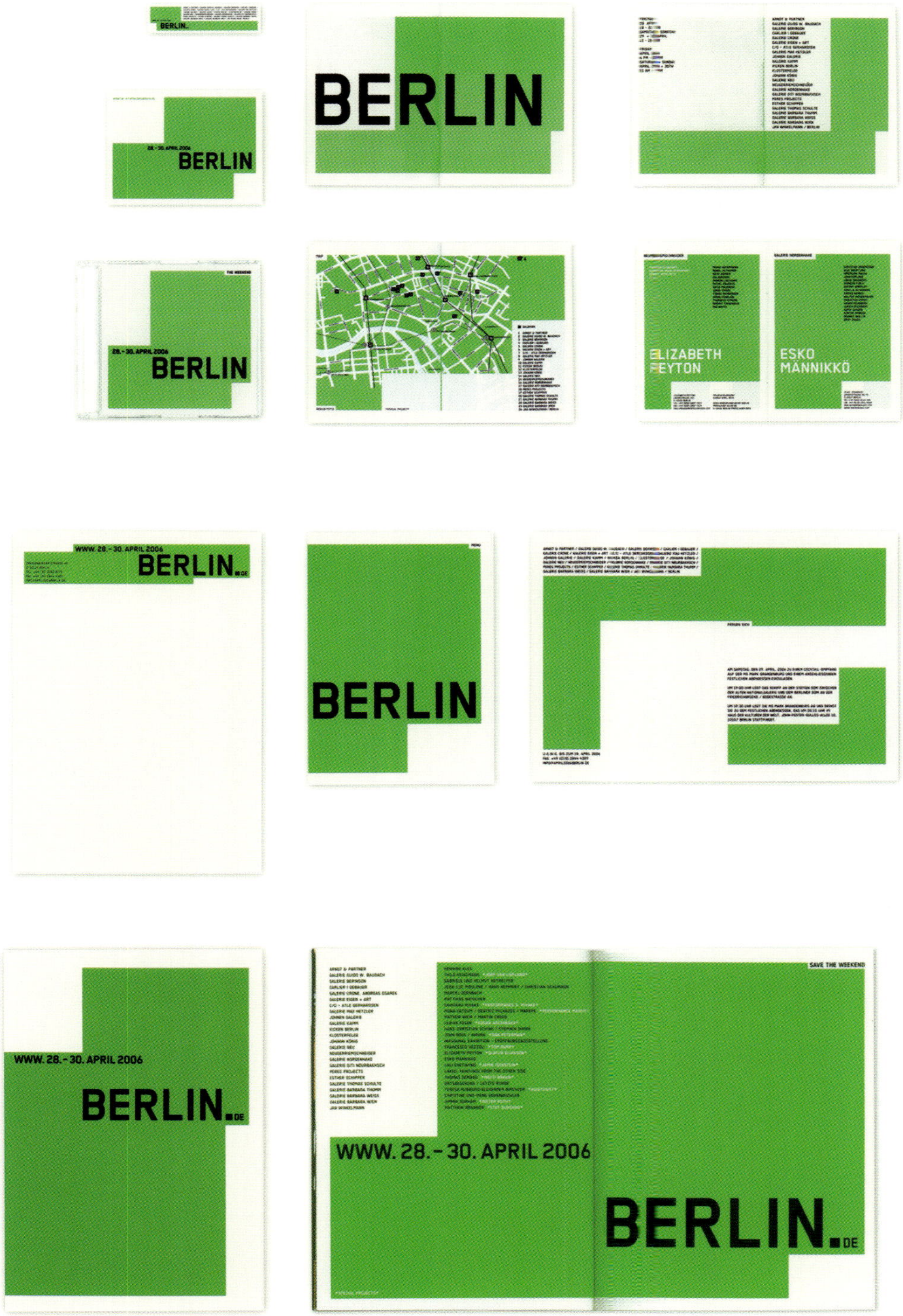

Auftraggeber Client
Berliner Galerienverbund
Berlin, Germany

Projekt Project
Berliner-Galerien-Wochenende
Verschiedene Kommunikationsmittel
Various communicative devices

Design
Christian Boros (CD)
Ingo Maak (AD)
Mathias Beyer (Design)
BOROS GmbH
Agentur für Kommunikation
Wuppertal, Germany

Verschiedene Kommunikationsmittel für den Zusammenschluss von 24 bedeutenden Berliner Galerien, die jährlich ein Wochenende der offenen Galerien veranstalten.

Different types of communicative devices for the union of 24 major galleries that organise an annual weekend of open galleries.

Auftraggeber Client
Württembergischer Kunstverein Stuttgart
Stuttgart, Germany

Projekt Project
Ausstellungsplakat Württembergischer Kunstverein Stuttgart
Plakat
Poster

Design
Sascha Lobe
Ina Bauer
L2M3 Kommunikationsdesign GmbH
Stuttgart, Germany

Für das Ankündigungsplakat zur Gruppenausstellung „Expanded Media. Medien im Raum" im Württembergischen Kunstverein Stuttgart wurde der Ausstellungstitel typografisch umgesetzt. Es entsteht ein einprägsamer visueller Code, der auch bei ausschnittartiger Abbildung, z. B. in gefalteten Flyern und Einladungskarten, sehr prägnant bleibt.
Durch das Beibehalten verschiedener Grundelemente und -positionen wie Adress- und Öffnungszeitenblock oder die Verwendung einer sehr feinen Typografie für den Infoblock fügt sich das Plakat trotz seiner prägnanten Gestaltung in die bestehende Serie von Ausstellungsplakaten des WKV, deren unterschiedliche Ansätze zwischen zurückhaltend typografierten Fotografien und expressivem Grafikdesign pendeln.

For the advertising poster for the group exhibition "Expanded Media. Media in Space" in the Württembergischer Kunstverein Stuttgart, the title of the exhibition was typographically transformed. A memorable visual code results that remains very concise even if represented in an extraction-like form, e.g. in folded flyers and invitations.
By keeping several basic elements and positions like the address and opening hours blocks, or the use of very dignified typography for the information block, the poster, despite its concise design, still blends into the existing series of WKV exhibition posters, whose various approaches range anywhere from subtle typographed photography to expressive graphic design.

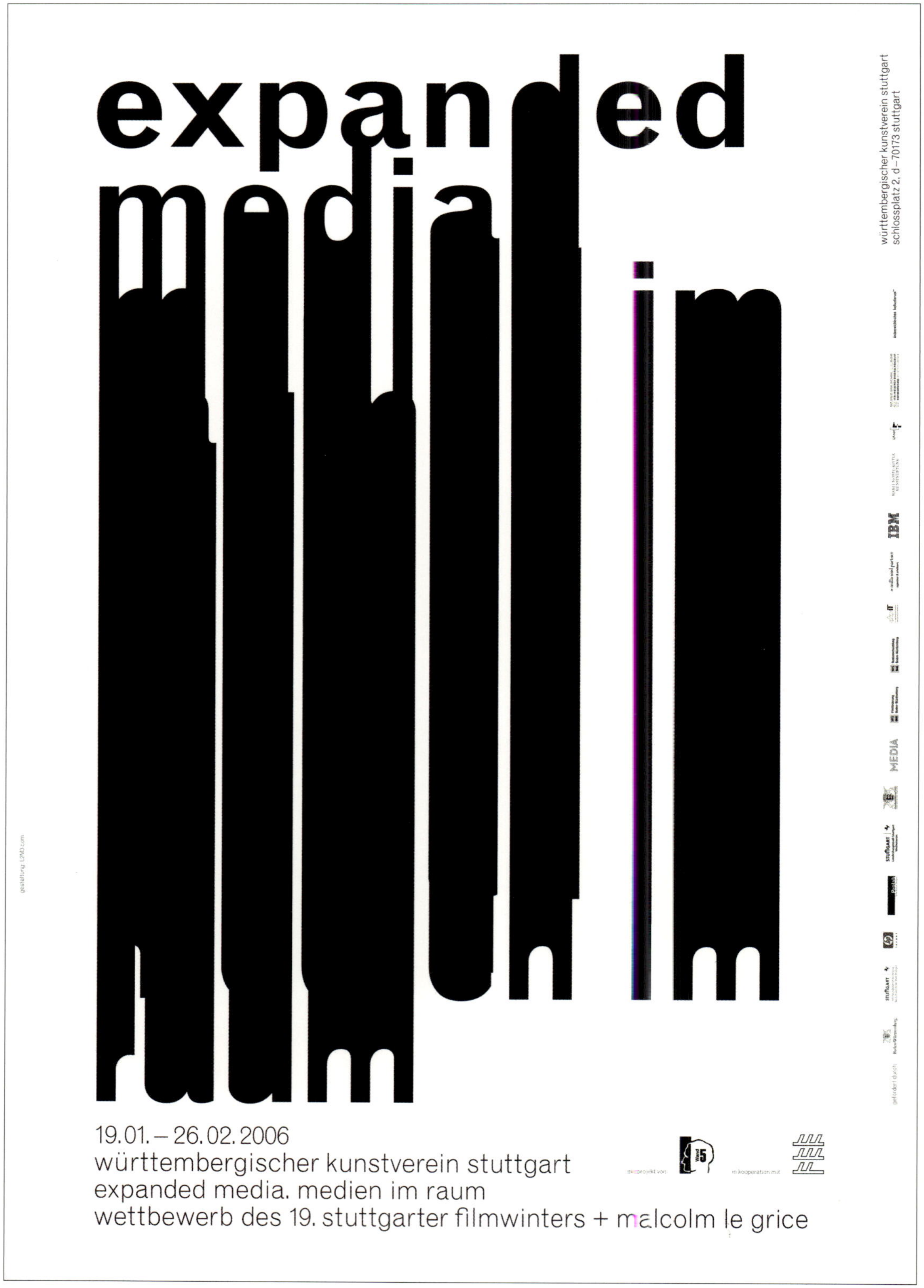

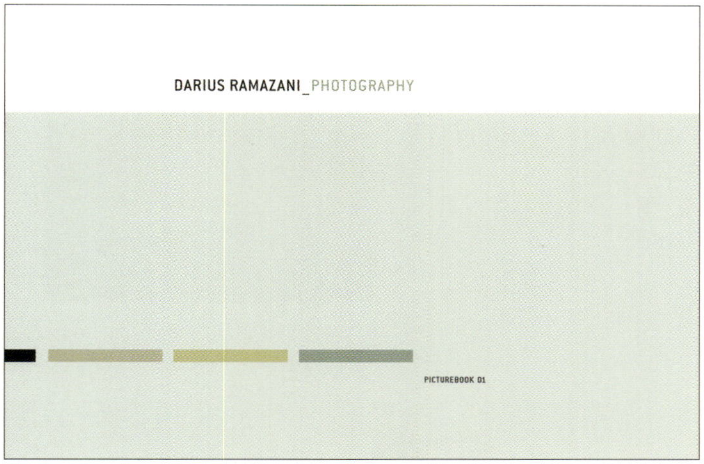

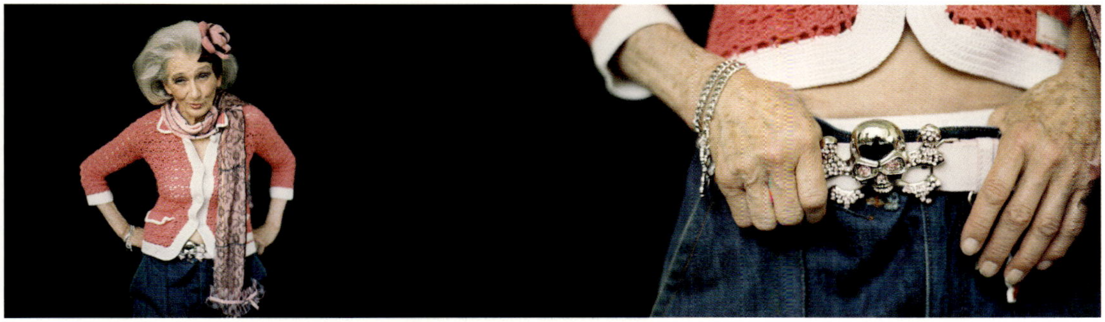

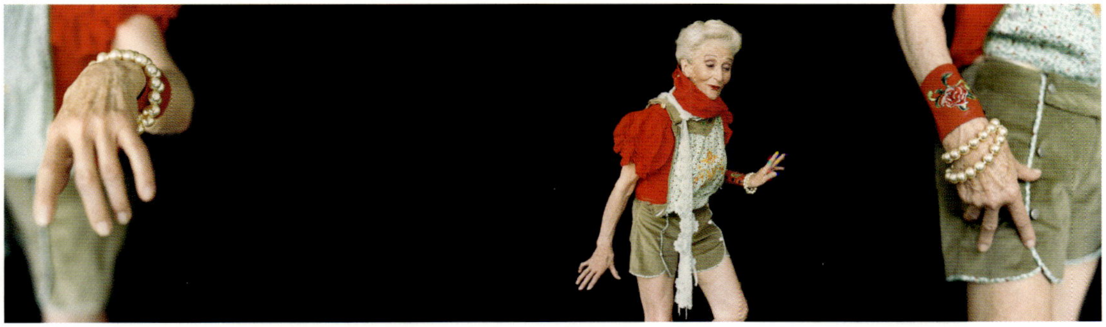

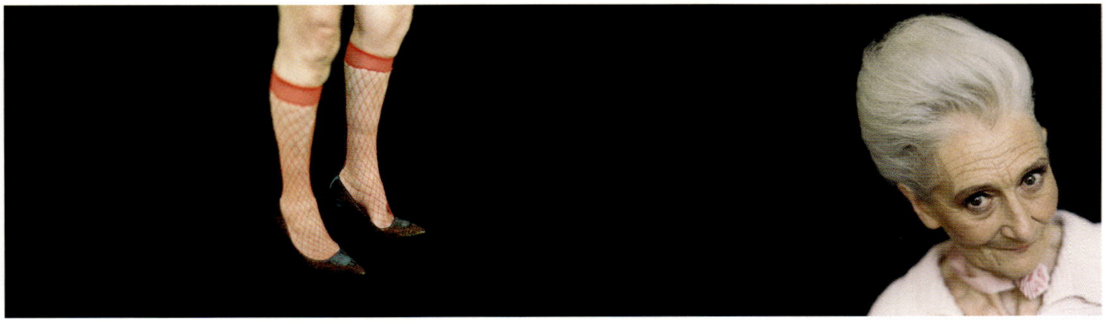

Auftraggeber Client
Darius Ramazani Photography
Berlin, Germany

Projekt Project
Darius Ramazani Photography Picturebook 01
Buch
Book

Design
Georg Andreas Suhr, Grafik Design
Berlin, Germany

Eigentlich war das Buch als Eigenwerbung gedacht, doch wer es in die Hand nimmt, mag es nicht mehr hergeben und stellt es prominent ins Regal: Der Fotoband des Fotografen Darius Ramazani enthält einen gelungenen Querschnitt aus freien und Auftragsarbeiten, darunter Modestrecken für die Modemesse BREAD & butter und Porträtserien. Porträts gelten als Stärke des 37-Jährigen, der zu seinen Kunden Konzerne wie Debitel, Nokia und Fortis zählt, aber auch für Projekte wie das Deutsche Down-Syndrom Center und den Deutschen Katholikentag fotografiert. Wirkungsvoll in Szene gesetzt wurden die Arbeiten im Fotobuch durch Grafiker Georg Andreas Suhr.

Initially the booklet of photographs by Darius Ramazani was intended for advertising purposes only. However customers and friends found that it was unputdownable. It shows a representative selection of commissioned work as well as free projects. Amongst other, readers will find a stretch of photographs taken at the Berlin fashion fair BREAD & butter as well as a series of portraits, the domain Ramazani is best-known for. At age 37, he works for companies like Debitel and Nokia and also enjoys projects such as taking pictures for the German Down-Syndrom Centre or the annual German Catholic Day. The unique layout was done by Georg Andreas Suhr.

Auftraggeber Client
Volkswagen AG
Wolfsburg, Germany

Projekt Project
GOLF LOVES EUROPE EUROPE LOVES GOLF
Katalog
Catalogue

Design
Johannes Plass
Paul Neulinger
Sven Ritterhoff
Julia Guther
Trine Skraastad
Trixi Rossi
Mutabor Design GmbH
Hamburg, Germany
Harm Clüver
Kay Dohnke
Uwe Pütz
G+J Corporate Media GmbH (Redaktion)
Hamburg, Germany
Rüdiger Schingale
Volkswagen AG (Auftraggeber)
Wolfsburg, Germany

Wie sieht Europa den Golf? Der Deutschen liebstes Kind ist ihr Auto. Doch was sagen Dänen und Tschechen, Niederländer und Schweizer dazu? Diese Frage stellte das Volkswagen Magazin zum 30. Geburtstag des Golf und richtete im vergangenen Jahr einen Fotowettbewerb aus. Die Studenten ausgewählter europäischer Fotoschulen suchten nach Antworten. Ihre Aufgabe war es, eine eigene, persönliche und landesspezifische Sichtweise zu finden. Nachwuchsfotografen aus sieben Ländern legten insgesamt 28 Arbeiten vor. Manche porträtieren den Golf verzückt und schwärmerisch ... andere können sich ein Augenzwinkern nicht verkneifen.

What does Europe think of the Golf? The Germans love their cars more than anything else. But what do the Danes and the Czechs, the Dutch and the Swiss think? Volkswagen Magazin asked this question on the occasion of the Golf's 30th birthday and held a photo competition on the subject last year. Students at selected European schools of photography tried to find some answers. Their task was to find their own personal and country-specific way of dealing with the question. Up-and-coming photographers from seven countries submitted a total of 28 works. Some portrayed the Golf effusively and adoringly ... Others couldn't resist a sly wink.

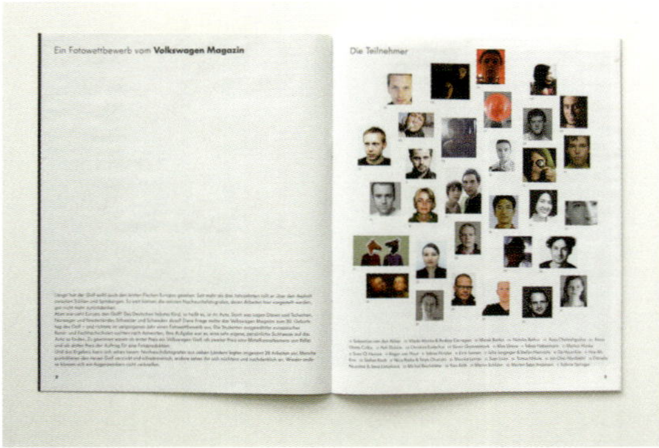

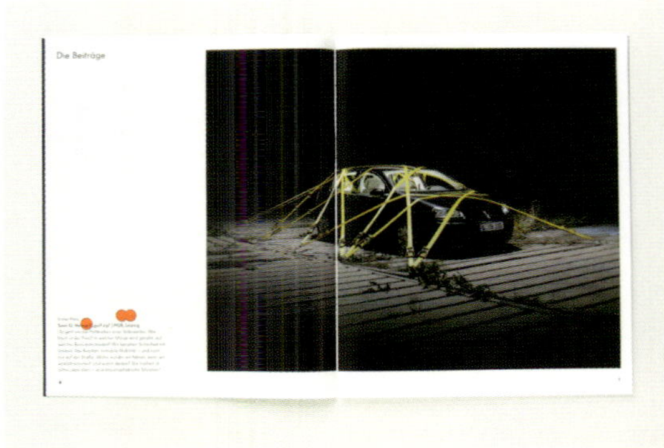

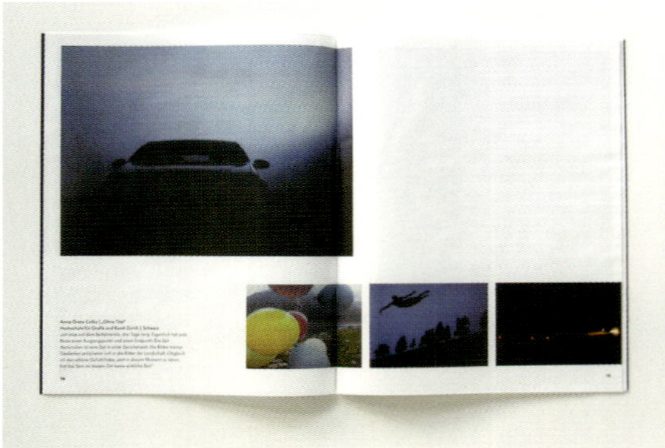

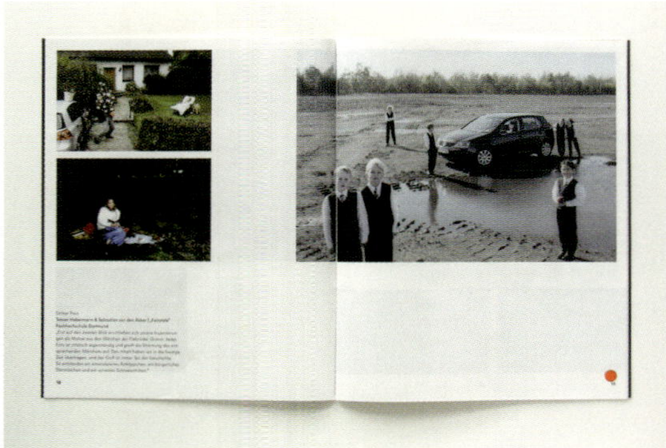

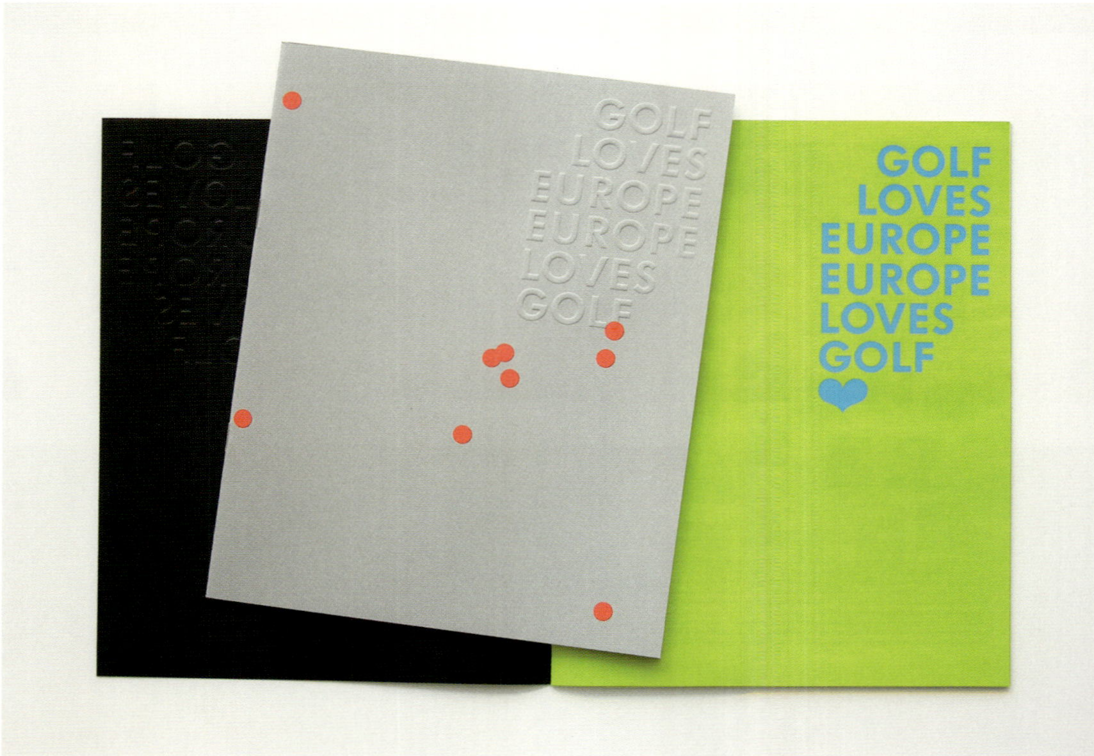

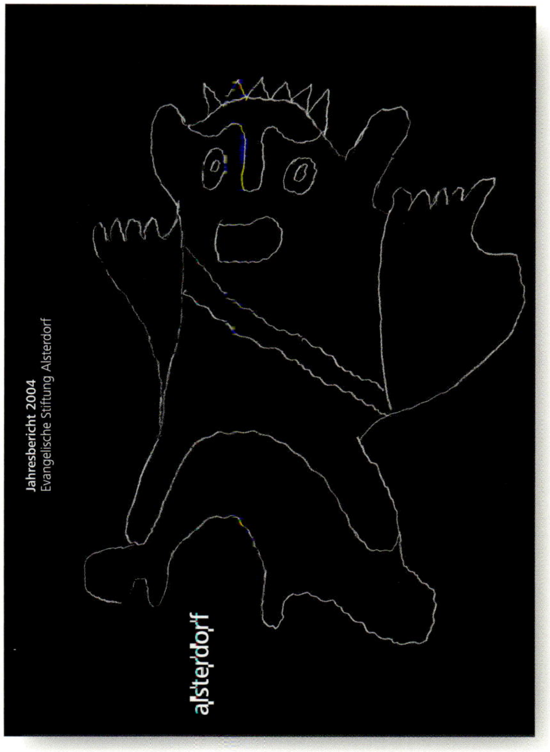

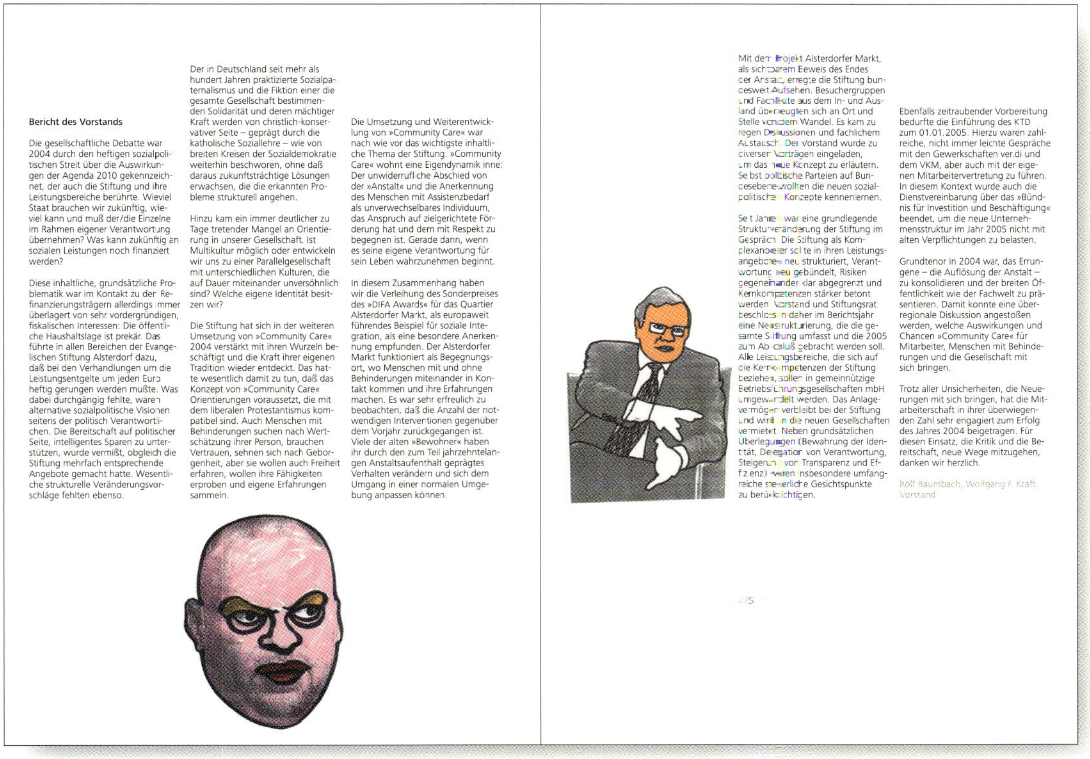

Bericht des Vorstands

Die gesellschaftliche Debatte war 2004 durch den heftigen sozialpolitischen Streit über die Auswirkungen der Agenda 2010 gekennzeichnet, der auch die Stiftung und ihre Leistungsbereiche berührte. Wieviel Staat brauchen wir zukünftig, wieviel kann und muß der/die Einzelne im Rahmen eigener Verantwortung übernehmen? Was kann zukünftig an sozialen Leistungen noch finanziert werden?

Diese inhaltliche, grundsätzliche Problematik war im Kontakt zu den Refinanzierungsträgern allerdings immer überlagert von sehr vordergründigen, fiskalischen Interessen: Die öffentliche Haushaltslage ist prekär. Das führte in allen Bereichen der Evangelischen Stiftung Alsterdorf dazu, daß bei den Verhandlungen um die Leistungsentgelte um jeden Euro heftig gerungen werden mußte. Was dabei durchgängig fehlte, waren alternative sozialpolitische Visionen seitens der politisch Verantwortlichen. Die Bereitschaft auf politischer Seite, intelligentes Sparen zu unterstützen, wurde vermißt, obgleich die Stiftung mehrfach entsprechende Angebote gemacht hatte. Wesentliche strukturelle Veränderungsvorschläge fehlten ebenso.

Der in Deutschland seit mehr als hundert Jahren praktizierte Sozialpaternalismus und die Fiktion einer die gesamte Gesellschaft bestimmenden Solidarität und deren mächtiger Kraft werden von christlich-konservativer Seite – geprägt durch die katholische Soziallehre – wie von breiten Kreisen der Sozialdemokratie weiterhin beschworen, ohne daß daraus zukunftsträchtige Lösungen erwachsen, die die erkannten Probleme strukturell angehen.

Hinzu kam ein immer deutlicher zu Tage tretender Mangel an Orientierung in unserer Gesellschaft. Ist Multikultur möglich oder entwickeln wir uns zu einer Parallelgesellschaft mit unterschiedlichen Kulturen, die auf Dauer miteinander unversöhnlich sind? Welche eigene Identität besitzen wir?

Die Stiftung hat sich in der weiteren Umsetzung von »Community Care« 2004 verstärkt mit ihren Wurzeln beschäftigt und die Kraft ihrer eigenen Tradition wieder entdeckt. Das hatte wesentlich damit zu tun, daß das Konzept von »Community Care« Orientierungen voraussetzt, die mit dem liberalen Protestantismus kompatibel sind. Auch Menschen mit Behinderungen suchen nach Wertschätzung ihrer Person, brauchen Vertrauen, sehnen sich nach Geborgenheit, aber sie wollen auch Freiheit erfahren, wollen ihre Fähigkeiten erproben und eigene Erfahrungen sammeln.

Die Umsetzung und Weiterentwicklung von »Community Care« war nach wie vor das wichtigste inhaltliche Thema der Stiftung. »Community Care« wohnt eine Eigendynamik inne: Der unwiderrufliche Abschied von der »Anstalt« und die Anerkennung des Menschen mit Assistenzbedarf als unverwechselbares Individuum, das Anspruch auf zielgerichtete Förderung hat und dem mit Respekt zu begegnen ist. Gerade dann, wenn es seine eigene Verantwortung für sein Leben wahrzunehmen beginnt.

In diesem Zusammenhang haben wir die Verleihung des Sonderpreises des »DIFA Awards« für das Quartier Alsterdorfer Markt, als europaweit führendes Beispiel für soziale Integration, als eine besondere Anerkennung empfunden. Der Alsterdorfer Markt funktioniert als Begegnungsort, wo Menschen mit und ohne Behinderungen miteinander in Kontakt kommen und ihre Erfahrungen machen. Es war sehr erfreulich zu beobachten, daß die Anzahl der notwendigen Interventionen gegenüber dem Vorjahr zurückgegangen ist. Viele der alten »Bewohner« haben ihr durch den zum Teil jahrzehntelangen Anstaltsaufenthalt geprägtes Verhalten verändern und sich dem Umgang in einer normalen Umgebung anpassen können.

Mit dem Projekt Alsterdorfer Markt, als sichtbarem Beweis des Endes der Anstalt, erregte die Stiftung bundesweit Aufsehen. Besuchergruppen und Fachleute aus dem In- und Ausland überzeugten sich an Ort und Stelle von dem Wandel. Es kam zu regen Diskussionen und fachlichem Austausch. Der Vorstand wurde zu diversen Vorträgen eingeladen, um das neue Konzept zu erläutern. Selbst politische Parteien auf Bundesebene wollten die neuen sozialpolitischen Konzepte kennenlernen.

Seit Jahren war eine grundlegende Strukturveränderung der Stiftung im Gespräch. Die Stiftung als Komplexanbieter sollte in ihren Leistungsangeboten neu strukturiert, Verantwortung neu gebündelt, Risiken gegeneinander klar abgegrenzt und Kernkompetenzen stärker betont werden. Vorstand und Stiftungsrat beschlossen daher im Berichtsjahr eine Neustrukturierung, die die gesamte Stiftung umfasst und die 2005 zum Abschluß gebracht werden soll. Alle Leistungsbereiche, die sich auf die Kernkompetenzen der Stiftung beziehen, sollen in gemeinnützige Betriebsführungsgesellschaften mbH umgewandelt werden. Das Anlagevermögen verbleibt bei der Stiftung und wird an die neuen Gesellschaften vermietet. Neben grundsätzlichen Überlegungen (Bewahrung der Identität, Delegation von Verantwortung, Steigerung von Transparenz und Effizienz) waren insbesondere umfangreiche steuerliche Gesichtspunkte zu berücksichtigen.

Ebenfalls zeitraubender Vorbereitung bedurfte die Einführung des KTD zum 01.01.2005. Hierzu waren zahlreiche, nicht immer leichte Gespräche mit den Gewerkschaften ver.di und dem VKM, aber auch mit der eigenen Mitarbeitervertretung zu führen. In diesem Kontext wurde auch die Dienstvereinbarung über das »Bündnis für Investition und Beschäftigung« beendet, um die neue Unternehmensstruktur im Jahr 2005 nicht mit alten Verpflichtungen zu belasten.

Grundtenor in 2004 war, das Errungene – die Auflösung der Anstalt – zu konsolidieren und der breiten Öffentlichkeit wie der Fachwelt zu präsentieren. Damit konnte eine überregionale Diskussion angestoßen werden, welche Auswirkungen und Chancen »Community Care« für Mitarbeiter, Menschen mit Behinderungen und die Gesellschaft mit sich bringen.

Trotz aller Unsicherheiten, die Neuerungen mit sich bringen, hat die Mitarbeiterschaft in ihrer überwiegenden Zahl sehr engagiert zum Erfolg des Jahres 2004 beigetragen. Für diesen Einsatz, die Kritik und die Bereitschaft, neue Wege mitzugehen, danken wir herzlich.

Rolf Baumbach, Wolfgang F. Kraft, Vorstand

4/5

Auftraggeber Client
Evangelische Stiftung Alsterdorf
Hamburg, Germany

Projekt Project
Jahresbericht Alsterdorf 2004
Jahresbericht
Annual report

Design
Prof. Andreas Uebele
Alexandra Busse
büro uebele visuelle kommunikation
Stuttgart, Germany

Das Unternehmen arbeitet mit behinderten Menschen. Die Vorstellung, dass die Kunden ihren eigenen Geschäftsbericht illustrieren, ist reizvoll. Die Künstlergruppe „Die Schlumper“, eine Einrichtung der Evangelischen Stiftung Alsterdorf, half uns bei der Umsetzung. Gelungen ist das Experiment bei den Porträts, die für den Bericht gefertigt wurden. Die Künstler reagieren auf den Inhalt. Bei anderen Seiten wird offenbar, dass sich Kunst dem Gebrauch entzieht. Die Abbildung von Kunst kann nur ihrem eigenen Zweck dienen.

The company works with disabled people. The idea of having customers illustrate their own annual report is fascinating. The group of artists "Die Schlumper" set up by the Alsterdorf Evangelical Foundation helped us put this into practice. Where the experiment has succeeded is in the portraits created for the report, where the artists respond to the contents. On other pages it becomes clear that art does not lend itself readily to other purposes. L'art, quite simply, is pour l'art.

Auftraggeber Client
BMW AG
München, Germany

Projekt Project
BMW Magazin special – Der neue 3er
Magazin
Magazine

Design
ringzwei
Hamburg, Germany
für/for
Hoffmann und Campe
Corporate Publishing
Hamburg, Germany

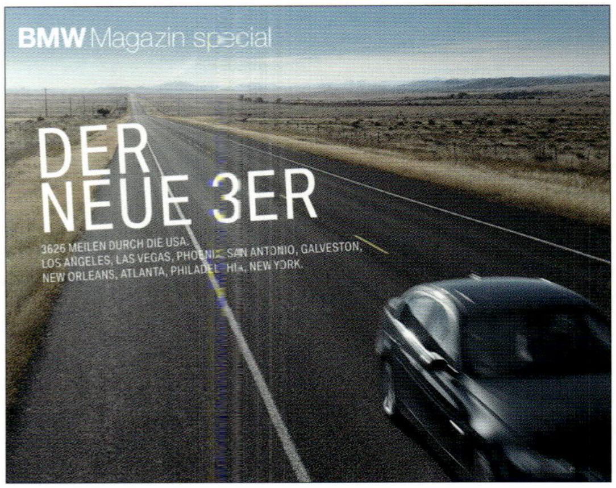

Das „BMW Magazin special – Der neue 3er" ist ein monothematisches Magazin, das zur Einführung des neuen BMW 3ers eine stimmige Markenwelt rund um das Fahrzeug kreiert und es auf einer Reise von Los Angeles nach New York mit seinen Besonderheiten vorstellt. Als Teil der BMW Magazin Markenfamilie nutzen die Specials eine hohe journalistische Glaubwürdigkeit und überdurchschnittlich lange Lesedauer, um Neukunden und BMW Fahrer zu begeistern. Sie sind exakt auf die Bedürfnisse von Interessenten zugeschnitten und werden weltweit distribuiert, über Händler, auf Messen und Events oder als Teil von Interessenten-Gewinnungsprogrammen.

"BMW Magazin special – Der neue 3er" is a magazine focusing solely on the new 3-Series BMW. It creates a coherent brand universe covering everything to do with the new car for its launch, introducing the vehicle and its special features on a journey from Los Angeles to New York. As part of the BMW magazine brand family, these specials have high journalistic credibility with a longer than average reading time to whet the appetite of new customers and BMW drivers. Tailored to the needs of potential buyers, the magazine is distributed worldwide via dealers, at trade fairs and events or during programs targeting potential buyers.

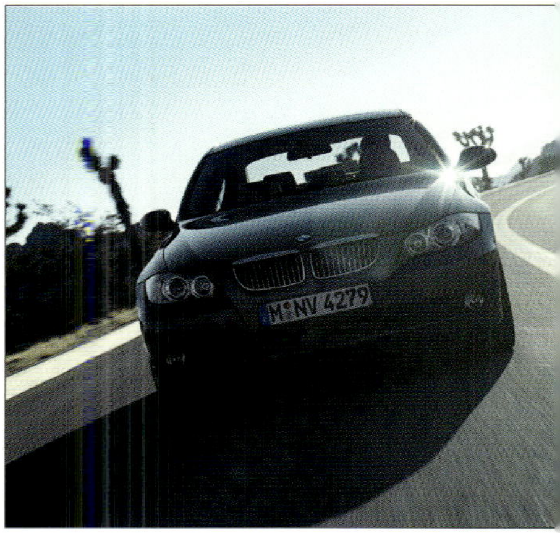

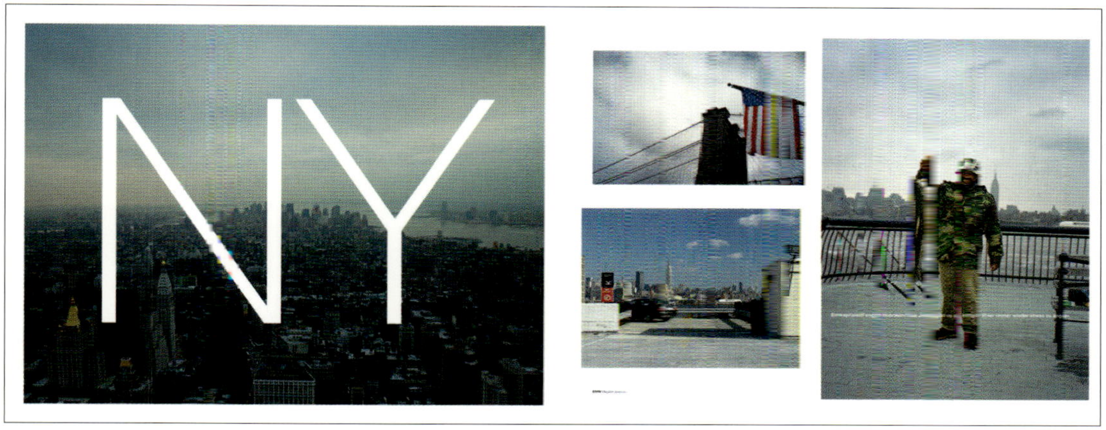
NY

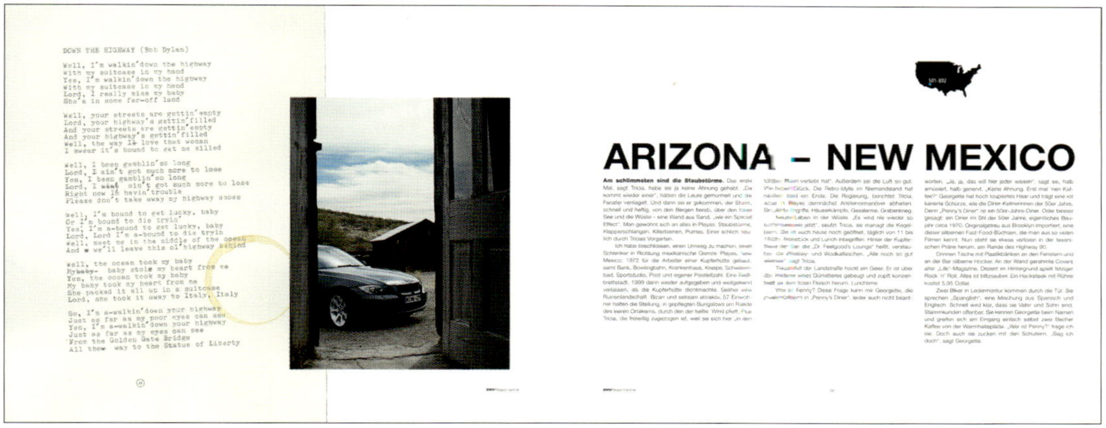
ARIZONA – NEW MEXICO

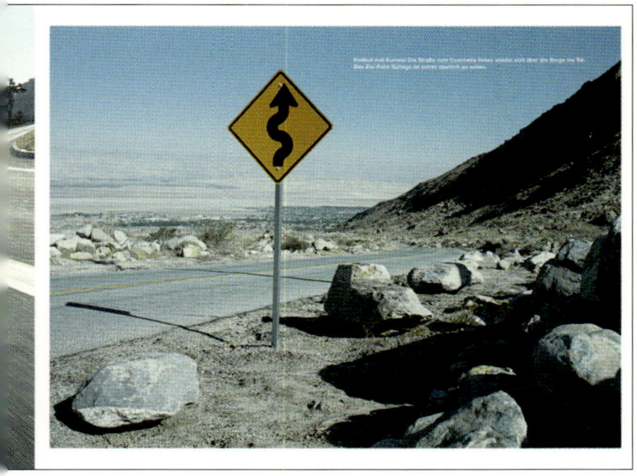

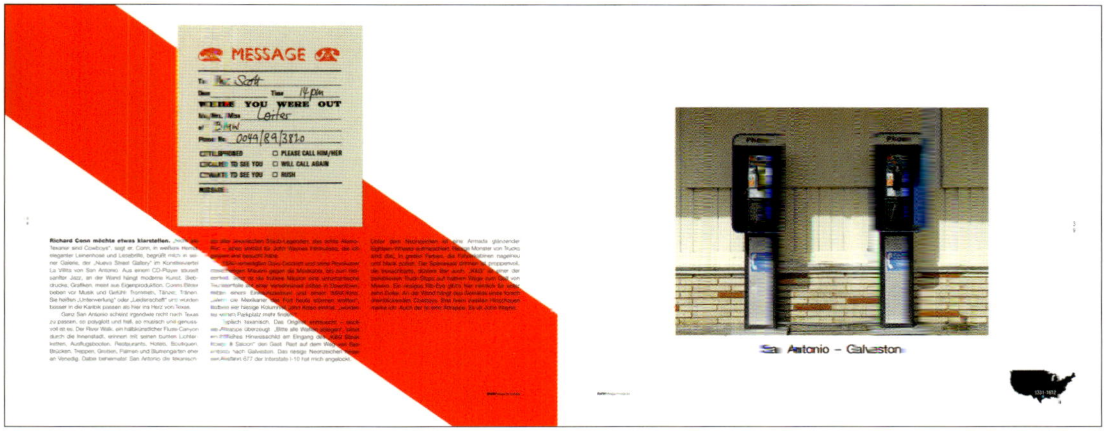
MESSAGE
YOU WERE OUT

Auftraggeber Client
ThyssenKrupp Elevator AG
Düsseldorf, Germany

Projekt Project
Airports Unlimited
Flughafenmagazin
Airport magazine

Design
Michaela Müller
Daniela Schäfer
RTS Rieger Team Werbeagentur GmbH
Düsseldorf, Germany

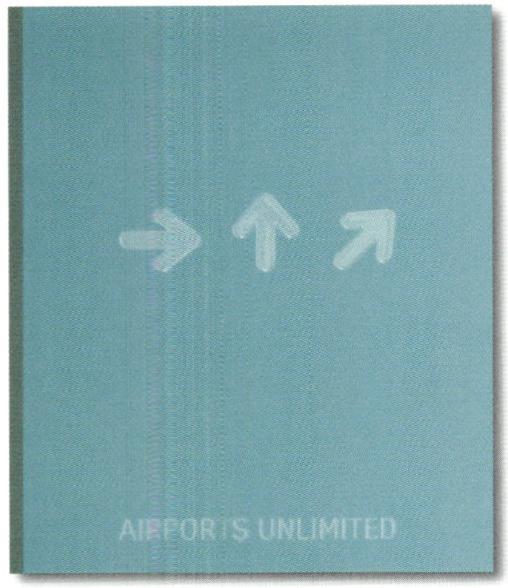

Was Menschen bewegt und was ThyssenKrupp Elevator alles dafür tut: Das sind die Schwerpunkte des Sonderbandes zum Thema Flughafen. Das Ganze fügt sich harmonisch zu einem Medium, das die umfassende Flughafenkompetenz und die weltweiten Referenzen des Unternehmens adäquat kommuniziert. Jeder Artikel folgt einer ganz eigenen, überraschenden Layout-Idee, die zum Teil durch interaktive Elemente unterstützt wird. Das Layout ist nicht uniform, sondern wird durch das Meta-Thema Flughafen geklammert. Nur Fliegen ist schöner.

The things that move people, and the various ways in which ThyssenKrupp Elevator supports that movement, provides the focus for the special magazine edition on airports. Everything flows together in a harmonious medium which appropriately communicates the company's extensive expertise in airports and its worldwide reference projects. Every article follows its own surprising layout idea, occasionally supported by interactive elements. Although the layout is not uniform, it is framed in by the meta-topic of "airports". Only flying could be more fun.

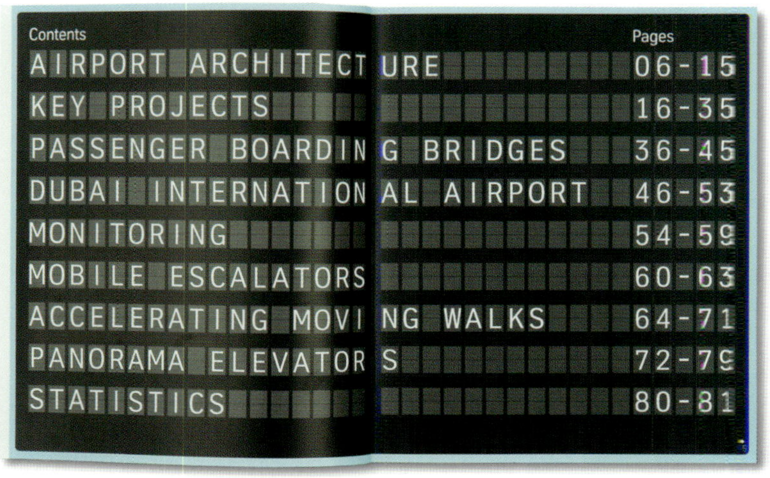
Contents
Pages
AIRPORT ARCHITECTURE 06-15
KEY PROJECTS 16-35
PASSENGER BOARDING BRIDGES 36-45
DUBAI INTERNATIONAL AIRPORT 46-53
MONITORING 54-59
MOBILE ESCALATORS 60-63
ACCELERATING MOVING WALKS 64-71
PANORAMA ELEVATORS 72-79
STATISTICS 80-81

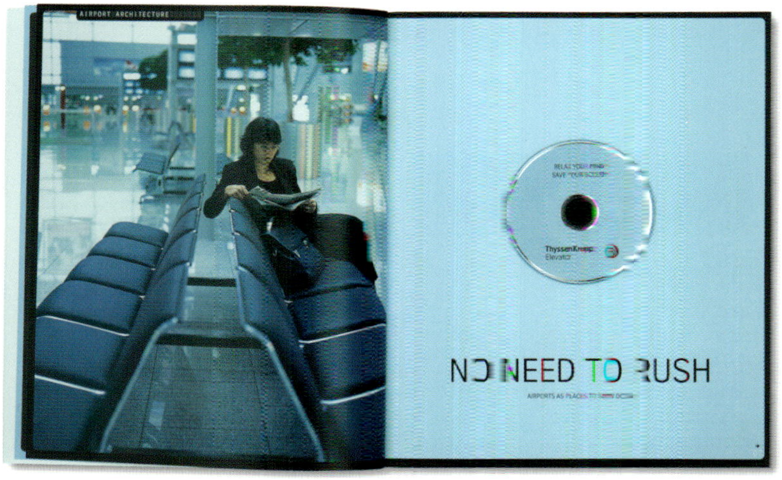
AIRPORT ARCHITECTURE
NO NEED TO RUSH

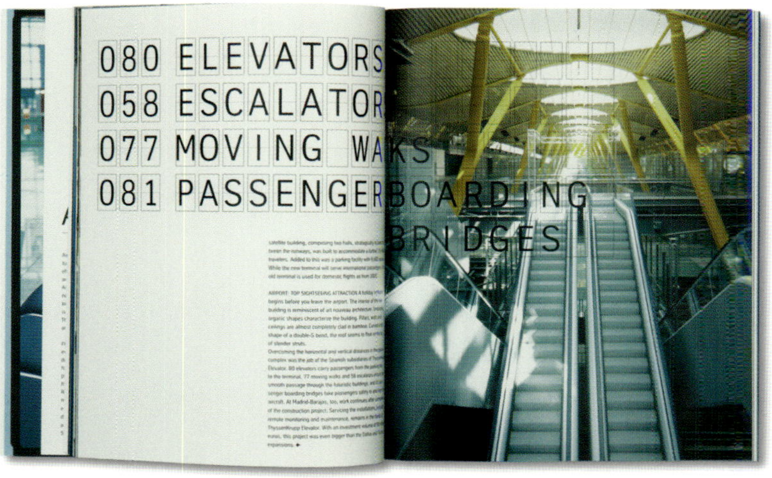
080 ELEVATORS
058 ESCALATORS
077 MOVING WALKS
081 PASSENGER BOARDING BRIDGES

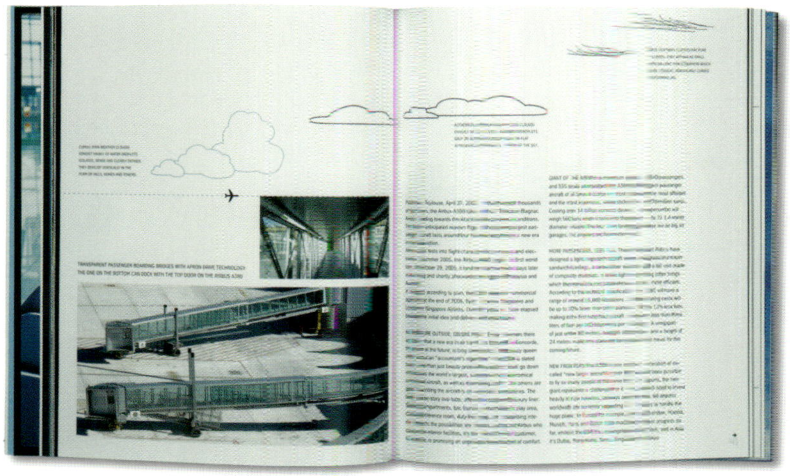

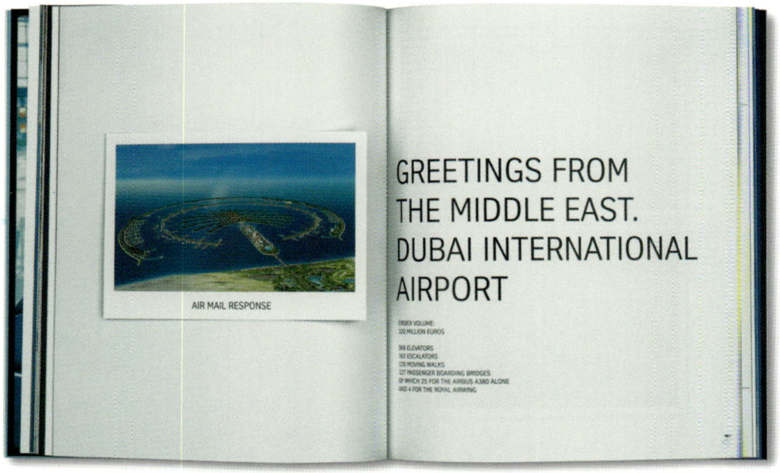
AIR MAIL RESPONSE
GREETINGS FROM
THE MIDDLE EAST.
DUBAI INTERNATIONAL
AIRPORT

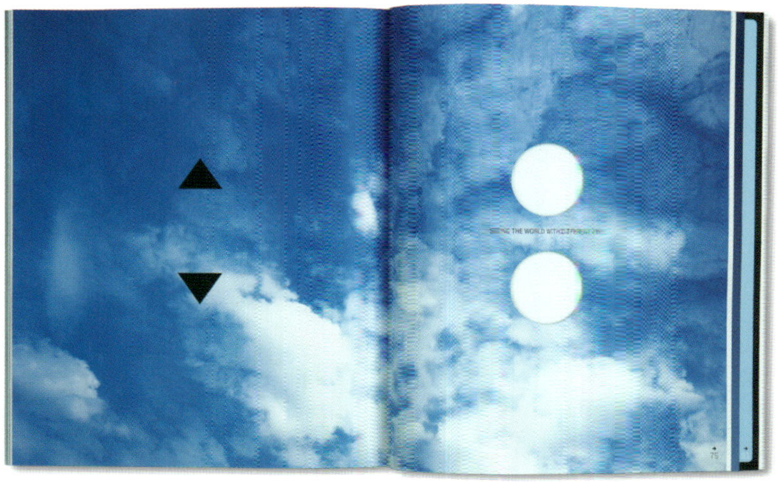

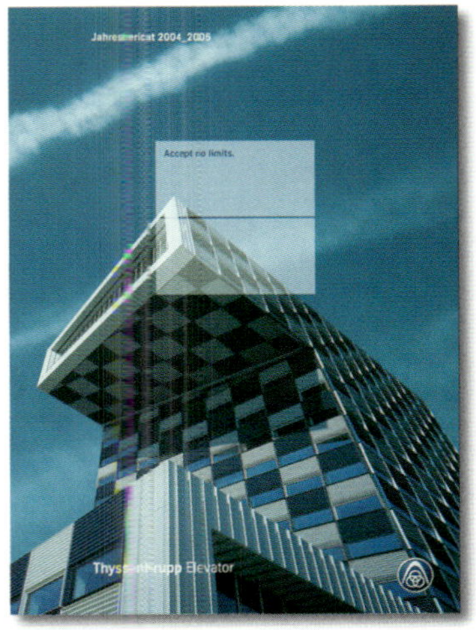

Auftraggeber Client
ThyssenKrupp Elevator AG
Düsseldorf, Germany

Projekt Project
Accept no limits
Jahresbericht
Annual report

Design
Michaela Müller
Verena Trurnit
RTS Rieger Team Werbeagentur GmbH
Düsseldorf, Germany

Der Jahresbericht bietet zum einen umfangreiche Informationen zum Unternehmen und zum Verlauf des Geschäftsjahres 2004/2005. Zum anderen werden in diesem Bericht die wichtigsten Referenzen von ThyssenKrupp Elevator anhand einer imaginären Weltreise präsentiert. Eine Darstellungsform, die die internationale Ausrichtung des Konzerns hervorhebt und den Leitsatz „Accept no limits“ emotional auflädt.

This annual report provides comprehensive information about ThyssenKrupp Elevator and its performance during the 2004/2005 financial year on the one hand, while showcasing examples of the company's main reference projects in the form of an imaginary trip around the world on the other. This is a method of representation that emphasizes the Group's international orientation and adds an emotive element to its guiding principle of "Accept no limits."

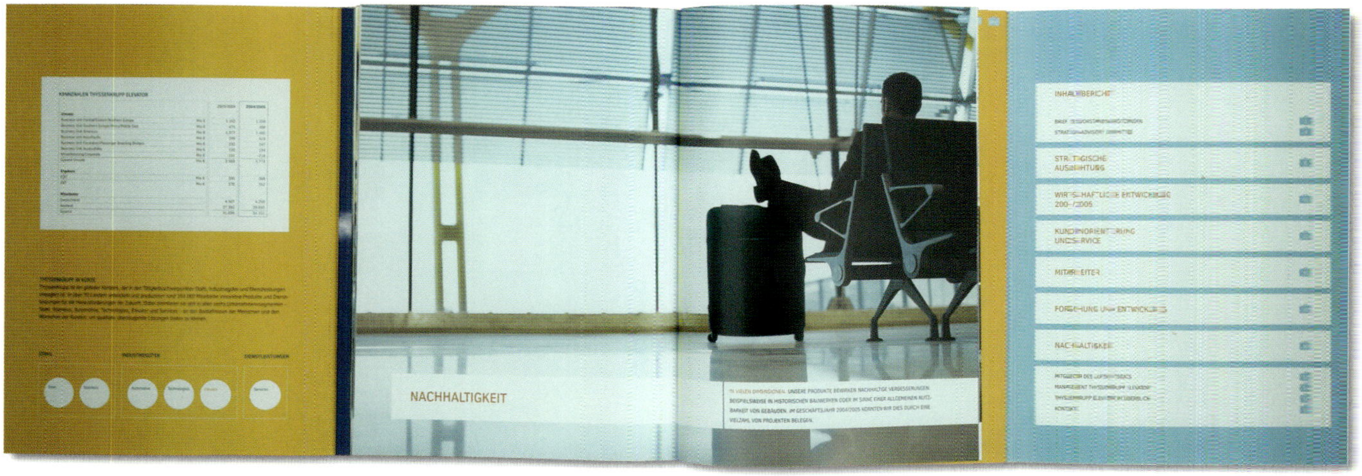
NACHHALTIGKEIT

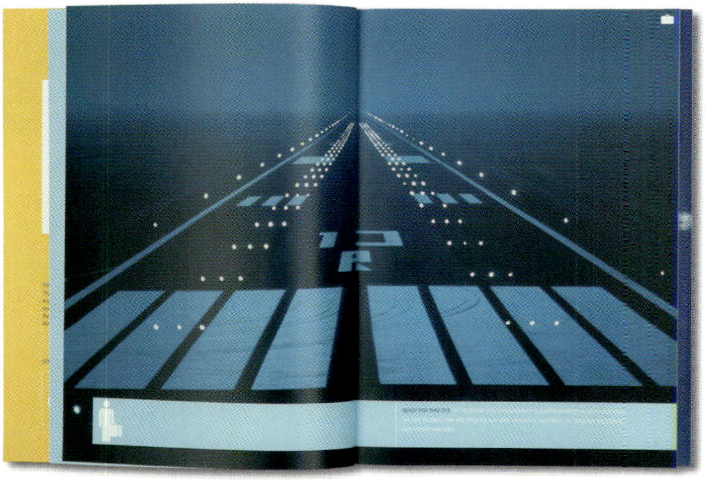

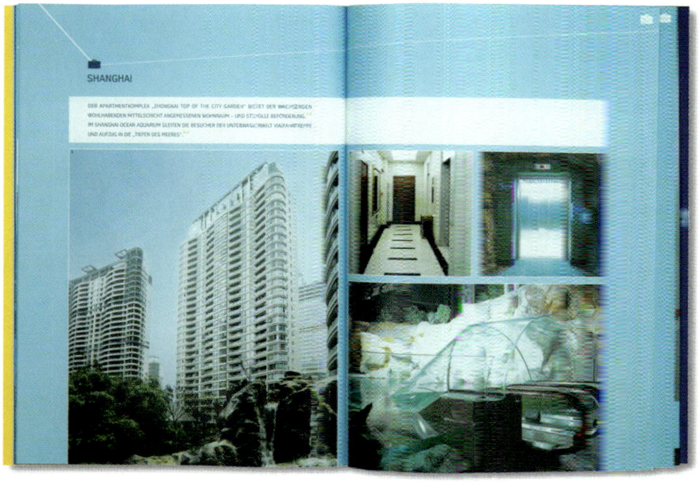
SHANGHAI

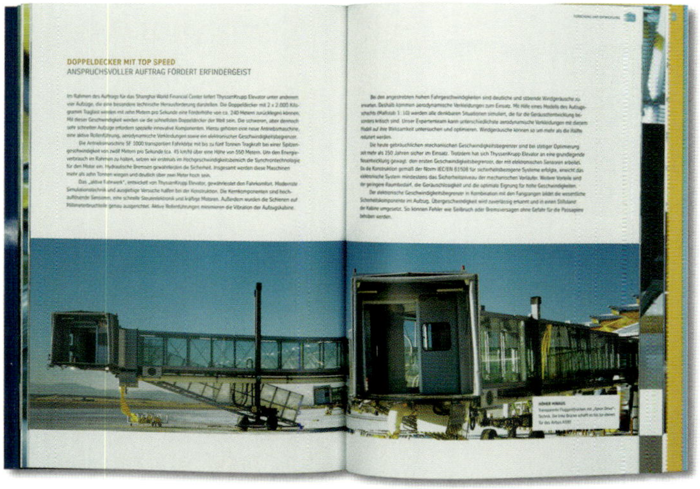
DOPPELDECKER MIT TOP SPEED
ANSPRUCHSVOLLER AUFTRAG FORDERT ERFINDERGEIST

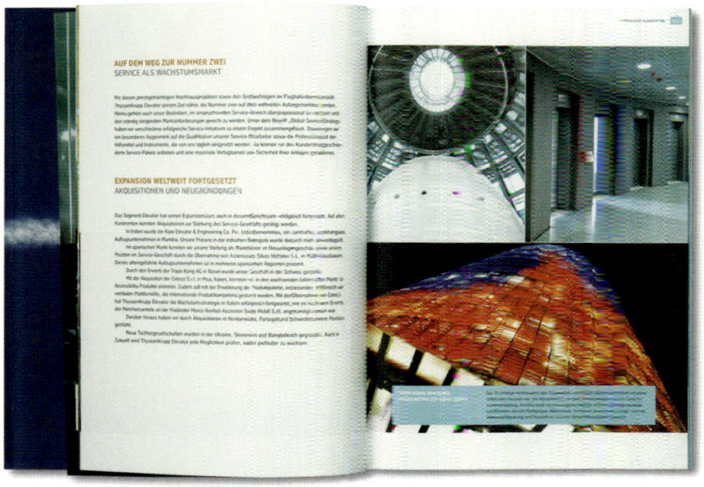
AUF DEM WEG ZUR NUMMER ZWEI
SERVICE ALS WACHSTUMSMARKT
EXPANSION WELTWEIT FORTGESETZT
AKQUISITIONEN UND NEUGRÜNDUNGEN

Auftraggeber Client
Viavai Winebar
Stuttgart, Germany

Projekt Project
Viavai Winebar
Visuelle Identität
Visual identity

Design
Prof. Andreas Uebele
Beate Gerner
büro uebele visuelle kommunikation
Stuttgart, Germany

Viavai. Der Name ist Rhythmus, Menschen spazieren vorbei. Ein Bild, das wir aus italienischen Gefilden kennen. Landschaften entstehen vor unserem inneren Auge, Farben und Formen, der Name erinnert an eine bekannte Welt und die grafische Form folgt dem Gedankenreisenden in eine typische Bar. Das Zeichen formt sich aus den Buchstaben des Namens. Das Gebilde ist abstrakt, aber vertraut. Es sagt: Ich bin eine italienische Bar. Die Farben sind Braun und Violett, Metapher der Eleganz, fa bella figura. Die Strenge des Zeichens bildet einen prägnanten Kontrast zu der weichen runden Schrift. Eine skurrile Form, die Grandezza beweist.

Viavai. The name itself conjures up the rhythm of people walking by. An image familiar from Italian climes. Whole landscapes arise before our inner eye, colors and shapes, the name reminds us of a familiar world and the design follows the mind-traveler into a typical local bar. The logo is created from the letters that make up the name. The shape is abstract, but familiar. It says: I am an Italian bar. The colors are brown and violet, symbolizing elegance, fa bella figura. The clear, clean design effectively offsets the soft curves of the typeface: a playful, striking shape exuding nothing less than grandezza.

VIAVAI

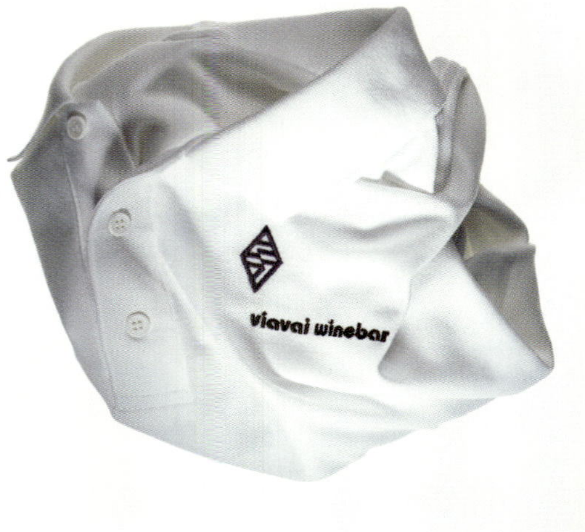

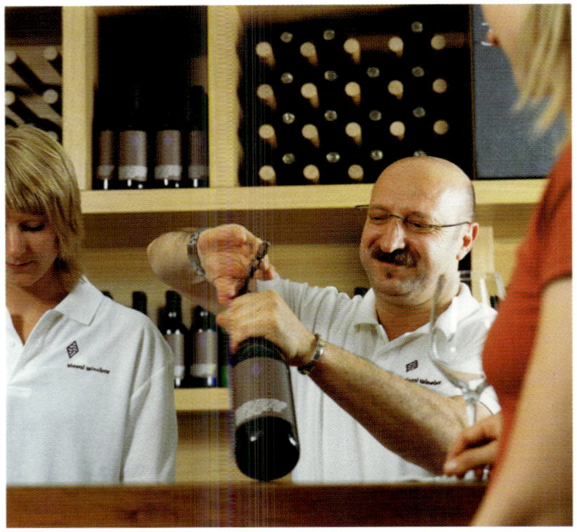

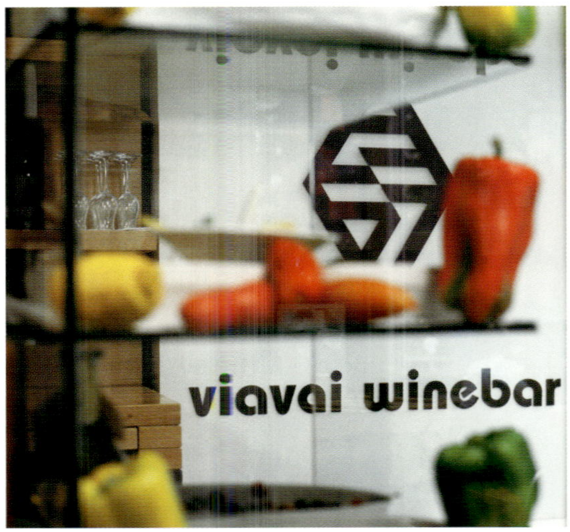

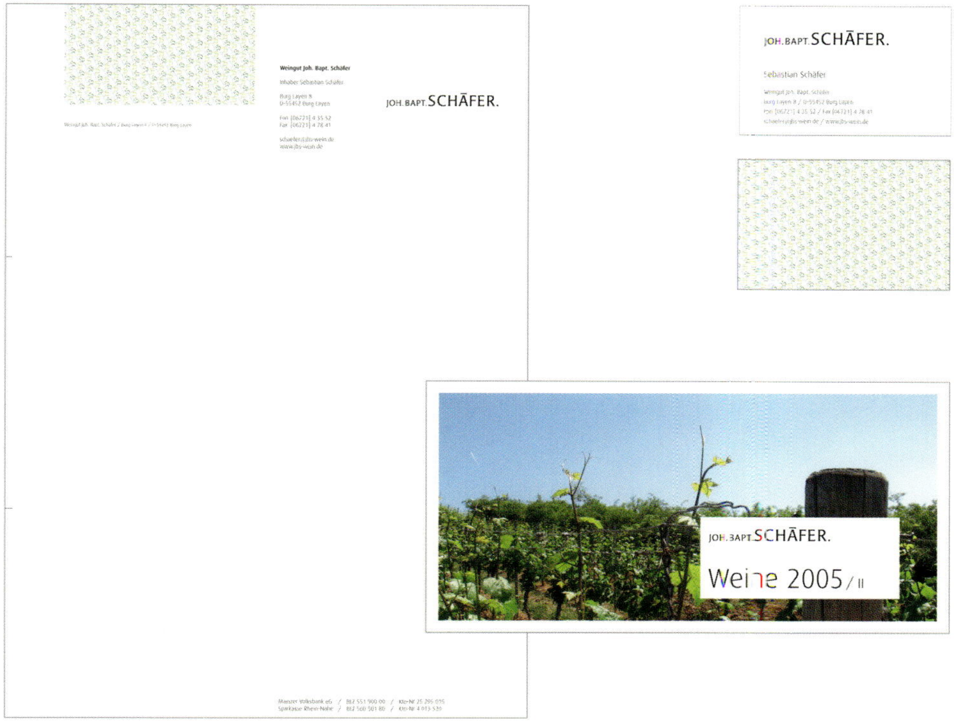

Auftraggeber Client
Weingut Joh. Bapt. Schäfer
Burg Layen, Germany

Projekt Project
CD Weingut Joh. Bapt. Schäfer
Corporate Design

Design
Fuenfwerken Design AG
Wiesbaden, Germany

Das Weingut Joh. Bapt. Schäfer hat eine lange Familientradition. Als Verweis auf die Historie haben wir für das neue Corporate Design die Initialen des originalen Schriftzugs in einen filigranen Rapport transferiert.

The Joh. Bapt. Schäfer vineyard enjoys a long family tradition. For the new corporate design, we referred to the enterprise's history by adopting the initials from the original lettering into a repeating filigree pattern.

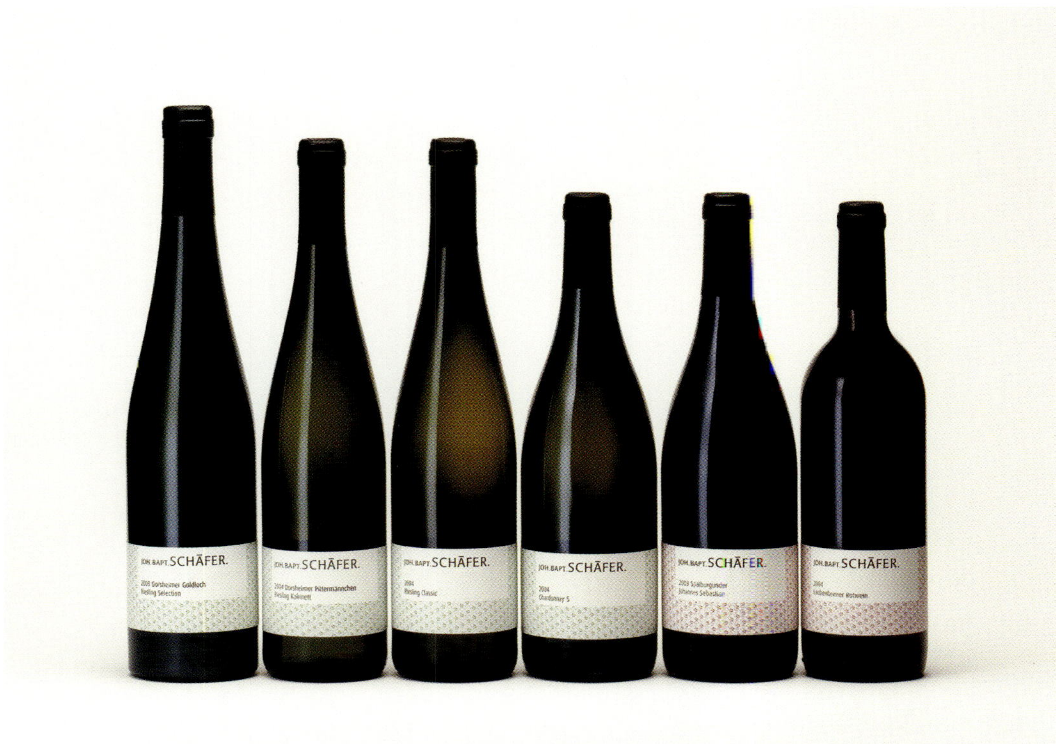

Auftraggeber Client
BMW Group
München, Germany

Projekt Project
Leistung bringen.
Geschäftsbericht
Annual report

Design
Frank Wagner
Thomas Tscherter
Markus Kirsch
Alexander Giandomenico
Vit Steinberger
häfelinger+wagner design
München, Germany

Ziel der BMW Group Geschäftsberichtkonzeption ist es – neben der Finanzberichterstattung –, die strategischen Ziele und Entscheidungen des Unternehmens magazinartig aufzubereiten und dem Leser anschaulich zu vermitteln. Es wurden drei Kapitel entwickelt, deren Einstieg jeweils mit einem mit Silber bedruckten Transparentpapier betont wurde. Der einseitige Silberschnitt des Imageteils steht im Kontext zum funktionalen, aber speziellen Register des Pflichtteils. Mit einer besonderen Fotokonzeption wurden alle 32 aktuellen Fahrzeuge der BMW Group Marken BMW, MINI und Rolls-Royce in einer einzigartigen Gesamtaufnahme vor Los Angeles gezeigt.

Besides financial reporting, the aim of the BMW Group Annual Report concept is to prepare and present the company's strategic objectives and decisions to the reader in an easy-to-read magazine format. Three chapters have been developed, the introduction to each of which is emphasised by a leaf of transparent paper with silver printing. The one-page silver of the image section is to be seen in the context of the functional yet special register of the mandatory section. A special photographic concept is used to show all 32 current models belonging to the BMW Group brands BMW, MINI and Rolls-Royce in a single, unique picture taken outside Los Angeles.

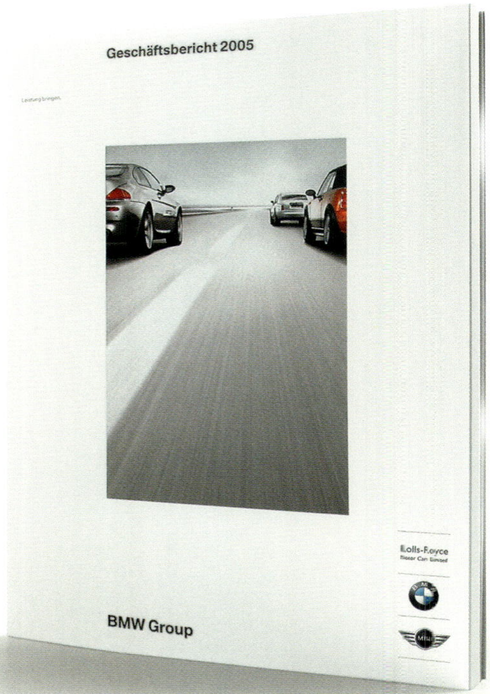

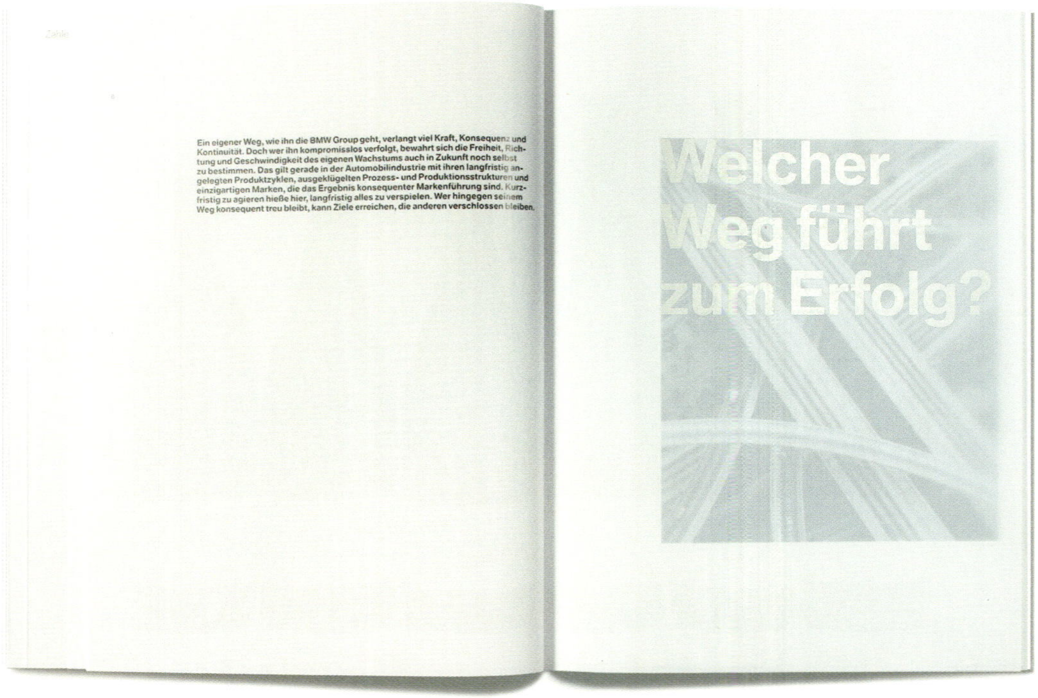

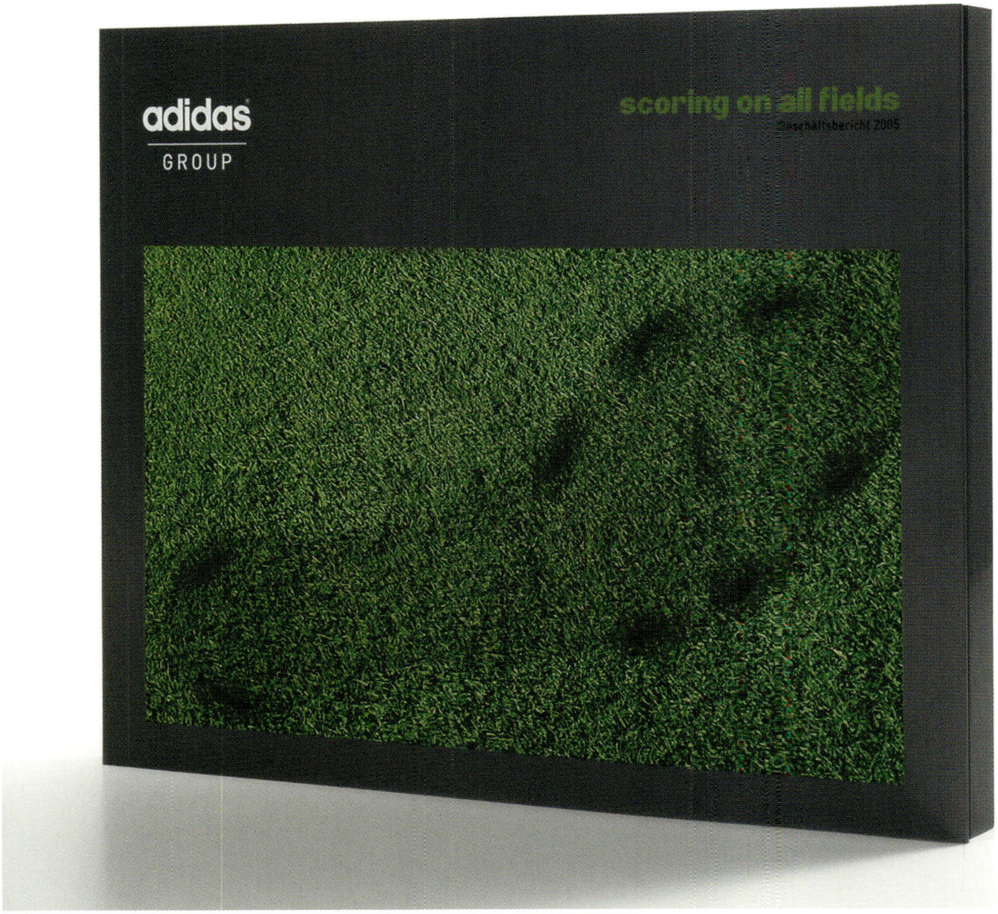

Auftraggeber Client
Adidas Group
Herzogenaurach, Germany

Projekt Project
scoring on all fields
Geschäftsbericht
Annual report

Design
Frank Wagner
Vit Steinberger
Thomas Tscherter
Christina Mayer
häfelinger+wagner design
München, Germany

Unter dem Motto „scoring on all fields" thematisiert adidas in seinem Geschäftsbericht 2005 das Potenzial und die Schlagfähigkeit des Unternehmens. Hintergrund ist der Zukauf von Reebok sowie die Fußball-Weltmeisterschaft im eigenen Land. Der Schuhabdruck auf dem Titel zeigt den Abdruck eines Fußballschuhs und verweist indirekt auf die Entwicklung des Stollenschuhs von Adi Dassler, mit dem der Erfolg von adidas begann. Grafisches Hauptthema ist ein Rasterfeld, das den Anzeigetafeln in Stadien entlehnt ist. Auch Grafiken und Tabellen oder auch das Vorstandsporträt stehen in grafischem Bezug zum Thema Fußball und einem Fußballspielfeld.

adidas uses the motto "scoring on all fields" in its annual report to illustrate the company's potential and its commercial vigour. The background is the recent purchase of Reebok, as well as Germany hosting the FIFA World Cup. The footprint on the cover shows the imprint of a football boot, an indirect reference to Adi Dassler's development of the boot with integral studs, which was the beginning of adidas' success story. The main graphic subject is a grid pattern modelled on the digital scoreboard screens in sports stadiums. Graphs, tables and even the Board portrait are graphically linked to the subject of football and a football pitch.

Auftraggeber Client
mannesmann plastics machinery GmbH
München, Germany

Projekt Project
mpm Bericht 04/05 „LÖSUNGEN INKL."
Jahresbericht
Annual report

Design
Frank Wagner
Markus Kirsch
Kerstin Weidemeyer
Andrea Mönch
häfelinger+wagner design
München, Germany

Die mpm ist eine Holding zur Produktion und zum Vertrieb von kunststoff- und plastikverarbeitenden Maschinen. Der jährlich neu konzipierte Jahresbericht stellt unter dem Motto „Lösungen inkl." die Innovationskraft des Unternehmens sowie seine Mitarbeiter in den Vordergrund. Jeweils eine bestimmte technische Innovation wird anhand einer Text-Bild-Strecke dargestellt. Die insgesamt fünf Imagestrecken zeigen die Vielfalt der Innovationen und die Menschen, die sie entwickelt haben. Das Fotokonzept lebt durch die reale Authentizität der Motive. Auf dem Titel wird das Thema Kunststoff durch eine partielle UV-Lackierung visualisiert.

mpm is a holding company concerned with the production and sale of machines for processing plastics. The Annual Report, which is redesigned every year, is subtitled "Solutions incl." and focuses its attention on the innovative power of the company and its employees. One particular technical innovation is presented in an individual series of text and pictures. A total of five image series shows the diversity of the innovations and the people who made them possible. The photographic concept is founded on the absolute authenticity of the subjects. The theme "plastics" is visualised on the title page by means of a partial UV lacquer finish.

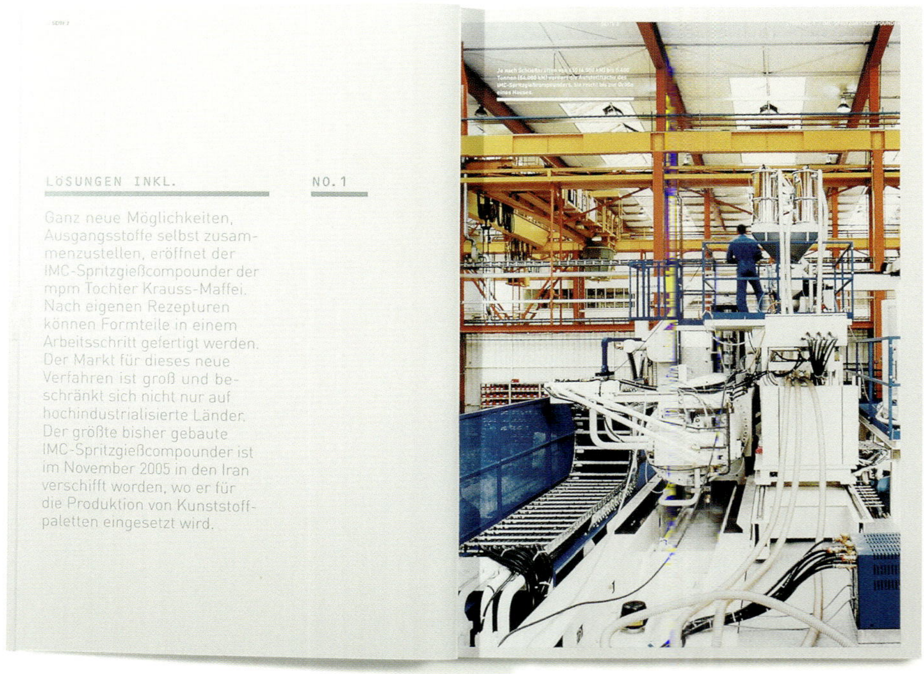

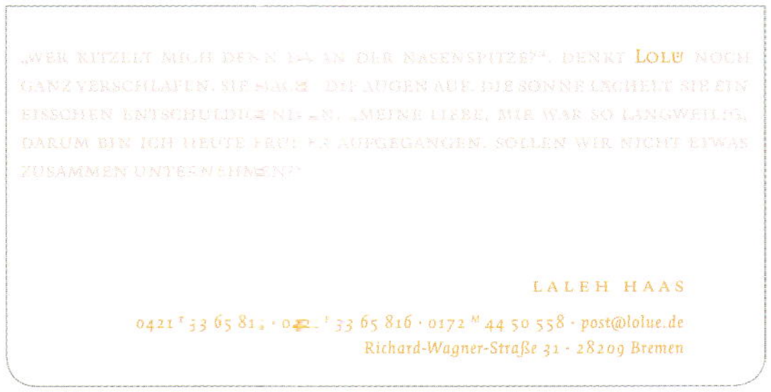

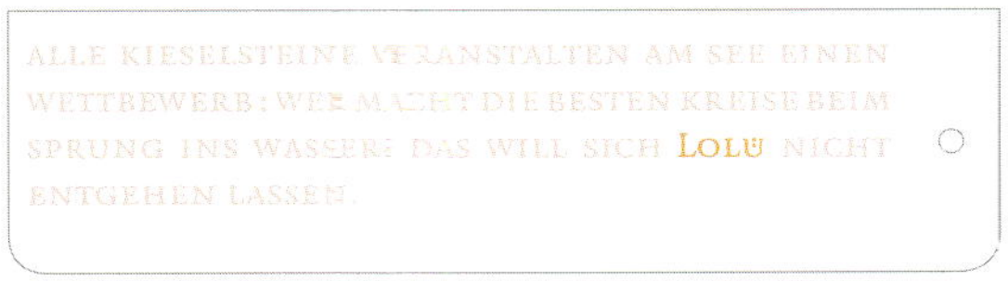

Auftraggeber Client
Lolü
Bremen, Germany

Projekt Project
Lolü
Corporate Identity

Design
Peter Bürnagel
Barbara Kotte
Anne-Lene Proff
scrollan
Berlin, Germany

Lolü macht liebevolle Kleidung und Accessoires für Kinder. Und die Welt der Kinder ist eine Welt voller Geschichten und Fantasie. Deshalb werden bei dem Corporate Design die Erlebnisse der kleinen Lolü erzählt, auf jedem Medium ein anderes.

Lolü makes lovable clothing and accessories for children. And a child's world is a world full of stories and imagination. This is why the corporate design tells stories about little Lolü's adventures – a different one for each medium.

Auftraggeber Client
DaimlerChrysler AG
Stuttgart, Germany

Projekt Project
360 GRAD
Umweltbericht
Environmental report

Design
Demian Bern
Jörg Dengler
Wolfram Schäffer
Tom Wahl
Vera Warter
Markus Wichmann
Alexander Stertzig
design hoch drei GmbH & Co.KG
Stuttgart, Germany

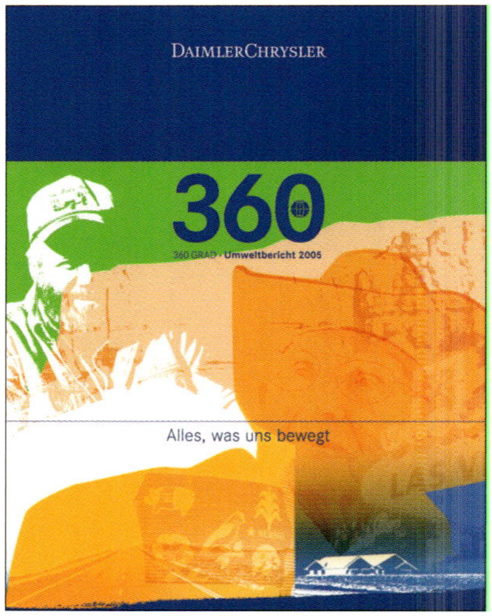

Bewegendes zum Thema Nachhaltigkeit. – Wie kommuniziert man als Unternehmen das oftmals technisch eingegrenzte Thema Umwelt wirksam? DaimlerChrysler setzt auf den emotional aufbereiteten Umweltbericht „360 GRAD“.
Um Energie und Emotionen kreist 2005 der Umweltbericht „Alles, was uns bewegt“. Als spannendes Magazin konzipiert und mit seinem Key Visual „360 GRAD“ zur eigenständigen Marke avanciert, unterhält das Unternehmensmedium seine Leser mit Qualitätsjournalismus, feinsten Illustrationen, bewegenden Fotostrecken und kreativem Layout.

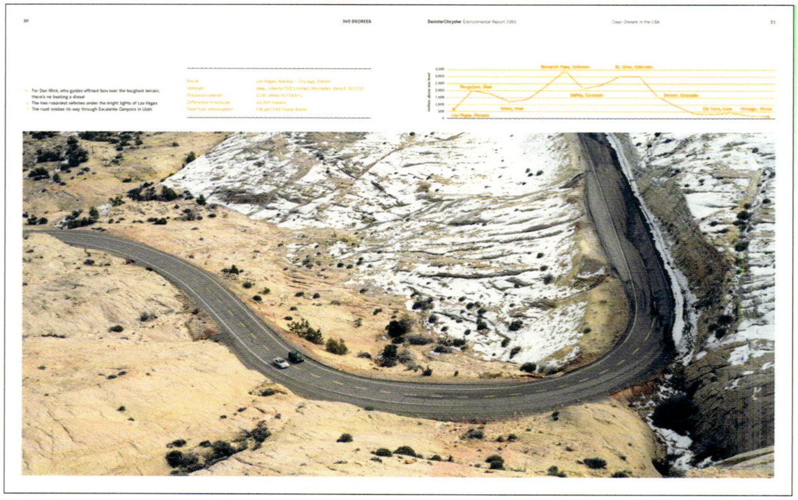

Moving forces on the issue of sustainability. – How can a company communicate technically contained environmental issues effectively? DaimlerChrysler focuses on the emotionally dressed environmental report “360 DEGREES”.
Energy and emotions – these are the topics the environmental report “The forces that move us” is concerned with in the year 2005. Conceived as an exciting magazine and with its key visual “360 DEGREES” advanced as an independent brand, the company medium entertains its readers with high-quality journalism, finest illustrations, emotional photography, and a creative layout.

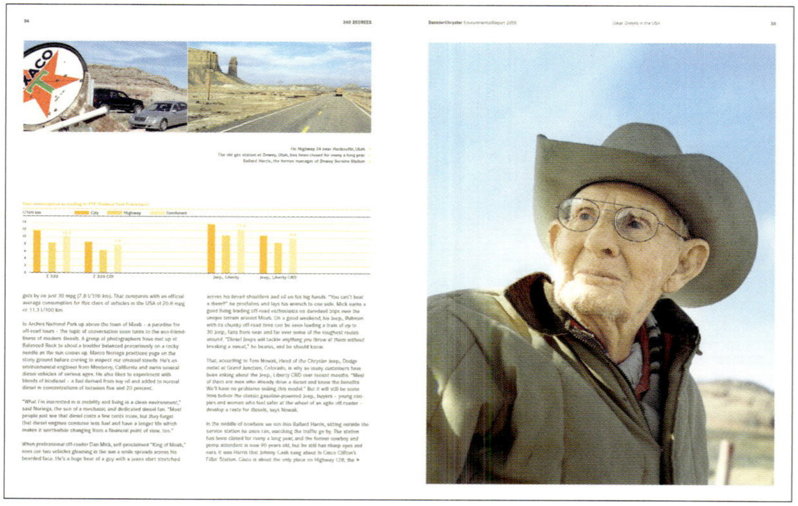

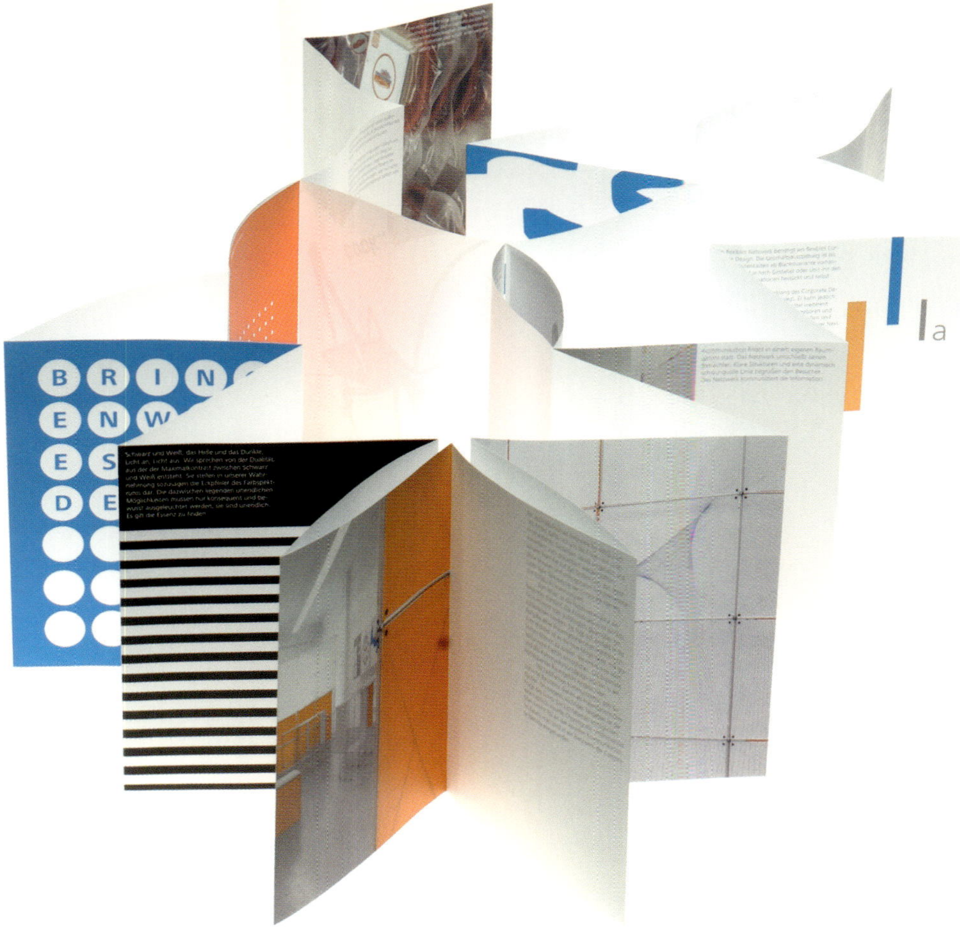

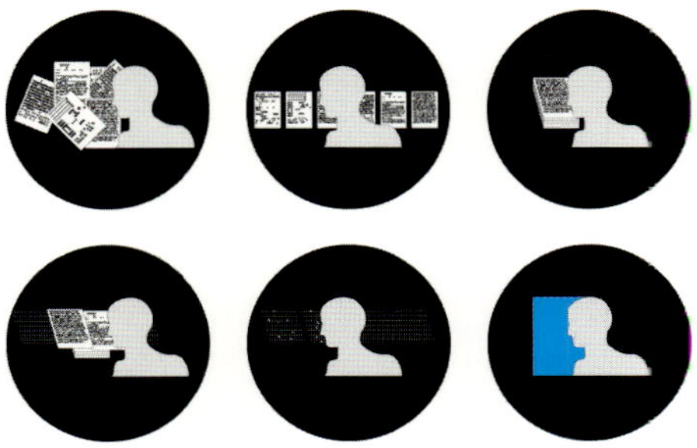

Auftraggeber Client
lichtpunkt // Netzwerk für Gestaltung
Heidenheim, Germany

Projekt Project
lichtpunkt // Corporate Identity & Design
Leporello
Accordion-folded brochure

Design
Marc Engenhart
Engenhart Visuelle Kommunikation
Stuttgart, Germany
Nico Hensel
Nico Hensel Gestaltung
Heidenheim, Germany

Diese Unternehmenskommunikation zeigt Aufgabe, Art und Beispiele der Dienstleistung des Netzwerks lichtpunkt. Die Aufgabe der Agentur ist es, verschiedenste Spezialisten aus unterschiedlichen Gestaltungsdisziplinen für komplexe Projekte zusammenzuführen und die Aufklärung gegenüber der Öffentlichkeit zu fördern. Der Gestaltungsbegriff beschränkt sich jedoch nicht nur auf Grafik-, Industrie- oder Produktdesign, sondern bietet ebenso Platz für alle anderen Arten der verbalen/auditiven und visuellen Gestaltung. Diese Medien transportieren die Corporate Identity und zeigen die Anwendung. lichtpunkt entwickelt erfahrbare visuelle Kommunikation.

This corporate communication shows challenge, character and samples of services of the network lichtpunkt. The vision of the agency is to bring together specialists from various branches of design to work jointly on complex projects while promoting the education of the public. The term "design", however, does not restrict itself to graphics to industrial or product design, but also provides room for all kinds of verbal/auditory and visual designs. These media transport the corporate identity and show its application. lichtpunkt develops visual communication which can be experienced.

Auftraggeber Client
Bravo Charlie
Stuttgart, Germany

Projekt Project
Bravo Charlie – Bar und Club
Corporate Design

Design
Oliver-A. Krimmel
Anja Osterwalder
Pia Bardesono
Christine Dorst
i-d buero
Stuttgart, Germany

Eine Bar und ein Club im ehemaligen Lufthansa-Terminal in Stuttgart. Die Anfangsbuchstaben von „Bar“ und „Club“ buchstabiert man im Pilotenalphabet „BRAVO“ und „CHARLIE“.
Ein Name wie ein Ausruf! Ein Logo wie eine Sprechblase. Ein Logo, das immer für direkte Kommunikation sorgt. Auf Gläsern, Visitenkarten, Speisekarten, T-Shirts oder auch als Street Art.

A bar and a club in the building of the former Lufthansa Terminal in Stuttgart. Spelled in the pilots' alphabet the initials of the two words "Bar" and "Club" are "BRAVO" and "CHARLIE".
A name like an interjection! A logo like a speech bubble! A logo that always arranges for direct communication. Printed on glasses, business cards, menues, T-shirts or reinterpreted as street art.

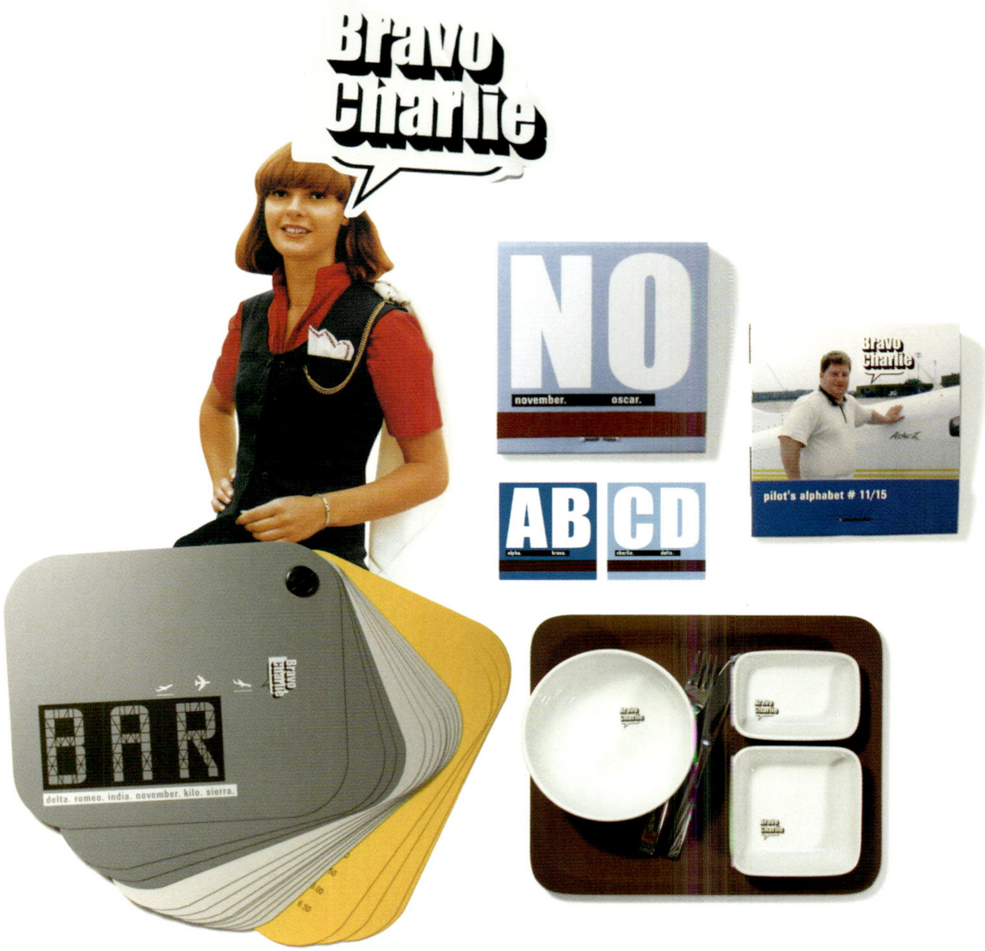

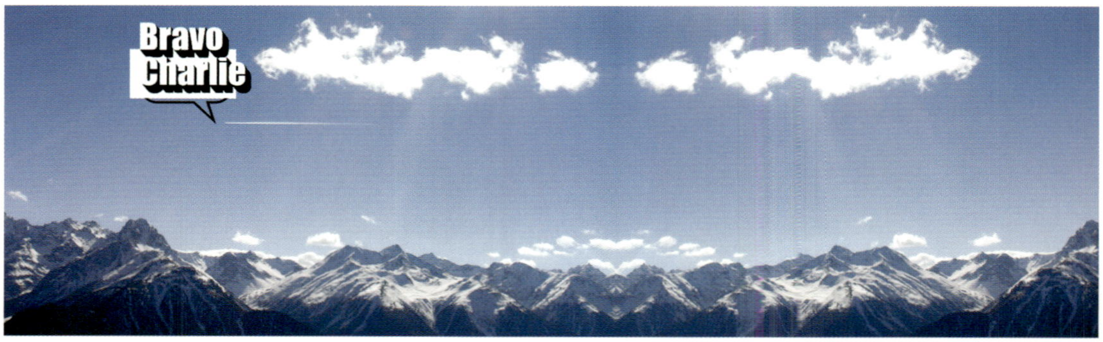

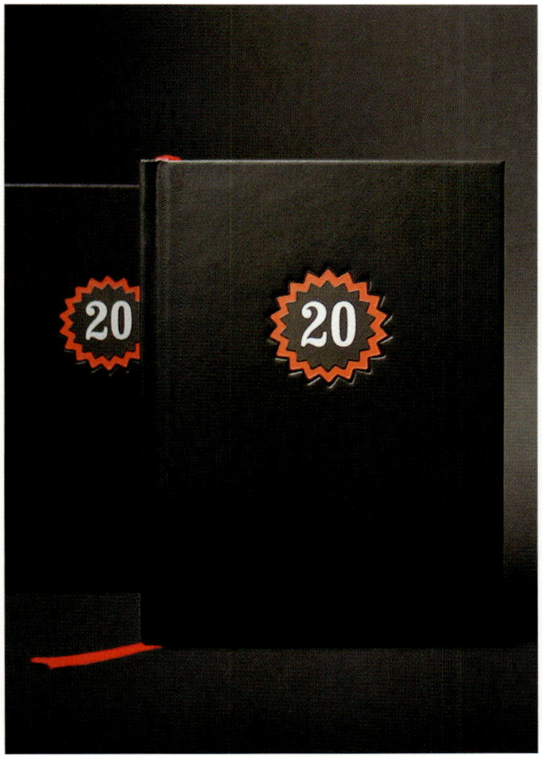

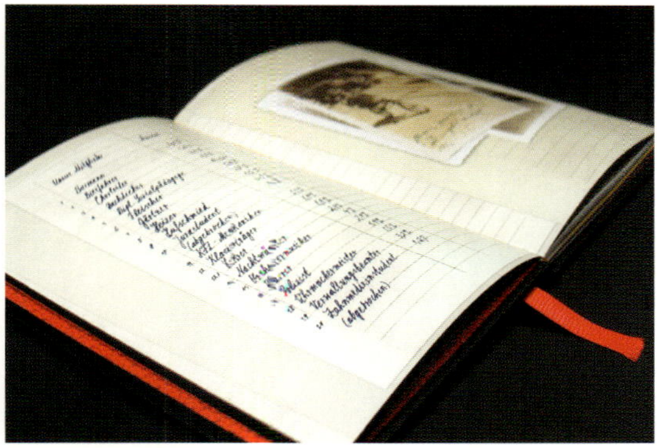

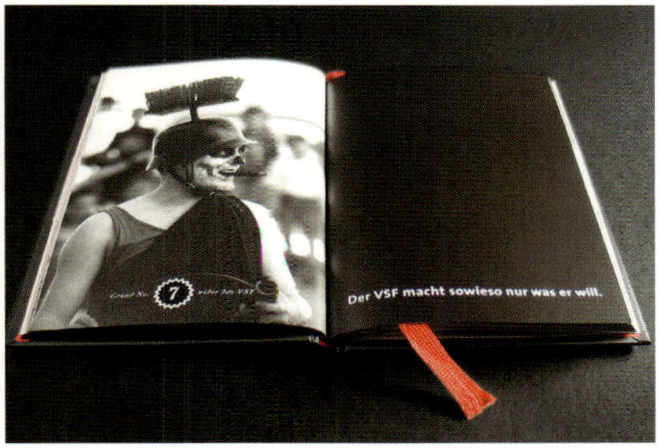

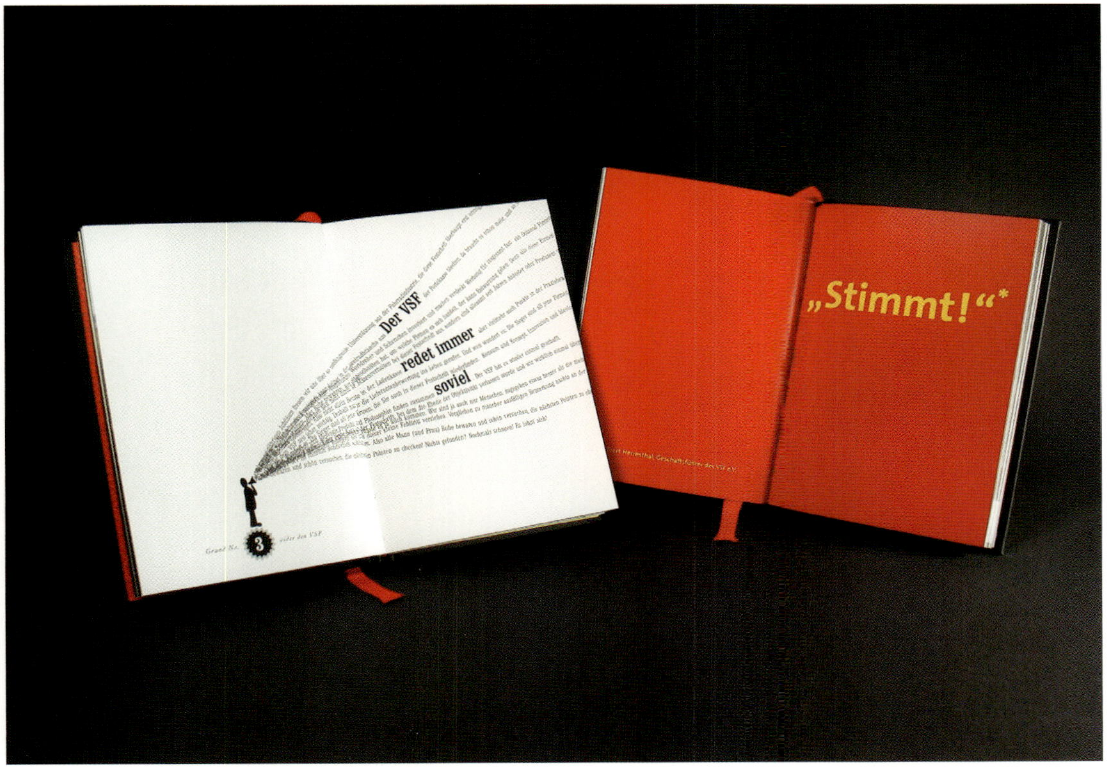

Auftraggeber Client
VSF – Verbund Selbstverwalteter Fahrradbetriebe e.V.
Aurich, Germany

Projekt Project
20 Gründe wider den VSF – Eine Festschrift
Buch
Book

Design
Sven Marten
Henning Parche
Barbara Grobe
echtweiß | Corporate Design
Heidelberg, Germany

„Klein und geheim“ war das Leitmotiv für Inhalt und Gestaltung der Festschrift zum 20-jährigen Bestehen des VSF (Verbund Selbstverwalteter Fahrradbetriebe e.V.), der sich seit seiner Gründung vom „alternativen Ökoverein“ zum professionellsten Verbund der Zweiradbranche in Deutschland entwickelt hat. Die Festschrift wurde bewusst als Gegenentwurf zu den Standards dieses Publikationsgenres entwickelt: Statt üblicher Selbstbeweihräucherung werden 20 gängige Vorurteile gegen den VSF gesammelt, in Schwarz-Weiß illustriert und auf jeweils folgenden farbigen Doppelseiten aufgelöst. Informatives und Historisches findet sich auf einem Beipackzettel.

"Small and covert" was the guiding theme for the contents and the layout of the Festschrift published on the occasion of the 20th anniversary of the VSF (network of self-governing bikeshops). Once founded as an "alternative ecological association" the VSF has grown to be the most professional biking sector network in Germany. The Festschrift is a clear counter to the regular publication of this type. Instead of typical self-adulation 20 prejudices against the VSF are taken, illustrated in black and white, and then rebutted in colour on the two following double pages. Informations and historical facts are placed on a package insert.

Auftraggeber Client
bayern design GmbH
Nürnberg, Germany

Projekt Project
DESIGNBERICHT BAYERN 2003-2005
Buch und Poster
Book and poster

Design
Frank Wagner
Thomas Tscherter
häfelinger+wagner design
München, Germany

bayern design fördert den Dialog zum Thema Design und baut aktiv an einer auf Bayern fokussierten Informationsplattform zum Thema Design. Der Tätigkeitsbericht – der zweite in einer über die nächsten Jahre angelegten Publikationsreihe – arbeitet mit ungewöhnlichen Typografieprinzipien wie beispielsweise den Ein- und Auszügen von Textabsätzen. Der Leser wird also mit Benutzen des Berichts indirekt aufgefordert, sich mit Gestaltung auseinander zu setzen. Neue Sichtweisen und Perspektiven freizulegen ist eine elementare Fähigkeit von Design. Der Bericht versucht dies mit Gestaltung deutlich zu machen, wie z.B. auch durch das den Umschlag umgebende Poster.

bayern design encourages dialogue on the subject of design and is actively working on a design information platform focussed on Bavaria. The progress report – the second in a series of publications planned for the next few years – makes use of unusual typographical principles such as the insertion or removal of text paragraphs. Reading the report therefore indirectly challenges us to concern ourselves with design and forms. Exposing new angles and perspectives is one of the fundamental abilities of design. The report attempts to do this by use of layout, such as the poster which surrounds the cover.

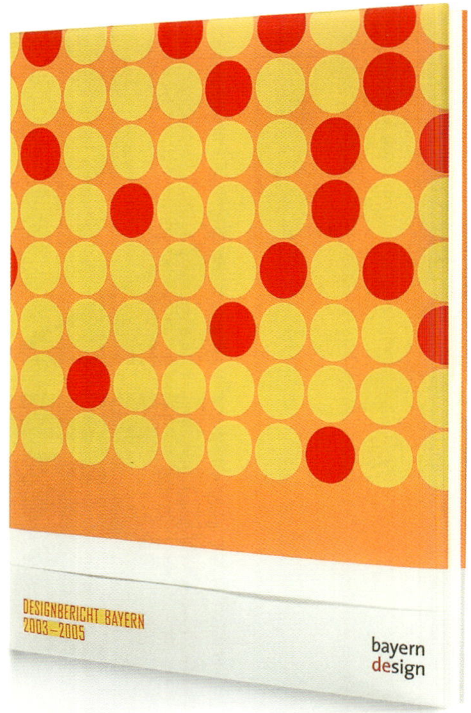

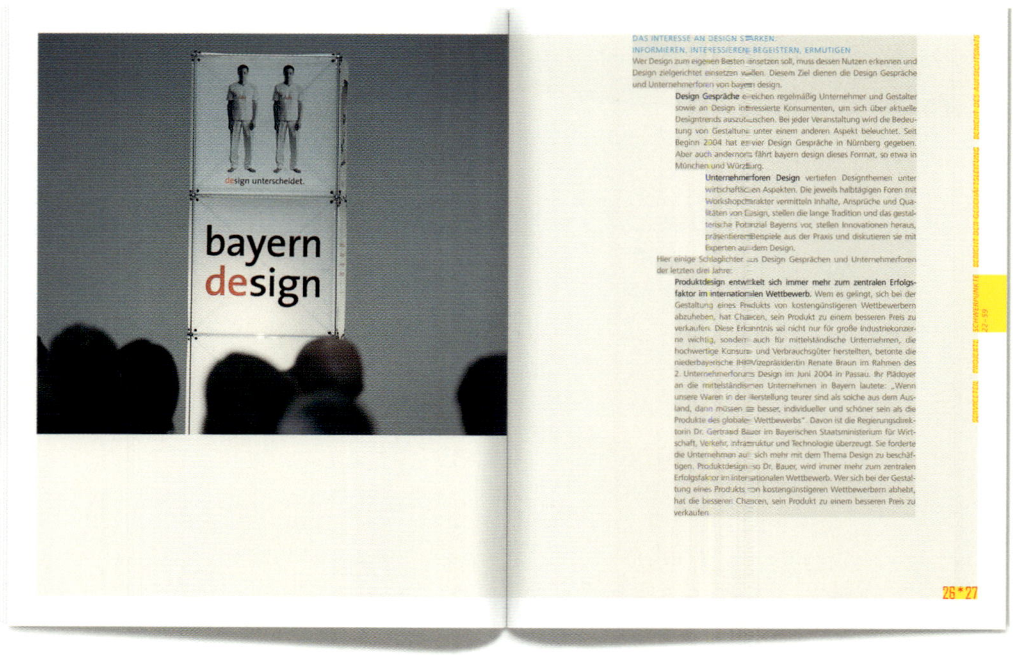

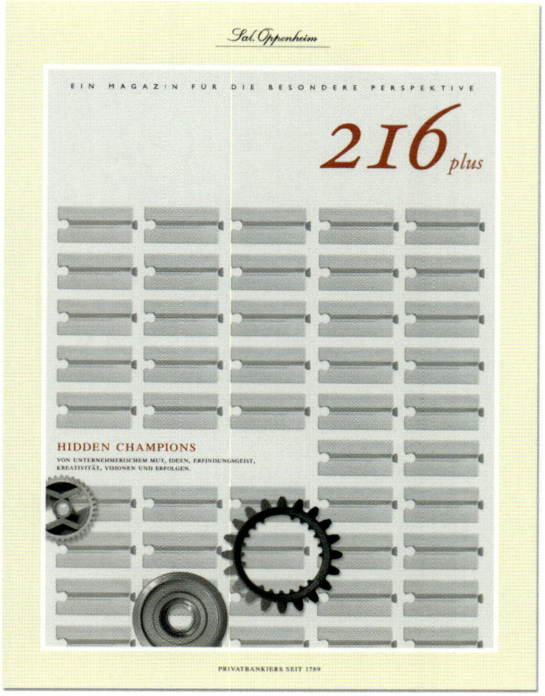

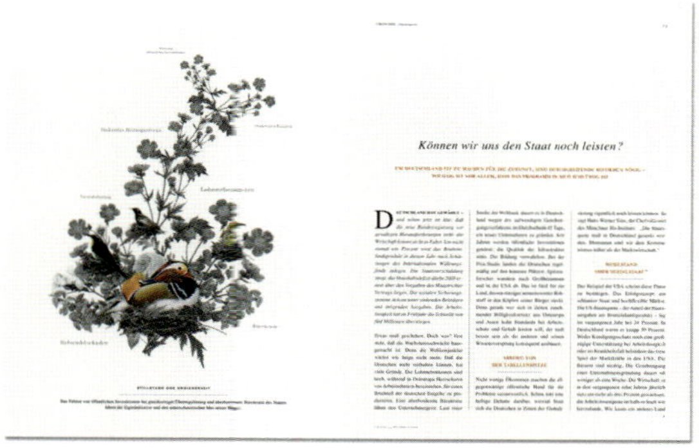

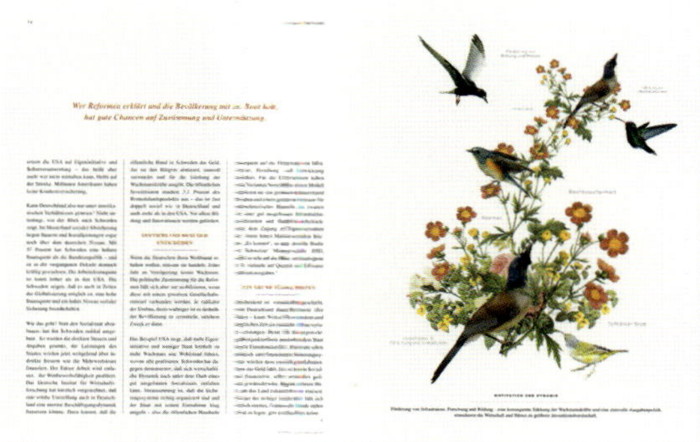

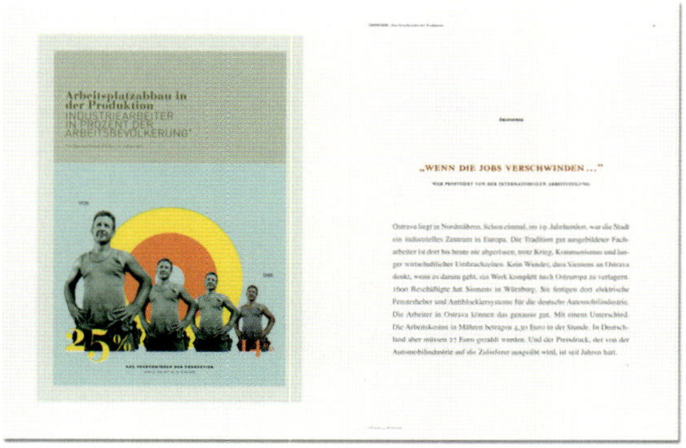

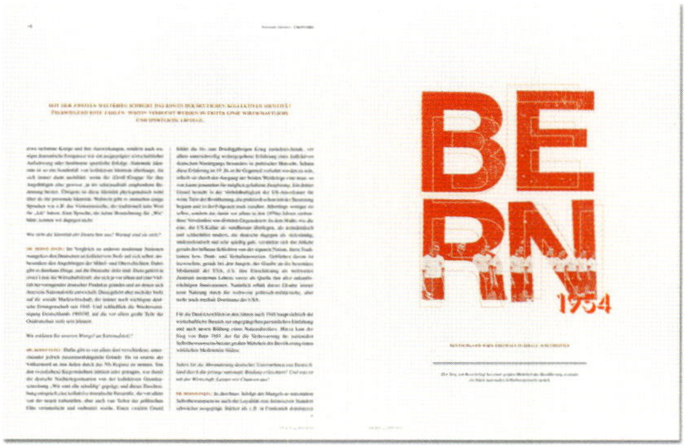

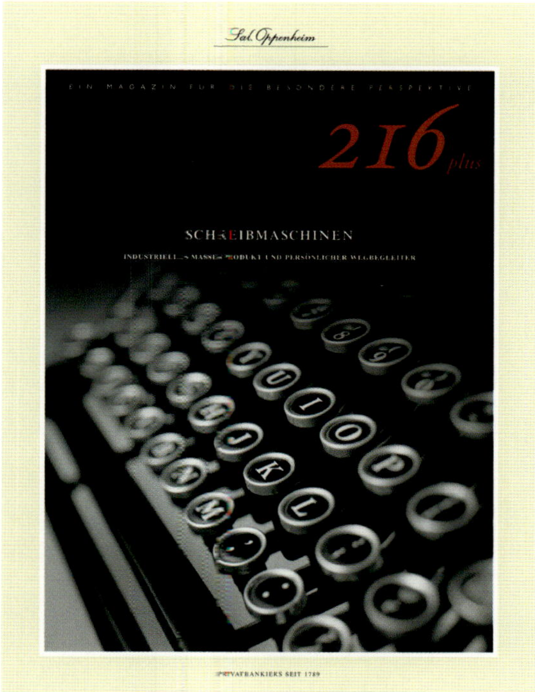

Auftraggeber Client
Sal. Oppenheim jr. & Cie. KGaA
Köln, Germany

Projekt Project
216plus – Ein Magazin für die besondere Perspektive
Kundenmagazin
Customer magazine

Design
Bernd Vollmöller
Simon & Goetz Design GmbH & Co.KG
Frankfurt am Main, Germany

Sal. Oppenheim ist eine der führenden Privatbanken Europas. Ziel war es, ein hochwertiges, aufmerksamkeitsstarkes und unterhaltendes Magazin zu schaffen, das dem anspruchsvollen Selbstverständnis des Hauses Sal. Oppenheim entspricht. Der Gestaltungsauftritt kommuniziert mit seinem klassischen, repräsentativen Charakter in unaufdringlicher Weise die im Hause Sal. Oppenheim bewusst gelebten Werte Tradition, Innovation und Exklusivität.
Die Themenbereiche Ökonomie, Gesellschaft und Kultur repräsentieren die Haltung und das vielseitige Engagement der Kölner Privatbankiers seit Gründung im Jahr 1789.

Sal. Oppenheim is one of the leading private banks in Europe. The task was to create a premium, attention-grabbing and entertaining magazine that communicates the sophisticated self-conception of Sal. Oppenheim. With its classic, representative character the magazine's design communicates the consciously lived values "tradition", "innovation" and "exclusiveness" in an unintrusive way.

Auftraggeber Client
Effenaar
Eindhoven, The Netherlands

Projekt Project
Effenaar
Visuelle Identität
Visual identity

Design
Serge Scheepers
René Toneman
Frank van Leeuwen
Emiel Rijshouwer
Eelko Groenewegen
Fabrique [design, communications and new media]
Delft, The Netherlands

Vor kurzem bezog der Eindhovener Poptempel Effenaar ein neues Domizil. Fabrique entwickelte eine passende visuelle Identität.
Weil der Veranstaltungskalender im Effenaar äußerst vielfältig ist, konzipierte Fabrique eine flexible Corporate Identity, die den verschiedenen kulturellen Strömungen entsprechend viel Raum lässt. Das „E“, das darin eine Schlüsselrolle spielt, erhält immer ein anderes Gesicht: Raster und Schraffierungen verleihen dem Buchstaben eine rohe, industrielle und stilistische Ausstrahlung. Durch die Verwendung von Farben entsteht ein kaleidoskopisches Bild.

The Eindhoven pop venue Effenaar recently moved to a new building. Fabrique developed a visual identity which is perfectly in line with the ambience of this building.
Because the programming at Effenaar is extremely varied, Fabrique therefore developed a flexible house style with room for various forms of cultural expression. The letter "E", which is given a central role, is given a different look each time: grids and shading give the letter a raw, industrial and stylistic look, while the use of colours creates a kaleidoscopic effect which is completely in line with Effenaar's unorthodox nature.

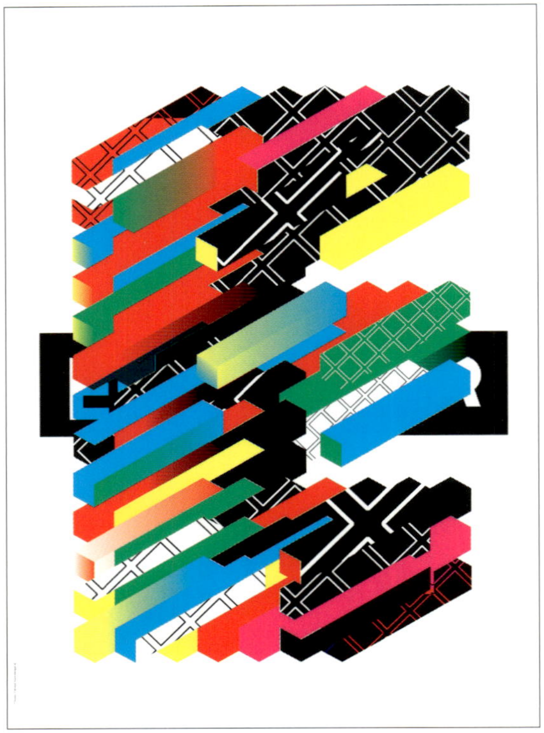

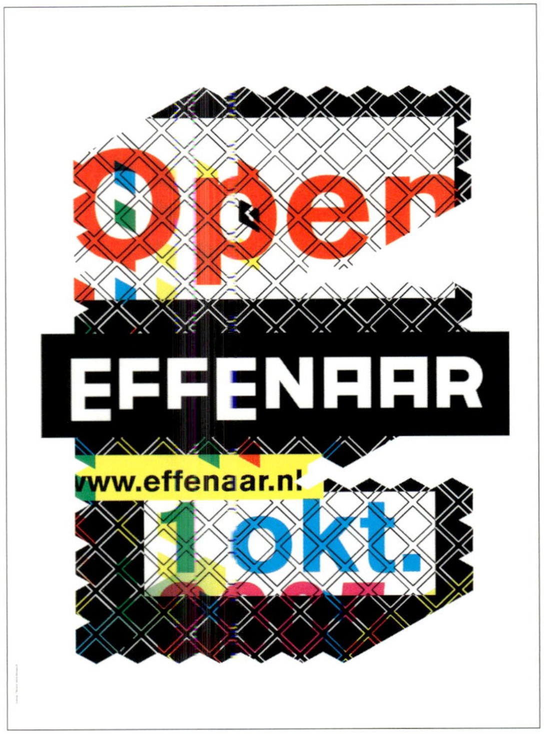

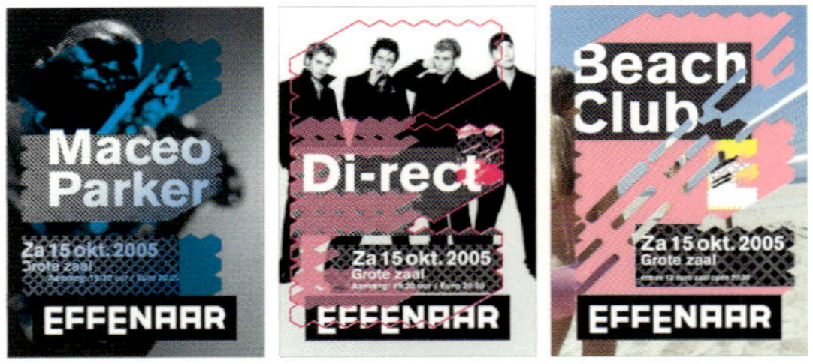

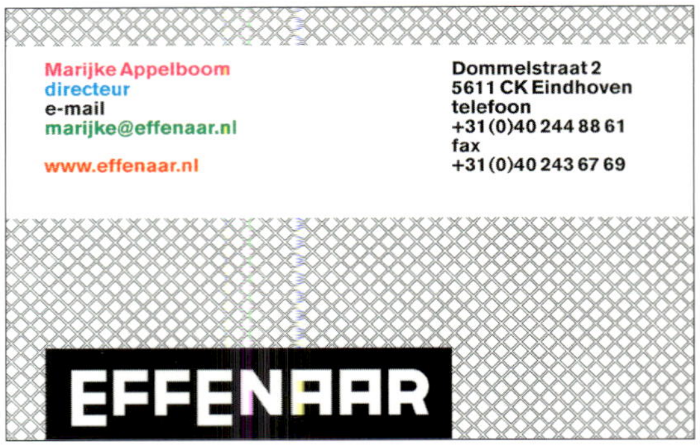

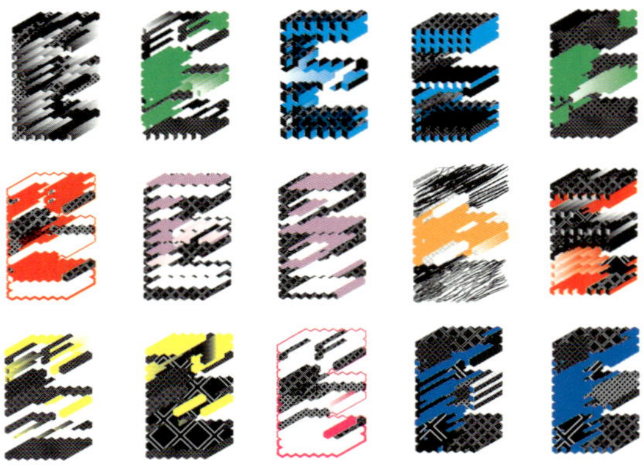

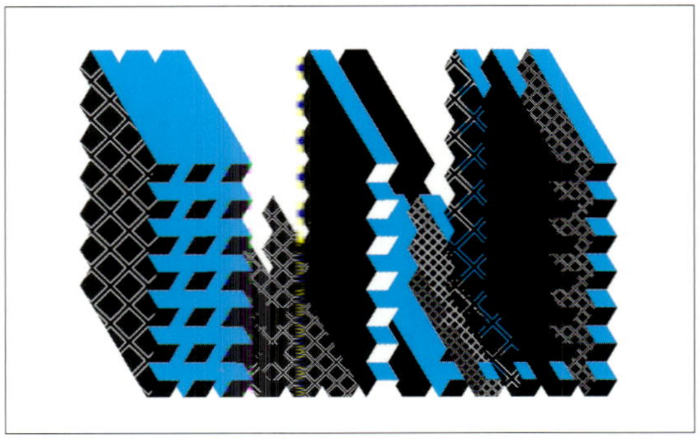

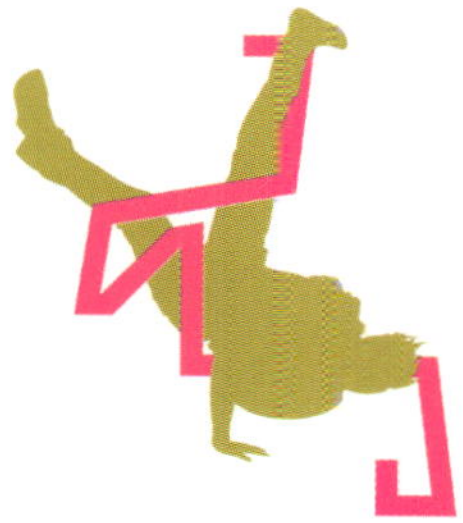

Auftraggeber Client
Holland Dance Festival
The Hague, The Netherlands

Projekt Project
Holland Dance Festival
Visuelle Identität
Visual identity

Design
Serge Scheepers
René Toneman
Maartje Wensing
Emiel Rijshouwer
Eelko Groenewegen
Fabrique [design, communications and new media]
Delft, The Netherlands

Für die 10. Auflage des Holland Dance Festivals entwickelte Fabrique die Corporate Identity und den Kampagnenstil rund um das Thema „dance approved". Ein Junge, der aufs Fahrrad springt – das immer wiederkehrende Bild in allen Kampagnenäußerungen –, steht sinnbildlich für das Thema Tanz im Alltag. Ausgangspunkt für das Design ist die These, dass Tanz immer und überall da ist. Weil Tanz nicht in nur einem einzigen Bild einzufangen ist, entwarf Fabrique eine Logosystematik: vier Bildsymbole und Schriftzüge, die durch ihre verschiedenartige Gestaltung jeweils einen anderen Tanzstil zeigen.

For the 10th edition of the Holland Dance Festival, Fabrique developed the identity and campaign style around the theme of "dance approved". A boy jumping on his bicycle, the recognisable image in all campaign communications, shows the dance of day-to-day life. The starting point for the design is the fact that dance is present everywhere and at all times. Because dance is not something that can be depicted in a single image, Fabrique designed a set of logos – four logos and word marks which, thanks to the varying design, each depict a different dance style.

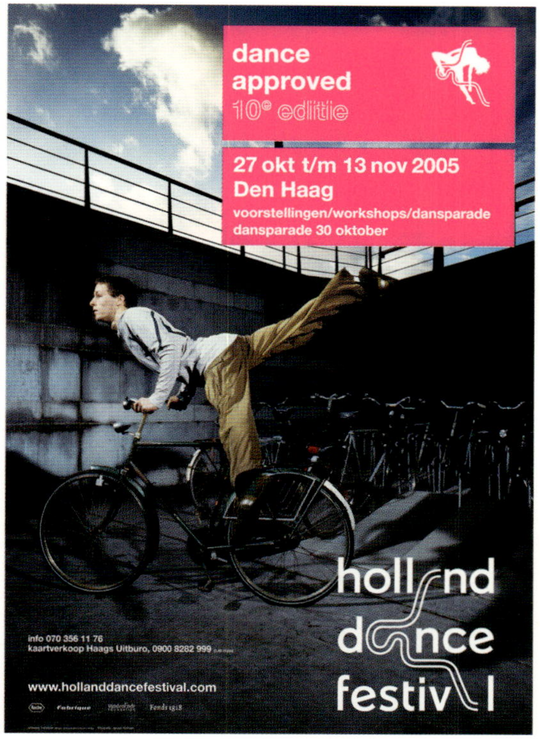

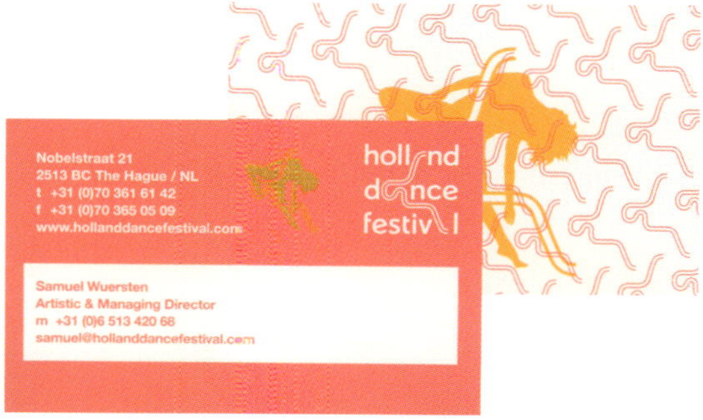

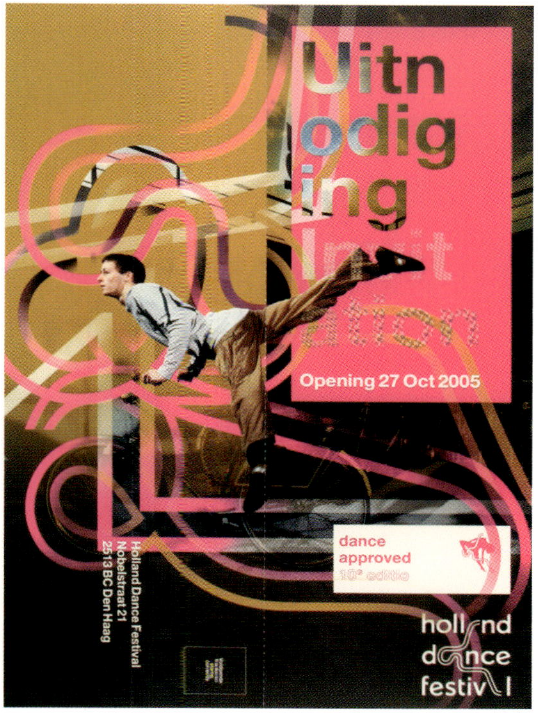

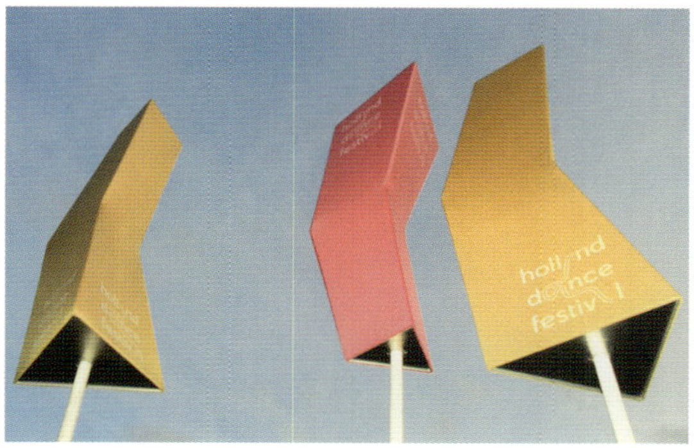

Auftraggeber Client
Jerini AG
Berlin, Germany

Projekt Project
Jerini AG Geschäftsbericht 2005
Geschäftsbericht
Company report

Design
Damm und Lindlar
Markenentwicklung und Corporate Design
Berlin, Germany

Jerini, ein biopharmazeutisches Unternehmen in Berlin, wurde vor zehn Jahren gegründet und ist im November 2005 an die Börse gegangen.
Bei dem Geschäftsbericht 2005 wird schon über das handliche Format Understatement, Teil des Selbstverständnisses von Jerini, vermittelt. Das Credo „We are Family" kommt in dem Kapitel „Jerini at Work" zum Tragen: In der Auswahl und Reihenfolge unserer Fotos wird zwischen Vorständen und Mitarbeitern nicht unterschieden. Die Entscheidung für den plakativen Einsatz eines Grasgrün im üblicherweise eher trockenen Finanzteil spricht für die Eigenwilligkeit des Unternehmens.

Jerini – a bio-pharmaceuticals concern in Berlin – was founded ten years ago and went public in November 2005.
The easy-to-handle format of the company report 2005 already conveys understatement, one aspect of the way in which Jerini sees itself. The credo "We are Family" is clearly operative in the chapter "Jerini at Work": there is no differentiation between board members and employees with regard to the selection and order of our photos. The decision in favour of a striking use of grass green in the generally rather arid financial section is an expression of the concern's unconventionality.

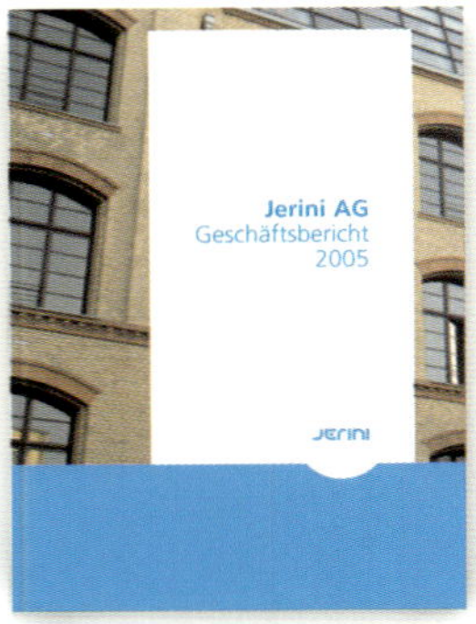

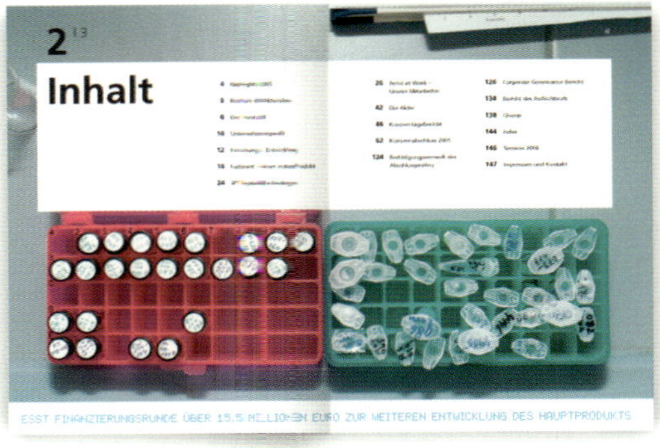

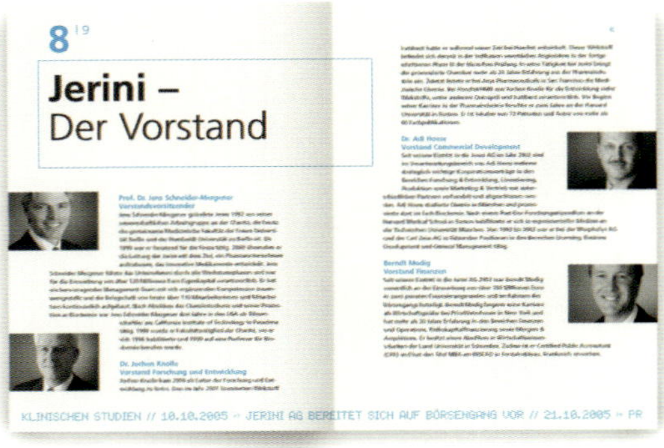

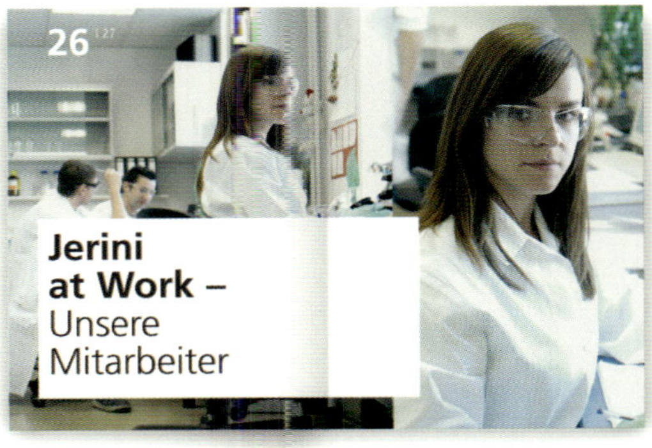

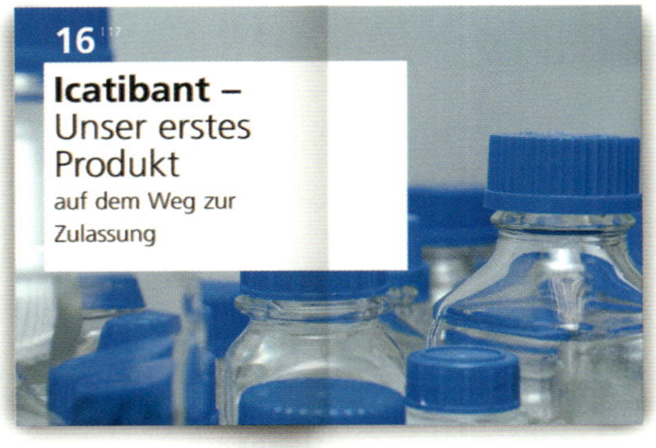

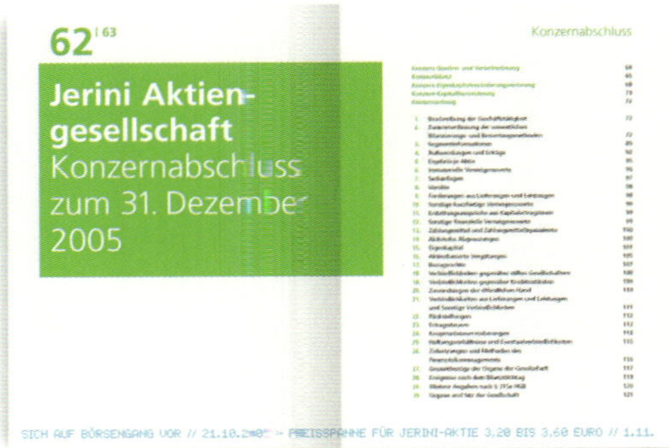

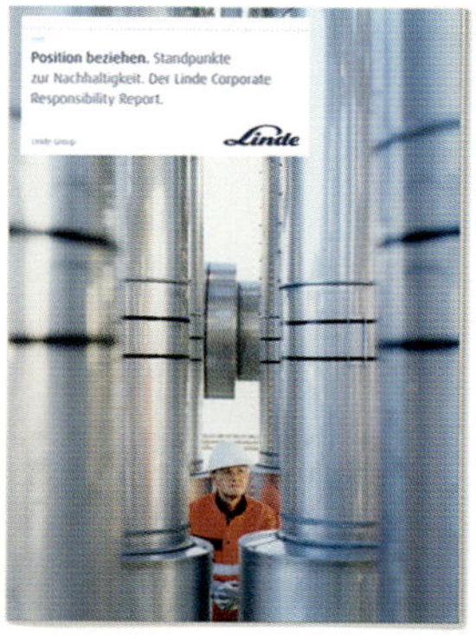

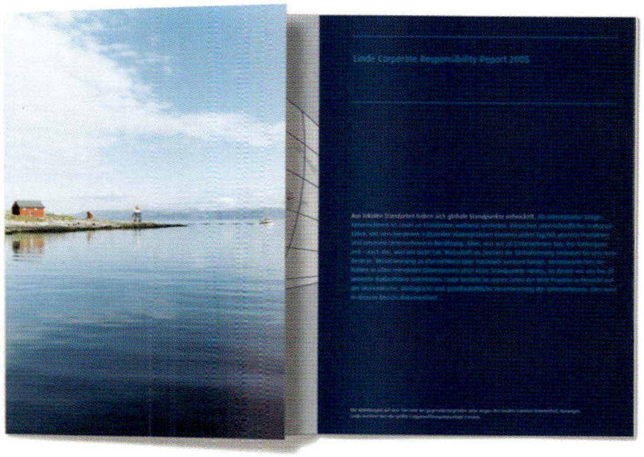

Auftraggeber Client
Linde AG
Wiesbaden, Germany

Projekt Project
Linde Corporate Responsibility Report
Broschüre
Brochure

Design
Peter Schmidt Group GmbH
Hamburg, Germany

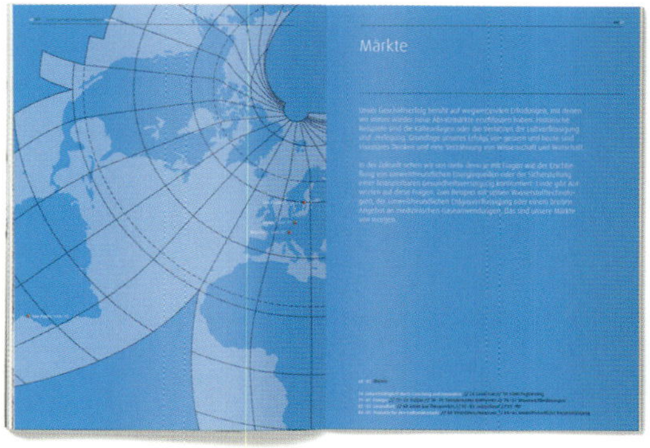

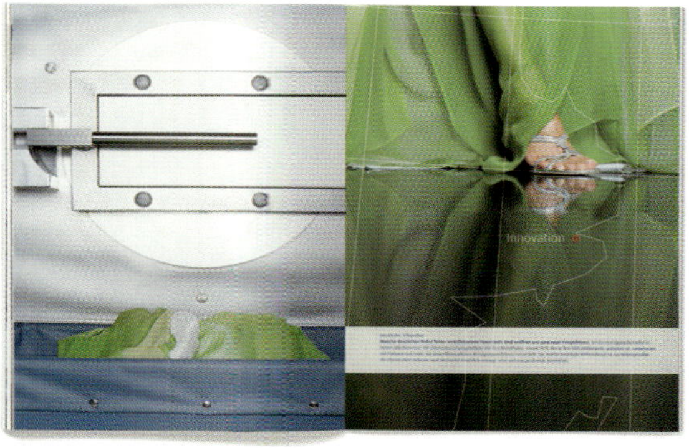

Mit dem CR-Report berichtet Linde 2005 erstmals konzernweit über sein Engagement in den vier strategischen Handlungsfeldern Umwelt, Mitarbeiter, Gesellschaft und Kapitalmarkt. Anhand von konkreten Beispielen wird beschrieben, wie z. B. Linde Technologien dazu beitragen, Umweltbelastungen zu reduzieren. Auf der Weltkarte im Titel-Ausklapper werden die geografischen Linde Standorte mit dem jeweiligen inhaltlichen Linde Standpunkt verbunden. Auf den Innenseiten werden alle aufgeführten Standorte und ihre Standpunkte bildlich und inhaltlich erläutert. Das Resultat: ein tiefer und authentischer Einblick in die Welt von Linde.

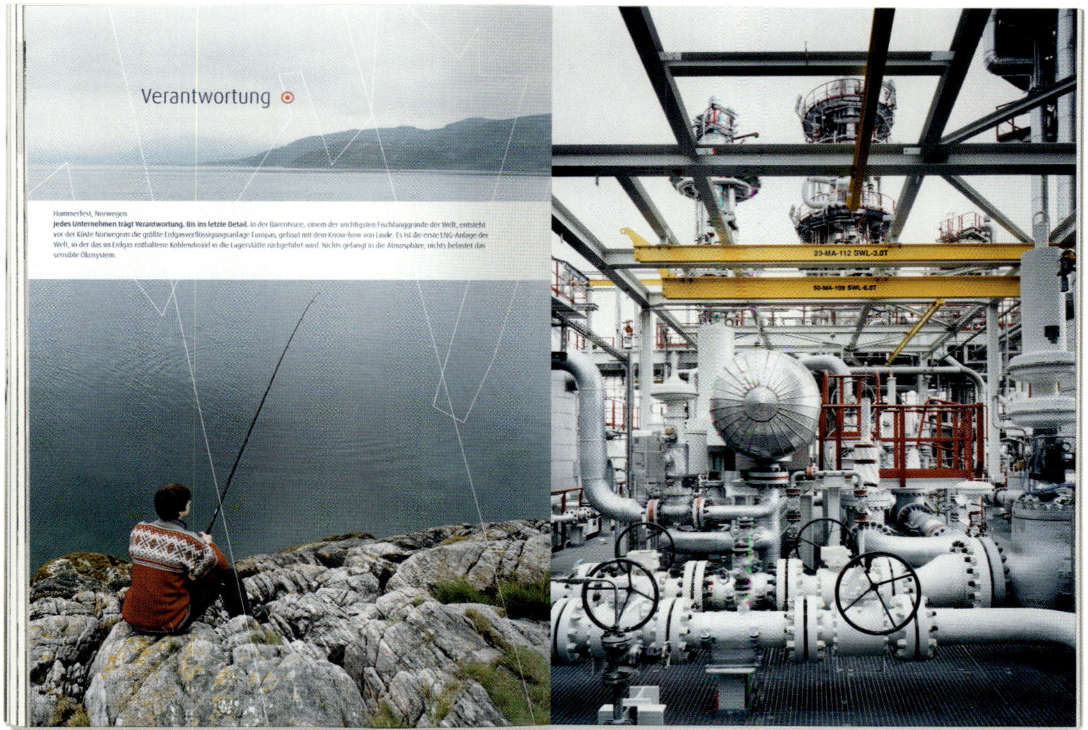

In its 2005 Corporate Responsibility Report, Linde reported group-wide for the first time on its commitment to four strategic action areas: the environment, human resources, society and capital markets. Focusing on specific cases, it describes how Linde is contributing to pollution-reducing technologies, to name just one example. The fold-out inside cover contains a map of the world linking the various Linde locations with content describing the respective Linde policy positions. The inside pages contain information about all the locations mentioned and their policy positions, including illustrations. The result is an in-depth and reliable insight into the world of Linde.

Auftraggeber Client
Staatstheater Mainz
Mainz, Germany

Projekt Project
CD Staatstheater Mainz
Corporate Design

Design
Fuenfwerken Design AG
Wiesbaden, Germany

Um über die identitätsstiftende Wirkung eines Logos hinaus einen Mehrfachnutzen zu schaffen, wurde für das Erscheinungsbild des Staatstheaters Mainz ein nicht statisch angeordneter Pfeil entwickelt, der auf Kerninformationen der verschiedenen Kommunikationsmedien hinweist. Eine transparente Farbfläche simuliert einen Theatervorhang, indem Spielszenen der Aufführungen teilweise oder vollständig abgedeckt werden. Die farbliche Kodierung orientiert sich an der programmatischen Ausrichtung.

To create a logo that would provide a sense of identity and serve a number of different purposes, a moveable arrow was developed for the Staatstheater Mainz's corporate design. The arrow points to key information in a variety of different communication media. A transparent colored area simulates a theater curtain either partially or fully concealing scenes from the performances. The color coding is based on the various program categories.

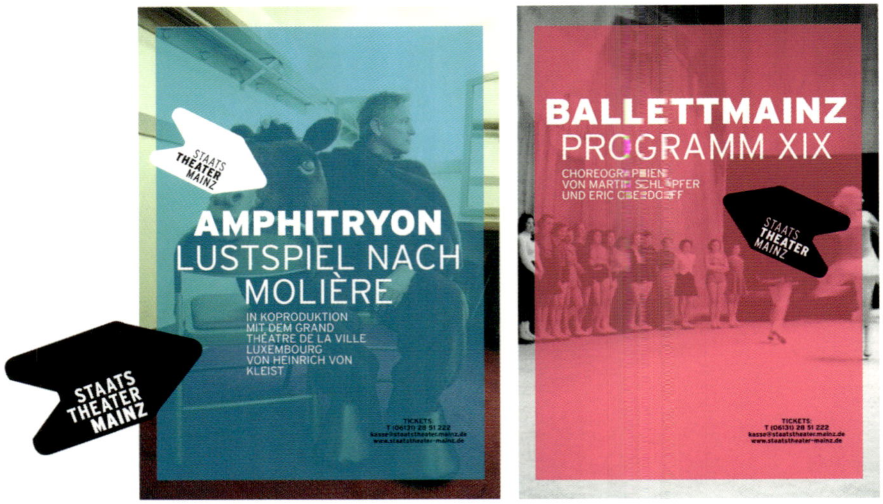

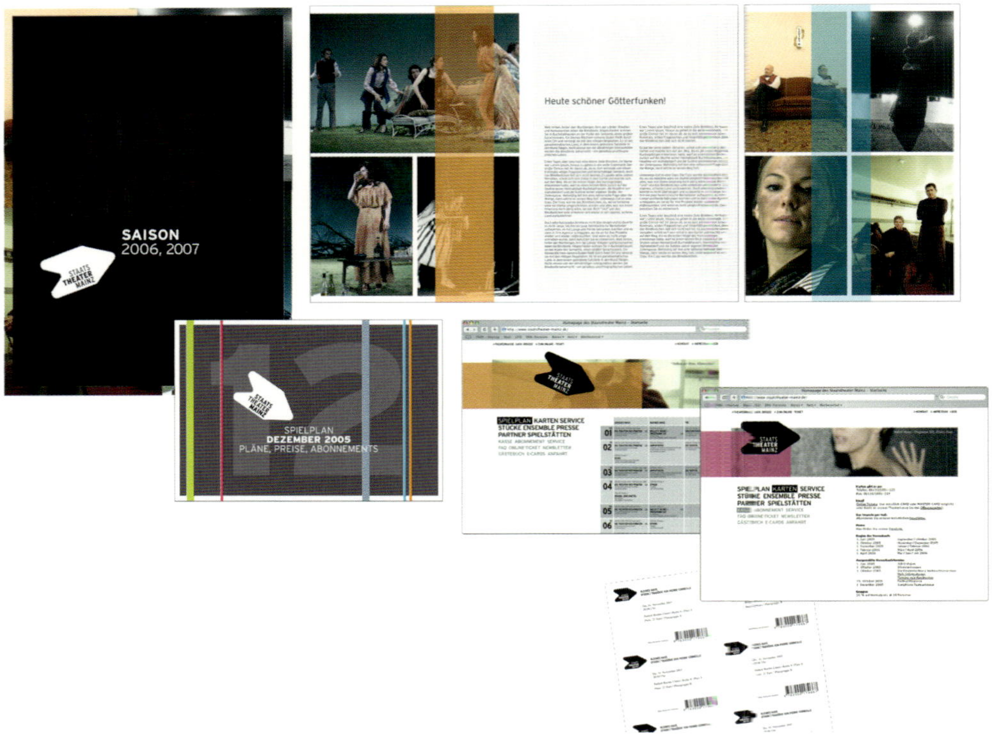

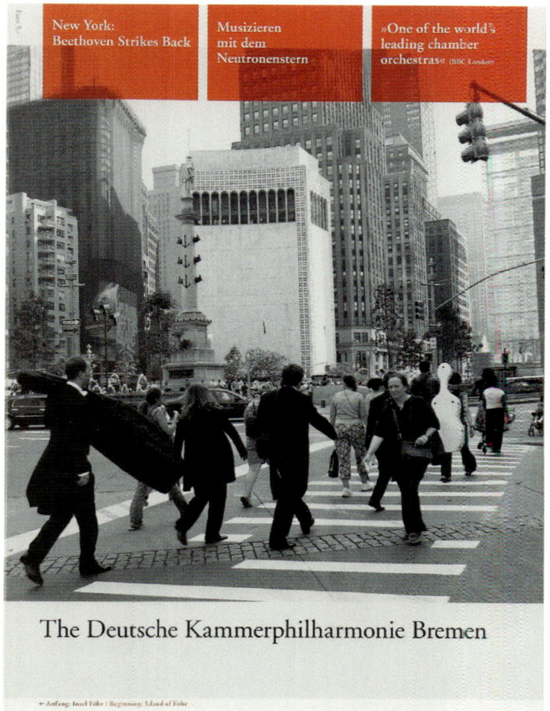

Auftraggeber Client
Die Deutsche Kammerphilharmonie Bremen
Bremen, Germany

Projekt Project
Kammerphilharmonie Bremen
Magazin
Magazine

Design
jung und pfeffer
visuelle kommunikation
Bremen, Germany

Die Deutsche Kammerphilharmonie Bremen zählt zu den internationalen Spitzenorchestern und tritt auf den weltweit renommiertesten Bühnen auf – von der Royal Albert Hall in London über die USA und Japan bis zu den Salzburger Festspielen. Das Magazin ist aus Anlass des 25-jährigen Orchester-Jubiläums erschienen und zeigt die Besonderheit und Einzigartigkeit der Kammerphilharmonie in vielen unterschiedlichen Facetten von der Geschichte über die Organisationsform und Dirigenten bis hin zur Entwicklung von neuen Präsentations- und Vermittlungsformen für klassische Musik.

The Deutsche Kammerphilharmonie Bremen is a leading international orchestra that performs in some of the world's most renowned venues – from Royal Albert Hall in London to many leading locations in the United States and Japan, all the way to the Salzburg Festival. This magazine was published for the orchestra's 25th anniversary and reveals the many different facets that make the Kammerphilharmonie so special and unique: from its history to its organization and conductors, all the way to its development of new modes of presenting and conveying classical music.

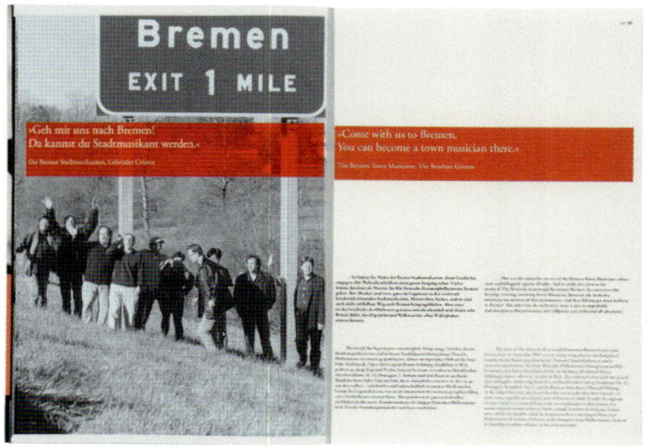

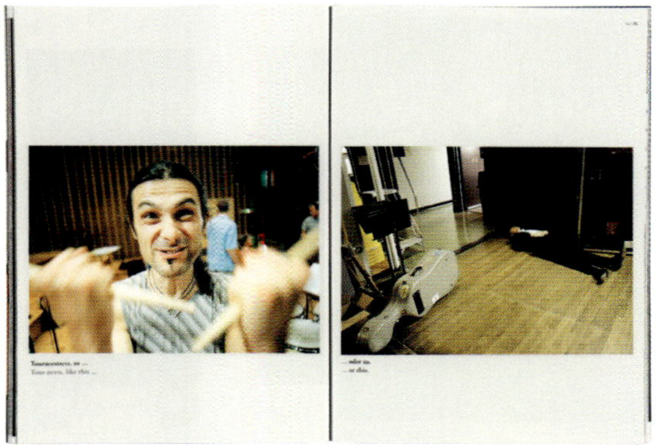

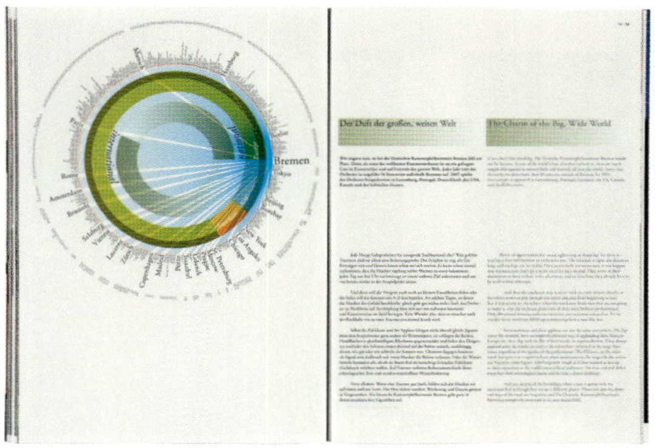

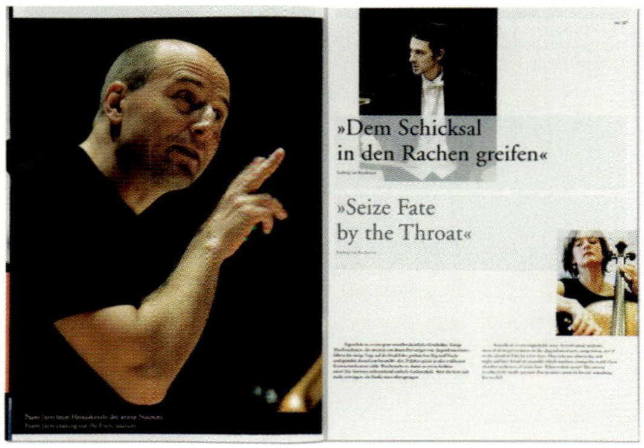

Auftraggeber Client
Podravka d.d.
Koprivnica, Croatia

Projekt Project
Feed Me / Ernähre mich
Jahresbericht
Annual report

Design
Bruketa & Zinic
Zagreb, Croatia

„Ernähre mich!“ ist ein Bericht, der die Firma Podravka, ihre Produkte und die Menschen, die für sie arbeiten, vorstellt. Podravka zeichnet sich aus durch Unternehmenswerte – Mission und Vision –, die wohlbekannt sind und in dem Bericht dargestellt werden. Da es sich um ein Unternehmen aus der Lebensmittelindustrie handelt, dreht sich der Bericht um Nahrung. Die Vorstandsmitglieder wurden als Köche fotografiert und tragen mit eigenen Rezepten zu dem Bericht bei. Ein Herz, das Firmenlogo, macht das Herz des Berichts aus. Das Titelblatt und die Eingangsfotos zeigen junge Menschen mit Löchern im Herzen, die sagen: „Ernähre mich!“ – und stellen so einen Bezug zu dem Firmenslogan her, der besagt, dass man das mit dem Herzen tun soll.

A report that presents the company, its people and products. Podravka has well-known company values, mission and vision, which are all portrayed in the report. Since it is a food company, the report revolves around food. Members of the management board were all photographed dressed as chefs, and they contributed a recipe. A heart, the logo of the company, is at the heart of the report. The cover and the introductory photos portray young people with holes in their hearts saying: "Feed me", as the company's slogan tells you to put your heart into everything you do. The report is interesting and fun to read, while providing important facts of the company.

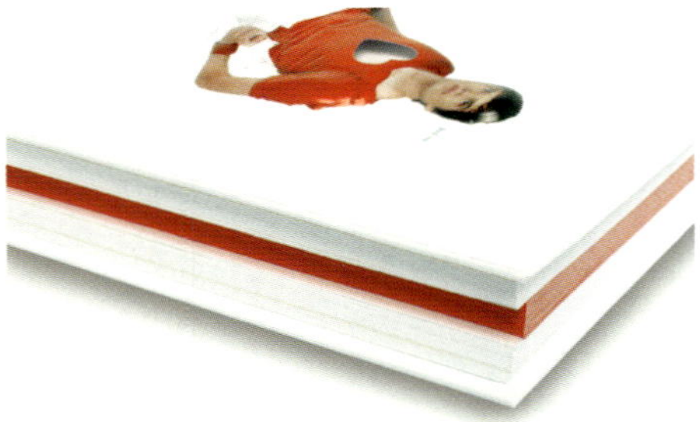

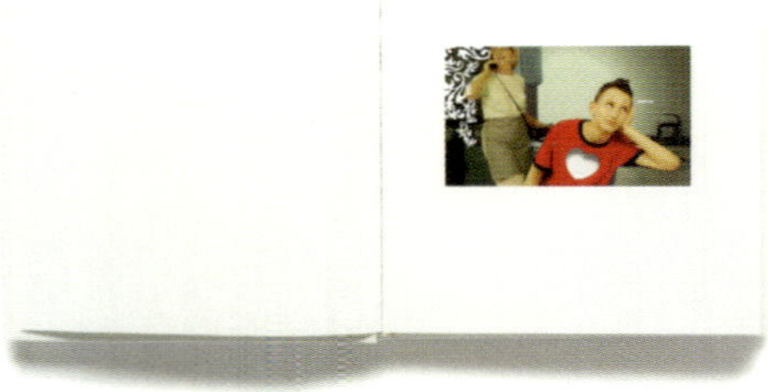

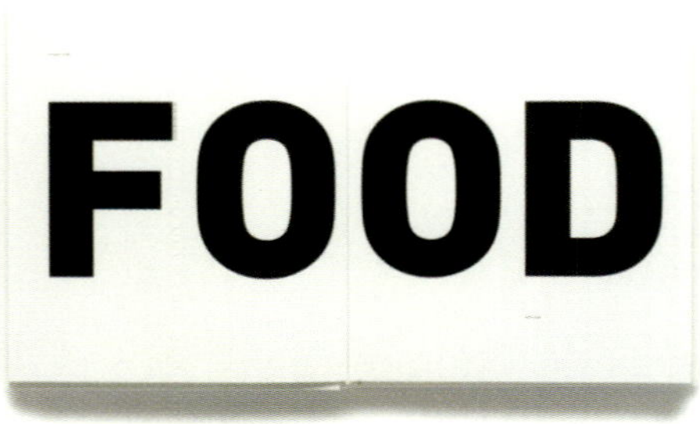

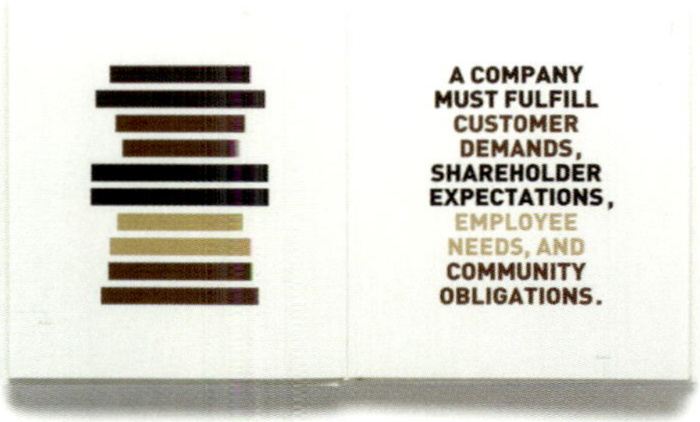

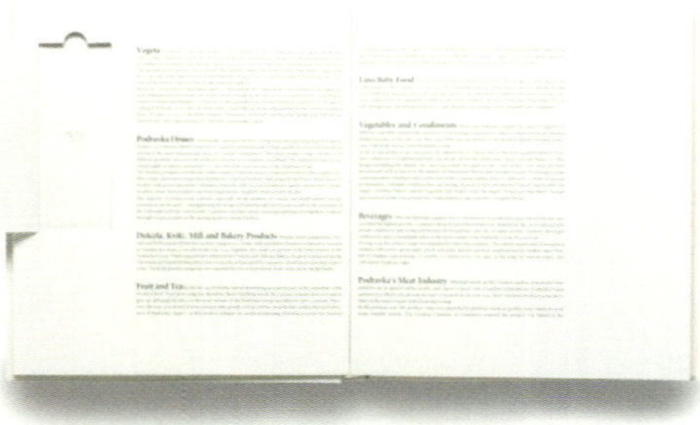

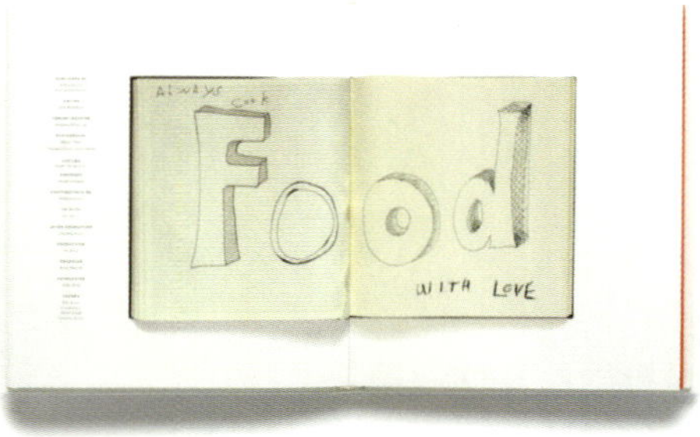

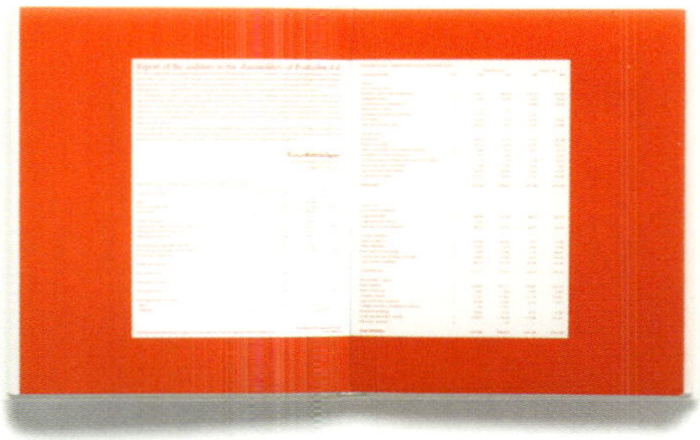

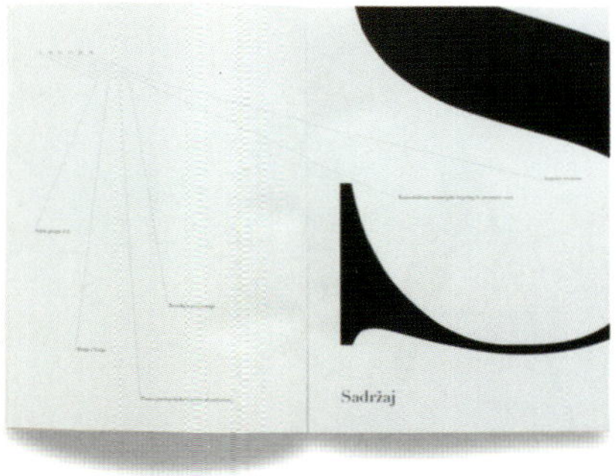

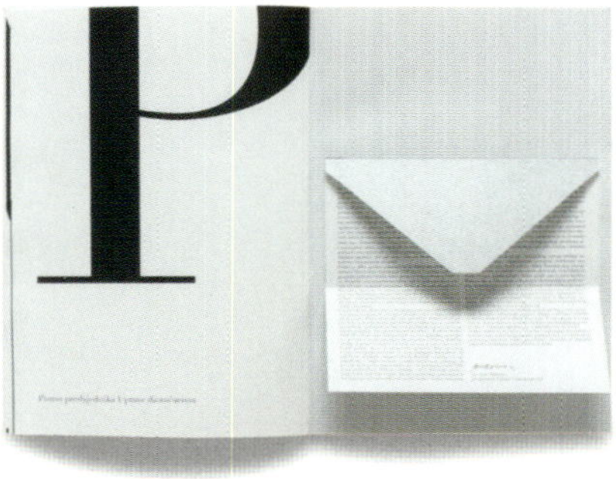

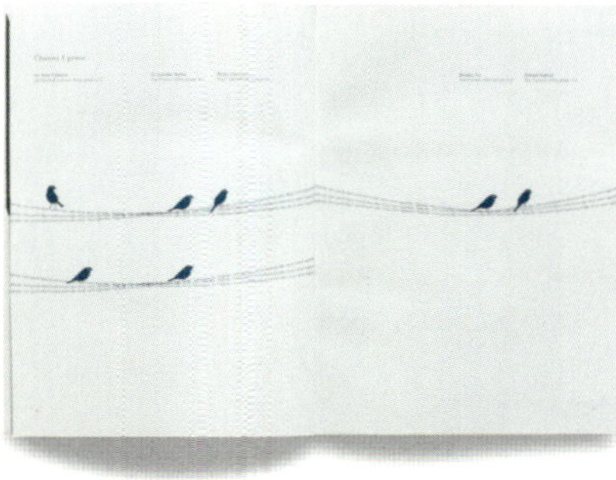

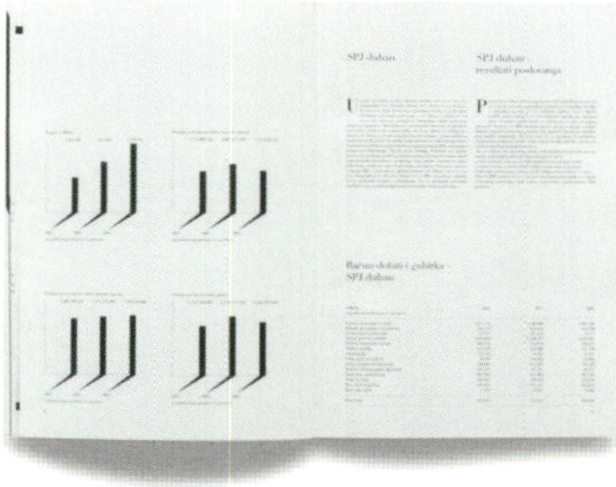

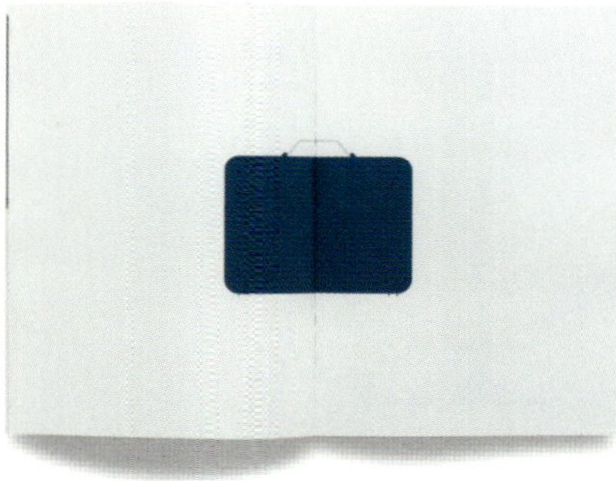

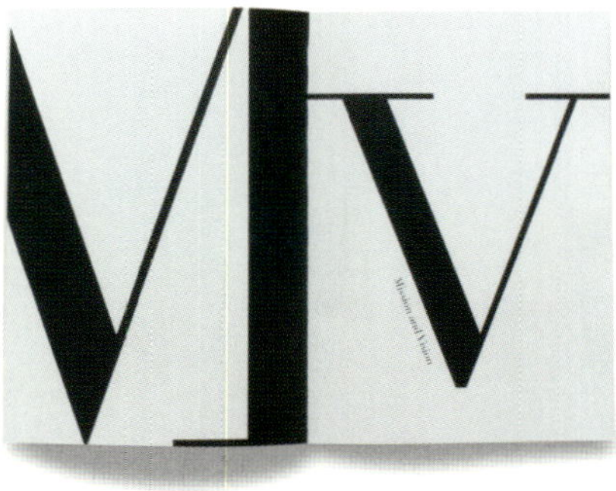

Auftraggeber Client
Adris Group
Rovinj, Croatia

Projekt Project
Positive Thoughts / Positive Gedanken
Jahresbericht
Annual report

Design
Bruketa & Zinic
Zagreb, Croatia

Das breite Spektrum der Aktivitäten der Adris Gruppe berücksichtigend, galt es, eine Lösung zu finden, die darstellt, dass es um ein Top-Unternehmer (oder auch: ein „1A“-Unternehmen) geht. Die Idee ging aus dem Firmenlogo hervor – dem Serifbuchstaben A im Didot-Font –, das schon für sich genommen herausstellt, dass es sich um eine starke Dachfirma handelt. Die Eingangsseiten der verschiedenen Teile des Berichts setzen sich aus Buchstaben zusammen, die aus dem Titel herausgelöst wurden und separate typografische Einheiten bilden. Außerdem enthält der Bericht Illustrationen, die die Idee der „Positiven Gedanken“ über bestimmte Motive vermitteln: Ein Bleistift steht z. B. für „Kreativität“, ein Fernglas für „Zukunft“.

Considering the wide range of Adris Group activities, the task was to find a solution that would represent the fact that it is an A1 company. The idea came from the logo, serif letter A in Didot font which itself tells that it is a powerful umbrella company. The introductory pages to different sections consist of letters extracted from the titles and form separate typographic units. Besides the titles based on typography, the report consists of simple illustrations based on the idea of "Positive Thoughts" with motifs like a bridge (for "idea"), a pencil ("creativity"), a briefcase ("entrepreneurship") and binoculars ("future").

Auftraggeber Client
Linde AG
Wiesbaden, Germany

Projekt Project
Linde Geschäftsbericht 2005
Geschäftsbericht
Annual report

Design
Peter Schmidt Group GmbH
Hamburg, Germany

Durch die Trennung von Finanzbericht und Magazin in zwei Publikationen beschreitet die Linde AG, einer der führenden Technologiekonzerne der Welt, mit dem Geschäftsbericht 2005 einen innovativen Weg im Bereich der Finanzkommunikation. Während der Finanzbericht alles Wesentliche zum Geschäftsjahr sachlich, kompakt und übersichtlich unter dem Motto „Zahlen, Daten, Fakten" bietet, bereitet das Linde Annual unter dem Titel „Richtungsweisend" Hintergrundinformationen zu den herausragenden Entwicklungen und Innovationen spannend und abwechslungsreich auf.

By splitting the financial report and magazine into two publications, Linde AG, one of the world's leading technology groups, has come up with an innovative method for communicating its financial results in its annual report for 2005. The Financial Report offers key data on the financial year in a factual, compact and clear fashion under the heading "Facts and Figures". The Linde Annual, entitled "Trend-setting", is an exciting and varied way of presenting background information on the group's outstanding developments and innovations.

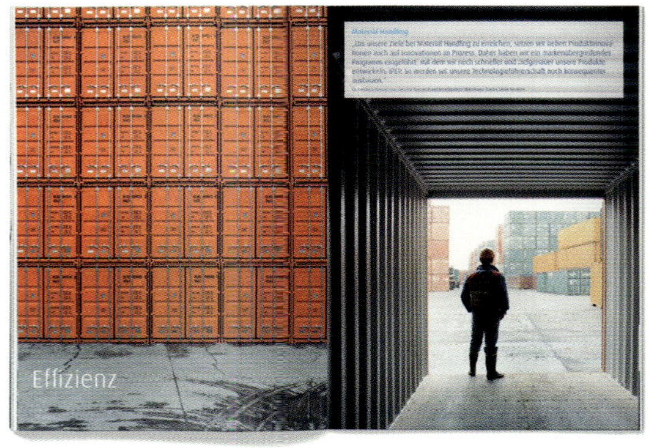

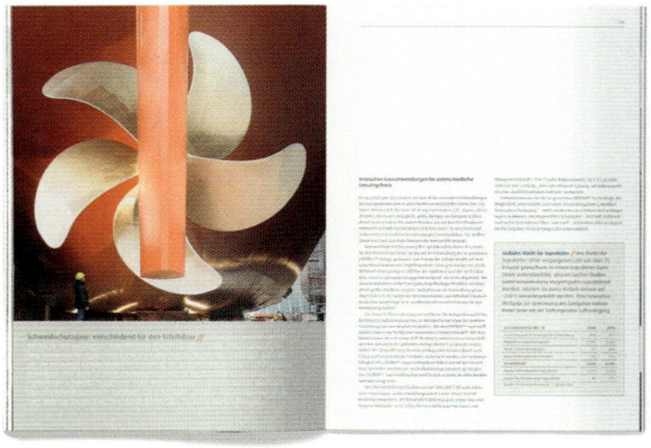

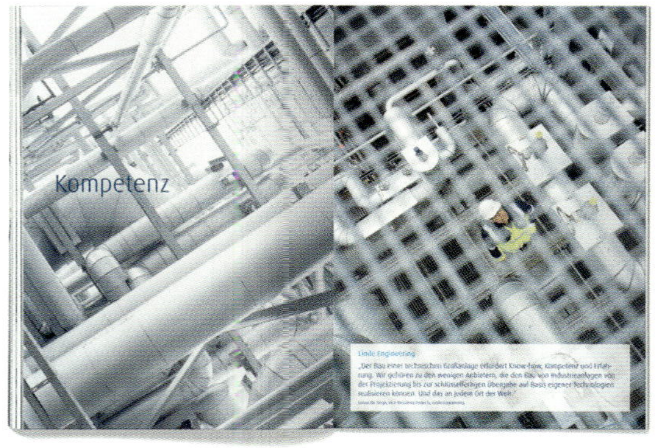

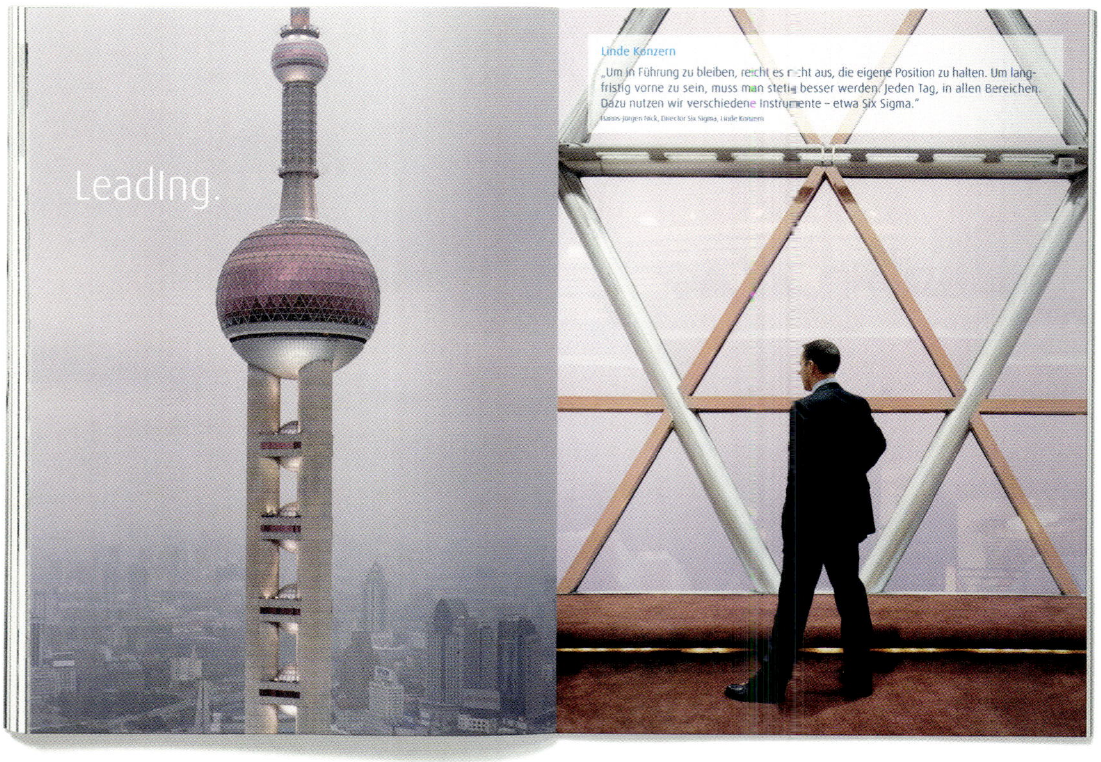

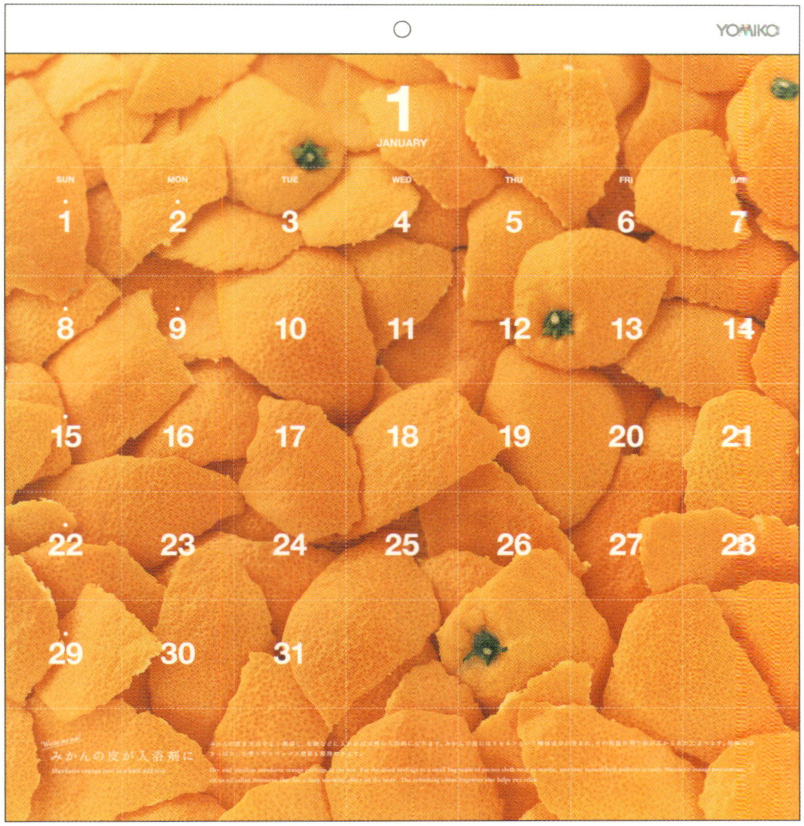

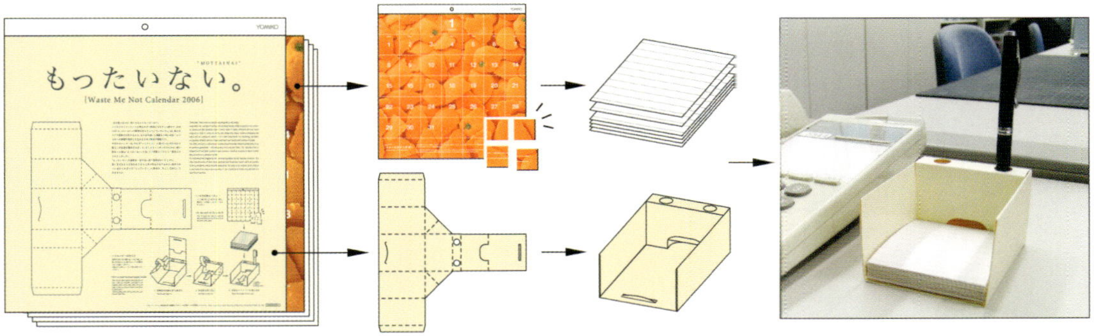

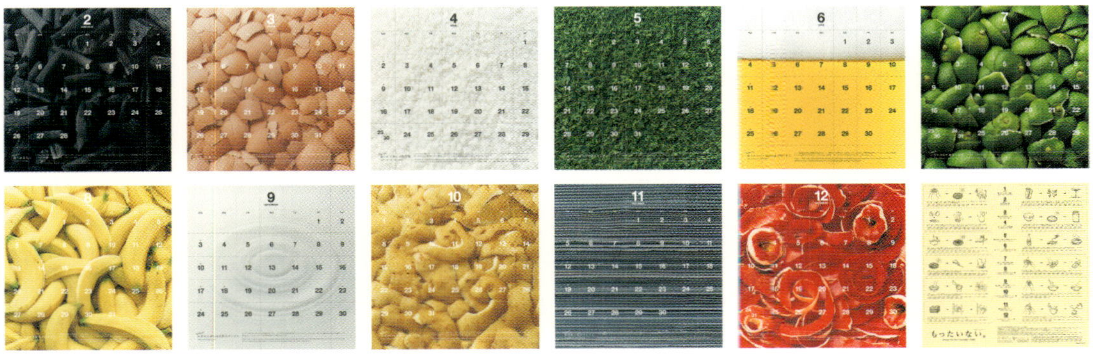

Auftraggeber Client
YOMIKO ADVERTISING INC
Tokyo, Japan

Projekt Project
Waste Me Not Calendar 2006 "Mottainai"
Kalender
Calendar

Design
YOMIKO ADVERTISING INC
Tokyo, Japan

Alle Fotos auf dem Kalender stellen Kunstwerke aus Abfällen dar. Gleichzeitig werden Hinweise gegeben, die in der japanischen Tradition seit Jahrhunderten Anwendung finden, wie diese Abfälle im Alltag nützlich wiederverwertet werden können. Neben den Fotos und Hinweisen spiegelt auch das Design des Kalenders unsere Bemühungen wider, diese Philosophie umzusetzen: Nach Gebrauch können die Kalenderseiten anhand von Perforationen in gleich große Teile zertrennt werden, deren Rückseiten als Notizblock benutzt werden können.
Wir haben damit aus einem Kalender, also aus einem „alten" Medium, ein neues Medium mit großer Interaktivität gemacht.

The calendar carries the message of "Mottainai" (Waste Me Not) to its users who are the source of mass products and advertising in Japan's consumer society.
All visuals in the calendar show artistically arranged garbage. In addition, interesting recycling tips for the "Waste Me Not" items are included that Japanese people have been using for centuries. True to the "Mottainai" spirit, each calendar page has been designed to be cut up into squares and re-used as memo paper.
The old medium "calendar" is thus transformed into a vital new medium, rich in interactivity. It is a completely new recycled and recycling calendar.

Auftraggeber Client
Axor Hansgrohe
Schiltach, Germany

Projekt Project
Axor WaterDream
Buch und Folder
Book and folder

Design
Hans-Jürgen Fein (CD/AD)
Andreas Sauer (Text)
Sabine Knapp (Grafik)
Werbung etc.
Stuttgart, Germany

Der „Axor WaterDream" wirft einen Blick in die Zukunft des Badezimmers. Entworfen von international renommierten Designern, wurden die visionären Badräume auf der Mailänder Möbelmesse präsentiert. Das für den Event und die begleitenden Werbemittel entwickelte Key Visual zeigt drei Wasserwellen, stellvertretend für die drei Designer und deren Arbeiten. Von der Einladungskarte bis zum abschließenden WaterDream-Buch sind die Wellen einer permanenten Veränderung unterworfen – genau wie unser Blick aufs Bad.

The "Axor WaterDream" takes a look into the future of the bathroom. Created by internationally renowned designers, these visionary bathrooms were presented at the Milan furniture fair. The key visual developed for the event and for the means of advertising shows three water waves, representing the three designers and their projects. From the first invitation card to the final WaterDream book, the waves are in constant flow – just like our view of the bathroom.

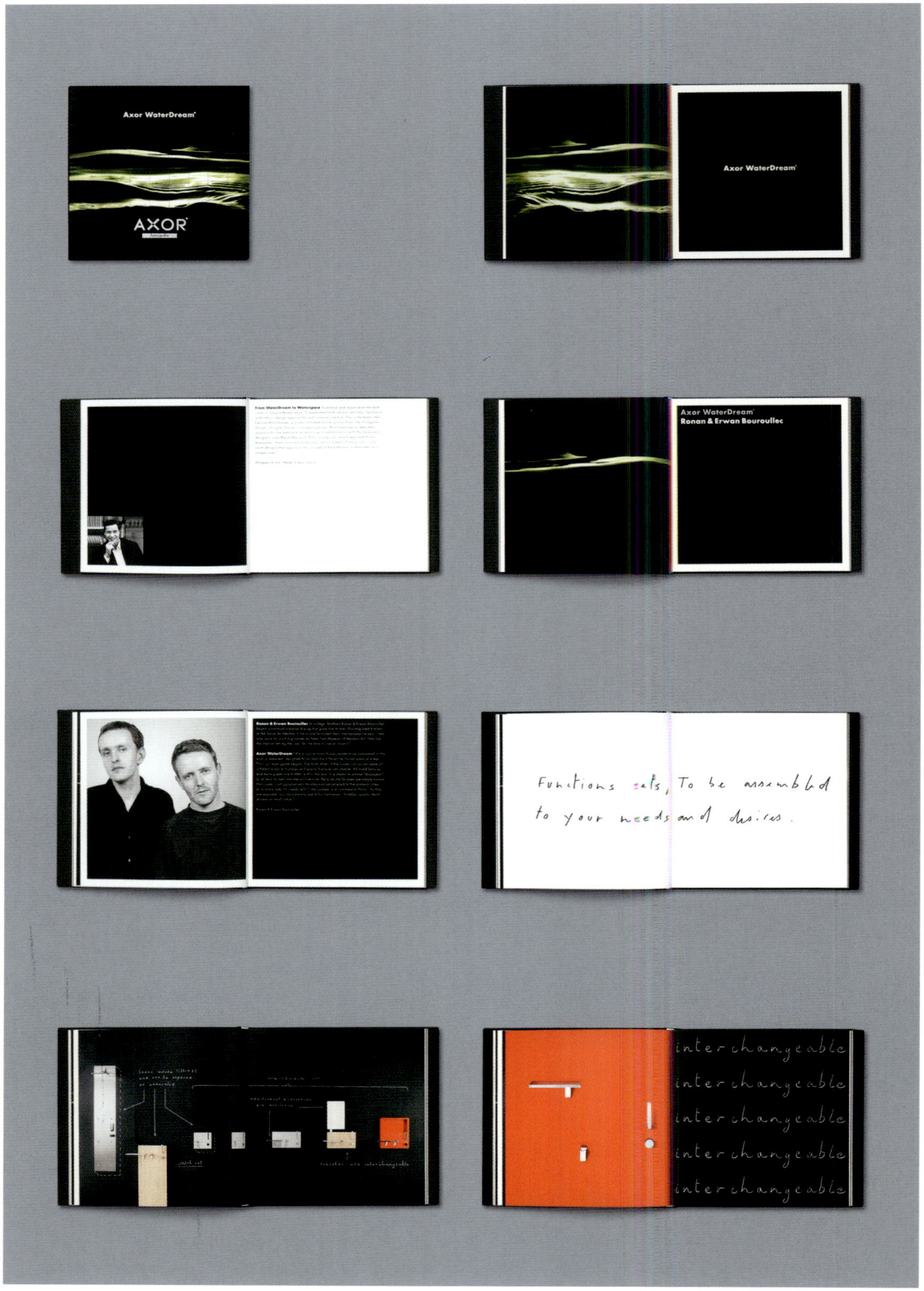

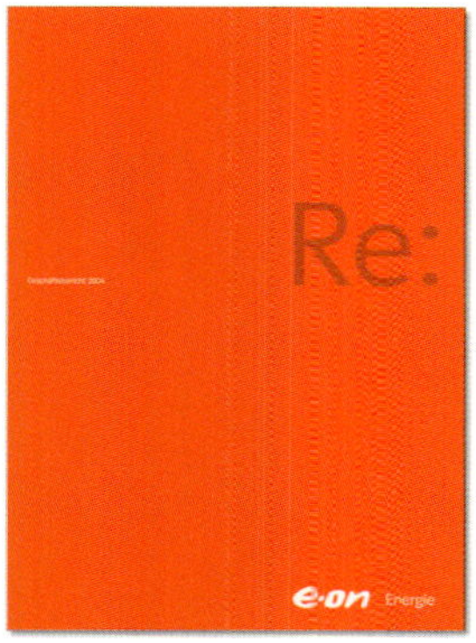

Auftraggeber Client
E.ON Energie AG
München, Germany

Projekt Project
Re:
Geschäftsbericht
Annual report

Design
Lesmo
Düsseldorf, Germany

Geschäftsbericht 2004 der E.ON Energie AG. – Die weiter wachsende Verantwortung des Unternehmens für Umwelt, Mitarbeiter und Gesellschaft steht im Vordergrund des Berichts: Nachhaltigkeit für die Zukunft kommender Generationen. Bilder von Kindern, die vordergründig naive Fragen stellen, regen zum Nachdenken an.

Annual Report 2004 of E.ON Energie AG – The report focuses on the company's growing responsibility for the environment, its employees and society. Sustainability will safeguard the future of the next generations. Photos of children asking ostensibly naïve questions provide food for thought.

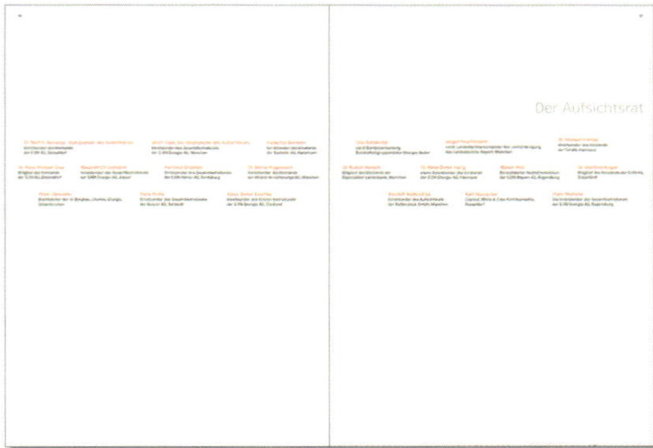

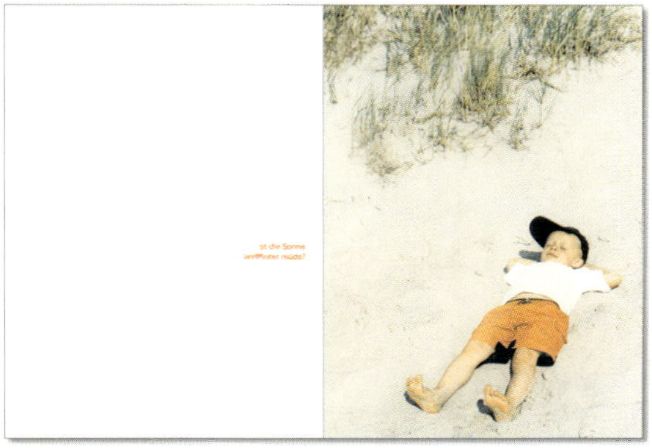

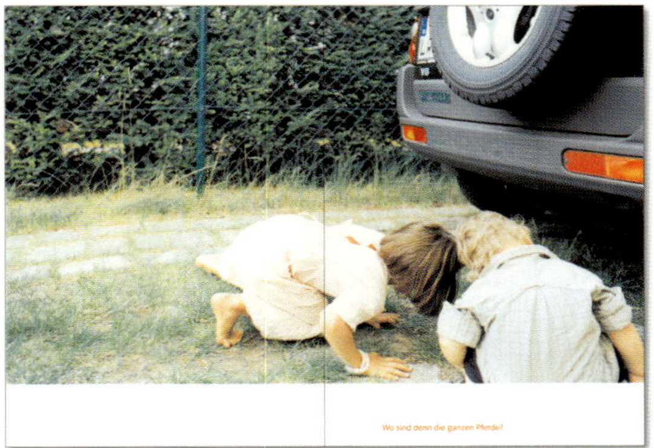

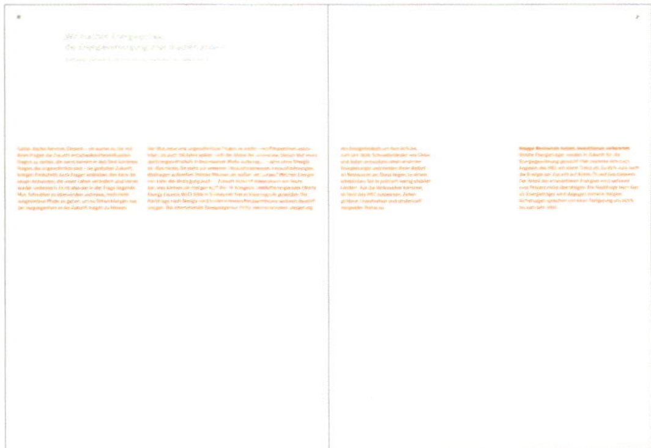

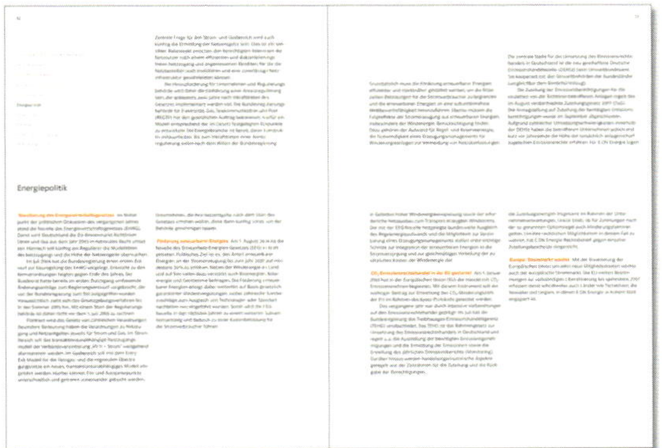

Auftraggeber Client
New Cat Orange
Wiesbaden, Germany

Projekt Project
Motiv-ator
Magazin
Magazine

Design
New Cat Orange
Wiesbaden, Germany

„Motiv-ator" ist etwas Besonderes. Es ist eine Agenturdarstellung, die ein Magazin ist, das wiederum ein Katalog ist. „Motiv-ator" schafft es, das Leistungsvermögen der Agentur zu demonstrieren, ohne darüber zu reden. Anstatt über die eigenen Fähigkeiten anzugeben, zeigt New Cat Orange seinen Partnern und (Neu-)Kunden in einem schön gestalteten und aufwändig produzierten Magazin, was man kann, zeigt in einer großen Foto-Strecke das Angebot des agentureigenen T-Shirt-Labels und liefert dazu noch eine Reihe lesenswerter Artikel zum Thema Motivation. So wird aus einem Sales Tool ein spannendes und schönes Stück Corporate Communication.

"Motiv-ator" is something very special. It is an agency presentation that is a magazine that is a catalogue. "Motiv-ator" is able to demonstrate the agency's capabilities without talking about them. Instead of bragging about its abilities, New Cat Orange shows its partners and (future) clients what they can do by providing them with a beautifully styled and expertly produced magazine, including an extensive photo feature on the agency's own t-shirt label, and a series of entertaining articles on motivation. In the end, a piece of sales communication is turned into an exciting and beautiful piece of corporate communication.

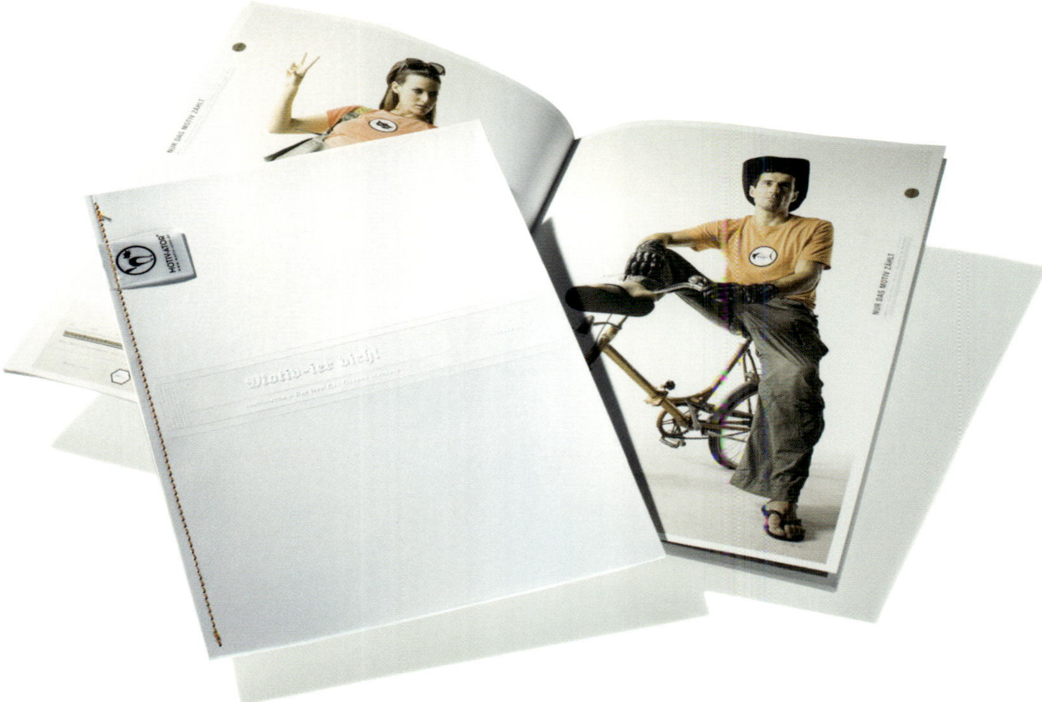

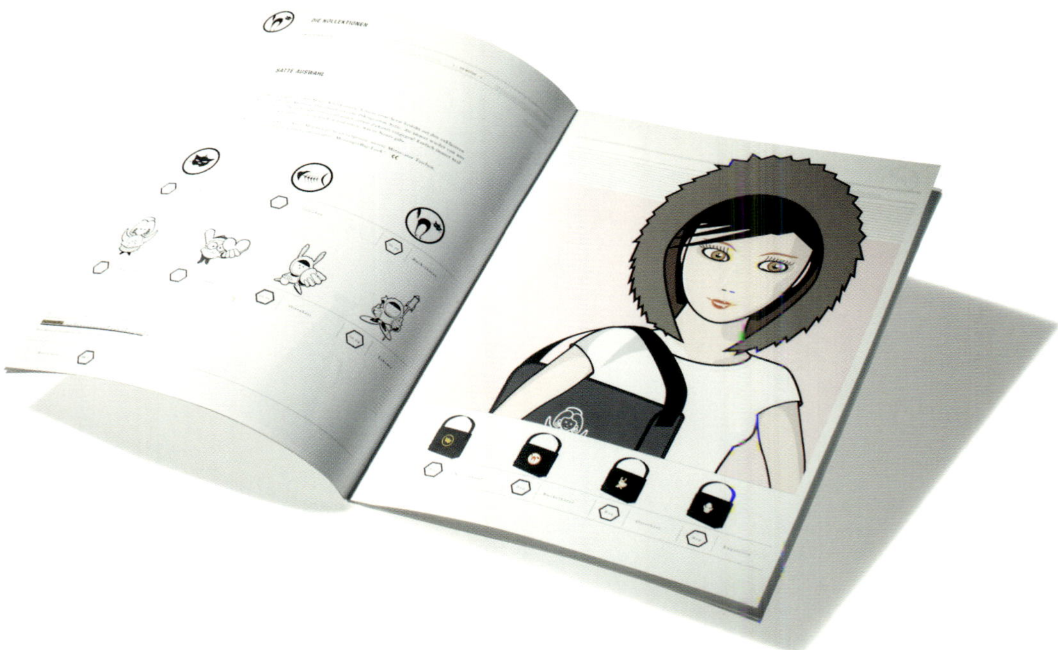

Auftraggeber Client
Laboratorium
Zagreb, Croatia

Projekt Project
Unendlicher selbstklebender Kalender
Everlasting adhesive calendar

Design
Ivana Vucic
Orsat Frankovic
Laboratorium
Zagreb, Croatia

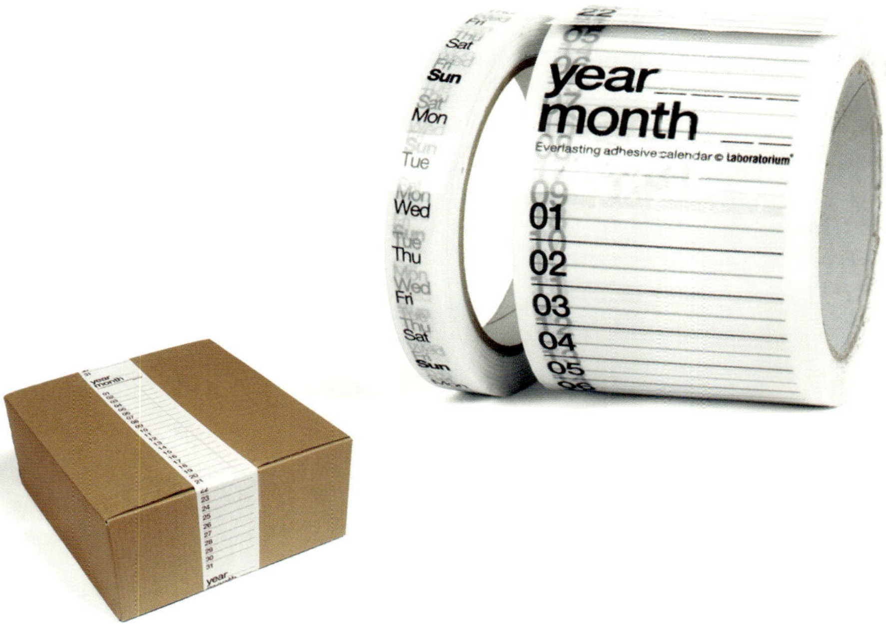

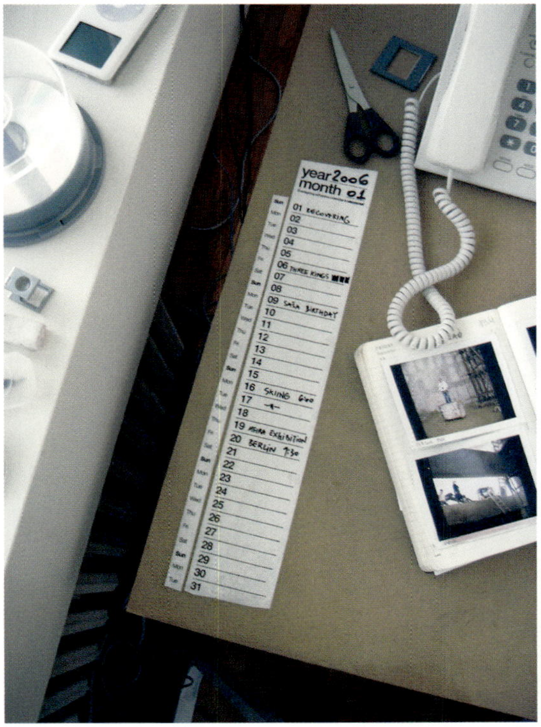

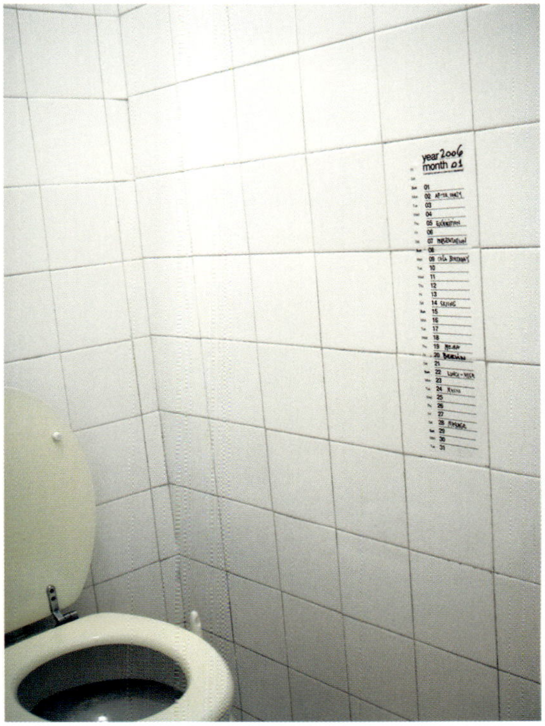

Der endlose Kalender ist ein multifunktionales Selbstklebeband, das man als Planer, Post-it, Klebeband, Aufkleber… nutzen kann. Nur die Länge des Bandes ist begrenzt. Der Benutzer kann jeden Monat zusammenstellen – egal welchen Jahres –, indem er zwei Bänder verschiedener Breiten in einer Tag-und-Datum-Kombination zusammensetzt und auf eine beliebige Oberfläche klebt. Der Kalender wurde als Promotions-Geschenk (Weihnachtskarte) an Kunden und Freunde von Laboratorium versandt.

The everlasting calendar is a multi-functional adhesive tape which functions as an organizer, planner, post-it, tape, sticker and suchlike at the same time. It is limited only by its own length. You can create each month in any year by a day-and-date combination of two separate rolls (different widths) and stick them together at the place where you need it, including different surfaces, walls, tables, toilet walls… The calendar was created as a promotional gift (Christmas card) for Laboratorium's clients and friends.

Auftraggeber Client
Carlsberg Deutschland Gruppe
Germany

Projekt Project
5,0 Original
Branding und Packaging
Branding and packaging

Design
feldmann+schultchen design studios
Hamburg, Germany

„Bier und gut!" – 5,0 Original ist eine neue Marke im Preiseinstiegssegment, die sich auf das Wesentliche konzentriert. Das einfache, zweifarbige Etikett und der simple Kronkorken verkörpern ein neues Selbstverständnis. Die Biermarke verzichtet auf Schnickschnack: kein kostspieliges Glanzgoldpapier, keine teure TV-Werbung und ein gebrauchter Kasten. 5,0 Original gibt es in den Sorten Pils, Weizen, Export und Lemon.

"Bier und gut!" ("Simply beer!") – 5,0 Original is a new brand in the low price segment that concentrates on the essentials. That simple two-colour label and that unassuming crown cork are the very embodiment of a new self-image. It's a beer brand that gets by without all those frills; there's no pricey, shiny gold paper, no expensive TV commercials and a used beer crate. 5,0 Original comes as a pilsner, a wheat beer, export and lemon.

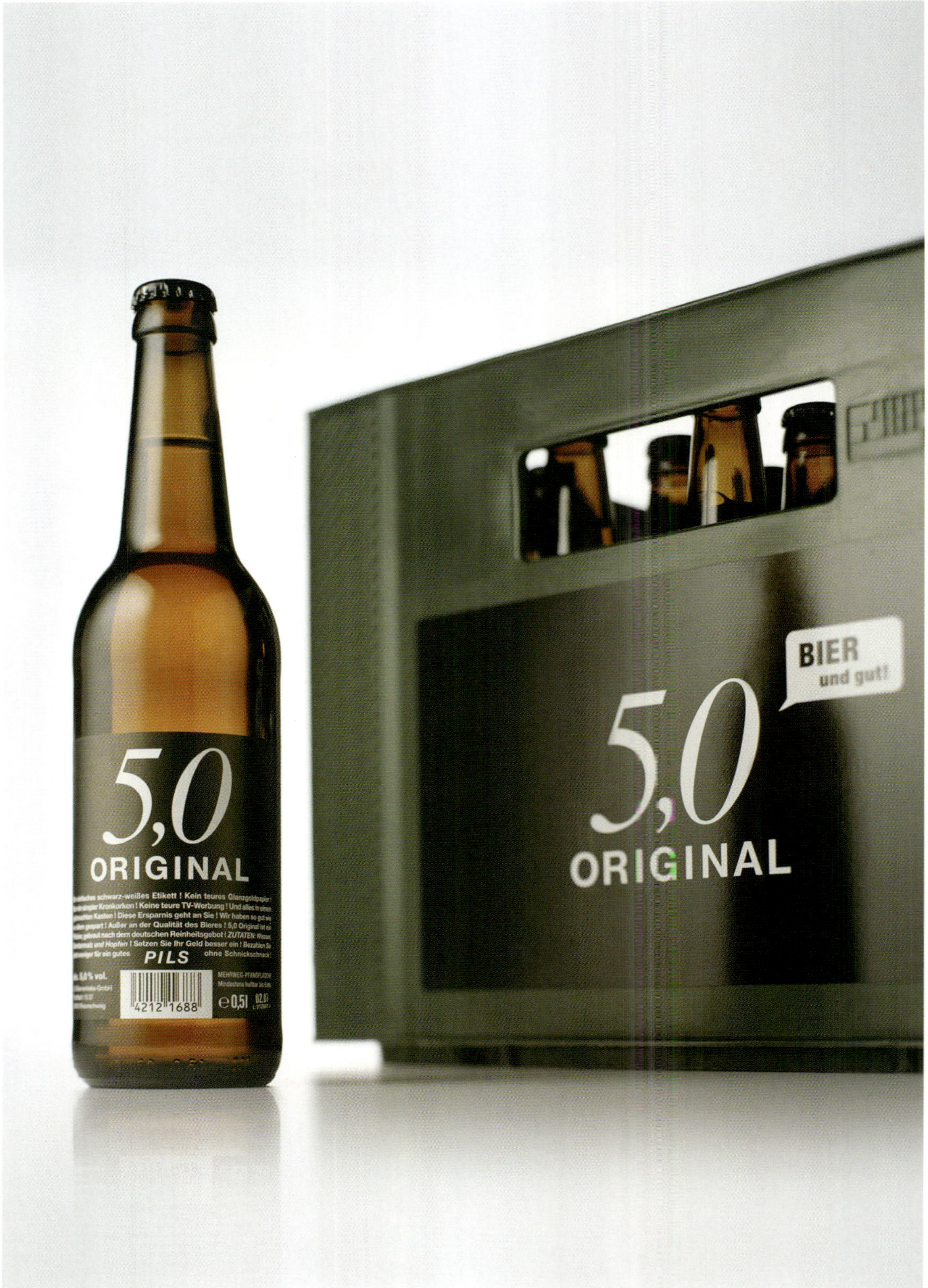

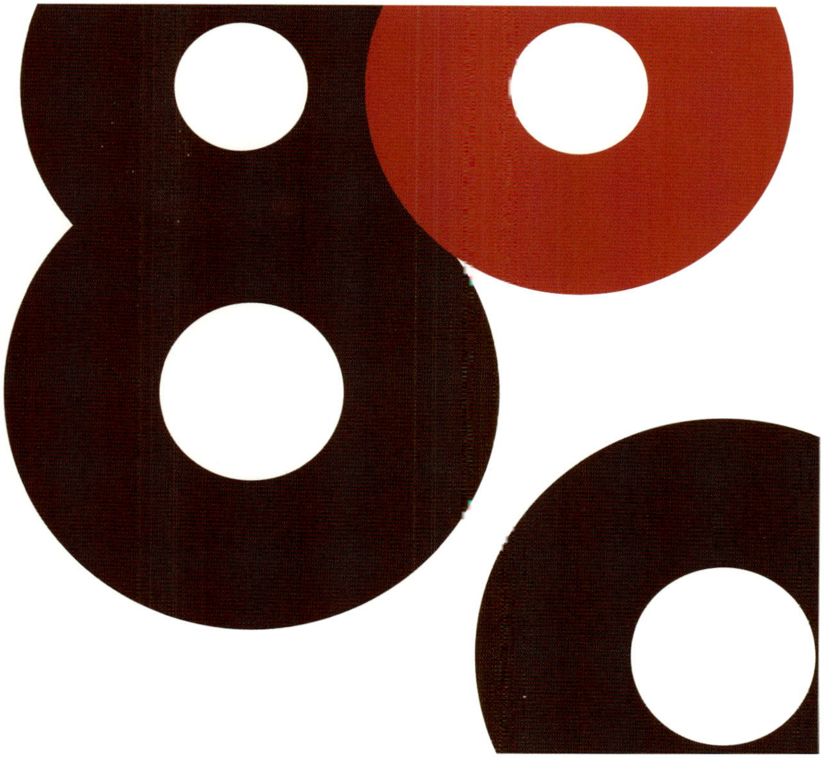

Auftraggeber Client
Q.Kom Archiv
Hannover, Germany

Projekt Project
SAKE 8
Buch und Label
Book and label

Design
Hanae Komachi
Hannover, Germany
Konstantin Grcic
KGID
München, Germany

Der Industriedesigner Konstantin Grcic hat für den 250 Jahre alten japanischen Sake-Hersteller Hachitsuru das Label SAKE 8 entworfen, das ohne herkömmliche japanische Zeichensprache auskommt. Dabei wurde die traditionelle Sake-Kultur mit der zeitgenössischen deutschen Designkultur zum ersten Mal verbunden. SAKE 8 ist ein deutsch-japanisches Kulturprojekt mit zwei Ausstellungen in beiden Ländern von Hanae Komachi. Dazu gehört das Bilderbuch SAKE 8, das zusammen mit dem Künstler Dieter Froelich entstanden ist – und den Entstehungsprozess des Labels und des Sakes darstellt und vom Künstler entwickelte Speisen zum Sake enthält.

Industrial designer Konstantin Grcic has come up with a SAKE 8 label design for the 250-year-old Japanese sake producer Hachitsuru that dispenses with the use of conventional Japanese characters. This is the first time traditional sake culture has joined forces with contemporary German design culture. SAKE 8 is a German-Japanese cultural project with two exhibitions in both countries by Hanae Komachi. It includes the illustrated book SAKE 8, created with artist Dieter Froelich, which depicts the creative process behind the label and how the sake is produced, and also features dishes created by the artist that go well with sake.

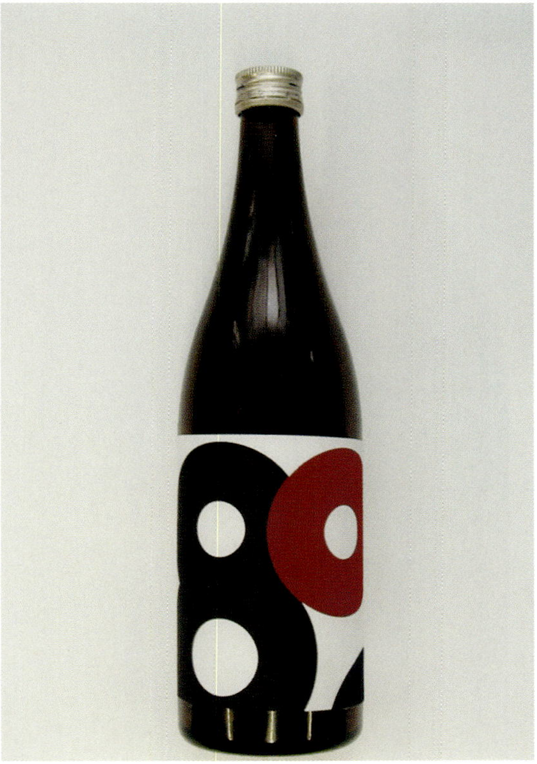

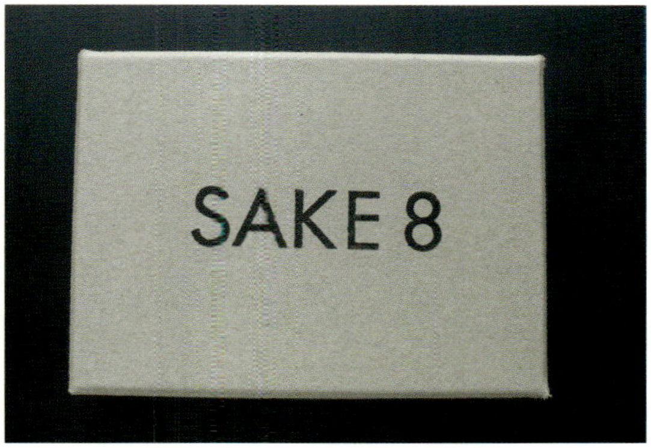

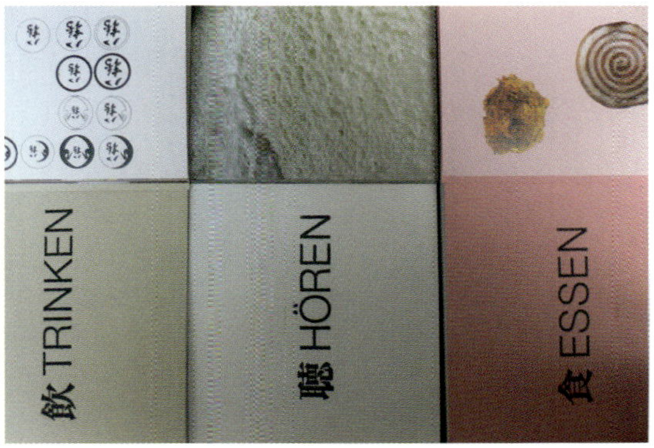

Auftraggeber Client
Realambient
Köln, Germany

Projekt Project
Realambient Vol. 03 – 05
CD / DVD-Cover
CD / DVD covers

Design
Klaus Untiet
Süleyman Kayaalp
Rob Fährmann
wppt:kommunikation GmbH
Gesellschaft für visuelle Kultur
Wuppertal, Germany

Die CDs „essen.momente“ und „La Défense – stage urbain“ und die DVD „Lisboa.reloaded“ von Michael Rüsenberg sind Soundscape-Kompositionen, Mischungen aus realen Klängen aus den Städten mit ihren Fußgängerzonen, Industrieanlagen und individuellen Geräuschteppichen.
Mit dem Ziel, eine Verpackung zu gestalten, die den Kompositionen gerecht wird, ist eine innovative Lösung entstanden, die sich vom Üblichen abhebt und einen Vorgeschmack auf die im Inneren zu erwartenden Erlebnisse bietet. Das Rohmaterial Pappe sowie die beiden im Siebdruck aufgetragenen Farben heben visuell und haptisch den authentischen, kraftvollen Eindruck der Klänge hervor.

The “essen.momente” and “La Défense – stage urbain” CDs and the “Lisboa.reloaded” DVD by Michael Rüsenberg are soundscape compositions, featuring blends of real urban sounds, including the relevant cities’ pedestrian zones, industrial plants and own unique tapestries of noise.
The objective was to design packaging that would do justice to the compositions, and an innovative solution was found which stands out from what is usually produced and offers a foretaste of the experiences that can be expected from its contents. The use of cardboard as a raw material as well as two screen-printed colors accentuates the authentic, powerful impression of the sounds both at the visual and tactile level.

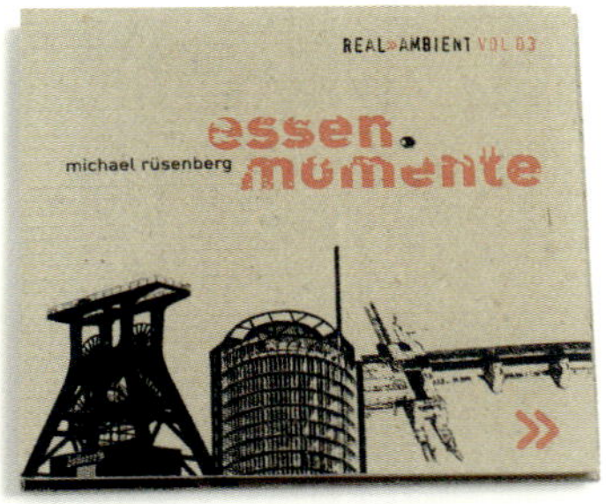

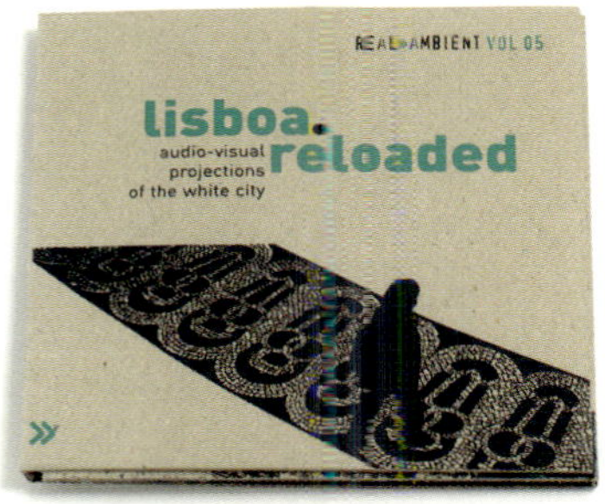

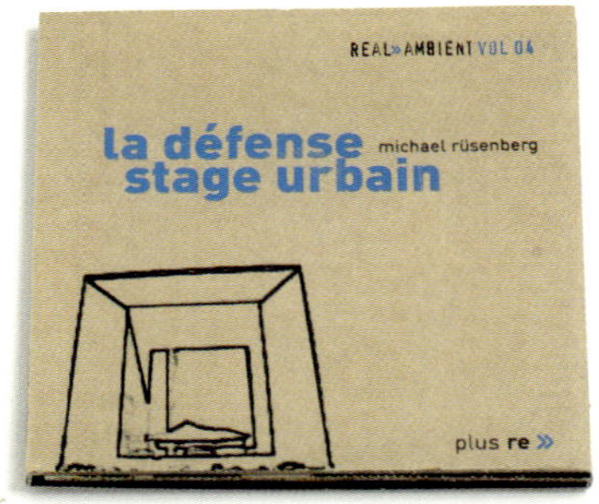

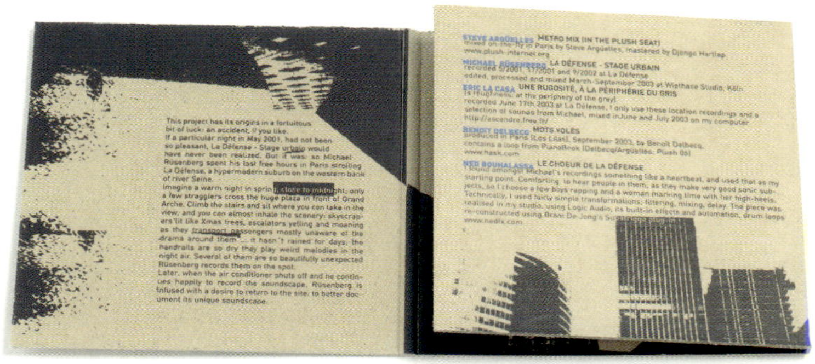

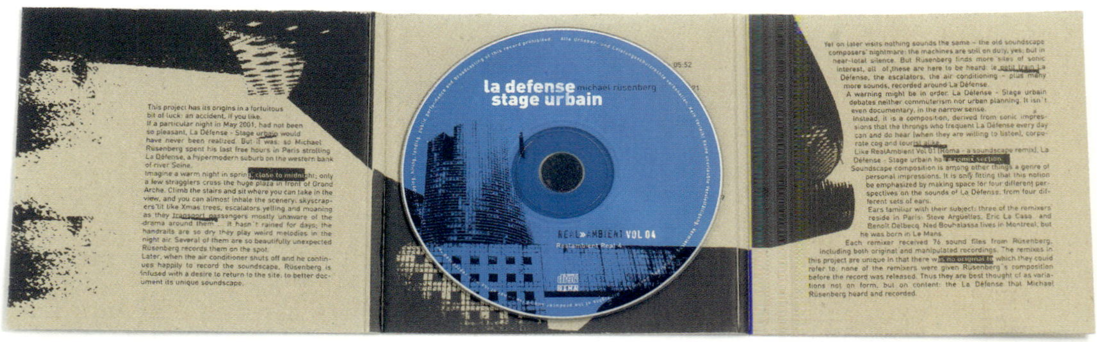

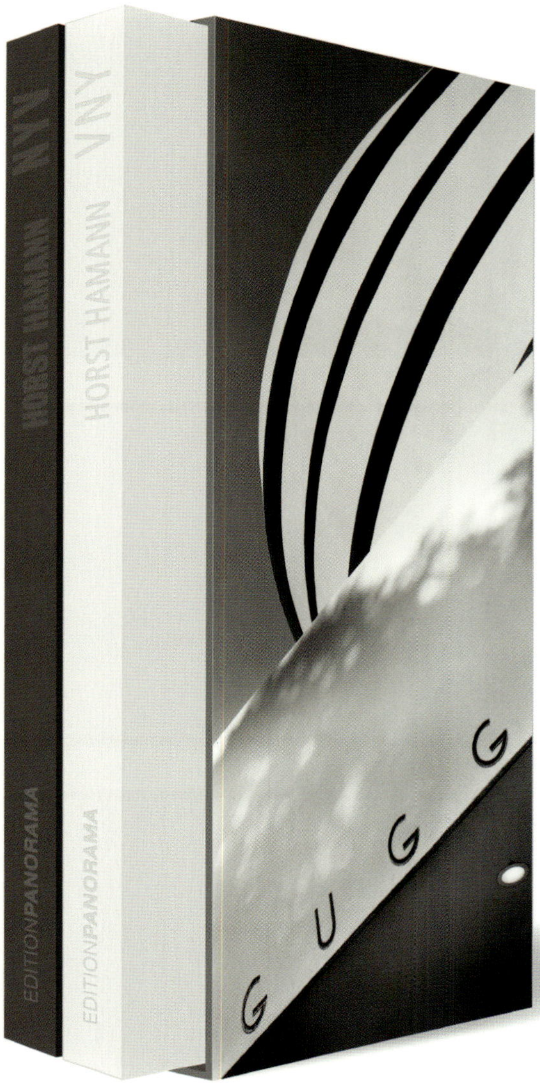

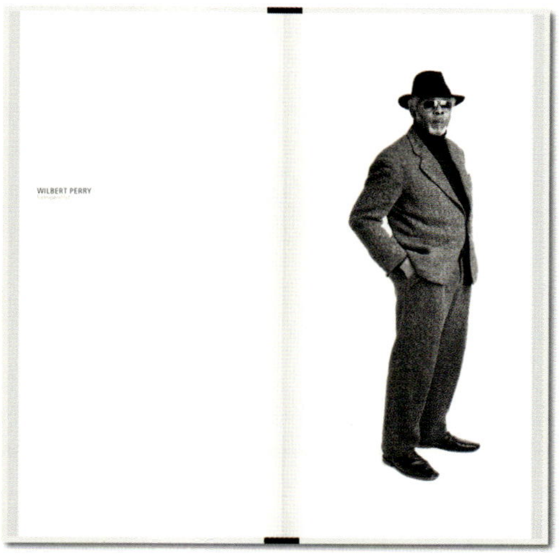

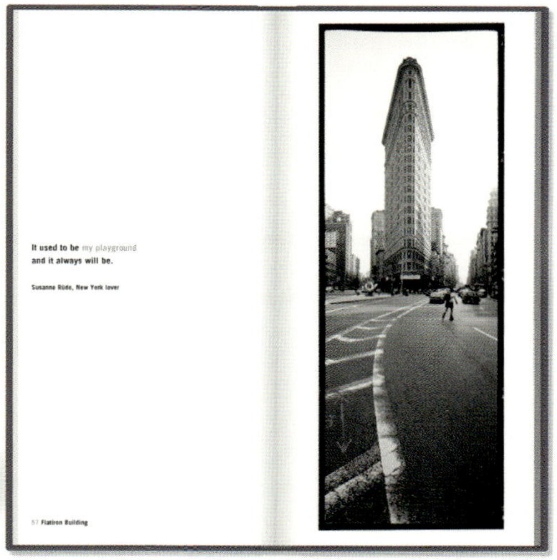

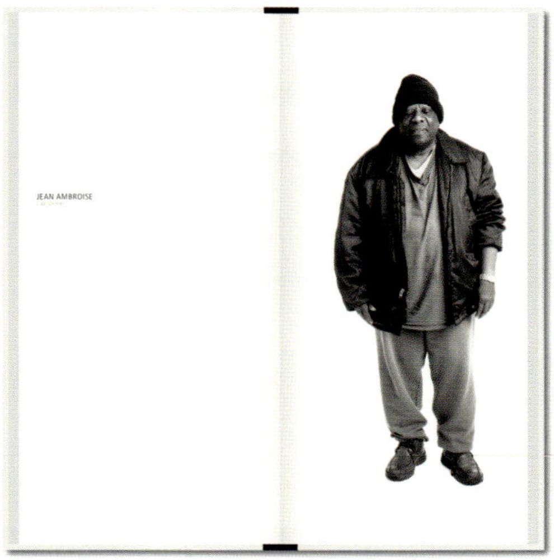

Auftraggeber Client
Edition Panorama
Mannheim, Germany

Projekt Project
New York Vertical / Vertical New Yorkers
Buch
Book

Design
Horst Hamann
Harald Vits
Sebastian Wiefler
Michael Gregoire
Edition Panorama
Mannheim, Germany

Wenn es jemals eine Stadt verdient hat, in vertikalen Panoramen fotografiert zu werden, dann ist und bleibt es New York. Horst Hamann hat mit diesen Aufnahmen nicht nur ein fantastisches Städtebild gezeichnet, sondern auch ein eigenes Genre geschaffen. Im zweiten Band finden sich Porträts von New Yorkern – ebenfalls im Hochformat, aber durch die geschlossenen Augen wirken die Porträts gelassen und konzentriert gleichermaßen.

If there is any city that has ever earned the right to be photographed in vertical panoramas it is New York and will always be New York. Horst Hamann has not only conjured a fantastic image of this city with these photographs, he has also established a new genre. In the second volume you will find portraits of New Yorkers – in vertical format as well, yet for the closed eyes of those being portrayed, the images exude utter calmness and concentration.

Auftraggeber Client
Edition Panorama
Mannheim, Germany

Projekt Project
Zeit des Wartens
Buch
Book

Design
Sebastian Wipfler
Edition Panorama
Mannheim, Germany

Der Fotograf Helmut Hirler hat mit der Kamera das Sterben seiner Großmutter begleitet. Dieses leise, bescheidene Buch dokumentiert das Sterben der alten Frau und legt damit Zeugnis vom Leben ab.

Photographer Helmut Hirler's camera accompanied his grandmother's passing away. This quiet, modest book documents the death of an aged woman and thereby testifies about life.

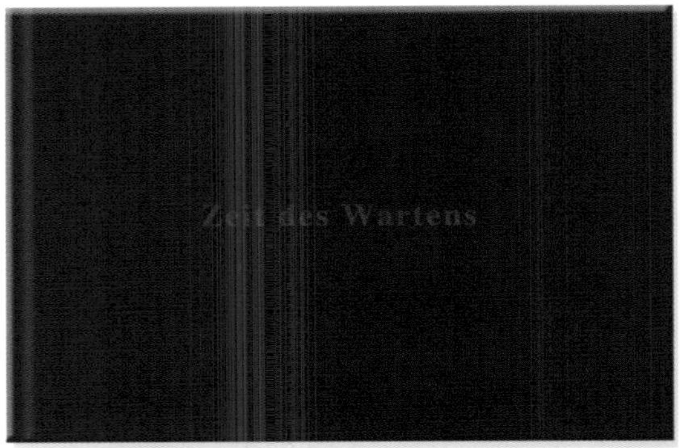

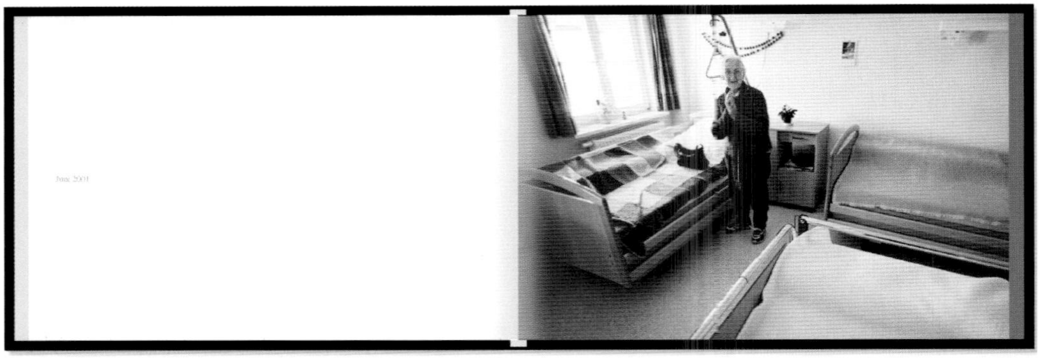

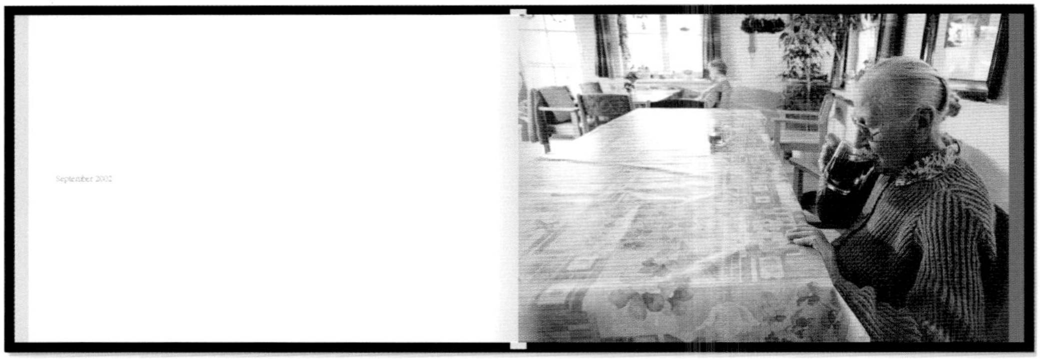

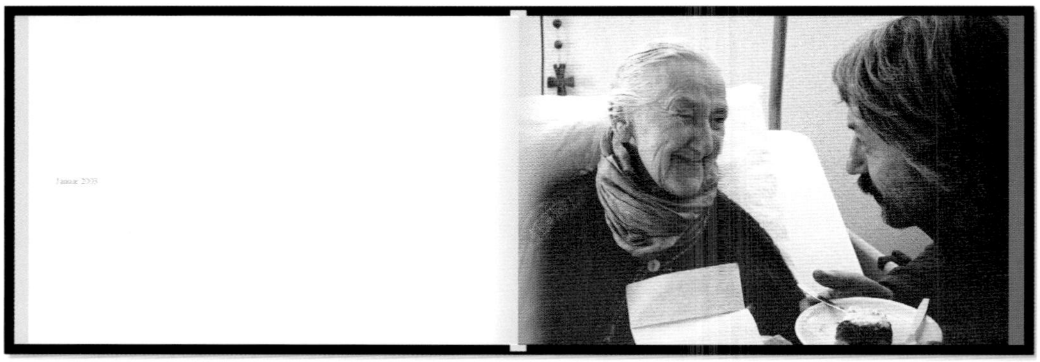

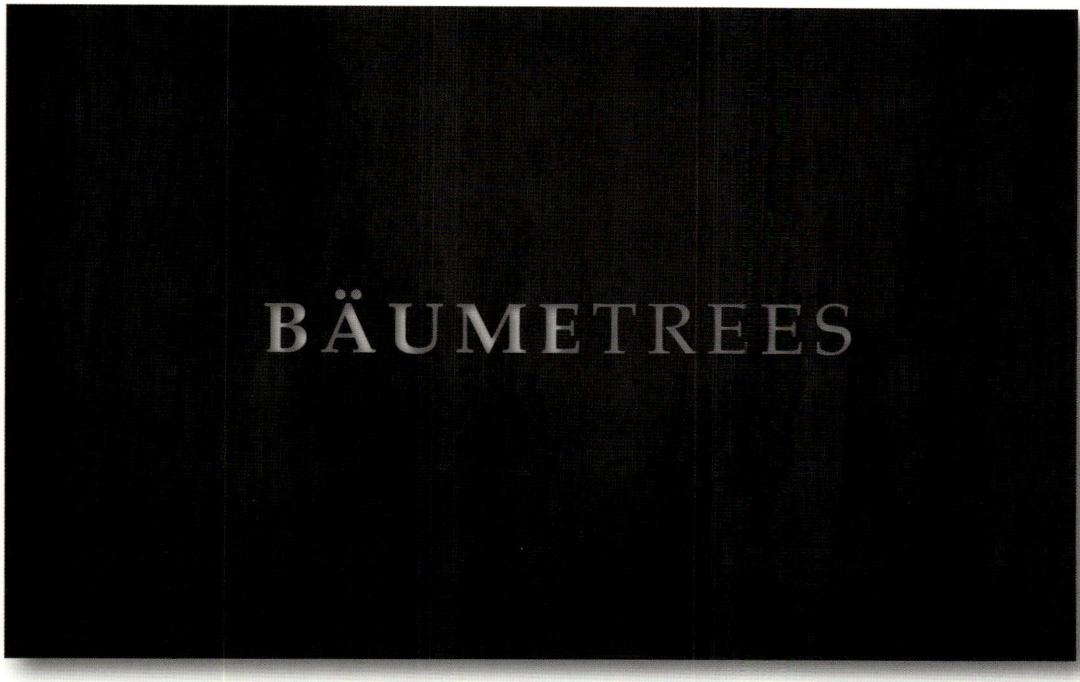

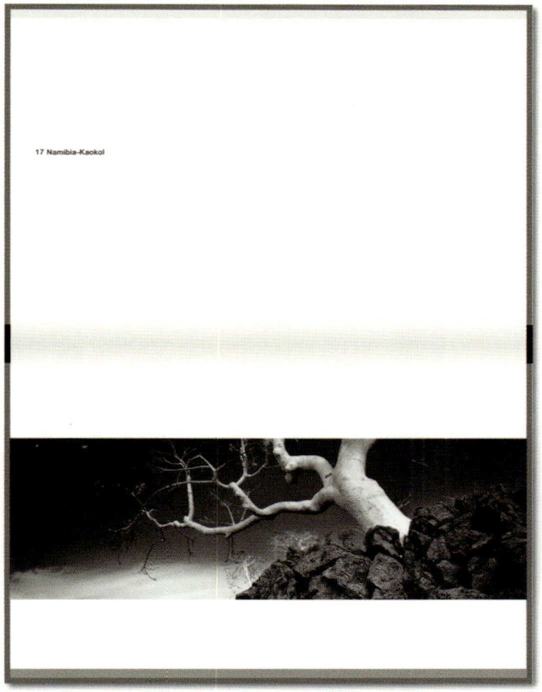

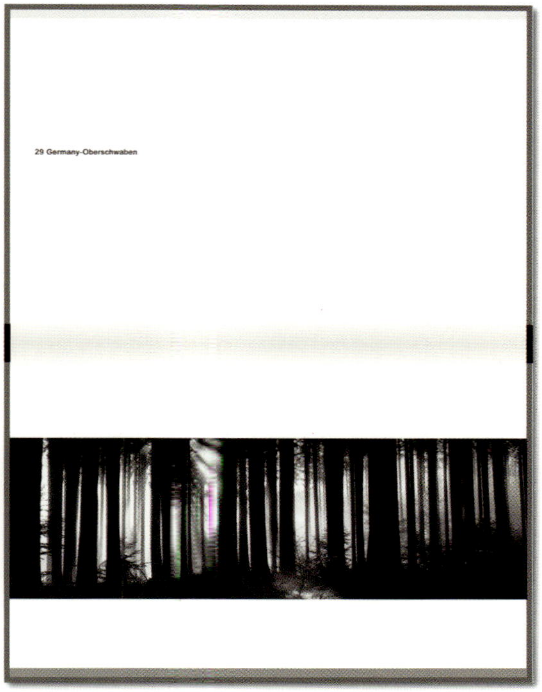

Auftraggeber/Client
Edition Panorama
Mannheim, Germany

Projekt/Project
Bäume/Trees
Buch
Book

Design
Sebastian Wipfler
Edition Panorama
Mannheim, Germany

Um Bäume ranken sich seit Menschengedenken Legenden, Märchen und Geschichten – mythisch erscheinen die Bäume und Wälder in diesem besonderen Buch auch in den schwarz-weißen Infrarot-Aufnahmen Helmut Hirlers, die die Himmel dunkel und das Blattwerk licht erscheinen lassen. Ein hochwertiger Schuber mit Laserstanzung macht das Holz der Bäume auch haptisch erlebbar.

From time immemorial, legends, fairy tales and stories have grown up around trees. The trees and forests of this monumental calendar reside in the realm of myth not least of all for the technique of black-and-white infrared photography used by Helmut Hirler. His photography renders the shimmering light of foliage against the background of darkly gleaming skies. A high-quality slipcase, engraved by laser technology, reproduces the tactile experience of the trees' wood.

Auftraggeber Client
Stadt Darmstadt
Darmstadt, Germany

Projekt Project
Armin Lindauer: Helmut Lortz – leicht sinnig
Buch
Book

Design
Armin Lindauer
Mannheim, Germany

Typografisch inszenierte Texte mit Zeichnungen.

Typographically arranged text with drawings.

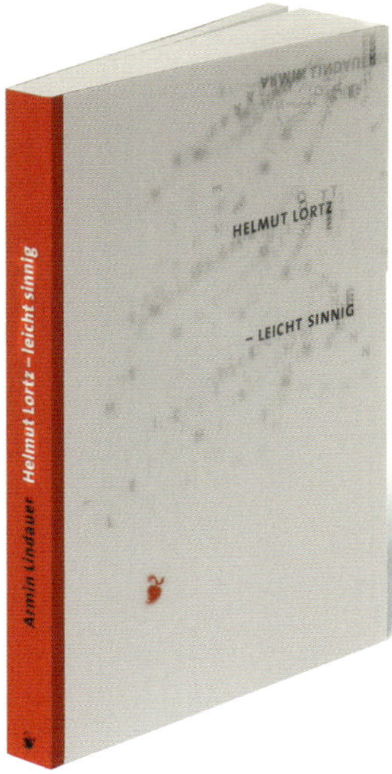

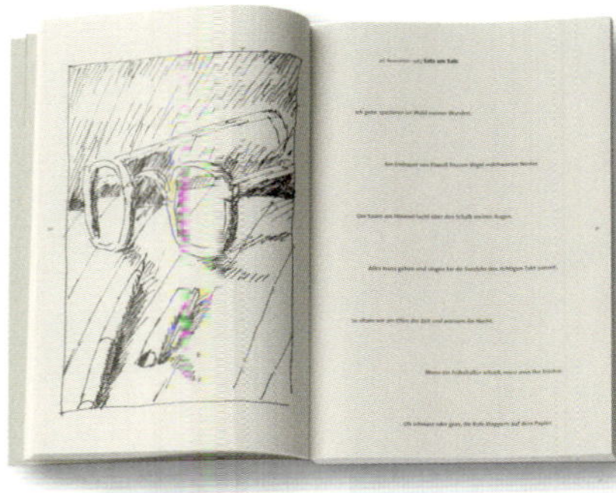

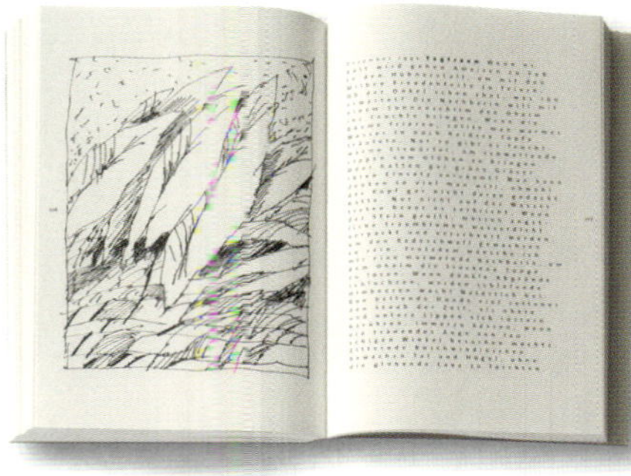

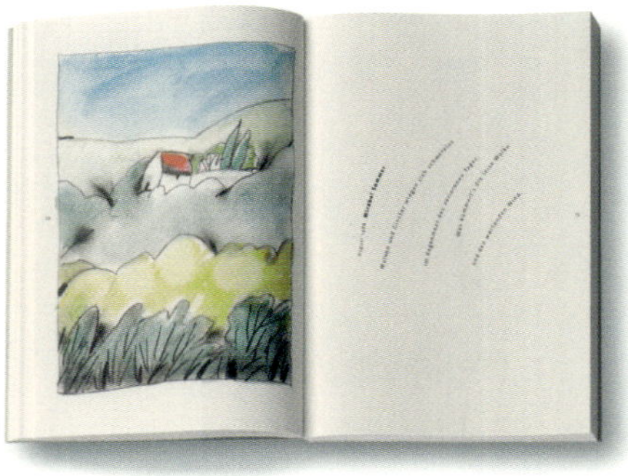

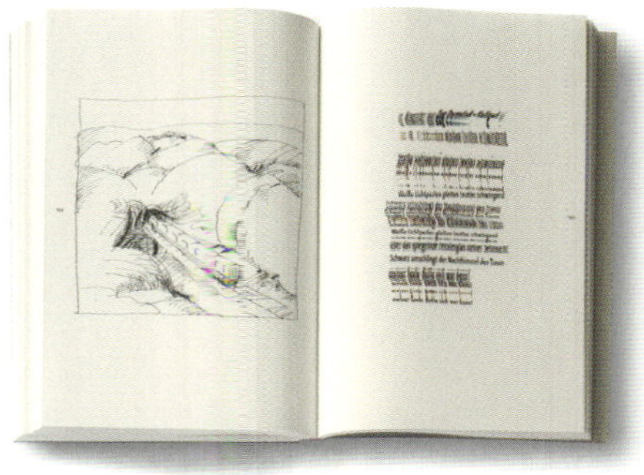

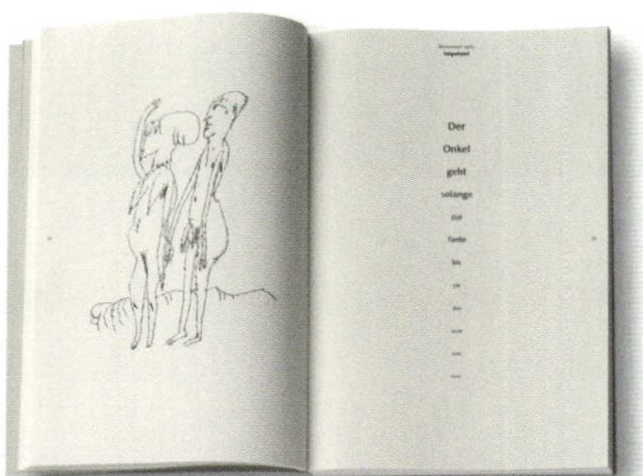

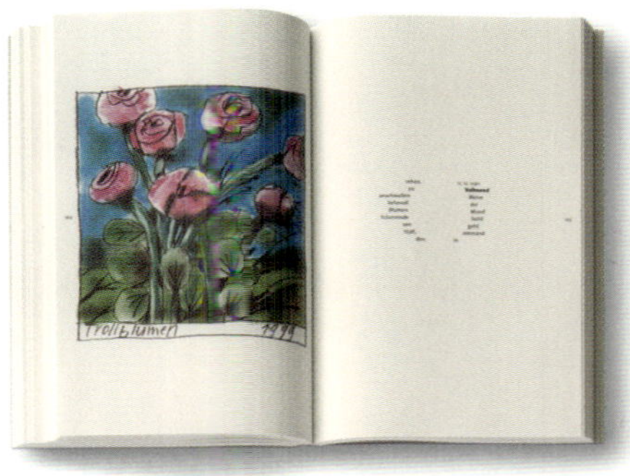

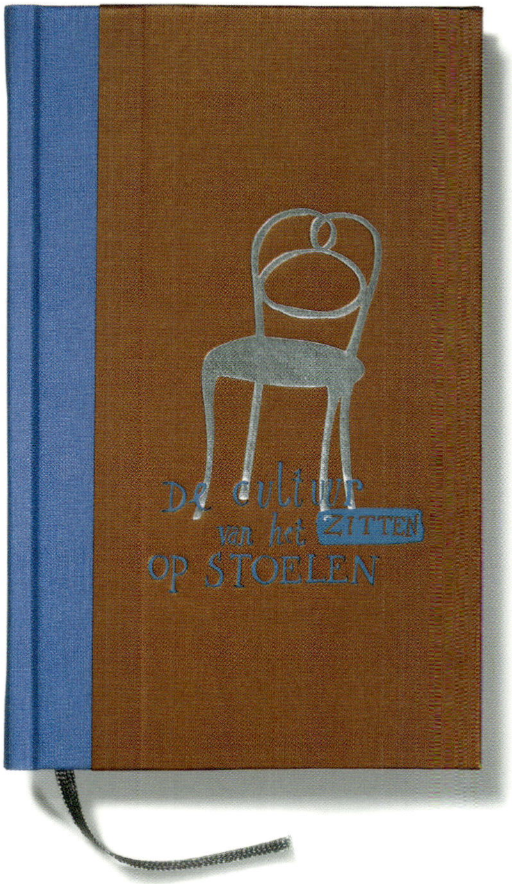

Auftraggeber Client
Bitter Culemborg BV, Thonet Importeur
AE Culemborg, The Netherlands

Projekt Project
De cultuur van het zitten op stoelen /
Die Kultur des Sitzens
Buch
Book

Design
Nicole Neubert
Heine/Lenz/Zizka
Frankfurt am Main, Germany
Dorine de Vos
Wich & de Vos bv
Rotterdam, The Netherlands

Entwicklung eines hochwertigen Breviers für Thonet Niederlande, das die kulturgeschichtliche Entwicklung des Stuhls von seinen Anfängen bis zur industriellen Entwicklung zum Massenobjekt beleuchtet.

Conception and design of a premium breviary for Thonet Netherlands, showing the cultural/historical evolution of the chair from its beginning to the industrial development as a mass product.

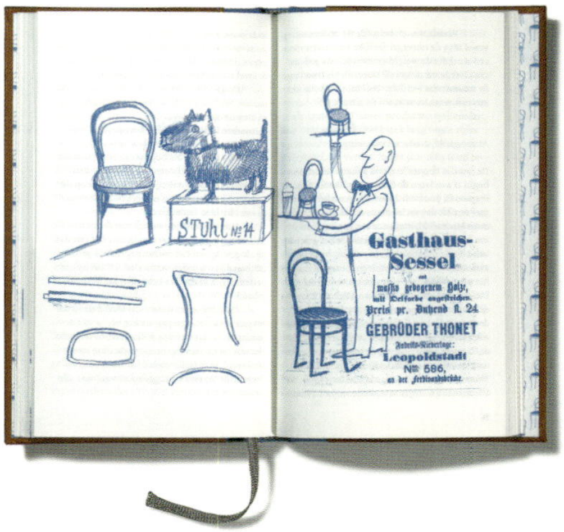

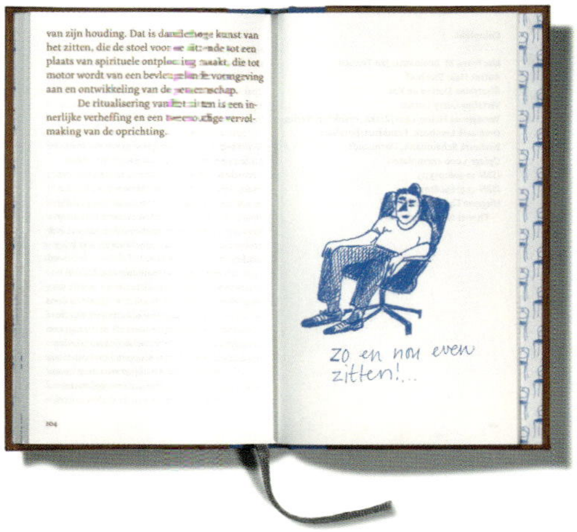

Auftraggeber Client
Papierfabrik Scheufelen GmbH + Co. KG
Lenningen, Germany

Projekt Project
The Book of Independence
Diary

Design
Anders Bergesen
Kirsten Dietz
Susanne Hörner
Anika Marquardsen
Jochen Rädeker
Felix Widmaier
Strichpunkt
Stuttgart, Germany

Unabhängigkeit bedeutet, sich frei entscheiden zu können. Mit PhoeniXmotion bietet die Papierfabrik Scheufelen ein Papier, das durch seine einzigartige Qualität einen Reichtum an Möglichkeiten eröffnet. „Be free" lautet daher das Leitmotiv, unter dem Strichpunkt das „Book of Independence" als Präsentationsmedium für PhoeniXmotion entwickelte. Eine Biografie der Möglichkeiten des Menschen, Freiheit und Unabhängigkeit zu leben. Fotografisch und typografisch illustriert das „Book of Independence" die verschiedenen Lebensstationen und zeigt nebenbei hochwertige Verarbeitungstechniken auf dem Premiumpapier. Ein besonderes Kalendarium, das beginnt, wann sein Nutzer es will, setzt das Thema „Unabhängigkeit" fort.

Independence means being free to make your own choices. The most recent example: PhoeniXmotion, a paper launched by Scheufelen whose one-of-a-kind quality offers a wealth of opportunities. Following the mantra "Be free!", Strichpunkt developed the Book of Independence as a presentation medium for PhoeniXmotion. A biography of the ways in which people experience freedom and independence. Through photographs and text, the Book of Independence illustrates the various stages of life, and shows high-quality processing methods on the premium paper. A special calendar explores the independence theme further: This calendar begins whenever you want it to – with stickers to highlight your own personal celebrations of freedom.

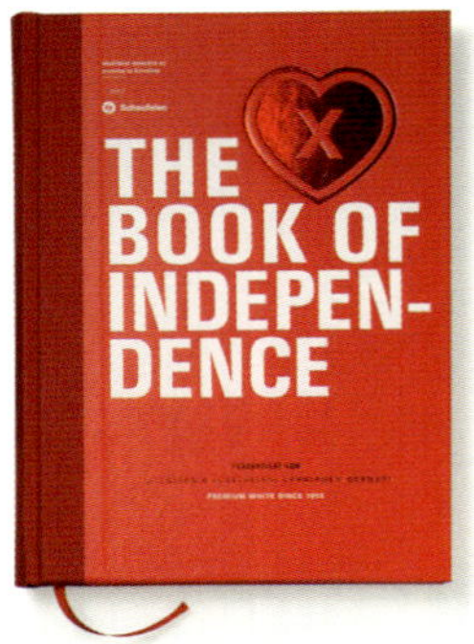

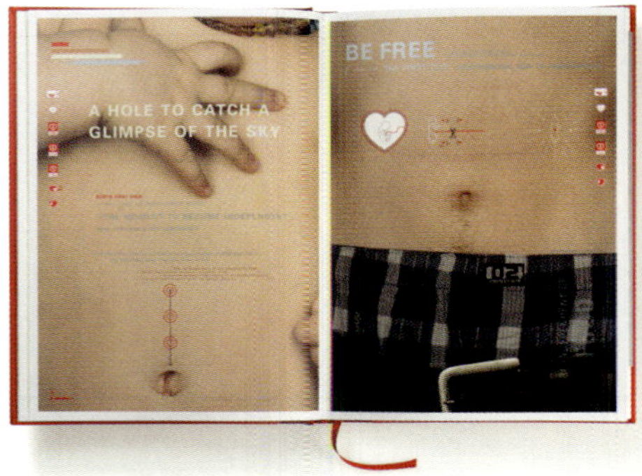

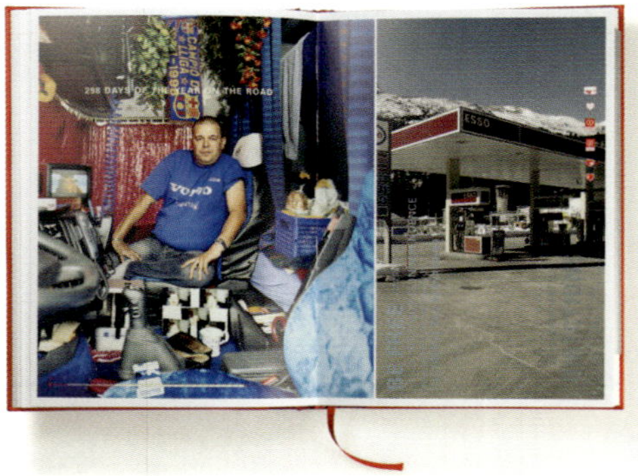

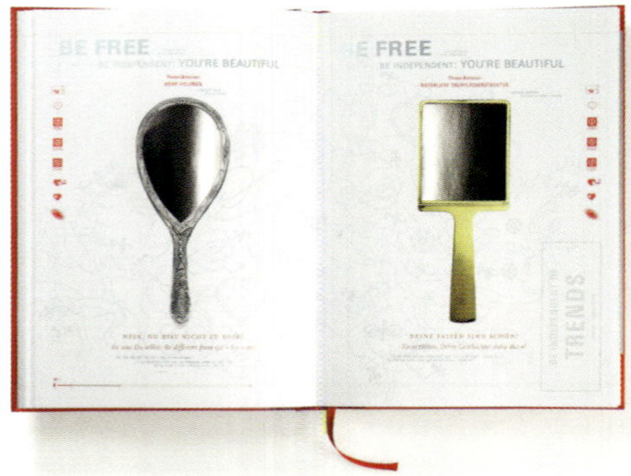

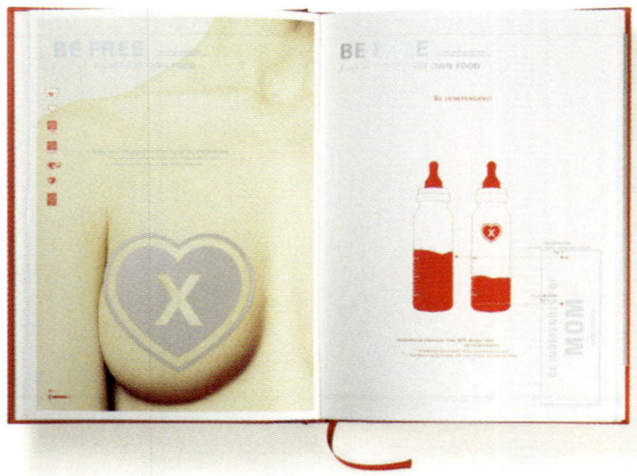

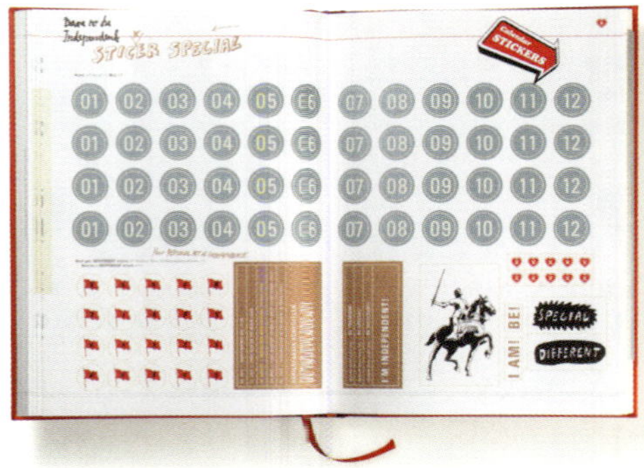

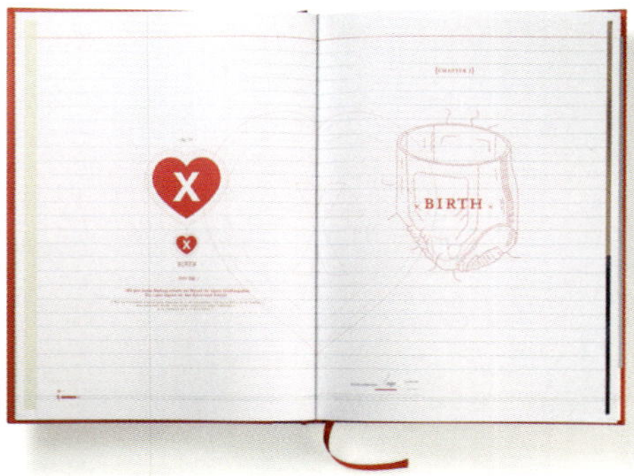

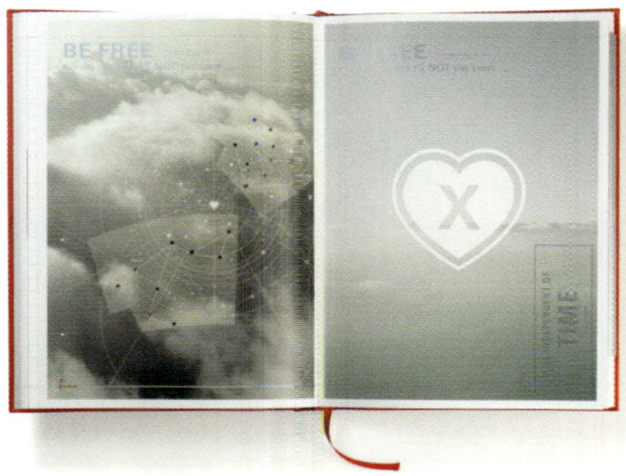

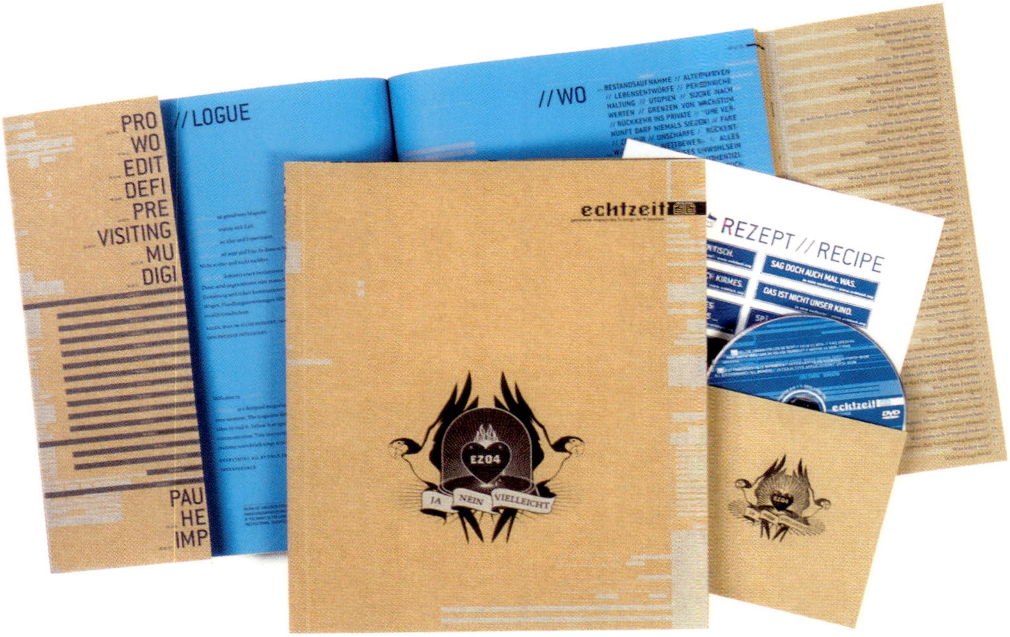

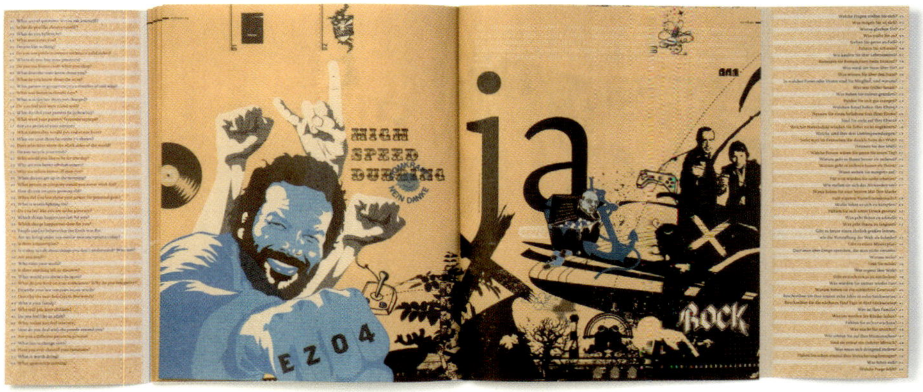

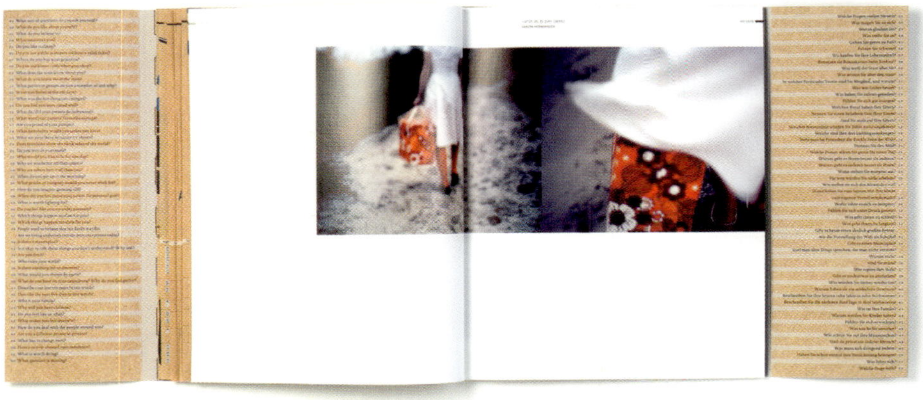

Auftraggeber Client
Redaktion „echtzeit" FH Potsdam
Potsdam, Germany

Projekt Project
„echtzeit04" Ja / Nein / Vielleicht
Magazin
Magazine

Design
Graham Ball
Svenja von Döhlen
Martin Eichhorn
Viola Eistert
Tim Finke
Carina Hinze
Thomas Kehr
David Löwe
Pablo Lütkenhaus
Sebastian Manger
Dennis Ratzlaff
Judith Schalansky
Jan Schütze
Henning Skibbe
Alex Thies
Steffen Wierer
Andreas Zecher
Redaktion „echtzeit" FH Potsdam
Potsdam, Germany

„echtzeit" ist gestaltetes Magazin, eigeninitiiert von Studenten der FH Potsdam. „echtzeit" ist Idee und Experiment, ist jetzt und hier in diesem Moment, nicht vorher und nicht nachher. In „echtzeit04" stellen wir 22 Menschen vor, deren Antworten uns Einblicke in Ansichten und Lebensentwürfe unterschiedlichster Art gewähren. Visuelle Extrakte hierzu formieren sich auf 52 Seiten zu teils skizzenhaften Porträts und opulenten Bildern. Der digitale Teil dieser Ausgabe bietet reichlich Raum, um Gelesenes zu vertiefen und zu vergleichen, weitere Antworten zu hören und den richtigen Ton zu treffen.

This is the fourth issue of "echtzeit". Some dust, last issue's theme, has found its way into "echtzeit04" – it accompanies us in the blueprints of countless forms and statutes from German civil offices. In contrast to the millions of printed and distributed applications, we present in this issue 22 people whose answers provide insight into individual lifestyles and very disparate, very private opinions. The digital section of this issue provides a wealth of additional space in which to extend the experience; to quantify and compare, discover additional answers and find the right point on our soundtrack.

Auftraggeber Client
jung und pfeffer
visuelle kommunikation
Bremen, Germany

Projekt Project
:output 08
Buch
Book

Design
jung und pfeffer
visuelle kommunikation
Bremen, Germany

:output ist der größte internationale Wettbewerb für Studenten der visuellen Kommunikation. Jedes Jahr schicken Studenten aus mehr als 40 Ländern über 1.000 Arbeiten zu dem Wettbewerb ein. Eine Jury aus international anerkannten Gestaltern wählt die Arbeiten aus, die in dem Jahrbuch :output veröffentlicht werden. :output fördert junge Talente und zeigt, was an Designhochschulen rund um den Globus passiert. jung und pfeffer haben das Konzept für :output ersonnen, sind Gründer der :output-Stiftung in Amsterdam (die das Projekt durchführt) und gestalten jedes Jahr das dazugehörige Jahrbuch.

:output is the largest international competition for students of visual communications. Every year students from more than 40 different nations submit over 1,000 entries to the competition, where a jury of internationally known designers selects the projects to be published in the :output yearbook. :output nurtures young talent and reveals the state of the art in design schools around the globe.
jung und pfeffer developed the concept for :output, established the :output foundation in Amsterdam (which organizes the project) and are responsible for the annual design of its yearbook.

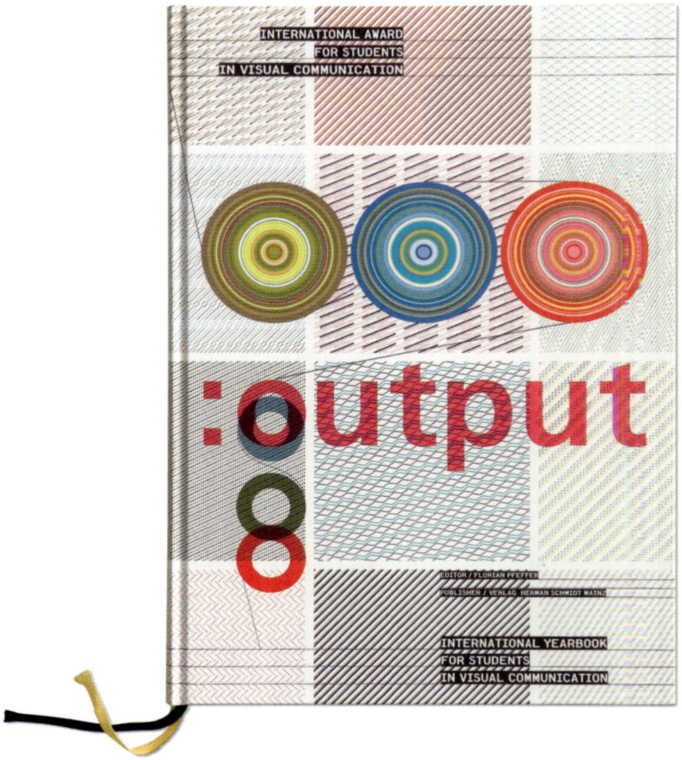

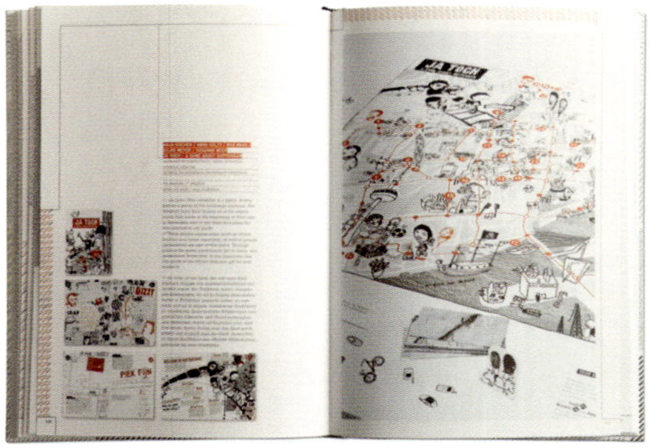

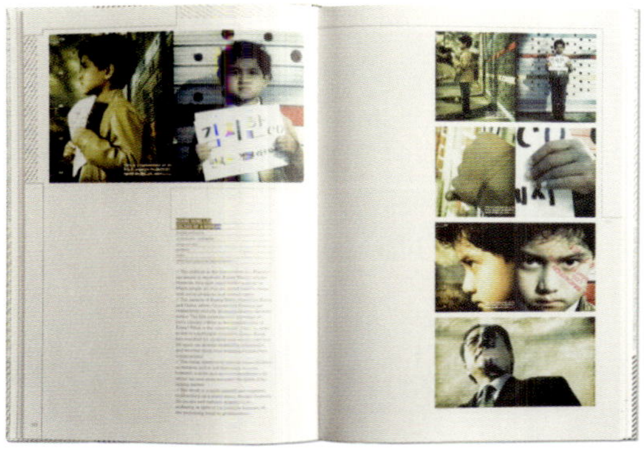

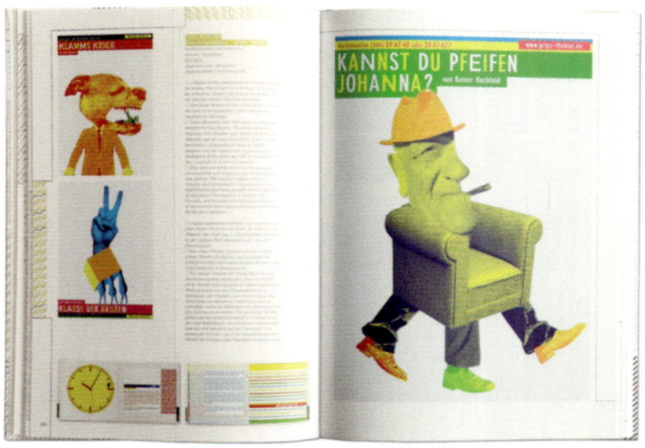

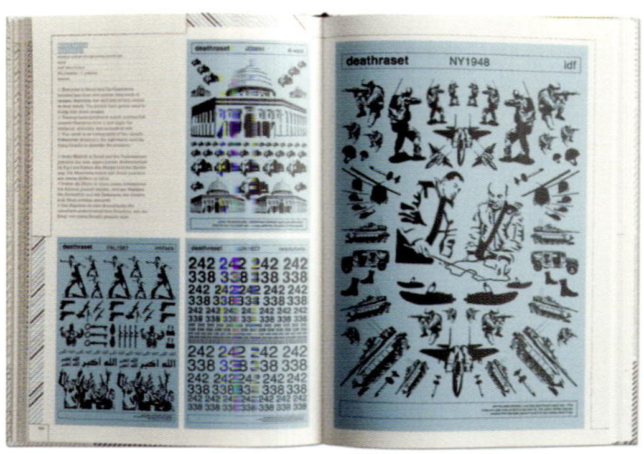

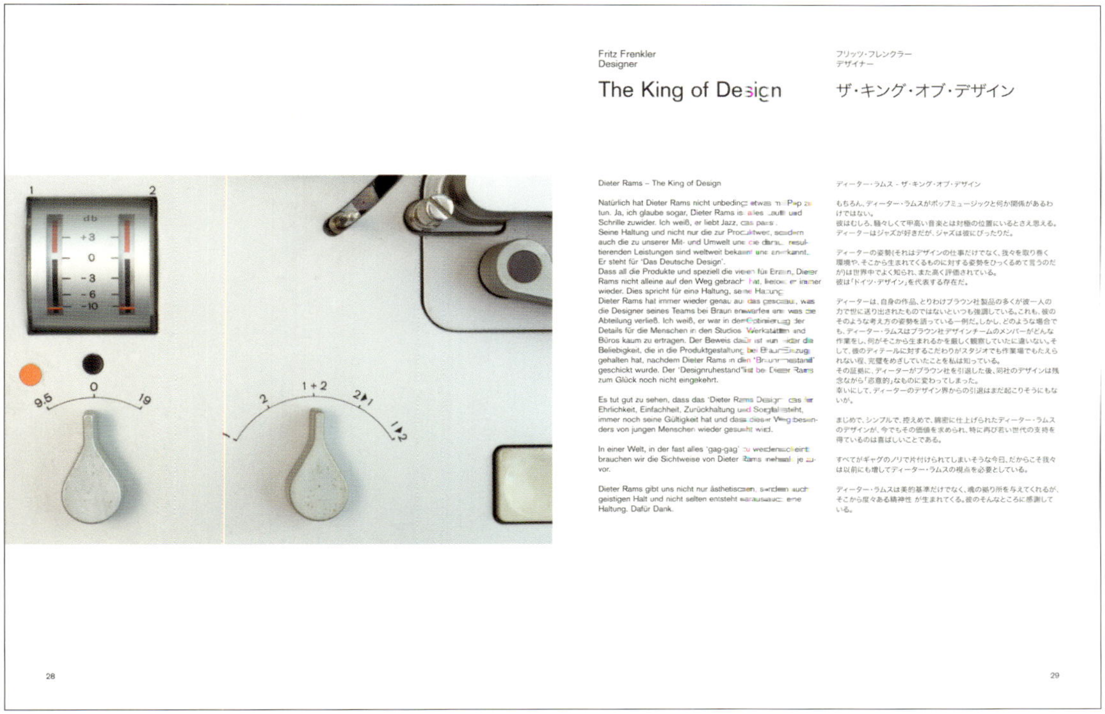

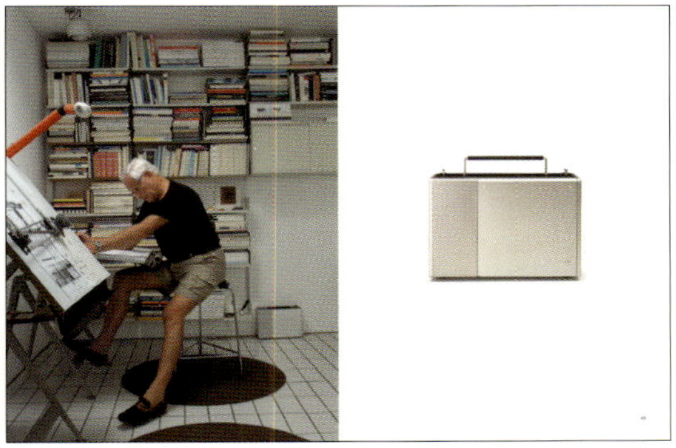

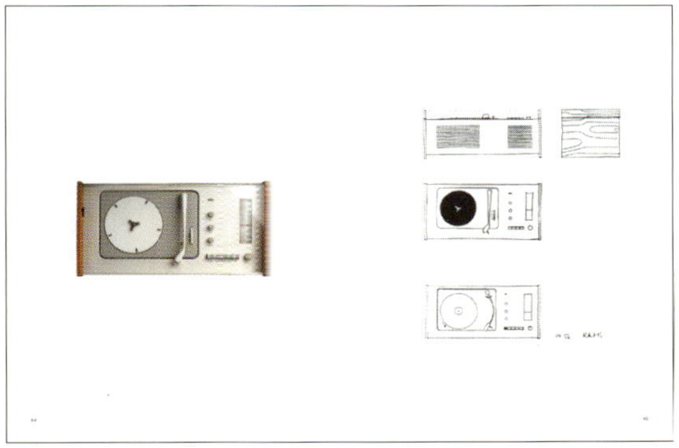

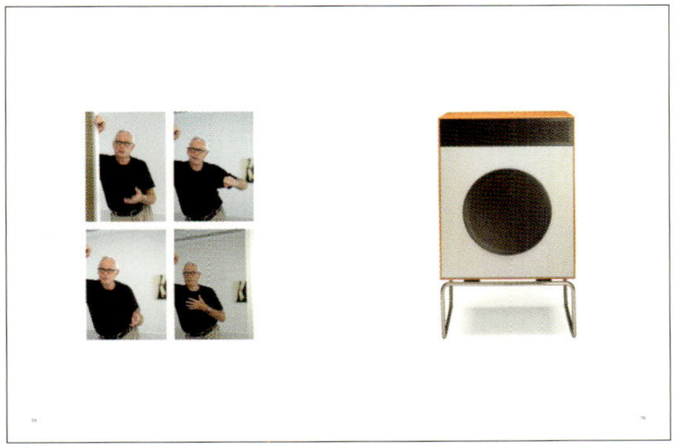

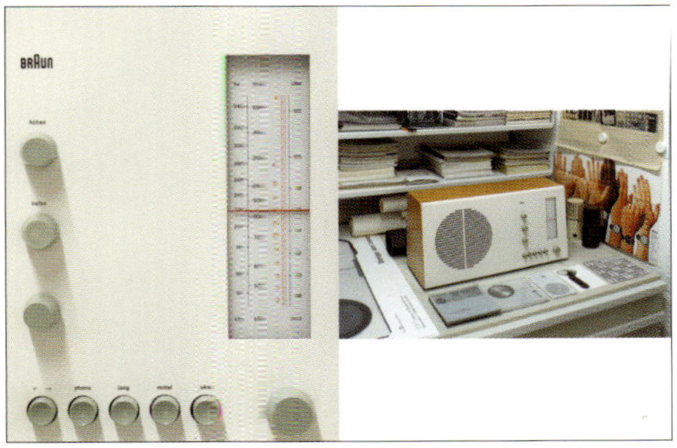

Auftraggeber Client
Miki Shimokawa
Frankfurt am Main, Germany

Projekt Project
„less but better – Die Welt von Dieter Rams"
Buch
Book

Design
f/p design gmbh
Frankfurt am Main, Germany
Miki Shimokawa (Konzept)
Frankfurt am Main, Germany

„less but better – weniger aber besser" lautete der Titel einer Ausstellung mit Arbeiten von Dieter Rams, die im Herbst 2005 im ältesten Zen-Tempel Japans, dem Kenninji Tempel in Kyoto, stattgefunden hat. Parallel dazu hat die Initiatorin der Ausstellung, Miki Shimokawa, mit f/p design das Buch „less but better – Die Welt von Dieter Rams" gestaltet und im Eigenverlag publiziert. Es geht in seiner Gestaltung auf die Persönlichkeit und die ästhetische Denkweise von Dieter Rams ein, wobei sein Alltagsleben in seinem Wohnhaus in Kronberg, fotografiert von Thomas Mayer, sowie Statements zu Dieter Rams und seinem Gesamtwerk im Vordergrund stehen.

"less but better – weniger aber besser!" is the title of an exhibition with products, designed by Dieter Rams, which was held in the autumn of 2005 in the oldest Zen temple in Japan, the Kenninji in Kyoto. To go along with this, the initiator of the exhibition, Miki Shimokawa, and f/p design have created the book "less but better – the world of Dieter Rams" and published it in a private publishing venture. It concerns the personality and the aesthetic mindset of Dieter Rams, and his everyday life in his residential house in Kronberg photographed by Thomas Mayer, and statements on Dieter Rams and his complete works are in the foreground here.

03 architecture & public spaces

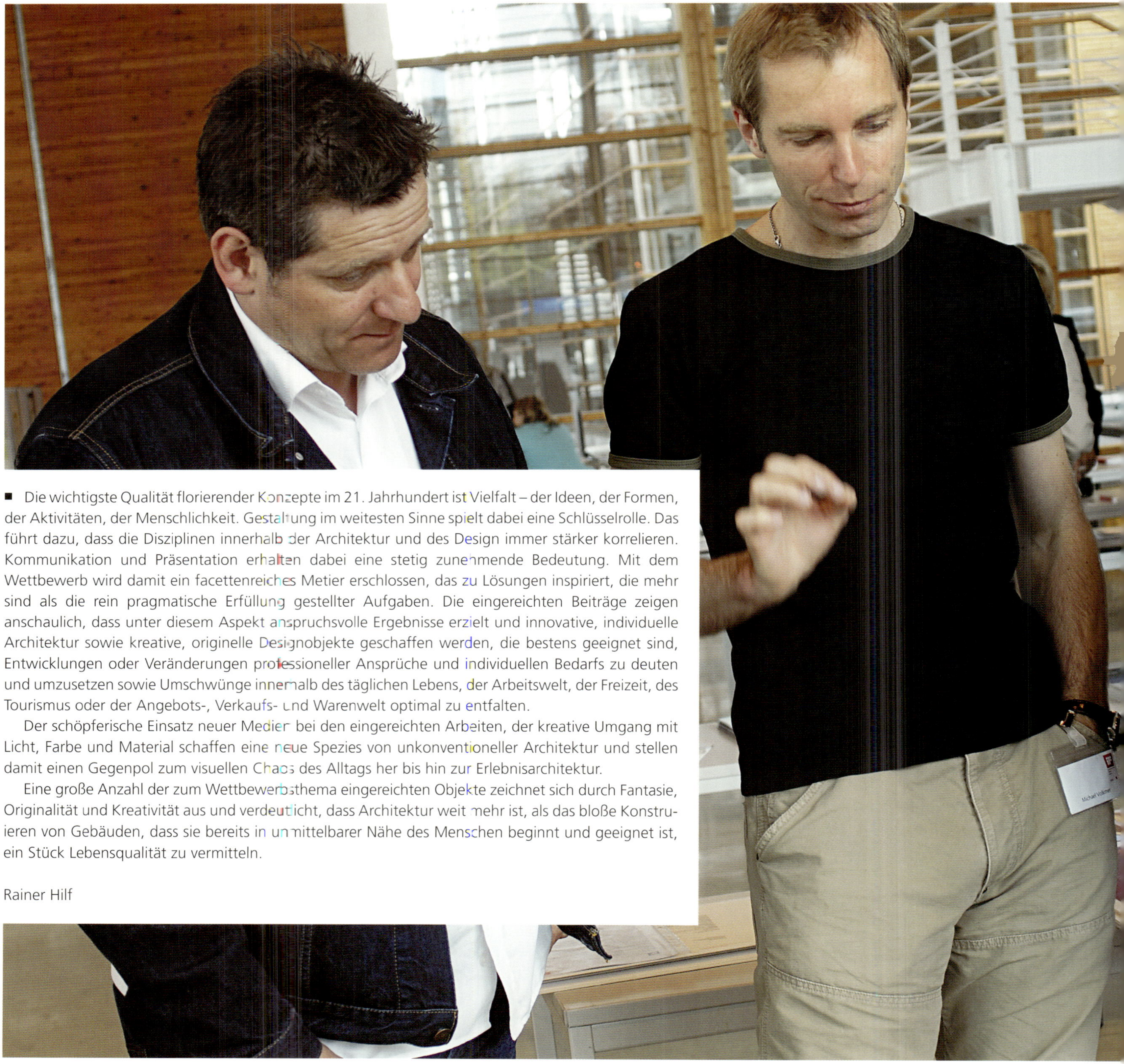

■ Die wichtigste Qualität florierender Konzepte im 21. Jahrhundert ist Vielfalt – der Ideen, der Formen, der Aktivitäten, der Menschlichkeit. Gestaltung im weitesten Sinne spielt dabei eine Schlüsselrolle. Das führt dazu, dass die Disziplinen innerhalb der Architektur und des Design immer stärker korrelieren. Kommunikation und Präsentation erhalten dabei eine stetig zunehmende Bedeutung. Mit dem Wettbewerb wird damit ein facettenreiches Metier erschlossen, das zu Lösungen inspiriert, die mehr sind als die rein pragmatische Erfüllung gestellter Aufgaben. Die eingereichten Beiträge zeigen anschaulich, dass unter diesem Aspekt anspruchsvolle Ergebnisse erzielt und innovative, individuelle Architektur sowie kreative, originelle Designobjekte geschaffen werden, die bestens geeignet sind, Entwicklungen oder Veränderungen professioneller Ansprüche und individuellen Bedarfs zu deuten und umzusetzen sowie Umschwünge innerhalb des täglichen Lebens, der Arbeitswelt, der Freizeit, des Tourismus oder der Angebots-, Verkaufs- und Warenwelt optimal zu entfalten.

Der schöpferische Einsatz neuer Medien bei den eingereichten Arbeiten, der kreative Umgang mit Licht, Farbe und Material schaffen eine neue Spezies von unkonventioneller Architektur und stellen damit einen Gegenpol zum visuellen Chaos des Alltags her bis hin zur Erlebnisarchitektur.

Eine große Anzahl der zum Wettbewerbsthema eingereichten Objekte zeichnet sich durch Fantasie, Originalität und Kreativität aus und verdeutlicht, dass Architektur weit mehr ist, als das bloße Konstruieren von Gebäuden, dass sie bereits in unmittelbarer Nähe des Menschen beginnt und geeignet ist, ein Stück Lebensqualität zu vermitteln.

Rainer Hilf

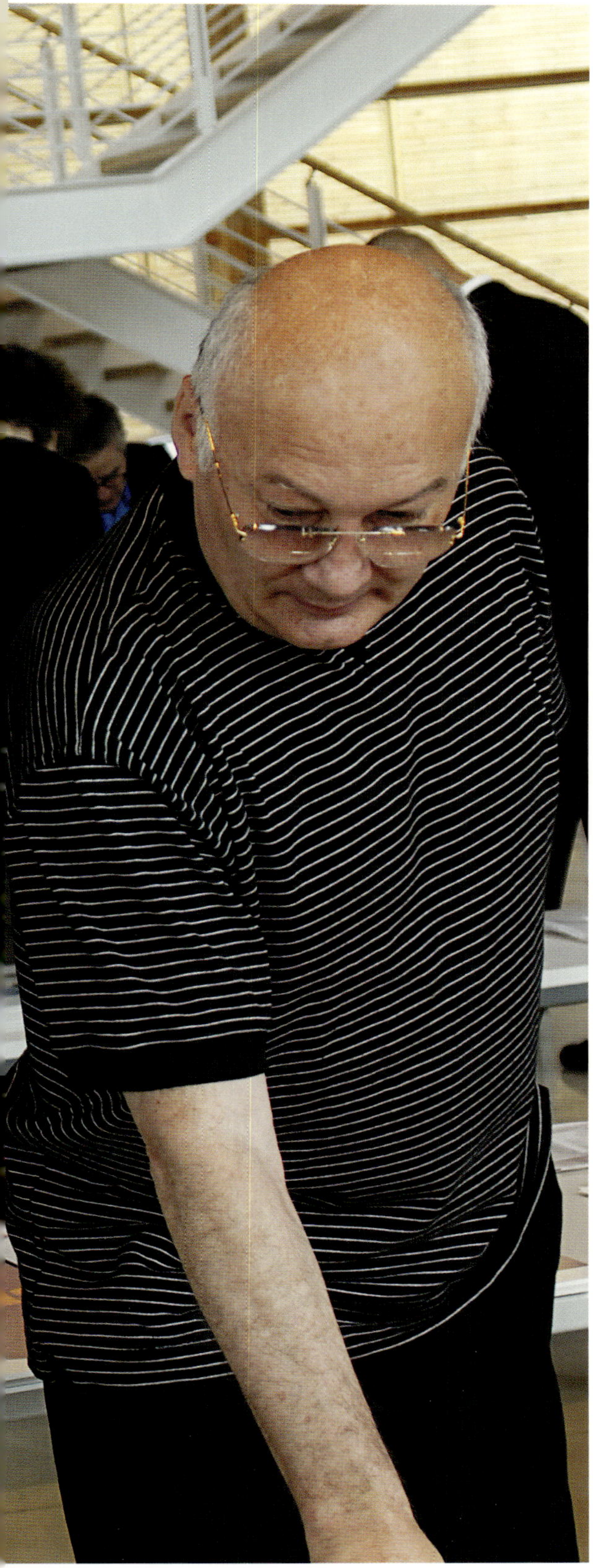

architecture & public spaces

■ The most important quality of flourishing concepts in the 21st century is their diversity – in terms of their ideas, shapes, activities and humanity. Design in the widest sense of the word has a key role to play in this trend. It also means that there is an increasing degree of correlation between the disciplines of architecture and design. As a result, communication and presentation are becoming more and more important. This competition encourages and inspires a multifaceted profession to come up with solutions that are more than just pragmatic ways of completing defined tasks. The projects submitted clearly illustrate that this process generates sophisticated results and innovative, individual architecture as well as creative and original designs. The works are ideal indicators of trends and changes in professional demands and individual needs, and highlight the best way to approach major changes in everyday life, the job market, recreation, tourism and product sales and marketing.

The creative use of new media in the projects entered in the competition and the innovative use of light, color and materials are indicative of a new era of unconventional architecture that acts as an antidote to the visual chaos of daily life and offers a new realm of architectural experience.

A large number of the projects entered in the competition displayed outstanding imagination, originality and creativity – indicating that architecture is much more than the mere designing of buildings. It begins with a good understanding of human nature and ideally ends up reflecting a particular quality of life.

Rainer Hilf

Rainer Hilf

Michael Volkmer

Ralf Höpfner

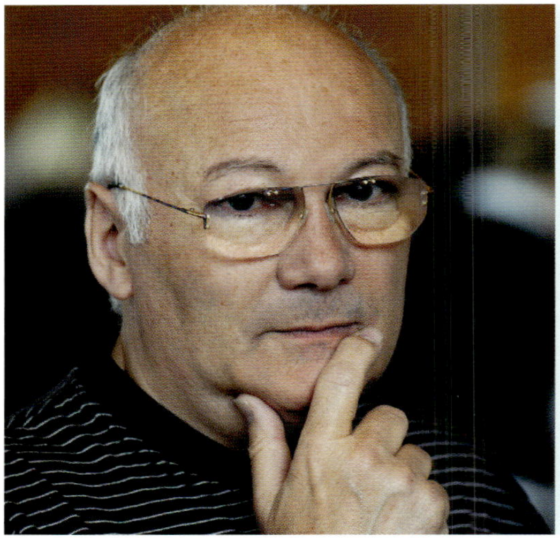

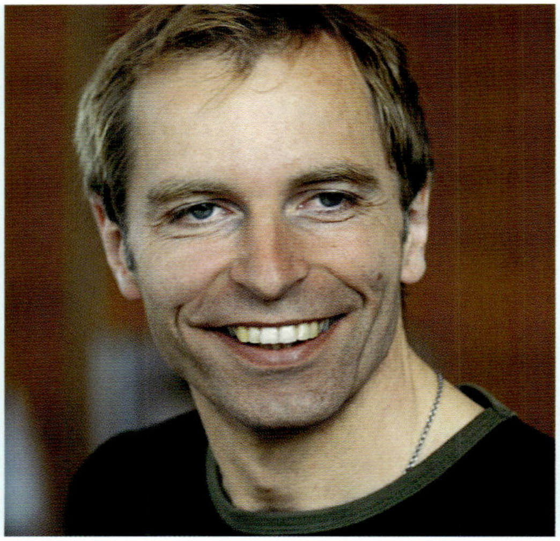

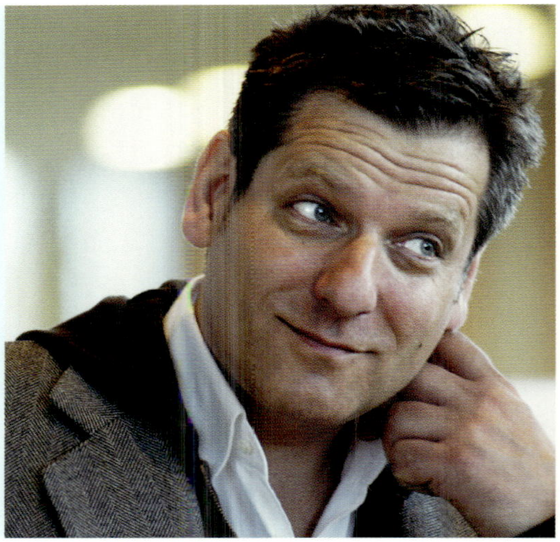

■ Rainer Hilf studierte Innenarchitektur an der Akademie der Bildenden Künste in Nürnberg und ist seit 1972 als freischaffender Innenarchitekt tätig. Er hat seit 1976 unterschiedliche Lehrtätigkeiten, u. a. eine Gastprofessur an der Hochschule für Kunst und Design Burg Giebichenstein in Halle/Saale, inne und ist seit 1995 Präsident des BDIA – Bund Deutscher Innenarchitekten. Rainer Hilf ist darüber hinaus Mitglied in zahlreichen Institutionen wie beispielsweise einer Lenkungsgruppe des Bundesministeriums für Verkehr, Bau- und Wohnungswesen.

■ Rainer Hilf studied Interior Design at the Academy of Fine Arts in Nuremberg and has been a freelance interior designer since 1972. He has held various teaching positions since 1976, including Visiting Professor at the Burg Giebichenstein University of Art and Design in Halle, Germany. He has served as President of the German Association of Interior Designers (BDIA) since 1995. Rainer Hilf is also a member of numerous institutions, including membership on a steering committee for the German Federal Ministry of Transportation, Building and Urban Development.

■ Michael Volkmer (*1965) studierte Kommunikationsdesign an der Fachhochschule Wiesbaden und gründete 1994 zusammen mit Anette Scholz die Agentur Scholz & Volkmer Intermediales Design GmbH in Wiesbaden. Seit 2003 ist er als alleiniger Geschäftsführer für die Planung, Gestaltung und Umsetzung von Projekten im Bereich der neuen Medien verantwortlich. Er lehrt an der Fachhochschule Darmstadt zum Thema „Darstellungsmethoden" und war bereits für nationale wie internationale Wettbewerbe als Jury-Mitglied tätig.

■ Michael Volkmer (*1965) studied Communication Design at the Wiesbaden University of Applied Sciences and, with Anette Scholz, set up the agency Scholz & Volkmer Intermediales Design GmbH in Wiesbaden in 1994. Since 2003 he has been responsible for planning, designing and implementing projects in new media. He teaches at Darmstadt's University of Applied Sciences on "methods of representation" and has served on a number of juries at national and international competitions.

■ Ralf Höpfner (*1959) war Berater bei BBDO Deutschland, Managing Director bei ID-Media und Elephant Seven sowie Geschäftsführer bei Heye & Partner. Seit 2005 ist er Gründer und Inhaber der markentechnisch orientierten Unternehmensberatung >jk< Markensteuer und zudem als Coach in den Konfliktbereichen Strukturen, Informationen, Beziehungen, Werte und Interessen sowie in der Unterstützung von unternehmerischen Positionierungs- und Change-Prozessen tätig.

■ Ralf Höpfner (*1959) has worked as a consultant at BBDO Deutschland, and has been Managing Director of ID-Media and Elephant Seven, as well as Managing Director of Heye & Partner. In 2005 he set up his own brand-oriented corporate consultancy, >jk< Markensteuer, also working as a coach in the conflict areas of structures, information, relationships, values and interests, as well as supporting corporate positioning and change processes.

Auftraggeber Client
Volkswagen AG
Wolfsburg, Germany

Projekt Project
Der Stil Passat
Presseveranstaltung
Press event

Design
Susanne Gocht
Joachim Hauser
Hajo Rehm
Hauser Gocht Meyer Kommunikation und Design GmbH
Frankfurt am Main, Germany
Christoph Schmuck
Kubix GmbH (Architektur)
Berlin, Germany

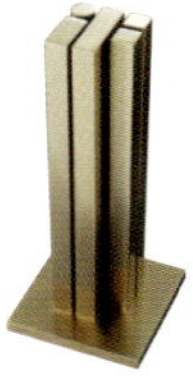

Februar 2005, Kreuzfahrtterminal Hamburg: Weltpremiere Passat vor 600 internationalen Pressevertretern. Ein 300 Meter langes, 25 Meter hohes Containerschiff legt als weithin sichtbares Landmark am Terminal an. Der Veranstaltung liegt ein festes Gestaltungs- und Farbkonzept zugrunde. Es greift die Rasterung der Architektur auf, variiert die Proportionen und füllt sie mit sinnlichen und menschlich-emotionalen Bildern. Eine 12 Meter breite Bühne bildet den Fokus des Raumes, ihre mediale Bühnenrückwand bespielt den Raum und überträgt die Live-Bilder der Redner, Künstler und Gastredner – von Susann Atwell bis Dr. Pischetsrieder, von Li Edelkoort bis Seal.

February 2005, Cruise Terminal Hamburg: World Premiere Passat with 600 international representatives of the press. A landing container ship – 300 meters long and 25 meters high – serves as eye-catching landmark. The event is based on a fixed design and color concept. It takes up the architectural raster, varies the proportions, and fills them with sensuous and humanely emotional pictures. A stage of 12 meters in width is the focus of the room. Its back plane is used as medium for projections and transmits the live pictures of the orators, the artists, and the guest speakers – from Susann Atwell to Dr. Pischetsrieder, and from Li Edelkoort to Seal.

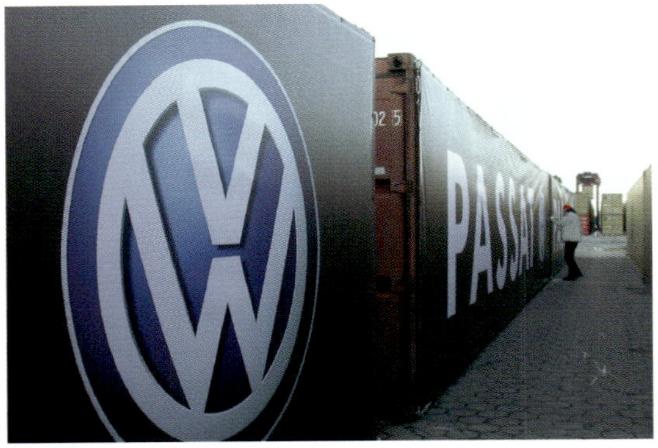

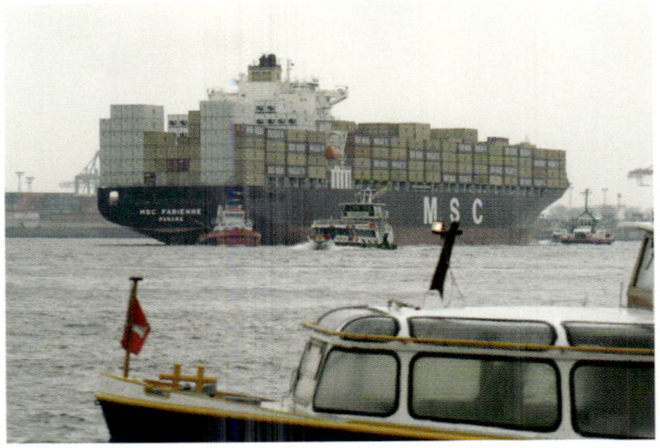

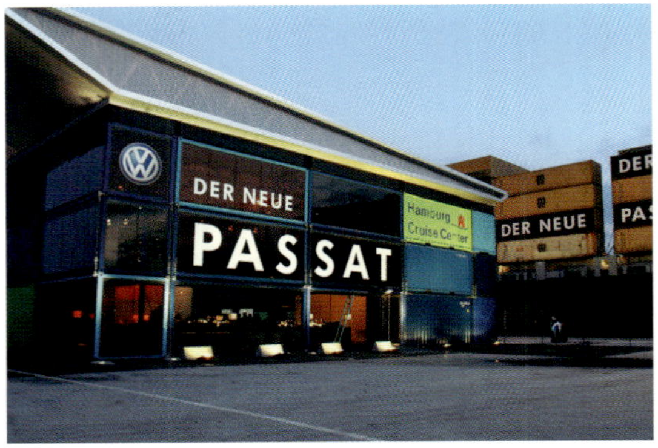

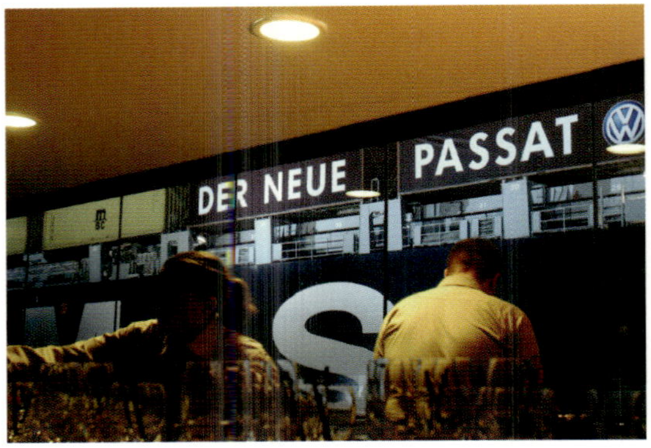

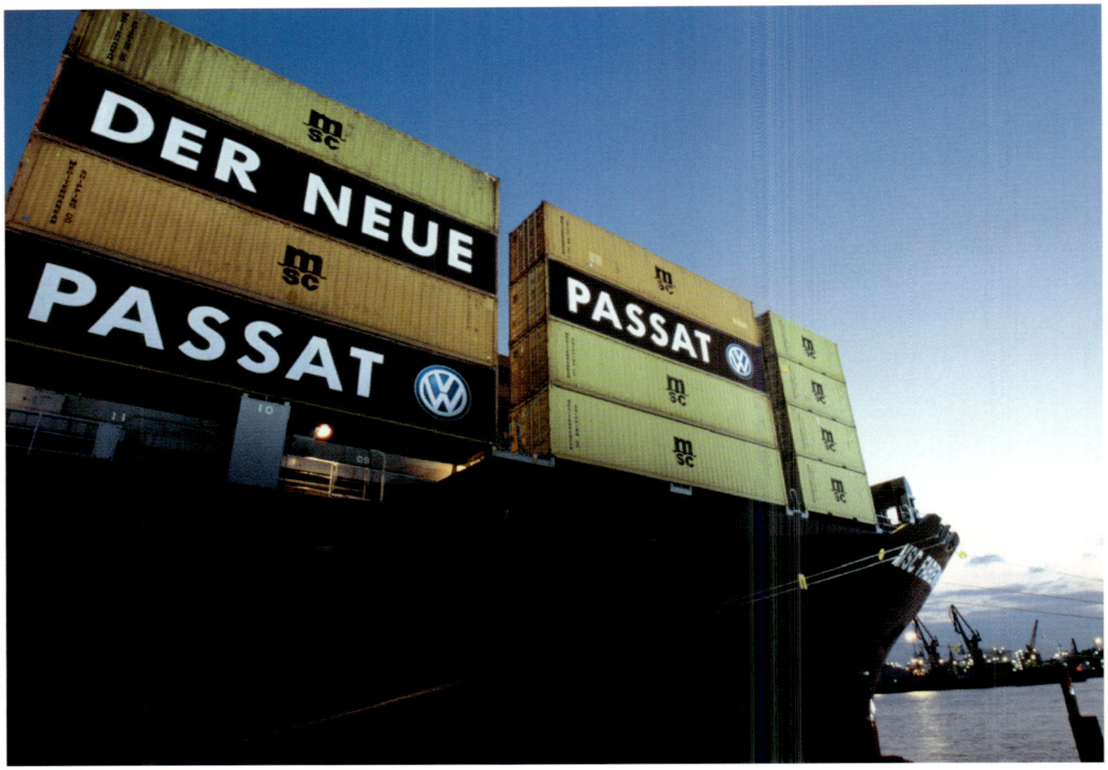

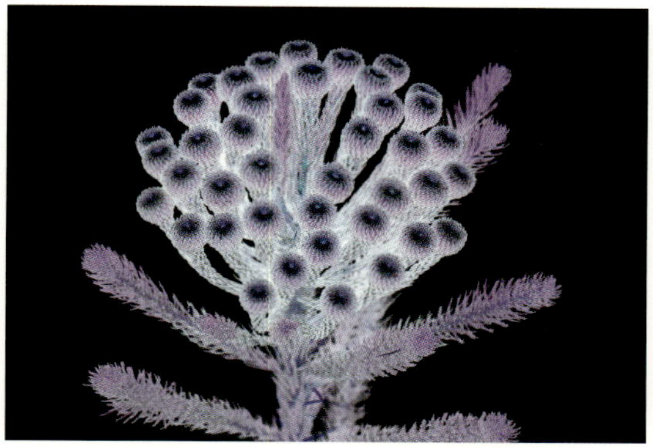
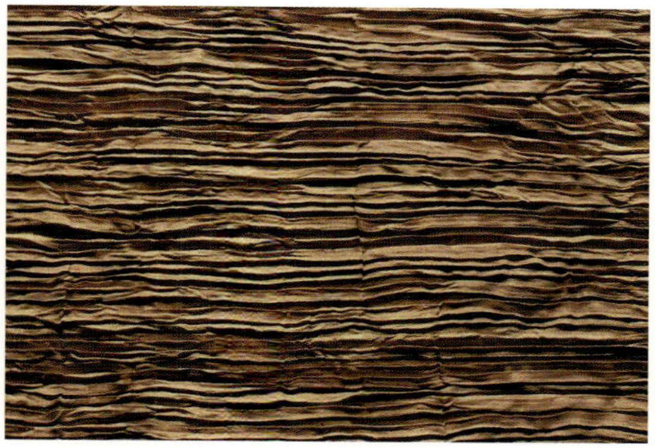
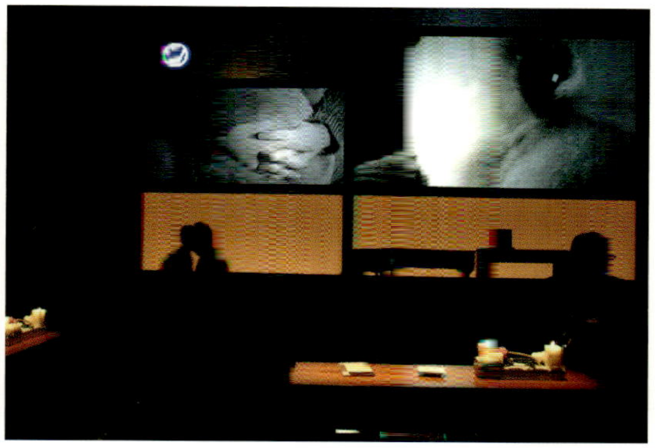
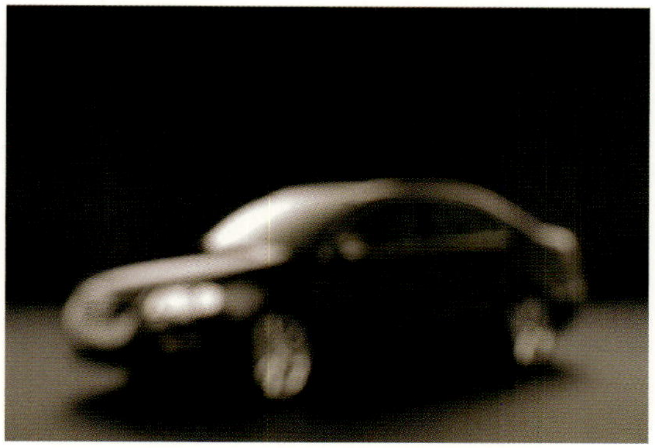
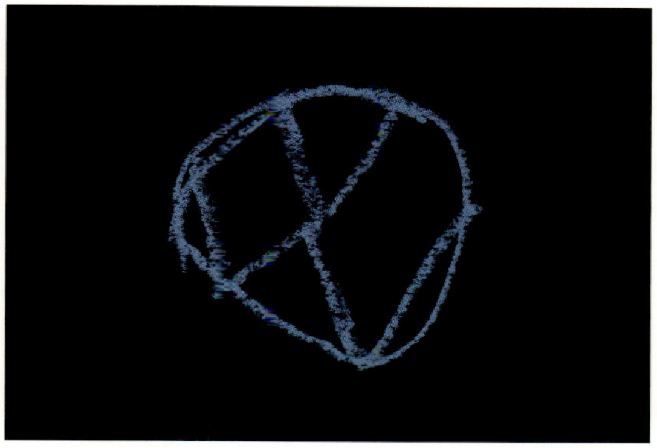
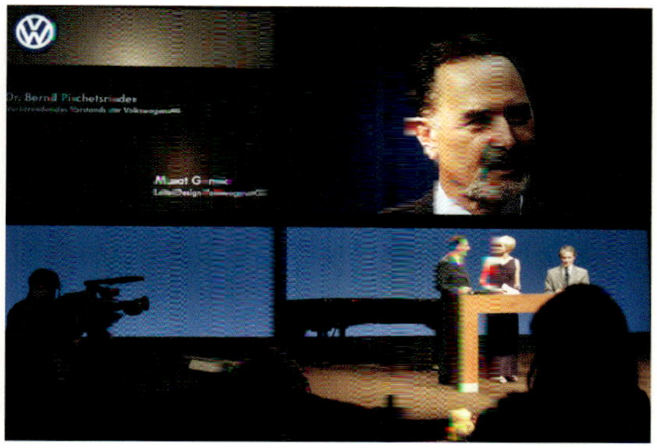
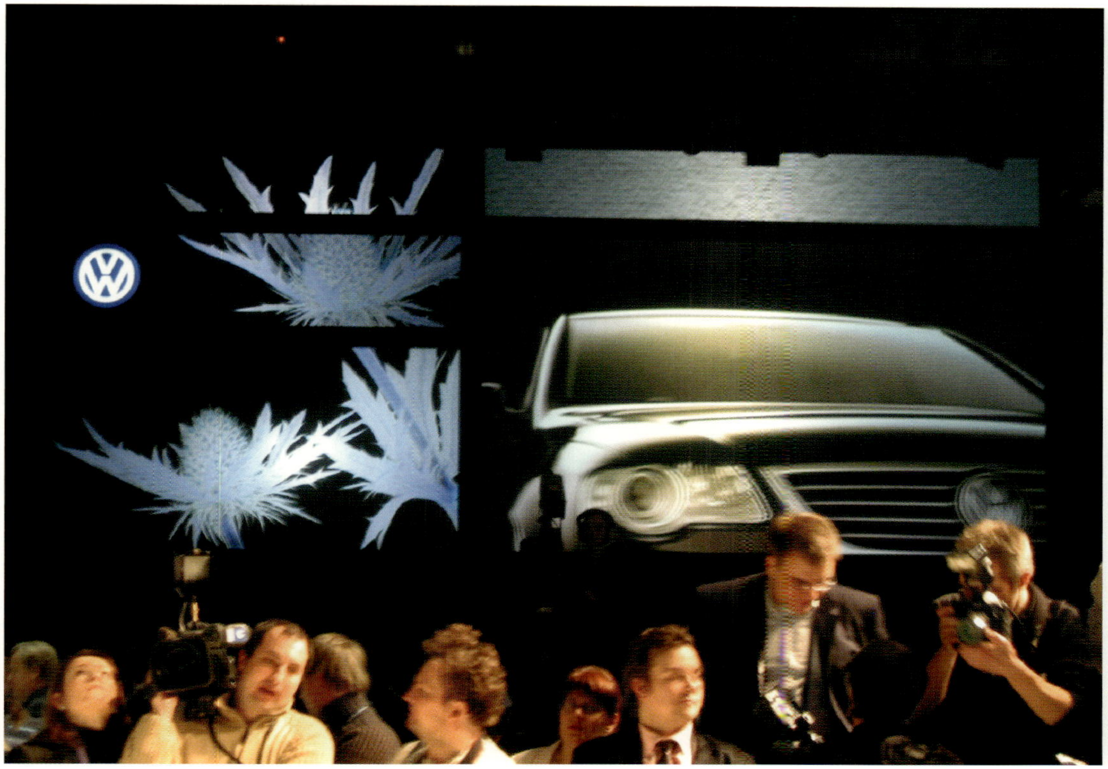
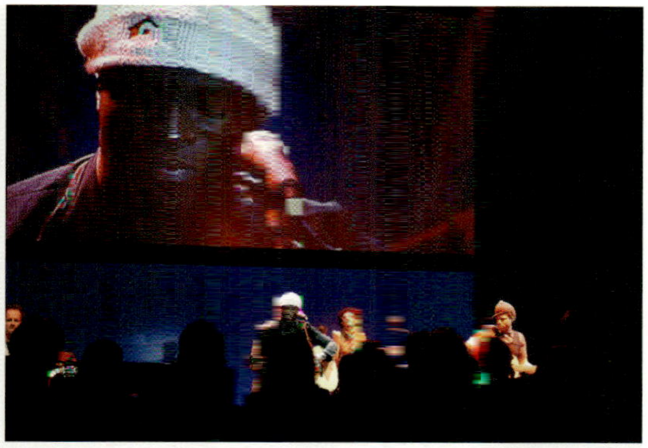
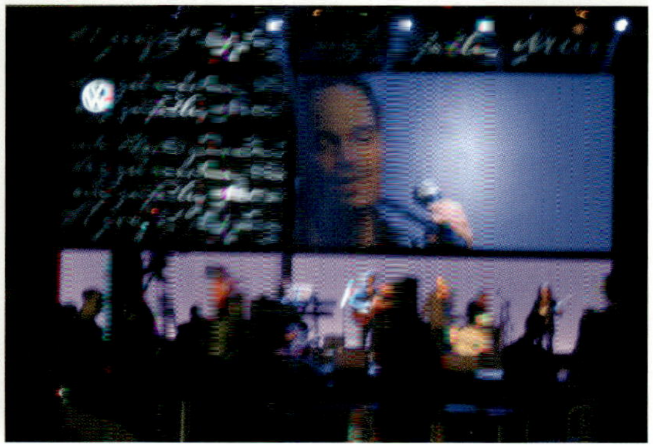

Auftraggeber Client
Sony Deutschland GmbH
Köln, Germany

Projekt Project
Sony auf der IFA 2005, Berlin
Messeauftritt
Trade fair appearance

Design
Design Company Agentur GmbH
München, Germany

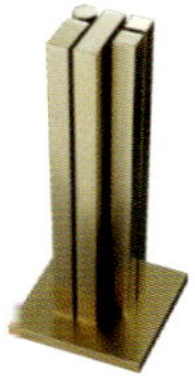

Ein schwarzer Kubus, bestehend aus abgehängten Stoffbahnen, füllt eine ganze Messehalle der IFA in Berlin. Musik- und Lichtreflexe dringen aus dem 2.300 m² großen Stoffkörper und hunderte von Hallenbesuchern schieben sich mit ihren bloßen Händen durch bewegende Räume. Die Entertainment-Botschaften der Marke Sony mit den Themen „Imaging", „Music", „Gaming" und „Picture" korrespondieren hier spielerisch mit den Wünschen und Sehnsüchten der Standbesucher. Innerhalb der Interaktion (Erlebnis) zwischen Marke und Raum entwickelt sich eine Identität zwischen Besucher und Sony. Der unvergleichbare Auftritt sorgte für ein großes Presseecho.

A black cube consisting of dangling lengths of fabric filled an entire hall at the IFA trade exhibition in Berlin. Reflections of music and light emanated from the 2,300 square meter cloth perimeter, and hundreds of visitors to the hall pushed their way through moving spaces with their bare hands. The entertainment messages coming from the Sony brand under the headings of "imaging", "music", "gaming" and "pictures" playfully matched the needs and desires of the stand's visitors. This interaction (adventure) between brand and space helped forge an identity between the visitors and Sony. This unparalleled trade fair appearance achieved a strong media impact.

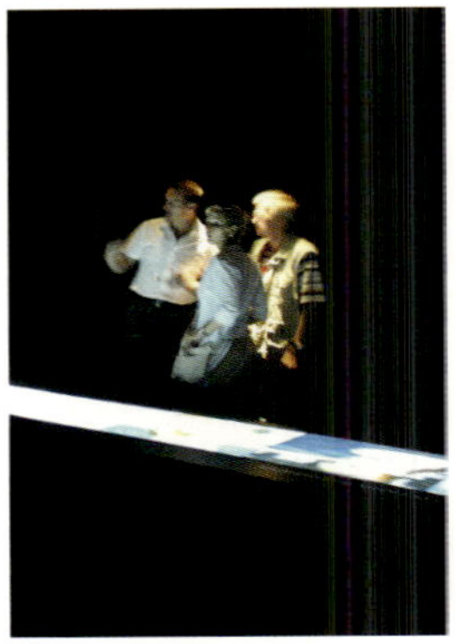

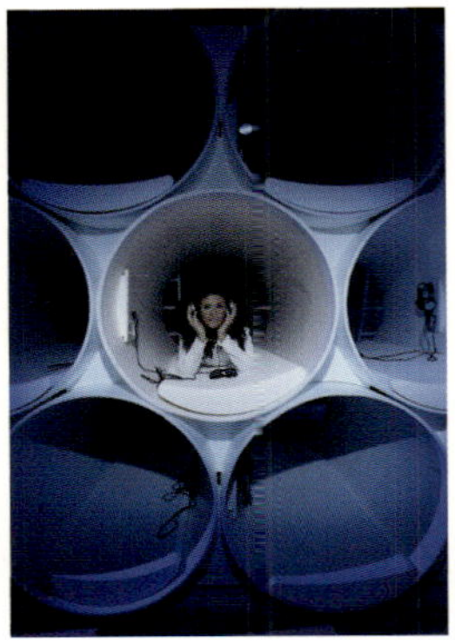

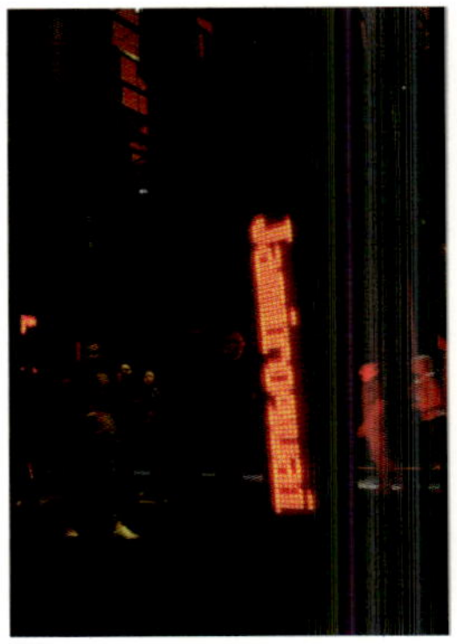

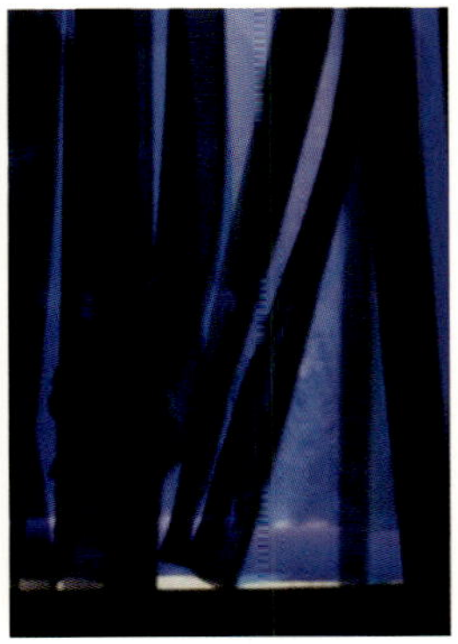

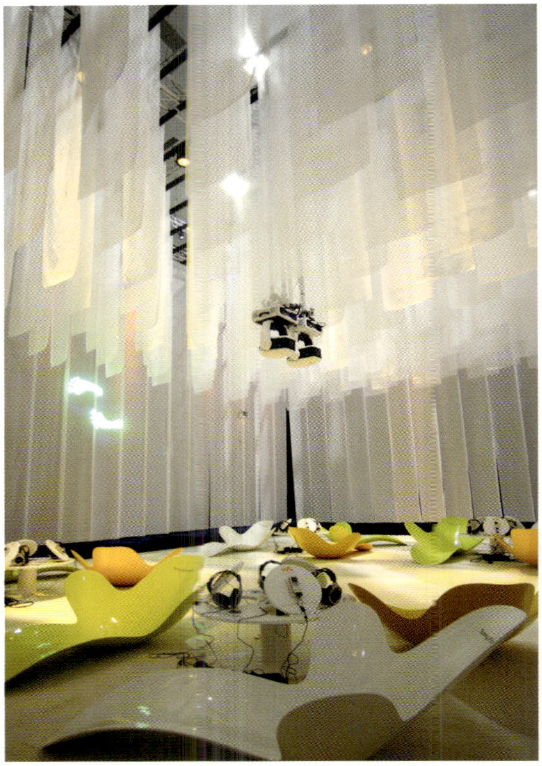

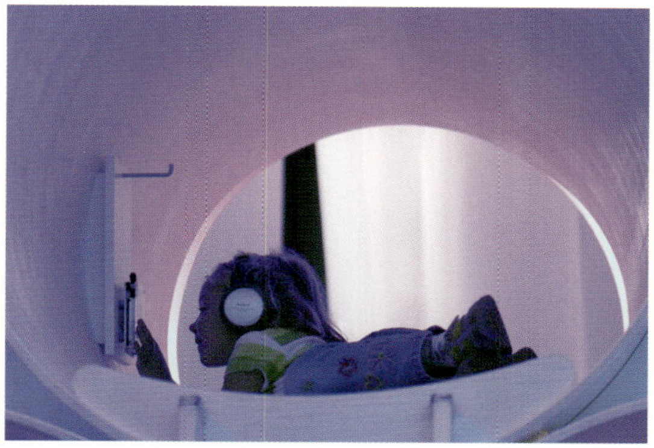

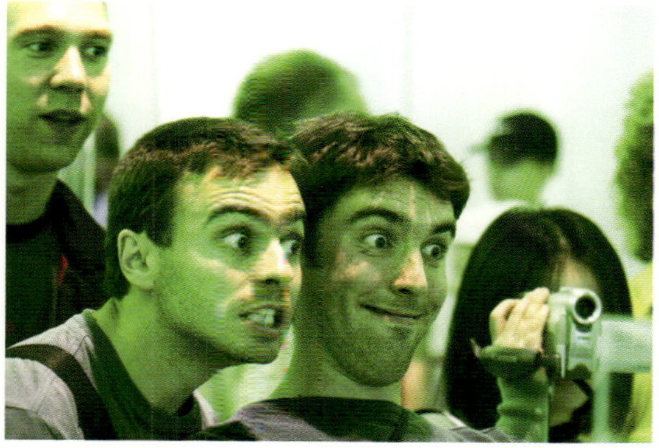

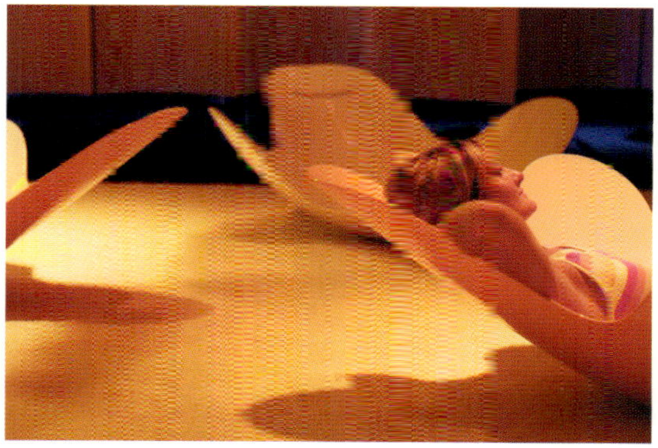

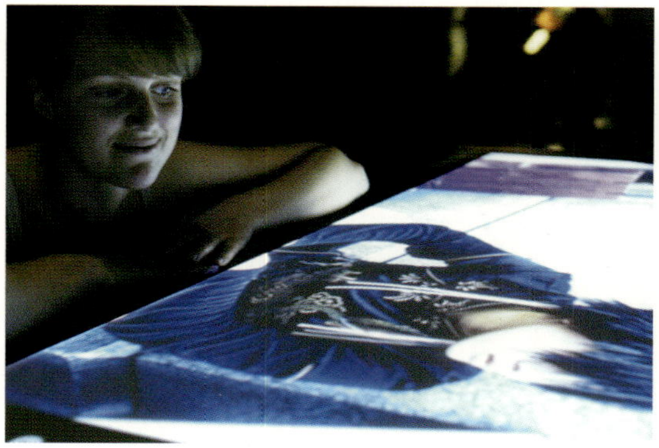

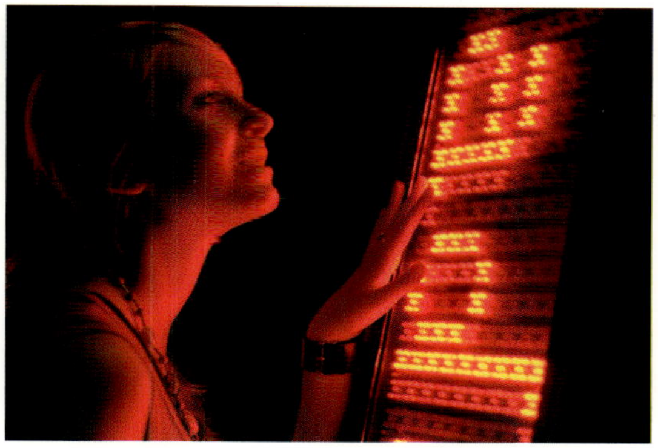

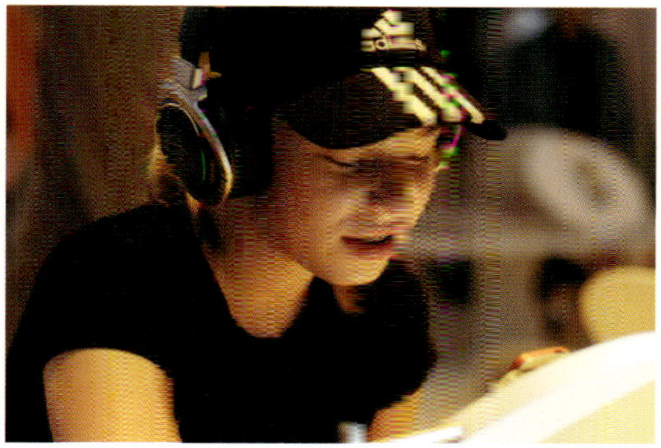

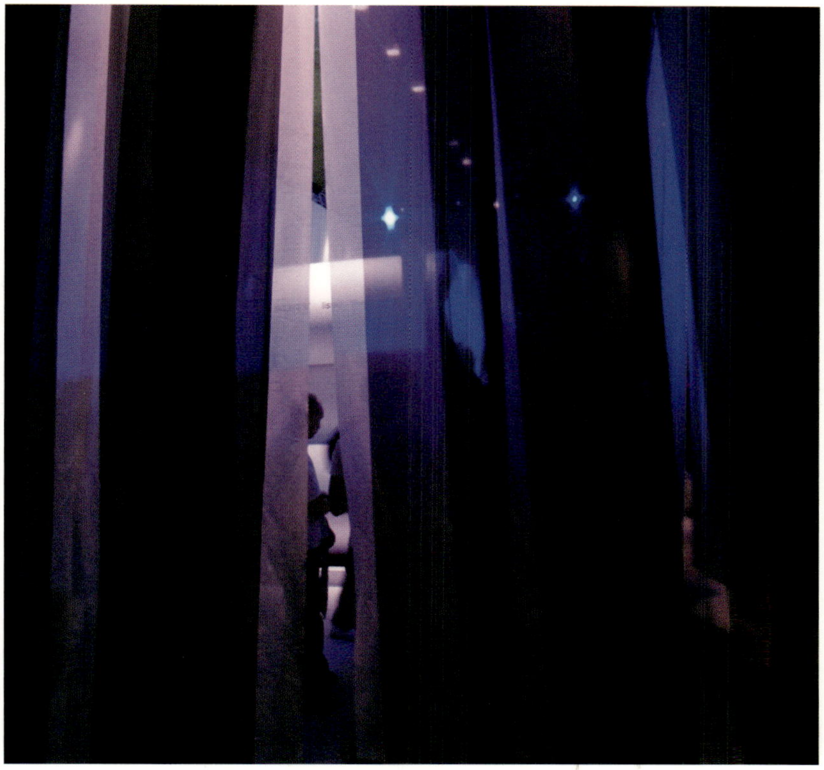

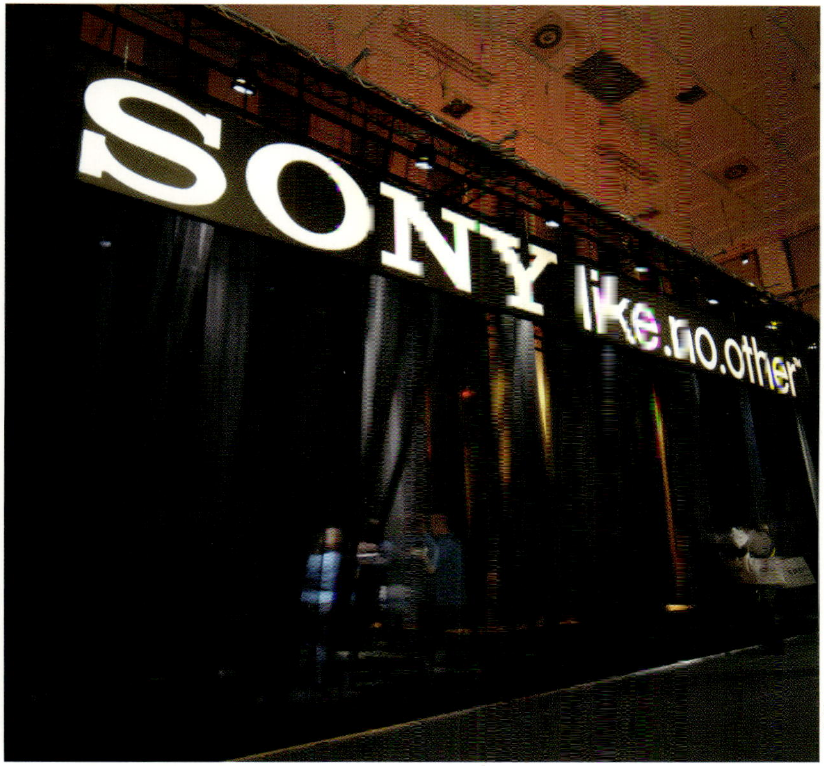
SONY like.no.other

Auftraggeber Client
Development Center for Biotechnology
Xizhi, Taiwan

Projekt Project
In Forum
Innenarchitektur
Interior design

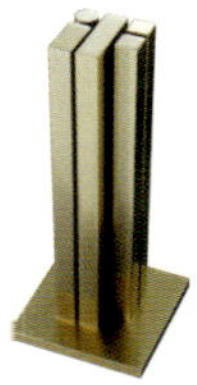

Design
Che-Hung Lu
JM Interior Architecture Design Inc.
Taipei, Taiwan

Die Aufgabe lautete, den rückwärtigen Bereich eines Forschungsgebäudes neu zu beleben. Das Problem war, ob die Leute diesen neu gestalteten und ehedem ungenutzten Ort annehmen würden. Als Lösung schufen wir von dort eine Verbindung zum vorderen, seitlichen Gebäudeteil, wodurch nicht nur ein imposanter Außeneingang entstand, sondern zudem die Grenzen dieses bis dato ungenutzten Raumes erweitert wurden. Zusätzlich entwarfen wir ein eigenes Logo für den Bereich und gestalteten den Verbindungsgang noch interessanter und individueller. Dieser neu gestaltete Bereich wird heute für Seminare und Pressekonferenzen genutzt, dient aber auch als Ort der Ruhe und Entspannung.

This case involved the rear corner of a research building. The problem was how to get people to approach this newly rearranged, previously unused space. To this end, we decided to extend this space by connecting it with the exterior corner in front of the building. In so doing, we not only added a conspicuous exterior entrance, but also extended the boundary of this formerly disused space. Furthermore, we created a CI image for this renewed space and were able to make the passageway more interesting and unique. This renewed space not only serves as a forum for seminars or press conferences, but can also serve as a place for rest or relaxation.

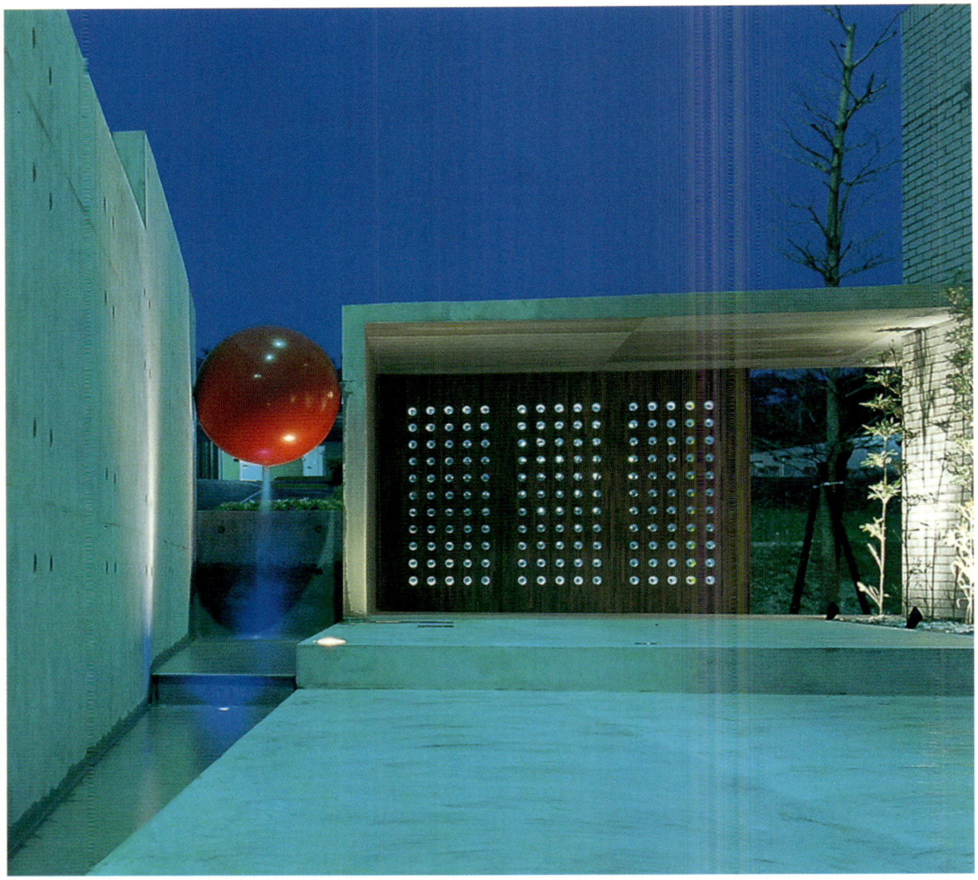

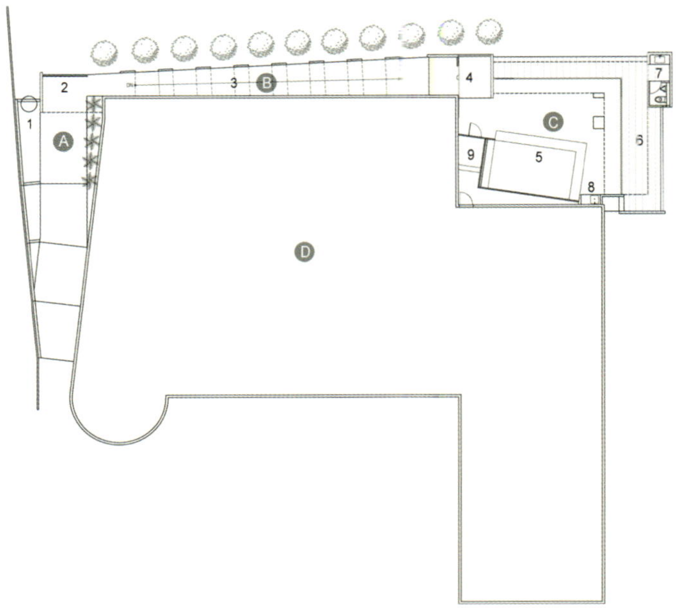

A. Foreground
1. spherical sculpture
2. tubes wall

B. The Connecting Corridor
3. ramp
4. lobby

C. The Forum
5. stepped grounding
6. balcony
7. restroom
8. cafe bar
9. preparing room

D. Researching Building

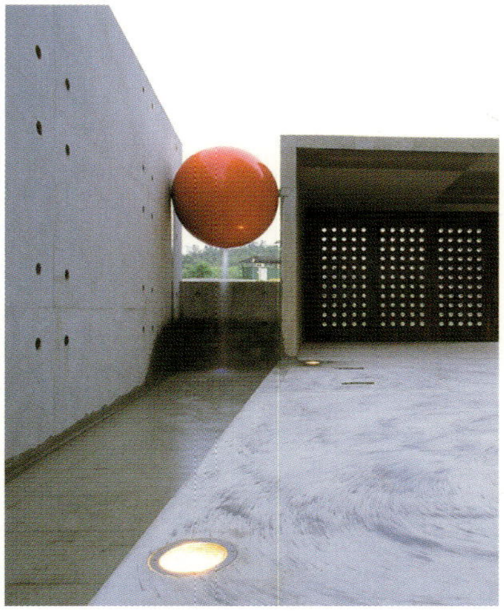

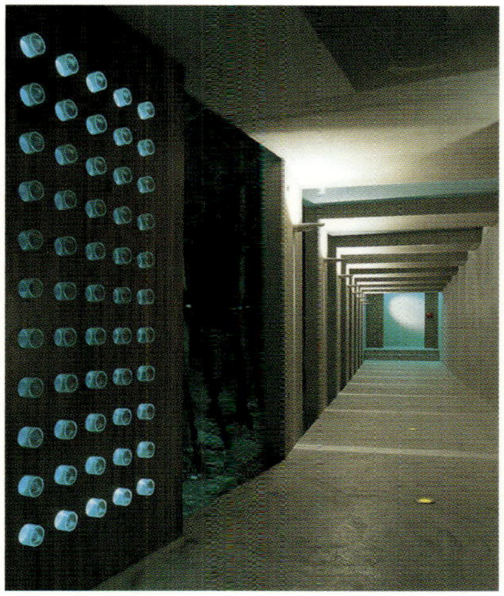

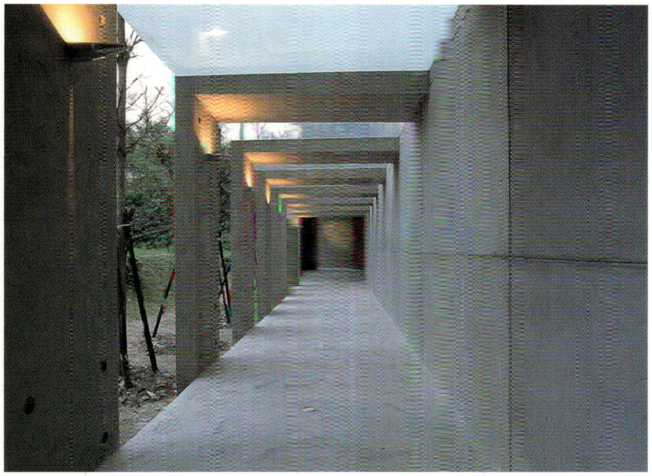

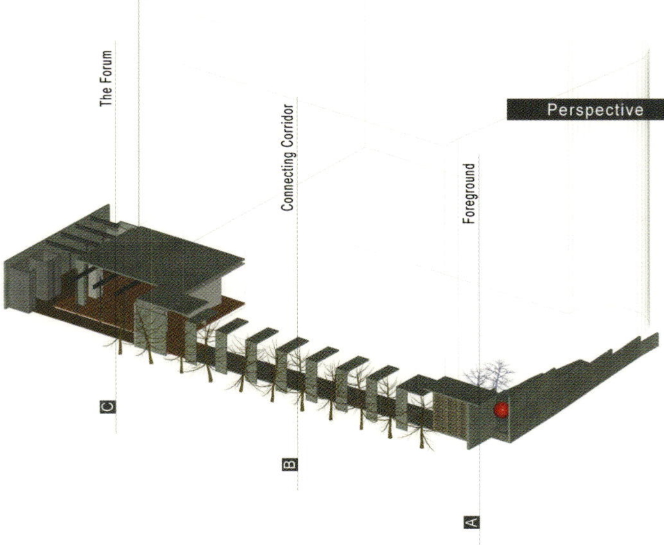

Perspective

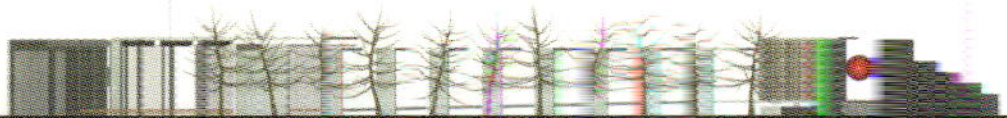

Facade

■ Space extension

The view will be felt like an unbounded extension both vertically and horizontally; produce the effect of space expansion

■ Divide and concatenate the space

Merging the original outdoor space and divided to three main spaces, then perfectly concatenate the divided space so there will be a close relationship between each space.

■ Create the level of space and enrich the variety

These three main spaces still keep their own level and independent, yet they still connect to each other along with the space variety.

■ Light and Shadow

Light and shadow interweaves just like the dancing fairy in the space, showing the lively difference between day and night.

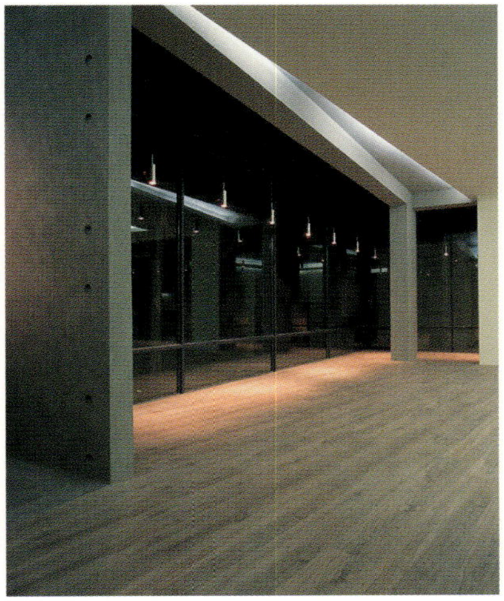

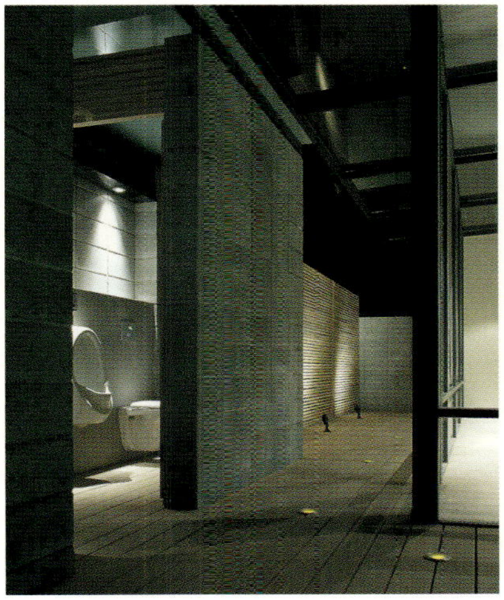

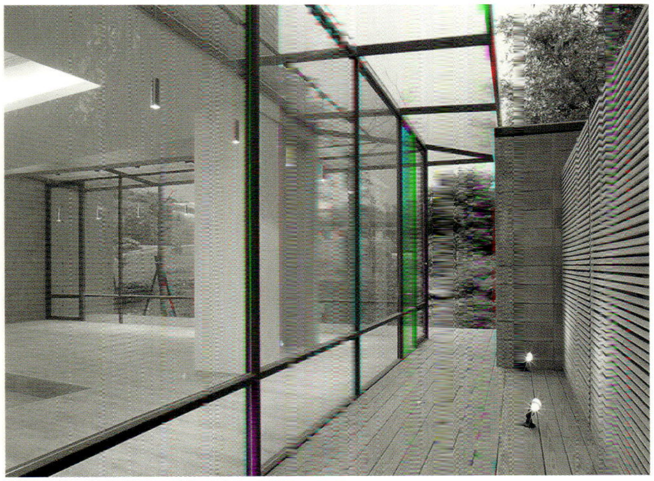

Auftraggeber Client
Österreichische Galerie Belvedere
Vienna, Austria

Projekt Project
Fahnenspur in Rot-Weiß-Rot
Interaktive Installationen
Interactive installations

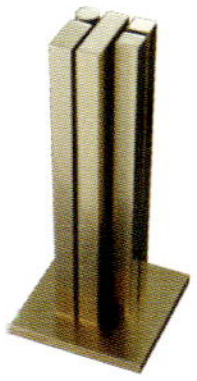

Design
Prof. Joachim Sauter
ART+COM AG
Berlin, Germany
Martin Kohlbauer
Vienna, Austria

ART+COM wurde mit der Mediengestaltung der Ausstellung „Das neue Österreich" beauftragt. Zentrales Element der Ausstellung ist neben historischen Exponaten und Kunstwerken die „Fahnenspur". Ein 250 m langes Band in den Farben der österreichischen Nationalflagge zieht sich durch die Ausstellungsräume. Mit interaktiven Installationen lädt es die Besucher zur Teilnahme am Ausstellungsgeschehen ein. Wenn der Besucher die Tasten eines Computerquiz betätigt, nimmt er an einem Fragespiel teil, wenn er einen Hörtrichter bedient, lauscht er in die Geschichte hinein; indem er sensitive Oberflächen berührt, löst er Begriffe aus. Die rot-weiß-rote Fahne wird zu einem vielseitigen Medium für die Geschichte Österreichs seit 1914.

ART+COM was commissioned to design the media for the exhibition "The New Austria". Its central element, alongside historical exhibits and selected art works, is the so called "flag trail". This is a 250 m long, winding banner in the colours of the Austrian national flag which passes through all the exhibition rooms. With countless interactive installations, it enables visitors to take an active part in the exhibition. The visitors can press the keys of a computer quiz to take part in a history quiz game, use an ear trumpet to listen in on history or expand key terms by means of touch-sensitive interfaces. The red/white/red flag is thus a versatile medium presenting Austria's eventful history since 1914.

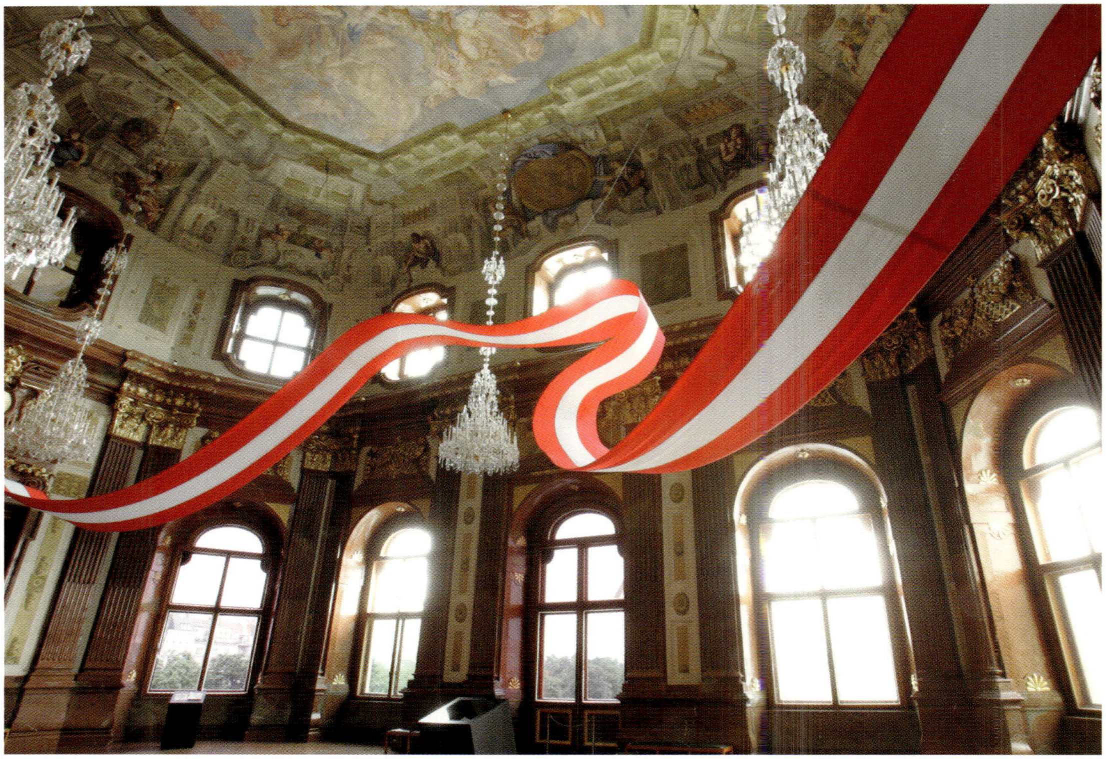

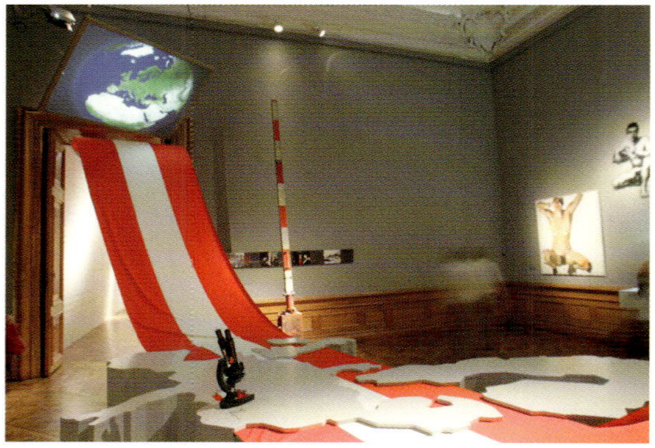

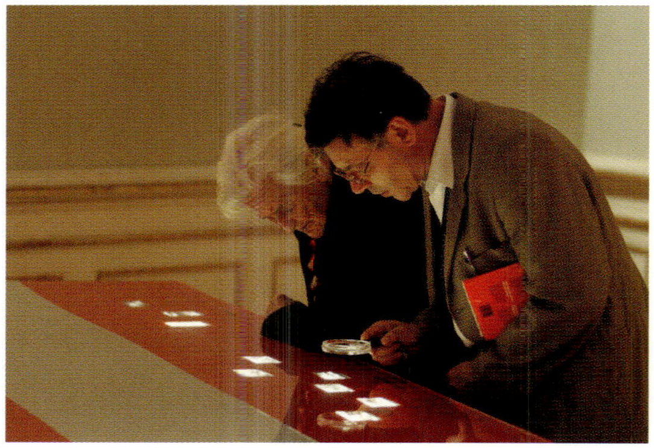

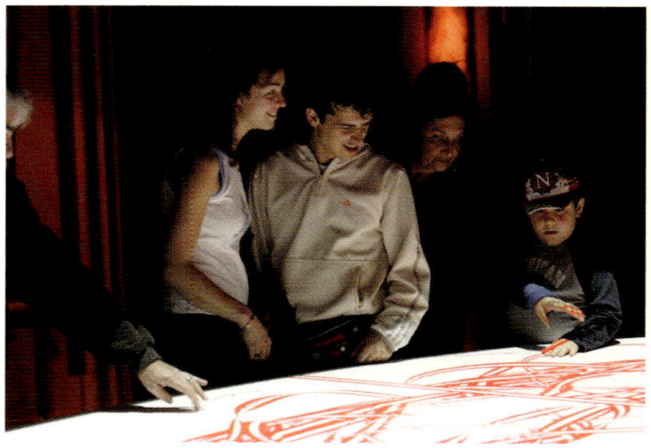

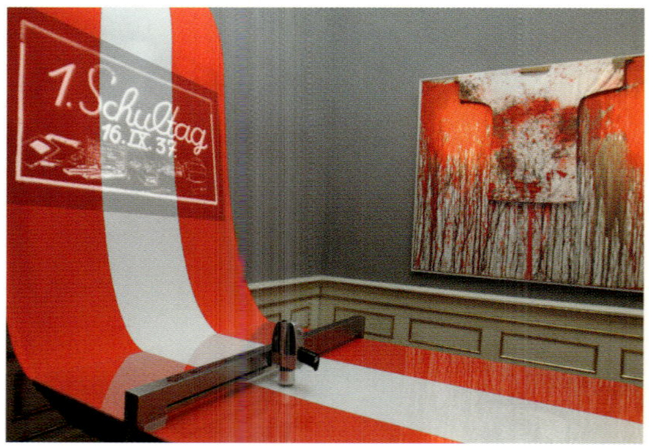

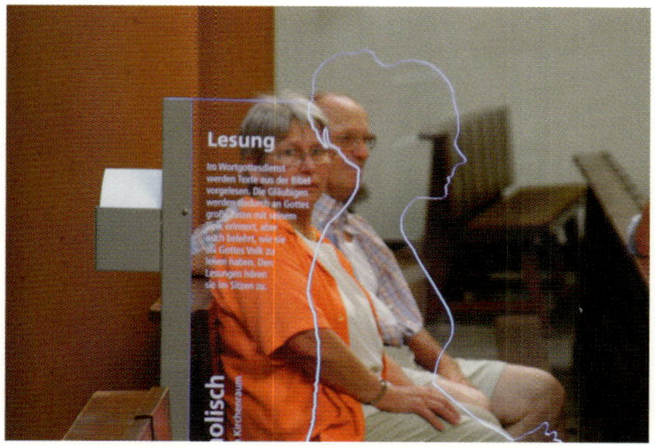

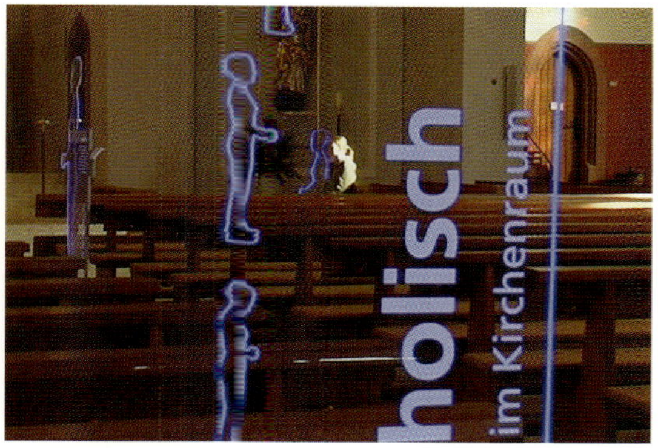

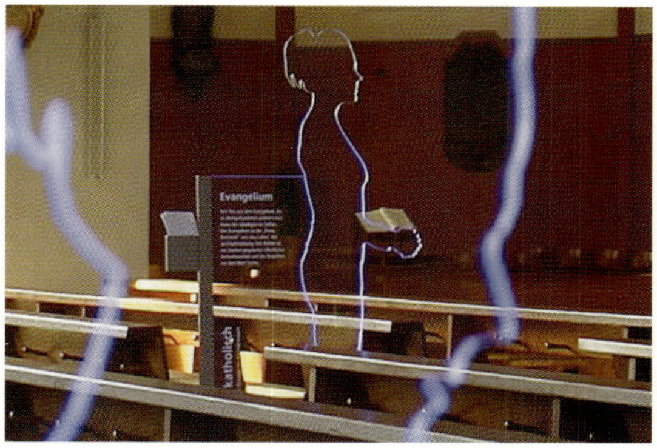

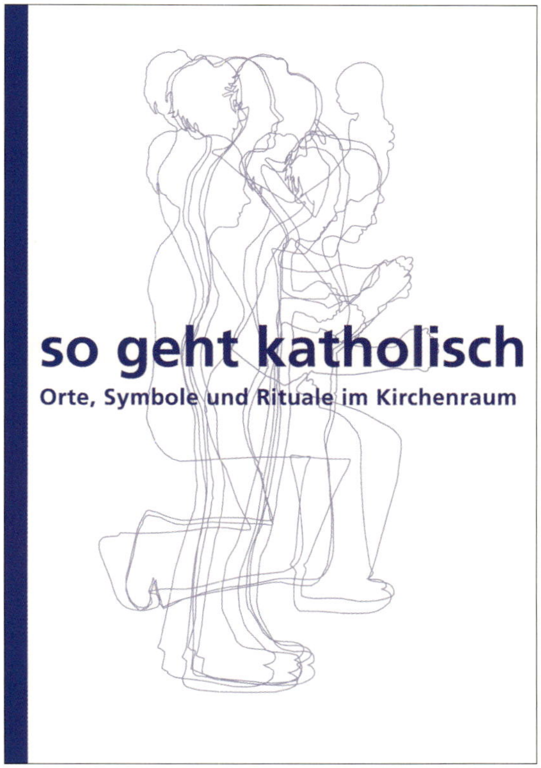

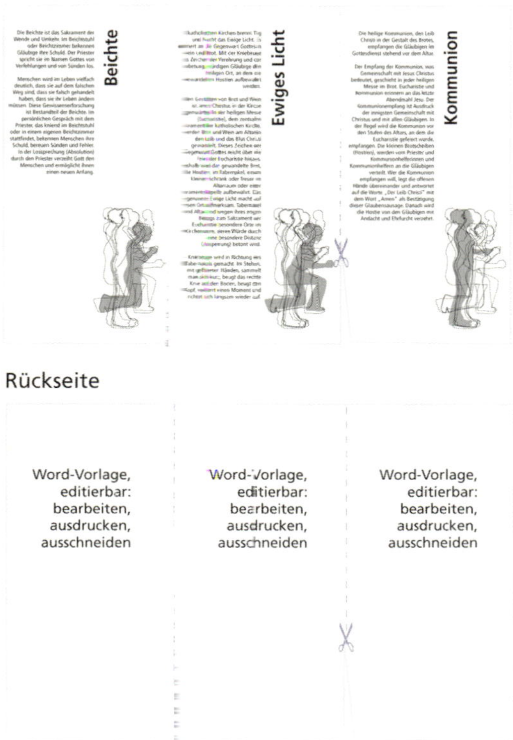

Auftraggeber Client
Bistum Limburg
Limburg an der Lahn, Germany

Projekt Project
so geht katholisch –
Ort, Symbole und Rituale im Kirchenraum
Ausstellung
Exhibition

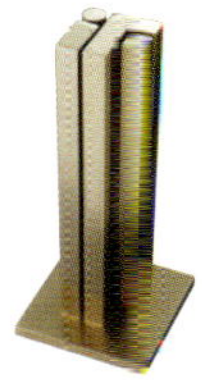

Design
einfallsreich marketing-
kommunikation gmbh & co. kg
Wiesbaden, Germany
SWID
Frankfurt am Main, Germany

„so geht katholisch" ist eine Wanderausstellung, die in katholischen Kirchen die Fragen zu grundlegenden Ausstattungsmerkmalen und zu auf den Ritus bezogenen Verhaltensformen der Kirchenbesucher beantwortet. Die Ausstellung besteht aus zehn Plexiglastafeln, auf denen jeweils eine menschliche Silhouette in einer typischen Haltung gezeigt wird. Diese Haltung wird in ihrer Bedeutung für den Glaubensvollzug erklärt. Das Objektsystem wurde in seiner Gestaltung und der Wahl des Materials so konzipiert, dass der sakrale Charakter der Kirche gewahrt bleibt.

"so geht katholisch" is a travelling exhibition, which tries to give answers to the questions people have when being in a catholic church. Questions about furniture and equipment of a church and the meaning of ritual behaviour. The exhibition consists of ten plexiglass tablets, each of them showing a human silhouette in a typical pose which is explained. Form and material of the exhibition system are chosen to suit the special sacral character of a church.

Auftraggeber Client
Volkswagen AG
Wolfsburg, Germany

Projekt Project
Project FOX
Fahrzeugpräsentation
Car presentation

Design
Fabian Tank (Creative Director)
Cedric Ebener (Project Manager)
Holger Pütting (Consulting and Finance)
eventlabs GmbH
Hamburg, Germany

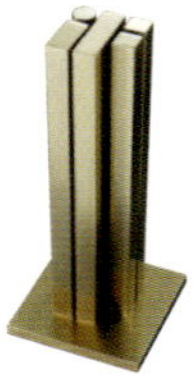

Project Fox

In Zusammenarbeit mit 21 internationalen Künstlergruppen und Designern / In cooperation with 21 international groups of artists and designers: AkimZasdBus, Antoine et Manuel, Benjamin Güdel, Birgit Amadori, Boris Hoppek, Container, E-Types, Freaklub, Friends With You , Geneviéve Gauckler, Hort, Kim Hiorthøy, Kinpro, Masa, Neasden Control Center, Pandarosa, Rinzen, Speto, tokidoki, Viagrafik, WK-Interact

Für die Fahrzeugpräsentation des Einsteigermodells Volkswagen Fox entwickelte die eventlabs GmbH im Frühjahr 2005 unter Einbeziehung der Fahrzeug-Zielgruppe ein dreiwöchiges Nachwuchs- und Kultur-Event in Kopenhagen: Project FOX.
Dafür wurde unter anderem das Hotel FOX geschaffen, in dem die eingeladenen Journalisten während der Präsentation übernachteten. Innerhalb von drei Monaten wurde ein altes 3-Sterne-Hotel durch 21 internationale Künstler zu einem der führenden Designhotels für „Young Urban Traveller" umgebaut. Anschließend wurde es dem ursprünglichen Eigner übergeben, der es unter dem Namen „Hotel FOX" im Sinne von Volkswagen weiterführen wird.
Neben dem Hotel FOX wurden zwei temporäre, öffentliche Orte geschaffen:
1. Studio FOX war sowohl „Open Workspace" für sieben Künstlergruppen, die während des Events Kunstinstallationen mit dem VW Fox entwickelten, als auch Relais-Station für die Journalisten und als Ausstellungsort zugänglich für die Kopenhagener Öffentlichkeit.
2. Club FOX war ein experimentelles Restaurant-Environment und gleichzeitig Club-Location für Kopenhagener und internationale Live-Acts.

For the dynamic car presentation of the Volkswagen Fox, the new youngster within the Volkswagen Family, eventlabs created in spring 2005 the Project FOX in Copenhagen, a cultural, three-week event with young artists, involving the car's target group. For this purpose, eventlabs created the Hotel FOX, in which the invited journalists stayed for the night. Within three months, 21 international artists had turned the old 3-star hotel into one of the leading design hotels for young urban travellers. After the event, it was resumed by the owners who continued running the Hotel FOX under the same name and according to the Volkswagen vision.
In addition to the Hotel FOX, two temporary, public venues were created:
1. Studio FOX was an open workspace for seven groups of artists, who set up artistic installations with the VW Fox. Studio FOX also served as a media workstation for the journalists and an exhibition hall open to the Copenhagen public.
2. Club FOX was an experimental gastronomic environment and also a club location for local and international live-acts.

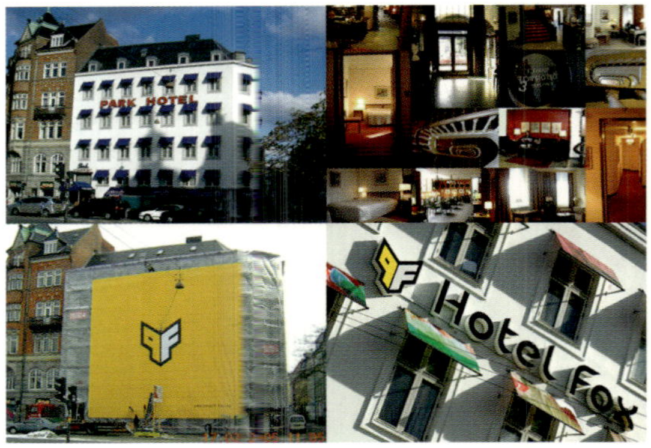

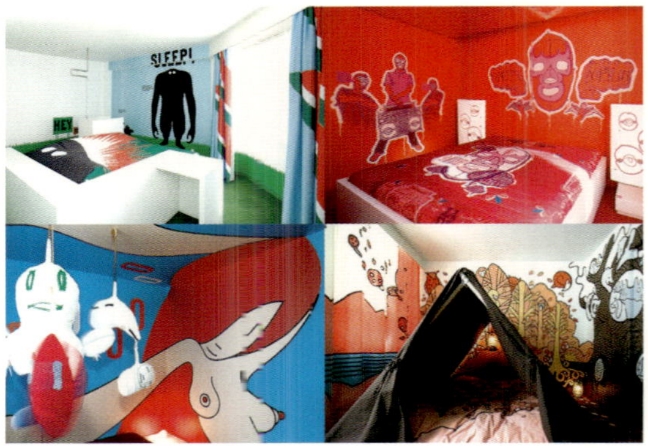

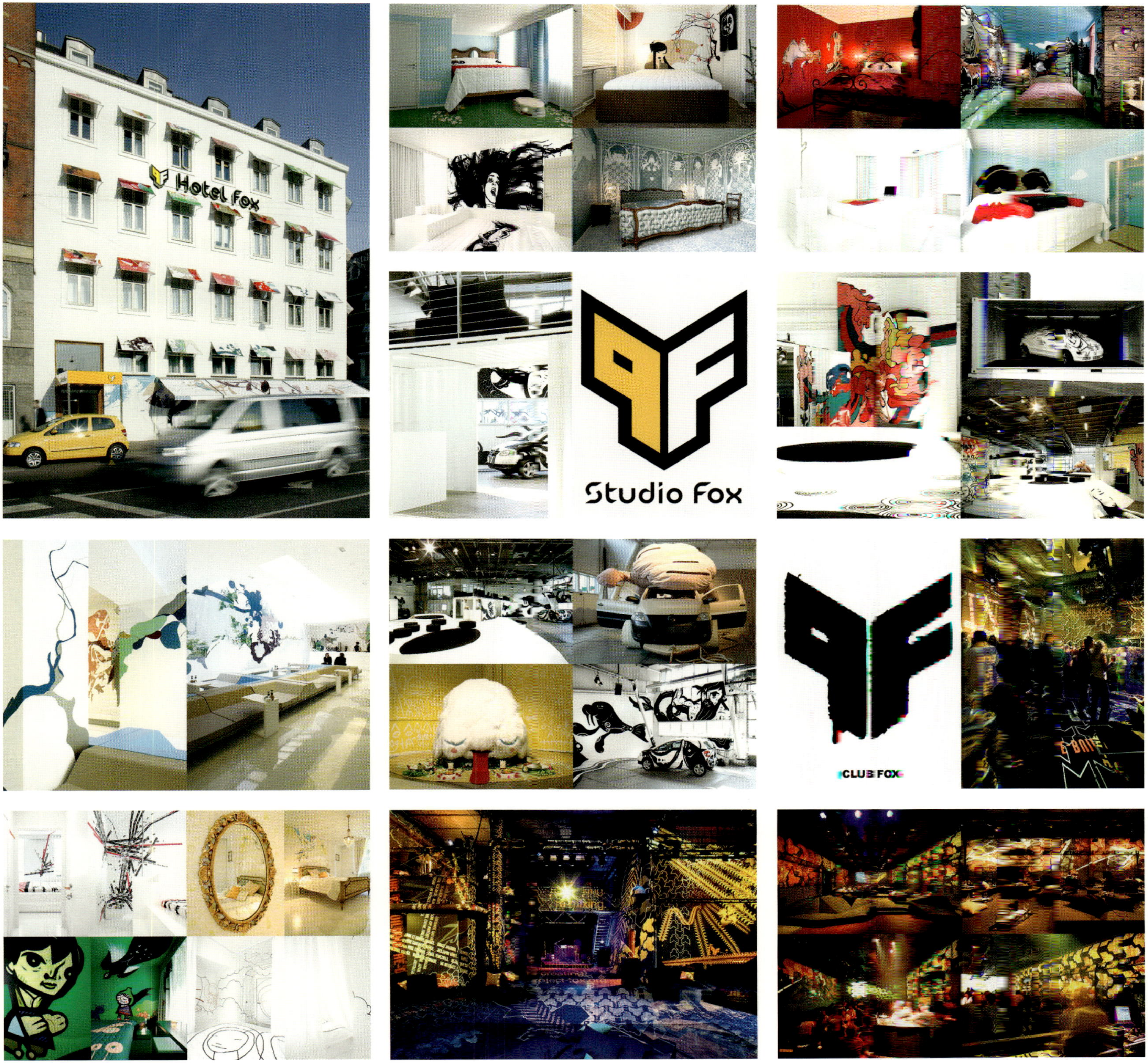
Hotel Fox
Studio Fox
CLUB FOX

Auftraggeber Client
Kunstmuseum Stuttgart
Stuttgart, Germany

Projekt Project
Max Bill. Eine Retrospektive
Ausstellungsdesign
Exhibition design

Design
Sascha Lobe
Jan Maier
Dirk Wachowiak
L2M3 Kommunikationsdesign GmbH
Stuttgart, Germany
Ulrich Zickler (Innenarchitektur)
Stuttgart, Germany

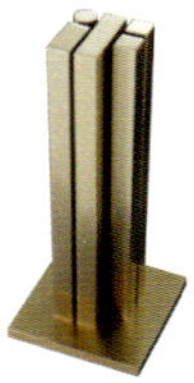

Eine Ausstellung, die allen Schaffensbereichen Bills gerecht werden möchte. Im räumlichen Zentrum der Ausstellung Max Bill, der Typograf: Wandgrafiken zeigen Schriftentwürfe und Zeichenentwicklungen sowie Bills grafische Hauptwerke. Von der Empore aus können in einer Pultvitrine ausgestellte Drucksachen mit den Wandgrafiken verglichen werden. In die angrenzenden Bereiche sind jeweils Blickbeziehungen möglich.

An exhibition designed to do justice to all areas of Bill's creations. At the physical center of the exhibition is Max Bill, the typographer: here, wall graphics show typography designs and the development of characters as well as Bill's main graphical works. Looking down from the gallery, printed materials laid out in a display case can be compared with the wall graphics. The exhibition layout also allows visitors to make visual connections between the different areas.

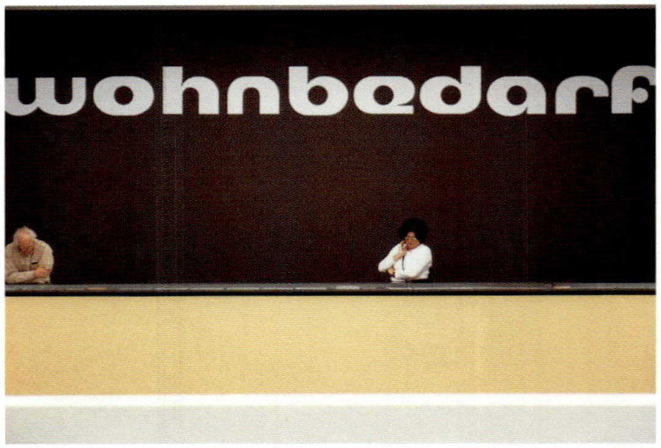

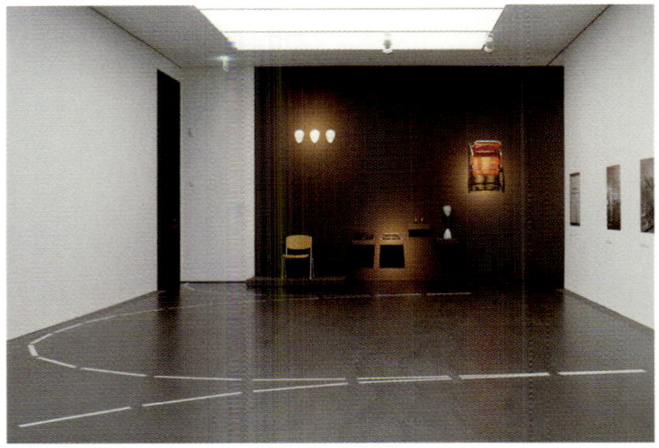

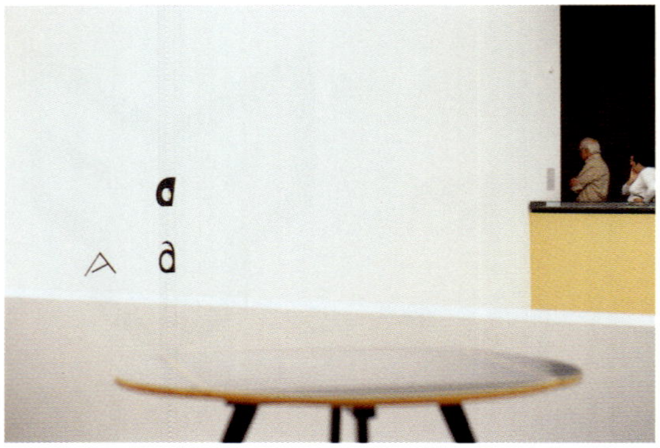

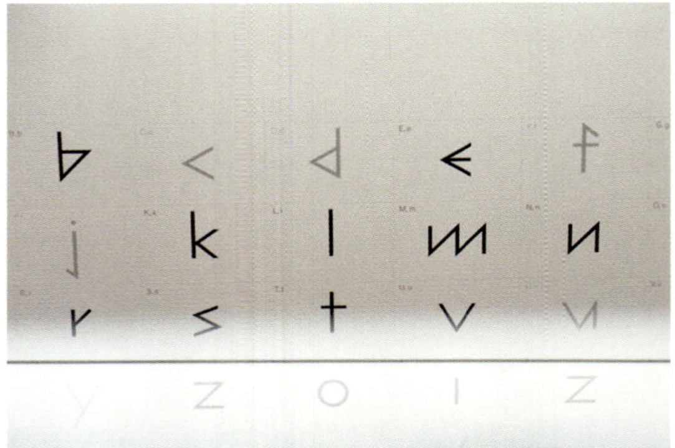

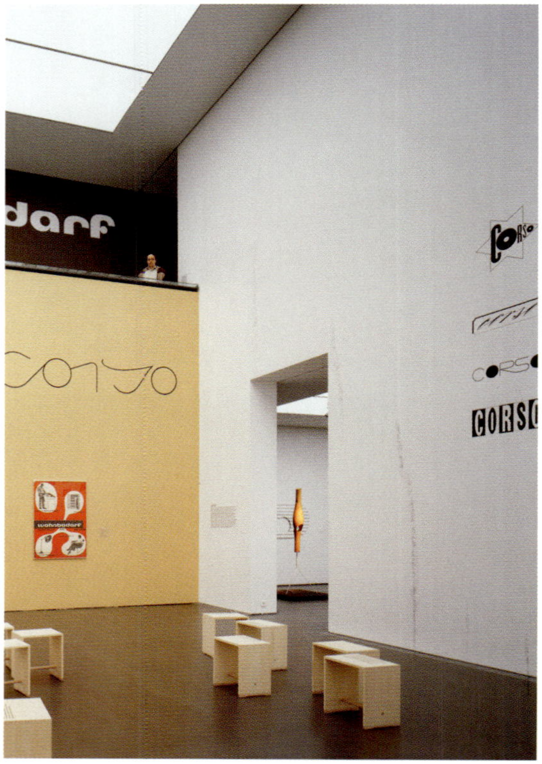

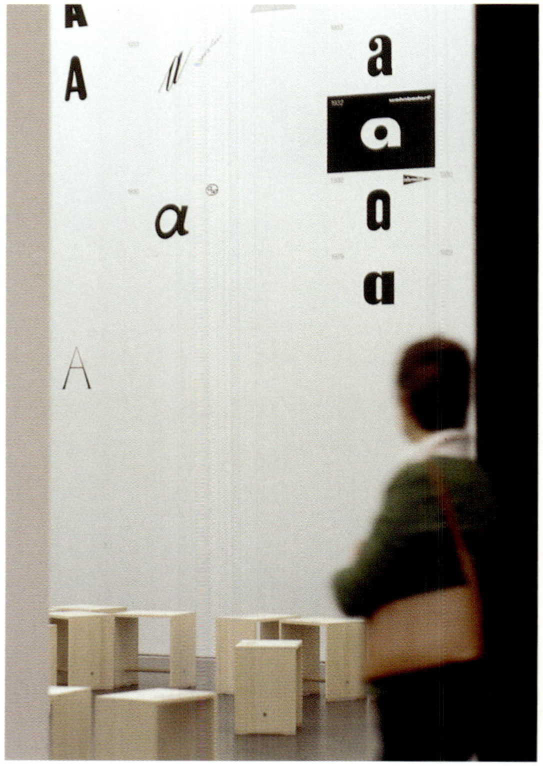

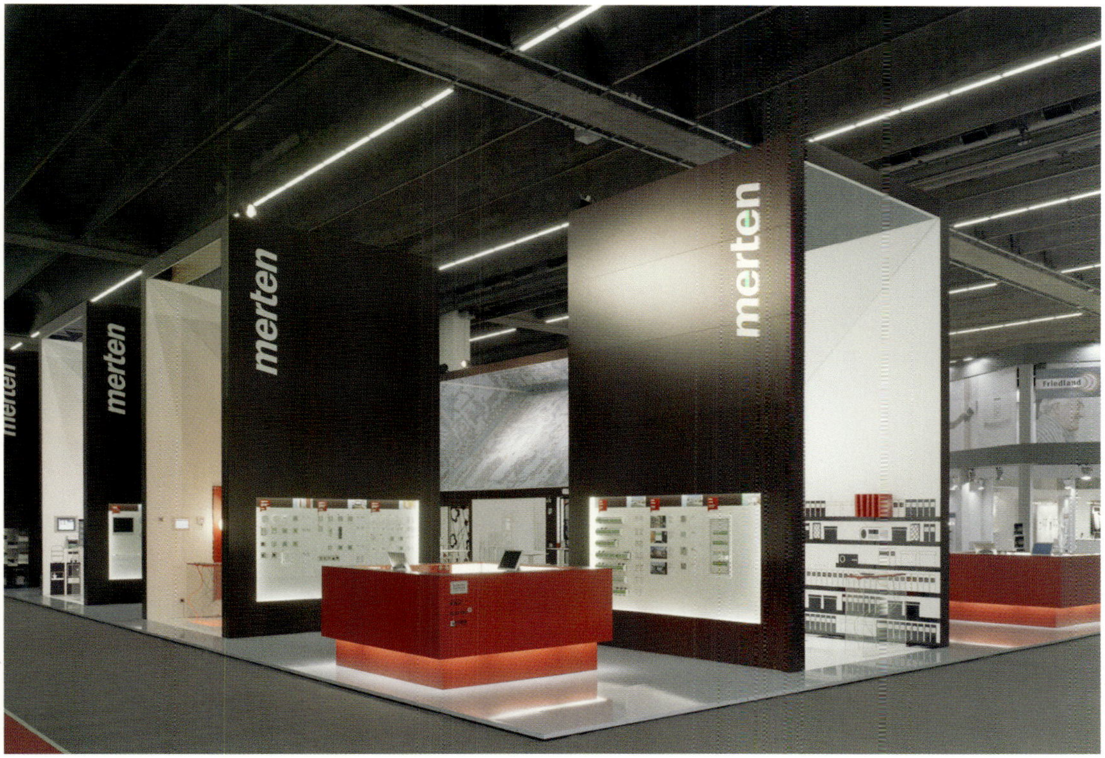

Auftraggeber Client
Merten GmbH & Co. KG
Wiehl, Germany

Projekt Project
Die Zukunft im Blick
Messestand auf der Light+Building 2006
Trade fair stand at the Light+Building 2006

Design
Martino Berghinz
Silvia Rossi
Studio Urquiola srl
Milan, Italy

Der Messestand zeigt innovative Gebäudesystemtechnik für Büro, Hotel und das eigene Zuhause in konkreter Anwendung. Der Mailänder Designer Martino Berghinz hat auf mehr als 400 m² fünf großvolumige Kuben geschaffen, in denen Raumsituationen modellhaft dargestellt sind. Die fantasievolle Inszenierung orientiert sich am Prinzip des Periskops: Nach oben hin sind die Kuben durch Spiegel begrenzt, die in einem Winkel von 45° zweidimensional reproduzieren, was auf der Grundfläche dreidimensional als Rauminszenierung vorhanden ist – ein Spiel mit Konkretem und Abstraktion, zwischen Idee und Realität. Die so projizierten Spiegel-Bilder entfalten eine starke visuelle Sogkraft.

The exhibition booth shows innovative building management systems for offices, hotels and private homes in concrete application. On more than 400 m² of exhibition space the Milan designer Martino Berghinz created five large cubes to exemplify living situations. The imaginative production is based on the principle of a periscope: the cubes are capped with mirrors, which reproduce in an angle of 45° and in two dimensions, what is three-dimensional on the floor – a play with facts and abstraction, between idea and reality. The projected mirror images evolve a strong visual appeal.

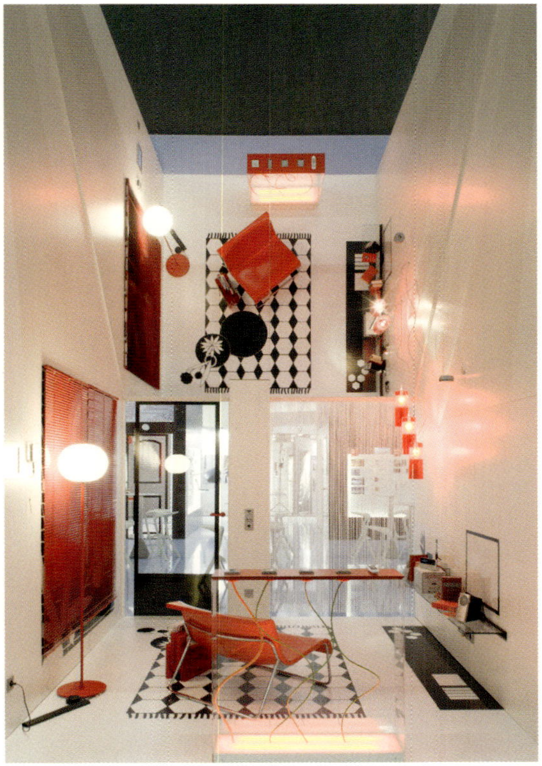

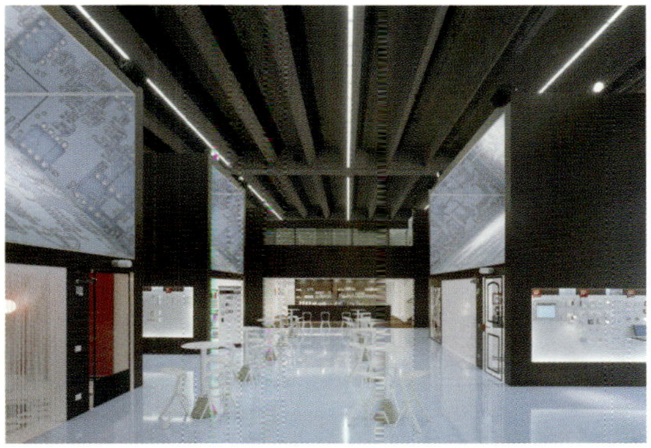

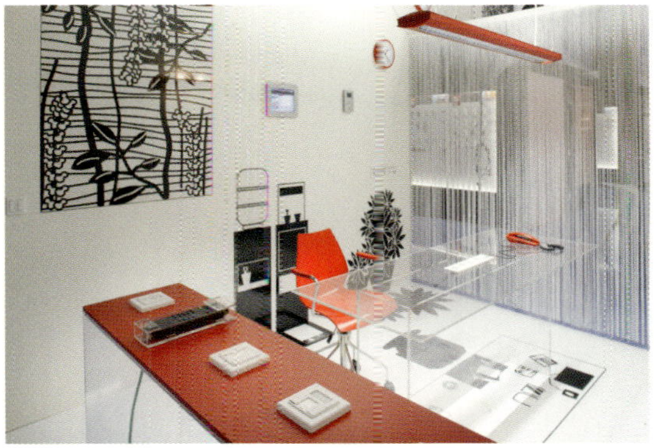

Auftraggeber Client
GM Europe
Rüsselsheim, Germany

Projekt Project
Präsentation des GM Produktdesigns der kommenden Jahre
Event-Architektur
Event architecture

Design
Simone Heckmann
Monika Kluthe
Norbert Fischbach
pmg profile marketing group
Wiesbaden, Germany

Pur, unverfälscht, glaubwürdig: die Designpräsentation des GM Produkt-Portfolios der kommenden Jahre. Die kommunikative Erlebniswelt ist geprägt von der strengen Architektur einer Industriebrache und lädt den Gast ein zur Konfrontation und Diskussion. Showplattformen mit schlichten Oberflächen und futuristischer Anmutung unterstreichen die ausgestellten Designstudien. Die patinierte Halle sorgt für die atmosphärische Trennschärfe.
Als Symbol für die verbindende Funktion der Dachmarke dient ein blauer Steg. Er führt und verbindet die Plattformen der Hersteller-Marken, deren Markenpersönlichkeiten für den Besucher medial und visuell erlebbar sind.
Das Design gilt als letzte große Chance, sich vom Wettbewerb zu unterscheiden. In Zeiten nahezu beliebig austauschbarer Plattformstrategien gehört die Zukunft dem Design und den Designern, der „Spirit of Creativity" als letzte große Differenzierungsmöglichkeit. Das Event folgte dieser Leitlinie und die durchweg positive Resonanz der Teilnehmer gab dem Konzept Recht.

Pure, unadulterated and plausible: the design presentation of the GM product portfolio for the next few years. The communicative style reflects the severe architecture of an abandoned industrial facility and invites confrontation and debate. Show platforms with simple lines and a futuristic appearance underscore the designs featured in the presentation. The hall, with its subtle patina, provides the right aura of exclusivity.
A blue catwalk serves as an underlying symbol of the GM umbrella brand. It leads and links the platforms of the production brands, while a sense of their separate brand identities is conveyed through a series of visual and multimedia effects.
Design is our last opportunity to stand out from the competition. In these times of practically interchangeable platform strategies, the future belongs to design and designers. A spirit of creativity is our last chance to establish a point of differentiation. The event was based on this principle, and the overwhelmingly positive feedback shows how right it was.

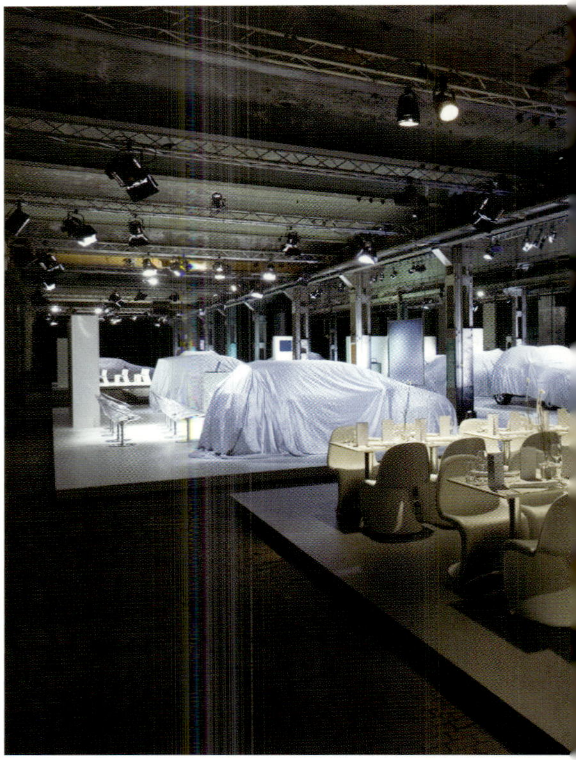

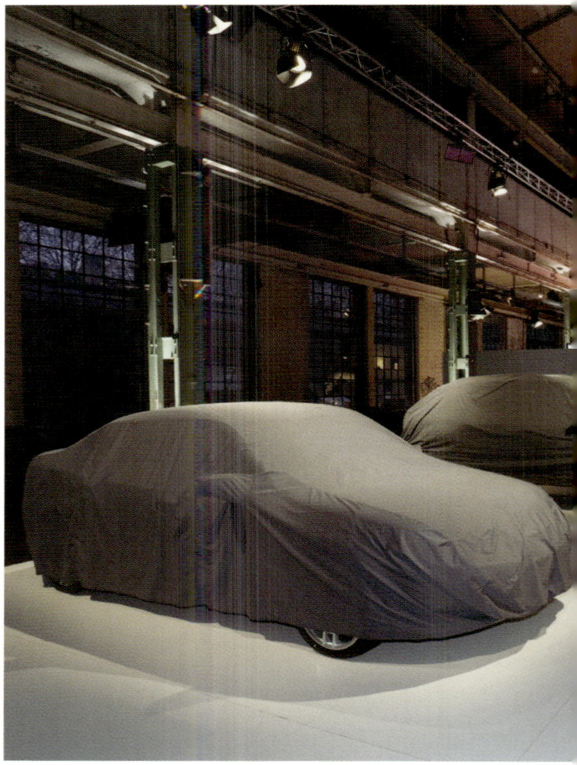

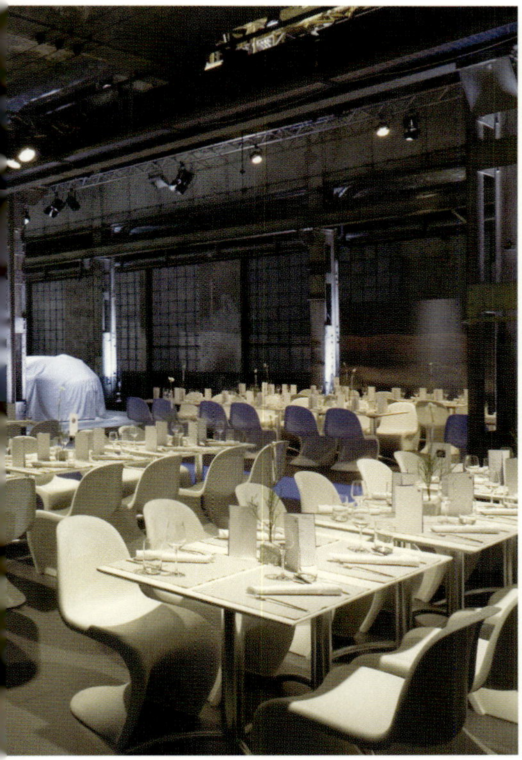

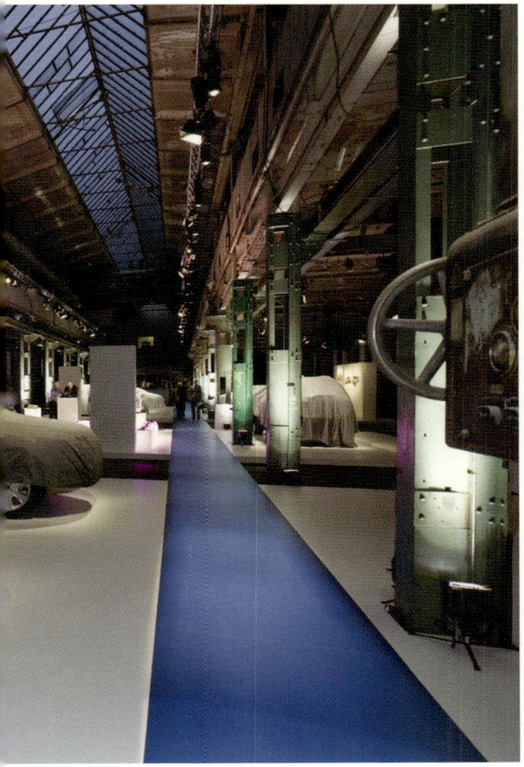

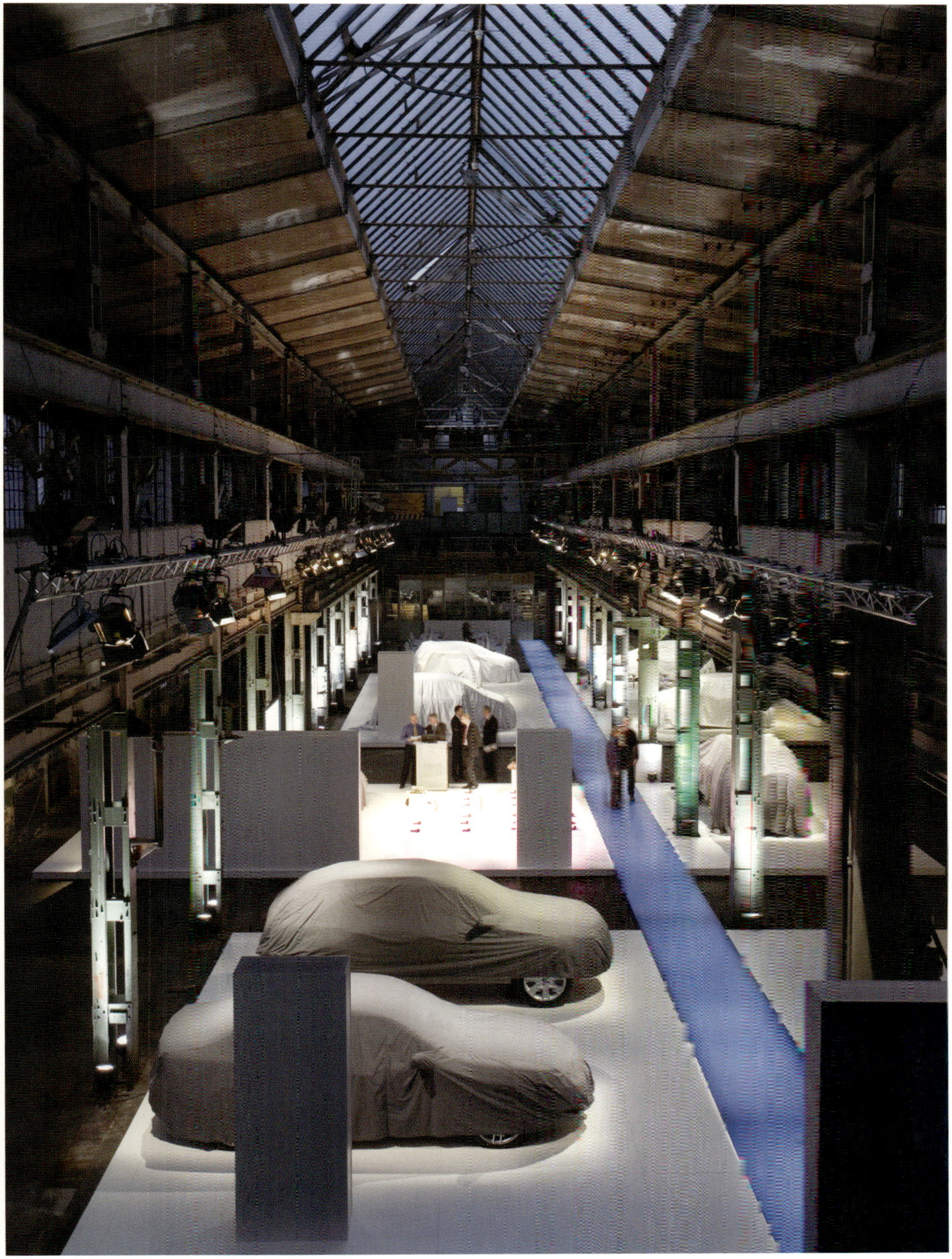

Auftraggeber Client
EnBW
Energie Baden-Württemberg AG
Karlsruhe, Germany

Projekt Project
EnBW Markenauftritt auf der HMI
Messestand
Trade fair stand

Design
Joseph Pölzelbauer (CD)
Jean Mierecke
Marcel Ermes
Meike Hoffmann
identis GmbH
Freiburg, Germany

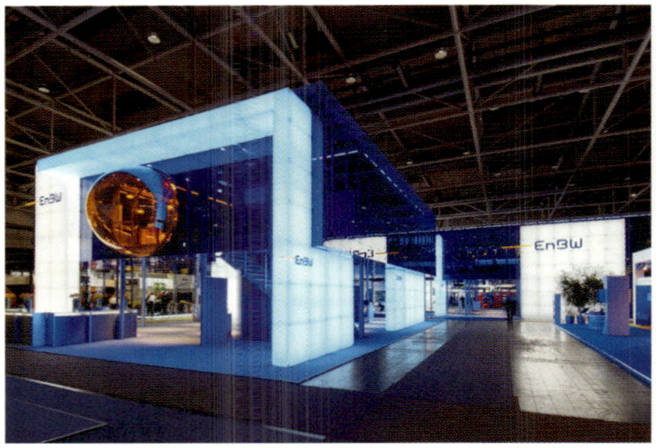

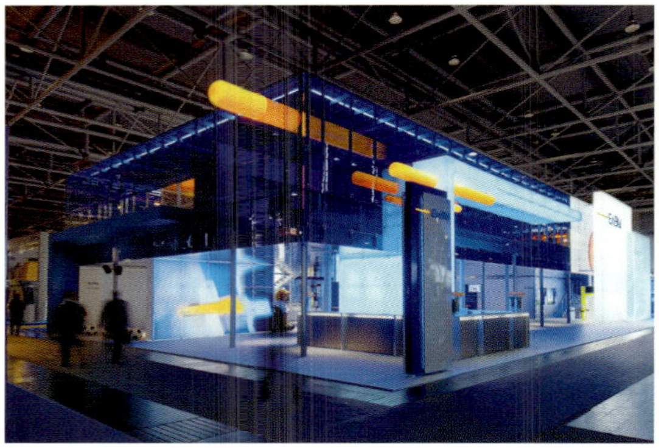

Auf der Hannover Messe 2006 trat die EnBW Energie Baden-Württemberg AG mit einem Messestand auf, der das Corporate Design und zugleich die Inhalte und Werte der Marke in die dritte Dimension umsetzt. Die Innovationskraft des Vordenkers und Wegbereiters EnBW kommt darin ebenso zum Ausdruck wie die Verbindung von Räumen, Welten und natürlich von Menschen.

Der Messestand ist die gebaute Idee der Marke EnBW. Er fungiert weniger als Gehäuse, mehr als Botschaft, als eine „Plastik der Energie". Eine der wesentlichen Konstanten des Corporate Designs ist die „Impulslinie". Dieses Element wird zu einer begehbaren Impulsröhre mit symbolischer und zugleich praktischer Funktion. Sie überbrückt die Distanz zwischen räumlichen Welten und bündelt die Verkehrsströme innerhalb des Messestandes.

Innerhalb der Marken-Welt EnBW bilden die Submarken des Konzerns ihre eigenen Räume und entfalten ihre visuellen Merkmale. Wo die Marke EnBW zur Kommunikation und Bewirtung einlädt, herrschen die Unternehmensfarben „Tiefenblau" und „Horizontorange" vor. Bei Yello Strom dagegen dominiert die Farbe Gelb, bei der Öko-Marke Naturenergie die Farbe Grün und bei der Mittelstands-Marke Watt die Farbe Rot. Die nach allen Seiten durchlässige Impulsröhre verbindet diese Marken-Welten.

EnBW Energie Baden-Württemberg AG appeared at Hannover Messe 2006 with a trade fair stand translating the company's corporate design as well as the brand's content and values into the third dimension. The stand expressed the innovative power of industry trailblazer EnBW just as much as it did the connection between realms, universes and naturally people.

The trade fair stand physically incorporated the idea of the EnBW brand. It functioned more as a message, a kind of "energy sculpture", than as a structure per se. One of the essential constants of the corporate design involved the "pulse line". This element became a walk-through "pulse pipe", exercising both a symbolic and practical function: It connected spatial realms and at the same time channeled traffic streams at the trade fair stand.

Within the EnBW brand universe, the Group's subbrands had their own individual realms where they were able to unfold their own visual characteristics. Wherever the EnBW brand invited guests in for communication or refreshments, the corporate colors of "deep blue" and "horizon orange" dominated. Within the Yello Strom sub-brand realm, on the other hand, the color yellow was dominant. For the ecological brand "Naturenergie", the color green was key and the SME brand "Watt" primarily deployed the color red. The "pulse pipe", which was open on all sides, connected these different brand universes.

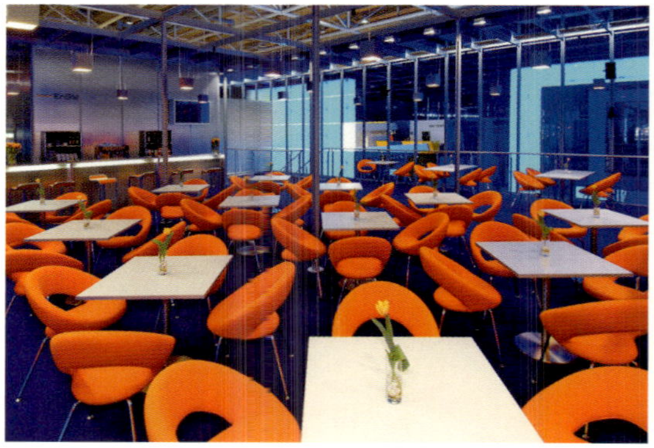

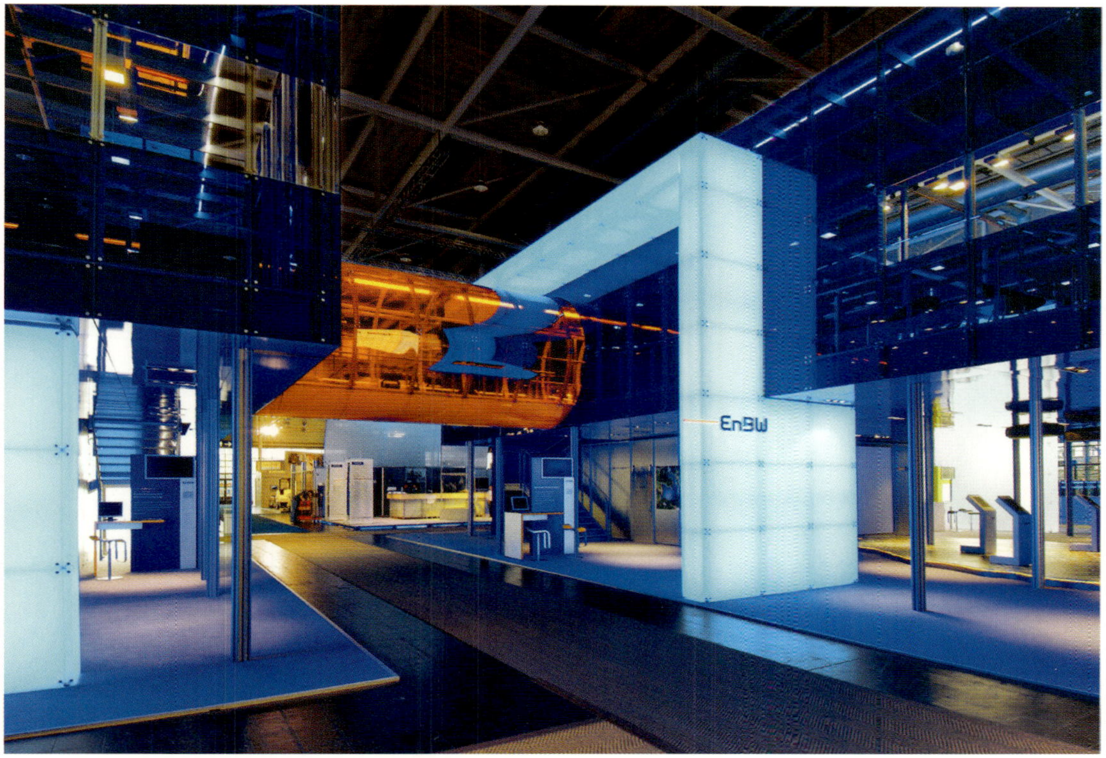
EnBW

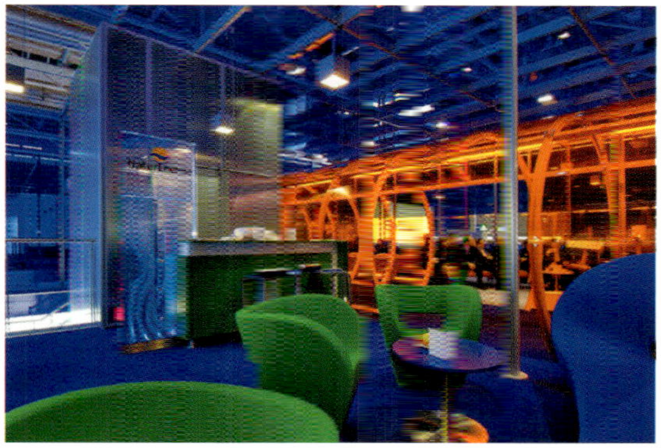

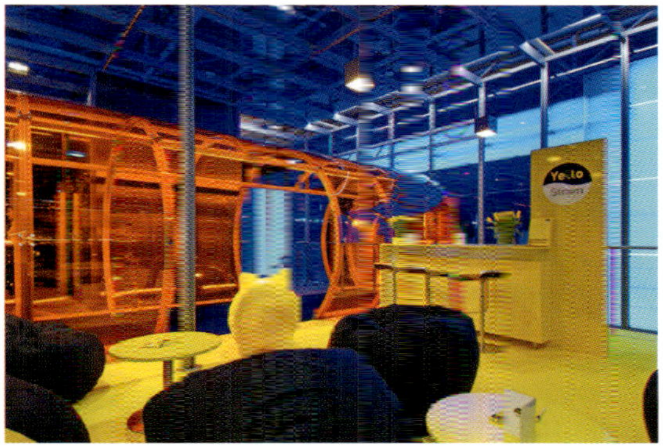

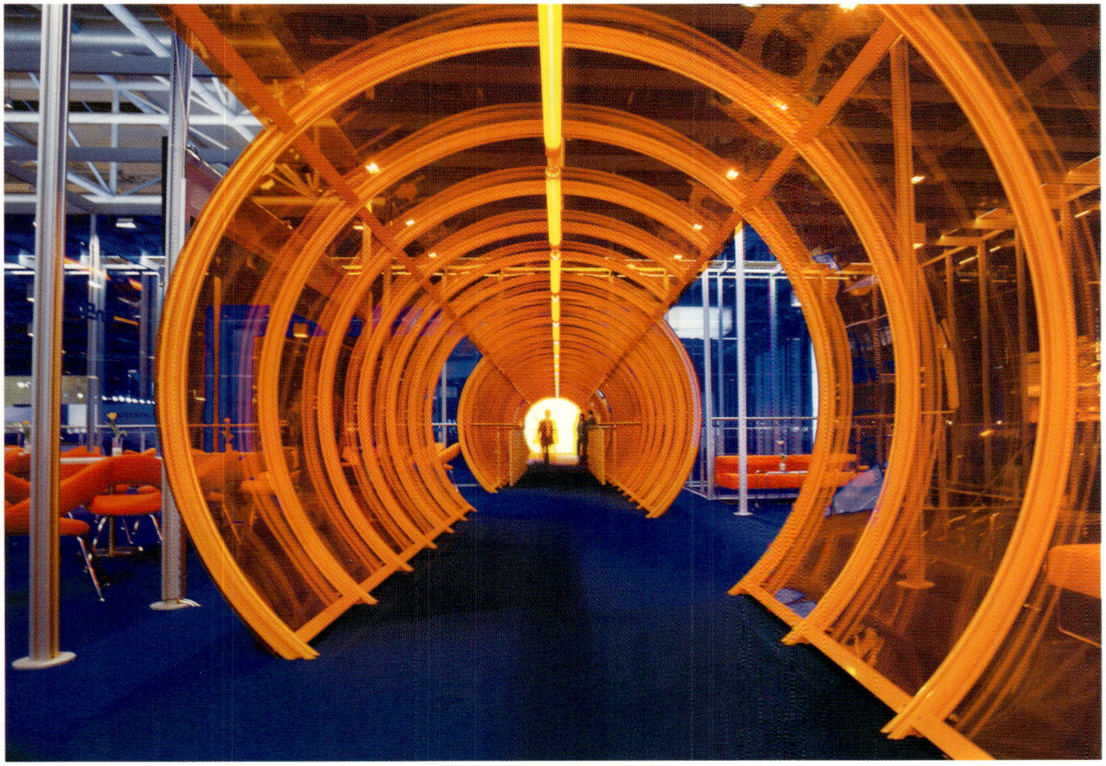

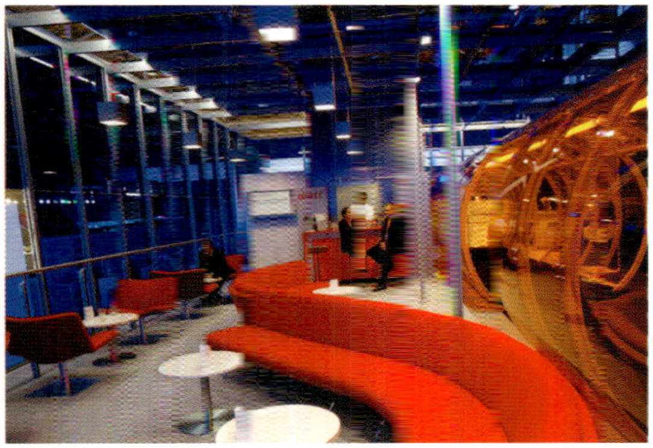

Auftraggeber Client
Bibliographisches Institut & F.A. Brockhaus AG
Mannheim, Germany

Projekt Project
Das Wissen der Welt
Markenraum, Presse- und Public-Event zur
Frankfurter Buchmesse
Brand space, press and public event
Frankfurt book fair

Design
Johannes Milla, Stefan Morgenstern, Martin Wagner,
Marion Kerckhoff
Milla und Partner
Stuttgart, Germany
Rainer Hellmann, Andreas Rell
BBDO Campaign GmbH Stuttgart
Stuttgart, Germany
Theo Bleckmann
EARrelevant Music
New York, USA
Marc Feigenspan
Ohrenkino.com
Stuttgart, Germany
WeimerSteinbergVega
Langenzenn, Germany

Installation von 30 dreieinhalb Meter großen Brockhaus-Bänden, die in einem weiten Kreis aufgestellt waren. Zu verstehen war dies als Hommage an die zeitlose Gültigkeit des Wissens in den Enzyklopädien von Brockhaus. Der Kreis konnte von allen Seiten betreten werden. Dazu drang eine permanente Klangperformance aus den verschiedenen Bänden ans Ohr des Besuchers: Das Wissen der Welt wurde hörbar. Eigens für das Enthüllungsevent wandelte sich der Buchkreis zur Bühne für eine Sprechoper, wobei die Künstler solo und im Chor singend durch die Installation schritten, bevor ihre Stimmen im Finale von oben aus den Büchern heraustönten.

The installation of 30 replicas of the Brockhaus Encyclopedia – each of them 3.40 m high – placed in a round formation was a tribute to the timeless validity of the knowledge in the encyclopedias of Brockhaus. The circle of enormous books could be entered from all sides. In addition, a permanent sound performance could be heard from each of the different volumes: the knowledge of the world became audible. For the unveiling of the installation, a phonetic opera for 30 + 1 voices was composed and performed, with the choir standing high above the audience in the Brockhaus-replicas, and the soloist in the middle of the circle.

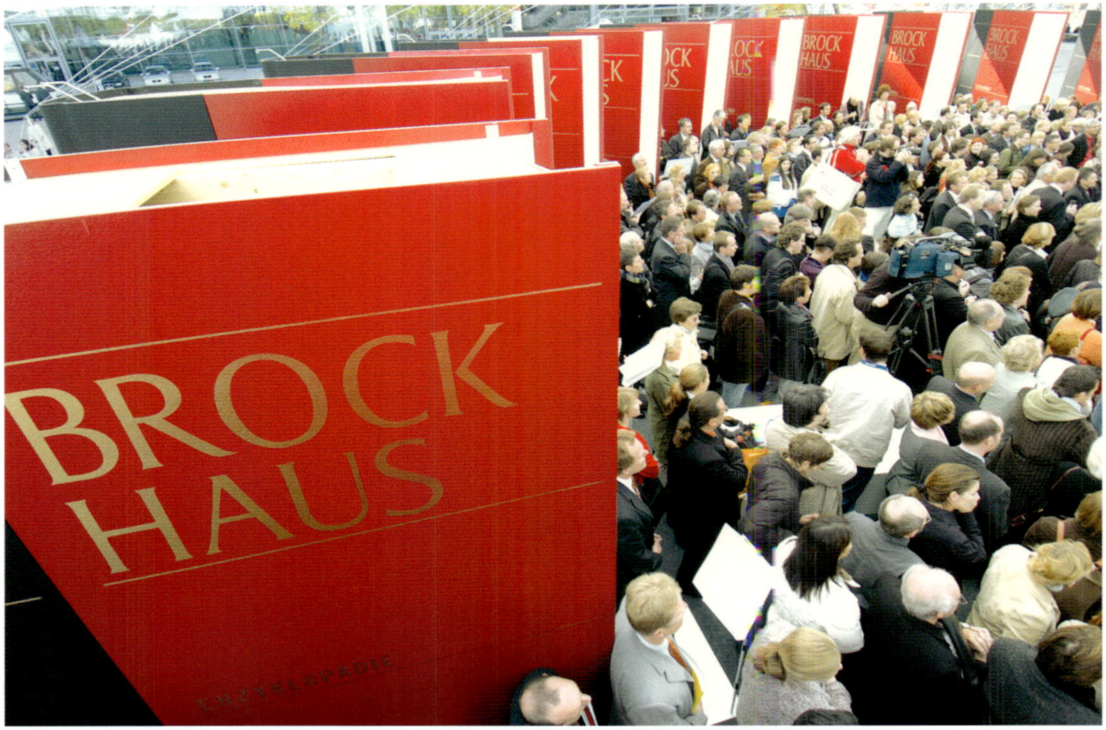

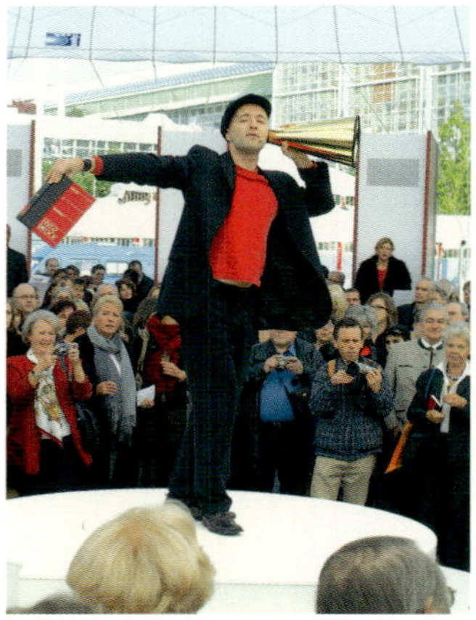

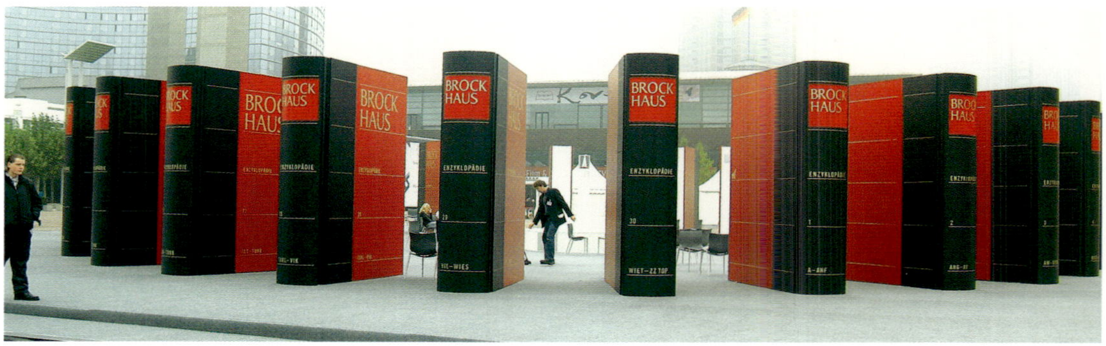

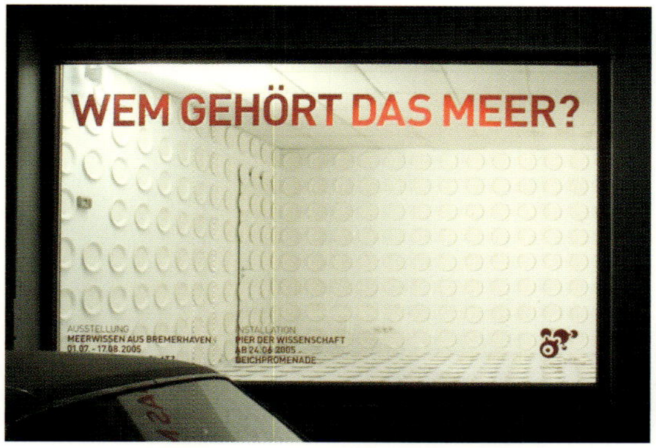

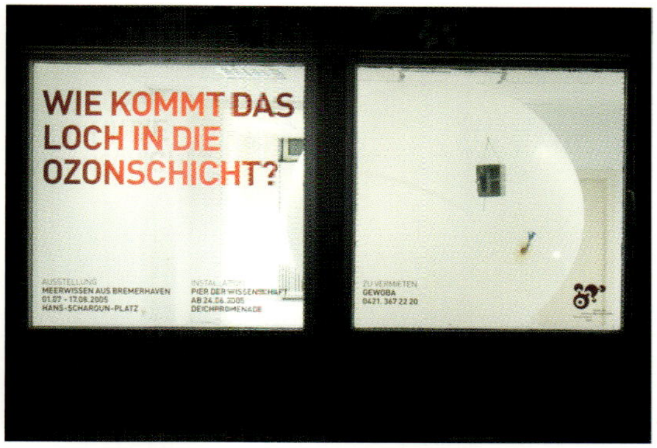

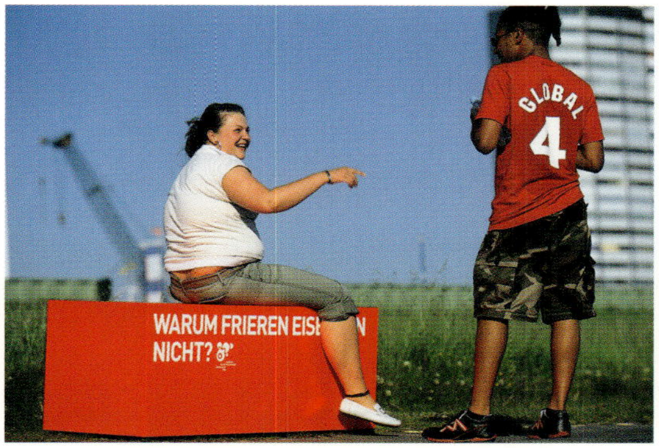

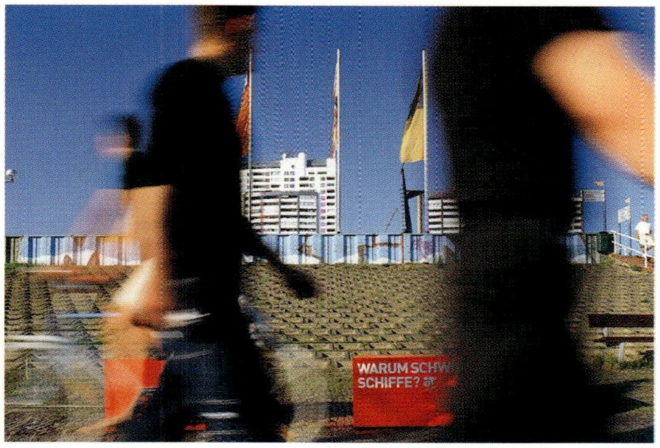

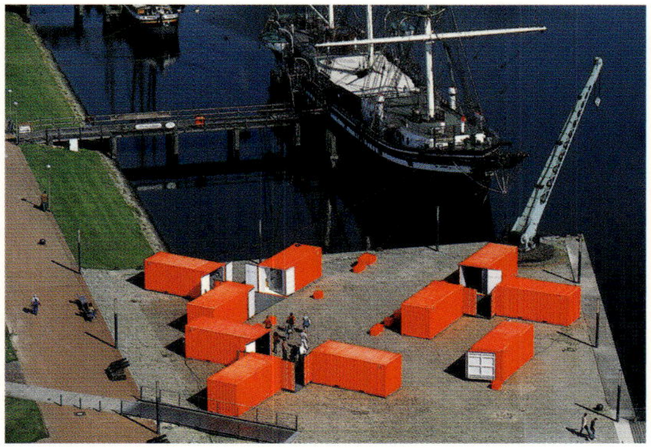

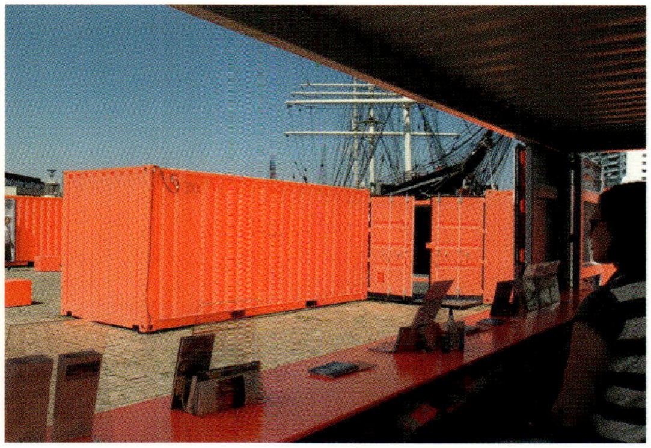

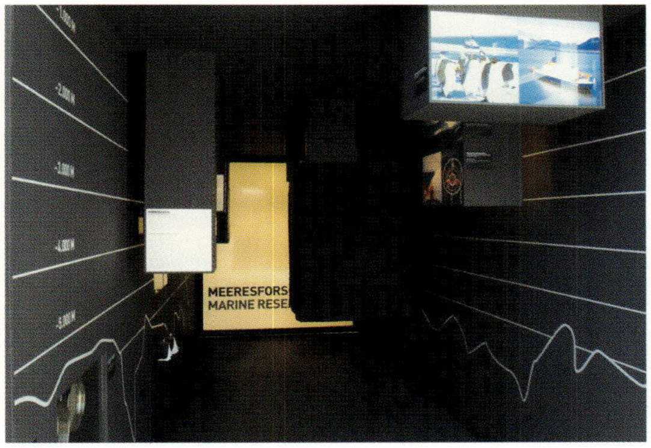

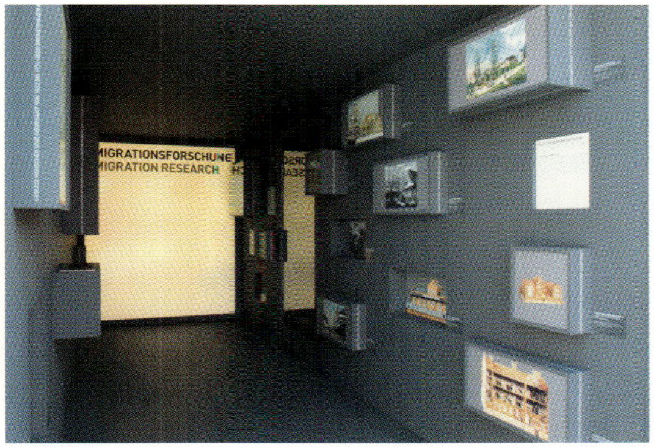

Auftraggeber Client
Magistrat Bremerhaven
Bremerhaven, Germany

Projekt Project
Pier der Wissenschaft 2005
Ausstellungen
Exhibitions

Design
Wilfried Kühn (Projektleitung)
Chris Rehberger (Projektleitung)
Tina Hage
Oki Kasajima
Verena Kern
Britta Knüppe
Nadja Riedel
Susana Soarez
Melanie Thron
Christoph Vaillant
Christina Kaufmann
Stipendium 2004/05
Designlabor Bremerhaven
Bremerhaven, Germany

Pier der Wissenschaft – der Ausstellungszyklus zur Stadt der Wissenschaft 2005 setzte auf variable Ausstellungsarchitektur, markantes Corporate Design und Kinderfragen. Hohe Präsenz und die sympathische Ansprache machten den Stadtraum zur Plattform für lebendige Wissenschaftsvermittlung. Auftaktkampagne, Informationspfad und Ausstellungen zeigten auf ungewöhnliche Weise, was in den Forschungseinrichtungen tagtäglich stattfindet. Verschiedene kreative Disziplinen wurden in einem Designkonzept zusammengeführt, um mit audiovisuellen Medien, Exponaten und Informationsgestaltung Wissenschaft für eine breite Öffentlichkeit zugänglich zu machen.

Science Pier – a series of exhibitions supporting the City of Science 2005 – made use of variable exhibition architecture, eye-catching corporate design and questions from children. Its high visibility and friendly way of talking to people turned the city into a platform for a lively conveyance of what science is all about. A launch campaign, information path and exhibitions revealed in an unconventional way what daily life is like at a research institute. Various creative disciplines flowed into a design concept comprising audiovisual media, exhibits and information design that aimed to make science accessible to the broad public.

Auftraggeber Client
Bertrandt AG
Ehningen, Germany

Projekt Project
Weitblick / Foresight
Messestand, IAA 2005 in Frankfurt am Main
Trade fair stand, International Motor Show (IAA) 2005 in Frankfurt am Main

Design
Ulrich Henzler
Carola Knoll
Ariane Raab
Totems I Communication & Architecture
Stuttgart, Germany

Die Bertrandt AG ist der Engineering-Partner für Fahrzeughersteller. Gebündelte Fach-Kompetenz – vom Serienprodukt bis zu den Zukunftsfeldern. Die Produkte sind streng geheim. Auf der Messe soll der Austausch mit Top-Entscheidern gefestigt, gefördert und erweitert werden.
Idee: Wir beweisen Weitblick. Entwicklung, Dynamik und Frische auf 160 m^2. Die Kommunikationsspuren werden nach unseren Plänen von Bertrandt selbst aus GFK gebaut. Exponate erzählen Geschichten. Die Bespielung teast an und reizt zur Vertiefung der Inhalte im persönlichen Gespräch. Die räumliche Wahrnehmung ist komfortabel, dienlich und erfrischend.

Bertrandt AG is the engineering partner for automobile manufacturers. Bundled specialist expertise – from standard products right through to future innovations. The products are highly secret. At the trade fair the interchange with top decision makers has to be firmly established, fostered and extended.
Idea: We demonstrate foresight. Development, dynamics and a fresh approach on 160 m^2. Communication tracks are built of GRP by Bertrandt themselves according to our plans. The complete presentation stimulates and encourages a deeper discussion of content in a personal conversation. The spatial perception is easy, expedient and refreshing.

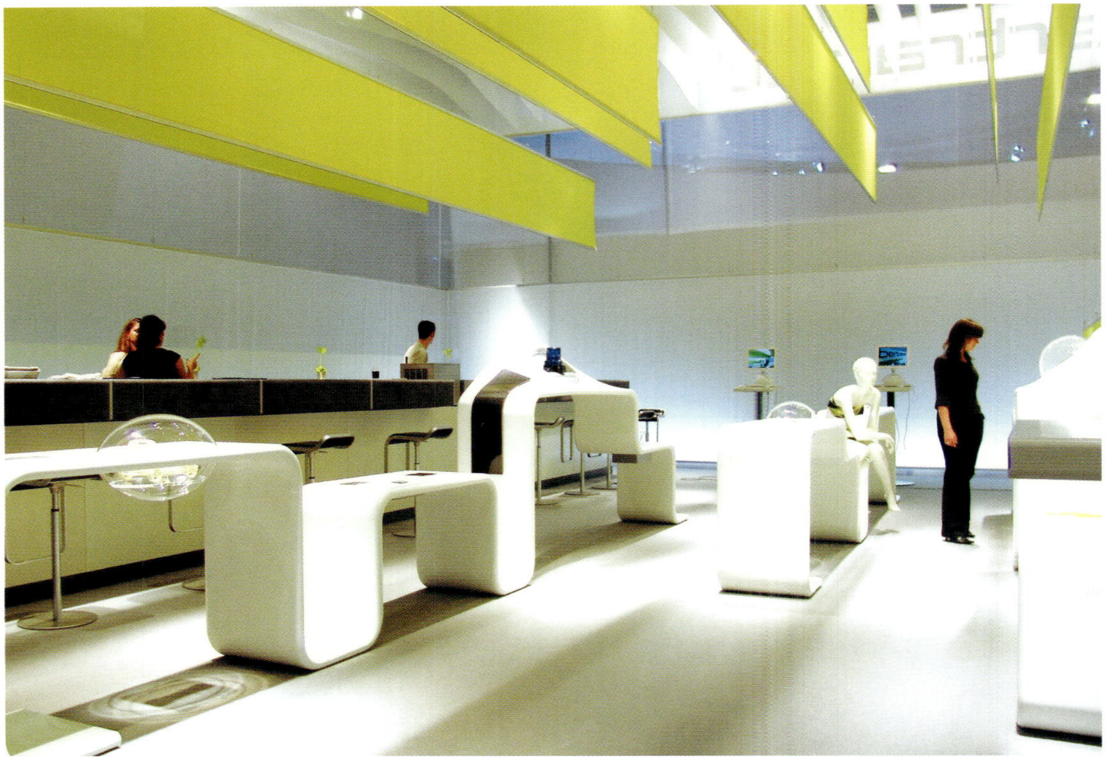

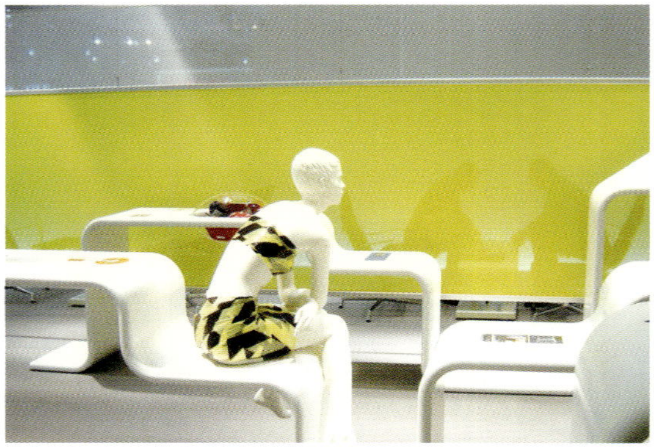

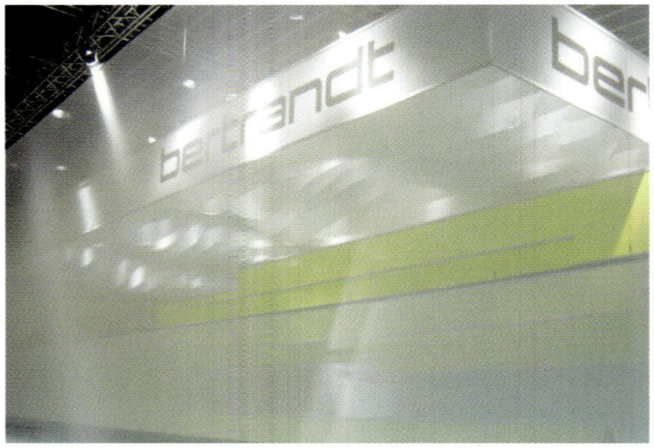

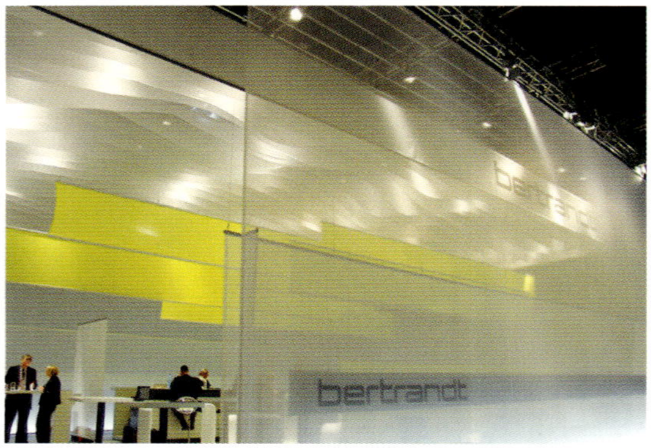

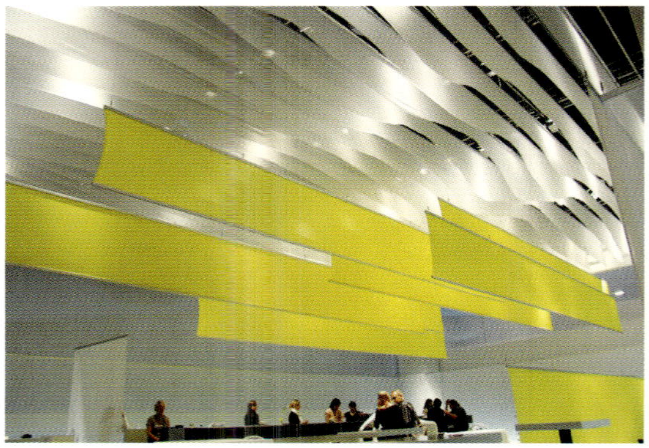

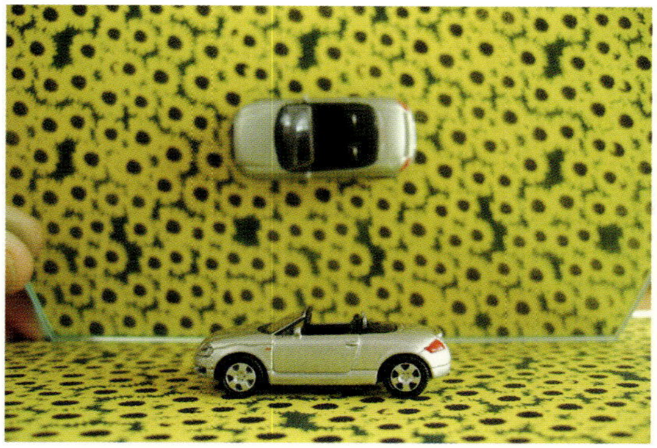

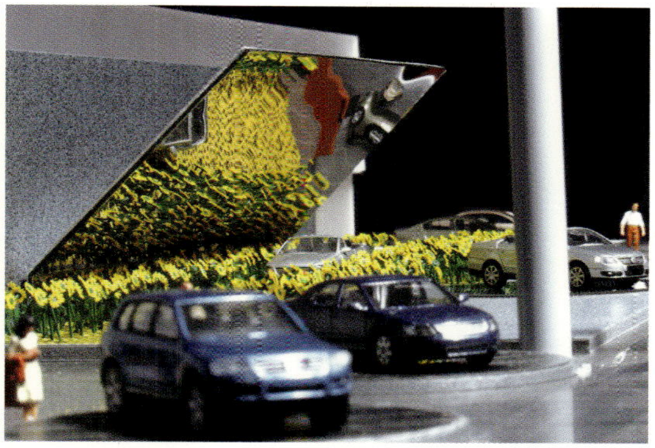

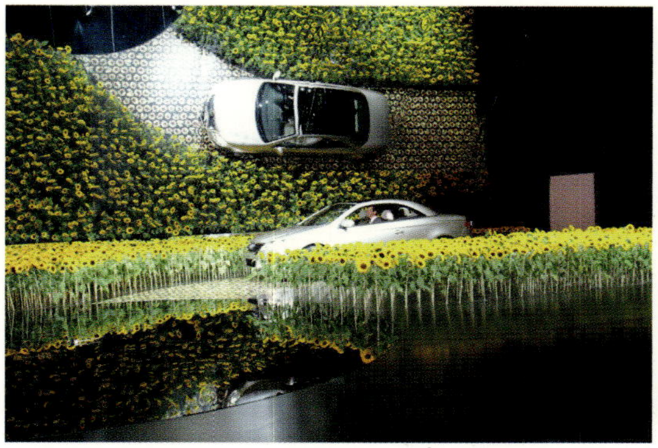

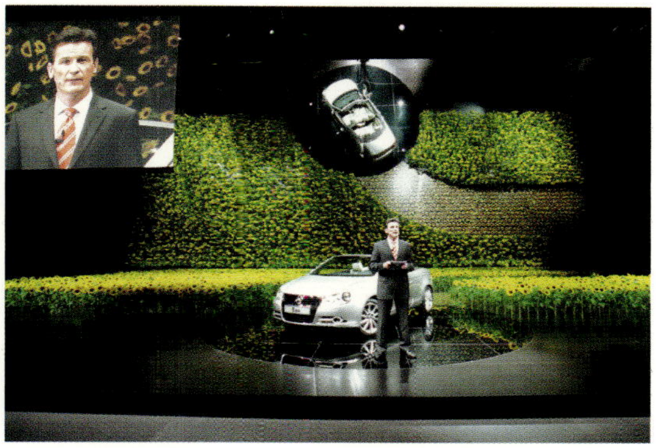

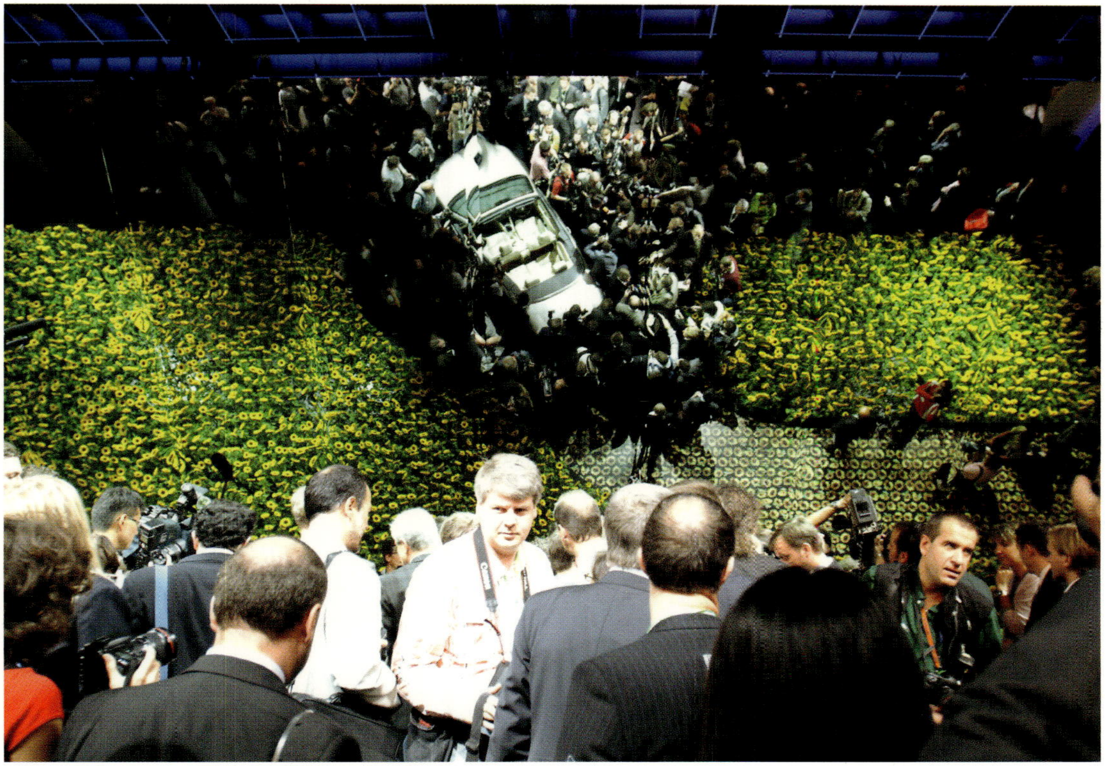

Auftraggeber Client
Volkswagen AG
Wolfsburg, Germany

Projekt Project
Sunflowers
Pressekonferenz
Press conference

Design
Susanne Gocht
Joachim Hauser
Hajo Rehm
Hauser Gocht Meyer Kommunikation und Design GmbH
Frankfurt am Main, Germany

Pressekonferenz von VW auf der IAA 2005 – erster großer Auftritt des Markenvorstandes Dr. Bernhard vor internationaler Presse und Weltpremiere des neuen Cabrios.
Ein reales Sonnenblumenfeld bildet das Key Visual. Um es weithin sichtbar zu machen, wird es in einem darüber befindlichen, um 45° geneigten Spiegel reflektiert. Ein einfaches, klares Bild für die Nähe zur Natur, Sommer, Cabriofahren. Zu Beginn ist das Bühnenbild verdeckt und filmisch bespielt. Zur Fahrzeugenthüllung hebt sich die mittlere Projektionsfläche und gibt das Sonnenblumenfeld frei. Ein überraschendes Bild, das die VW Markenwerte verkörpert: sympathisch, freundlich, nahbar.

Volkswagen Press Conference at the IAA 2005 – first big appearance of the board member Dr. Bernhard before the international press, and world premiere of the new Cabrio.
A real field of sunflowers as key visual. To be visible from afar it is reflected in a mirror, positioned above and tilted by 45°. A clear picture depicting closeness to nature, summer, driving a cabrio. In the beginning, the stage design is covered and used as screen for projections. The middle screen is lifted to unveil the car; the picture of the sunflowers is seen. A surprising picture that embodies the Volkswagen brand values – likeable, friendly and accessible.

Auftraggeber Client
o2 Germany GmbH & Co. OHG
München, Germany

Projekt Project
o2 Germany Messeauftritt CeBIT 2006
Messestand
Trade fair stand

Design
Michael Keller
Birgit Vogel
Christoph Rohrer
Wahan Mechitarian
Julia Romeiß
Andreas Koch
Melanie Sauer
KMS Team GmbH
München, Germany
Susanne Schmidhuber
Kerstin Arleth
Schmidhuber+Partner
München, Germany

Die visuelle Präsenz von o2 Germany auf der CeBIT 2006 wurde durch die konzeptionelle Weiterentwicklung des Messeauftritts gegenüber dem Vorjahr noch verstärkt. Erneut bildete die Deckenkonstruktion „Medienwolke" den Blickfang des Standes: ein über 1.000 m^2 großer Farbbildschirm aus rund 28.000 Pixeln, über den Videobilder und Textsequenzen liefen. Er schwebte in Form einer Wolke über den Köpfen der Besucher und bildete ein harmonisches Ensemble mit der Standlandschaft darunter.

The visual presence of o2 Germany at CeBIT 2006 was enhanced by a further development of the concept of the previous year's exhibit. Again, the eye catcher of the exhibition stand was the "media cloud" ceiling structure, a 1,000 m^2 color display composed of about 28,000 pixels, showing video images and textual sequences. The cloud hovered above the heads of visitors, forming a harmonic ensemble together with the landscape of the exhibition stand below.

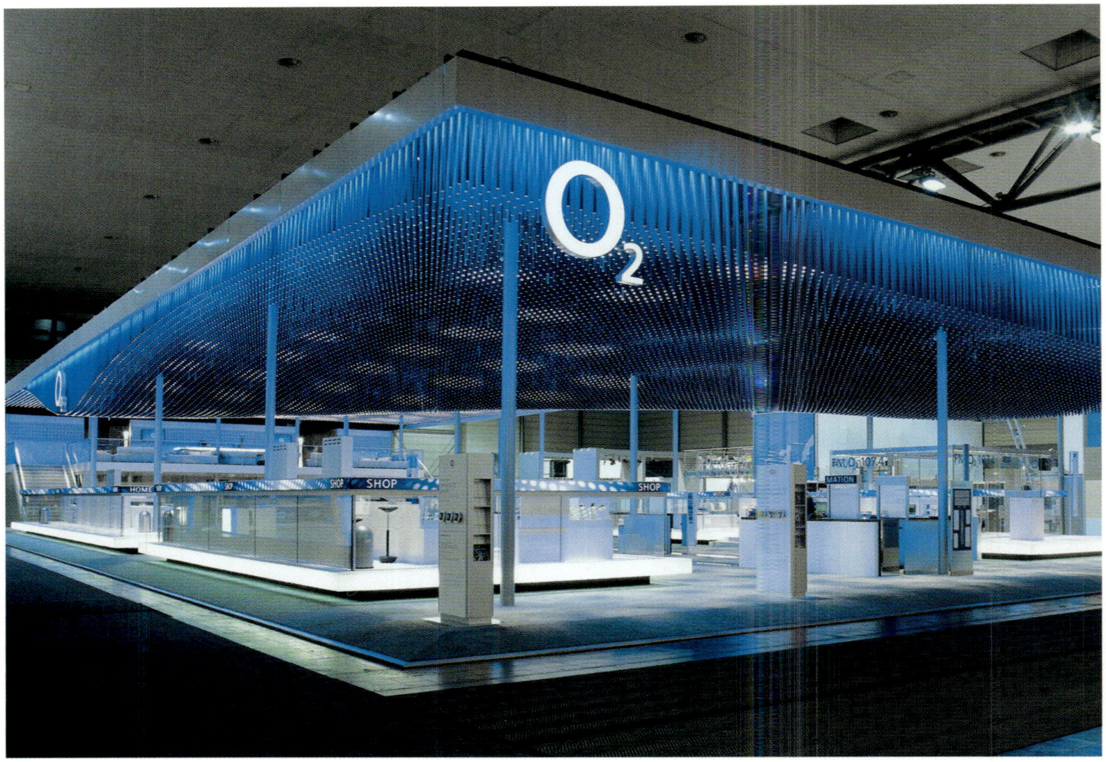

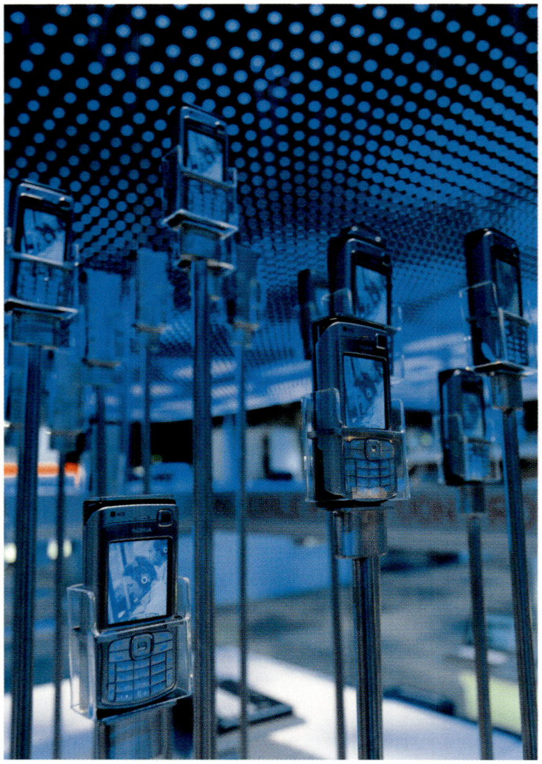

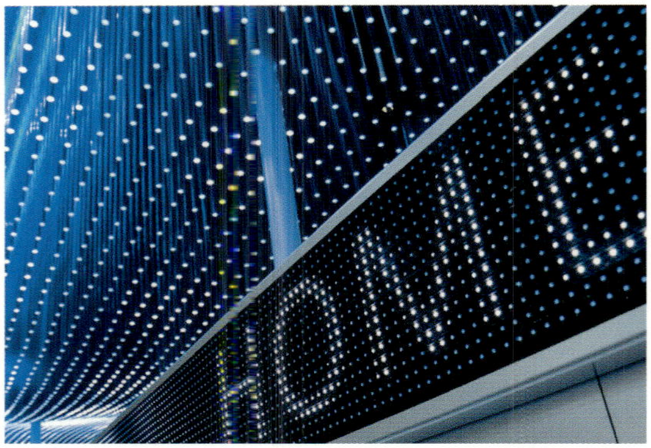

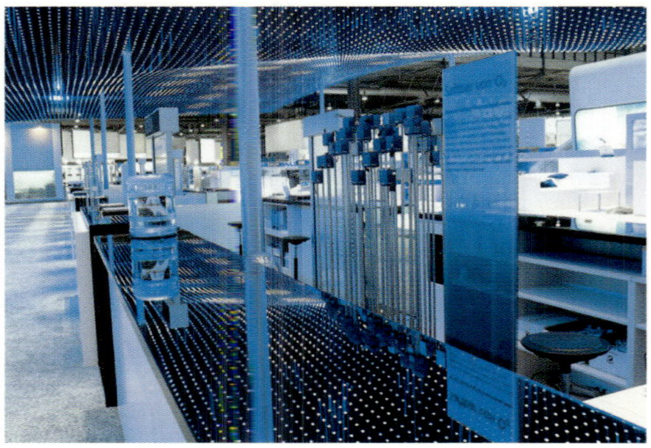

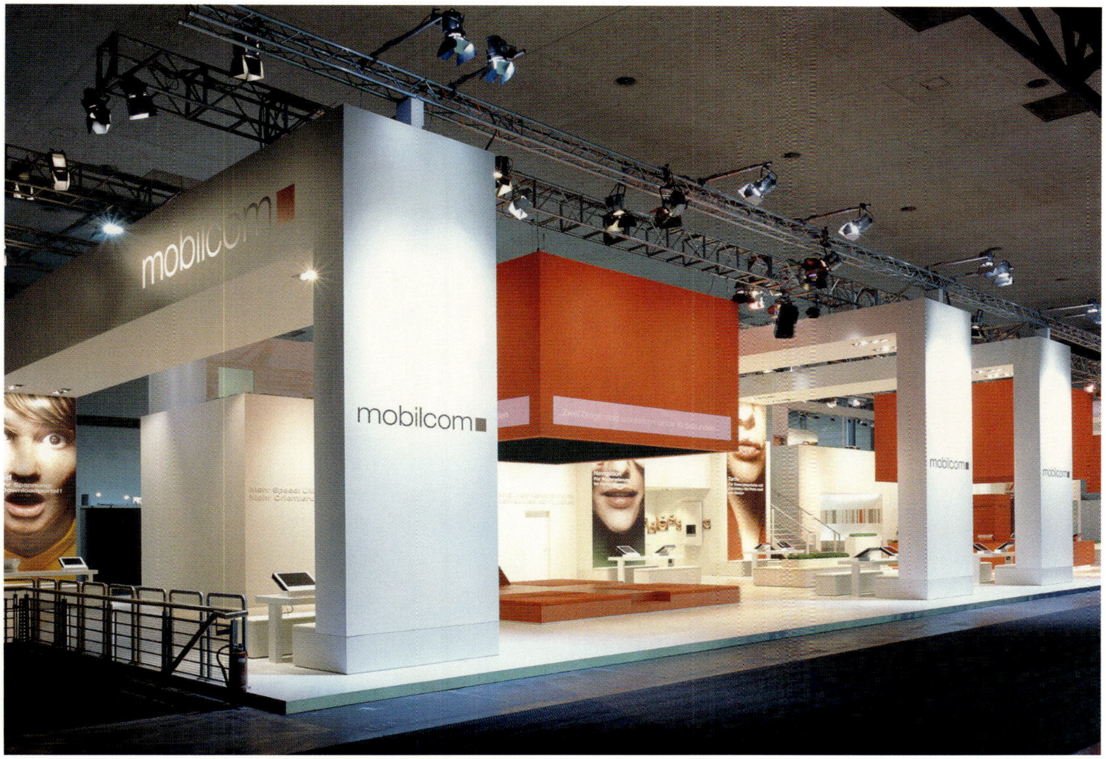

Auftraggeber Client
mobilcom Communicationstechnik GmbH
Büdelsdorf, Germany

Projekt Project
mobilcom | CeBIT 2005
Messestand
Trade fair stand

Design
Triad Berlin Projektgesellschaft mbH
Berlin, Germany

Triad Berlin gestaltete den Messeauftritt von mobilcom auf der CeBIT 2005 als intuitiven Erlebnisraum. Die klare Designsprache trägt nicht nur der CI von mobilcom Rechnung. Sie transportiert auch das Messethema, das transparente und übersichtliche Tarifsystem des Mobilfunkproviders zu kommunizieren. Der Kunde erschließt sich die CI und das Messethema durch die offene und klare Architektur bereits bei der Annäherung. Auf dem Stand selbst wird dieser Eindruck durch großformatige Bilder mit integrierten Textbotschaften verstärkt. Eine übersichtliche Anordnung der Produktpräsentations- und Beratungsbereiche rundet das Gesamtkonzept ab.

Triad Berlin designed mobilcom's appearance at the CeBIT 2005 as an intuitive realm of experience. The clear design of the booth not only does justice to the mobilcom CI, it also conveys the focus topic of the convention by communicating the wireless carrier's clear and transparent system of rates. The open, clear architecture transmits the CI and the convention topic to customers even as they approach the stand. At the booth itself, this impression is reinforced with large-scale pictures featuring integrated text messages. An uncluttered array of product presentation and consultation areas rounds out the overall concept.

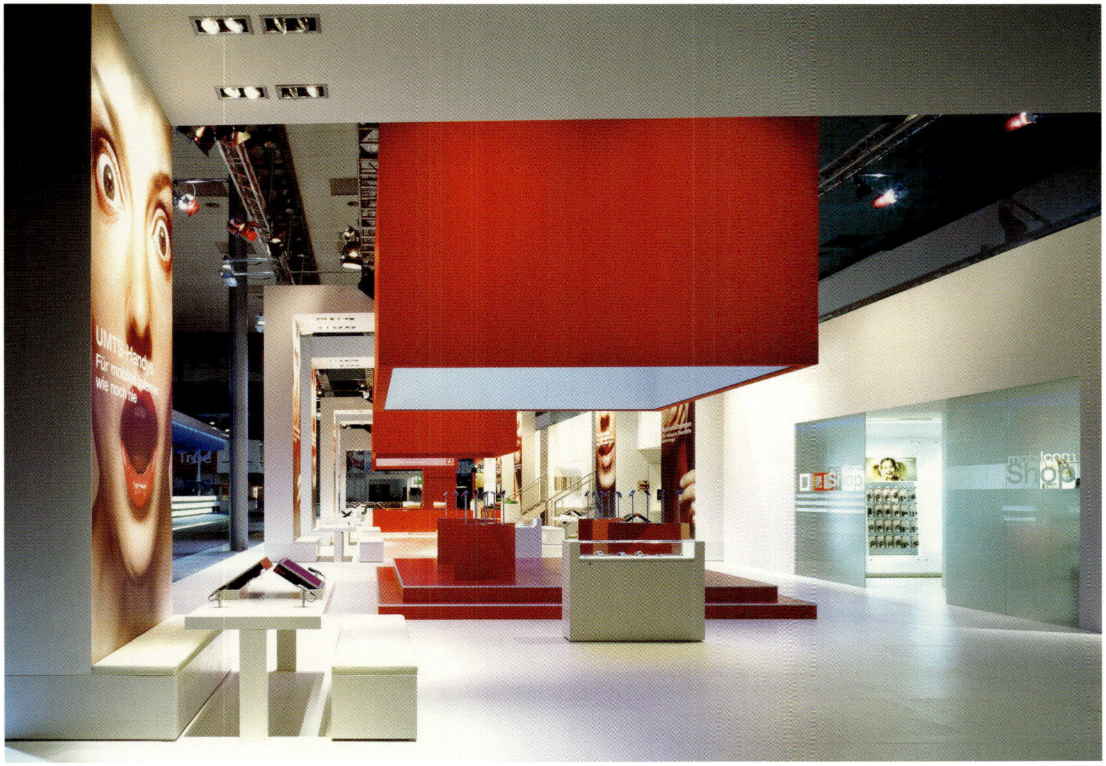

Auftraggeber Client
DaimlerChrysler AG
Stuttgart, Germany

Projekt Project
Voraus denken. Mercedes-Benz auf der IAA Frankfurt 2005.
Messestand
Trade fair stand

Design
Atelier Markgraph GmbH (Kommunikation)
Frankfurt am Main, Germany
Kauffmann Theilig & Partner Freie Architekten BDA (Architektur)
Ostfildern, Germany
design hoch drei GmbH & Co.KG (Grafikdesign)
Stuttgart, Germany
TLD Planungsgruppe GmbH (Lichtdesign)
Wendlingen, Germany

7.000 Quadratmeter Ausstellung, 900.000 Besucher, rund 45 Minuten Besuchsdauer – und eine mit Spannung erwartete Weltpremiere. Der Auftritt von Mercedes-Benz auf der IAA 2005 verlangt eine klare Dramaturgie. 15 Minuten Flanieren durch die vielfältige Produktpalette von Mercedes-Benz, 30 Minuten Erleben der neuen S-Klasse: Der dramaturgische Ansatz legt den Fokus auf ein inszeniertes „Rendezvous" der Besucher mit dem Flaggschiff von Mercedes-Benz. Am Beispiel der Luxuslimousine, die alle Markenwerte in Reinform in sich vereint, werden Innovationskraft, Leidenschaft und Qualität von Mercedes-Benz konzentriert kommuniziert.

7,000 square meters of floor space, 900,000 visitors, each of whom spent an average 45 minutes on the booth, a long-awaited world premiere – the Mercedes-Benz presentation at the IAA 2005 demanded a clear dramatic structure. Fifteen minutes to look around, thirty for the S-Klasse – in a meticulously planned rendezvous with the brand flagship, visitors experience the innovative power, passion and quality of Mercedes-Benz in their most concentrated form.

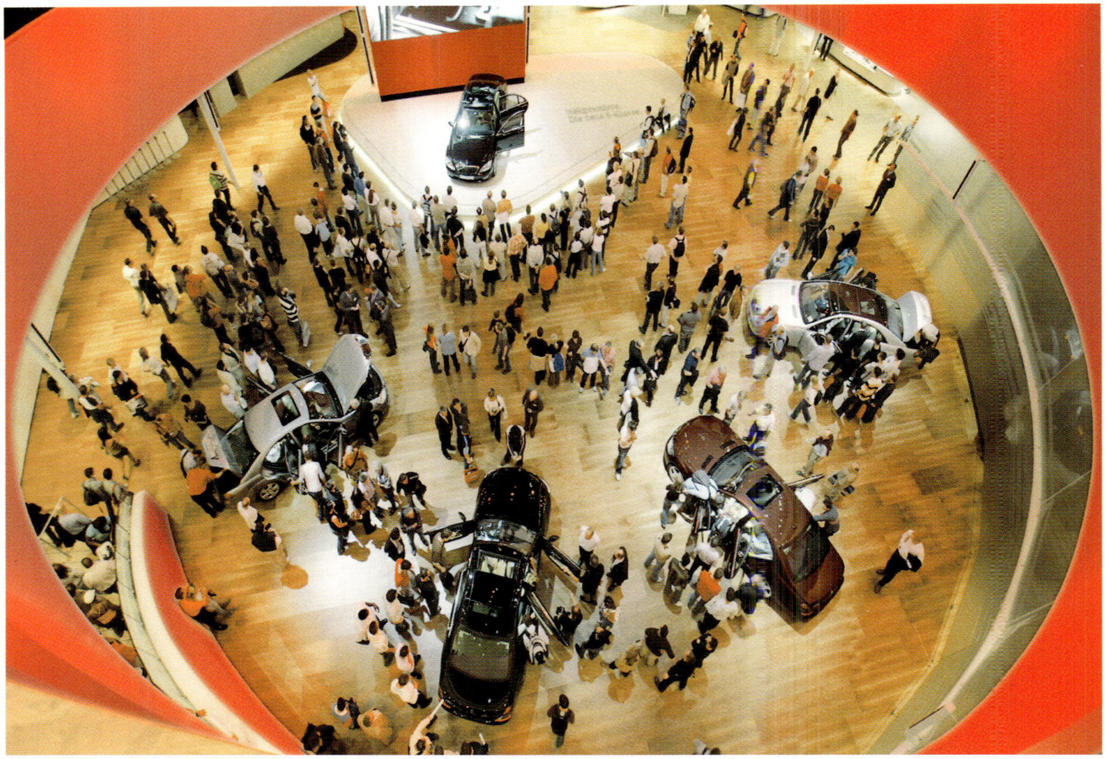

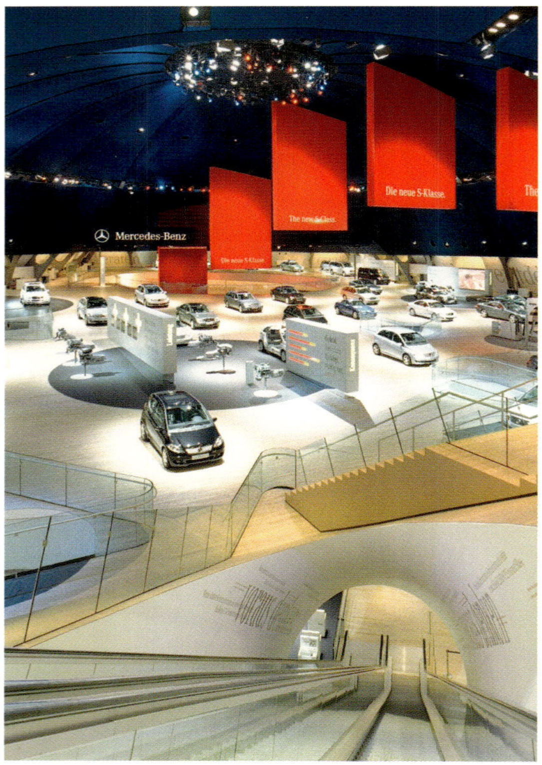

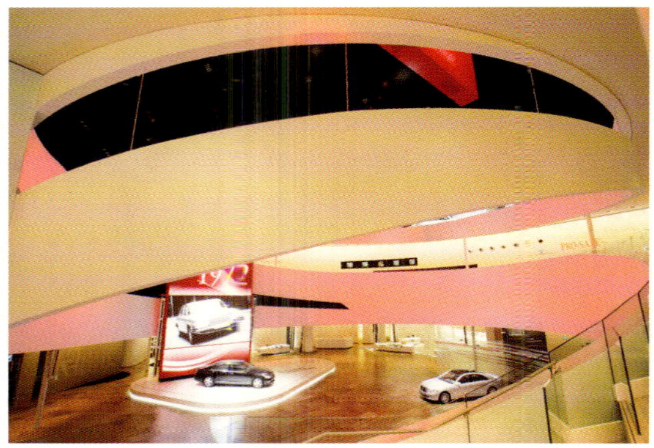

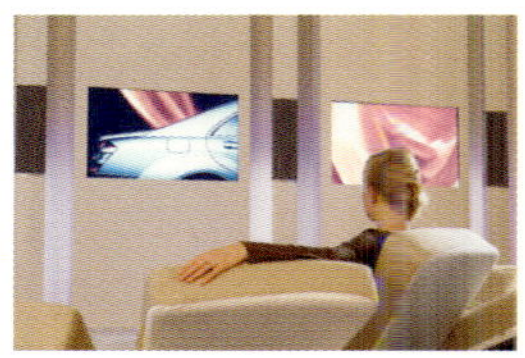

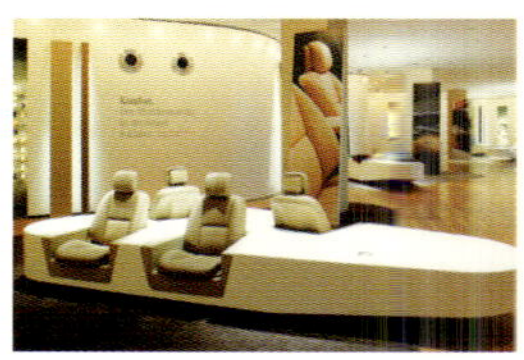

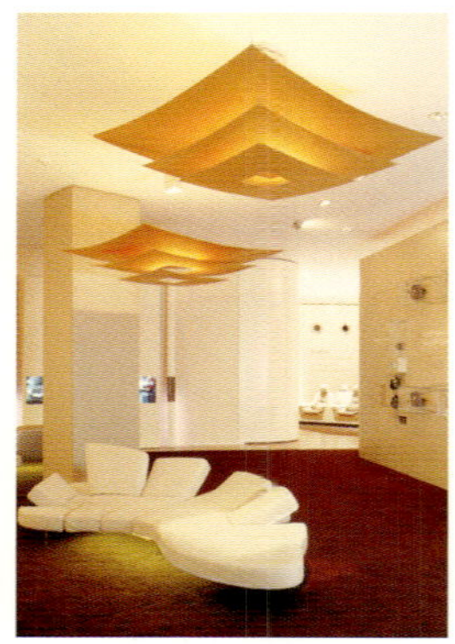

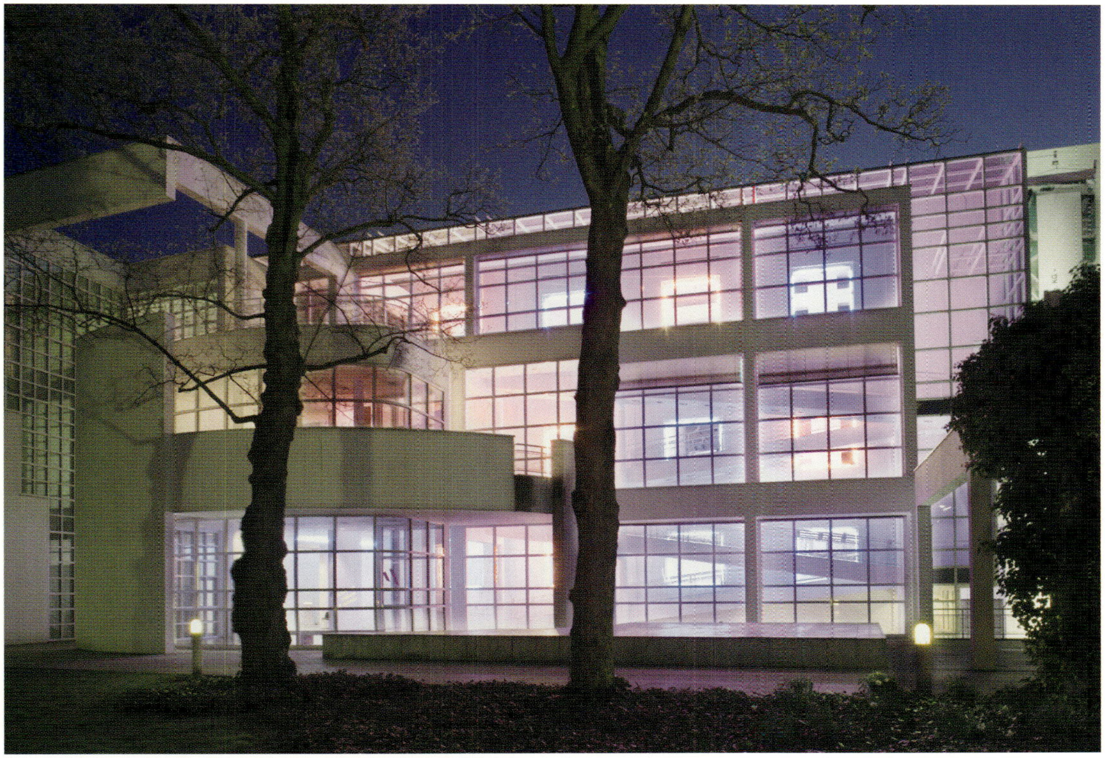

Auftraggeber Client
Merten GmbH & Co. KG
Wiehl, Germany

Projekt Project
Switch – Intuition im Raum
Ausstellung anlässlich der Luminale 2006
Exhibition on the occasion of the Luminale 2006

Design
Marc Krusin
Milan, Italy
Stylepark AG
Frankfurt am Main, Germany

Die von Stylepark kuratierte Ausstellung dokumentierte den Findungsprozess einer neuen Designsprache für die Firma Merten. Ob Schalter, Sicherheitsprogramme, Gebäudesystemtechnik in öffentlichen Gebäuden, Hotels oder Gewerbe: Technologische Herausforderungen benötigen Visionen für die Bedienung von morgen. Bekannte Vertreter aus der Architektur- und Designszene haben für Merten neu und zukunftsweisend gedacht. Unter der Ausstellungsregie des Mailänder Designers Marc Krusin waren im Museum für Angewandte Kunst Frankfurt u.a. Konzeptionen von Konstantin Grcic, Jürgen Mayer H., Antonia Henschel, Reed Kram und Max Ratjen zu sehen.

The exhibition curated by Stylepark documented the search of Merten's new design language. Whether switches, safety programs, building system technology in public buildings, hotels or trades: Technological challenges need visions for tomorrow. Well-known representatives from the fields of architecture and design thought anew and found trend-setting ideas for Merten. Under the exhibition direction of the Milan designer Marc Krusin the Museum of Applied Art in Frankfurt showed conceptions of Konstantin Grcic, Jürgen Mayer H., Antonia Henschel, Reed Kram and Max Ratjen.

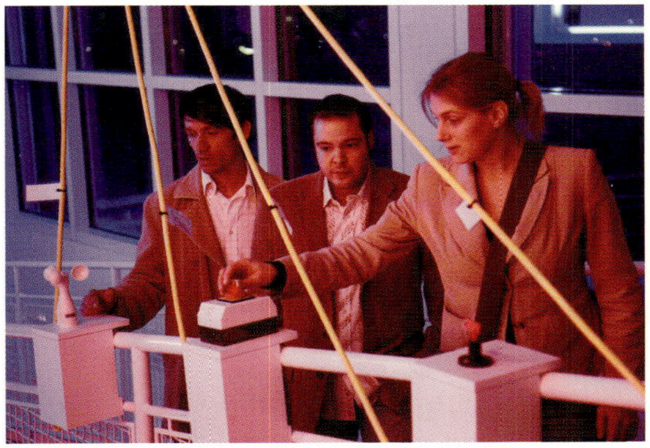

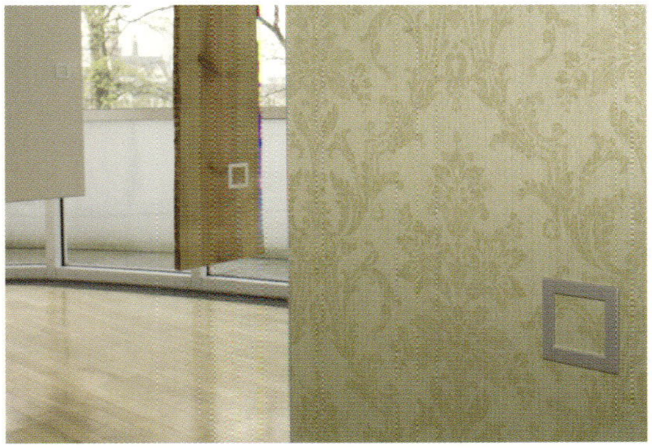

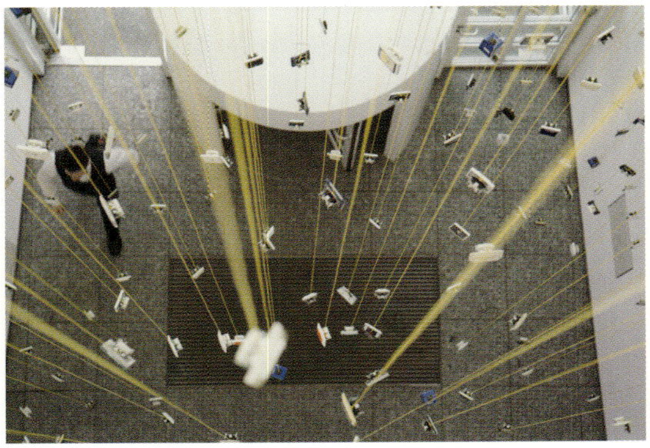

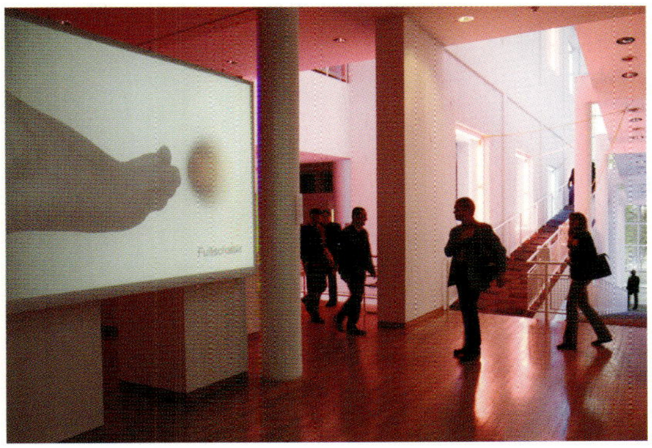

Auftraggeber Client
BMW AG
München, Germany

Projekt Project
BMW/MINI Messeauftritt Tokyo Motor Show 2005
Messestand
Trade fair stand

Design
Bernhard Franken
Frank Brammer
Thomas Remdisch
Berthold Scharrer
Fang Tian
Christoph Trenner
Franken Architekten GmbH
Frankfurt am Main, Germany
Interbrand, Zintzmeyer & Lux
Zurich, Switzerland

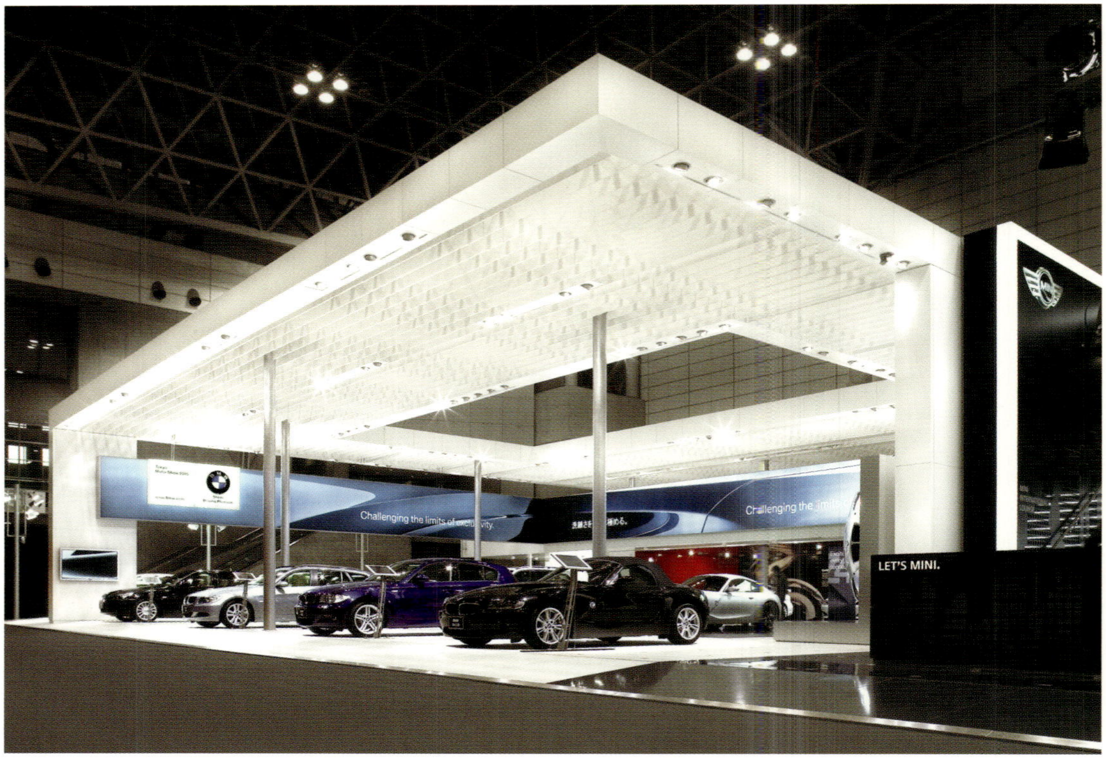

Die Präsentation der beiden Marken erfolgt Seite an Seite auf einer Ausstellungsfläche.
Das Prinzip der markentypischen Architektur von BMW basiert auf der Schichtung von raumbildenden Scheiben, in denen durch das Besuchererlebnis ein fließender Raum entsteht. Das Leitmotiv „Reflektionen" unterstreicht den technologisch innovativen Anspruch, die kultivierte Eleganz, Präzision und Dynamik der Fahrzeuge.
Der Messeauftritt von MINI setzt sich durch seine schwarze kubische Architektur deutlich von BMW ab. Farbige Frames rahmen Clubräume ein, in denen die MINI Specials „Seven", „Checkmate" und „Park Lane" in eigenen Raumwelten inszeniert werden.

The two brands are being presented side by side on one single exhibition stand.
The BMW brand architecture is based on the layering of slabs that create a floating space for the visitor. The graphical leitmotif "reflections" emphasizes the innovative technological claim, the cultivated elegance, precision and dynamics of the vehicles.
The MINI presentation contrasts clearly with BMW through its cubic black architecture. Coloured "Frames" enclose "Clubs", in which the MINI Specials "Seven", "Checkmate" and "Park Lane" are put on stage in specific thematic environments.

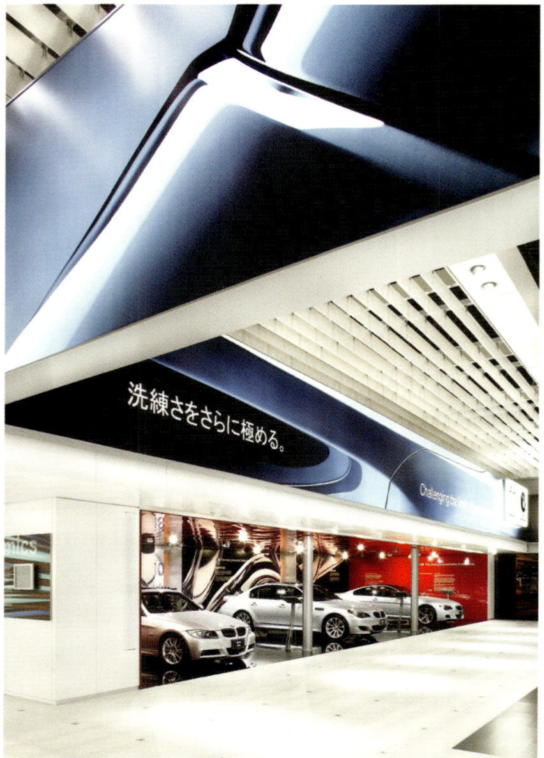

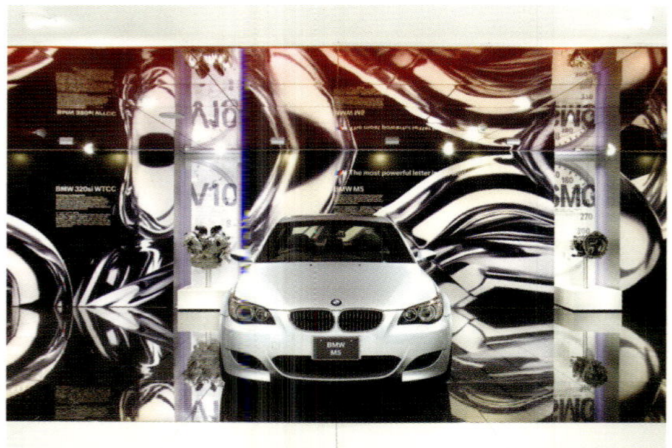

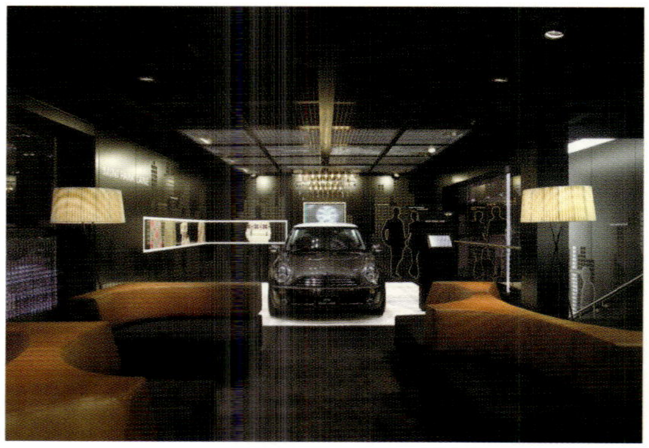

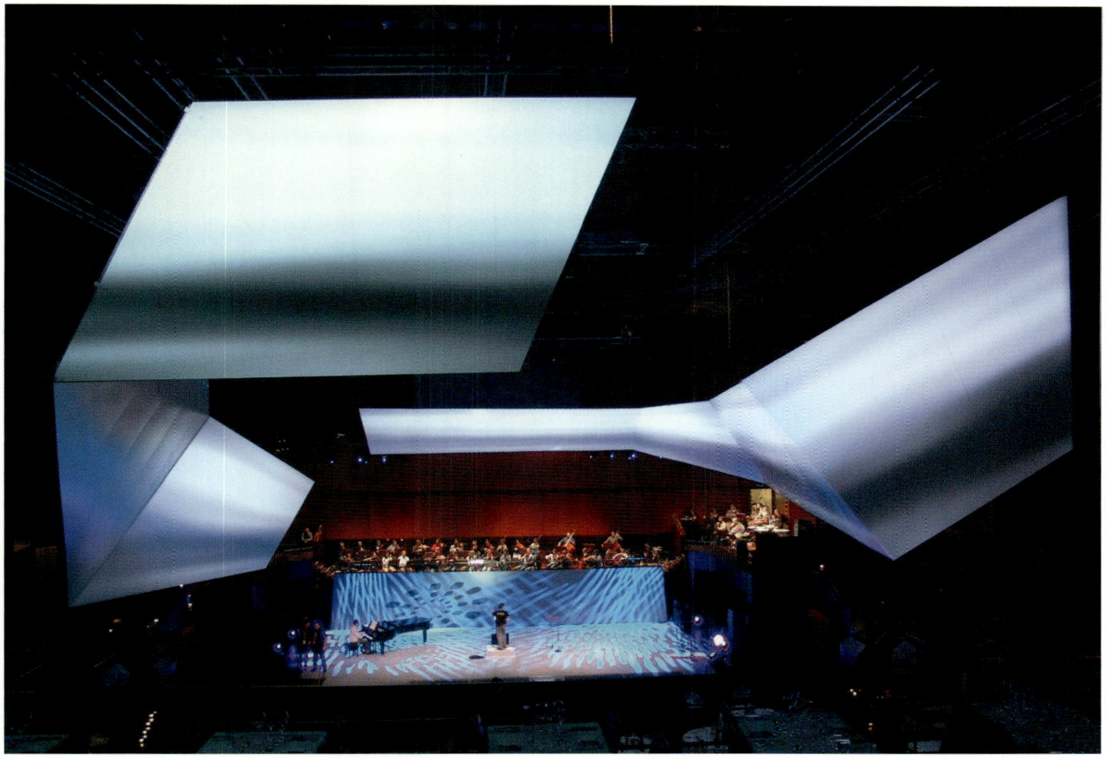

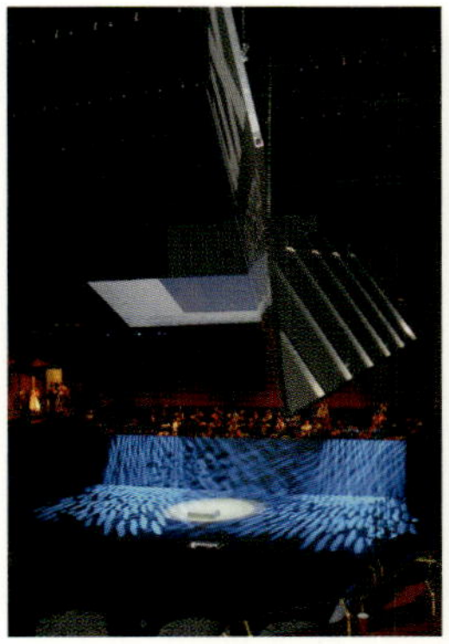

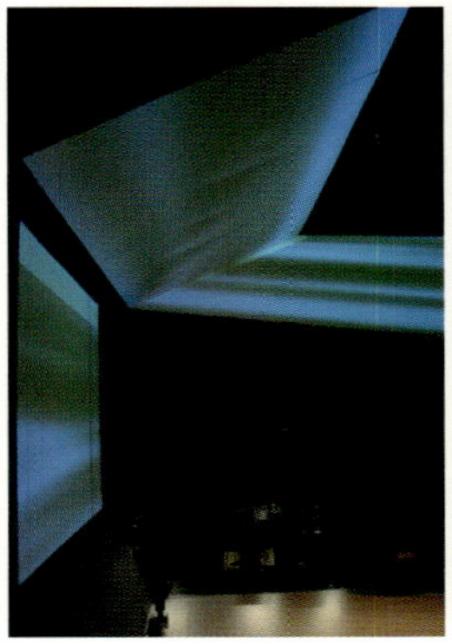

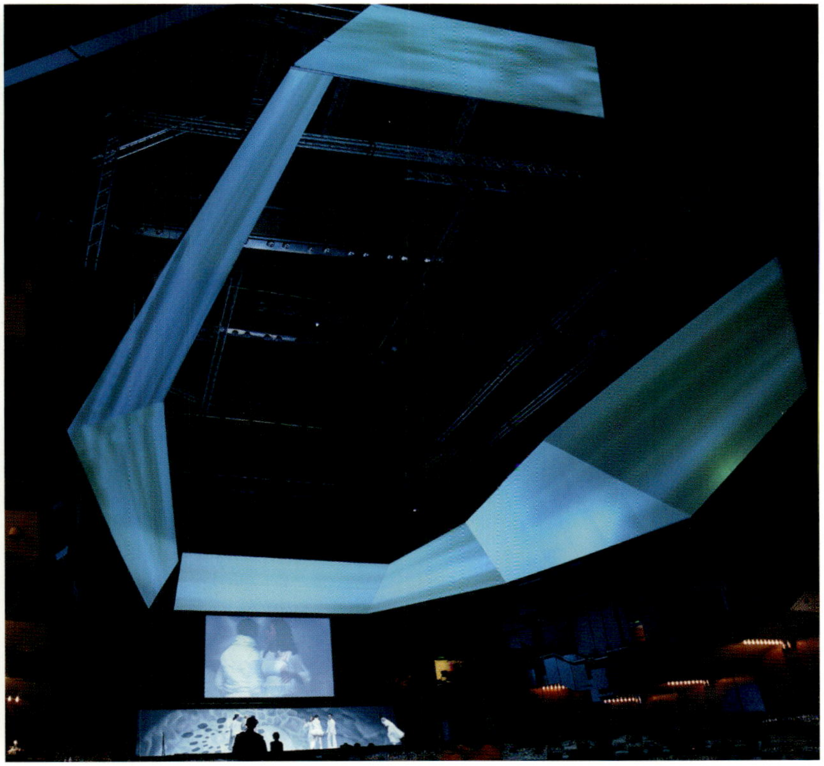

Auftraggeber Client
Quasar Communication GmbH
Wiesbaden, Germany

Projekt Project
50 Jahre dit – Allianz Dresdner Global Investors – Jubiläum
Bühnenarchitektur
Stage design

Design
dreizueins architektur designgrafik
Frankfurt am Main, Germany

Im Konzertsaal der Alten Oper Frankfurt rückt eine architektonische Inszenierung unter dem Motto „Aufbruch Wachstum Zukunft" den Gast in den Mittelpunkt der Jubiläumsfeier zum 50-jährigen Bestehen des Finanzunternehmens dit.
Ausgehend von der Bühne bilden medial bespielte Bänder eine dreidimensionale Skulptur. Sie löst die Distanz zur Bühne auf, der Gast wird Teil des Geschehens, das Bühnenbild entfaltet sich im gesamten Raum.
Auf den Projektionsbändern, deren Form einem Börsendiagramm entspringt, zeichnen sich ein halbes Jahrhundert Erfolgsgeschichte, vor allem aber die Zukunftsfähigkeit und Innovationskraft des Unternehmens raumgreifend medial ab.

In the concert hall of the Alte Oper Frankfurt an architectonic staging, themed with the headline "departure growth future", made the guest be the heart of the 50th anniversary of the financial business dit.
Starting from the stage, ribbons – serving as media projection surfaces – form a three-dimensional sculpture. The distance to the stage dissolves, the guest becomes part of the game – the stage setting unfolds into the entire space.
Stock exchange diagrams give the form for the ribbons onto which half a century's story of success is projected, most notably the sustainability and innovative strength of the company.

Auftraggeber Client
Messe Frankfurt Venue GmbH&Co KG
Frankfurt am Main, Germany

Projekt Project
Café City
Innenarchitektur / Gastronomie
Interior design / Catering

Design
DasProjekt GmbH
Hofheim am Taunus, Germany

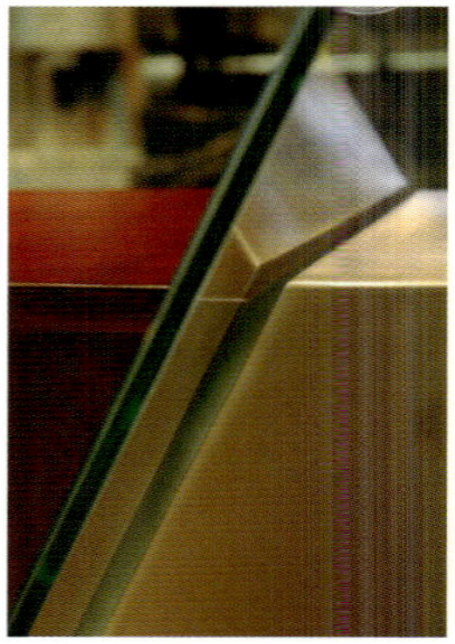

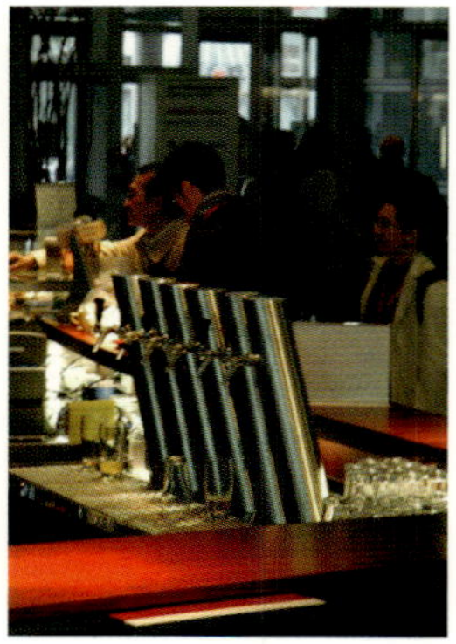

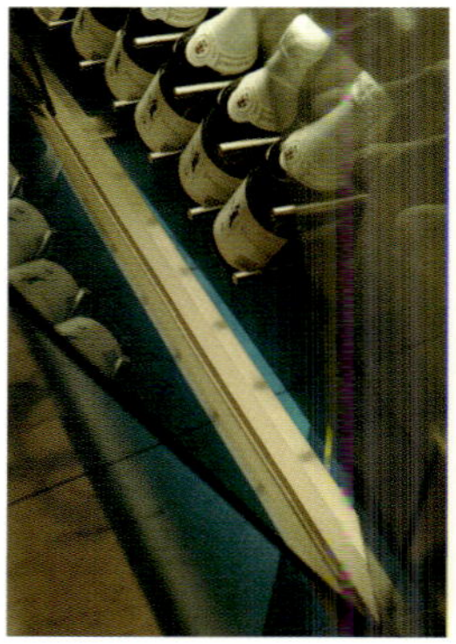

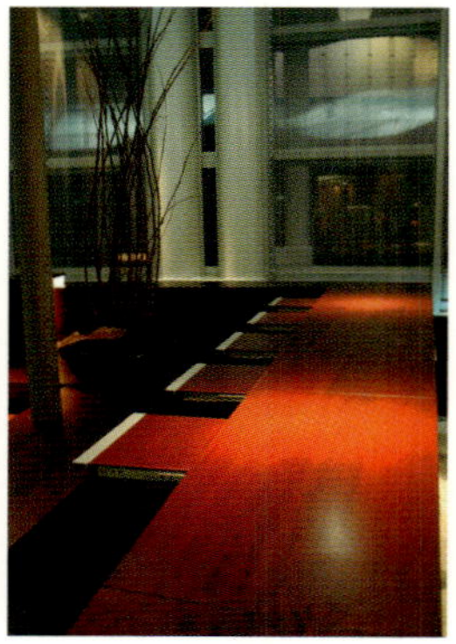

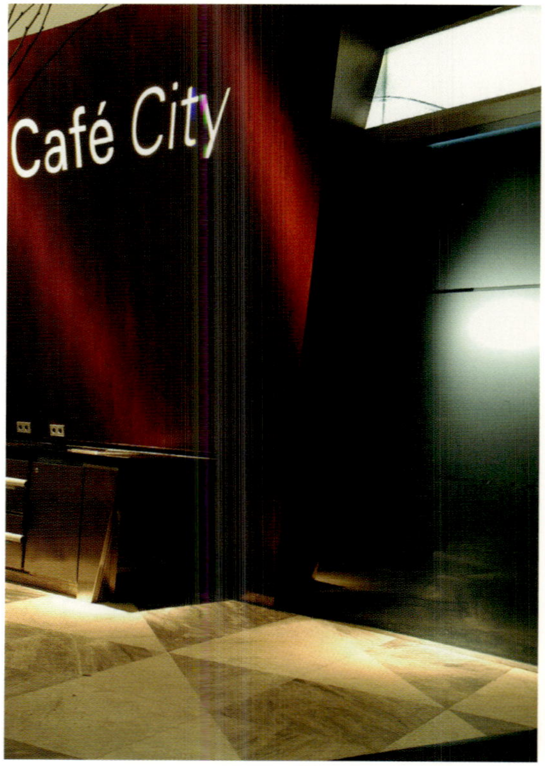

Eingang City_Messe Frankfurt. In dem Rundbau aus den 80er Jahren erscheint das Café City in einem „neuen Licht". Ziel war ein Corporate Design, das Bezug nimmt auf die Messe Frankfurt und gleichzeitig ihrem Anspruch auf Innovation und Modernität gerecht wird.
Die Bewegungsströme aufnehmend neigt sich das neue Café dem Messebesucher einladend entgegen. Die bewegte Form bildet den Kontrapunkt zu der strengen Gestaltung und der zweckmäßigen Atmosphäre im Eingangsfoyer. Sämtliche Holzeinbauten folgen der vorgegebenen Bewegung. Sie neigen sich 14° zum Raummittelpunkt und 14° in die Bewegungsrichtung.
Material- und Farbkonzept sind inspiriert von der klassischen Farbgebung früherer Kaffeehäuser. Rotbraunes Holz (Kirsche) dominiert die Einbauten. Es unterstreicht mit seiner starken Maserung den Charakter des monolithischen Baukörpers. Sitzpodest, Theke und Wand wirken wie aus einem Holz geschnitten.
Am Abend wird das neue Café zur Bar. Die stimmungsvolle Illumination lockt Besucher, sich vor oder nach einem Event in entspannter Atmosphäre niederzulassen. Eine Leuchtblende mit Bildern architektonischer Highlights der Messe verbindet das Café thematisch mit dem Ort.

City Entrance_Messe Frankfurt. Housed in a round building dating from the eighties, Café City now appears in a whole "new look". The objective was to create a corporate design that makes reference to Messe Frankfurt while living up to its image of innovation and modernity.
Taking up the theme of movement and flow, the new café leans invitingly toward the influx of trade fair visitors. The curving, fluid forms create a welcome counterpoint to the severe lines and functional atmosphere of the entrance foyer. All wooden fixtures follow a prescribed line of movement. They lean at a 14° angle toward the room's centre and 14° in the direction of movement.
Materials and colour concept were inspired by classic coffeehouse interiors. The reserved, abstract design concentrates on just a few materials and a clear language of form. Reddish brown cherry wood is the predominant material used for the interior fixtures. Its prominent grain underscores the monolithic character of the structural components. Seating platform, counter and wall seem to be carved out of a single piece of wood.
In the evening, the new café becomes a bar. Lighted screens encircling the space show highlights of the trade show architecture, connecting the café thematically with its location.

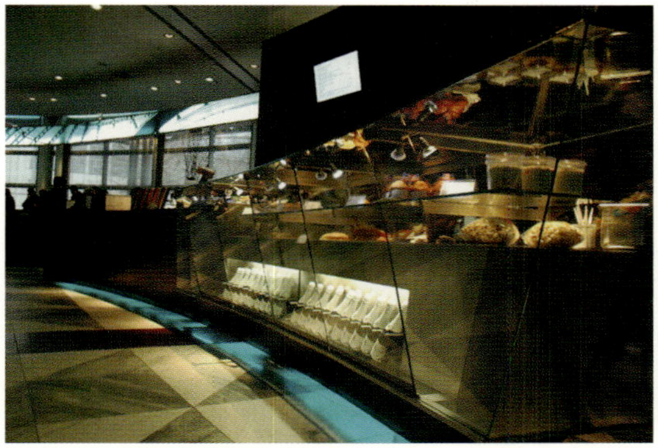

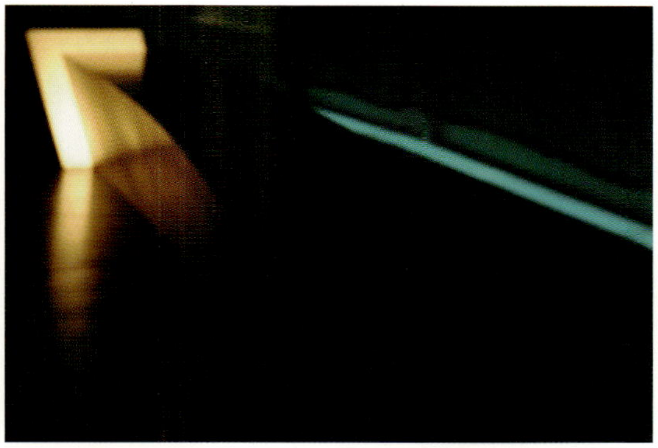

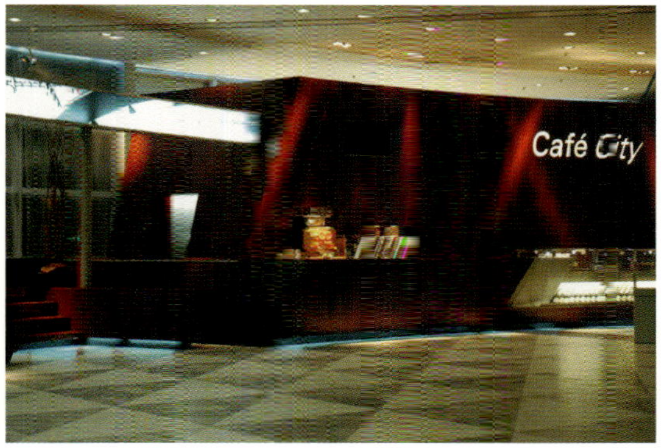
Café City

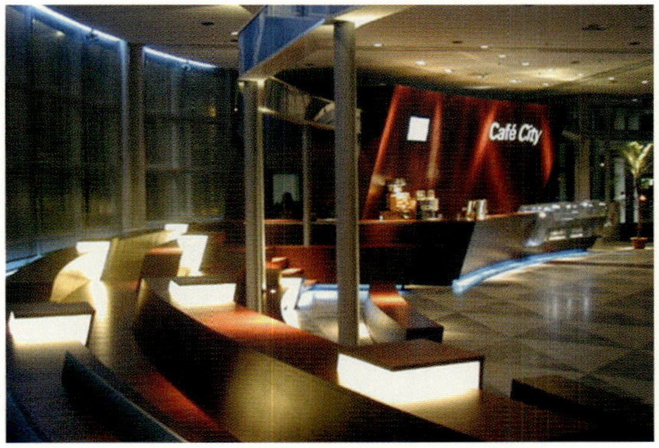
Café City

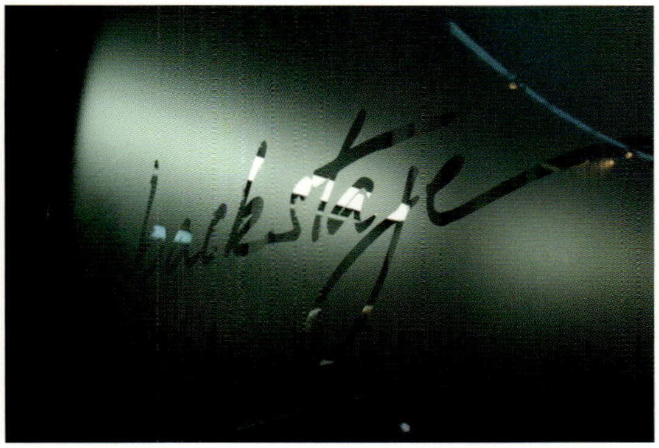
backstage

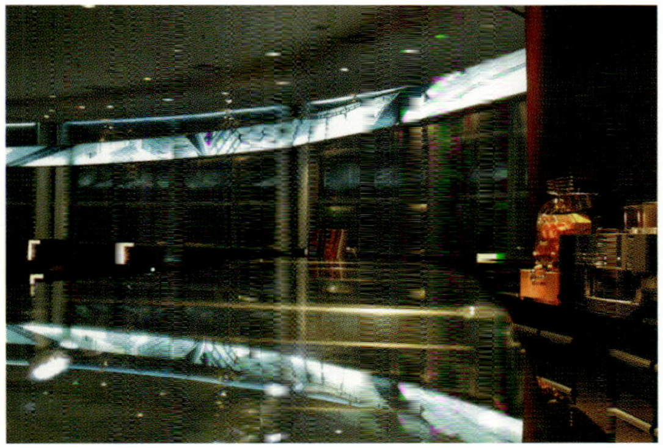

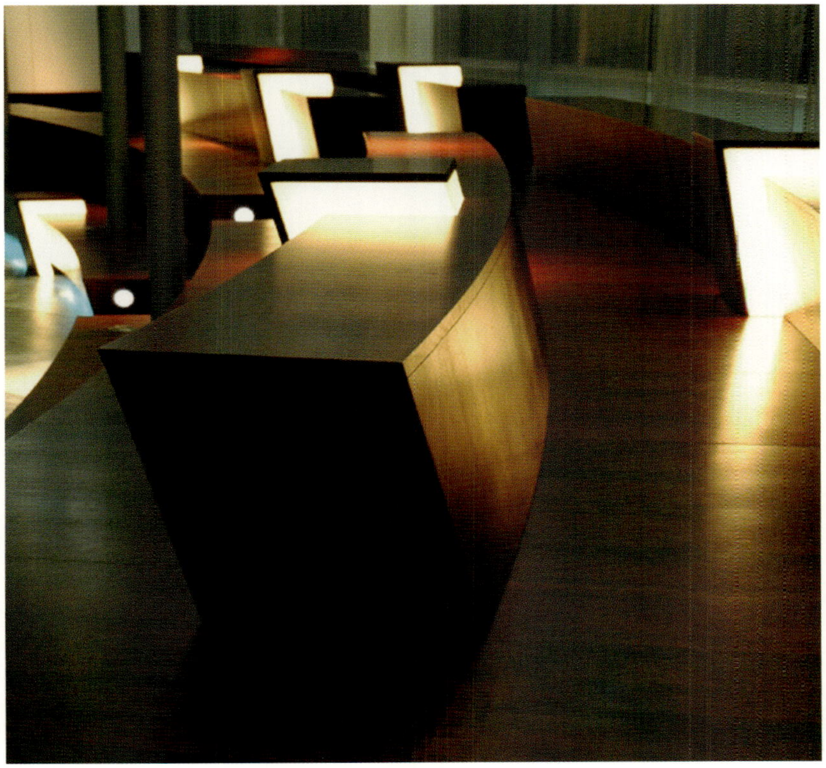

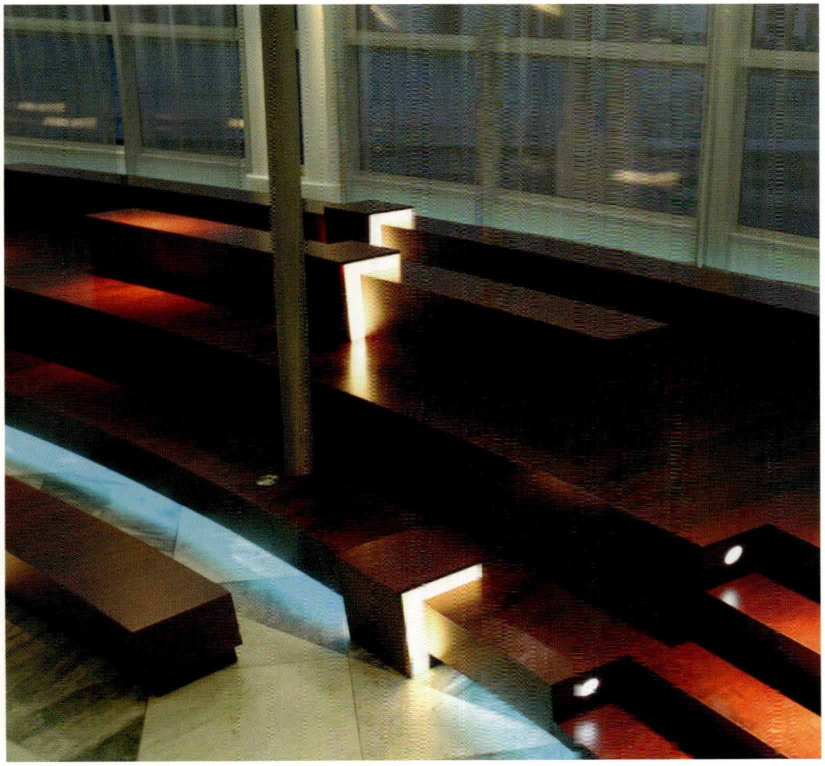

Auftraggeber Client
Infospace Mobile – Elkware Studio
Hamburg, Germany

Projekt Project
Infospace
Corporate Interior

Design
feldmann+schultchen design studios
Hamburg, Germany

Das Interior Design für die Deutschland-Dependance von Infospace Inc., dem führenden Anbieter von „Mobile Media", variiert mit einem System aus mobilen Möbelelementen das Motiv minimalistischer Spielelandschaften. Hier arbeiten junge Spiele- und Computerfreaks an der Entwicklung von Mobile-Games. Wie aus dem Boden herausgedrückt erheben sich Flächen zum Sitzen, Liegen, Loungen und Spielen, öffnen sich Räume zum Stauen, Hängen und Ablegen. Die Elemente scheinen in jeder Position mit dem Boden zu verfließen. Eine Spielwiese zum Verstecken und Toben für große Jungs im grauen Büroalltag.

The interior design for the offices of the German branch of Infospace Inc., leading provider of "mobile media", is a skillful mix of mobile furniture elements to generate a motif of minimalist game landscapes. The rooms are populated by young computer and game geeks developing games for the mobile phone platform. Areas for sitting, lounging and gaming rise up as if pressed straight out of the floor, opening up spaces for storing and hanging up things. All elements appear to blend in ideally with the floor from any position in the room. A playground for big boys to hide and romp about in the everyday grey office world.

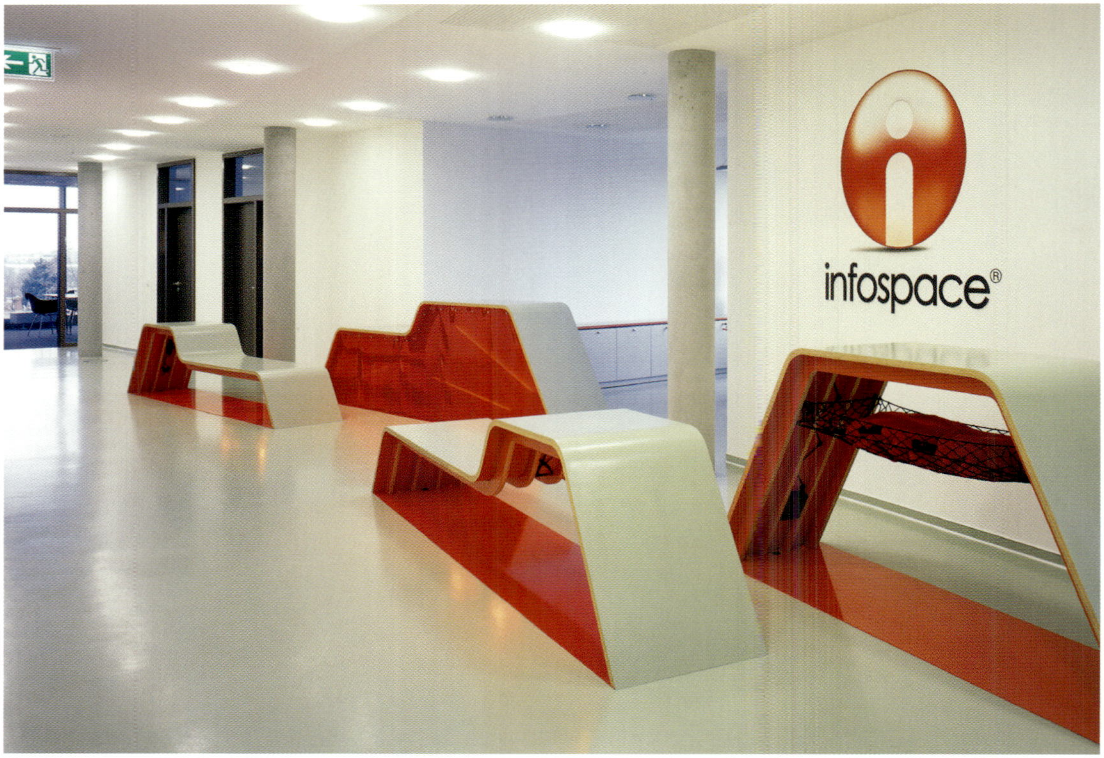

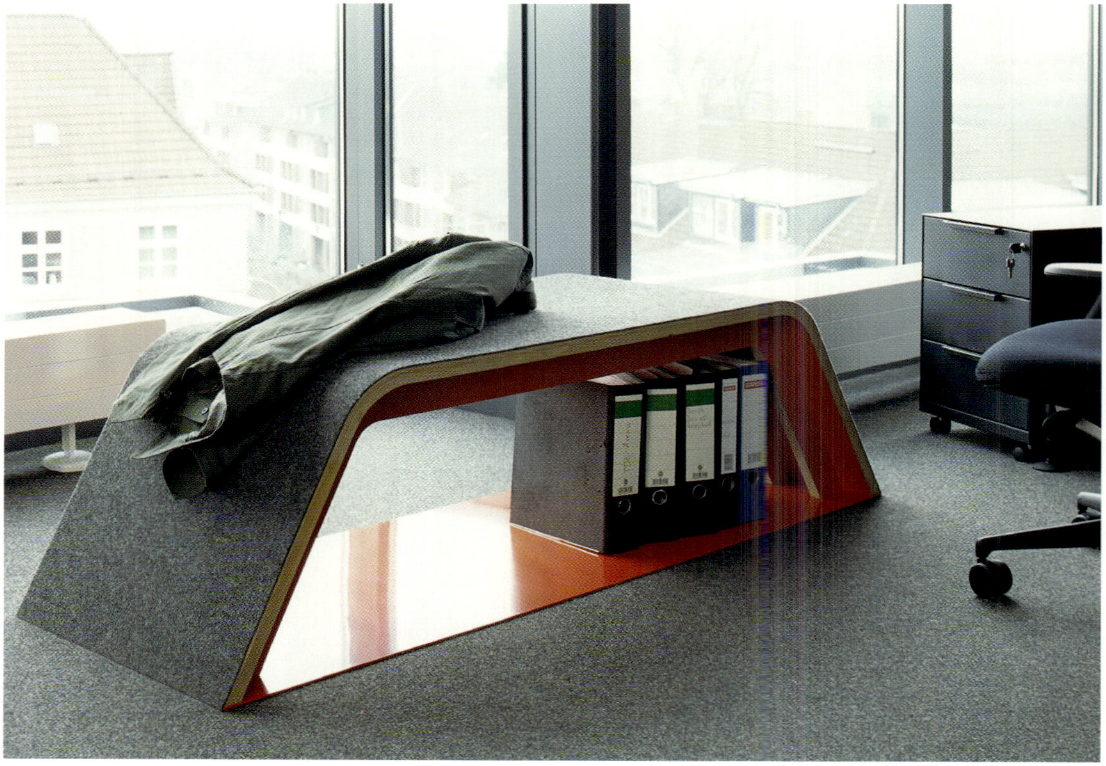

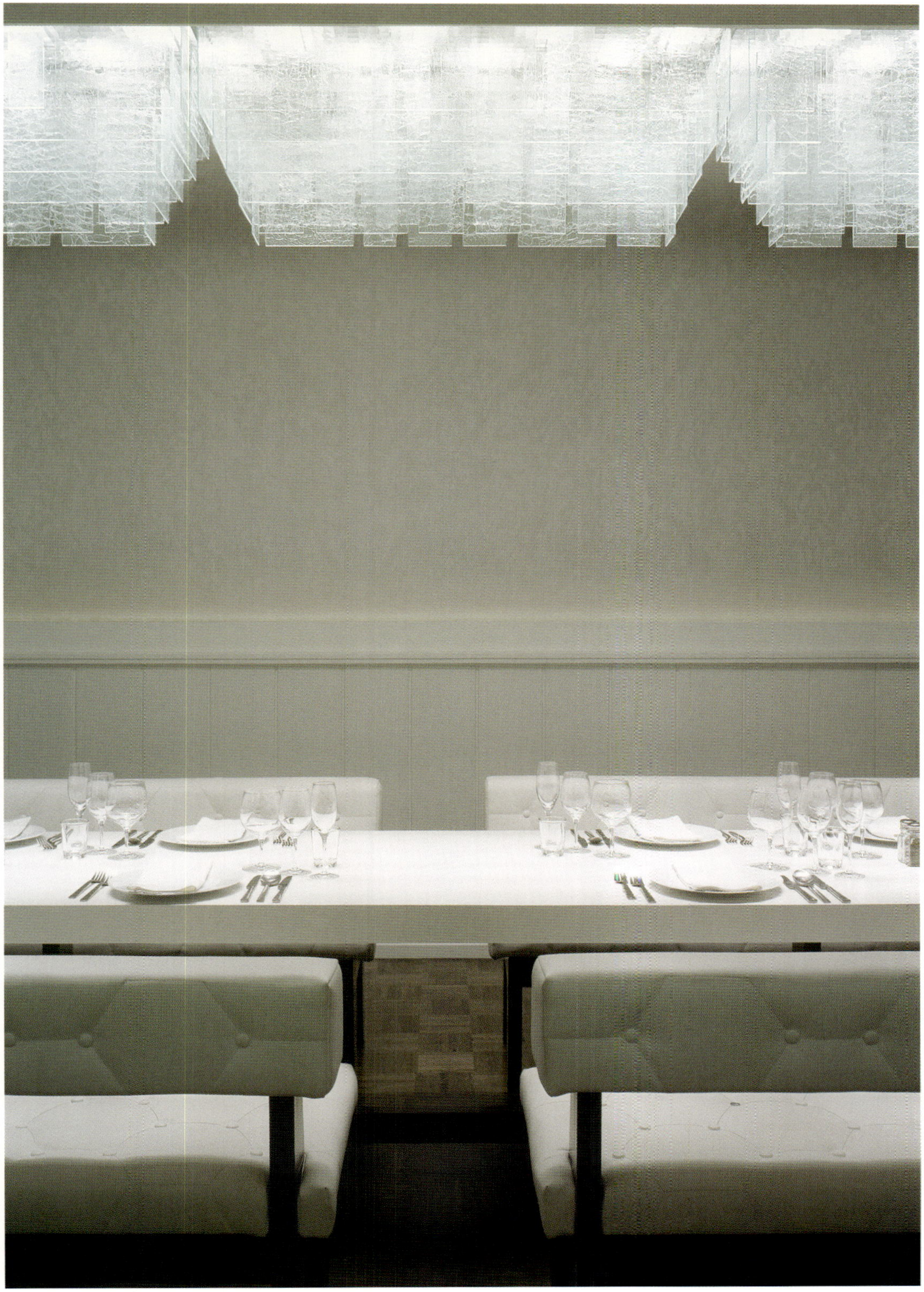

Auftraggeber Client
Kern, Radiogatec GbR
Restaurant Schneeweiß
Berlin, Germany

Projekt Project
Restaurant Schneeweiß
Restaurantarchitektur
Restaurant interior design

Design
Hinnerk Dedecke
Heike Dettmann
Thilo Fuchs
unit-berlin
Berlin, Germany

Weiß strahlend präsentiert sich das Restaurant. Der Name „Schneeweiß" ist Programm: Die alpenländische Küche wird in Räumen serviert, die wie mit einer sanften Schneedecke überzogen scheinen, ohne kühl zu wirken. Die Kombination verschiedener weißer Materialien, die mit dunklem Holz kontrastiert werden, erzeugt eine zauberhafte Verbindung von großstädtischem Szene-Lokal und Berghüttengemütlichkeit. Das Hexagon als abstrahiertes Motiv eines Schneekristalls zieht sich vom Logo bis zur Polsterung durch das medienübergreifende Gestaltungskonzept. Architektur, Möbel, Kronleuchter und grafische Gestaltung Print/Online ergänzen sich spannungsvoll.

In the restaurant "Schneeweiß" (Snow White), the "Snow White" theme extends to more than just the name: Swiss specialities are served in an interior that appears to be covered with a soft blanket of snow but without the chill of winter. The combination of different white-coloured materials with dark wooden elements creates an enchanting liaison between a trendy, cosmopolitan restaurant and that cosy, uniquely Alpine feel. The abstract hexagon motif inspired by the form of a snow crystal is an integral part of the entire design concept, and is featured everywhere from the logo to the leather upholstery. The restaurant's architecture, furniture, chandelier and graphic elements complement each other thrillingly.

Auftraggeber Client
Duravit AG
Hornberg, Germany

Projekt Project
Duravit Design Center „Leben im Bad"
Firmenmuseum und Showroom
Company museum and showroom

Design
Prof. Uwe R. Brückner
Dominik Hegemann
atelier brückner GmbH
Stuttgart, Germany

Duravit Design Center „Leben im Bad" – Am Firmenstandort Hornberg im Schwarzwald wird, in einem von Philippe Starck gestalteten Gebäude, ein Showroom mit musealem Anspruch eingerichtet, der die Marke Duravit als designorientiertes, weltweit agierendes Unternehmen darstellt.
Auf fünf Geschossen bekommt der Besucher Einblick in den Entstehungsprozess der anspruchsvollen Produkte und somit eine unbekannte Perspektive aus der Sicht der Produktion. Nach allen Regeln der Präsentationskunst wird das Thema „Leben im Bad" auf allen Etagen in eigenen Raumbildern neu inszeniert.

Duravit Design Centre – Company Museum and Showroom, 2004. The concern Duravit is located in Hornberg in the Black Forest in a new building designed by Philippe Starck. Here an ambitious showroom with a demanding, museum-like attitude was arranged to present the brand as a design-oriented company operating globally.
In the five-floor building visitors gain an insight into the manufacturing process of the high-value products, thus from the view of the flow of production. With all the tricks of the trade Duravit presents a new take on the theme "Living Bathrooms" on each of the five floors.

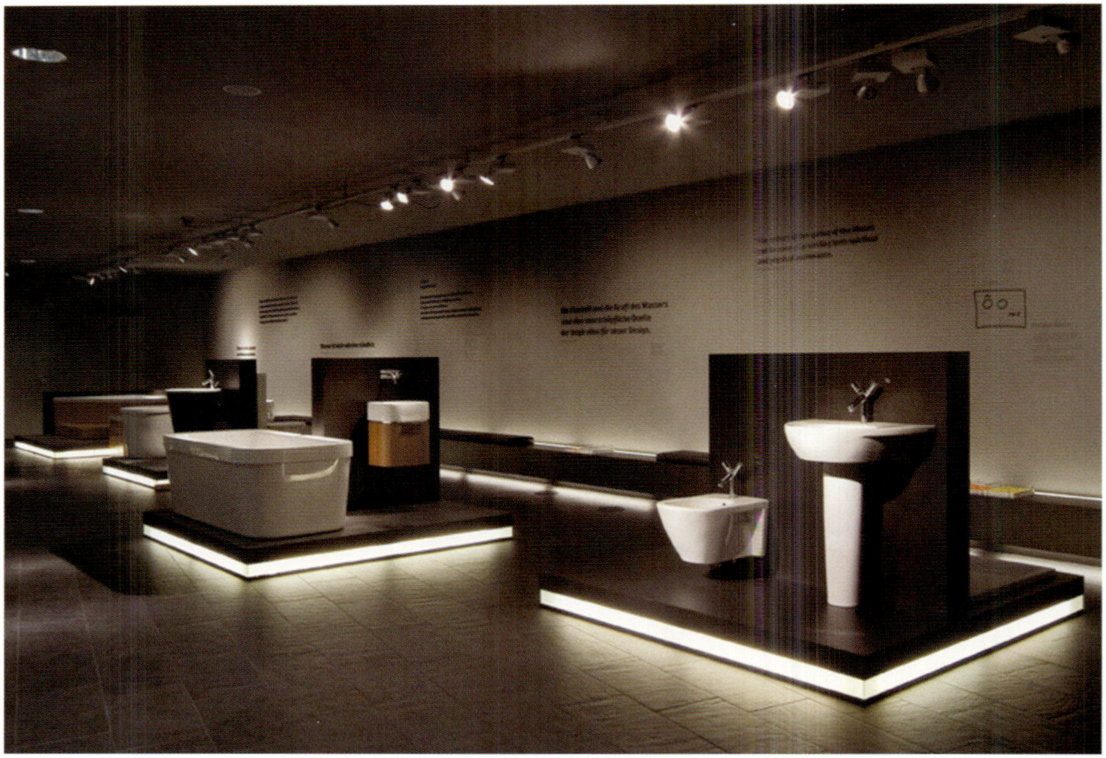

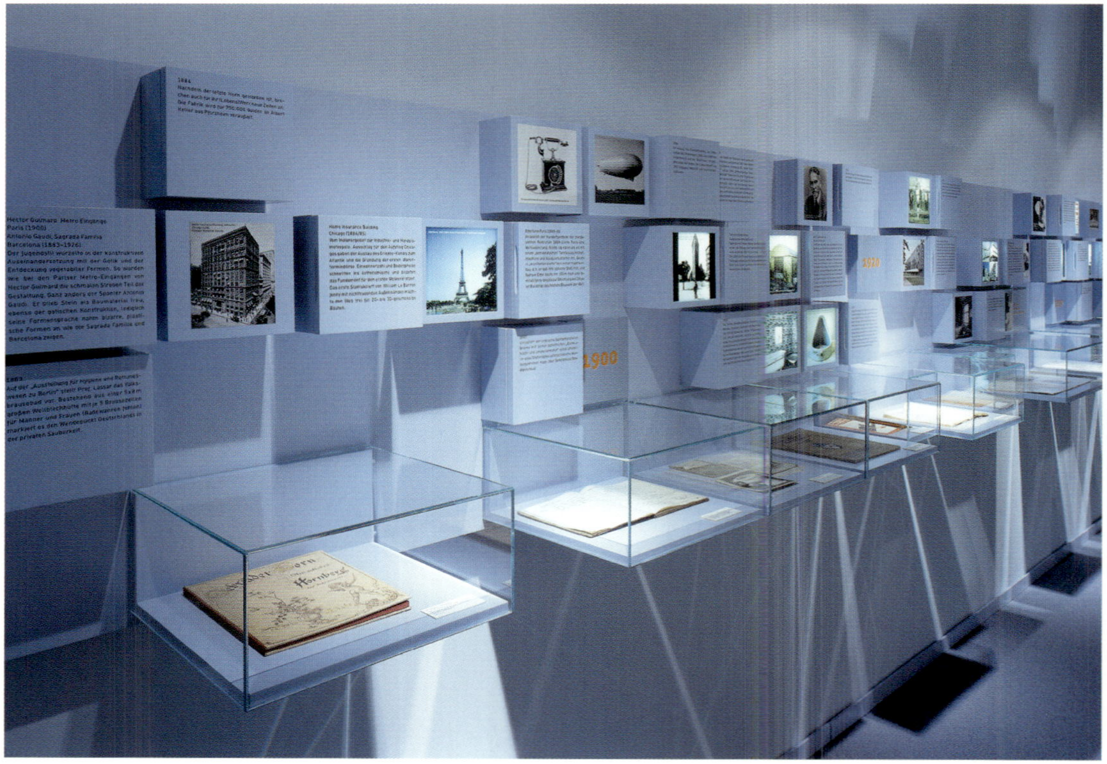

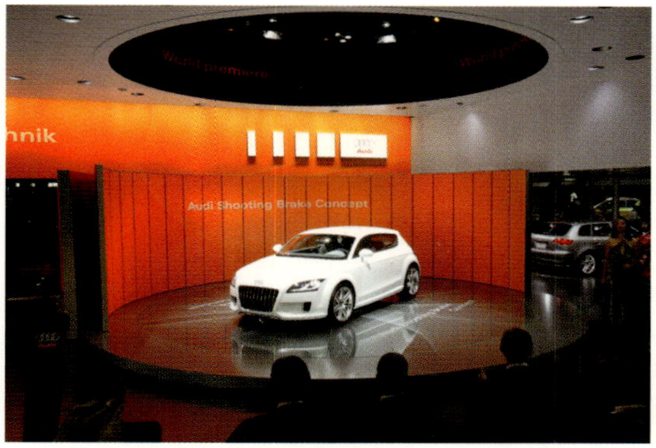

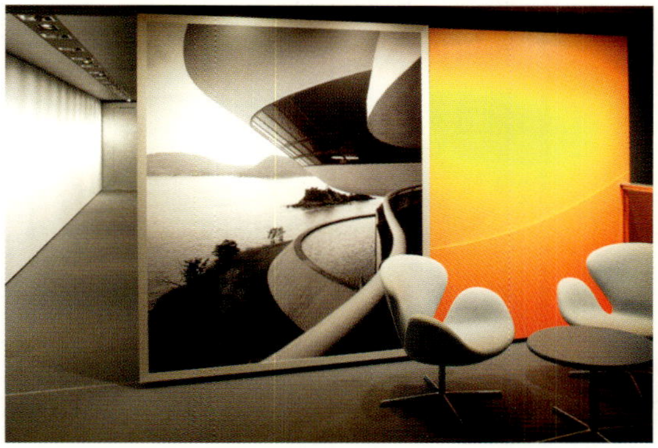

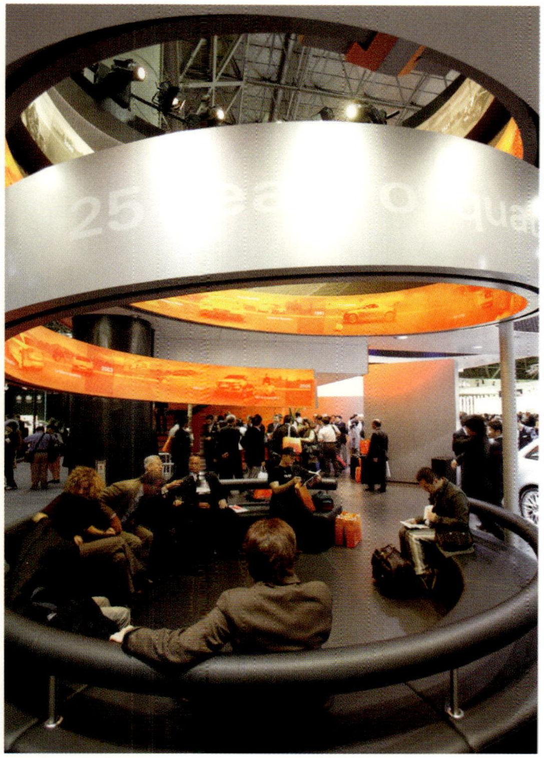

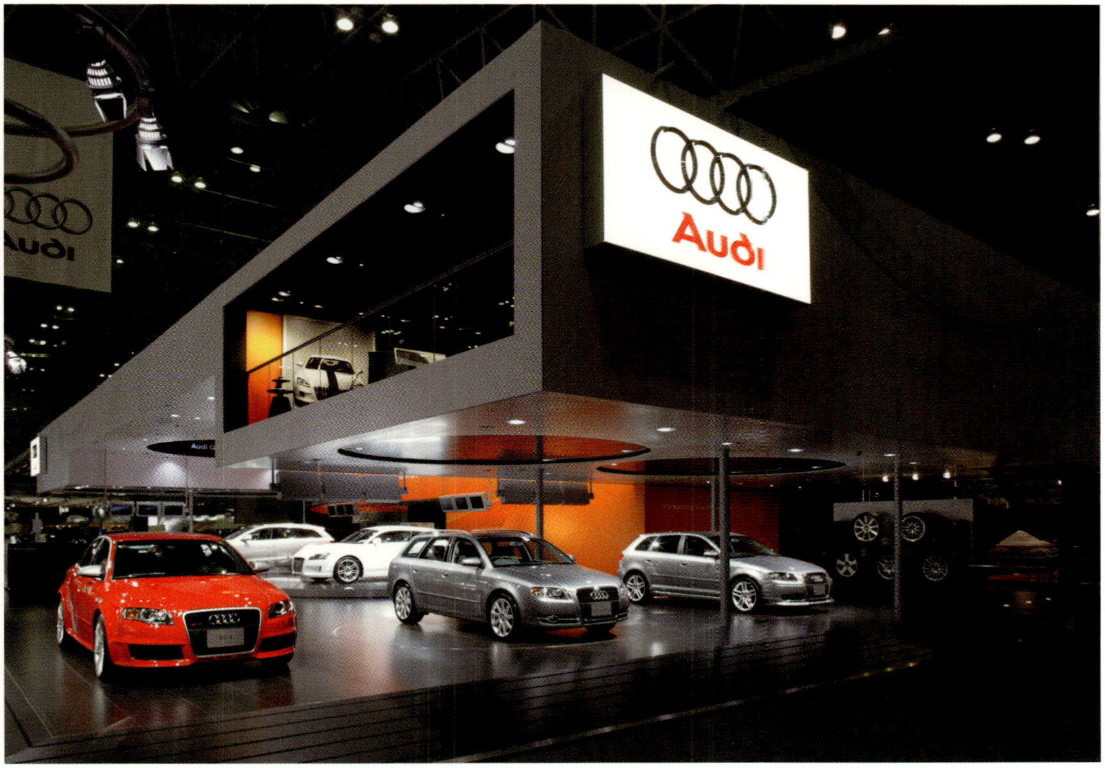

Auftraggeber Client
Audi AG
Ingolstadt, Germany

Projekt Project
Audi Messeauftritt Tokyo 2005
Messestand
Trade fair stand

Design
Johannes Plass
Axel Domke
Marc Antosch
Patrick Molinari
Kai Riemland
Mutabor Design GmbH
Hamburg, Germany
Schmidhuber + Partner GbR (Architektur)
München, Germany
Bernhard Neumann
Audi AG (Auftraggeber)
Ingolstadt, Germany

Beim Messeauftritt des Automobilherstellers bei der Tokyo Motor Show 2005 verschmelzen die Gene von Bauhaus und Tadao Ando zu einer progressiven und zugleich puristischen Gesamtinszenierung nach einem Prinzip: Die Architektur ist Quadrat, die Kommunikation ist Kreis. Teil davon ist ein schwebendes Gebäude. Wie ein unsichtbar gehaltenes Dach ist es über den Fahrzeugen installiert. Es konnte erst durch eine besonders ausgeklügelte Ingenieurleistung in der Statik umgesetzt werden. Im Zentrum des Messestandes steht eine weltweit erstmals eingesetzte, neu entwickelte Enthüllungsinszenierung aus kreisförmig angeordneten, beweglichen Lamellen.

At the car manufacturer's stand at the Tokyo Motor Show 2005, the genes of Bauhaus and Tadao Ando are merging to form a simultaneously progressive and puristic overall performance true to one principle: The architecture is square, the communication is circular. Part of this is a suspended building. It is installed above the vehicles like an invisibly suspended roof. It took a particularly ingenious piece of structural engineering to bring it to fruition. The centre of the fair stand is occupied by the global premiere of a newly developed unveiling production comprising a circular formation of movable slats.

Auftraggeber Client
Kästner Optik
Stuttgart, Germany

Projekt Project
Kästner Optik
Innenarchitektur / Shop-Design
Interior design / shop design

Design
ippolito fleitz group
Stuttgart, Germany
Textkunst
Bruno Nagel
Berlin, Germany

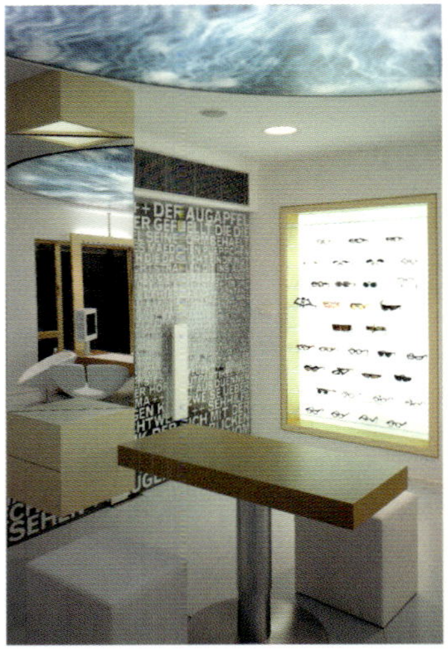

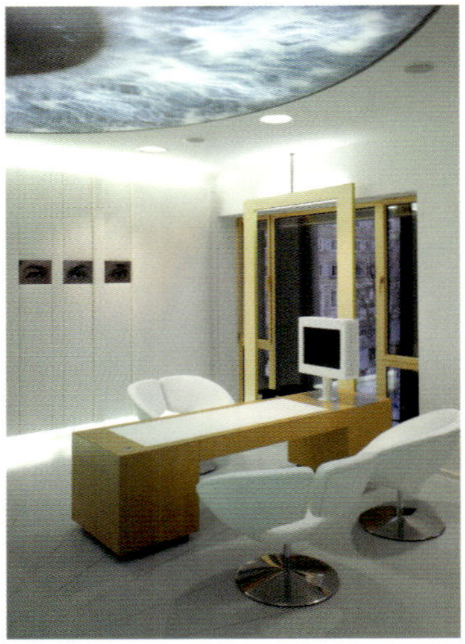

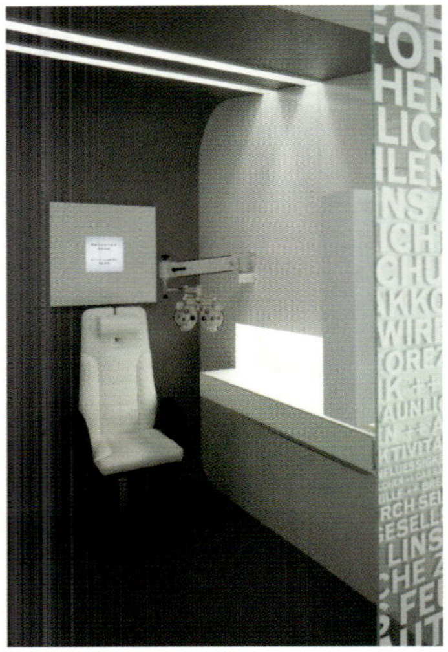

Kästner Optik ist ein alteingesessenes und erfolgreiches Optikergeschäft im Herzen Stuttgarts im Königsbau am Schlossplatz. Aufgabe war es, durch den Umbau des ersten Obergeschosses der Kernkompetenz eines Optikers, also der Arbeit rund um das menschliche Auge, einen neuen Rahmen zu geben. Zu der zunehmenden Wahrnehmung des Optikers als Style-Berater bei der Wahl der Brillenfassung soll hier ein klares Gegengewicht gesetzt werden. Im neuen Sehzentrum soll in einer klaren, Präzision vermittelnden Atmosphäre die optometrische Arbeit stattfinden sowie in entspannter Atmosphäre des After-Sale-Gesprächs eine noch stärkere Kundenbindung erreicht werden.

Kästner Optik is an optometrist's shop with a long tradition of excellence, situated in the very heart of Stuttgart. The task was to position an optometrist's core competence, everything relating to the human eye, in a new framework by redeveloping the first floor. The intention was to set a clear counterweight to the increasingly prevalent perception of an optometrist as a mere style advisor for selecting the right spectacle frames. In the new Optical Center, optometric work will take place in an atmosphere that exudes clarity and precision while at the same time creating a relaxed atmosphere for the after sales dialog.

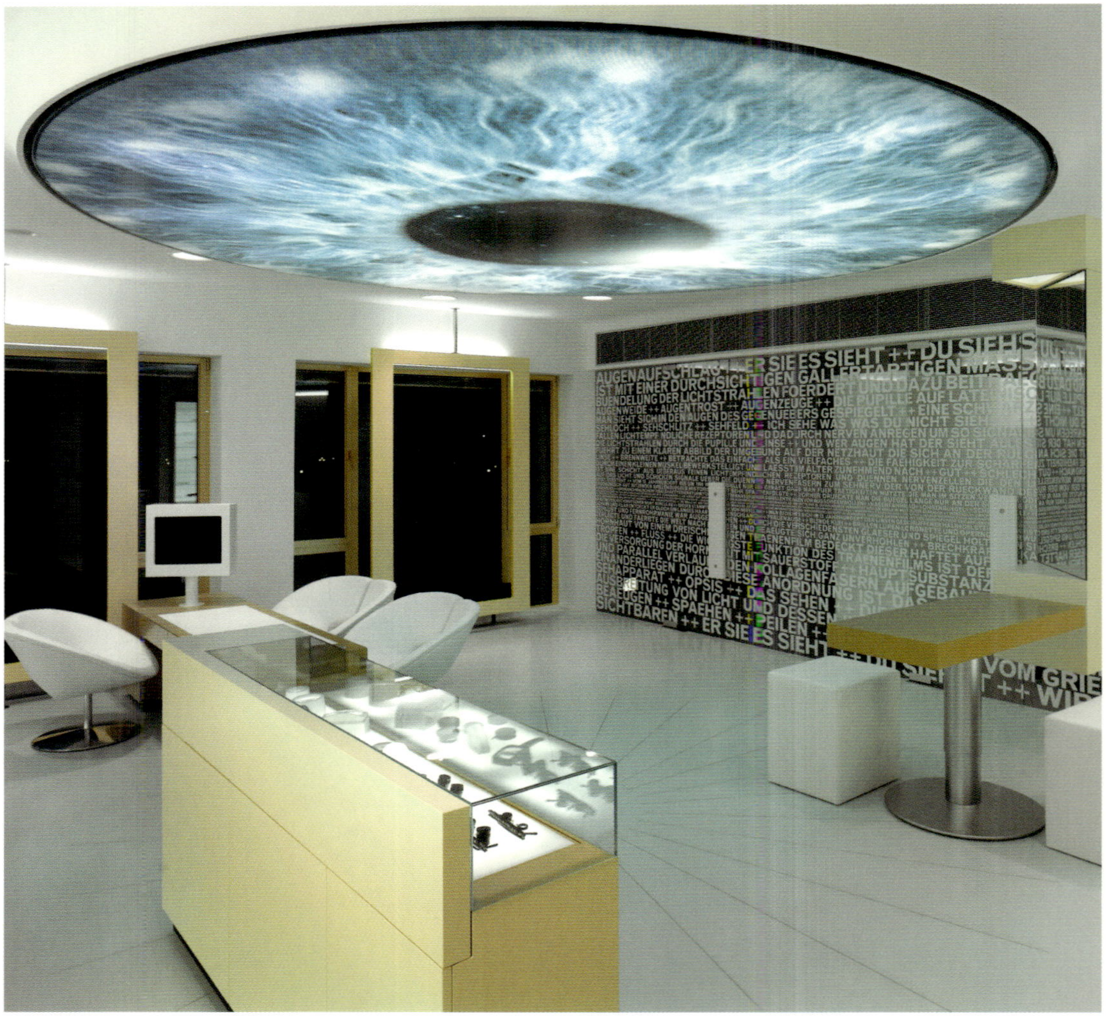

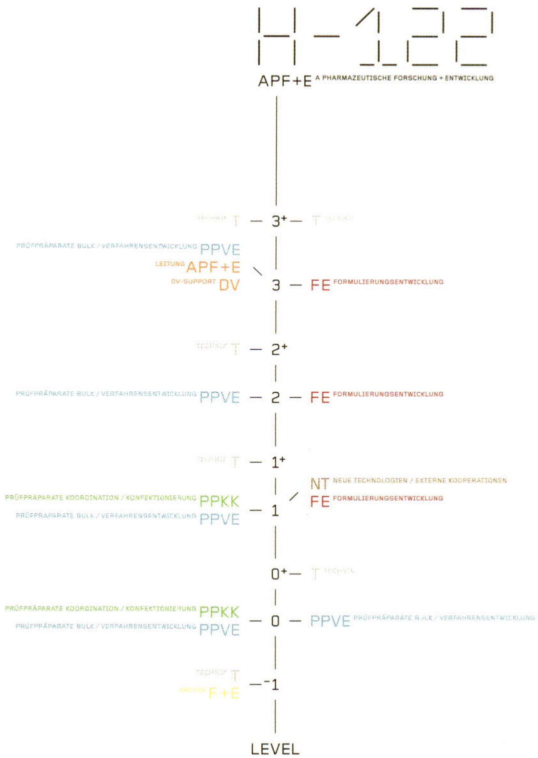

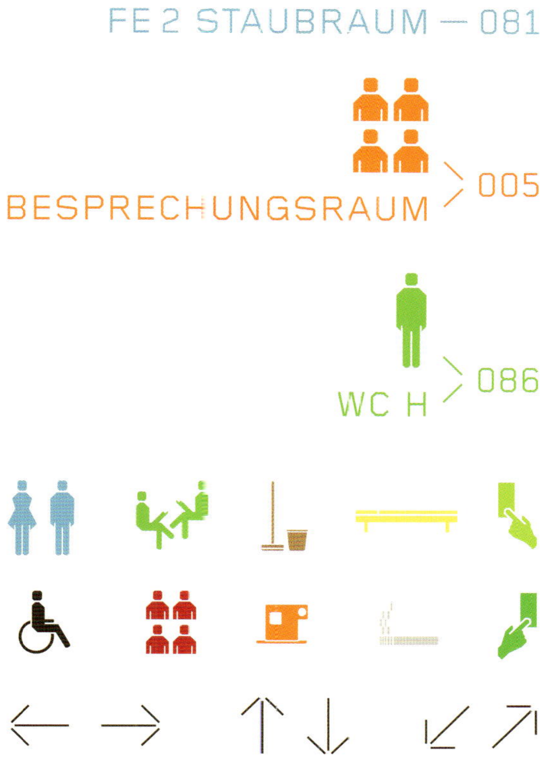

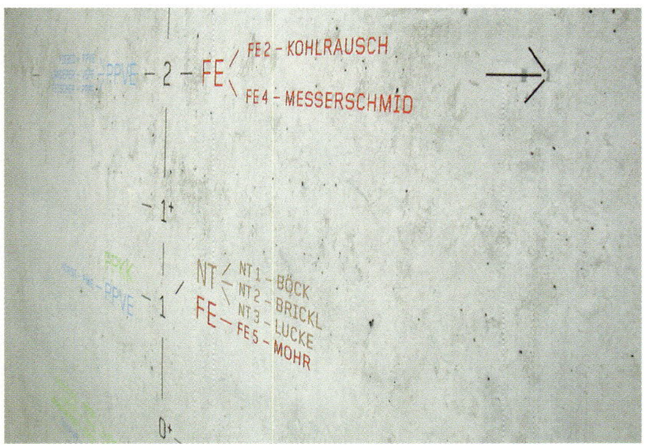

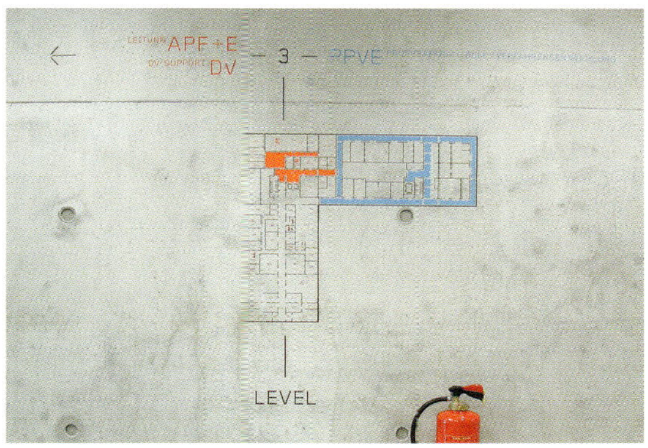

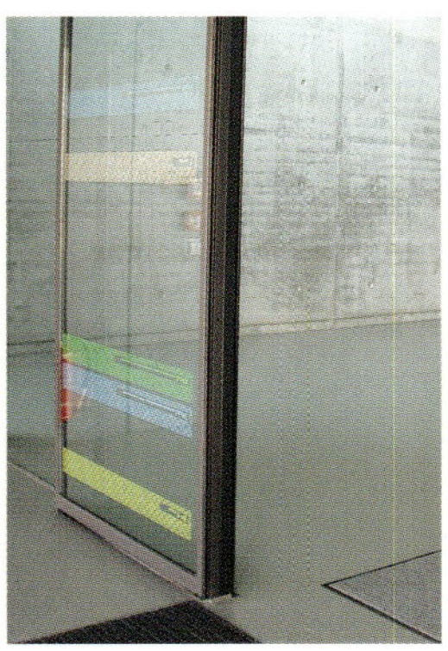

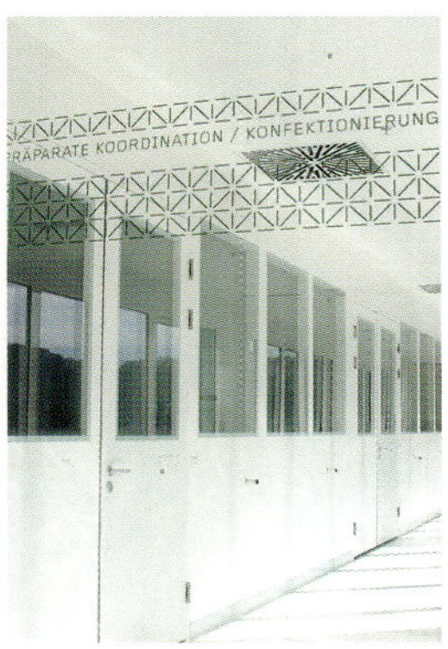

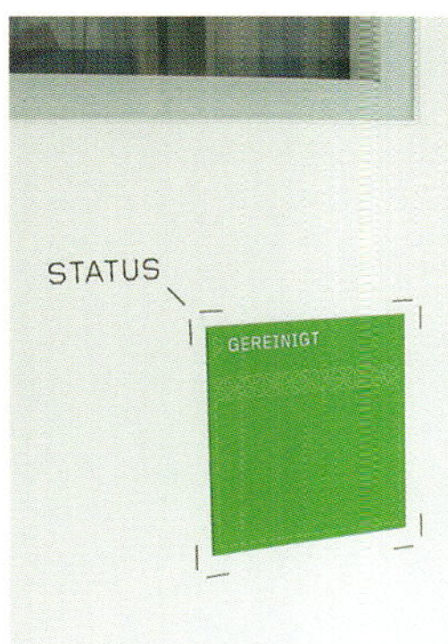

Auftraggeber Client
Boehringer Ingelheim
Pharma GmbH & Co. KG
Biberach / Riss, Germany

Projekt Project
Neubau H122
Leit- und Orientierungssystem
Signage and orientation system

Design
Designklinik
Büro für Design und Kommunikation
Stuttgart, Germany

Erfolgreiche Forschungsarbeit liegt in der Kombination von Konzentration, Kommunikation und experimenteller Arbeit. Das Gebäude greift diese Nutzung auf und stellt mit seinen zwei Baukörpern und verschiedenen Geschosshöhen eine hohe Anforderung an das Orientierungssystem. Die Grundlage für das grafische Konzept bildet das chemische Verkettungsprinzip einer Formel. Die verbindenden Elemente von Gebäude, Etagen, Raumnummern, Namen der Mitarbeiter, den Bereichen, Farben und der Schriftanwendung lassen die kommunikative Achse von H122 entstehen und sorgen für ein einheitliches, spielerisches Konzept mit verständlicher und einprägsamer Orientierung.

Successful research is a combination of concentration, communication and experimentation. This diversity is reflected in the design of the building, whose two distinct edifices and varying heights between floors placed great demands on the newly commissioned orientation system. The basis for this graphic concept comes from the chemical chain principle used in formulas. The connective elements of buildings, floors, room numbers, employee names, areas, colors and typeface create the communicative axis of H122 and ensure an integrated, playful concept in an orientation system that is easy to grasp and remember.

Auftraggeber Client
Silverstein Properties Inc.
New York, USA

Projekt Project
Virtual Window
Interaktives Informationssystem
Interactive information system

Design
Michael Keller
Christoph Rohrer
Andreas Niessner
Sven Sonnendorfer
Julia Peglow-Peters
Cecil Rustemeyer
KMS Team GmbH
München, Germany
Christian Arkay-Leliever
Skidmore, Owings & Merrill LLP
New York, USA

Das „Virtual Window" besteht aus neun Plasmabildschirmen, die in die Glasfassade des Raums integriert sind. Live-Kameras übertragen den Blick aus der Fensterfront auf die Bildschirme. Sobald die Anwendung aktiviert wird – per Handbewegung durch einen mit Sensoren bestückten Rahmen –, lässt sich die Kamera in alle Richtungen steuern, sodass Details von Ground Zero auf dem Groß-Monitor erscheinen; wird ein Gebäude ausgewählt, verwandelt sich die reale Aufnahme in eine interaktive Animation, die sämtliche verfügbaren Baudaten in visuelle Darstellungen umsetzt.

The central element are nine plasma screens integrated into the inner side of the glass façade. Initially they present the same view one would enjoy looking through the windowed front of the building – only that it is in fact captured by an outdoor camera and simultaneously rendered as an image on the screens. When the application is activated, the camera can be steered in all directions such that details of Ground Zero appear on the giant display. If a building or site is selected, the actual camera shot morphs into a virtual view. This is an interactive animation that converts all available construction data into visual representations.

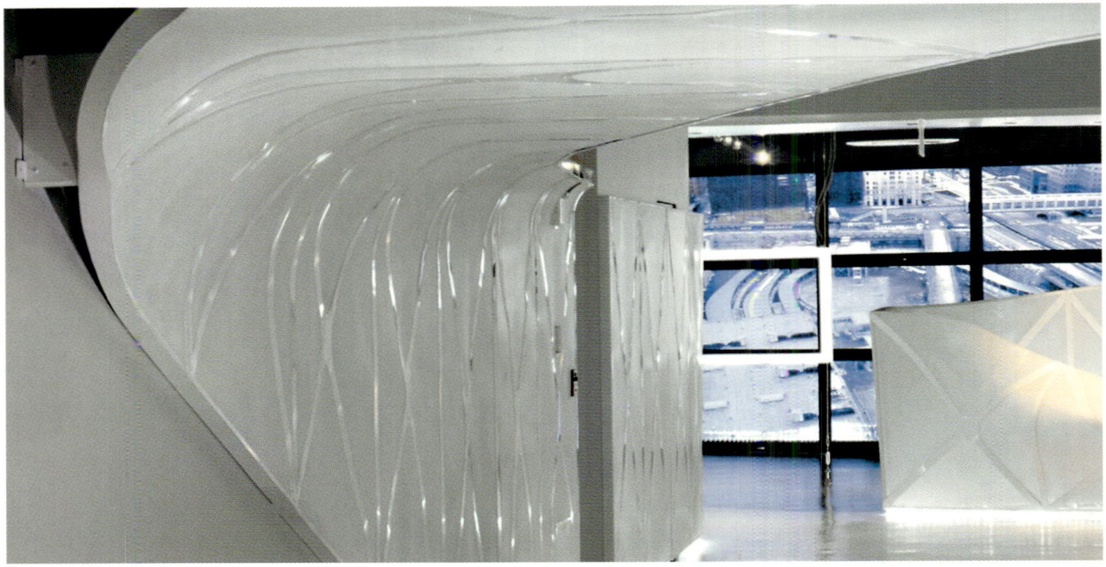

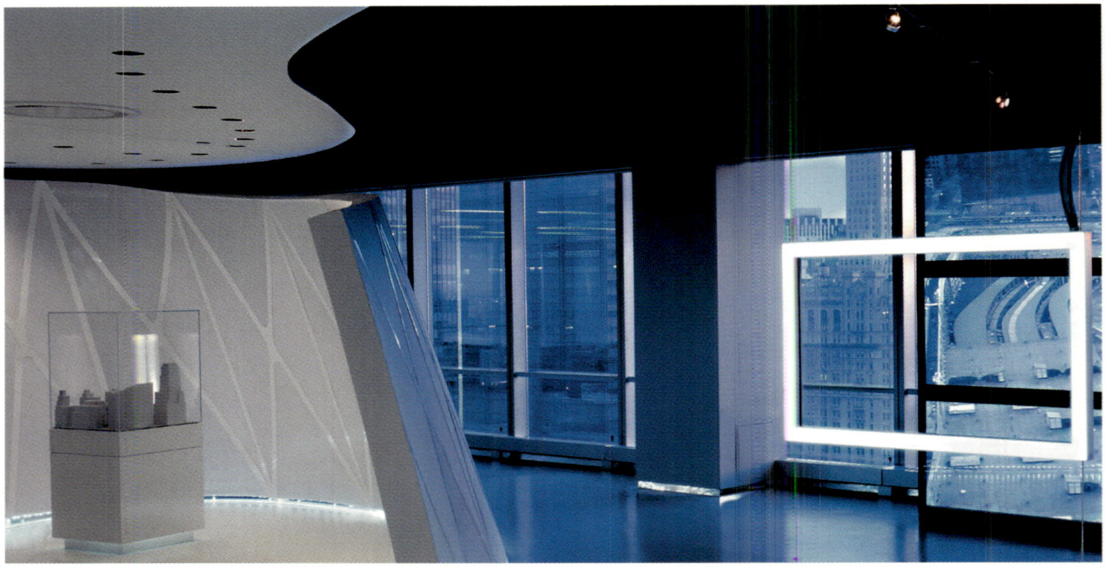

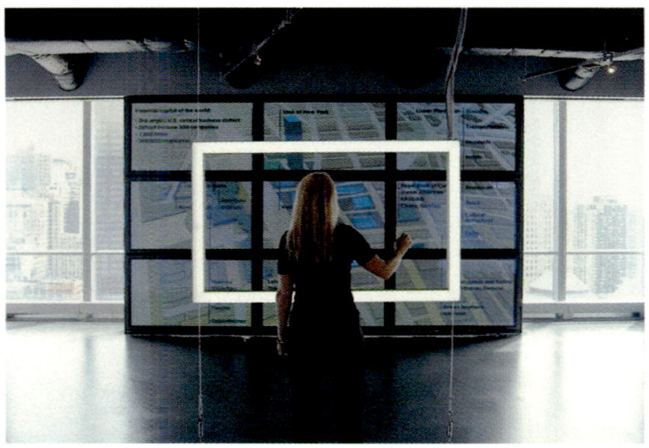

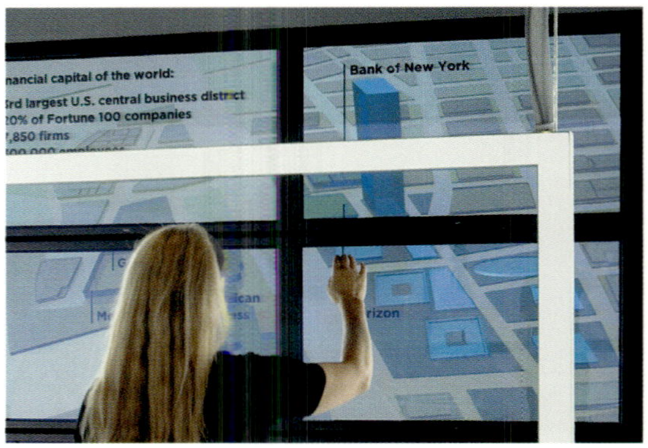

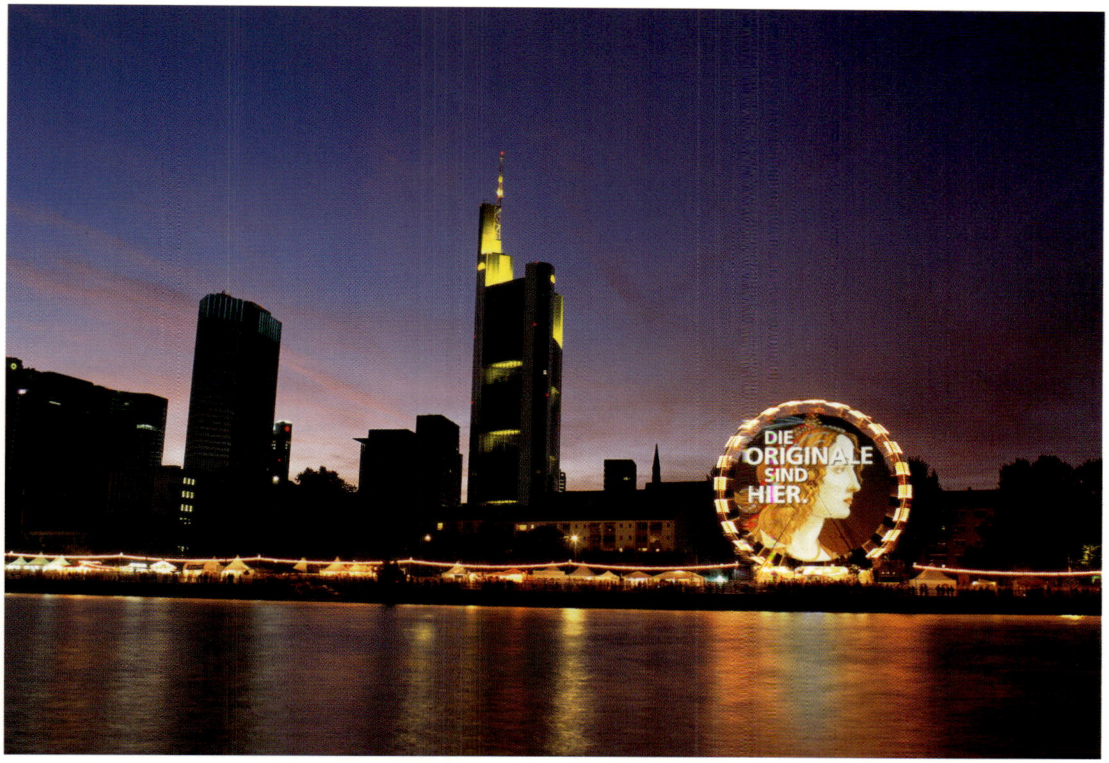

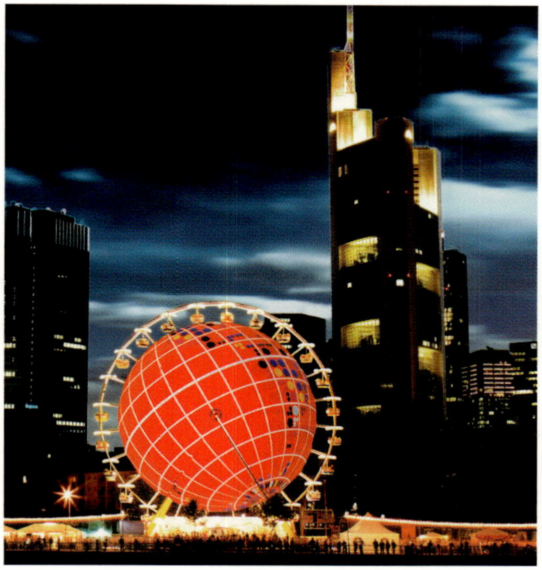

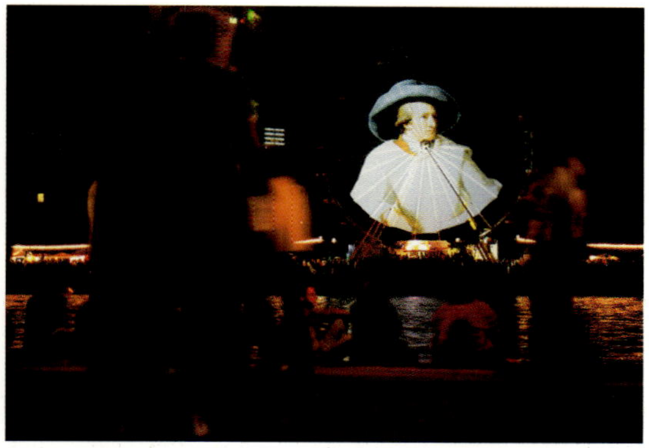

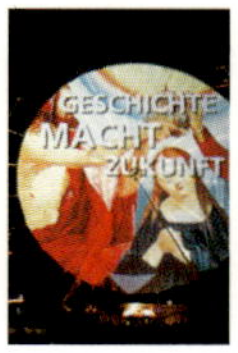

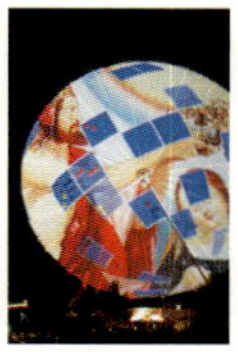

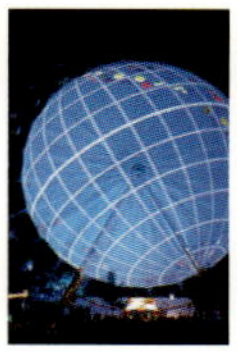

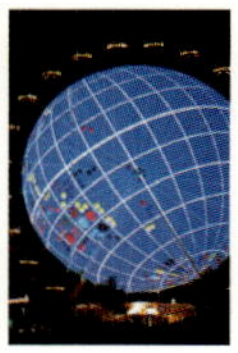

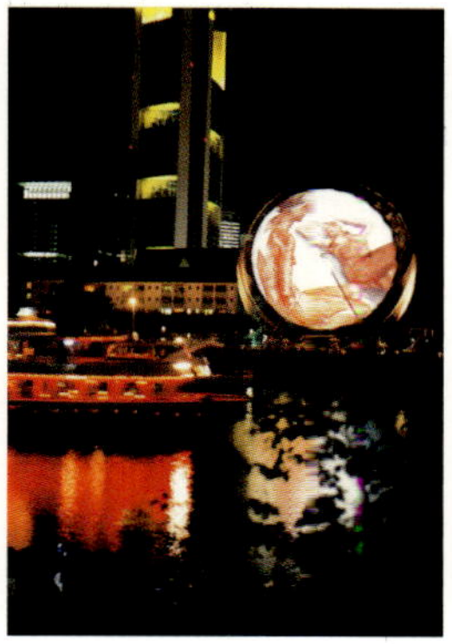

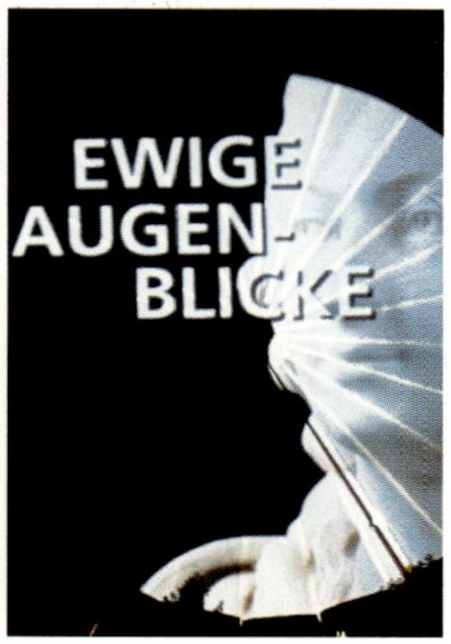

Auftraggeber Client
Tourismus+Congress GmbH Frankfurt am Main
Frankfurt am Main, Germany

Projekt Project
KUNSTIRAD
Licht- und Medieninszenierung
Light and media sculpture

Design
Roland Lambrette
Stefan Weil
Kristin Trümper
Jan Schmelter
Andreas Behl
Atelier Markgraph GmbH (Kommunikationskonzept und -design)
Frankfurt am Main, Germany
Sebastian Oschatz
meso | digital media systems design (Media Design)
Frankfurt am Main, Germany
Ole Schulte
Aufwärts Projektagentur (Sound Design)
Oberursel, Germany
Dietrich Körner (Lichtdesign)
Köln, Germany

Museumsuferfest Frankfurt 2005: Ein Riesenrad wird zur Bühne für die Museen. Das KUNSTIRAD bringt internationale Meisterwerke aus 16 Ausstellungshäusern am Main auf die große Leinwand. Jede Gondel repräsentiert ein Museum. Hat der Zeiger eine Museumsgondel ausgewählt, startet die Bespielung. Die Fahrt führt durch Epochen, Stile, Farben und Formen. Über Sensoren am Riesenrad ist die Radbewegung mit der digitalen Bespielung verkoppelt. Die Besucher selbst bestimmen so den Rhythmus der Projektion. Das KUNSTIRAD vereint auf zauberhafte Weise romantischen Jahrmarkt und Großstadtkulisse. Und setzt ein Zeichen für die Frankfurter Kulturlandschaft.

Museumsuferfest Frankfurt 2005: A Ferris Wheel becomes a stage for the museums. ARTIWHEEL brings international masterpieces from 16 Frankfurt museums out onto the big screen. Each museum has its own gondola; as soon as the pointer selects it, the projection begins its fascinating trip through the styles, colours and shapes of different ages. The visitors themselves determine the projection's rhythm, thanks to sensors in the wheel that synchronise its movements to the digital media feeds. ARTIWHEEL is a magical mix of fairground romance and city skyline – and a symbol of Frankfurt's cultural landscape.

04 crossmedia

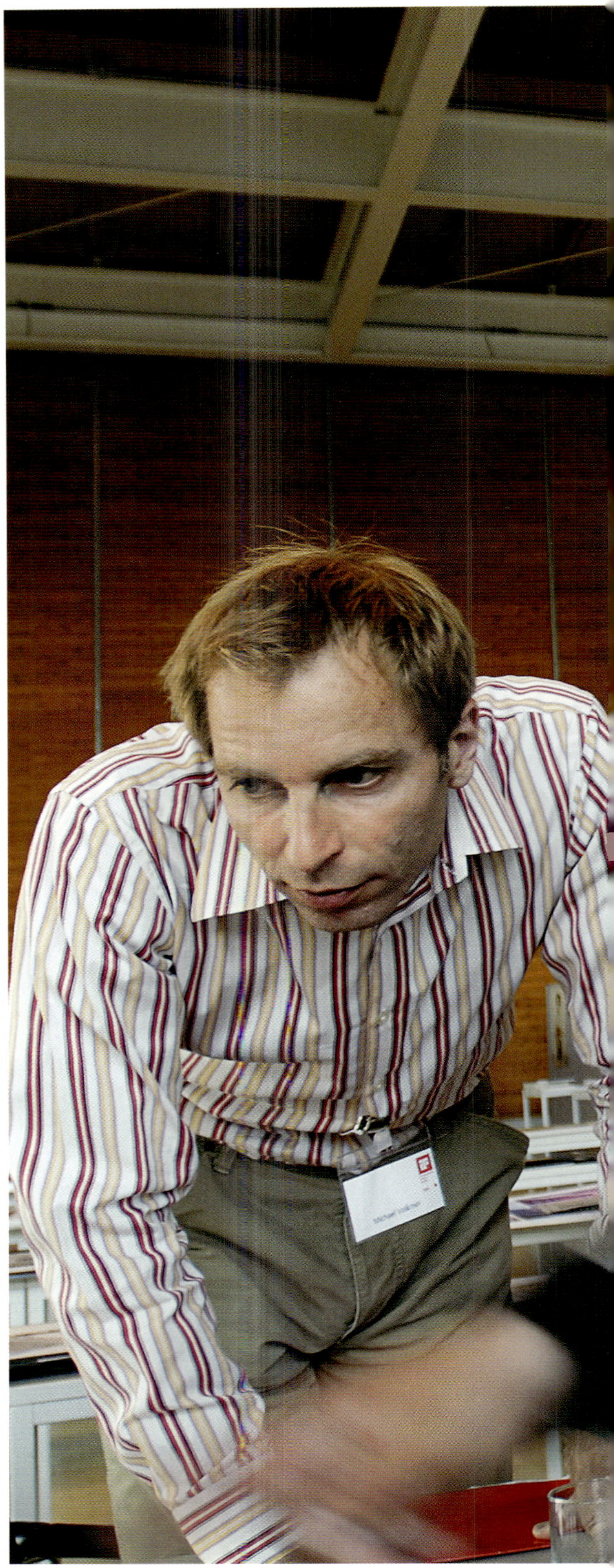

■ Was ist crossmedial? Multitasking. Der moderne Mensch rezipiert quer durch die Medien. Am Schreibtisch werden fast simultan Telefon, Computer und Akten im Auge behalten. Beim Autofahren nimmt der Fahrer parallel den Verkehr, das Gespräch mit dem Beifahrer, Musik aus dem Radio und die Werbebotschaft an der Litfasssäule wahr. Physikalisch betrachtet ist das mit Sicherheit crossmedial. Aber ist es auch aus Sicht des Marketings als crossmedial zu bezeichnen? Mit Sicherheit nein. Es fehlt das verbindende Glied, die Zusammengehörigkeit, das aufeinander aufbauende Moment.

Während der Begriff Crossmedia in den Marketing-Standardwerken von Esch, Meffert oder Kroeber-Riel gar nicht auftaucht, ist er in der Werbebranche in aller Munde. Nur definiert ihn jeder ein wenig anders. Die treffendste Definition findet sich bei Wikipedia. Es sollen „mindestens zwei Medien vom Nutzer genutzt werden, damit die Werbebotschaft inhaltlich komplett übermittelt werden kann". Klingt einfach und plausibel. Dennoch waren einige Einreichungen im iF communication design award 2006 in dieser Kategorie unzureichend. So ist es nach Einschätzung der Jury nicht crossmedial, wenn ein Corporate Design einmal auf einem Briefbogen und ein zweites Mal auf einem LKW angewandt wird. Zwar handelt es sich hier durchaus um zwei unterschiedliche Medien, aber das aufeinander aufbauende Moment fehlt. Andererseits kann es durchaus crossmedial sein, wenn eine Botschaft via Banner angerissen und dann auf einer Website vertieft wird: physikalisch nur ein Medium – inhaltlich aber zwei unterschiedliche Werbeformen. Wenn man sich in der Branche umhört, gibt es fast keine größere Kampagne, die ihre Dramaturgie nicht innerhalb unterschiedlicher Medien aufbauen möchte. Leider wird oft nur auf formaler Ebene stimmig oder „integriert" kommuniziert, anstatt Botschaften inhaltlich „crossmedial" zu vernetzen oder gar zu steigern. Nämlich vorne im Sales Funnel Aufmerksamkeit zu erzielen, sprich den Verbraucher „abzuholen". Ihm dann die gewünschten Informationen zu geben, um ihn schließlich weiter hinten im Sales Funnel via Dialog etc. zu einer Transaktion zu führen. Das geht oft nur über verschiedene Medien – eben crossmedial.

Michael Volkmer

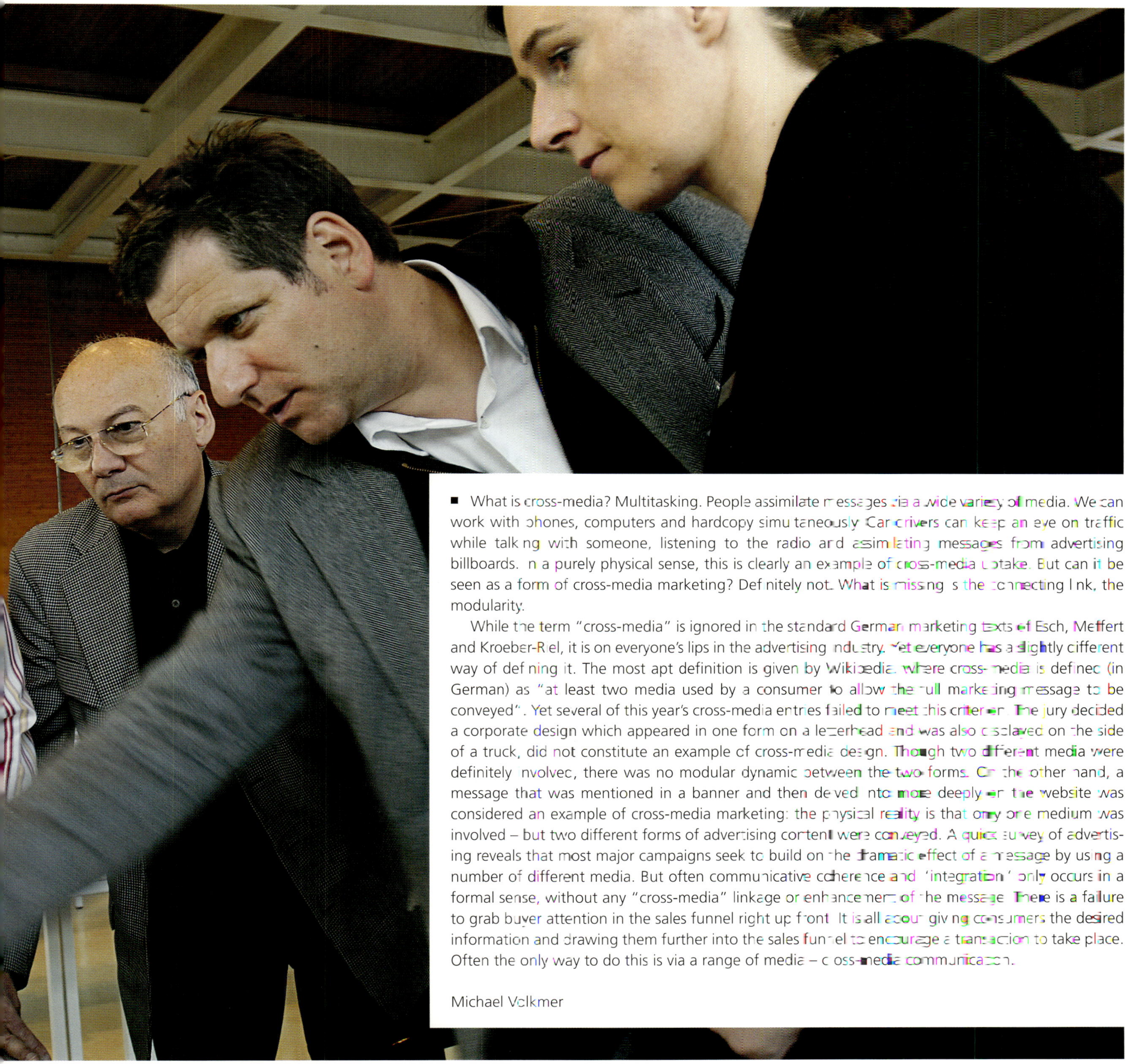

■ What is cross-media? Multitasking. People assimilate messages via a wide variety of media. We can work with phones, computers and hardcopy simultaneously. Car drivers can keep an eye on traffic while talking with someone, listening to the radio and assimilating messages from advertising billboards. In a purely physical sense, this is clearly an example of cross-media uptake. But can it be seen as a form of cross-media marketing? Definitely not. What is missing is the connecting link, the modularity.

While the term "cross-media" is ignored in the standard German marketing texts of Esch, Meffert and Kroeber-Riel, it is on everyone's lips in the advertising industry. Yet everyone has a slightly different way of defining it. The most apt definition is given by Wikipedia, where cross-media is defined (in German) as "at least two media used by a consumer to allow the full marketing message to be conveyed". Yet several of this year's cross-media entries failed to meet this criterion. The jury decided a corporate design which appeared in one form on a letterhead and was also displayed on the side of a truck, did not constitute an example of cross-media design. Though two different media were definitely involved, there was no modular dynamic between the two forms. On the other hand, a message that was mentioned in a banner and then delved into more deeply on the website was considered an example of cross-media marketing: the physical reality is that only one medium was involved – but two different forms of advertising content were conveyed. A quick survey of advertising reveals that most major campaigns seek to build on the dramatic effect of a message by using a number of different media. But often communicative coherence and 'integration' only occurs in a formal sense, without any "cross-media" linkage or enhancement of the message. There is a failure to grab buyer attention in the sales funnel right up front. It is all about giving consumers the desired information and drawing them further into the sales funnel to encourage a transaction to take place. Often the only way to do this is via a range of media – cross-media communication.

Michael Volkmer

■ Michael Volkmer (*1965) studierte Kommunikationsdesign an der Fachhochschule Wiesbaden und gründete 1994 zusammen mit Anette Scholz die Agentur Scholz & Volkmer Intermediales Design GmbH in Wiesbaden. Seit 2003 ist er als alleiniger Geschäftsführer für die Planung, Gestaltung und Umsetzung von Projekten im Bereich der neuen Medien verantwortlich. Er lehrt an der Fachhochschule Darmstadt zum Thema „Darstellungsmethoden" und war bereits für nationale wie internationale Wettbewerbe als Jury-Mitglied tätig.

■ Michael Volkmer (*1965) studied Communication Design at the Wiesbaden University of Applied Sciences and, with Anette Scholz, set up the agency Scholz & Volkmer Intermediales Design GmbH in Wiesbaden in 1994. Since 2003 he has been responsible for planning, designing and implementing projects in new media. He teaches at Darmstadt's University of Applied Sciences on "methods of representation" and has served on a number of juries at national and international competitions.

■ Ralf Höpfner (*1959) war Berater bei BBDO Deutschland, Managing Director bei ID-Media und Elephant Seven sowie Geschäftsführer bei Heye & Partner. Seit 2005 ist er Gründer und Inhaber der markentechnisch orientierten Unternehmensberatung >¡< Markenfeuer und zudem als Coach in den Konfliktbereichen Strukturen, Informationen, Beziehungen, Werte und Interessen sowie in der Unterstützung von unternehmerischen Positionierungs- und Change-Prozessen tätig.

■ Ralf Höpfner (*1959) has worked as a consultant at BBDO Deutschland, and has been Managing Director of ID-Media and Elephant Seven, as well as Managing Director of Heye & Partner. In 2005 he set up his own brand-oriented corporate consultancy, >¡< Markenfeuer, also working as a coach in the conflict areas of structures, information, relationships, values and interests, as well as supporting corporate positioning and change processes.

■ Rainer Hilf studierte Innenarchitektur an der Akademie der Bildenden Künste in Nürnberg und ist seit 1972 als freischaffender Innenarchitekt tätig. Er hat seit 1976 unterschiedliche Lehrtätigkeiten, u. a. eine Gastprofessur an der Hochschule für Kunst und Design Burg Giebichenstein in Halle/Saale, inne und ist seit 1995 Präsident des BDIA – Bund Deutscher Innenarchitekten. Rainer Hilf ist darüber hinaus Mitglied in zahlreichen Institutionen wie beispielsweise einer Lenkungsgruppe des Bundesministeriums für Verkehr, Bau- und Wohnungswesen.

■ Rainer Hilf studied Interior Design at the Academy of Fine Arts in Nuremberg and has been a freelance interior designer since 1972. He has held various teaching positions since 1976, including Visiting Professor at the Burg Giebichenstein University of Art and Design in Halle, Germany. He has served as President of the German Association of Interior Designers (BDIA) since 1995. Rainer Hilf is also a member of numerous institutions, including membership on a steering committee for the German Federal Ministry of Transportation, Building and Urban Development.

Michael Volkmer

Ralf Höpfner

Rainer Hilf

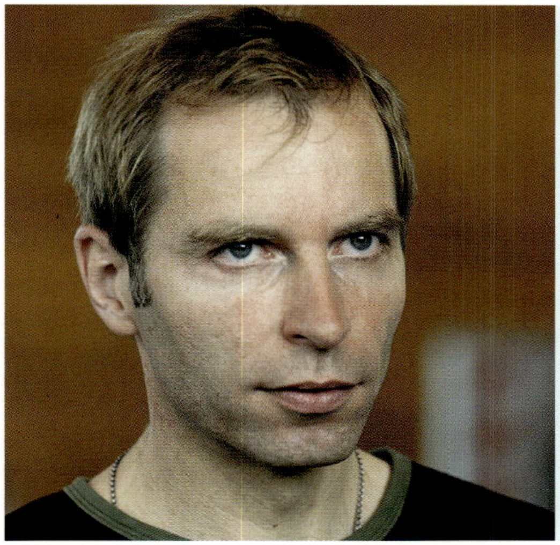

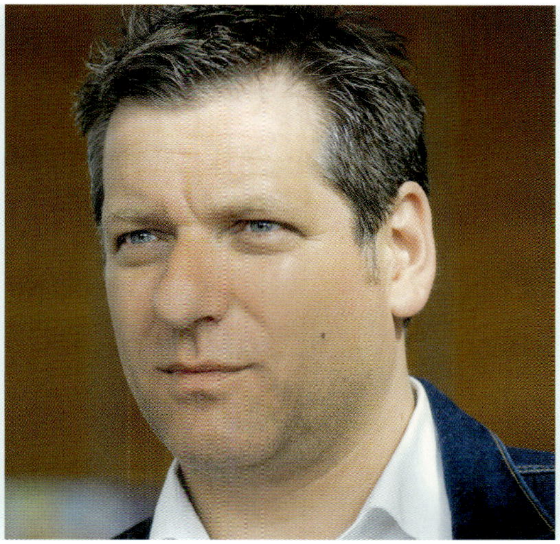

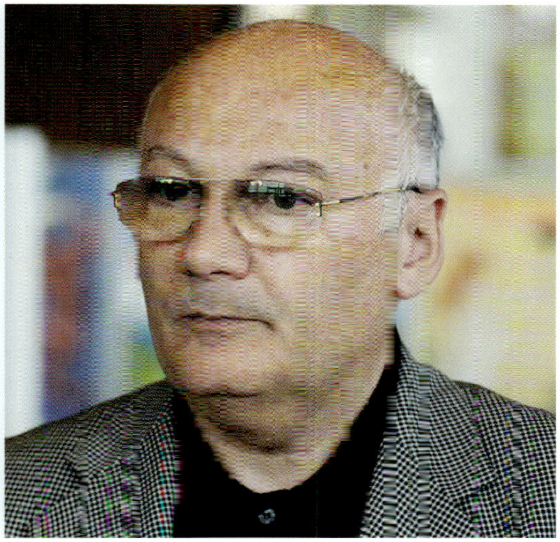

gold award: crossmedia

Auftraggeber Client
ProChrist e.V.
Kassel, Germany

Projekt Project
Die kleinste Kirche der Welt
Vorkampagne zur Veranstaltung ProChrist 2006
Campaign for the event ProChrist 2006

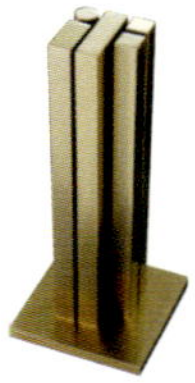

Design
Michael Buttgereit
Wolfram Heidenreich
Heiko Rafflenbeul
Buttgereit und Heidenreich GmbH
Haltern am See, Germany

70 Smarts. 40 Wochen. 70 europäische Städte. 3.600 Fahrer. Über 2 Millionen Kilometer. Das ist die Bilanz einer Kampagne, die von uns mit Unterstützung des EKD Ratsvorsitzenden Bischof Huber und Daimler-Chrysler für die überkonfessionelle Bewegung Pro-Christ konzipiert wurde. Als „Die kleinste Kirche der Welt" war sie über zehn Monate auf dem Weg zu den Menschen. 60.000 Gespräche über Alltagssorgen und Nöte, über Glauben und Zweifeln wurden in diesen mobilen Minikirchen geführt. Die Kirche auf dem Weg zu den Menschen – das wird auch von den Medien positiv aufgenommen: 17 Agenturmeldungen, 93 TV-Beiträge, über 1.500 Berichte in Print, Radio und Internet.

70 Smarts. 40 weeks. 70 European cities. 3,600 drivers. Over 2 million kilometers. These are the figures of a campaign that we conceived with support of Bishop Huber of the EKD (Evangelische Kirche Deutschlands) and DaimlerChrysler, a campaign for the supradenominational movement "ProChrist". As "The smallest church of the world" it was on the way to the people over a period of ten months. There were 60,000 conversations about worries and needs, about faith and doubts in these mobile mini-churches. The church on the way to the people was also present in the media: 17 agency notices, 93 reports on TV, over 1,500 reports in print, radio and on the Internet.

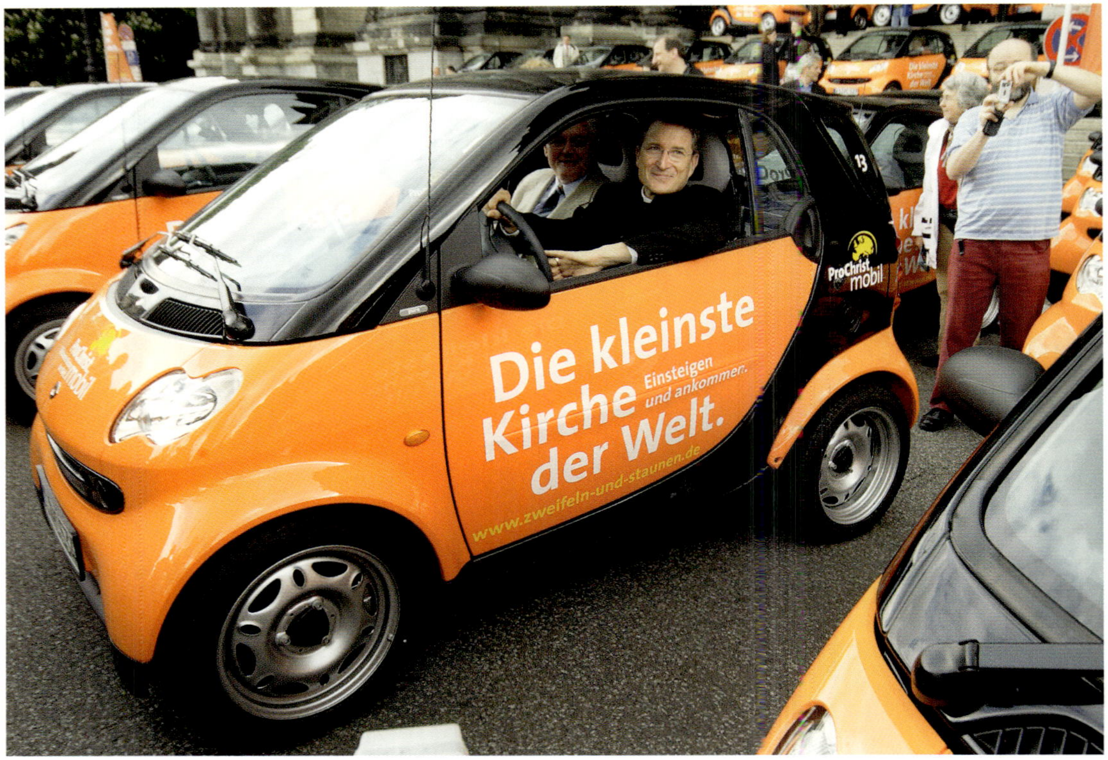

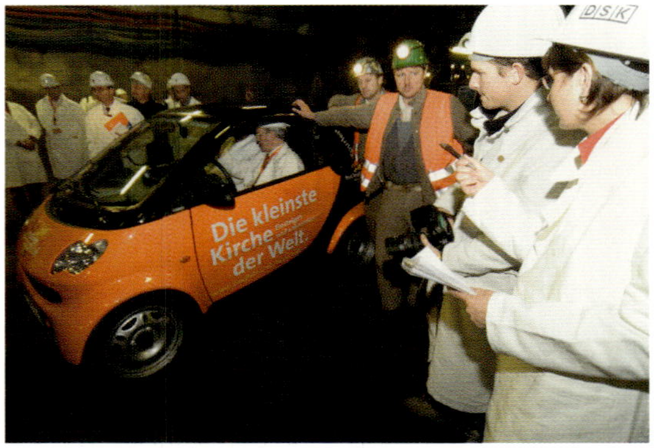

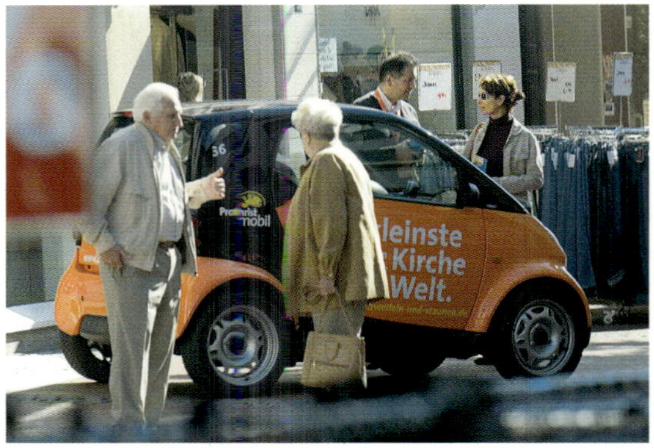

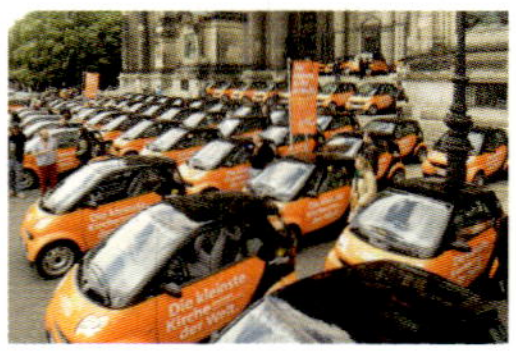

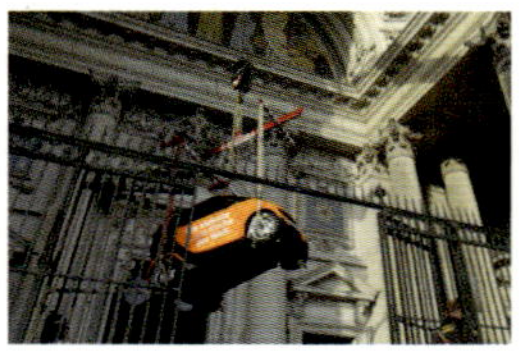

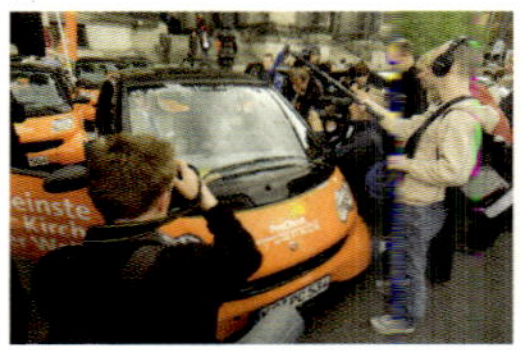

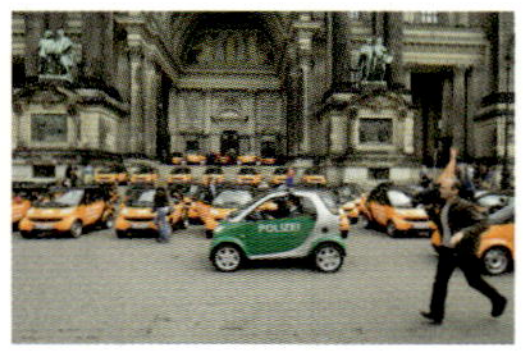

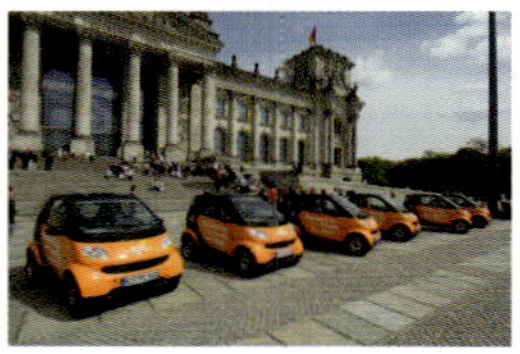

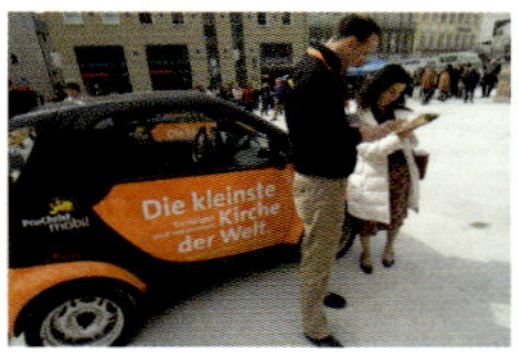

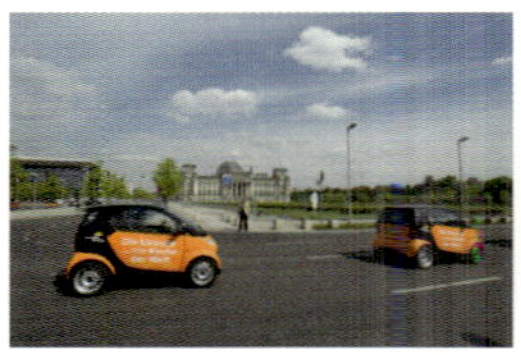

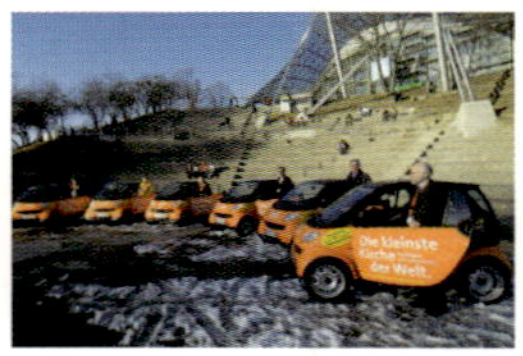

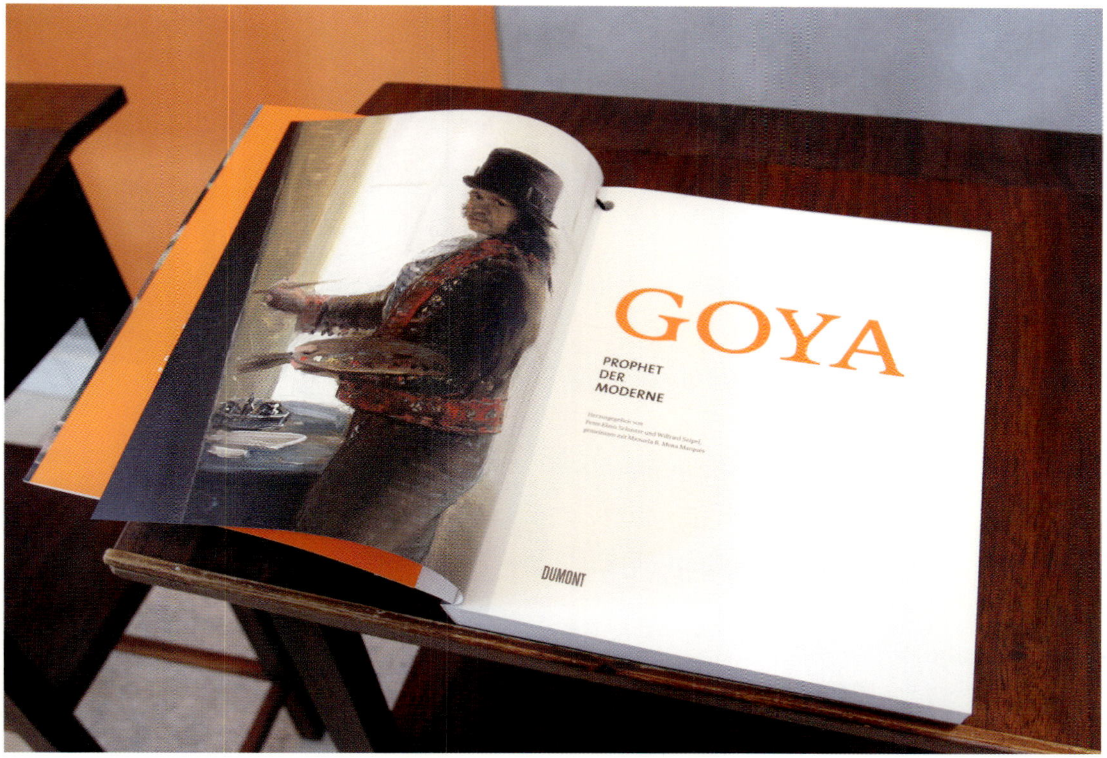

Auftraggeber Client
Verein der Freunde der Nationalgalerie e.V.
Berlin, Germany

Projekt Project
Goya – Prophet der Moderne
Corporate Design

Design
MetaDesign AG
Berlin, Germany

Der Name „Goya“ steht im Kunst- und Kulturkontext für einen Wegbereiter der Moderne. Vor diesem Hintergrund fand 2005 die bislang wichtigste Ausstellung der Werke Francisco de Goyas in der Alten Nationalgalerie, Berlin, statt. Der Name des Künstlers steht wie ein Markenzeichen prägnant im Zentrum der Kommunikation. Der Schriftzug „GOYA“ wurde in leuchtendem Orange auf einzelne Werke appliziert. Die auffällige Farbwahl und die typografische Gestaltung schafften eine aufmerksamkeitsstarke Kulturkampagne, der 230.000 Besucher folgten.

In the art and cultural world, Goya is considered one of the fathers of modernity. In 2005, the Alte Nationalgalerie in Berlin presented the most important exhibition of the artist's work to date. Goya's name was the focal point of communication activities, used like a distinct trademark. A "GOYA" word mark was printed over reproductions of individual works in glowing orange. The eye-catching color and the typographical design formed the heart of a striking cultural campaign that drew 230,000 visitors.

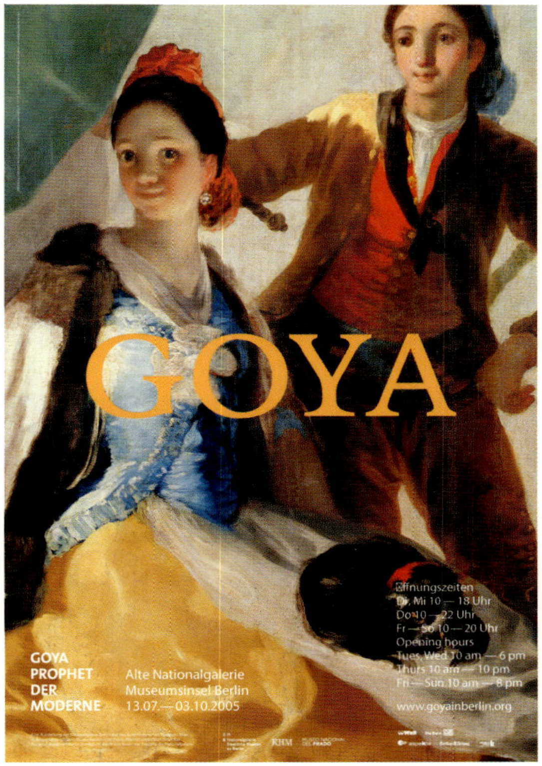

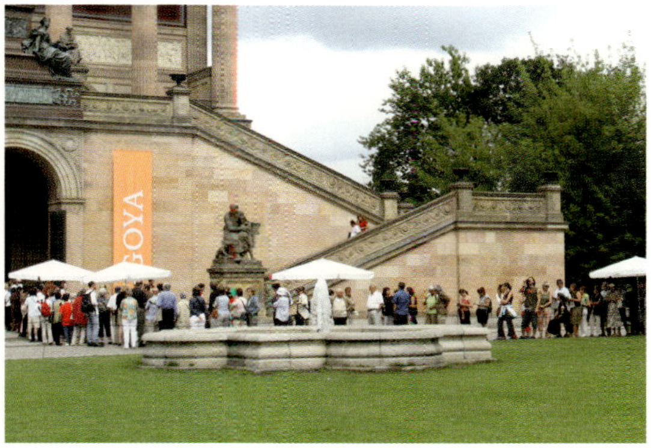

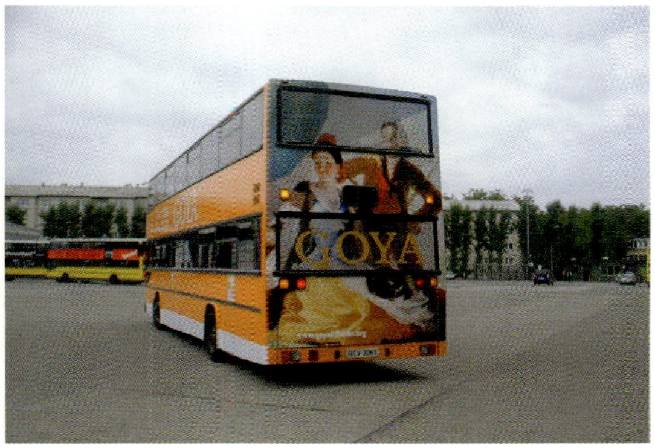

Auftraggeber Client
Deutsche Telekom AG
T-Com Zentrale
Bonn, Germany

Projekt Project
T-Com Haus
Integrierte Marketingkampagne
Integrated marketing campaign

Design
Regine Haschka-Helmer (Strategie/CD)
Birgit Frenzel, Volker Krön (Projektleiter)
Thomas Allgeier, Marco Massier, Ingo Floren,
Christoph Hammer (Projektmanager)
Robert Novakovic (Art Director)
Marcus Thormälen, Hervé Cecchi (Design)
I-D Media AG
Berlin, Germany

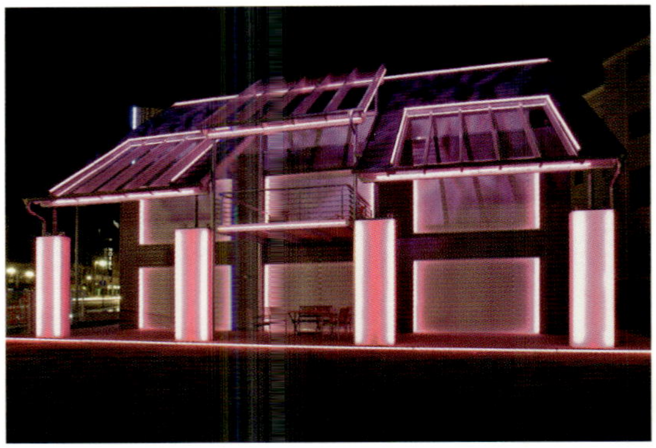

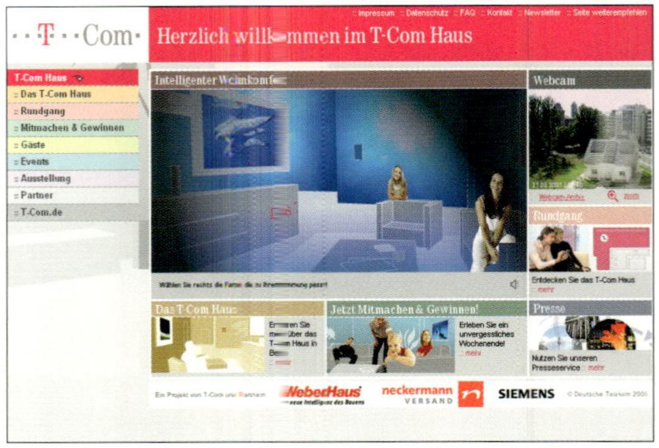

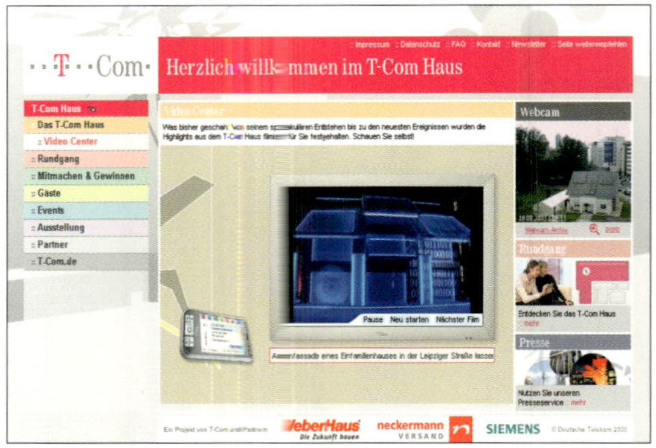

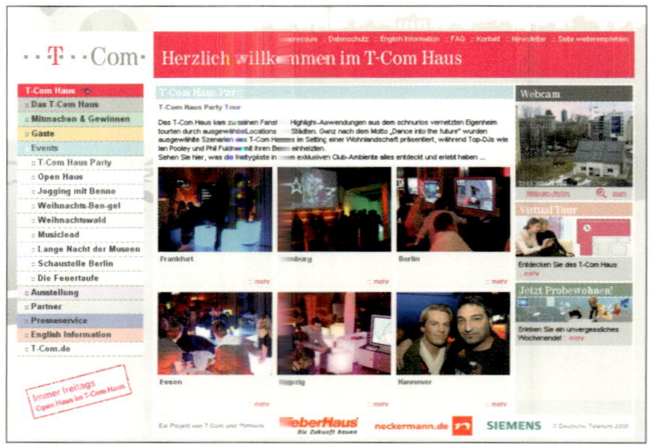

Im Zentrum des integrierten Marketingkonzepts stand das T-Com Haus, das physisch inmitten Berlins errichtet wurde. Modernste technische Einrichtungen der Kommunikations-, Entertainment-, Sicherheits- und Haustechnologie wurden im T-Com Haus erlebbar.
Wöchentlich sind Bewerber ins T-Com Haus eingeladen worden, um dort ein Wochenende zu verbringen. Sie wurden über verschiedenste Medienkanäle aufgerufen, sich für den Aufenthalt im Haus zu bewerben (Broschüren, Banner-Kampagne, Litfasssäulen, Events). Zentrale Plattform dabei war der Internetauftritt www.t-com-haus.de.
Im Zusammenspiel von illustrativen Elementen und realem Fotomaterial wurden die Formen und Farbgebung der T-Com CI in sehr moderner Anmutung umgesetzt. Bereits das Gesamtdesign des Projekts ist so konzipiert worden, dass eine ideale datenkomprimierte Adaption für den Online-Auftritt und für Online-Werbemittel möglich war und gleichzeitig stark emotionale Gestaltungen für den Print-Bereich entstanden sind.

The T-Com Haus offered consumers and journalists insights into the present and near-future of media use as well as communication technologies within the private household.
At the center of the integrated marketing concept was the T-Com Haus, which has been physically constructed in the middle of Berlin. Within the scope of a consumer program, each week T-Com raffled off a long-weekend stay in the T-Com Haus to a different team.
The central platform of the integrated marketing campaign was the website www.t-com-haus.de. Numerous other communication channels (brochures, posters, banner campaign, advertising columns etc) assured the positive image impact of the project with a broad consumer base.
The overall design of the project was conceived in such a way as to enable an ideal adaptation for both online site and online advertising materials in terms of data compression. At the same time, an emotional design for the print sphere has been accomplished.

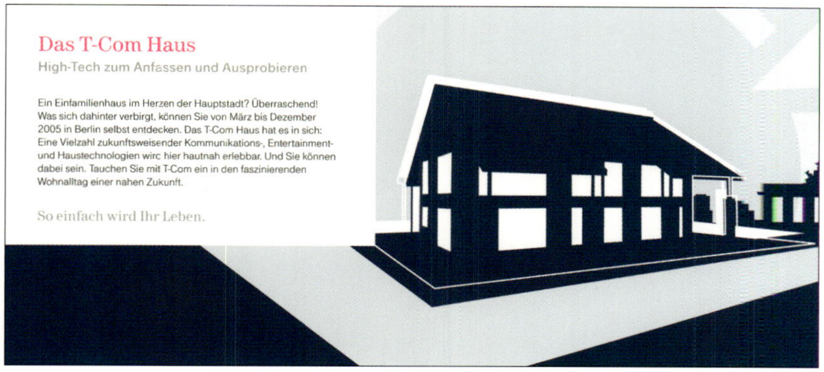
Das T-Com Haus
High-Tech zum Anfassen und Ausprobieren
Ein Einfamilienhaus im Herzen der Hauptstadt? Überraschend! Was sich dahinter verbirgt, können Sie von März bis Dezember 2005 in Berlin selbst entdecken. Das T-Com Haus hat es in sich: Eine Vielzahl zukunftsweisender Kommunikations-, Entertainment- und Haustechnologien wirc hier hautnah erlebbar. Und Sie können dabei sein. Tauchen Sie mit T-Com ein in den faszinierenden Wohnalltag einer nahen Zukunft.
So einfach wird Ihr Leben.

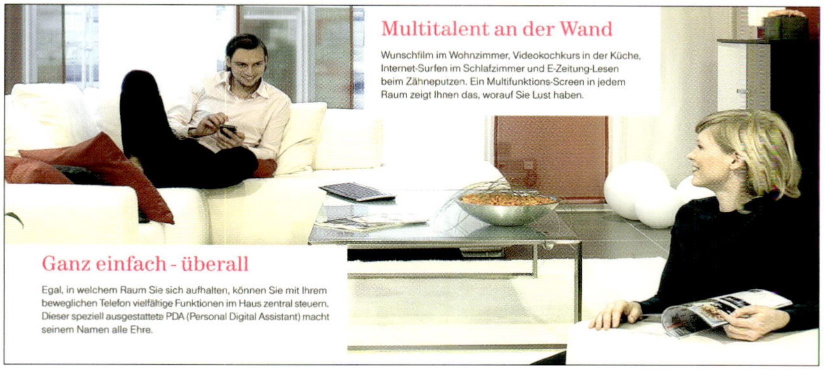
Multitalent an der Wand
Wunschfilm im Wohnzimmer, Videokochkurs in der Küche, Internet-Surfen im Schlafzimmer und E-Zeitung-Lesen beim Zähneputzen. Ein Multifunktions-Screen in jedem Raum zeigt Ihnen das, worauf Sie Lust haben.
Ganz einfach - überall
Egal, in welchem Raum Sie sich aufhalten, können Sie mit Ihrem beweglichen Telefon vielfältige Funktionen im Haus zentral steuern. Dieser speziell ausgestattete PDA (Personal Digital Assistant) macht seinem Namen alle Ehre.

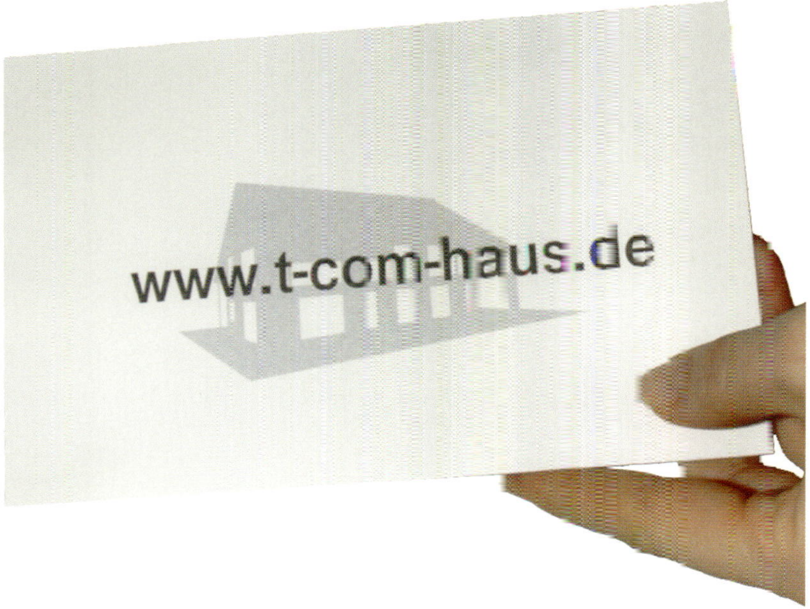
www.t-com-haus.de

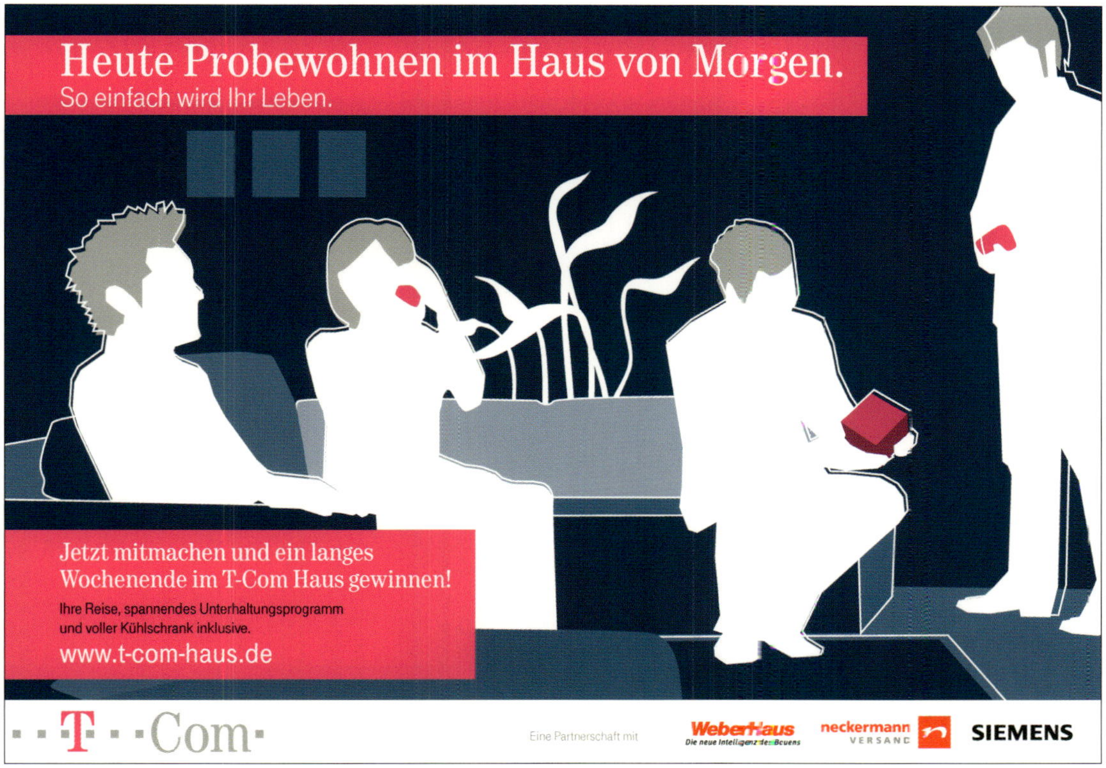
Heute Probewohnen im Haus von Morgen.
So einfach wird Ihr Leben.
Jetzt mitmachen und ein langes Wochenende im T-Com Haus gewinnen!
Ihre Reise, spannendes Unterhaltungsprogramm und voller Kühlschrank inklusive.
www.t-com-haus.de
T · · Com ·
Eine Partnerschaft mit
WeberHaus
Die neue Intelligenz des Bauens
neckermann
VERSAND
SIEMENS

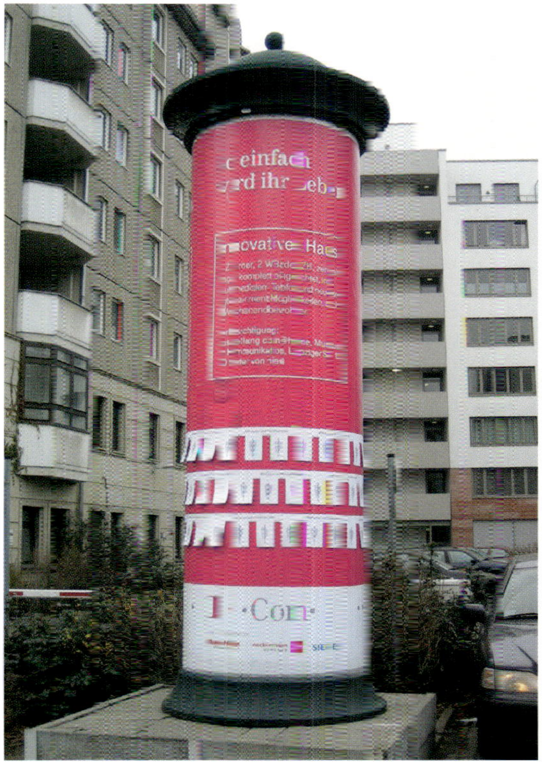

crossmedia

Auftraggeber Client
Hochschule für Gestaltung und Kunst Zürich hgkz
Hochschule Musik und Theater Zürich hmt
Zurich, Switzerland

Projekt Project
hgkz + hmtz = zhdk
Corporate Design

Design
Richard Feurer
Matthias Michel
Gisèle Schindler
feurer network ag
Zurich, Switzerland

Projekt: Gemeinsames Corporate Design im Hinblick auf den Zusammenschluss zur Zürcher Hochschule der Künste (zhdk).
Anspruch: Kontinuität und Wandel: Das CD schreibt die Geschichte einer modern-konkreten Tradition fort und kommuniziert zugleich eine offene Zukunft kreativer Vielfalt. Institution und Individuation: Ein einheitlicher Auftritt nach außen, möglichst große Kreativspielräume für unterschiedlichste Forschungs- und Studienbereiche nach innen.
Kernkonzept: hgkz und hmt (die künftige zhdk) werden als „open space" begriffen, in dem eine Vielzahl von Teilbereichen mit jeweils eigenen Identitäten integriert ist. Dieser „offene Raum" wird zum formalen Prinzip. Die „Einheit" ist nicht in einem Logo oder Schriftzug repräsentiert, sondern im prägnanten gestalterischen Umgang mit der „Vielfalt".

Project: Joint corporate design for the fusion spawning the Zurich School of the Arts (zhdk).
Brief: Continuity and change: The corporate design updates the history of a concrete, modern tradition, while at the same time communicating a future that is wide open to creative diversity. Institution and individuation: An integrated external appearance and as much scope as possible for internal creativity in widely divergent areas of research and study.
Core concept: hgkz and hmt (the future zhdk) are seen as an open space integrating a large number of departments, each with their own identity. This "open space" becomes the design principle. "Unity" is not represented in a logo or lettering, but in a concise creative treatment of "diversity".

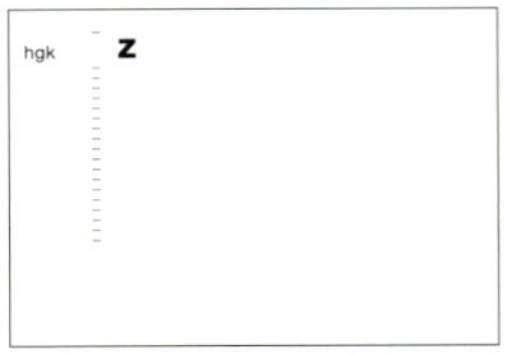

a

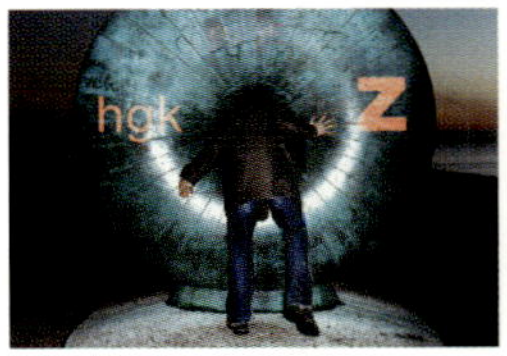

b

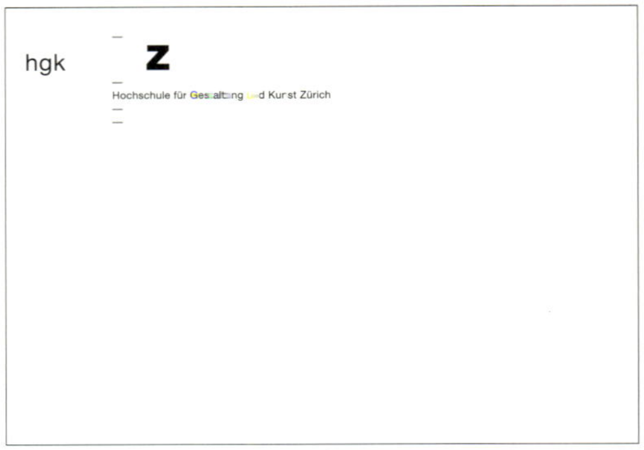

c

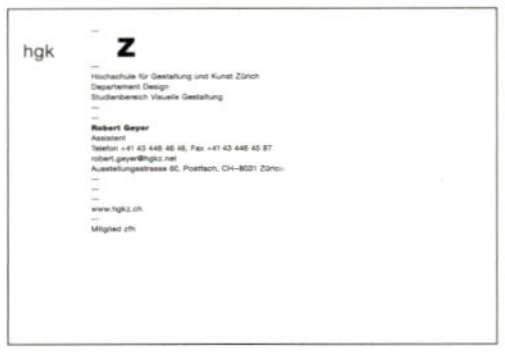

d

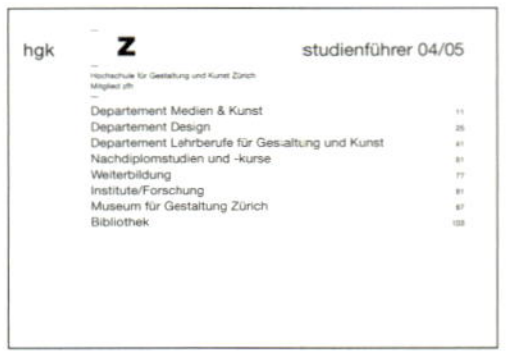

e

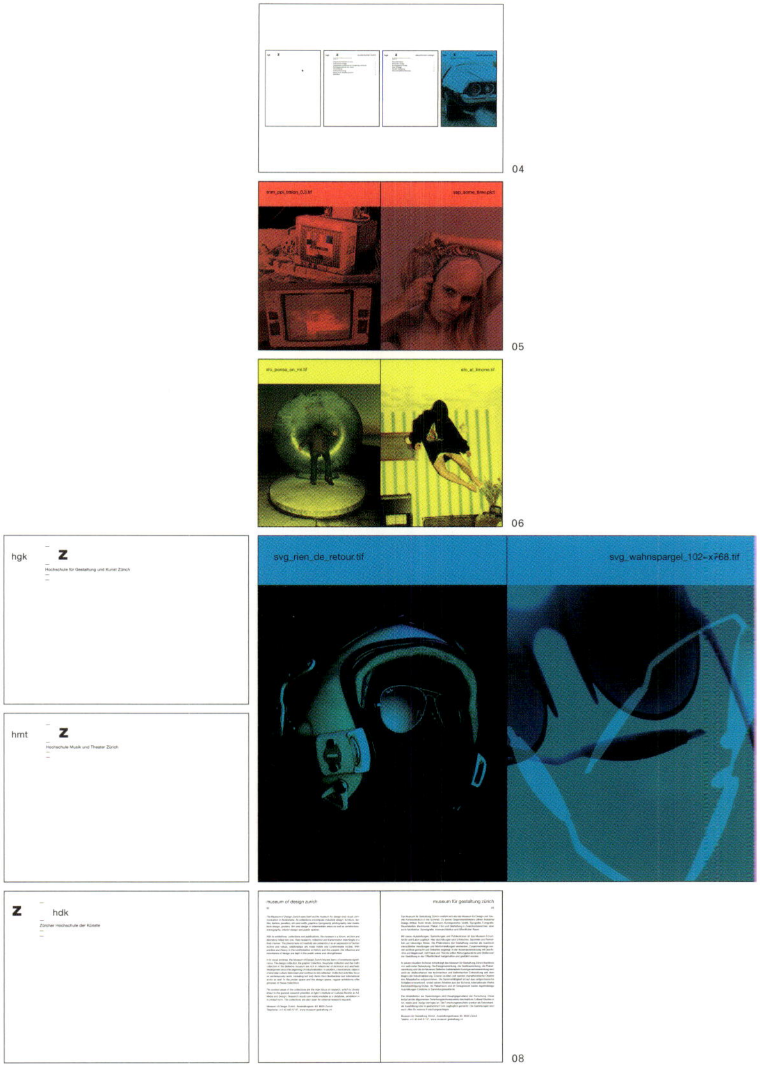

04 05 06 01 07 02 03 08

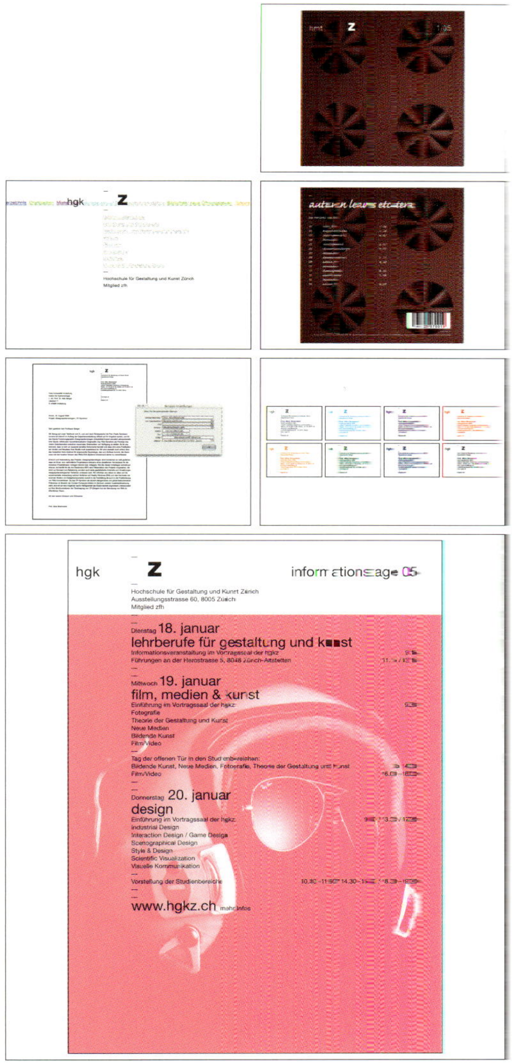

09 10 11

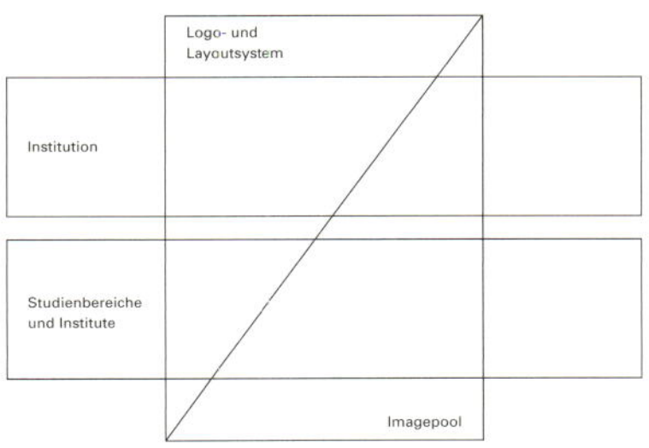

- a Ästhetisch-rh[illegible]orische Gru[illegible]
- b Bild-[illegible] Erl[illegible]br[illegible]elt
- c Schr[illegible]tzu[illegible]
- d Log[illegible]system
- e Lay[illegible]tsy[illegible]tem

01 S[illegible]d L[illegible]system
02 S[illegible]d L[illegible]system
03 S[illegible]d L[illegible]system
04 S[illegible]ernr[illegible]en
05 B[illegible]elt[illegible]medien
06 B[illegible]elt[illegible]medien
07 B[illegible]elt[illegible]medien
08 L[illegible]ut K[illegible]edien
09 V[illegible]
10 K[illegible]figur[illegible] Briefbogen
11 V[illegible]nstal[illegible]splakat
12 [illegible] Se[illegible]hmtz records
13 [illegible] Se[illegible]hmtz records
14 V[illegible]enka[illegible]

Auftraggeber Client
mamamoto
Hamburg, Germany

Projekt Project
mamamoto
Geschäftsausstattung, Produkte, Event, Website, Buch
Business equipment, products, event, website, book

Design
Claudia Fischer-Appelt
Petra Matouschek
Thomas Kappes
Hana Sedelmayer
Arne Schmidt
Ligalux
Hamburg, Germany

Die Welt ist in Bewegung. Nur das Familienbild, das uns täglich kommuniziert wird, trotzt jeder Neuerung. mamamoto fordert eine „Neuauflage".
Die Marke bedient sich auf allen Kommunikationskanälen eines zeitgemäßen, klaren Designs – jenseits von Blümchen und Bärchen. Gestartet ist mamamoto mit Geschäftsausstattung, Website, Manifest und dem „T-Shirt-Tausch-Tag" (Event- und Presseausstattung). Weitere Projekte sind: Studien zum Familienbild, eine Kampagne, das Buch „Family Business" sowie weitere Produkte. Damit Familie wieder einen Platz im Herzen der Gesellschaft findet.

The world is changing. Only the image of the family, which we are inundated with daily, is still digging its heels in. mamamoto calls for a makeover.
Throughout all channels of communication, the brand steers clear of the tritely snug world of teddy bears and flowers, opting for a pure, contemporary design. The initiative began with business equipment, a website, the manifesto and the "T-shirt exchange day". Alongside studies on the image of the family, a campaign ("Praise the parents"), and a book ("Family Business") are scheduled. The aim is to return the family to its rightful place at the heart of society.

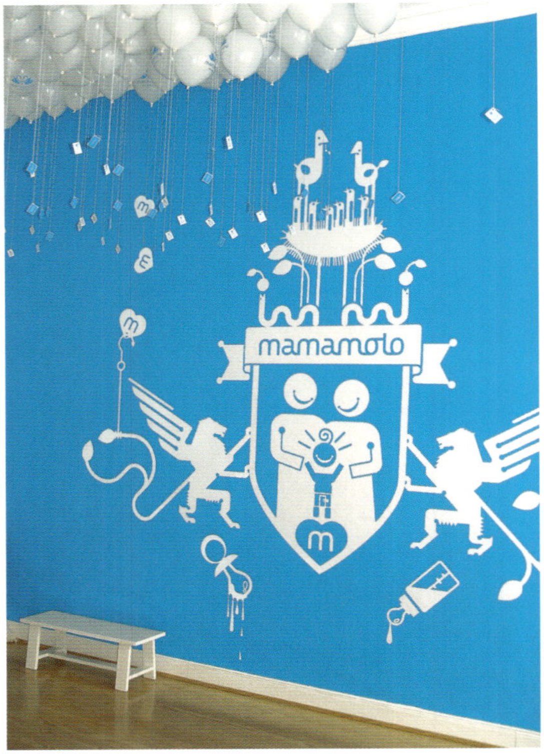

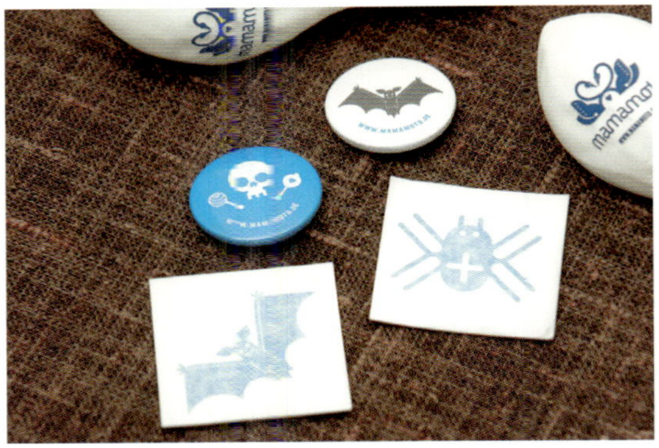

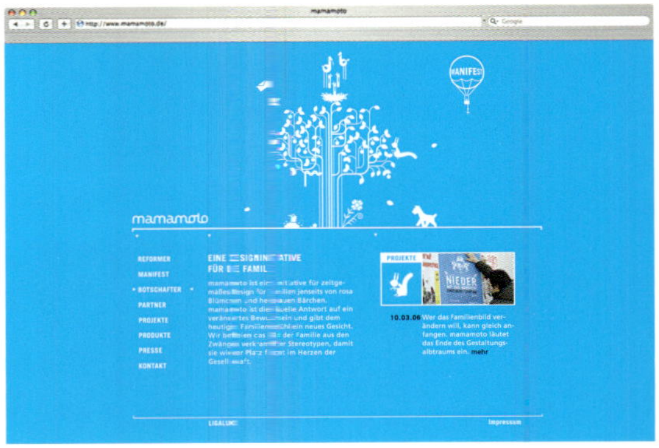

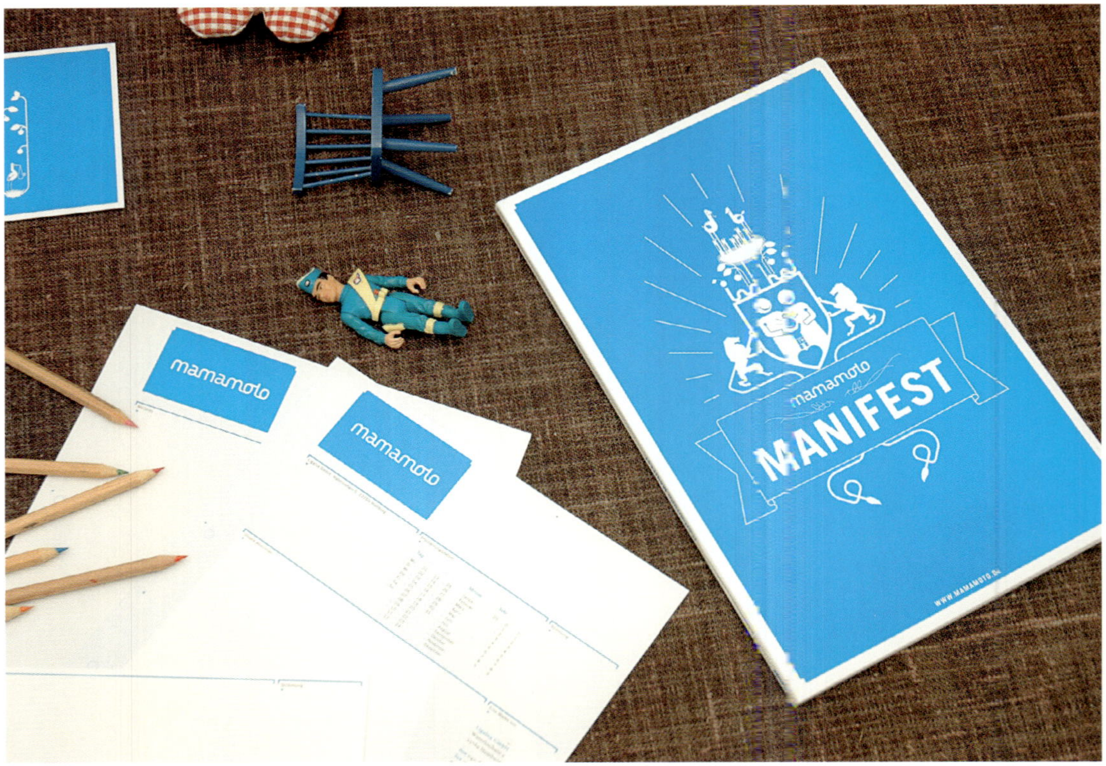

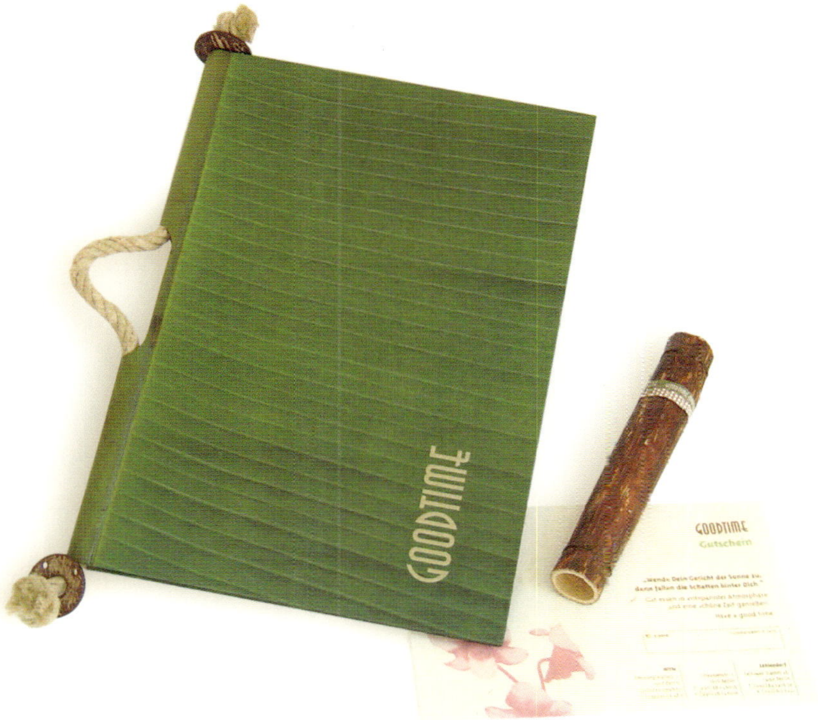

Auftraggeber Client
Goodtime GbR
Berlin, Germany

Project Project
Goodtime
Speisekarte, Gutschein, Internet
Menu, voucher, website

Design
Florian Braun
Holger Klitzke
Andreas Berthel
terralisto
Berlin, Germany

„Appetit anregen, ohne Essen darzustellen.“ – In der traditionellen thailändischen Küche vereinen sich Einflüsse aus Indien, Asien, Europa und Lateinamerika. Die südostasiatische Küche kann man demzufolge auch als das erste „Fusion Food“ der Geschichte bezeichnen.
Bei der Übersetzung dieser besonderen Auszeichnung in ein Corporate Design diente uns die Collage als das treffsicherste Gestaltungsmerkmal. Ganz nah am gastronomischen Konzept der Goodtime-Restaurants gestalteten wir seitenweise explorative Welten in Analogie zu den verschiedenen Speiserichtungen.

"Whetting the appetite without presenting food." – Traditional Thai cooking has Indian, Asian, European and Latin American influences. South-East Asian cuisine can thus be described as the first "fusion food" in history. Collage was the most effective way of conveying this distinctive feature in corporate design. Working closely with the gastronomic concept of Goodtime Restaurant, we designed explorative worlds in line with the range of different foods.

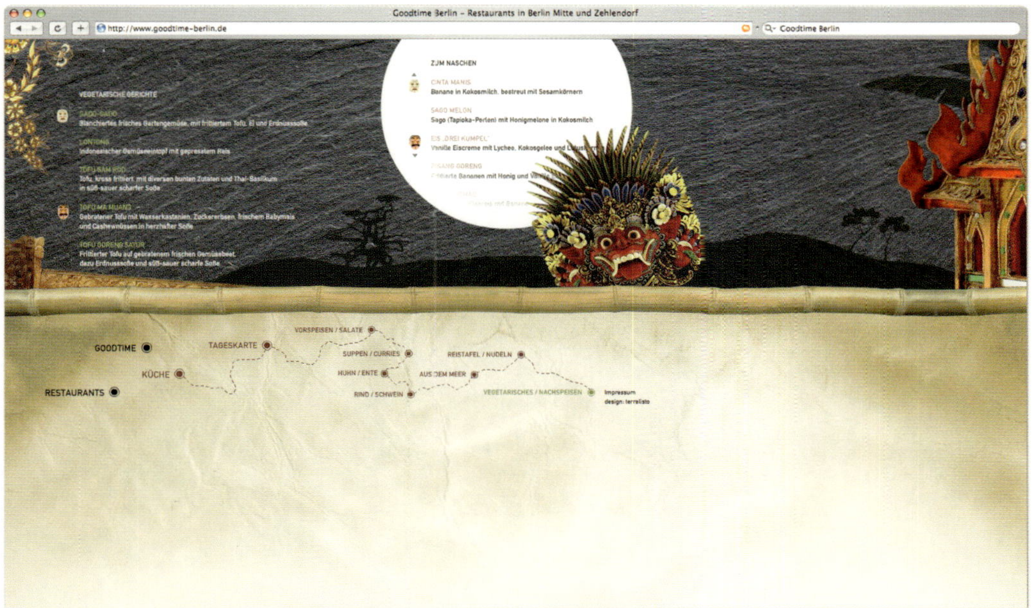

crossmedia

Auftraggeber Client
Axelmeiselicht
München, Germany

Projekt Project
Occhio Sento von Axelmeiselicht
Marken-Launch und Marken-Implementierung
Brand launch and implementation

Design
Peter Martin, Daniel Karczinski, Birgit Fuhrmeister, Daniela Scharf, Corinna Rödel, Jaana Zammer, Mike Rabensteiner, Birgit Pfisterer
(Design/Konzeption)
Nicola Sacher
(Text/PR)
Robert Sprang, Matthias Garvelmann
(Fotografie)
Peter Martin, Jürgen Drändle (Drändle 70 l 30), Axel Meise
(3D/Messe)
Martin et Karczinski GmbH
München, Germany

Aufgabe war es, das neue Leuchtensystem Occhio Sento an die bereits von Martin et Karczinski entwickelte Markenstrategie von Occhio zu koppeln und in die Markenkommunikation zu integrieren.
Um die systemischen Qualitäten effektiv zu kommunizieren, wird Occhio Sento als eigenständiger Brand im Markt eingeführt. Im Namen erkenntlich bleibt dabei Herkunft und funktionale Kompatibilität zum eingeführten Leuchtensystem Occhio.
Umsetzung: Definition der Brandrelation, Produktname, Claim, Wortmarke.
Medien: Produkt- und Anwendungskataloge, Preislisten, PoS-Displays, Verpackungen, Standkonzeption zur Light+Building 2006, Öffentlichkeitsarbeit.

The task was to tie in the new Occhio Sento lighting system with the Occhio brand strategy already developed by Martin et Karczinski, and to integrate it into the corporate communication.
For effective communication of the systemic qualities of Occhio Sento, it was launched as a separate brand. Its name reveals its origins and indicates its functional compatibility with the Occhio lighting system already available on the market.
Realisation: Definition of brand relation, product name, claim, word-marque.
Media: Product and application catalogues, price lists, POS displays, packaging, stand design for Light+Building 2006, PR work.

axelmeiselicht

Occhio | sento

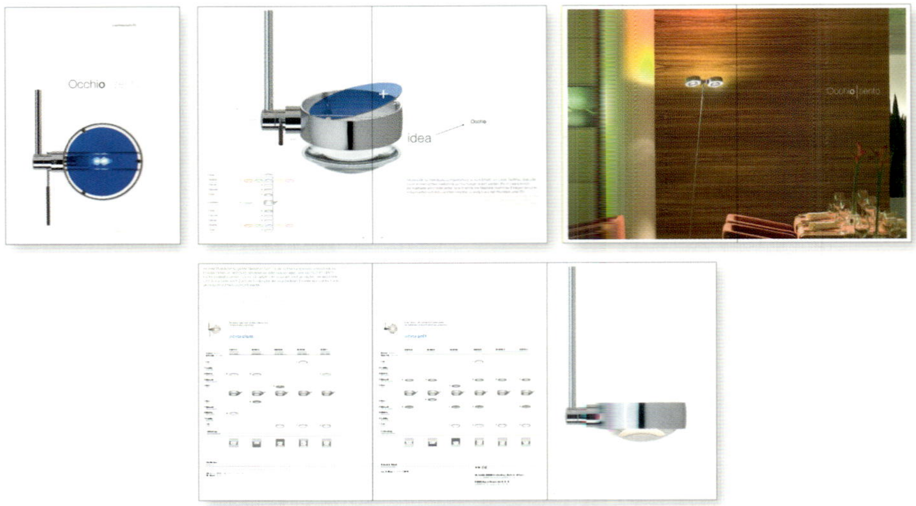

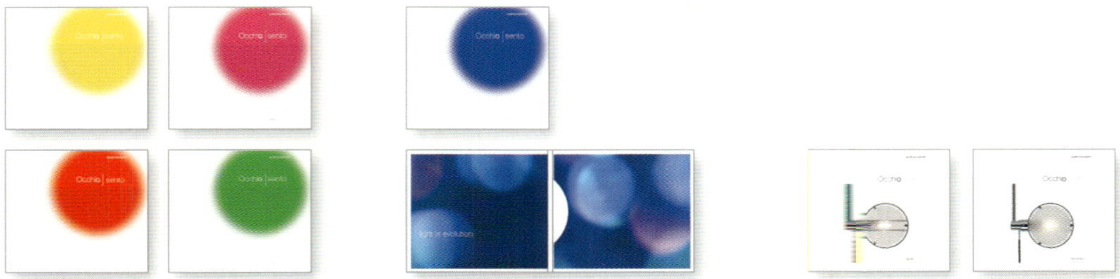

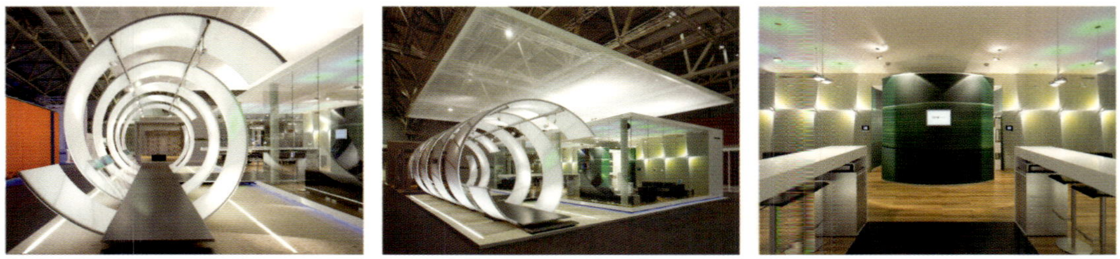

ARCHITEKTUR ADMINISTRATION ART

Raum und Atmosphäre

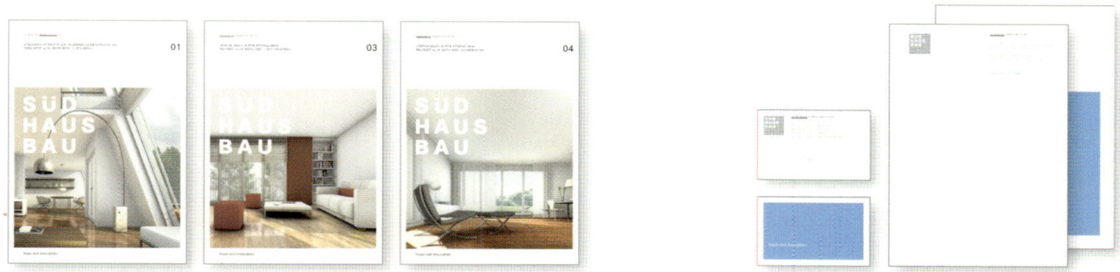

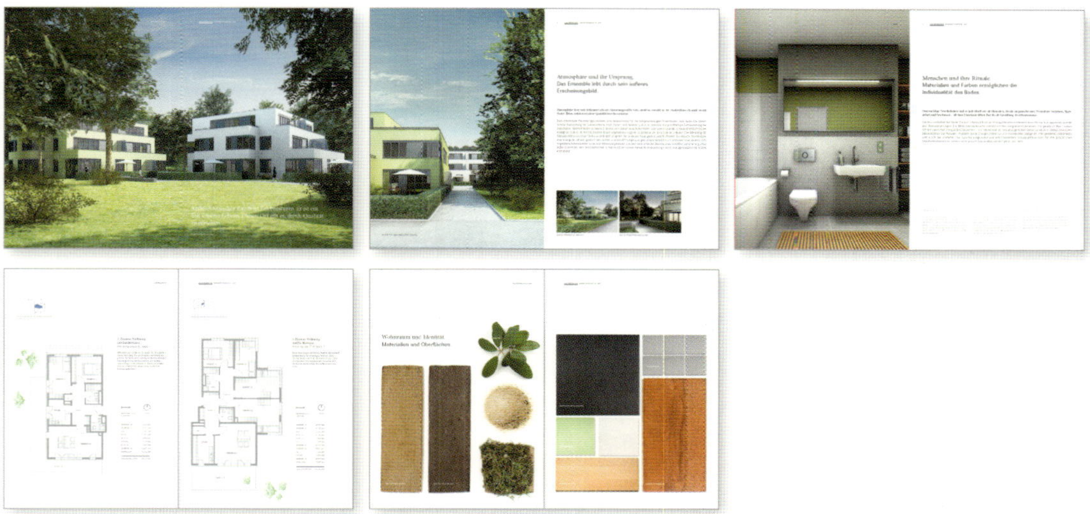

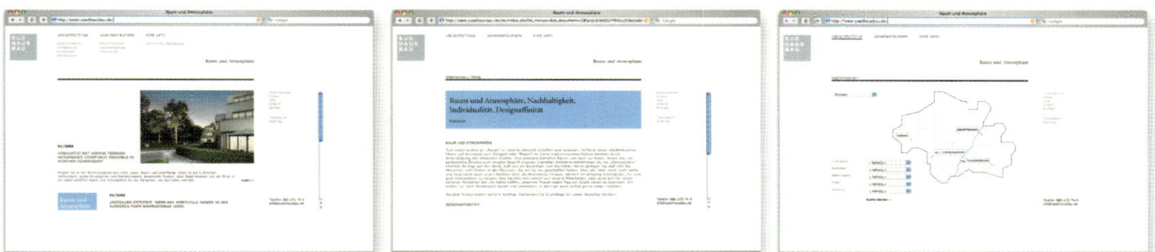

Auftraggeber Client
Ottmann GmbH & Co Südhausbau KG
München, Germany

Projekt Project
Südhausbau
Markenrelaunch
Rebrand

Design
Peter Martin, Daniel Karczinski, Nicola Sacher
(Strategie/Positionierung)
Peter Martin, Daniela Scharf, Petra Fössel,
Nicola Bernhard, Mike Rabensteiner
(Design)
Daniel Karczinski, Nicola Sacher, Matthias Höofl
(Text/PR)
Katrin Aser, Martin Aigner, Torsten Melcherczyk
(Web)
Martin et Karczinski GmbH
München, Germany
Ralph Rieger
VisualServices
Hamburg, Germany

Der Münchner Bauträger Südhausbau verwaltet allein in Bayern mehr als 7.000 Wohnungen und Gewerbeeinheiten. Neben einem auf allen modernen Medien anwendbaren Erscheinungsbild und einer Neupositionierung am Markt ist der Qualitätsaufbau der Renderings und Grundrisse als Differenzierungsmerkmal zentraler Teil der Corporate Communication.
Umsetzung: Unternehmens- und Marktanalyse, Neupositionierung, Namensfindung zur Benennung der Geschäftsfelder, Corporate Design, Re-Design der Wort-/Bildmarke, Claim, Bildsprache.
Medien: Geschäftsausstattung, Exposés für Neu-/Bestandsobjekte, Imagemappen, Bauschilder, Fahnen, Website, Anzeigen, Messeauftritt.

Within Bavaria alone Südhausbau, a Munich-based building developer, manages over 7,000 apartments and business units. Alongside a corporate image usable in all modern media and a repositioning on the market, the quality of the renderings and the ground plans are a distinctive feature and core component in the corporate communication.
Realisation: Company and market analysis, repositioning, name-finding for the business divisions, corporate design, redesign of the word/picture-marque, claim, imagery.
Media: Business papers, exposés for new/existing buildings, image folders, site signs, flags, website, adverts, fair presentation.

Auftraggeber Client
Thyssen Polymer GmbH
Bogen, Germany

Projekt Project
Thyssen Polymer wird zu INOUTIC
Markenrelaunch
Rebrand

Design
Peter Martin, Daniel Karczinski
(Strategie / Positionierung / Naming)
Peter Martin, Daniela Scharf, Birgit Pfisterer,
Patrick Wachner
(Design)
Daniel Karczinski, Nicola Sacher, Matthias Höpfl
(Text / PR)
Martin Aigner, Katrin Asen, Torsten Malcherczyk
(Web)
Simon Hoegsberg
(Fotografie)
Peter Martin, Jürgen Drändle (Drändle 70 I 30)
(3D / Messe)
Martin et Karczinski GmbH
München, Germany

Der Fenster-Profilhersteller Thyssen Polymer wurde an den belgischen Konzern Deceuninck verkauft, was zum Erlöschen der Namenslizenz führte. Der neue Name INOUTIC (in I out I ic) setzt sich zusammen aus „inside" und „outside" und ist definiert als die Wissenschaft von den architektonischen und wirtschaftlichen Zusammenhängen zwischen Innen und Außen. Umsetzung: Markt-/ Unternehmensanalyse, Neupositionierung, internationale Markenstrategie, Namensfindung („tell the logo"), Claim, Wort-/ Bildmarke. Medien: Imagebroschüre, Produktprospekte, Internetauftritt, Imagefilm, Messeauftritt, Corporate Sound, Mitentwicklung neuer Produkte, PR, Anzeigenkampagne.

Thyssen Polymer, a manufacturer of window profiles, was sold to the Belgian group Deceuninck, a move which led to the termination of the licence on the company name. The new name, INOUTIC (in I out I ic), created out of the words "inside" and "outside", is defined as the science of the architectural and economic relations between inside and outside. Realisation: Market / company analysis, repositioning, international brand strategy, name-finding ("tell the logo"), claim, word / picture-marque. Media: Image brochure, product brochures, website, image film, trade-fair presentation, corporate sound, co-development of new products, PR, advertising campaign.

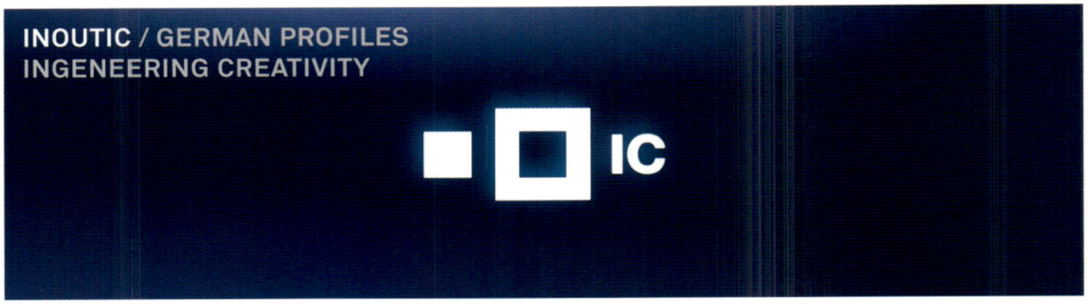

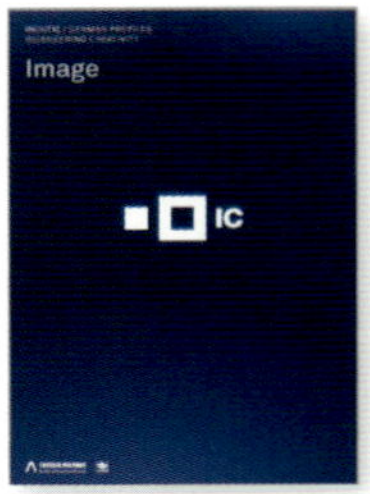

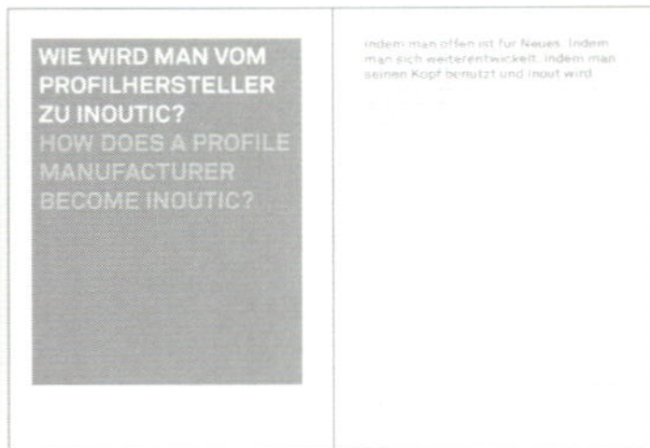

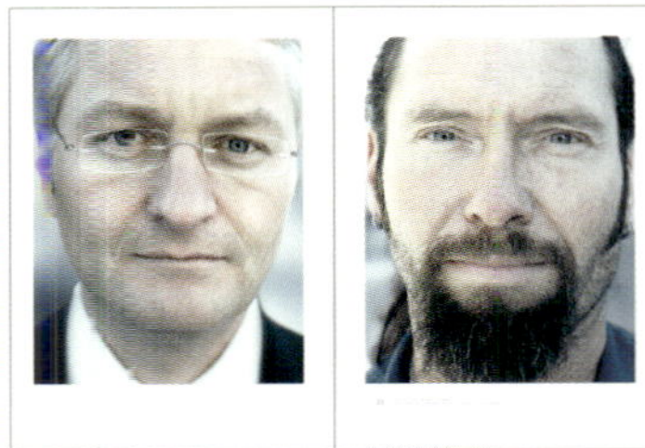

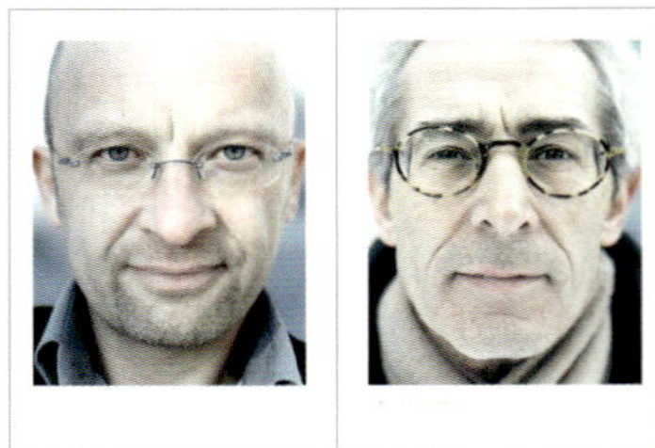

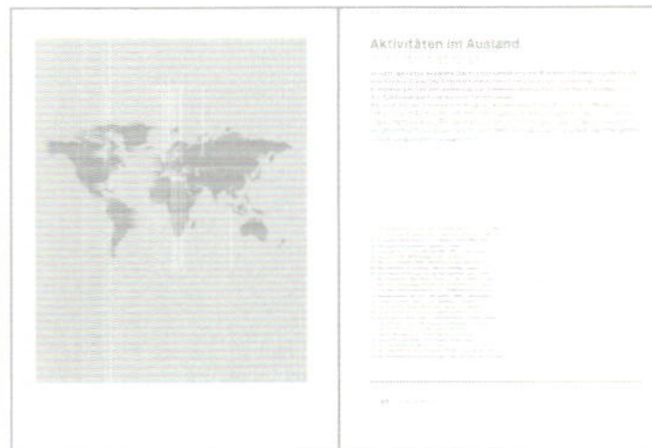

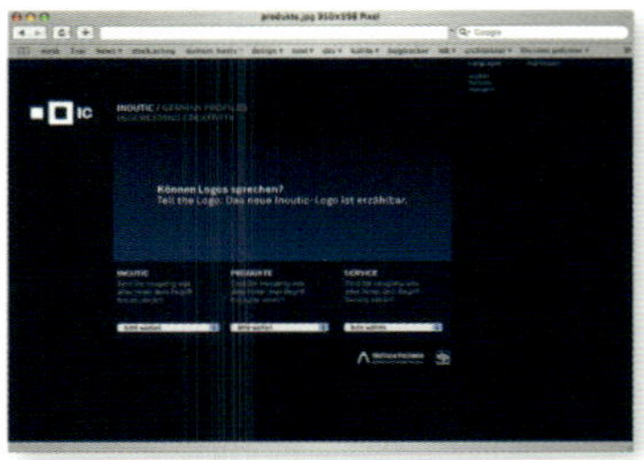

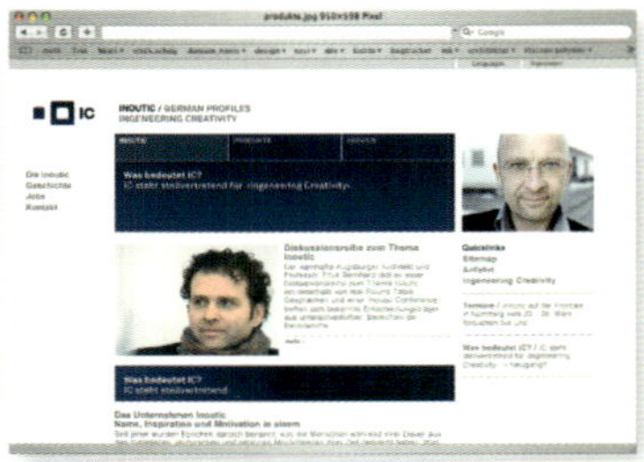

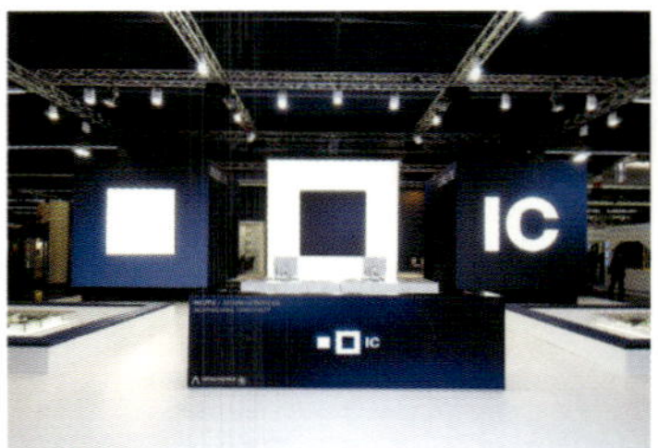

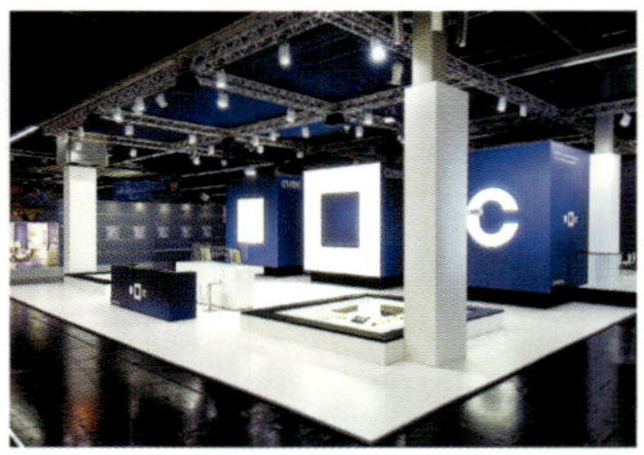

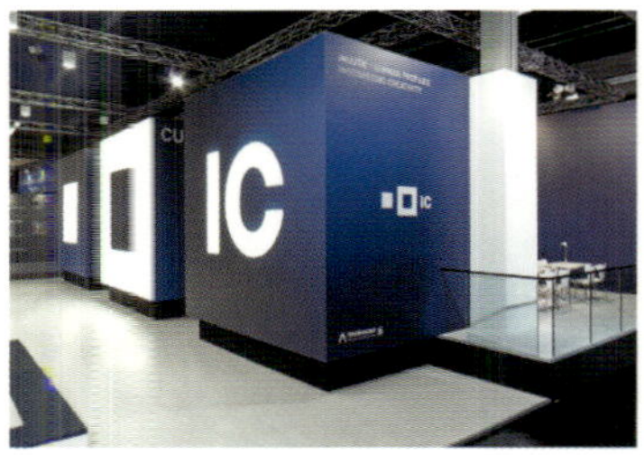

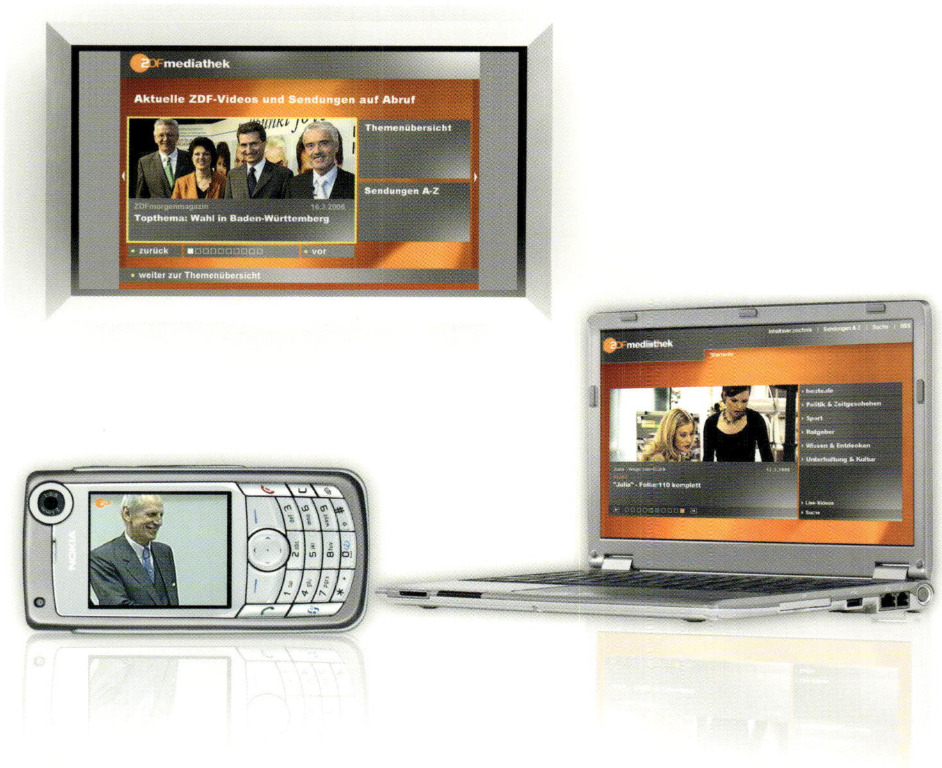

Auftraggeber Client
ZDF
Mainz, Germany

Projekt Project
ZDFmediathek
Video-on-Demand-Angebot
Video on demand offer

Design
Jochen A. Zeitler
Pixelpark AG
Köln, Germany

Die Mediathek mit einem Video-Angebot aus aktuellen Sendungen des ZDF ist eines der ersten umfassenden IPTV-Angebote in Europa und in ihrer stringenten Konvergenz aus Web, Mobile und IPTV derzeit einzigartig. Dabei ist jedes einzelne Frontend von der Bedienung her konsequent an das Eingabegerät und das Nutzungsszenario angepasst (TV-Fernbedienung, PC-Tastatur, Handy). Die Anbindung an den digitalen Workflow des ZDF erlaubt es, alle Kanäle ohne kanalspezifischen Mehraufwand zu befüllen. Die einzelnen Angebote selbst wurden konsequent in auf der IP-Plattform etablierten Standards programmiert (XHTML, Javascript, Flash), um nicht an einzelne, proprietäre Plattformen gebunden zu sein.

The ZDFmediathek is one of the most extensive IPTV-offers in Europe. Presently the convergence of web, mobile and IPTV is of a unique kind. The usability of every single frontend is adapted consequently to the input device and the user scenario (remote control, keyboard, mobile phone). The look and feel and the brand of the ZDF persists across all distribution channels. The connection to the digital workflow of the ZDF enables the filling of the channels without additional work. The coding of the several offers is based on established standards of the IP-platform (like XHTML, Javascript, Flash) to be independent of proprietary platforms.

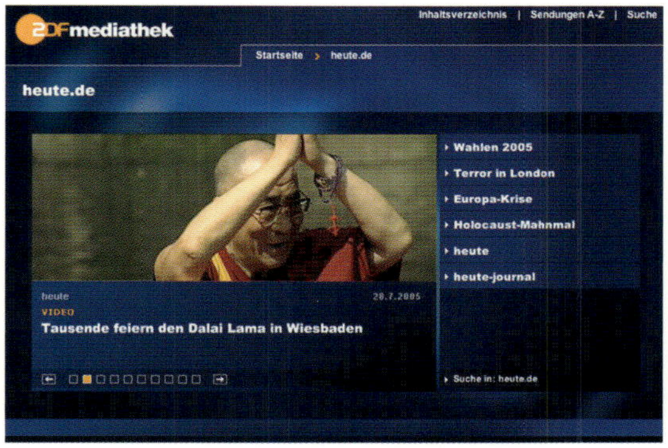

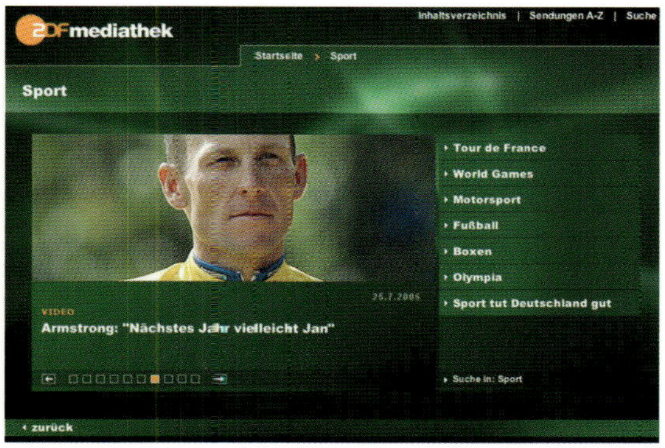

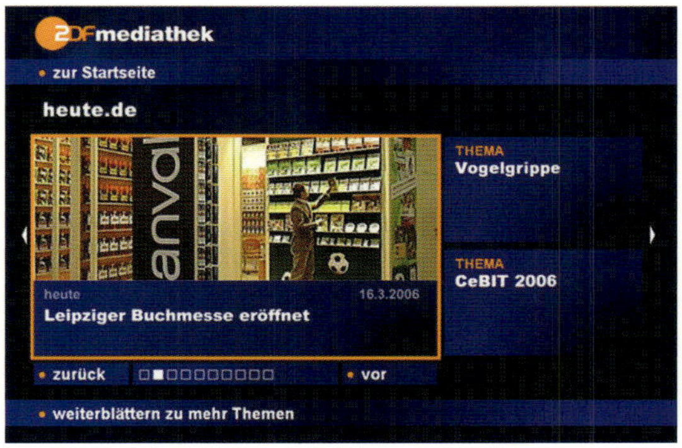

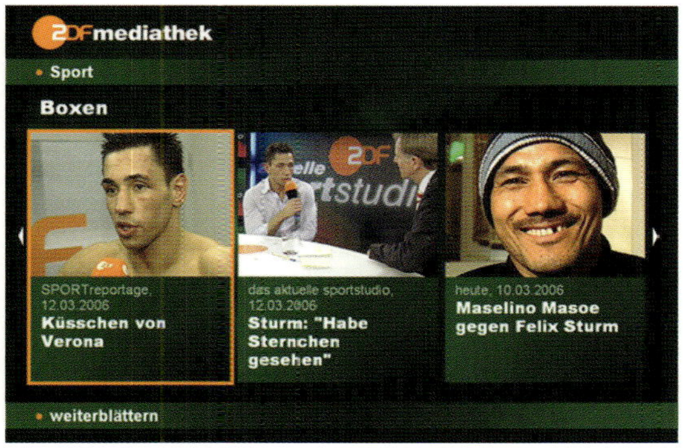

05 too good to be true

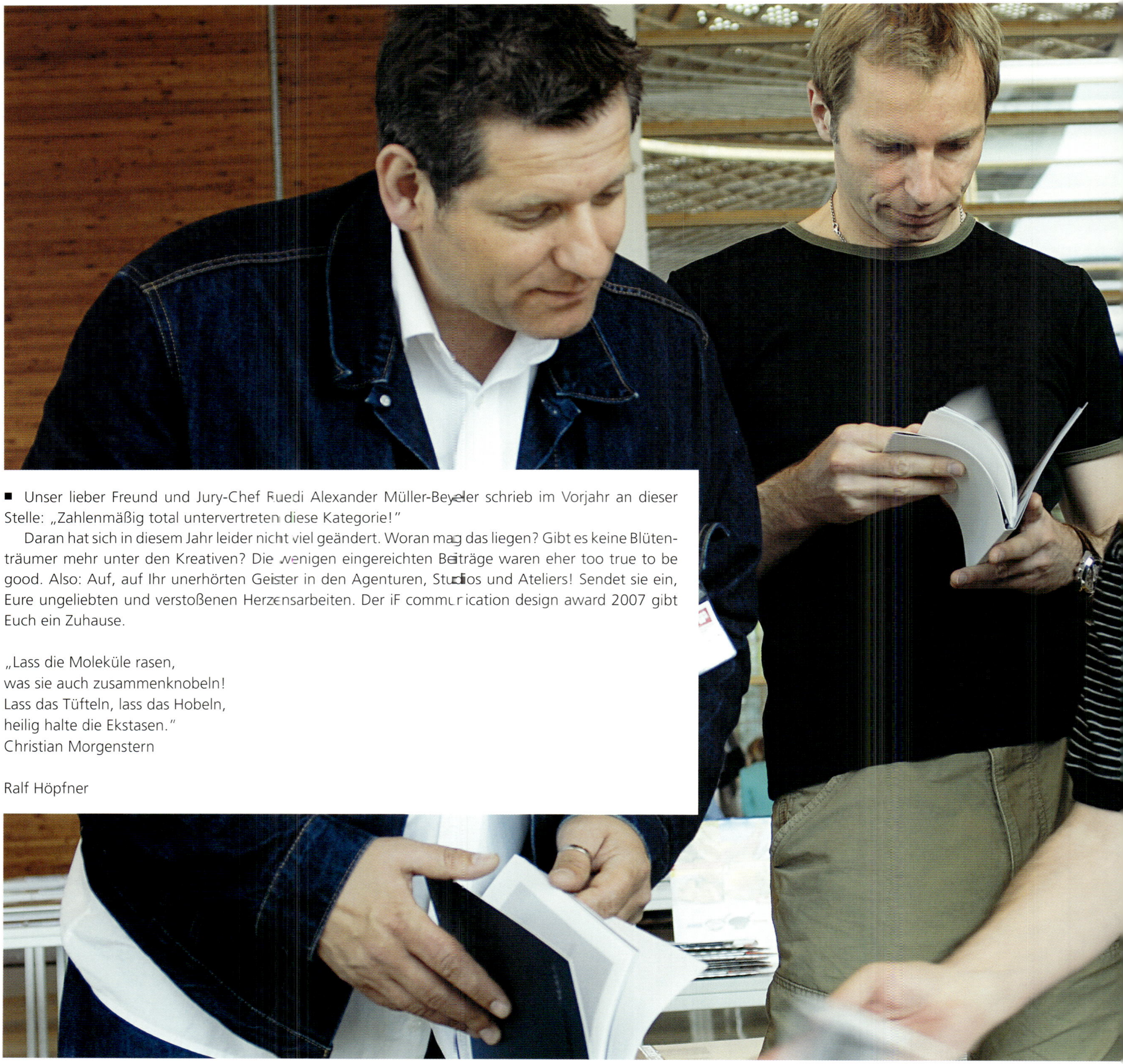

■ Unser lieber Freund und Jury-Chef Ruedi Alexander Müller-Beyeler schrieb im Vorjahr an dieser Stelle: „Zahlenmäßig total untervertreten diese Kategorie!"

Daran hat sich in diesem Jahr leider nicht viel geändert. Woran mag das liegen? Gibt es keine Blütenträumer mehr unter den Kreativen? Die wenigen eingereichten Beiträge waren eher too true to be good. Also: Auf, auf Ihr unerhörten Geister in den Agenturen, Studios und Ateliers! Sendet sie ein, Eure ungeliebten und verstoßenen Herzensarbeiten. Der iF communication design award 2007 gibt Euch ein Zuhause.

„Lass die Moleküle rasen,
was sie auch zusammenknobeln!
Lass das Tüfteln, lass das Hobeln,
heilig halte die Ekstasen."
Christian Morgenstern

Ralf Höpfner

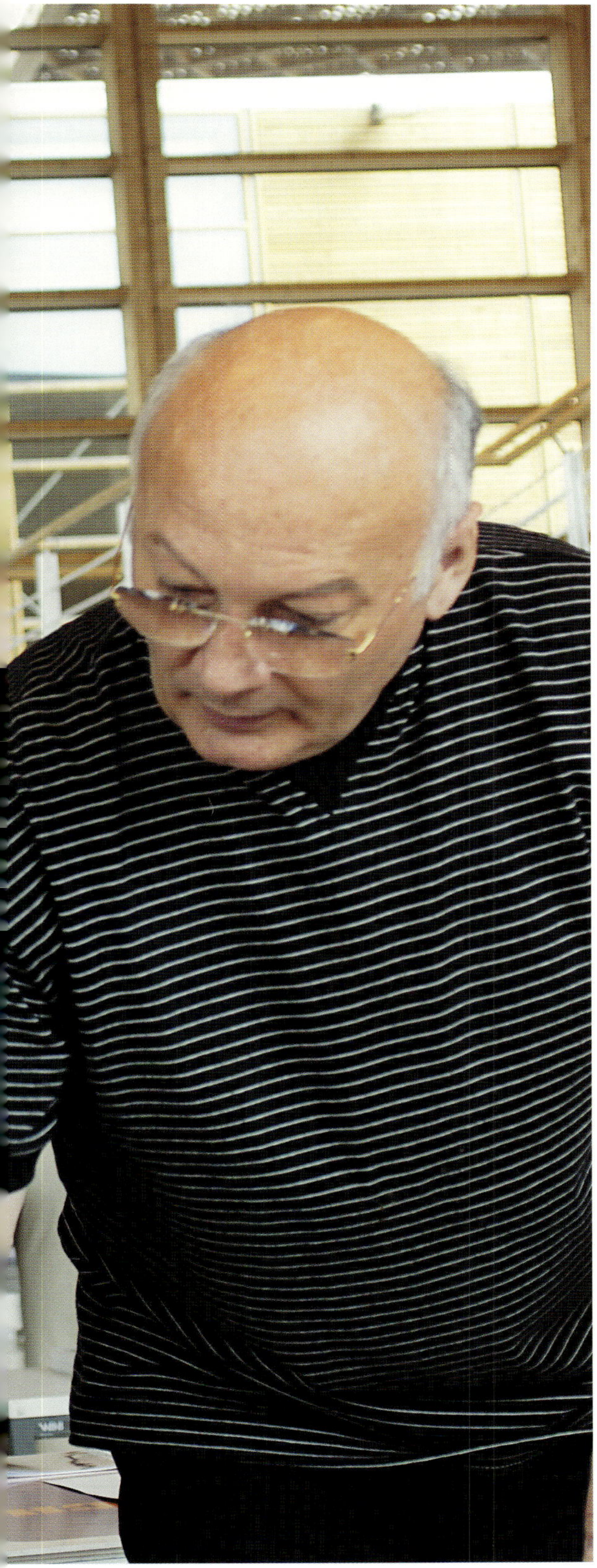

too good to be true

■ Our dear friend and Jury Chairperson Ruedi Alexander Müller-Beyeler wrote the following words in this space last year: "Totally underrepresented, this category was!"

Unfortunately, not much has changed over the last 12 months. Why not? Aren't there any more wild dreamers out there among all the creative minds? The handful of projects that were submitted tended to be rather "too true to be good". Therefore: Stand up and be heard, you untamed souls in the agencies, studios and ateliers! Send us your unpopular, scorned pet projects. The iF communication design award 2007 will provide you with a home.

"Let the molecules race free,
however much they dash and spring!
Stop your meddling, stop your modeling
save this sacred ecstasy!"
Christian Morgenstern (trans.: Katja Moorwood)

Ralf Höpfner

Ralf Höpfner

Rainer Hilf

Michael Volkmer

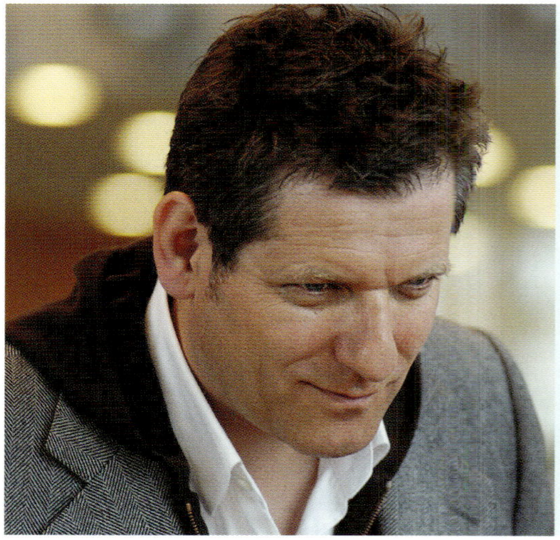

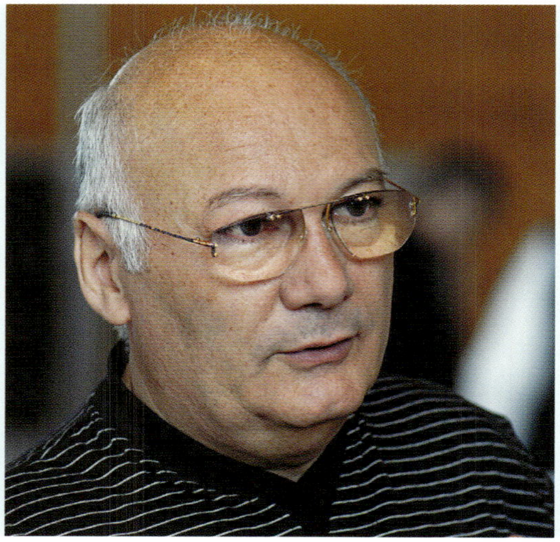

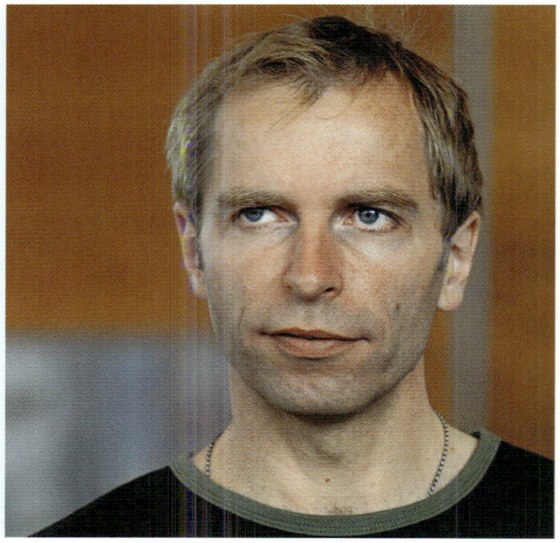

■ Ralf Höpfner (*1959) war Berater bei BBDO Deutschland, Managing Director bei ID-Media und Elephant Seven sowie Geschäftsführer bei Heye & Partner. Seit 2005 ist er Gründer und Inhaber der markentechnisch orientierten Unternehmensberatung >¡< Markenfeuer und zudem als Coach in den Konfliktbereichen Strukturen, Informationen, Beziehungen, Werte und Interessen sowie in der Unterstützung von unternehmerischen Positionierungs- und Change-Prozessen tätig.

■ Ralf Höpfner (*1959) has worked as a consultant at BBDO Deutschland, and has been Managing Director of ID-Media and Elephant Seven, as well as Managing Director of Heye & Partner. In 2005 he set up his own brand-oriented corporate consultancy, >¡< Markenfeuer, also working as a coach in the conflict areas of structures, information, relationships, values and interests, as well as supporting corporate positioning and change processes.

■ Rainer Hilf studierte Innenarchitektur an der Akademie der Bildenden Künste in Nürnberg und ist seit 1972 als freischaffender Innenarchitekt tätig. Er hat seit 1976 unterschiedliche Lehrtätigkeiten, u. a. eine Gastprofessur an der Hochschule für Kunst und Design Burg Giebichenstein in Halle/Saale, inne und ist seit 1995 Präsident des BDIA – Bund Deutscher Innenarchitekten. Rainer Hilf ist darüber hinaus Mitglied in zahlreichen Institutionen wie beispielsweise einer Lenkungsgruppe des Bundesministeriums für Verkehr, Bau- und Wohnungswesen.

■ Rainer Hilf studied Interior Design at the Academy of Fine Arts in Nuremberg and has been a freelance interior designer since 1972. He has held various teaching positions since 1976, including Visiting Professor at the Burg Giebichenstein University of Art and Design in Halle, Germany. He has served as President of the German Association of Interior Designers (BDIA) since 1995. Rainer Hilf is also a member of numerous institutions, including membership on a steering committee for the German Federal Ministry of Transportation, Building and Urban Development.

■ Michael Volkmer (*1965) studierte Kommunikationsdesign an der Fachhochschule Wiesbaden und gründete 1994 zusammen mit Anette Scholz die Agentur Scholz & Volkmer Intermediales Design GmbH in Wiesbaden. Seit 2003 ist er als alleiniger Geschäftsführer für die Planung, Gestaltung und Umsetzung von Projekten im Bereich der neuen Medien verantwortlich. Er lehrt an der Fachhochschule Darmstadt zum Thema „Darstellungsmethoden" und war bereits für nationale wie internationale Wettbewerbe als Jury-Mitglied tätig.

■ Michael Volkmer (*1965) studied Communication Design at the Wiesbaden University of Applied Sciences and, with Anette Scholz, set up the agency Scholz & Volkmer Intermediales Design GmbH in Wiesbaden in 1994. Since 2003 he has been responsible for planning, designing and implementing projects in new media. He teaches at Darmstadt's University of Applied Sciences on "methods of representation" and has served on a number of juries at national and international competitions.

Auftraggeber Client
MTV Networks GmbH & Co. OHG
On Air
Berlin, Germany

Projekt Project
Paradies-Spot
Firmenpräsentation
Company presentation

Design
Benjamin Brettschneider
MTV Networks GmbH & Co. OHG
On Air
Berlin, Germany

special award

Zusammen, was zusammen gehört! MTV und VIVA: ihre Konkurrenz, ihre Zusammenführung und ihre Weiterentwicklung zu der Einheit „VIACOM Brand Solution".

Together what belongs together! MTV and VIVA: their competition, their unification and their further evolvement into the unity of "VIACOM Brand Solution".

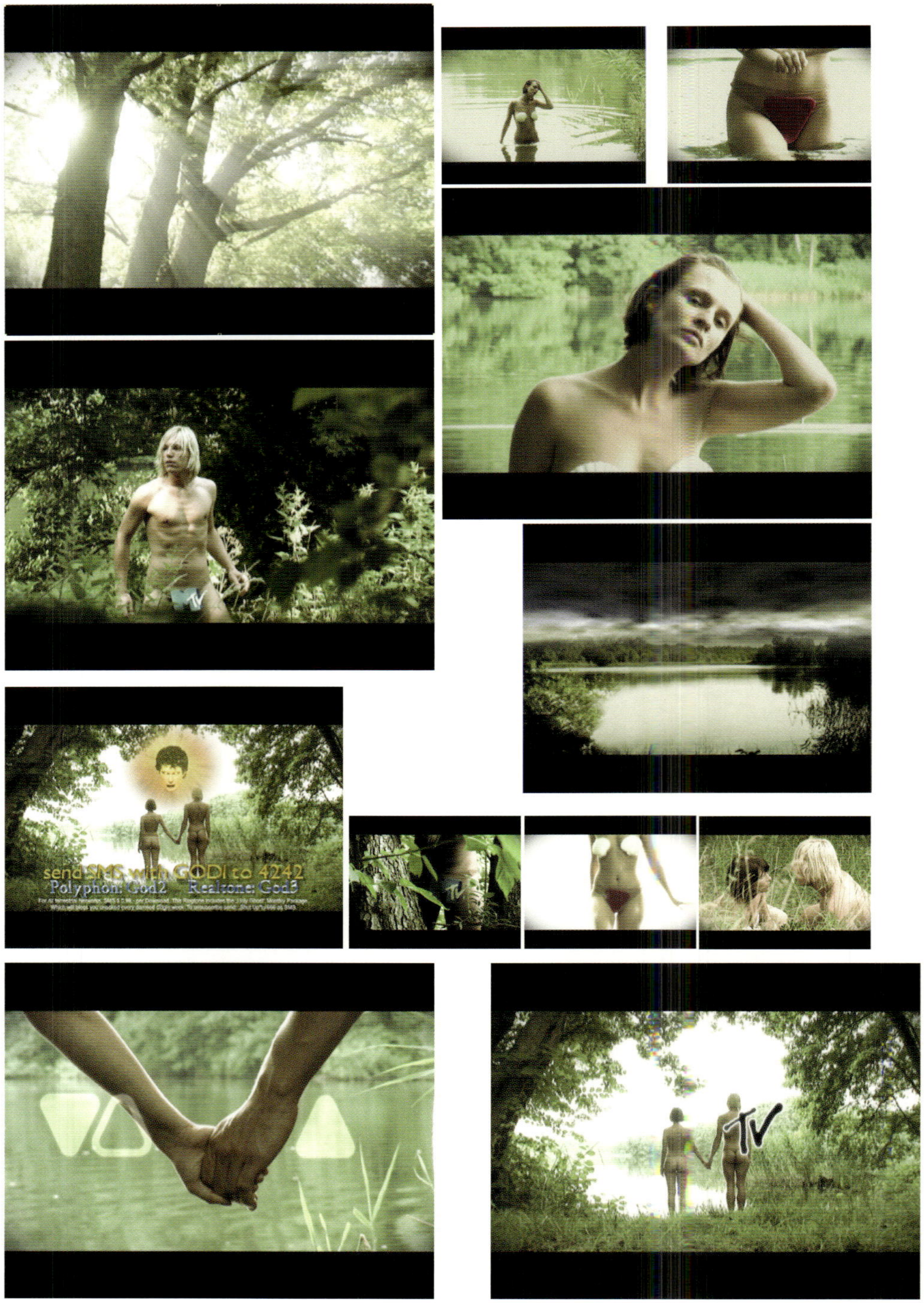

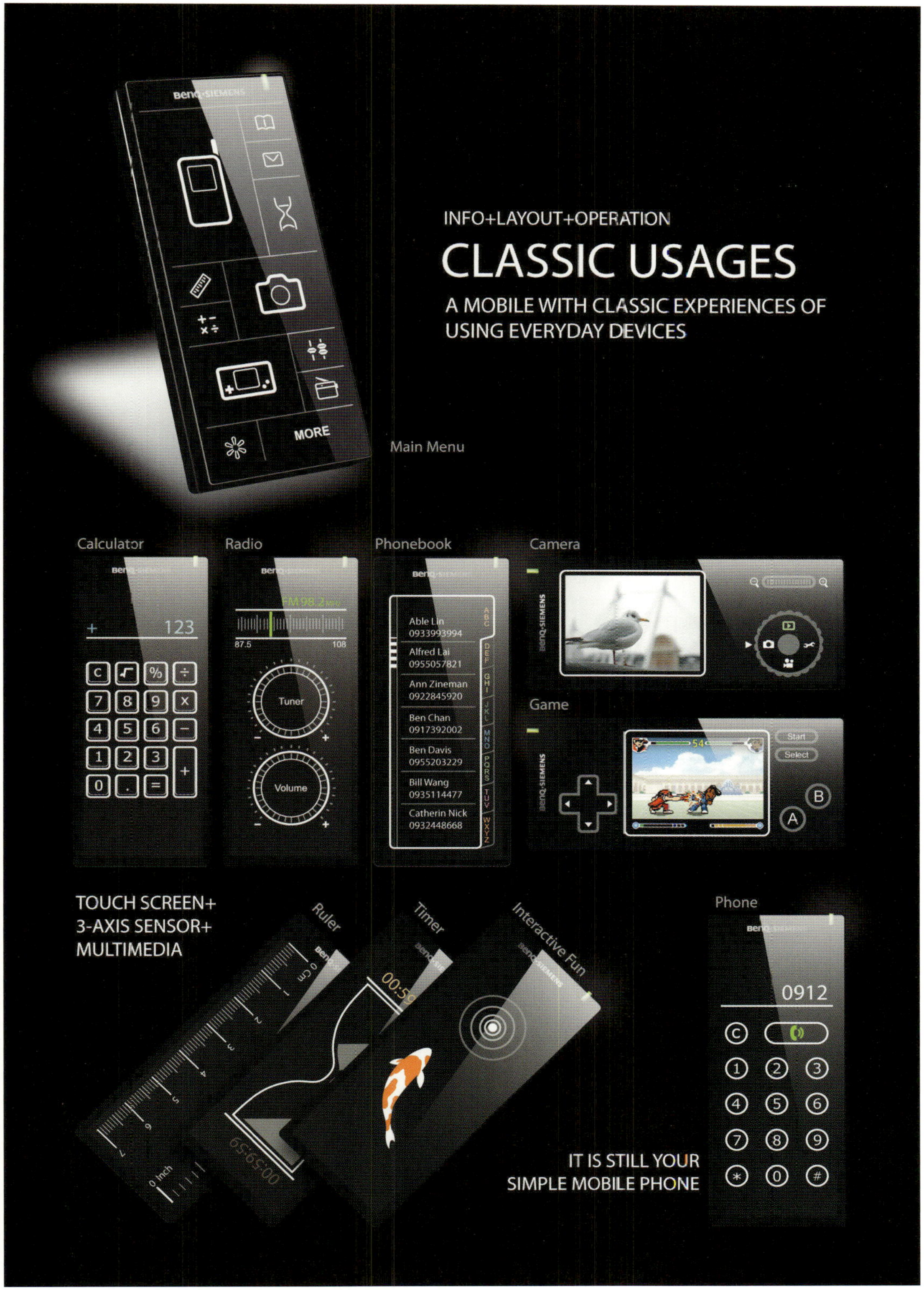

Auftraggeber Client
BenQ Corporation
Taipei, Taiwan

Projekt Project
Classic Usages
Mobiltelefon
Mobile phone

Design
BenQ Corporation
Taipei, Taiwan

special award

Mobiltelefone, die quasi alles können und daher mit immer mehr Funktionen in einer Gerätelösung überfrachtet werden, erweisen sich in der Anwendung als zu komplex. Ihre Handhabung korrespondiert nicht mehr mit den Gewohnheiten der Benutzer und schreckt ab. Das Classic Usages Konzept ist eine Rückbesinnung auf klassische Konventionen, die wieder Geltung erhalten. Endgeräte sollen sich im Hinblick auf ihren Informationsgehalt, ihr Layout und ihre Anwendung an der tatsächlichen Erlebenswelt des Benutzers orientieren. Es wird nur eine minimale Anzahl wesentlicher Bedienungsmöglichkeiten beibehalten, die aber wiederum den Großteil der Benutzerwünsche abdeckt.
Rückbesinnung auf Klassisches. Zurück zu den Wurzeln.

When an increasing number of functions need to be packed into a single device, when a mobile phone tries to do everything, the solutions are often too complex. The functionality no longer communicates with users in a familiar way, making usage more difficult and compromising the user experience. The Classic Usages concept involves recalling and respecting the classic, long-standing conventions and real-world experience of using various common tools and devices in terms of their information, layout, and operation. It retains only the meaningful and minimal interface elements to fulfill the maximum number of user desires. It appears to be what it used to be, and nothing more.
Back to the classics. Back to basics.

too good to be true

Projekt Project
26 Letters, www.26-letters.de
Website / Microsite

Design
Frank Gürgens
(Konzept, Design, Programmierung)
u-double-u
Berlin, Germany

special award

Die internationalen Patent- und Copyrightgesetze behindern die Entwicklung von Kultur und Technologie teilweise mehr, als sie diese fördern. Der Schutz der Verwertungsrechte wird vor den Allgemeinnutz der Erkenntnisse gestellt. Weitere Verschärfungen der gesetzlichen Regelungen sind geplant.
Die vorgestellte Arbeit leistet Lobbyarbeit für eine Sensibilisierung der Öffentlichkeit und der politisch Verantwortlichen zu diesem Thema, um eine entsprechende öffentliche Diskussion zu unterstützen bzw. in Gang zu setzen.

International patent and copyright laws may restrict cultural and technical development more than they promote it. The microsite www.26-letters.de aims to sensitize the public for this topic and hopes to encourage a public discussion in a surprising and playful manner.

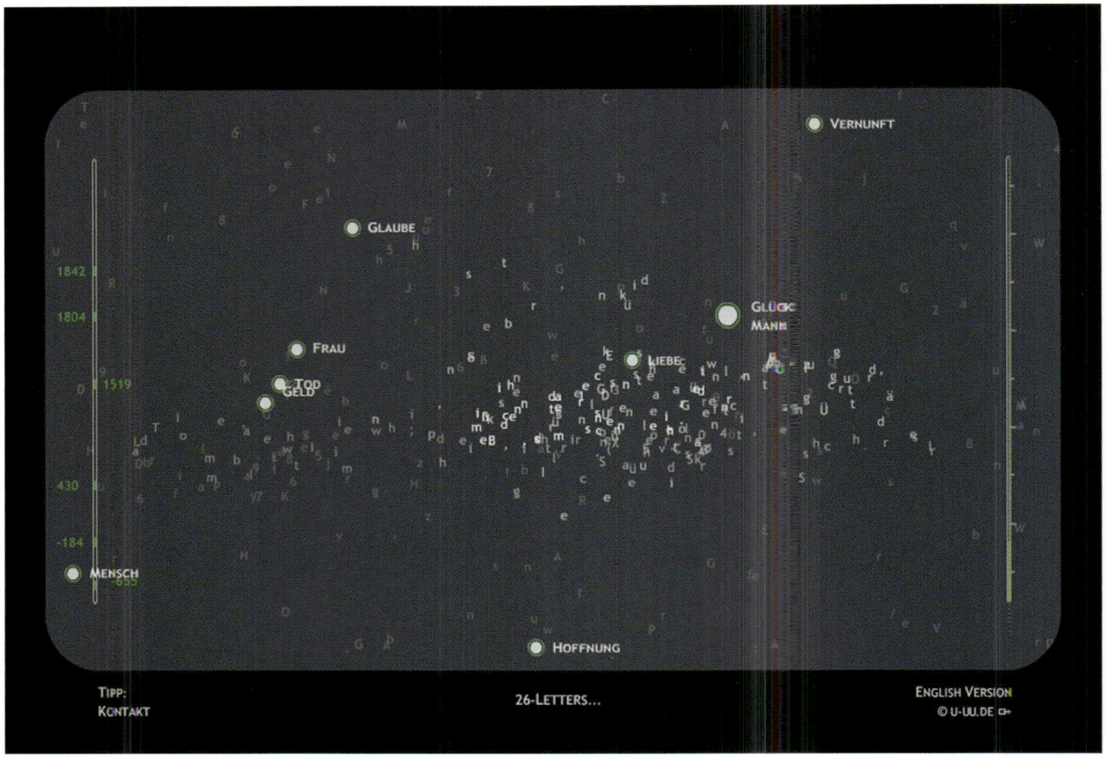

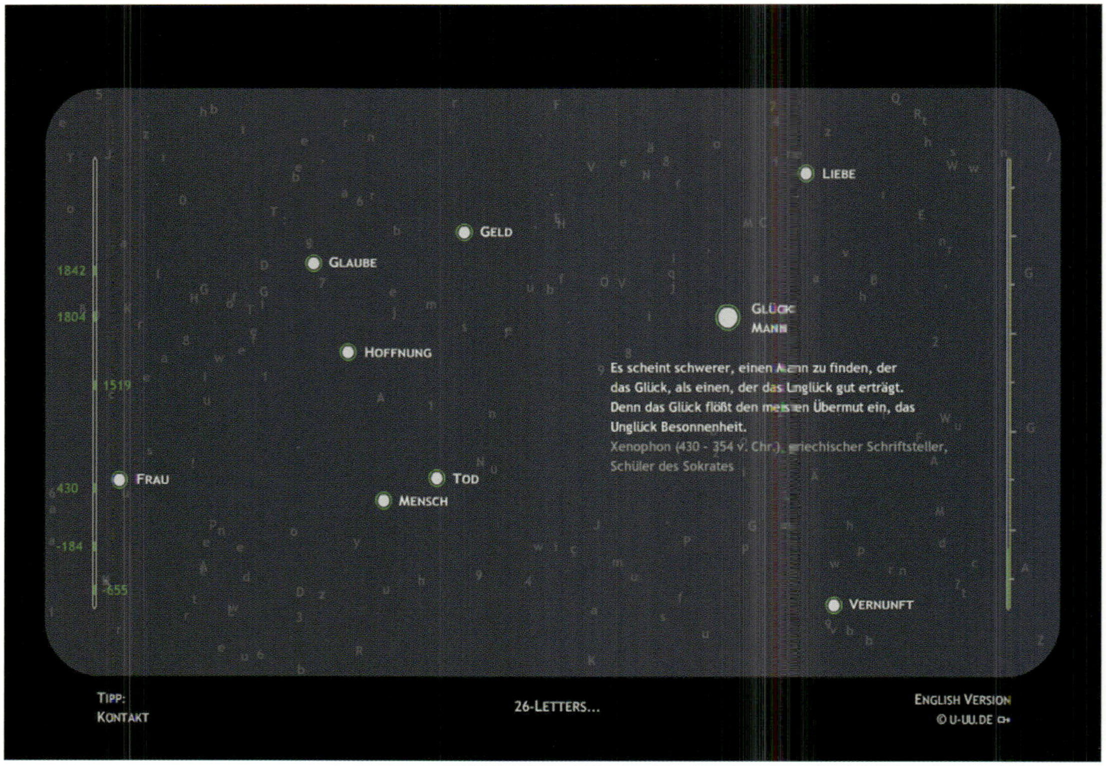

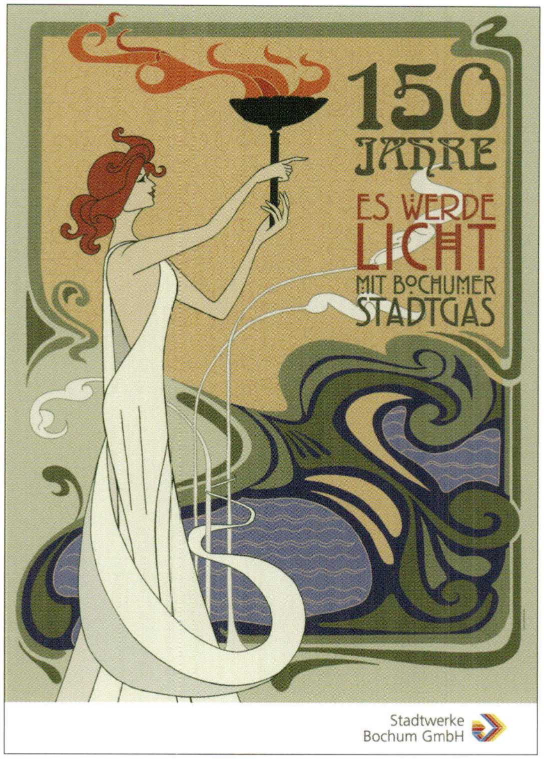

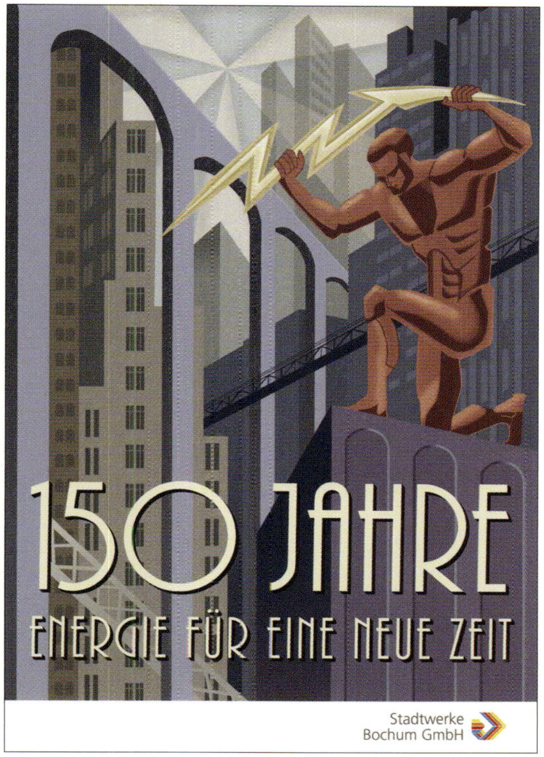

Auftraggeber Client
Stadtwerke Bochum GmbH
Bochum, Germany

Projekt Project
150 Jahre Stadtwerke Bochum
Plakatserie
Poster series

Design
René Wynands
Silke Löhmann
Orlando Hoetzel
Oktober Kommunikationsdesign GmbH
Bochum, Germany

Eine Serie aus vier Plakaten zum Jubiläum der Stadtwerke Bochum. Ihre Gestaltung entspricht den Stilen wichtiger Epochen der letzten 150 Jahre.

A series of four posters for the 150th anniversary of the Stadtwerke Bochum. The design of the posters corresponds to the styles of important eras in the last 150 years.

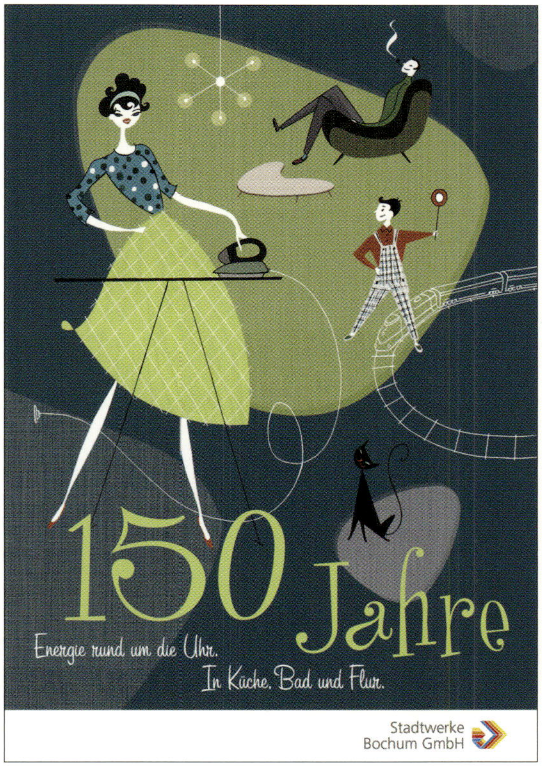

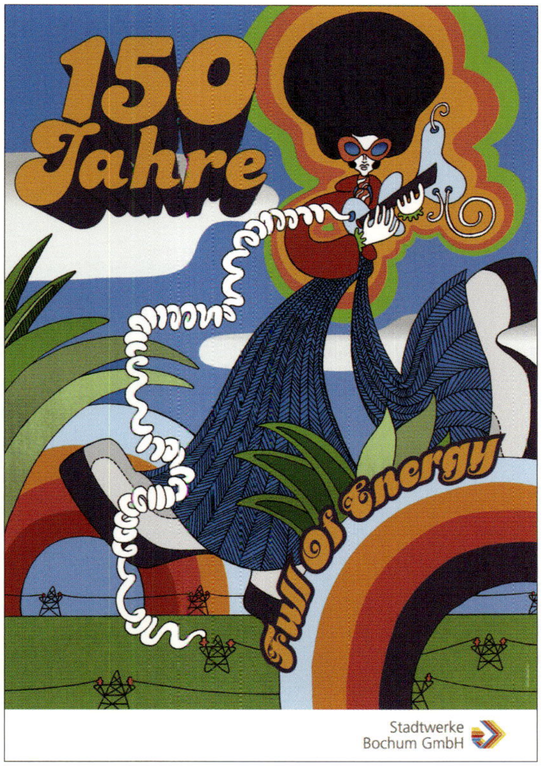

06 iF concept award communication

iF concept award communication

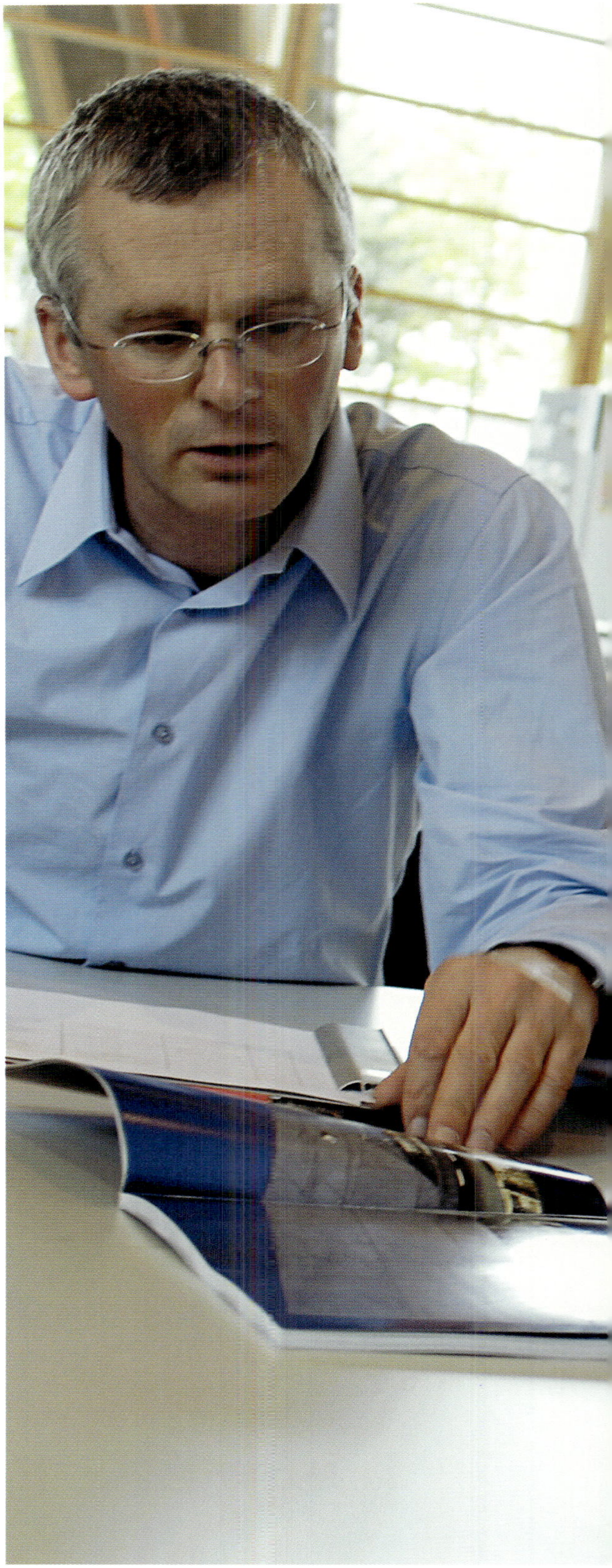

■ Im Gegensatz zu Arbeiten, die im Bezugsfeld der realen Auftragssituation entstehen, sind Projekte und Konzepte von Studenten an Designhochschulen oftmals frei von einschränkenden Rahmenbedingungen. Daher können studentische Arbeiten sogar höhere Qualitäten oder innovativere Ansätze aufweisen als vergleichbare Projekte der Praxis. Der diesjährige iF concept award communication hat diese These bestätigt, obwohl die hohe Erwartung der Jury, die sich sehr auf die Begutachtung der überaus vielfältigen Arbeiten freute, in diesem Jahr nur teilweise erfüllt werden konnte.

Viele eingereichte Projekte verfügen über sehr interessante Ansätze oder Ideen, die ein hohes Potenzial für überzeugende Ergebnisse in sich getragen hätten. So manche Arbeit ist aber an der Oberfläche haften geblieben, die derzeit scheinbar „angesagte" visuelle Sprache zieht sich weltweit fast durch alle Designschulen. Das gibt zu denken. Eine größere Bandbreite, Eigenständigkeit und Differenzierung in der Nutzung der visuellen Mittel und Medien ist zu erhoffen. Den Lehrenden und Studierenden an Hochschulen ist zu wünschen, souverän zu bleiben (oder wieder zu werden) und inhaltliche mit gestalterischer Arbeit sorgfältig zu verknüpfen. Denn: Die Basis jeder guten Arbeit ist ein überzeugendes Konzept, das mit adäquaten Mitteln in der Ganzheit wie im Detail anspruchsvoll und sorgfältig umgesetzt ist; egal, ob es sich dabei um eine eher persönliche illustrative Äußerung, eine einfache Wortmarke oder eine komplexe wissenschaftliche Darstellung handelt.

Die ausgezeichneten Arbeiten, und insbesondere die Top 5, weisen sowohl eine hohe konzeptionelle und kommunikative Qualität als auch eine konsequente und im Detail überzeugende Umsetzung auf. Dazu herzlichen Glückwunsch! (Und in Zukunft mehr davon.)

Prof. Detlef Rahe

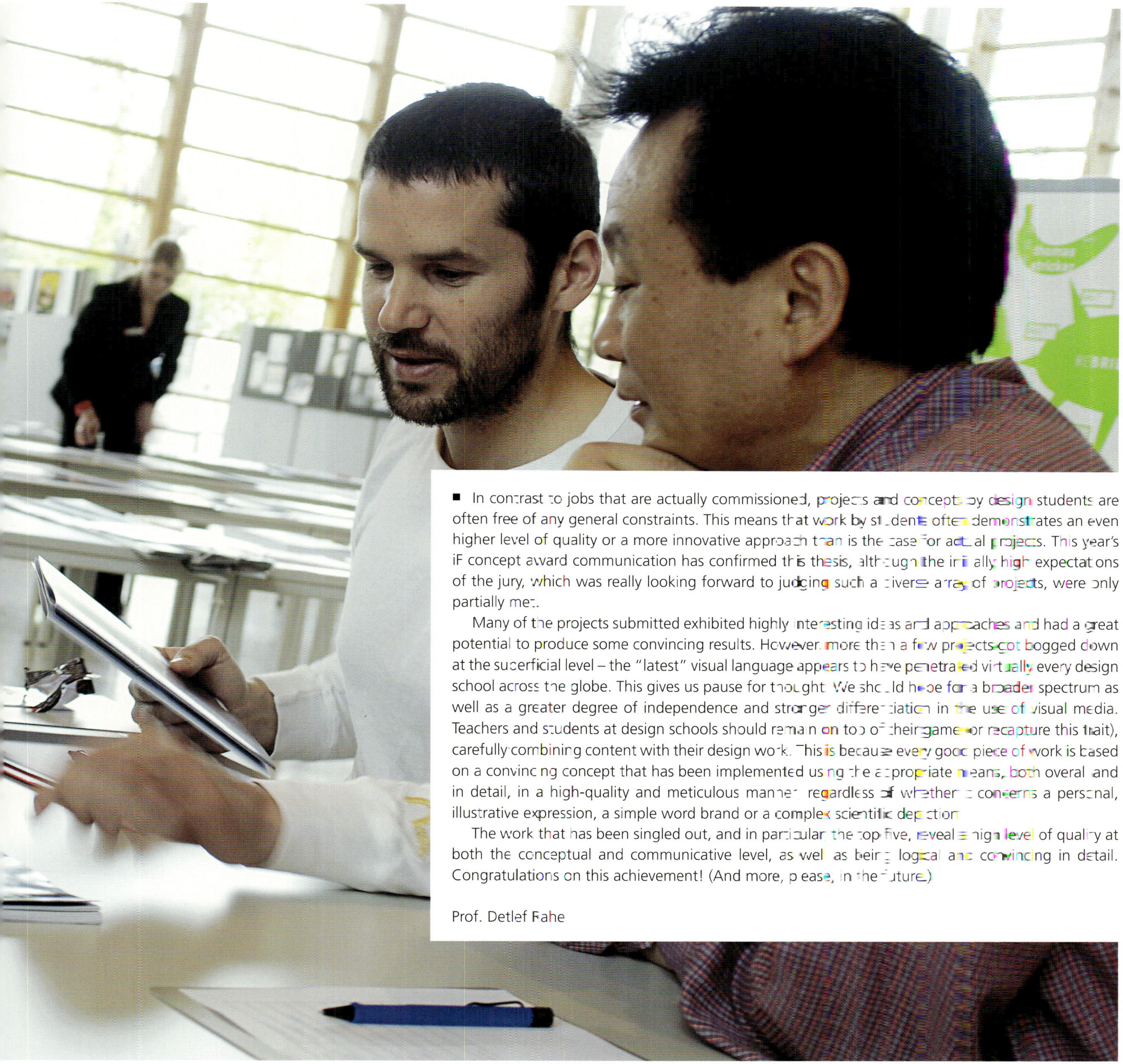

■ In contrast to jobs that are actually commissioned, projects and concepts by design students are often free of any general constraints. This means that work by students often demonstrates an even higher level of quality or a more innovative approach than is the case for actual projects. This year's iF concept award communication has confirmed this thesis, although the initially high expectations of the jury, which was really looking forward to judging such a diverse array of projects, were only partially met.

Many of the projects submitted exhibited highly interesting ideas and approaches and had a great potential to produce some convincing results. However, more than a few projects got bogged down at the superficial level – the "latest" visual language appears to have penetrated virtually every design school across the globe. This gives us pause for thought. We should hope for a broader spectrum as well as a greater degree of independence and stronger differentiation in the use of visual media. Teachers and students at design schools should remain on top of their game (or recapture this trait), carefully combining content with their design work. This is because every good piece of work is based on a convincing concept that has been implemented using the appropriate means, both overall and in detail, in a high-quality and meticulous manner, regardless of whether it concerns a personal, illustrative expression, a simple word brand or a complex scientific depiction.

The work that has been singled out, and in particular the top five, reveal a high level of quality at both the conceptual and communicative level, as well as being logical and convincing in detail. Congratulations on this achievement! (And more, please, in the future.)

Prof. Detlef Rahe

■ Prof. Detlef Rahe (*1964) ist Diplom-Designer/ Master of Fine Arts und gründete 1989 „rahe+rahe design" gemeinsam mit seiner Frau Ulrike Rahe. Er war u. a. Professor für Produkt- und Systemdesign an der Hochschule Anhalt (FH) im Fachbereich Design in Dessau sowie dessen Dekan. Detlef Rahe hat seit 1998 eine Professur für 3-D-Design an der Hochschule für Künste Bremen inne und ist für den Aufbau und die Leitung des wirtschaftsnahen An-Instituts, dem i/i/d Institut für Integriertes Design, zuständig.

■ Prof. Detlef Rahe (*1964) studied Design and has an MA in Fine Arts. In 1989 he co-founded "rahe+rahe design" with his wife Ulrike Rahe. He was professor of Product and Systems Design within the Design Faculty at the Anhalt University of Applied Sciences in Dessau, Germany, where he also served as the Dean. Since 1998 Detlef Rahe has been professor of 3D-Design at the University for the Arts in Bremen and is responsible for developing and managing the Institute for Integrated Design (i/i/d), a research institution with close links to business.

■ James Auger war von März 2002 bis Januar 2005 Forschungsassistent am Media Lab Europe. Im Anschluss arbeitete er in Japan für Issey Miyake. Er absolvierte bei Rolls-Royce eine Ingenieursausbildung im Bereich Flugzeugantriebe und erwarb einen Masterabschluss in Produktdesign am Londoner Royal College of Art (RCA). James Auger ist gegenwärtig in der Forschungsabteilung für Interaction Design des RCA tätig, die mit Philips Research kooperiert. Außerdem ist er Lehrkraft am Fachbereich für Interaction Design des RCA.

■ James Auger was a research associate at Media Lab Europe between March 2002 and January 2005. After that he moved to Japan to work for Issey Miyake. He has an Engineering apprenticeship from Rolls-Royce (aero engines) and an MA in Design Products from the Royal College of Art in London. James Auger is currently a research fellow within the Interaction Design department at the RCA collaborating with Philips research and teaches in the Interaction Design MA course.

■ Jae-Jin Shim studierte Industrial Design (BFA) an der Hongik University, Korea. Er war u.a. Managing Director des LG Electronics Europe Design Center und ist seit 2002 Vice President des LG Electronics DDM (Digital Display & Media) Design Research Lab. Er ist Vice President der Korea Federation of Design Associations, Director der Korea Society of Design Science, Invited Designer, Korea Industrial Design Exhibition sowie Director der Design Brand Management Society.

■ Jae-Jin Shim studied Industrial Design (BFA) at Hongik University in Korea. He has held a number of positions at LG Electronics, including Managing Director of the LG Electronics Europe Design Center, and since 2002 has served as Vice President of the LG Electronics DDM (Digital Display & Media) Design Research Lab. He is Vice President of the Korea Federation of Design Associations, Director of the Korea Society of Design Science, an Invited Designer at the Korea Industrial Design Exhibition as well as Director of the Design Brand Management Society.

Prof. Detlef Rahe

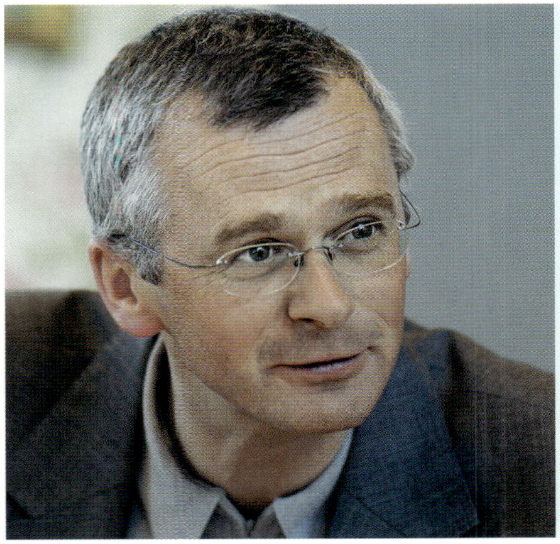

James Auger

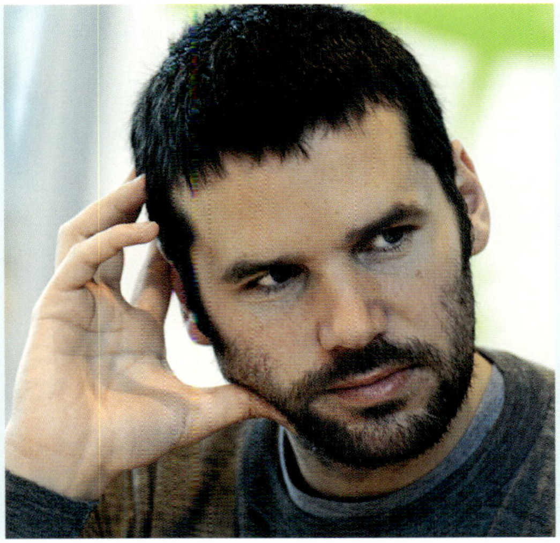

Jae-Jin Shim

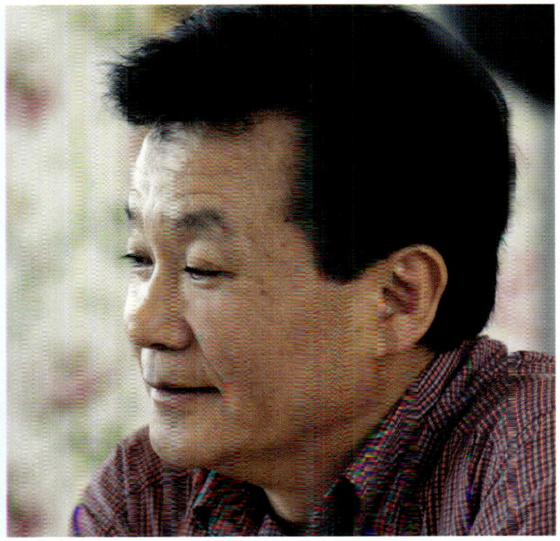

Auftraggeber Client
HTWG Konstanz
Kommunikationsdesign
Konstanz, Germany

Projekt Project
Babbel – Der Entwurf einer Bildersprache
Zeichen
Signs

Design
Daniel Lorch
Berlin, Germany

Mit Lichtgeschwindigkeit surfen wir inzwischen um den Globus – dank Internet. Kulturelle Unterschiede werden durch multinationale Konzerne geebnet, und an das Stolpern über Sprachbarrieren haben wir uns längst gewöhnt. Kurzum: Noch nie zuvor war die Notwendigkeit einer Sprache, die alle sprechen und verstehen, größer als heute!
Die Babbel ist eine Kunstsprache aus 650, überwiegend selbsterklärenden Zeichen und acht Grammatikregeln. Die Sprache lässt sich in wenigen Tagen erlernen, und das Vokabular reicht aus, um das Wichtigste im Alltag zu formulieren. Anwendungsgebiete: internationales Marketing, Rehabilitation, Signaletik, Internet.

Nowadays we surf around the globe at the speed of light – thanks to the Internet. Cultural differences are smoothed over by multinational companies and we are used to stumbling over language barriers. In short: Never before has the need for one language spoken and understood by all been more important than it is today.
Babbel is an artificial language with 650, mainly self-explanatory signs and eight rules of grammar. The language can be learned in a few days and the vocabulary is sufficient to describe most things in everyday life. Fields of application: international marketing, rehabilitation, signage systems, Internet.

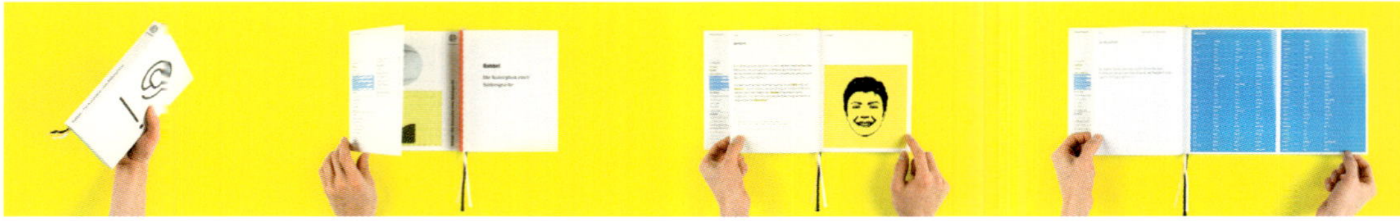

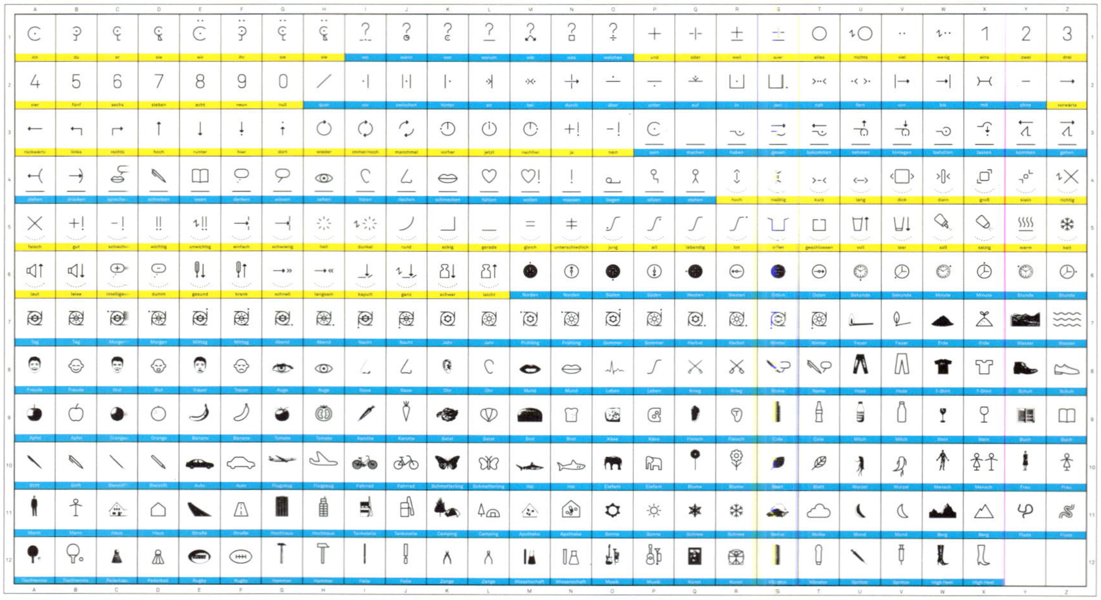

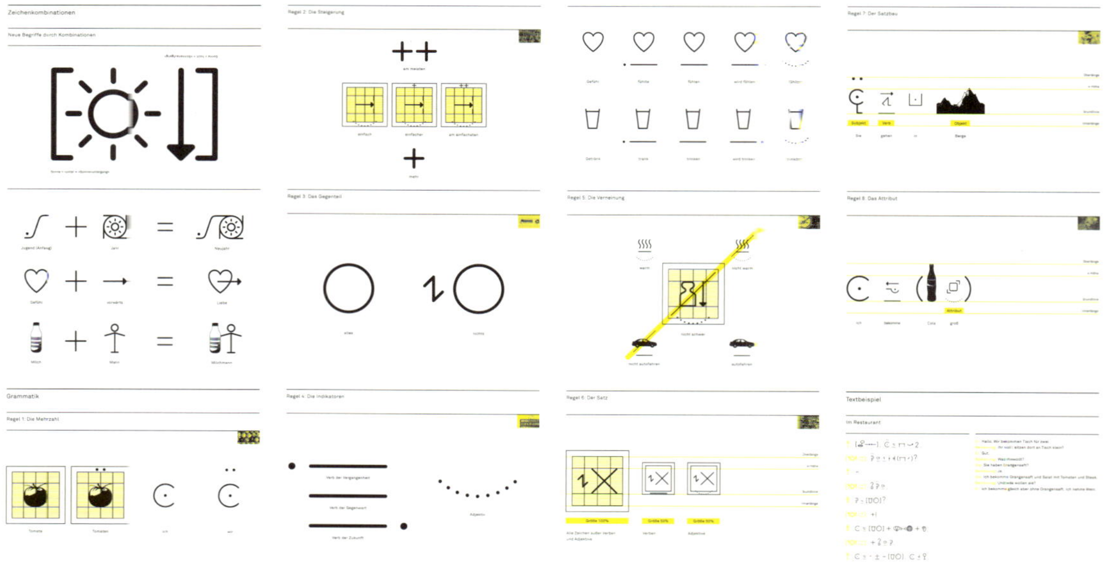

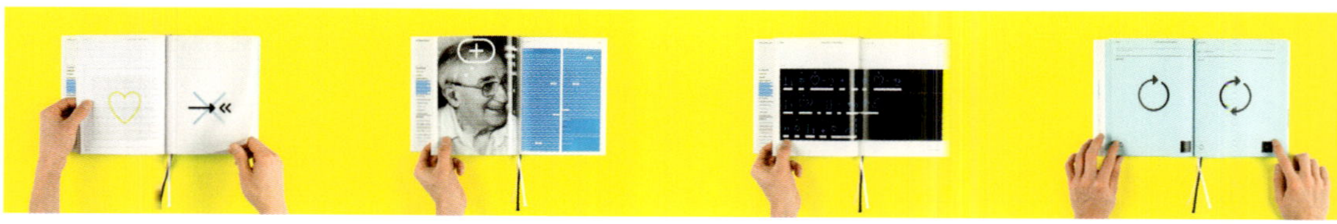

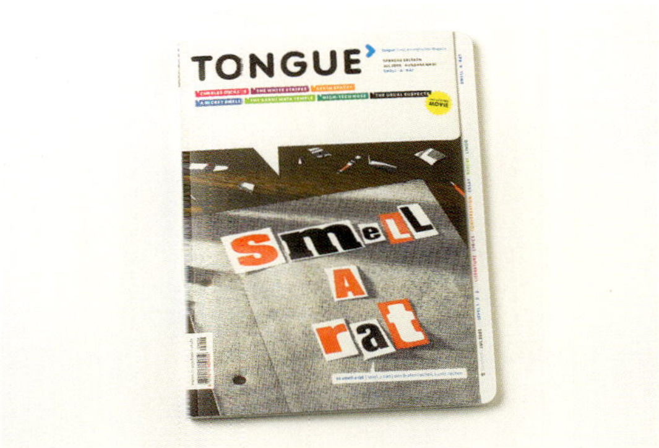

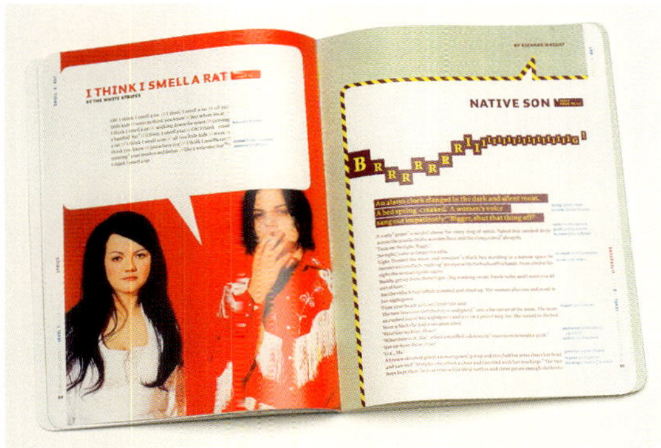

Auftraggeber Client
Muthesius Kunsthochschule
Kommunikationsdesign
Kiel, Germany

Projekt Project
TONGUE
Englisches Sprachmagazin für deutsche Leser
English language magazine for German readers

Design
Hanna Sedelmayer
Hamburg, Germany

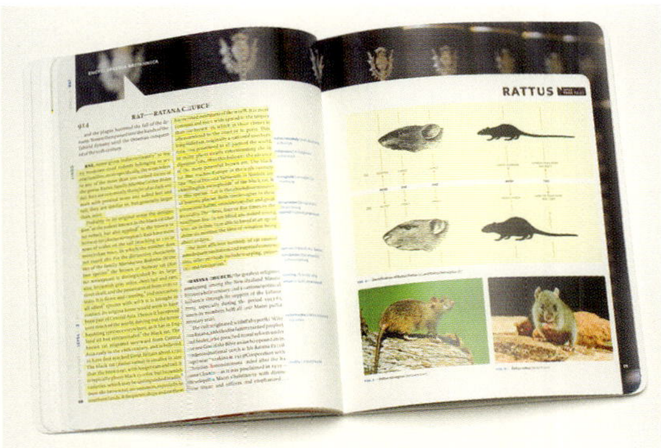

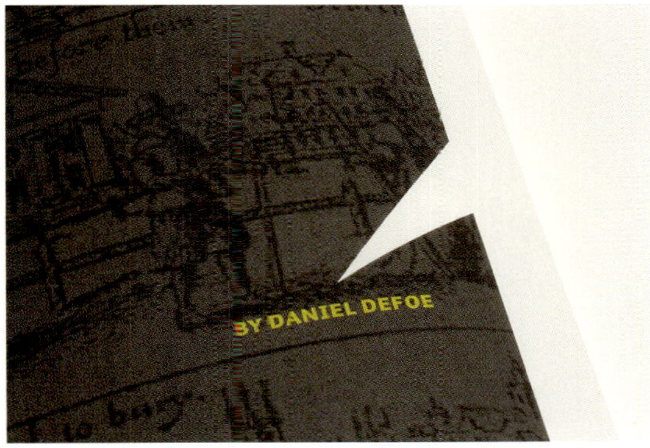

Schluss mit Grammatik Pauken! Das Magazin TONGUE – eine Collage aus original englischen Texten versehen mit deutschen Vokabelhilfen – motiviert seine Leser mit jeder Ausgabe von neuem, sich mit der Fremdsprache zu befassen. Unter einem bestimmten Oberthema vermittelt ein breites Spektrum an Texten u. a. aus Literatur, Comic, Musik, Film, Presse und Wissenschaft den Lesern ein Gefühl für die vielen Facetten der englischen Sprache. Funktional und emotional gestaltete Layouts wecken die Neugier – und mit dem beigefügten Spielfilm auf DVD (Originalton mit Untertiteln) wird das moderne Sprachtraining auch für die Ohren zum Erlebnis.

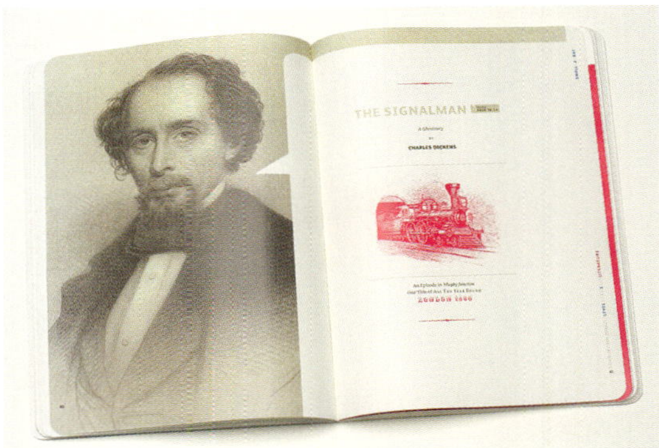

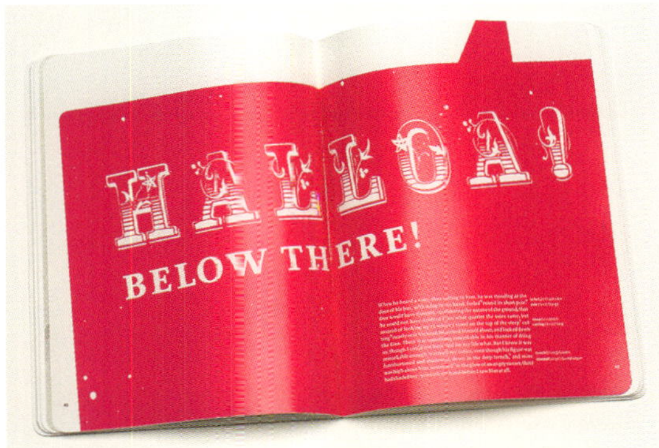

No more boring grammar lessons! The magazine TONGUE is a thrilling blend of original English texts and German vocabulary hints. Each issue arouses a new interest in the foreign language. A wide range of texts (from the fields of music, literature, movies, press, science and many others) dealing with a certain subject gives a sense for the different aspects of the English language. Layouts with functional and emotional designs awaken the reader's curiosity. The attached feature film on DVD (original version with subtitles) makes the modern language training also an exciting hearing experience.

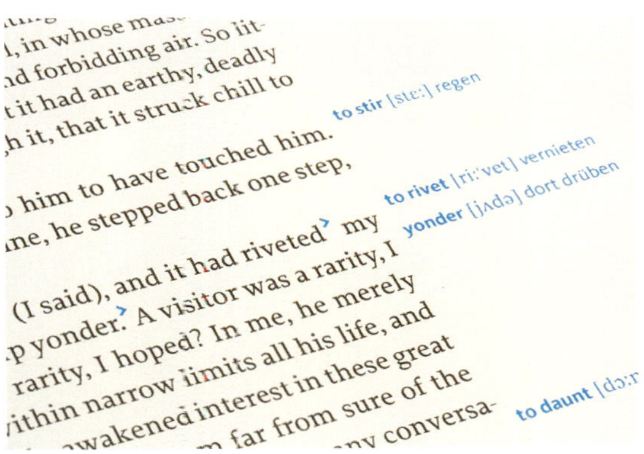

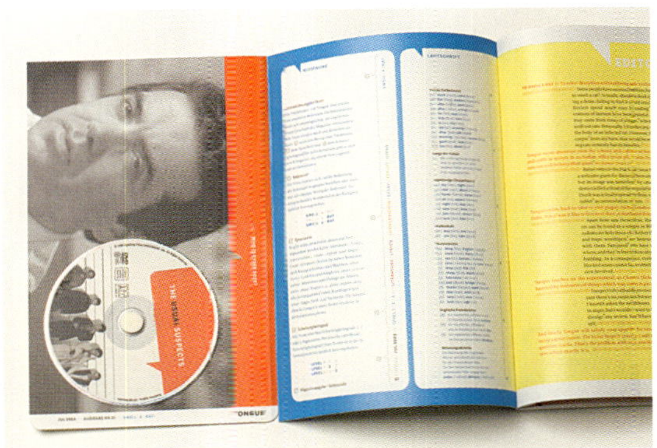

iF concept award communication prize money 2,000 euros

Auftraggeber Client
Universität Duisburg-Essen
Essen, Germany

Projekt Project
SUSHI de luxe
Corporate Design für ein Sushi-Restaurant
Corporate design for a Sushi restaurant

Design
Katrin Schmuck
Essen, Germany

Der Gestaltungsansatz liegt in der Verbindung des Ostasiatischen mit dem Europäischen und setzt charakteristische Merkmale wie Strenge, Reduktion und Einfachheit als Gestaltungsmittel ein. Ausgangspunkt der Arbeit ist die Fragestellung, wie sich ein Sushi-Restaurant im übersättigten Markt abheben und sich die hohe Qualität von Sushi in der hohen Gestaltungsqualität des Erscheinungsbildes widerspiegeln kann.
Wichtiger Ansatz der Arbeit ist es, das Thema Sushi auf sensible, aber auch spielerische Art visuell zu transportieren.

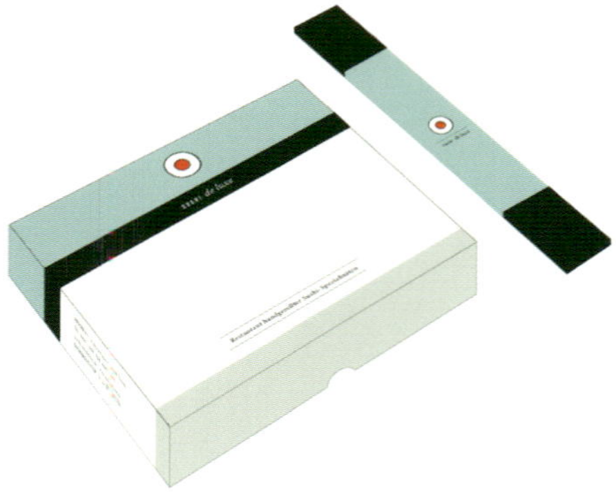

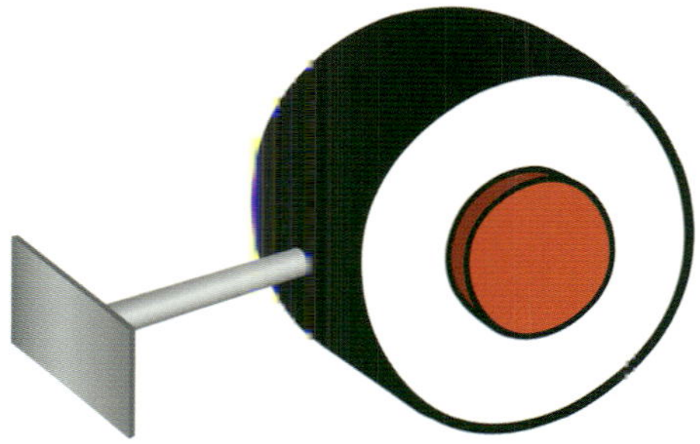

The design approach is based on the connection of the East Asian style with the European one, deploying characteristic features such as severity, reduction and simplicity as design elements. Starting point of the concept was the idea that in a saturated market a Sushi restaurant can stand out from the crowd by reflecting the high quality of the Sushi food in the restaurant's high-class corporate design.
In order to visually communicate the subject "Sushi" in the corporate design, the concept pursues a very detailed and playful but at the same time appropriately sensitive approach.

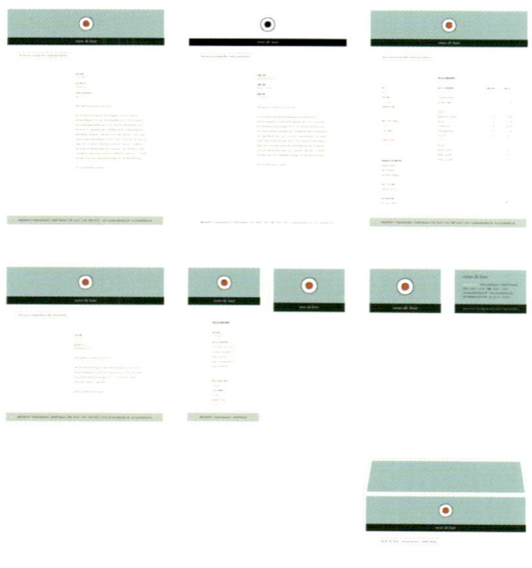

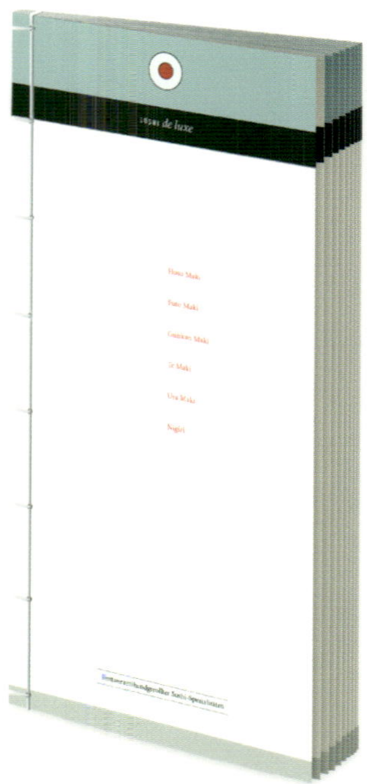

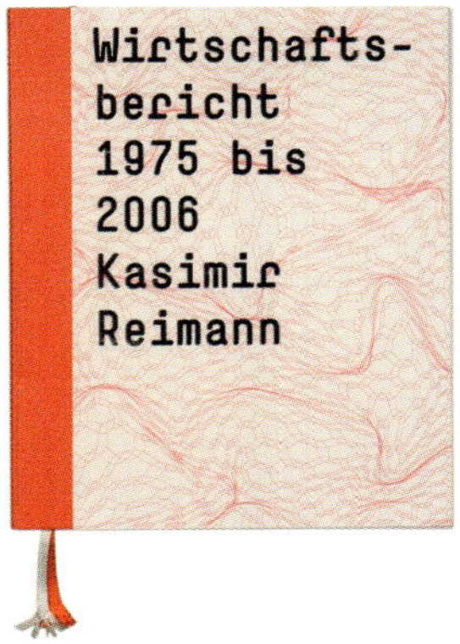

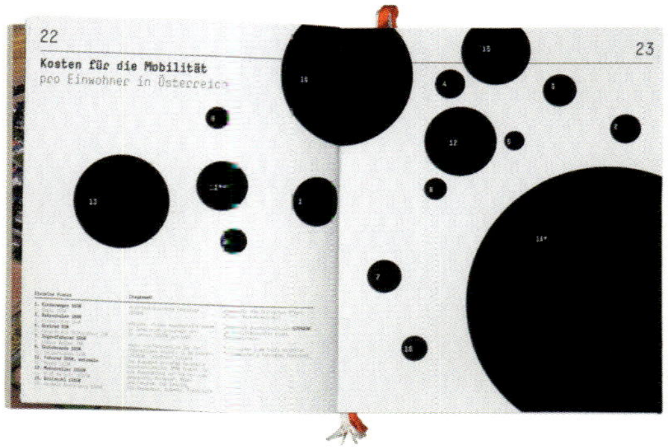

Auftraggeber Client
Universität für angewandte Kunst Wien
Institut für Design, Grafik Design
Vienna, Austria

Projekt Project
Wirtschaftsbericht 1975-2006 Kasimir Reimann
Wirtschaftsbericht / Buch
Economic review / book

Design
Kasimir Reimann
Vienna, Austria

Am Übergang seiner schulischen Laufbahn zur Erwerbstätigkeit legt Kasimir Reimann ein finanzielles Resümee seines Lebenslaufs vor. Durch die monetäre Bewertung seiner individuellen Verhältnisse werden Themen wie Human-, Sozial- und Sachkapital exemplarisch im Kontext allgemeiner Statistiken berechnet und als ökonomische Kategorien ins Bewusstsein gerufen.

At the change-over from his educational career to gainful employment, Kasimir Reimann presents a financial summary of his CV. Through the monetary assessment of his individual circumstances, subjects such as human, social and real capital are exemplarily calculated in the context of general statistics and brought into awareness as economic categories.

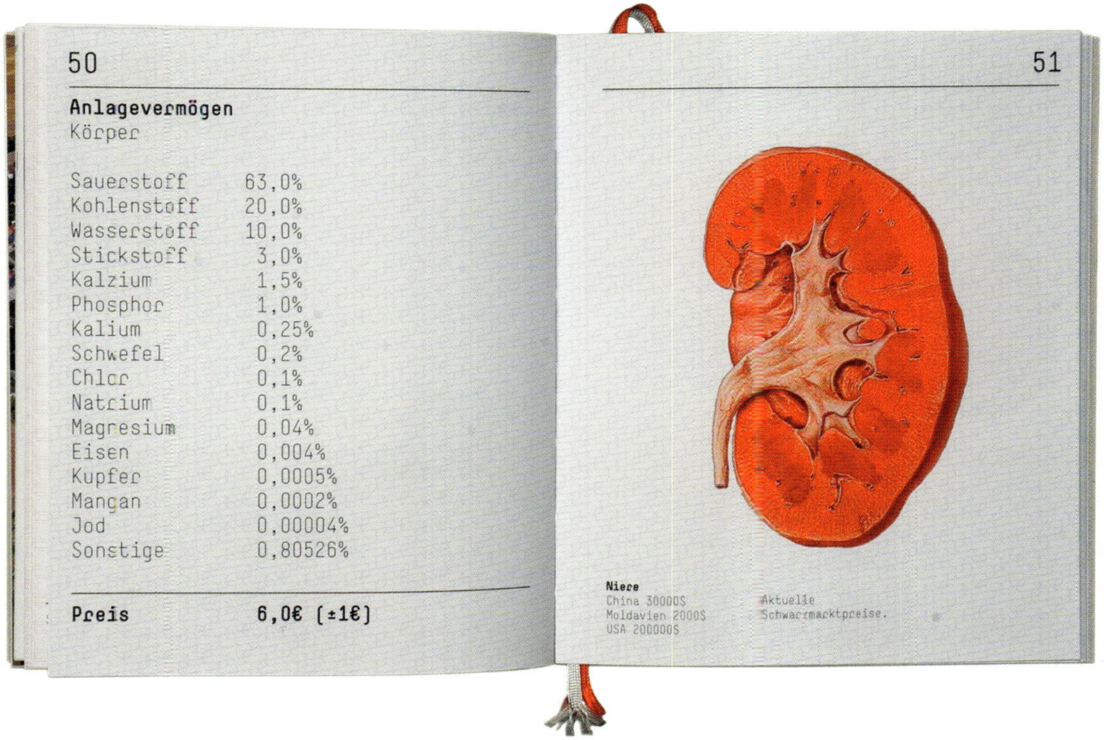

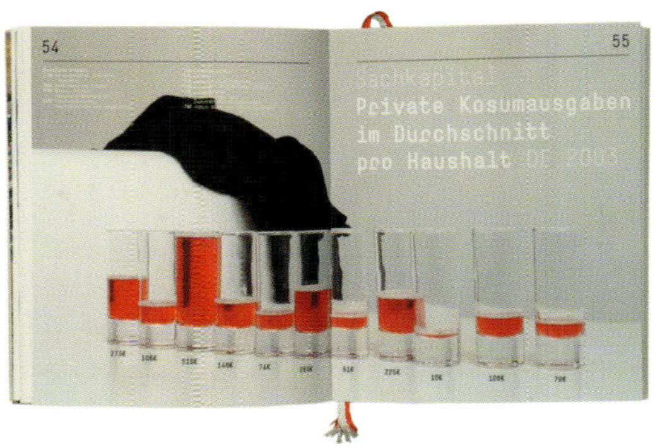

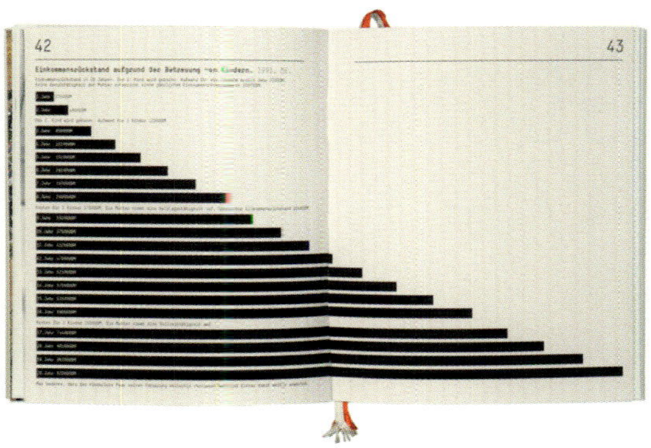

Auftraggeber Client
Hochschule für Bildende Künste Braunschweig
Braunschweig, Germany

Projekt Project
concrete dreams / zaha hadid's phaeno
Buch
Book

Design
Jan Grothklags
Wolfsburg, Germany

Das Phaeno Science Center in Wolfsburg, ein Gebäude von Zaha Hadid, ist ein interaktives Wissenschaftsmuseum. Daher lautet auch das Motto des Museums: „Phaeno – ich entdecke!"
In den Ansätzen des Buchprojekts geht es um den Betrachter der Fotos, der die gleiche Stellung wie der Besucher des Phaeno einnimmt. Er soll das gezeigte Bild betrachten, hinterfragen und zu einem eigenen Ergebnis kommen, wodurch die Interaktivität zwischen dem Betrachter und den Fotos gefördert wird. Das Interesse für das eigentliche Bauwerk Phaeno wird somit auf die gleiche Art und Weise geweckt, mit der das fertige Gebäude auf den Besucher wirkt.

The Phaeno Science Center is Germany's unique experiment landscape for science and technology laid out within spectacular architecture by Zaha Hadid. The architecture as well as the exhibition concept offer discovery and adventure to the visitor.
The building concept has been transferred into a photographic project, which shows the developing process of the Phaeno over the last two years. The final medium is a book, which builds up a plot between text and image level inviting the user of the book on a journey of discovery. In opposition to the viewer who only sees the Phaeno building from the outside and who only enters it after its completion, the book shows the development of the building's space from the inside.

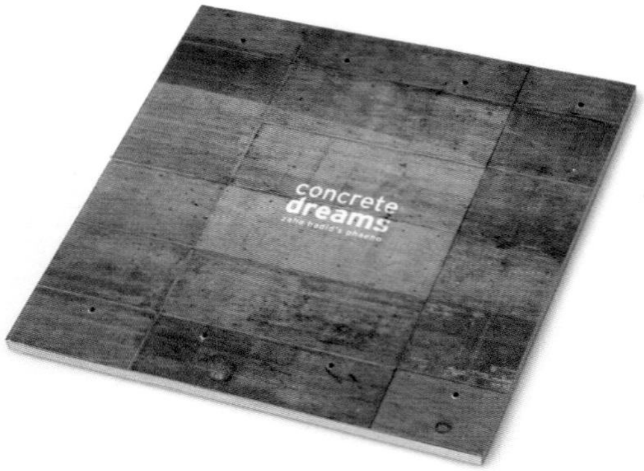

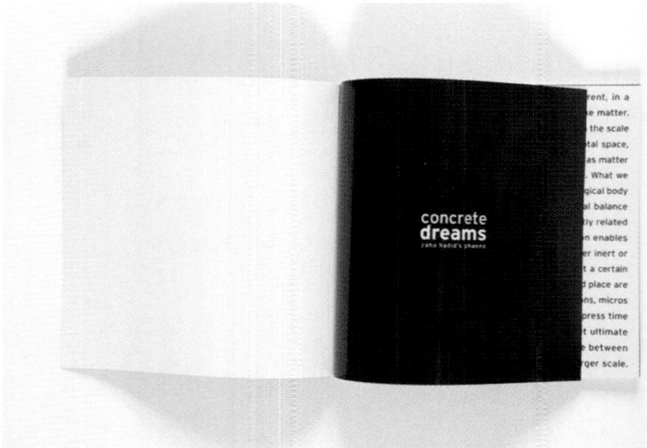

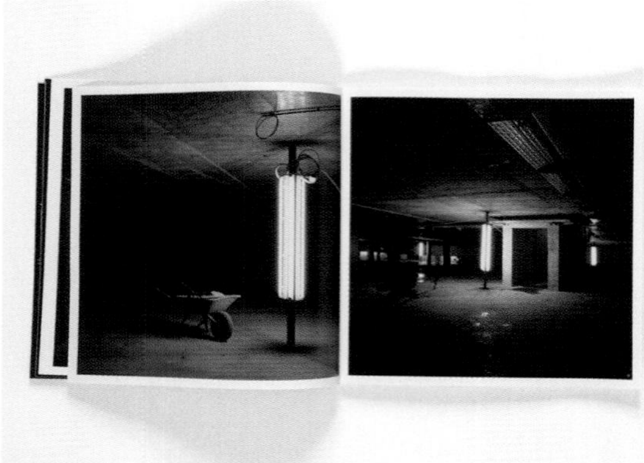

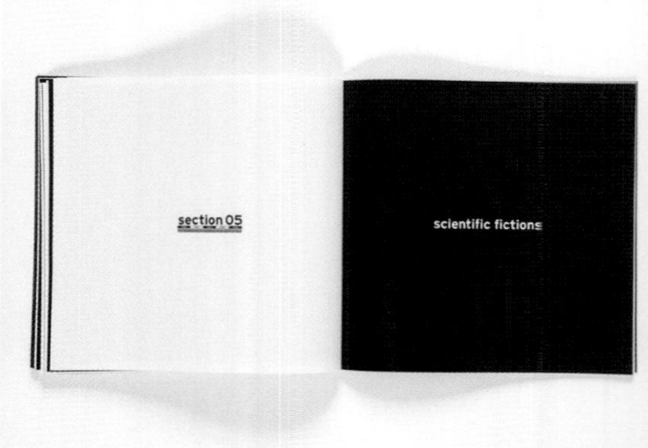

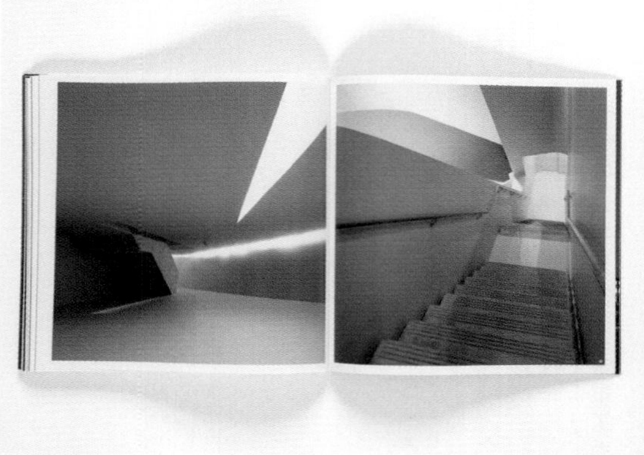

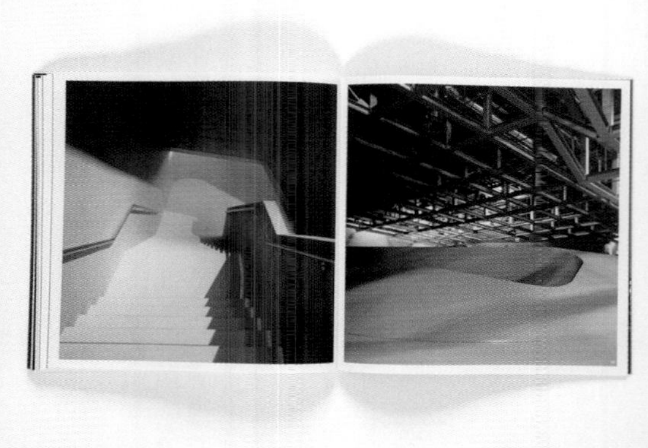

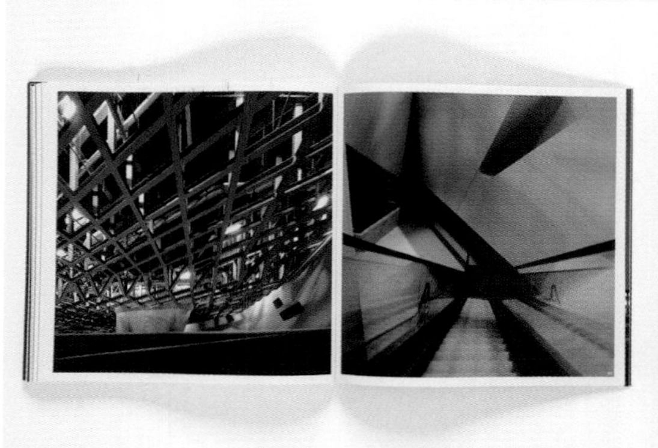

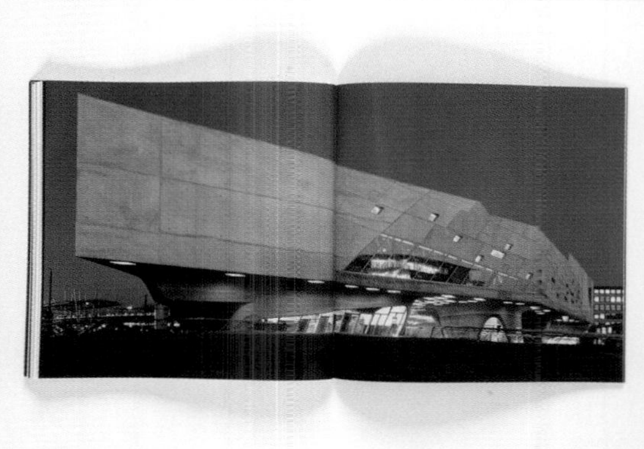

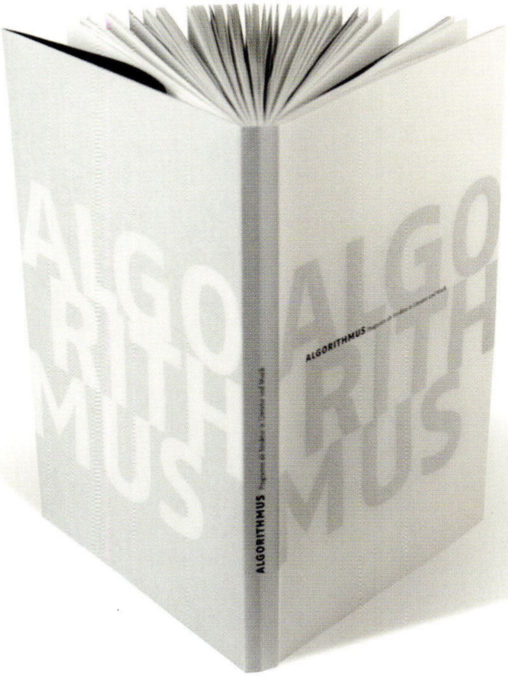

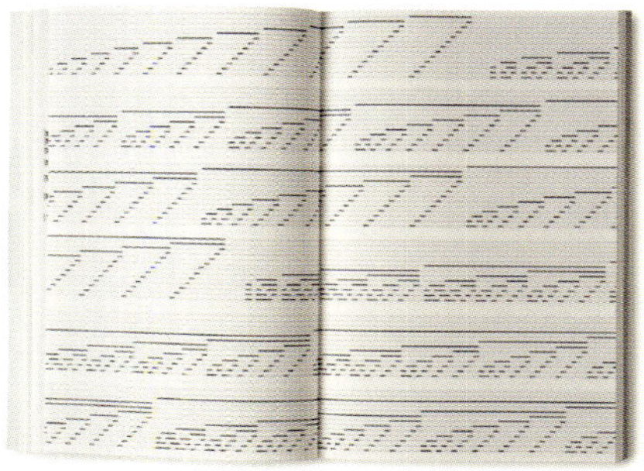

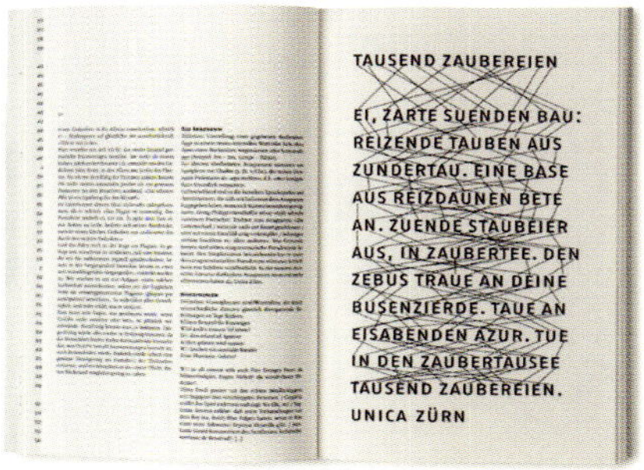

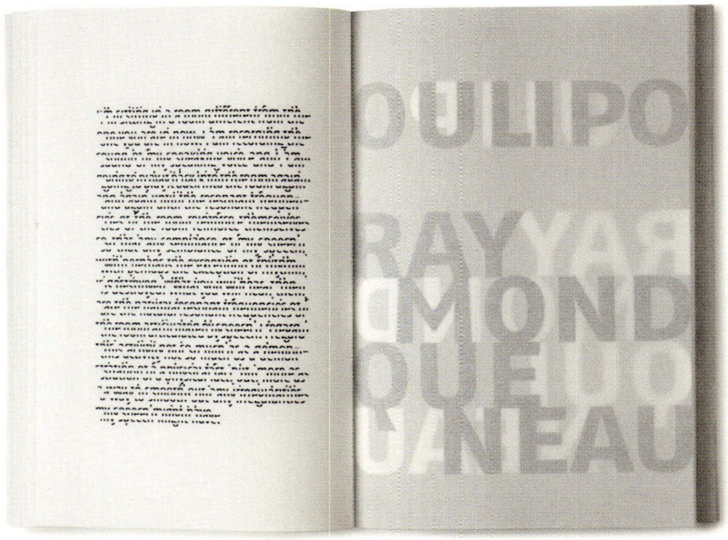

Auftraggeber Client
Muthesius Kunsthochschule
Kommunikationsdesign
Kiel, Germany

Projekt Project
Algorithmus – Programm als Struktur in Literatur und Musik
Buch
Book

Design
Simone Bräutigam
Hamburg, Germany

Meine Diplomarbeit ist von einer Ausstellung im ZKM in Karlsruhe „Die algorithmische Revolution" inspiriert worden. Algorithmische Strukturen treten in den vielseitigsten Erscheinungsformen auf: von der mathematischen Gesetzmäßigkeit über Programme in der Informatik bis hin zu Arbeitsanweisungen und mathematischen Strukturen in Literatur, Kunst, Musik, Architektur etc.
Das Medium Buch diente als experimenteller Raum zur typografischen und illustrativen Interpretation von algorithmischen Prozessen aus Literatur und Musik. In Form von rhythmischen Partituren sind musische und sprachliche Elemente neu kombiniert.

My dissertation was inspired by the exhibition "Die algorithmische Revolution" at the ZKM Karlsruhe. Algorithmic structures appear in very different forms: in mathematical laws, in computer programs, and even in literature, art and music.
The book is used as a medium which gives room for typographic illustrations of algorithmic processes from the fields of literature and music in an experimental way. The results are rhythmic scores as a combination of musical and linguistic elements.

Auftraggeber Client
Hochschule für Künste Bremen
Bremen, Germany

Projekt Project
VIER
Das Magazin der Hochschule für Künste Bremen
Magazine of the Bremen University of the Arts

Design
Thorsten Konrad
Thomas Schrott
Florian Holstein
Alexander Böll
Lisa Kreutzer
Judith Gärtner
Esther-Marie Nink
Britta Isenrath
Bremen, Germany

Identität sichtbar machen, Identifikation erfahren. Das Magazin VIER ist monothematisch – „Identität" Schwerpunkt der ersten Ausgabe. VIER hat als Informationsinstrument die Aufgabe, die komplexe und übergreifende Lehre und Forschung der HfK Bremen deutlich zu machen, dem Status und der Kultur der Hochschule ein Gesicht zu geben. 100 Seiten Inhalt und ein achtseitiger Umschlag zeigen die interdisziplinäre, integrative und internationale Ausrichtung der HfK. Raffiniert spiegelt die ausgefeilte Ästhetik des Editorial Design künstlerische, wissenschaftliche und praxisorientierte Zusammenhänge und zeigt weite Räume für richtungsweisende Arbeiten.

Visualize identity, experience identification! Current topic of the Bremen University of the Arts (HfK Bremen) magazine VIER is "identity". The interesting platform VIER gives detailed information about the complex and cross functional research and teaching at the HfK Bremen, its cultural role and mission. About 100 pages of content and an eight-sided cover visualize the interdisciplinary, integrative and international focus of the HfK. The aesthetics of the editorial design sophistically reflect the joint endeavors and links between art, music, theory and design. VIER underlines the broad and open methods in trendsetting workings.

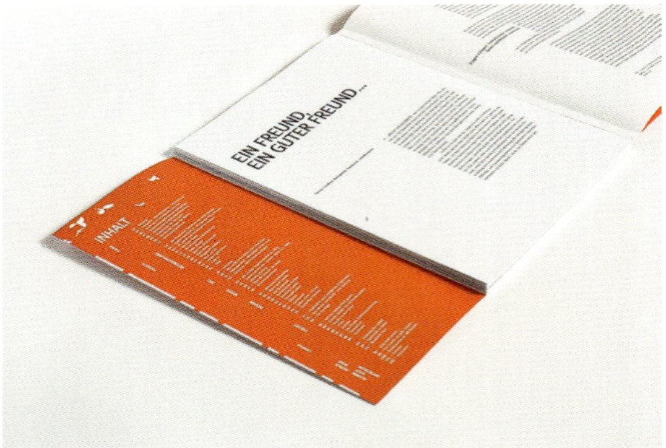

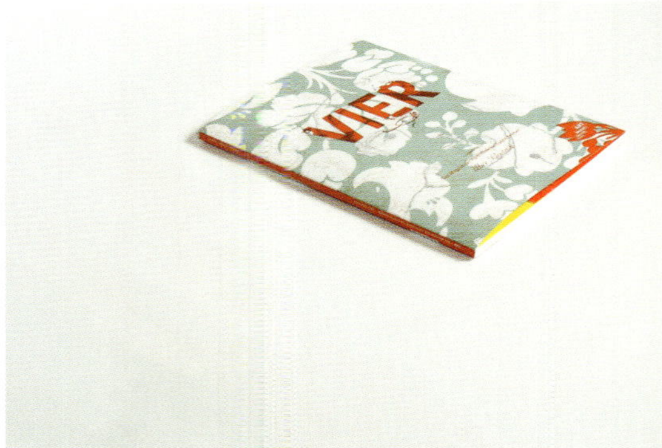

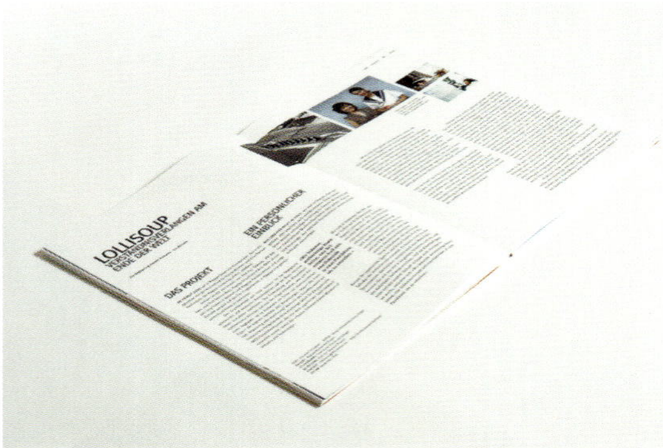

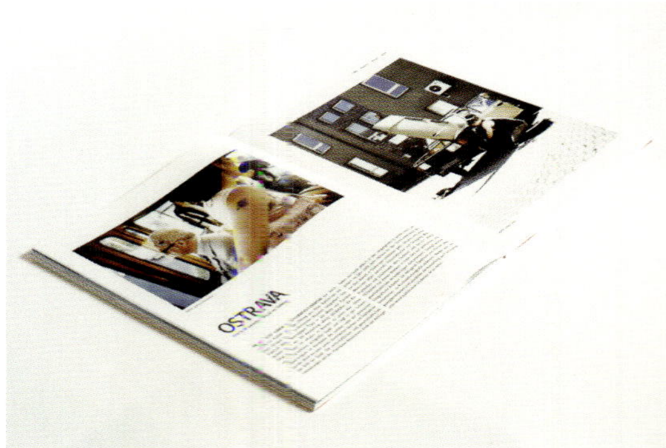

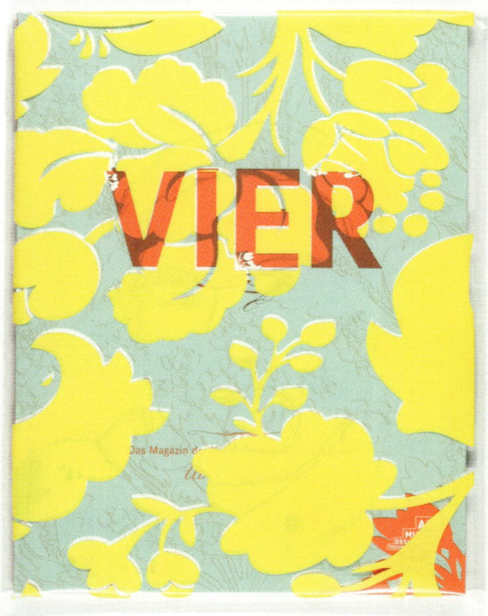

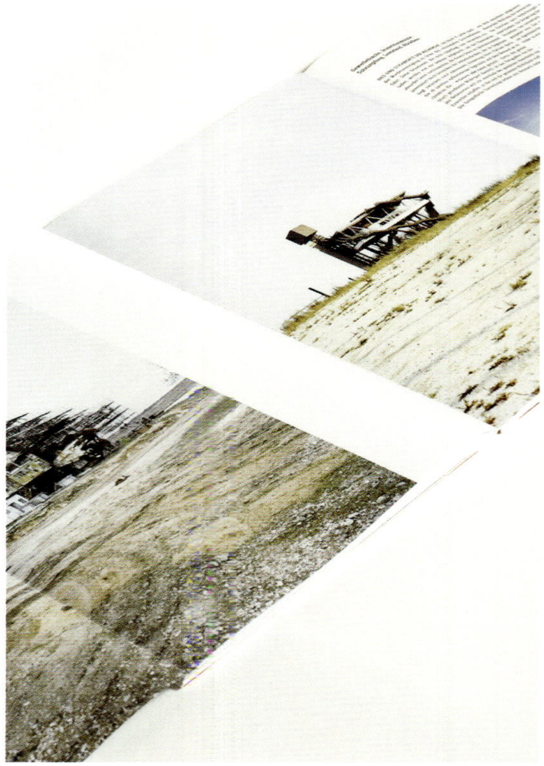

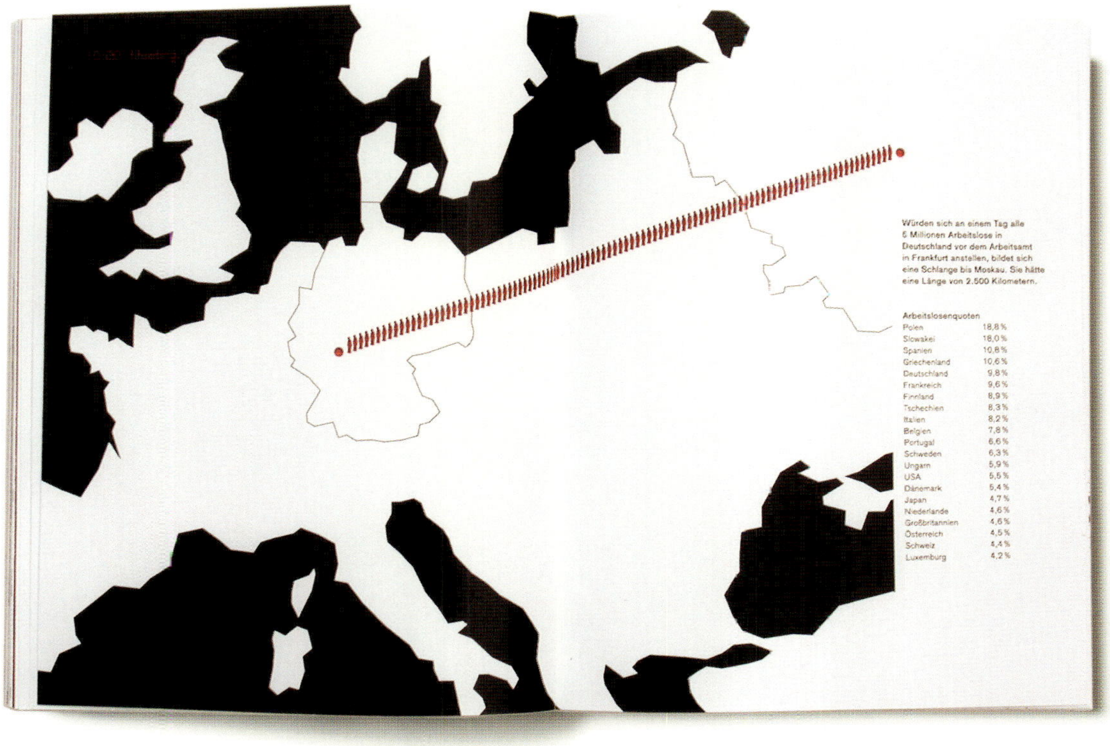

Auftraggeber Client
Fachhochschule Düsseldorf
Fachbereich Design
Düsseldorf, Germany

Projekt Project
Ein statistischer Tag
Buch
Book

Design
Ariane Bayer
Essen, Germany

Statistik – nur trockene Zahlen? Das Buch „Ein statistischer Tag" vermittelt die Daten der Amtlichen Statistik auf unterhaltsame, einprägsame und verständliche Weise. Als Basis und Kapiteleinteilung dient die Zeitverwendungsstatistik der Deutschen. Zusätzlich führt eine Kurzgeschichte, die einen gewöhnlichen Alltag widerspiegelt, durch das Buch und schafft so eine Verbindung der unterschiedlichen Statistiken. Diese werden in Form prägnanter bildstatistischer Illustrationen dargestellt. Die bildhafte Übertragung erleichtert dabei das Verständnis der Inhalte, ermöglicht auch große Zahlen zu begreifen und fördert die Einprägsamkeit.

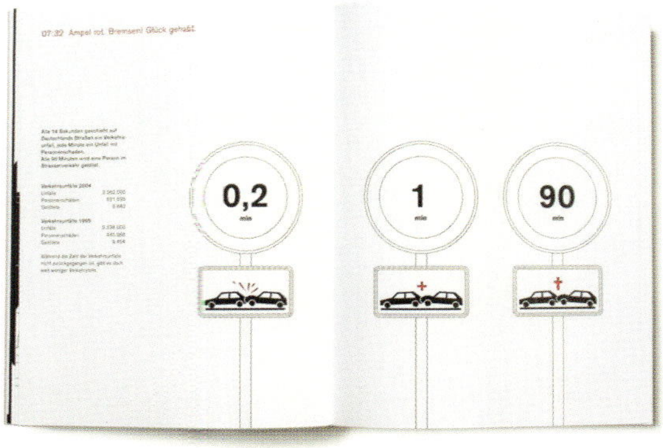

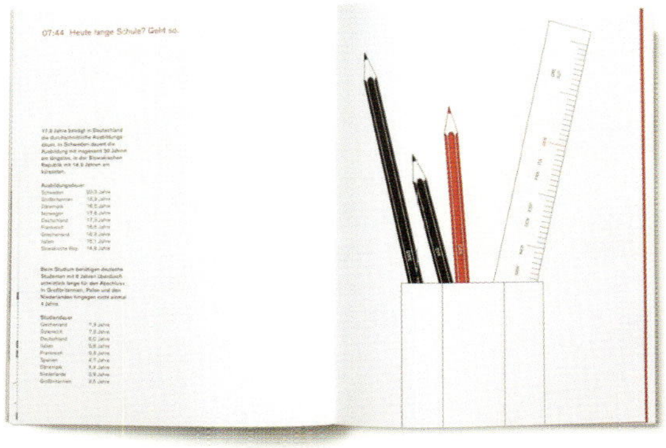

Statistics – just dry numbers? The book "Ein statistischer Tag" ("A Statistical Day") conveys the data of official statistics in an entertaining, memorable and comprehendible way. A statistic concerning the time usage in Germany serves as the underlying basis and chapter allocation device. A short story about everyday life guides through the book and thus connects the various statistics. These are displayed by concise picture-statistical illustrations. The transfer from numbers into pictures facilitates comprehension of content – even if big numbers are dealt with – and therefore enhances memorability.

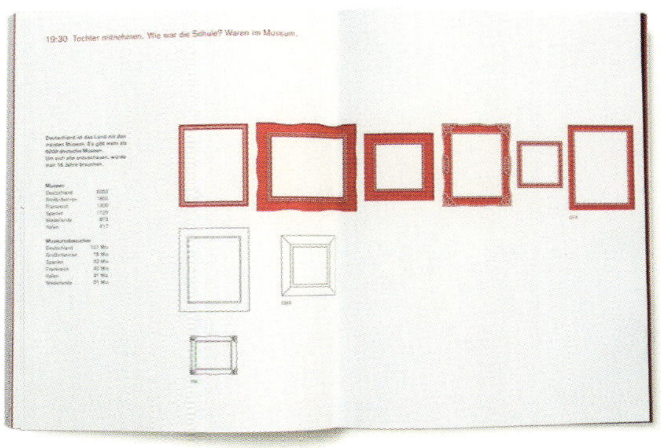

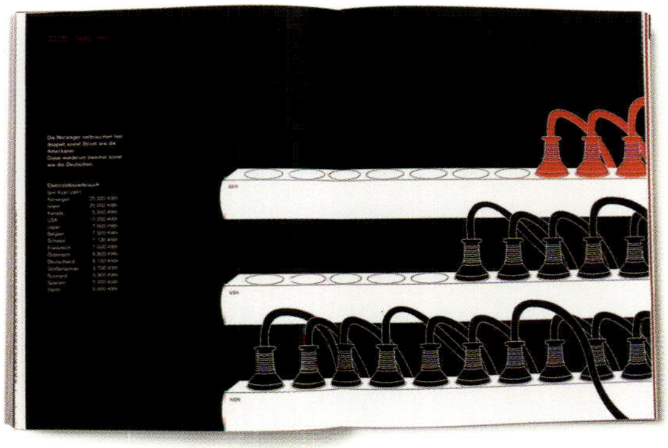

Auftraggeber Client
Royal College of Art
London, United Kingdom

Projekt Project
ARC issue 01-03
Magazin
Magazine

Design
Birgit Pfisterer
Matilda Saxow
Tom Lucas
ARC, Royal College of Art
London, United Kingdom

ARC ist das Hochschulmagazin des Royal College of Art (RCA). ARC untersucht zeitgenössische Kunst und Design inner- und außerhalb des RCA. Der bold gesetzte Titel fungiert als Poster, wenn ARC in der Öffentlichkeit gelesen wird. Das Format ändert sich mit jeder Ausgabe. Durch die Kombination von Ein- und Zweifarbdruck, die Verwendung unterschiedlicher Papiersorten und die Suche nach alternativen Druckverfahren versucht ARC das Maximale aus seinem kleinen Budget herauszuholen. Mit Interviews, Kurzgeschichten, Fotografien, Illustrationen, Artikeln, Kritiken ist ARC eine Momentaufnahme der Vielfalt zeitgenössischer Kultur und Kreativität.

ARC is the magazine of the Royal College of Art. ARC aims to represent the college from the angle of its students, creating a diverse portrait of contemporary creative culture. Maximising a limited printing budget, the designers explored one and two colour printing, the use of different paper stocks, and a variety of unbound formats for each issue. Through its bold and clear typography and layout, ARC is accessible, stimulating and very readable, ensuring full enjoyment of both content and design. This is reflected in the way the front and back covers are only seen to their full advantage when the magazine is actually being held up and read.

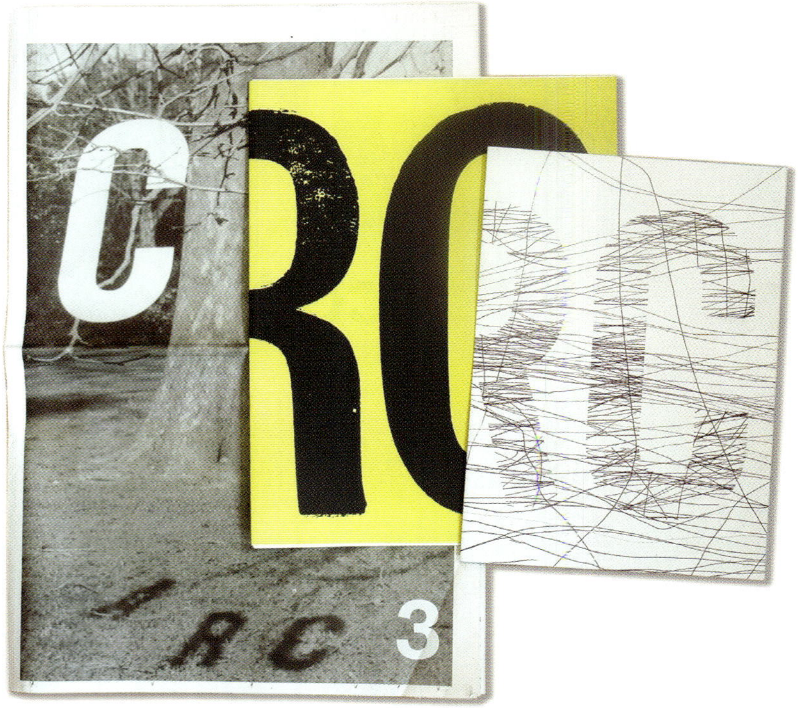

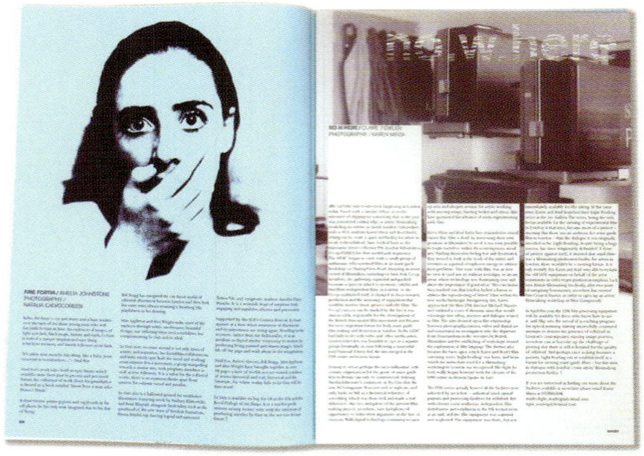

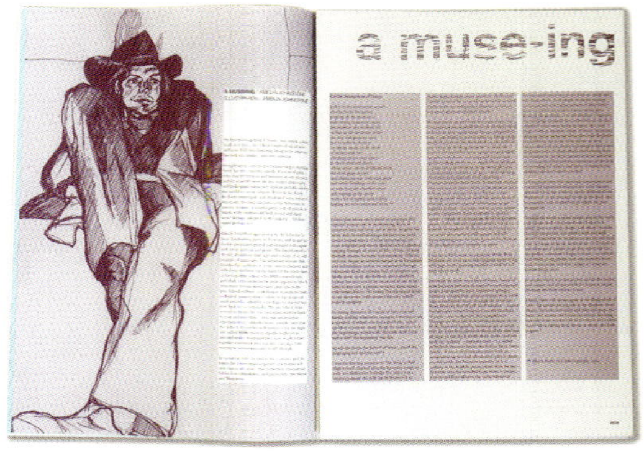

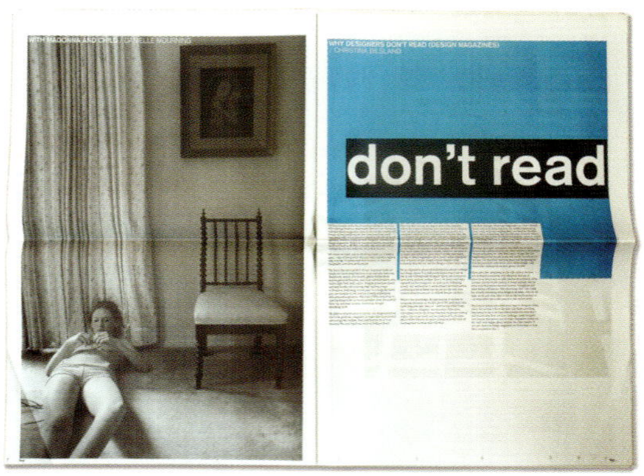

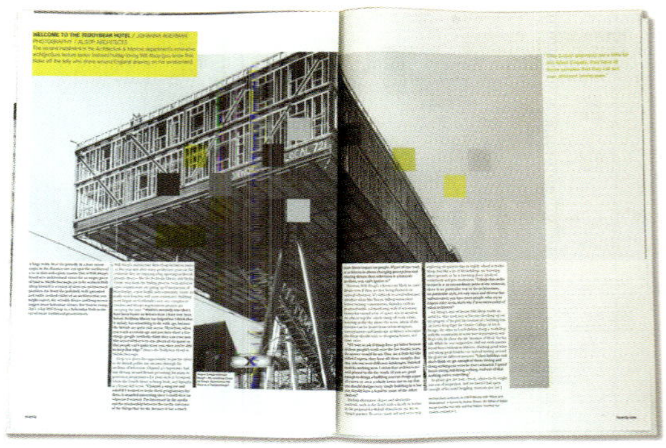

Auftraggeber Client
Hochschule für Gestaltung Offenbach am Main
Visuelle Kommunikation
Offenbach, Germany

Projekt Project
Hier + Jetzt. 64 Anregungen zur Zeitverschwendung
Zeitung
Newspaper

Design
Catrin Schnabel
Offenbach, Germany

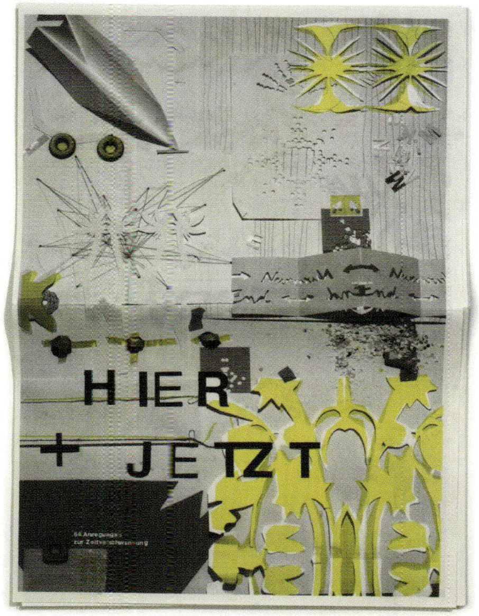

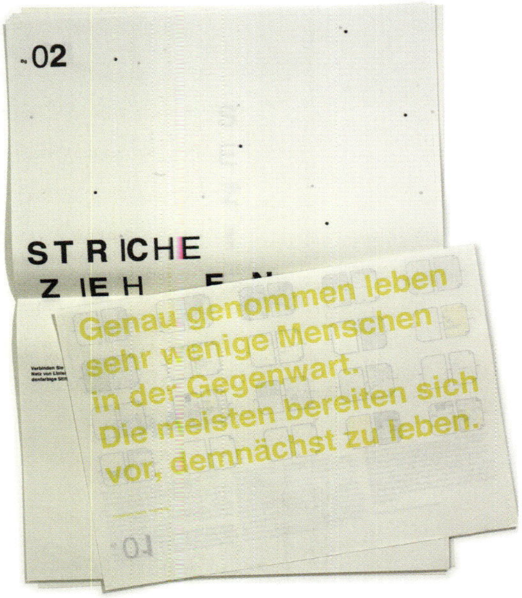

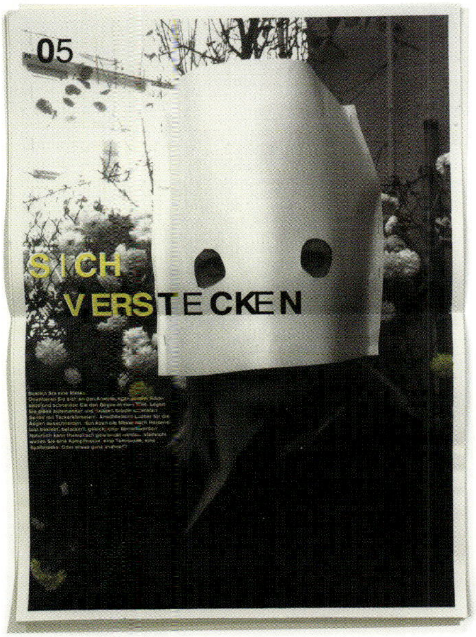

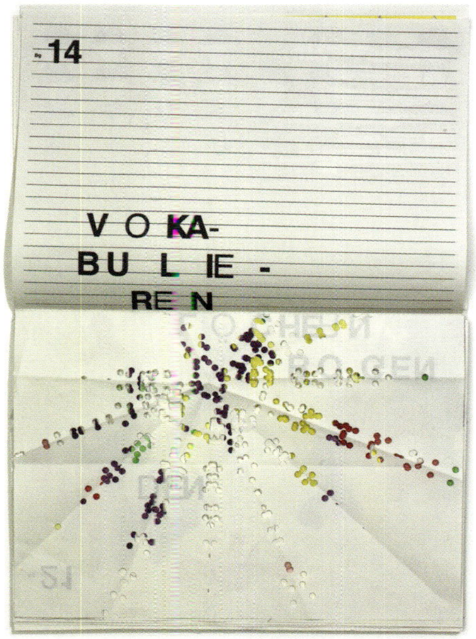

„Hier + Jetzt" ist eine Zeitung für Erwachsene, die keine Zeit haben. Sie besteht aus einer losen Bogensammlung mit Übungen, die zum Verschwenden eines Augenblicks anregen. Teils finden die Übungen direkt auf dem Papier statt, teils fordern sie auf, etwas mit der Umgebung oder der Situation, in der man sich gerade befindet, zu tun. Beispielsweise soll der Bogen mit dem Locher zerfressen oder komplett ausgemalt werden. Die Übungen können überall und mit den gerade zur Verfügung stehenden Mitteln gemacht werden. „Hier + Jetzt" ist als eine Art Lockerungs- oder Entschleunigungsprogramm gedacht.

"Here + Now" is a newspaper for stressed adults who regard themselves as having no spare time at all. It consists of several sheets of paper filled with ideas and exercises how to "waste" moments in time. There is one exercise on each sheet. Some exercises are meant to be done using one of the newspaper sheets directly, other instructions how to waste time encourage the readers to reflect on their current situation and their environment. The exercises can be done everywhere and one can use everything that is available at that very moment. "Here + Now" is meant to be a relaxation or distraction program.

Auftraggeber Client
Fachhochschule Dortmund
Fachbereich Design
Dortmund, Germany

Projekt Project
Leichtkraft
Fotografie
Photography

Design
Ivo Mayr
Dortmund, Germany

Konzept: Die Bildserie „Leichtkraft" ist ein Spiel mit den Sehgewohnheiten des Betrachters. Kernidee ist es, Fotos durch eine einfache Drehung aus ihrem eigentlichen Kontext zu lösen. Die dadurch hervorgerufene Irritation zwingt den Betrachter genauer hinzusehen, um das ungewohnte Bild zu verstehen. Umsetzung: Ich fotografiere Menschen im öffentlich-urbanen Raum mit viel Beton und flächigen Elementen. Der Ausschnitt der Fotos ist sehr reduziert und abstrakt gehalten. An diesen Plätzen lasse ich Menschen agieren in gestellten Posen, die Bewegungen und Haltungen andeuten, die in Wirklichkeit so nicht möglich wären. Die Schwerkraft scheint aufgehoben. Wo die Schwerkraft tatsächlich wirkt, ist erst auf den zweiten Blick erkennbar: Es sind z. B. fallende Haare bzw. Kleidung, die den entscheidenden Hinweis geben.

Concept: The picture series "Leichtkraft" plays with the viewing habits of the observer. The main idea is to separate photos from their actual context through a simple rotation. This creates confusion for the observer and forces him to have a closer look in order to understand the "strange" picture.
Implementation: I take pictures of people in open urban spaces with a lot of concrete and flat elements. The frame of the photos is very reduced and abstract. In these locations I then have people act in artificial poses that suggest movements and postures which would not be possible like that in real life. Gravitation seems to be abolished. Where gravitation really acts can only be seen at a second glance: hanging hair or falling clothes give a clue for example.

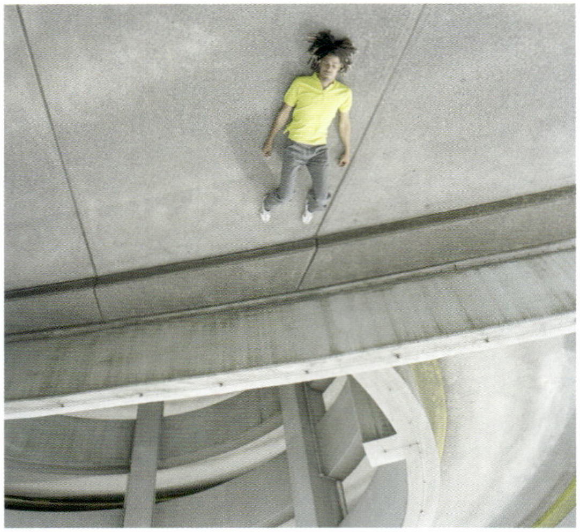

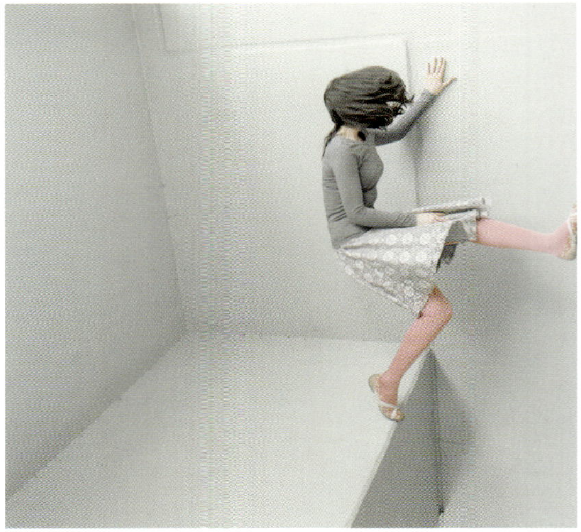

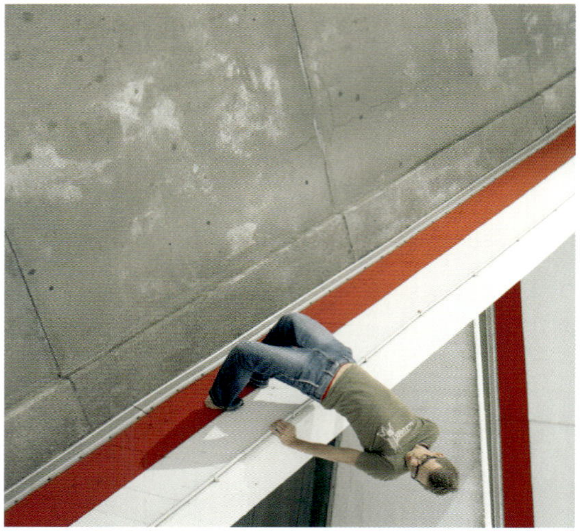

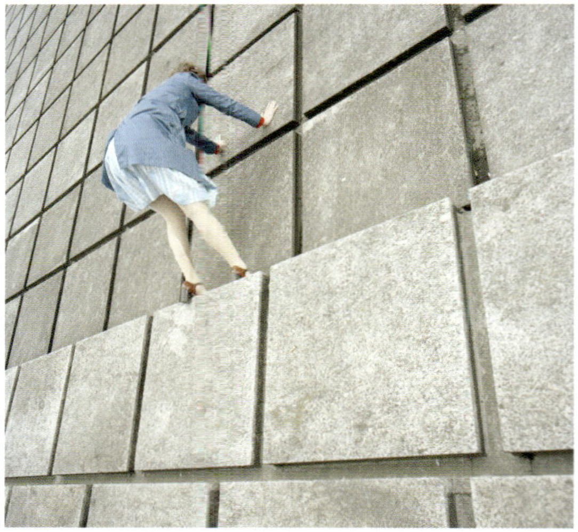

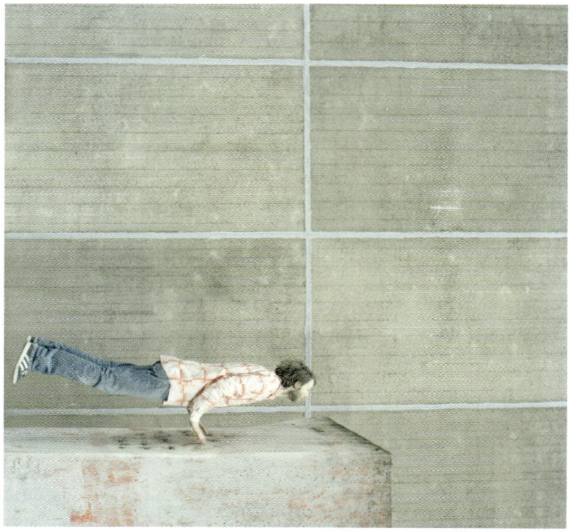

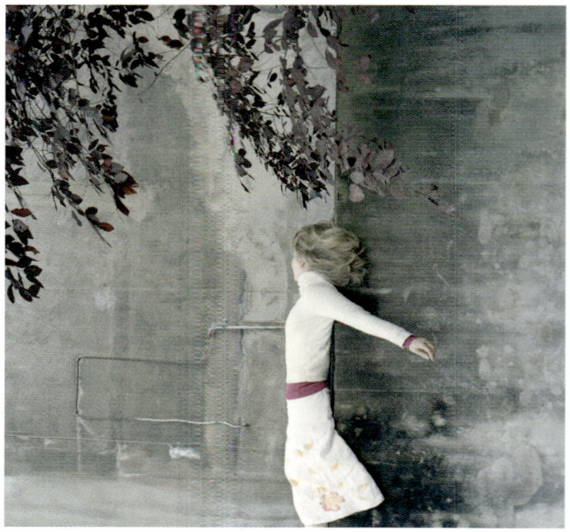

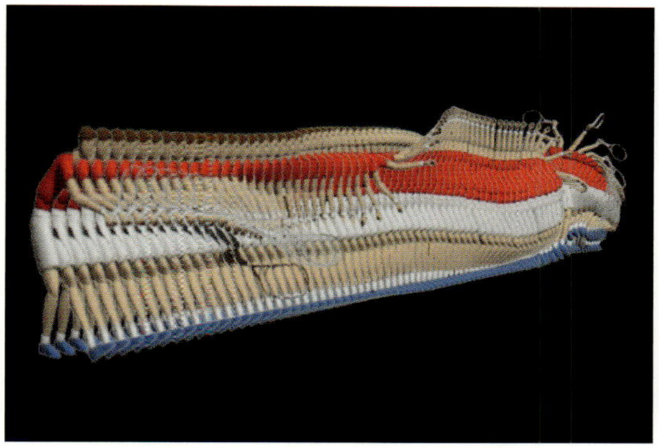

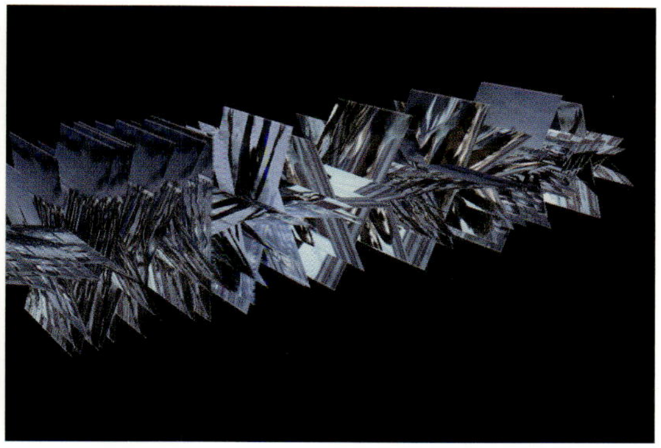

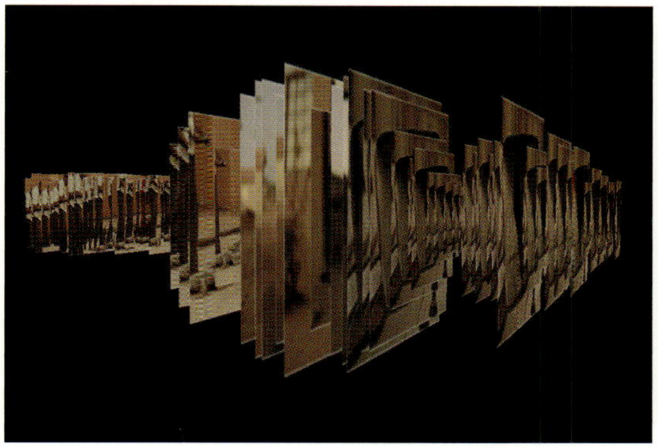

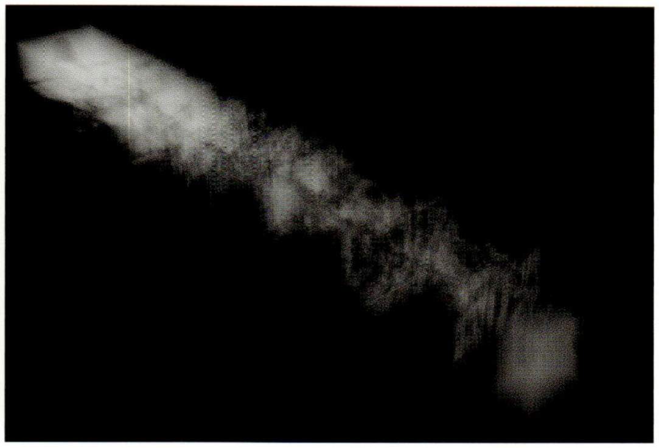

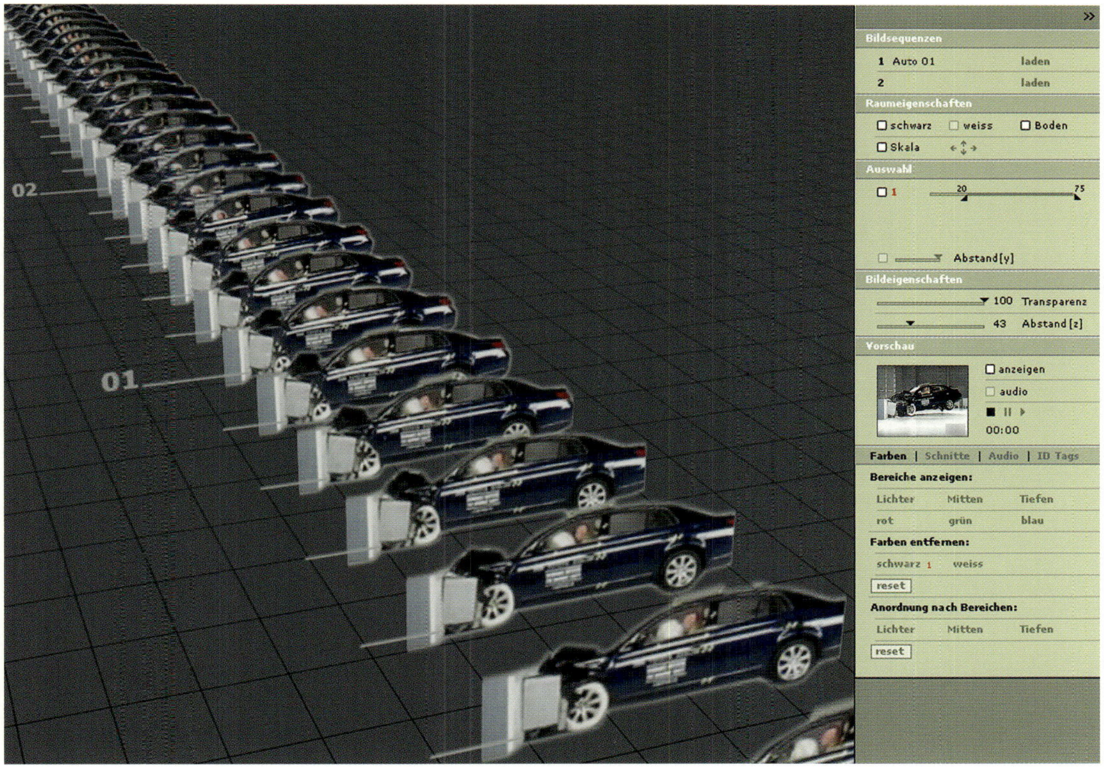

Auftraggeber Client
Fachhochschule Würzburg-Schweinfurt
Fachbereich Gestaltung
Würzburg, Germany

Projekt Project
Recreating Movement
Interaktives Werkzeug
Interactive tool

Design
Martin Hilpoltsteiner
Bremen, Germany

„Recreating Movement" ist ein interaktives Werkzeug, das neue Möglichkeiten in der Betrachtung von filmischem Material aufzeigt. Die Einzelbilder einer Filmsequenz werden im virtuellen Raum hintereinander angeordnet. Es entsteht ein dreidimensionaler Körper, der den Ablauf der einzelnen Filmbilder darstellt. Raum, Zeit und Bewegung werden auf diese Weise in einem virtuellen Block eingefroren und können anhand unterschiedlicher Filter und Einstellungsoptionen genauer untersucht bzw. miteinander verglichen werden.

"Recreating Movement" is an experimental tool for analysing film sequences. Single frames of a film are extracted and arranged one after another in a three-dimensional space. The frames – which are normally visible for only a split second during a film – are rendered into a tube-like frameset that "freezes" a particular time span in a film. Various filters and settings can be applied to the film sequence with a displayable menu bar.

Auftraggeber Client
Fachhochschule Hannover
Fachbereich Design und Medien
Hannover, Germany

Projekt Project
DIET RIOT – Eat War!
www.diet-riot.net
Website

Design
Philipp Zurmöhle
Hannover, Germany

Während meines Aufenthalts in Japan musste ich manchmal mit dem Essen kämpfen, das ich bekam. Nun rächen sich die Speisen ... Vielleicht kämpfen sie gegen Diäten, gegen sich selbst oder einfach nur gegen alles Schlechte auf Erden.
Auf der animierten Website www.diet-riot.net kann man sehen, wie Maki-San, das militante Sushi, Breakdance tanzt, das Ei Atamago im „DJ Style" kocht oder wie die Squid Squad einfach nur rumhängt.
Außerdem habe ich Poster und Sticker entworfen, die zusammen mit der Website in einer Ausstellung in Hiroshima präsentiert wurden.

During my stay in Japan I sometimes had to fight with the strange food served there. Now this food strikes back ... Maybe it fights against diets, against itself or just against all the bad in the world.
On the animated website www.diet-riot.net you can watch Maki-San, the militant Sushi, doing the breakdance, the egg Atamago cooking in "DJ style" or the Squid Squad just hanging around.
I also designed posters and stickers, which were presented with the website in an exhibition in Hiroshima.

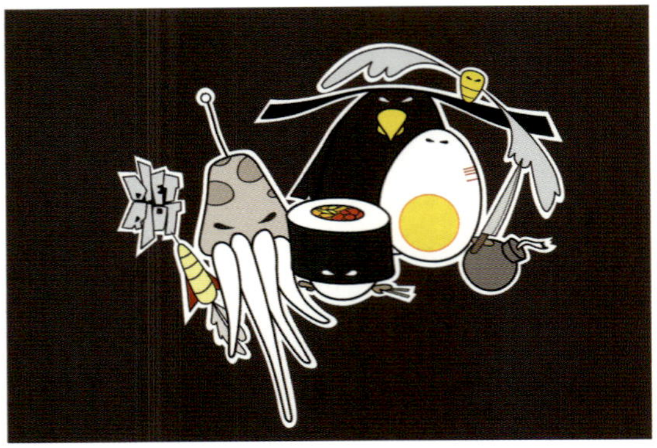

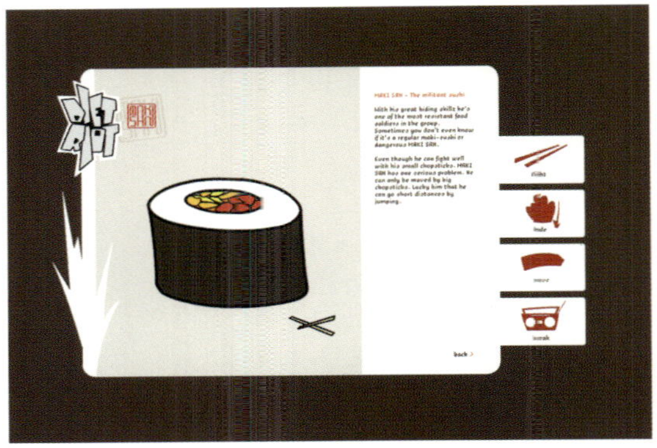

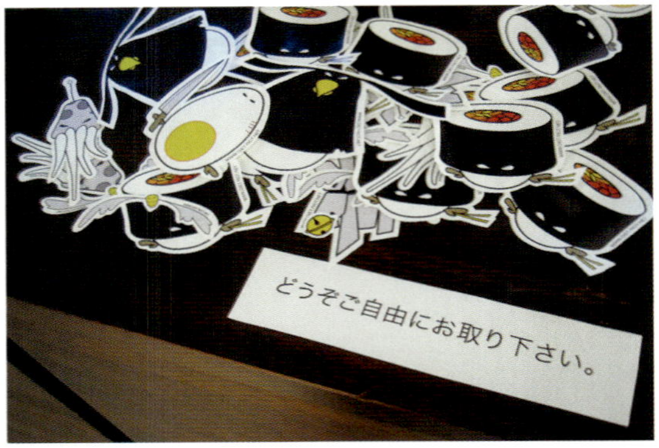

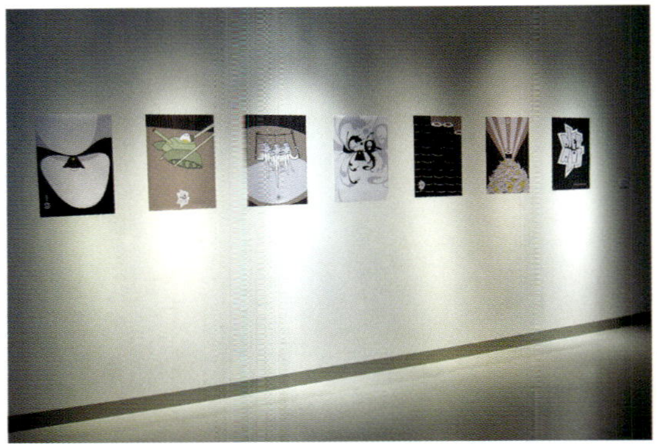

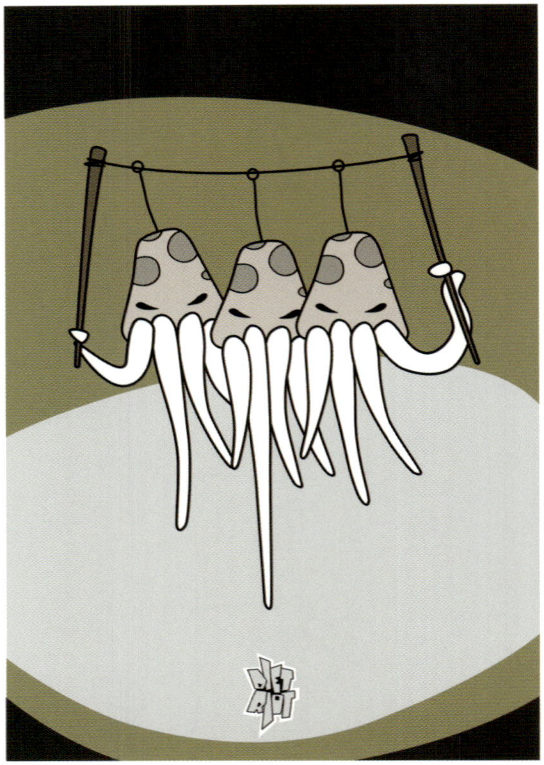

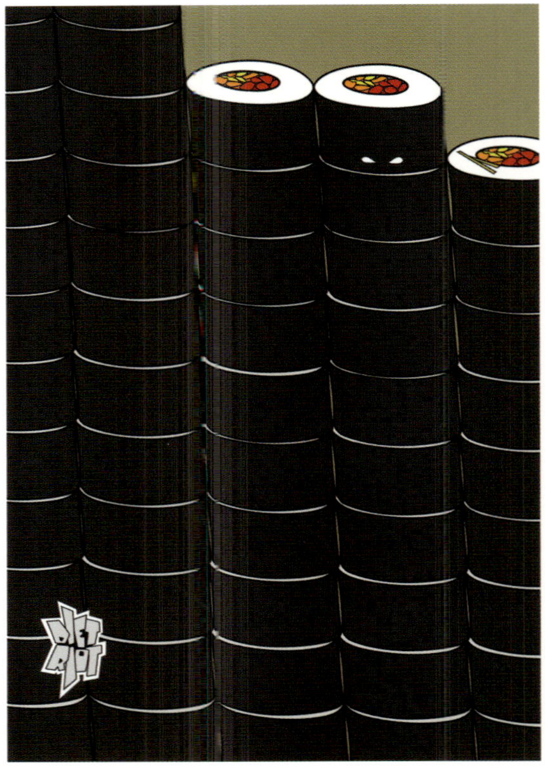

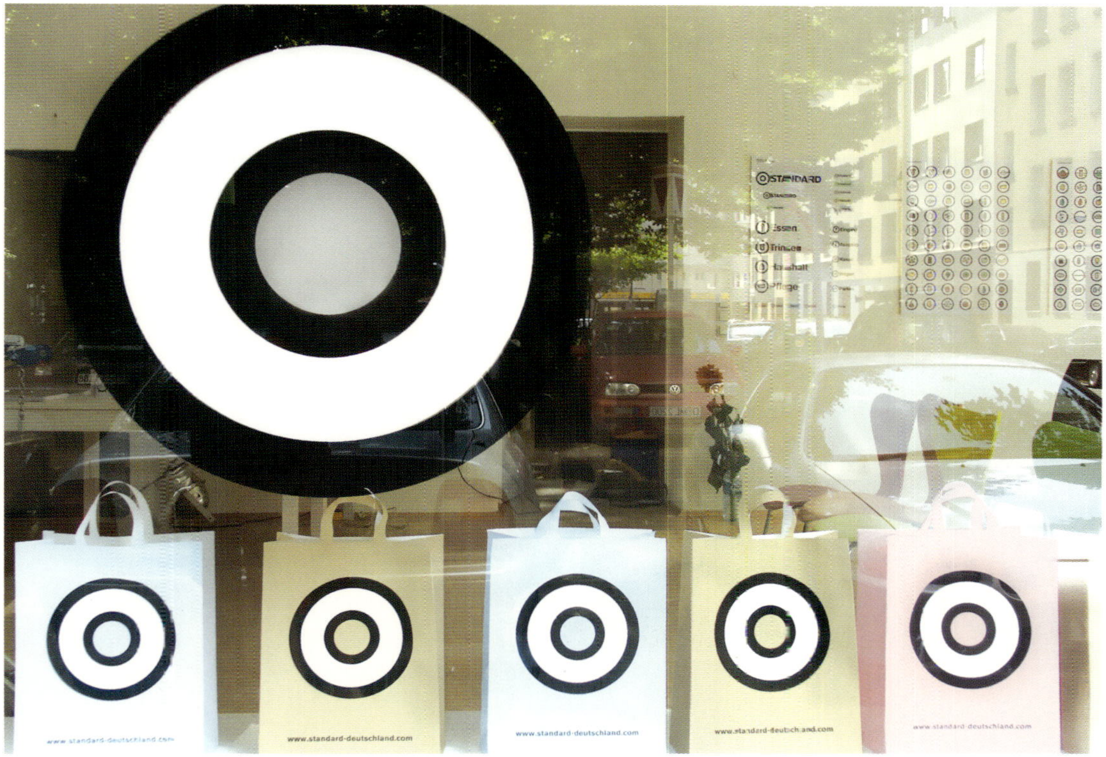

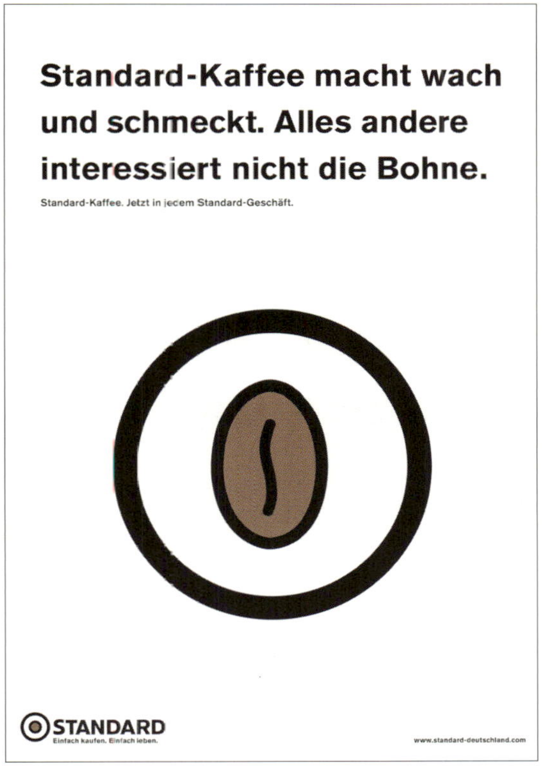

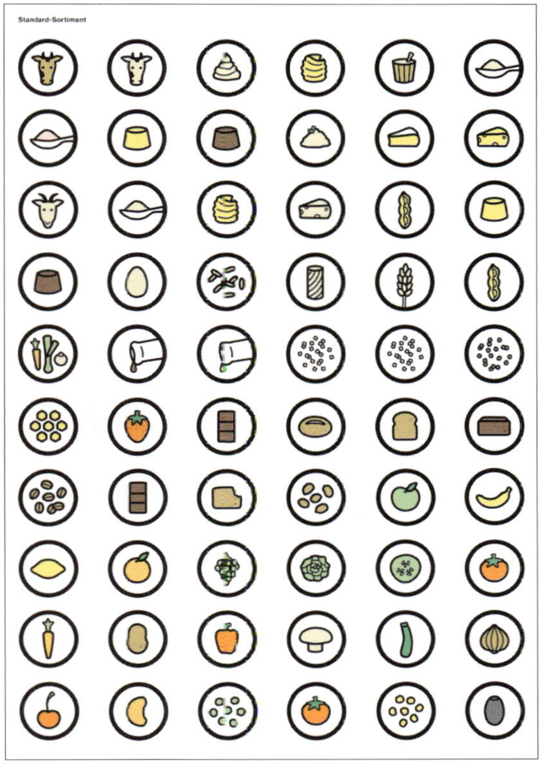

Auftraggeber Client
Fachhochschule Düsseldorf
Fachbereich Design / Kommunikationsdesign
Düsseldorf, Germany

Projekt Project
Standard
Konzeption und Entwurf einer markenästhetischen Utopie
Conception and design of a brand utopia

Design
Sarah Heidelberg
Berlin, Germany

Standard ist ein neuartiges Einkaufsangebot im alltäglichen Verbrauchsgütermarkt. Es bietet mit seinem begrenzten Sortiment und der visuellen Beschränkung auf das Notwendige eine Alternative zu bestehenden Konsum-Formen auf dem Markt.
Standard will den Verbraucher in der immer unüberschaubareren Einkaufslandschaft entlasten und damit Orientierung schaffen. Statt unsozialer Discount-Mentalität setzt Standard auf nachhaltigen Handel, regionale Produktion und logofreie Packungen. Damit ermöglicht Standard seinen Kunden zukunftsfähigen Konsum unabhängig von leeren Markenimages. Dies drückt auch der Claim aus: „Standard. Einfach kaufen. Einfach leben."

Standard is a new approach to shopping for daily needs. With its limited range of products and the restriction of visual elements to the essentials it offers an alternative to current forms of consumerism.
Standard aims to relieve the customer from the confusing shopping landscape and to create orientation. It differentiates itself from the antisocial mentality of discount brands, as well as it stresses values such as fair trade, regional produce and trademark-free packaging. All this provides the customer with a sustainable approach to consumerism, free of meaningless brand images. This is expressed by the claim: "Standard. Shop simple. Live simple."

Auftraggeber Client
Fachhochschule Düsseldorf
Fachbereich Design
Düsseldorf, Germany

Projekt Project
Zeichenstern
Kompendium der visuellen Zeichen und Markenzeichen
Compendium of visual signs and brands

Design
Nadine W. Wilms
Köln, Germany

Zeichenstern ist ein Kompendium der visuellen Zeichen und Markenzeichen, ein Nachschlagewerk, eine Orientierungshilfe und Inspirationsquelle sowie eine Argumentationsstütze für Gestalter. Es ist eine Sammlung von über 1.200 Zeichen, die formal-grafisch nach der Kategorisierung der Wiener Klassifikation des Deutschen Marken- und Patentamts zusammengestellt sind.
Der erste Teil des Buchs beschäftigt sich mit Fragen der Identität, des Namens, der visuellen Form, des Re-Designs, der Markenanmeldung und des Plagiats. Neben der Zeichensammlung findet der Leser ein Kapitel, das sich der umfangreichen Terminologie der Zeichen widmet.

Zeichenstern is a compendium of visual signs and brands, a reference book, a guideline and source of inspiration as well as an argumentation support for designers. It is a collection of over 1,200 signs which are arranged graphically and formally referring to the categorization of the Viennese classification of the German Patent and Trade Mark Office.
The first part of the book is concerned with questions of identity, names, visual form, re-design, brand registration and plagiarism. In addition to the collection of signs, the reader finds a chapter which is dedicated to the extensive terminology of the signs.

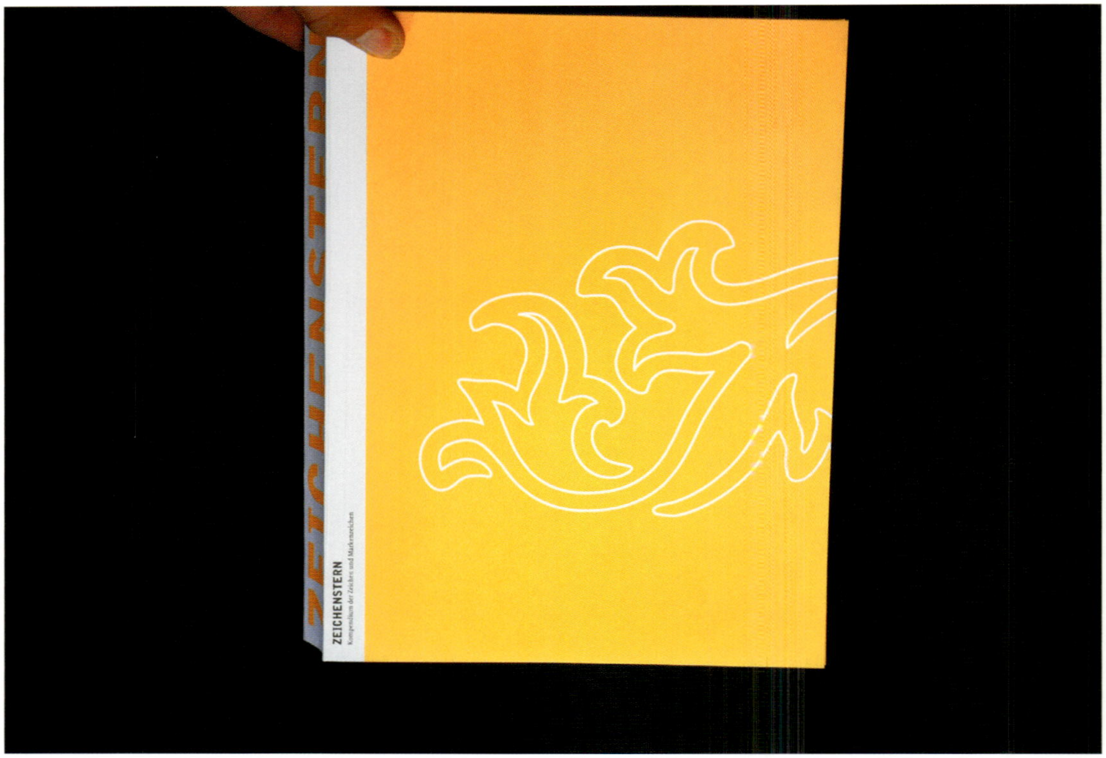

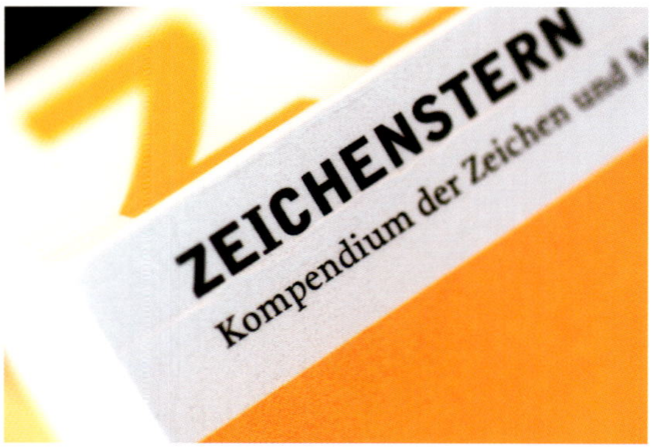

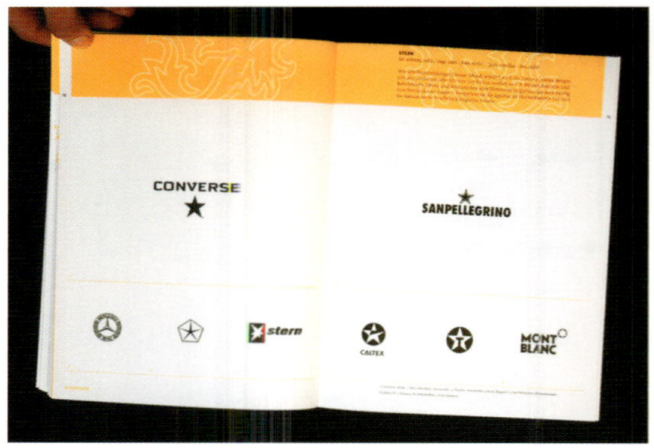

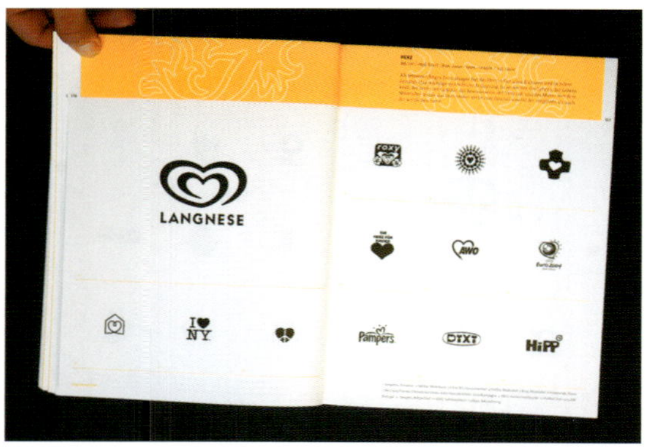

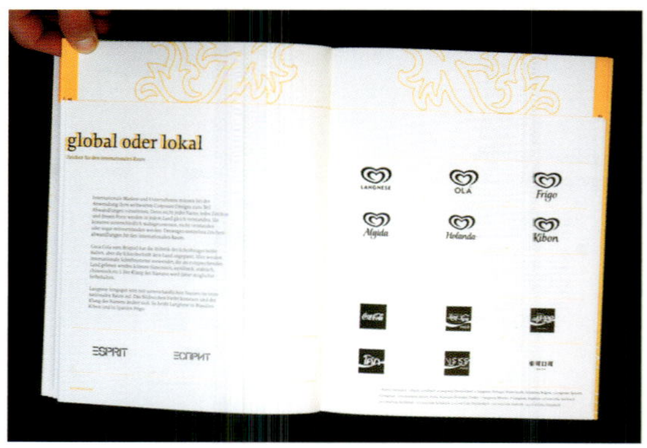

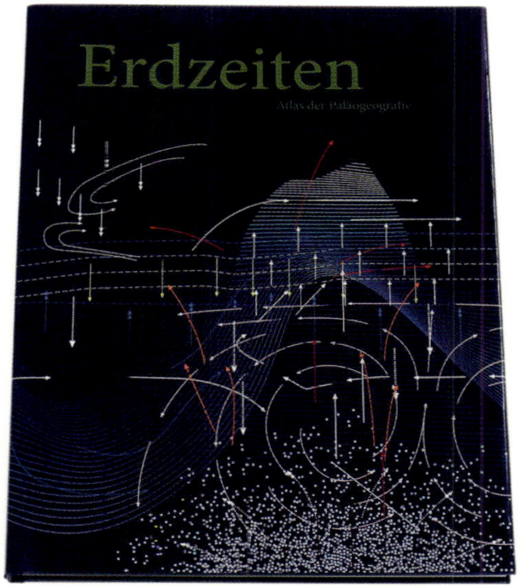

Auftraggeber Client
Senckenbergmuseum
Frankfurt am Main, Germany
Hochschule für Gestaltung Offenbach am Main
Offenbach, Germany

Projekt Project
Erdzeiten – Atlas der Paläogeografie
Atlas

Design
Martin Gorka
Offenbach, Germany
Cornelia Vogt
Osnabrück, Germany
Violetta Walter
Neu-Isenburg, Germany

Der Atlas der Erdzeitalter entstand an der HfG Offenbach unter Betreuung von Prof. Klaus Hesse in Zusammenarbeit mit dem Senckenbergmuseum Frankfurt. Der Atlas ist der erste seiner Art und zeigt Plattentektonik, Verteilung von Land und Meer, klimatische Verhältnisse, Vulkanismus, Tier- und Pflanzenwelt sowie Meeresströmungen vor vor 290 Millionen Jahren bis heute.
In vier Monaten intensiver Arbeit entwickelten wir Konzept und Gestaltung. Diese vermitteln unkompliziert die Fülle an Informationen und zeigen die Erde als einen lebendigen Planeten.

The atlas of geological eras was developed at HfG Offenbach under the supervision of Prof. Klaus Hesse in cooperation with the Senckenbergmuseum Frankfurt. With this project the museum intends to increase the awareness for paleogeography amongst pupils and the general public.
The atlas "Erdzeiten" is the first one to cover in this manner continental drift, coastal changes, climatic conditions, volcanic activity, flora and fauna as well as ocean currents over the last 290 million years.
Within four months of intensive work we developed the book's concept and design. They both make the versatility of information accessible and bring our Earth out as a lively planet.

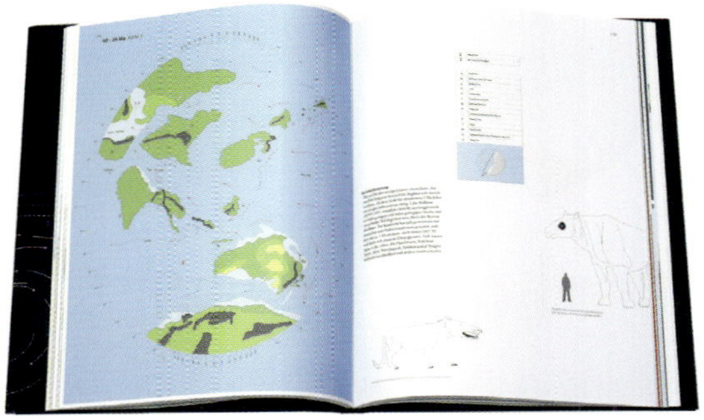

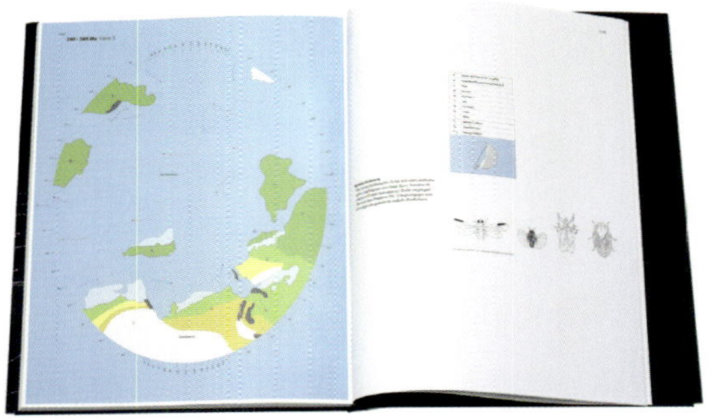

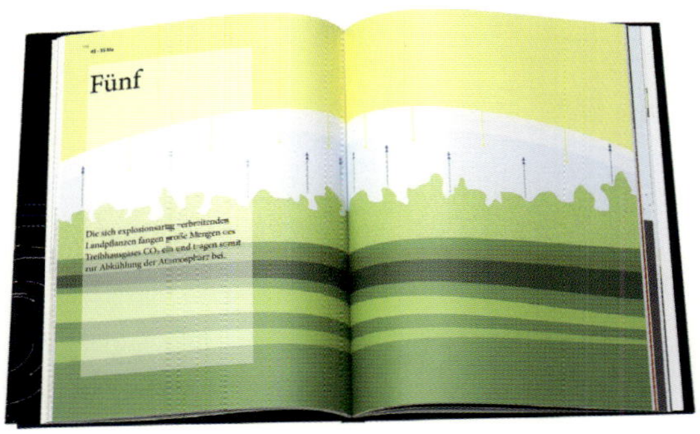

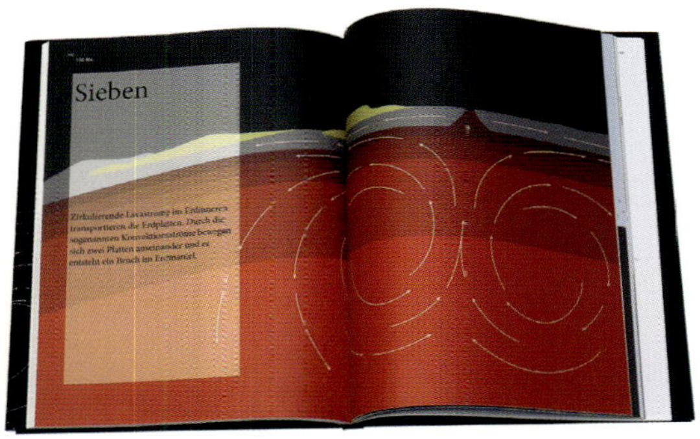

Auftraggeber Client
Hochschule für Angewandte Wissenschaften (HAW) Hamburg
Fakultät Design, Medien und Information
Studiendepartment Design
Hamburg, Germany

Projekt Project
NON – Phantome, Pseudonyme, Generische Personen
Sachbuch
Non-fiction book

Design
Stefan Schröter
Hamburg, Germany

„NON" ist ein Sachbuch über Phantome, Pseudonyme und fiktive Personen aus Medien, Kultur und Politik. Die einzelnen Figuren werden vorgestellt, untersucht und visualisiert – das Phänomen des Personenmythos wird von verschiedensten Seiten beleuchtet.
Das Buch ist eine Reise durch die Welt der Phantome, auf der mit Absicht am Ende nicht alle Geheimnisse gelüftet werden.
Featuring: der Yeti, das Ungeheuer von Loch Ness, Zorro, Osama Bin Laden, der Otto Normalverbraucher und viele andere.

"NON" is a non-fiction book about phantoms, pseudonyms and fictitious individuals from the spheres of media, culture and politics. The different characters are introduced, analysed and visualised. In doing so the phenomenon of the myth of the person is viewed from different perspectives.
The book is a journey through the world of phantoms, where intentionally not all secrets will be disclosed. Featuring: the Abominable Snowman, the Loch Ness Monster, Zorro, Osama Bin Laden, Alan Smithee and many others.

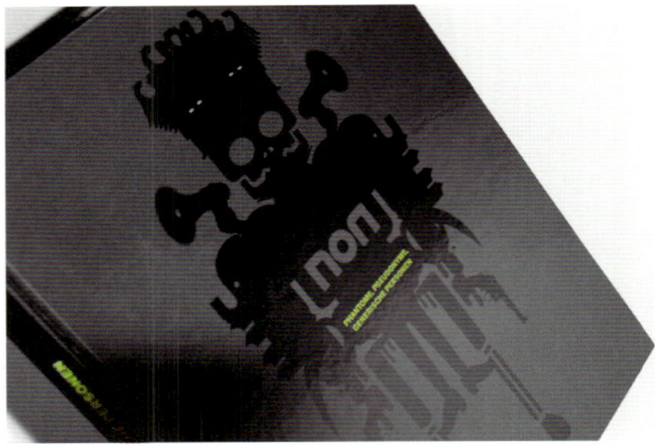

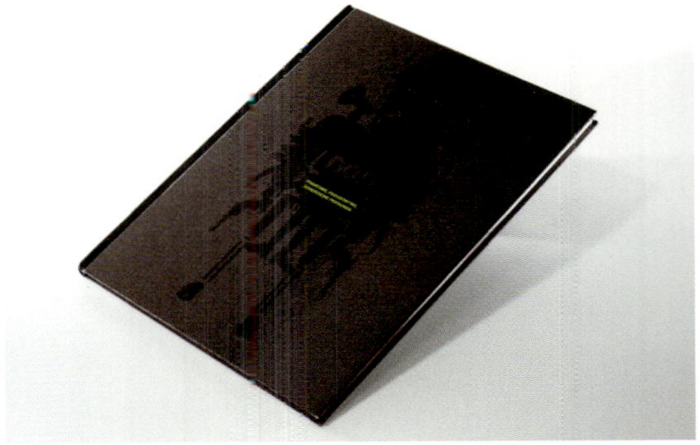

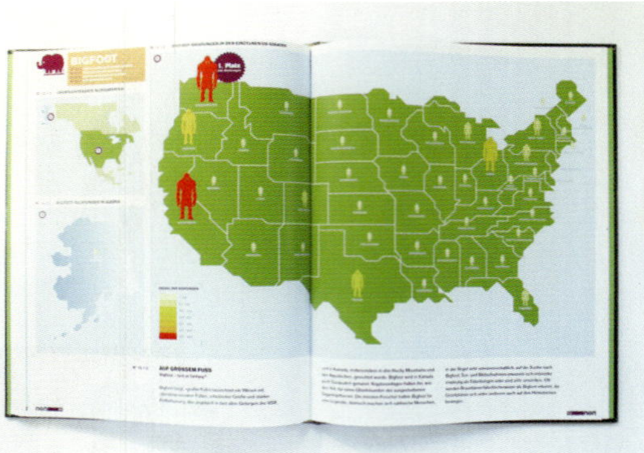

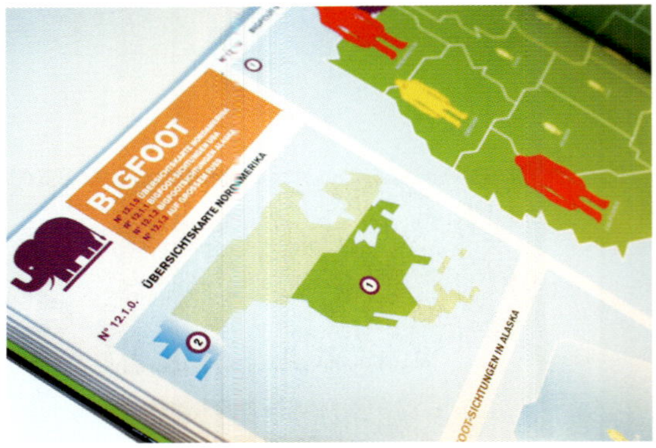

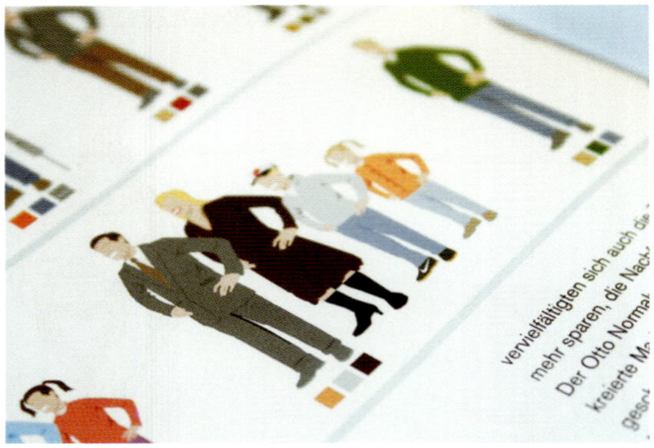

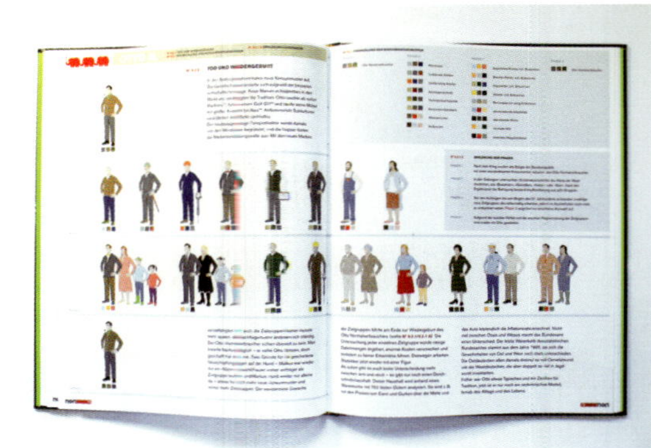

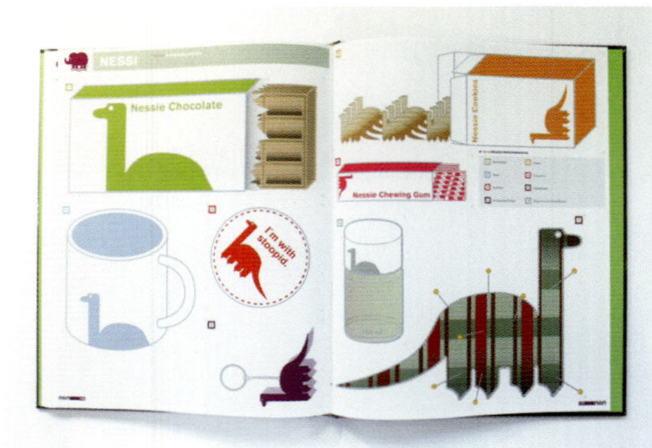

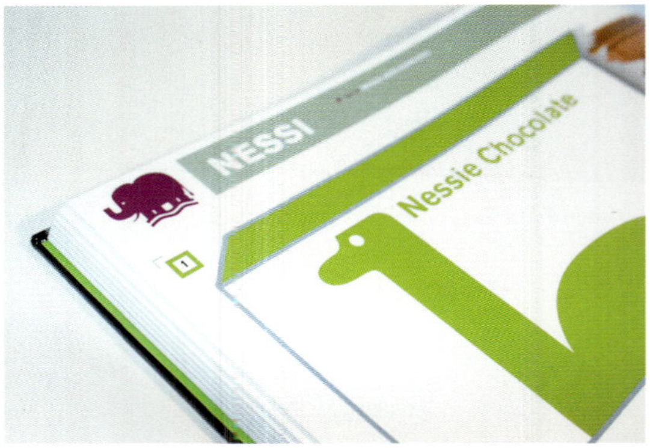

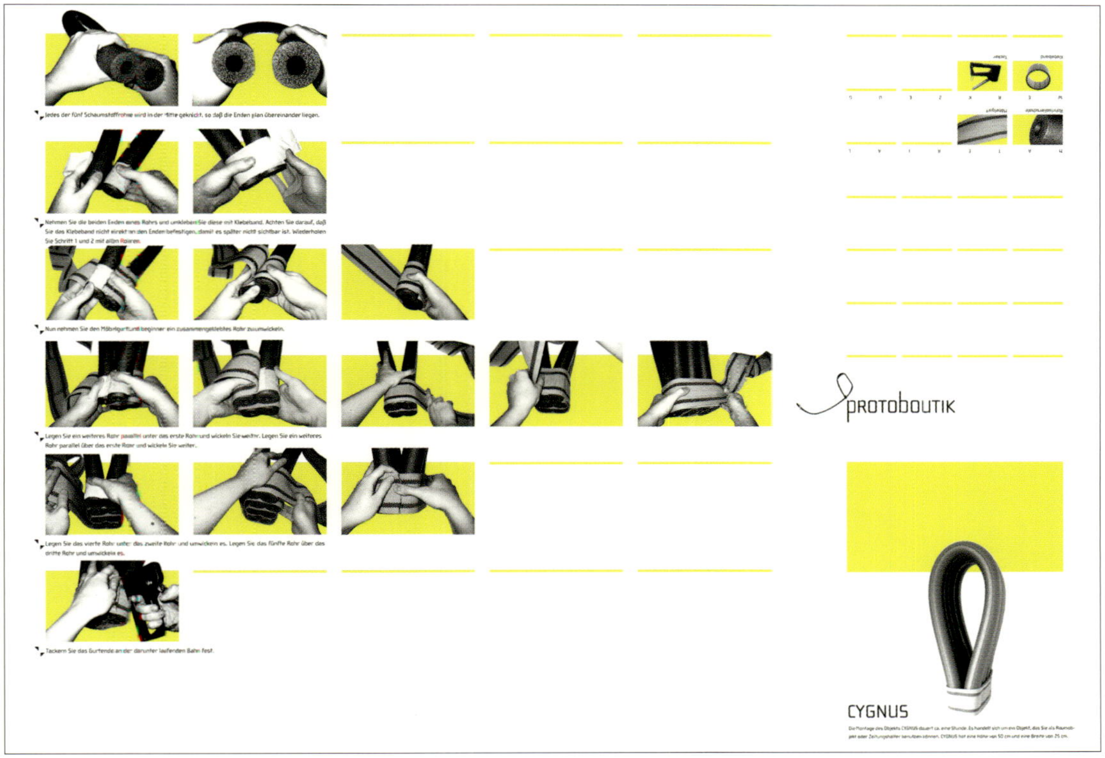

Auftraggeber Client
Fachhochschule Mainz
Design
Mainz, Germany

Projekt Project
Protoboutik
Bauanleitungen, Katalog, Objekte
Construction manual, catalogue, objects

Design
Nadine Bauer
Wiesbaden, Germany

Protoboutik ist ein Designkonzept und dient gleichzeitig als Werbekampagne für einen Baumarkt. Designinteressierte Kunden sollen anhand von Bauanleitungen angereizt werden, Designobjekte aus rohen Baumarktmaterialien für wenig Geld nachzubauen – eine neue Inspiration abseits von Garten- und Wohnungsgestaltung. Werkausstellungen im Baumarkt und Plakatserien dienen dem Projekt. Die Bauanleitungen befinden sich zusammengefaltet in durchsichtigen Tüten mit einer Papierverkleidung. Diese Tüten liegen in Kartenständern kostenlos aus. Dazu existiert ein Katalog, der gegen eine Schutzgebühr einen konkreten Überblick über Protoboutik verschafft.

Protoboutik on the one hand serves as a design concept and on the other hand as an advertising campaign for a DIY superstore. On the basis of a construction manual customers interested in design shall be attracted to build design objects out of raw materials for little money – a new inspiration apart from garden and home styling! Varying expositions in the DIY superstore and advertising bills support the project. The construction manuals are folded in transparent bags and are distributed free of charge. In addition, a catalogue which illustrates a complete picture of the project is available for a nominal fee.

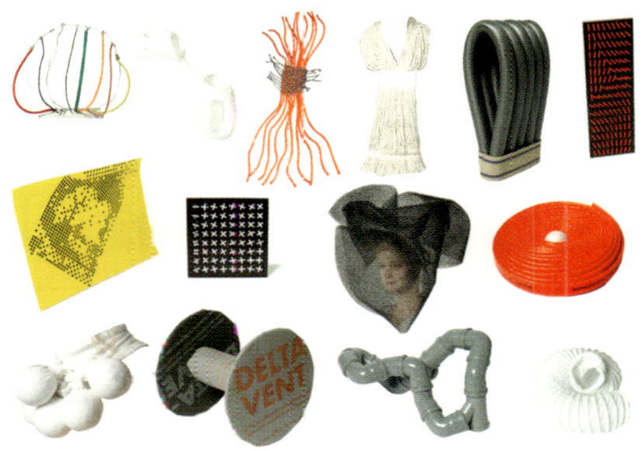

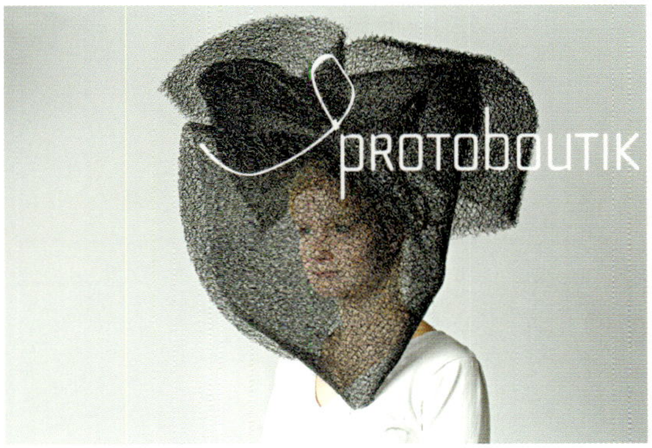

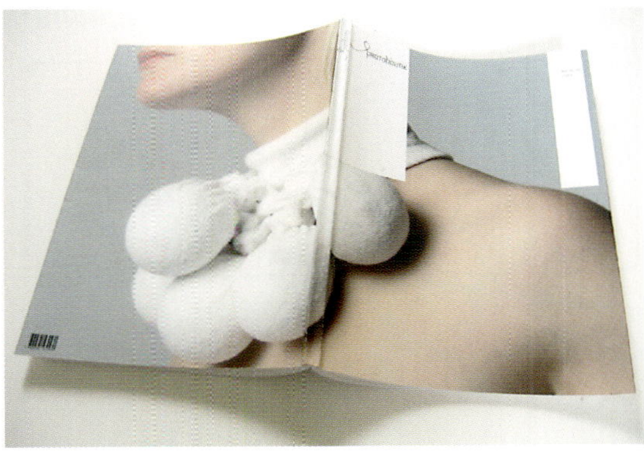

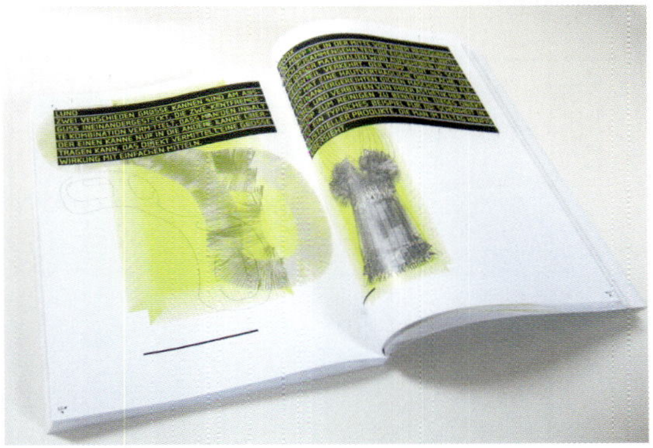

iF concept award communication

Auftraggeber Client
Bauhaus-Universität Weimar
Weimar, Germany

Projekt Project
Truck & Type – Transit Grafik in Europa
Buch
Book

Design
Michael Diebold
Weimar, Germany

Große Schrift, klare Farben, Pfeile, Kreise, Balken – 18 Meter Werbefläche, Kilometer für Kilometer. LKW-Planen sind Visitenkarten im Überformat. Gestaltet werden sie selten von Designern, sondern meist von den Fuhrunternehmern selbst. Und doch, oder gerade deshalb, faszinieren die meist ungewöhnlichen Lösungen.
„Truck & Type" ist eine Bestandsaufnahme aktueller Grafiken auf LKWs. Neben einer Website unter www.truck-and-type.net nähert sich das Buch „Truck & Type – Transit Grafik in Europa" dem Thema theoretisch, bildlich und analytisch. Eingeschlagen in ausrangierte LKW-Planen erzählt jedes Buch seine ganz eigene Geschichte der Straße.

Big type, clear colors, arrows, circles, bold lines – 18 meters of advertising space, kilometer by kilometer. Truck tarps are oversized business cards. They are rarely designed by graphic designers, but mostly by the carter himself. Nevertheless – or just because of that – they are fascinating due to their unusual solutions.
"Truck & Type" is an inventory of current graphics on trucks. In addition to a website on www.truck-and-type.net, the book "Truck & Type – Transit Graphics in Europe" contains theory, photo gallery and facts and figures on the topic. The book jacket is made from old truck tarps. So every book has its own character by telling its own road story.

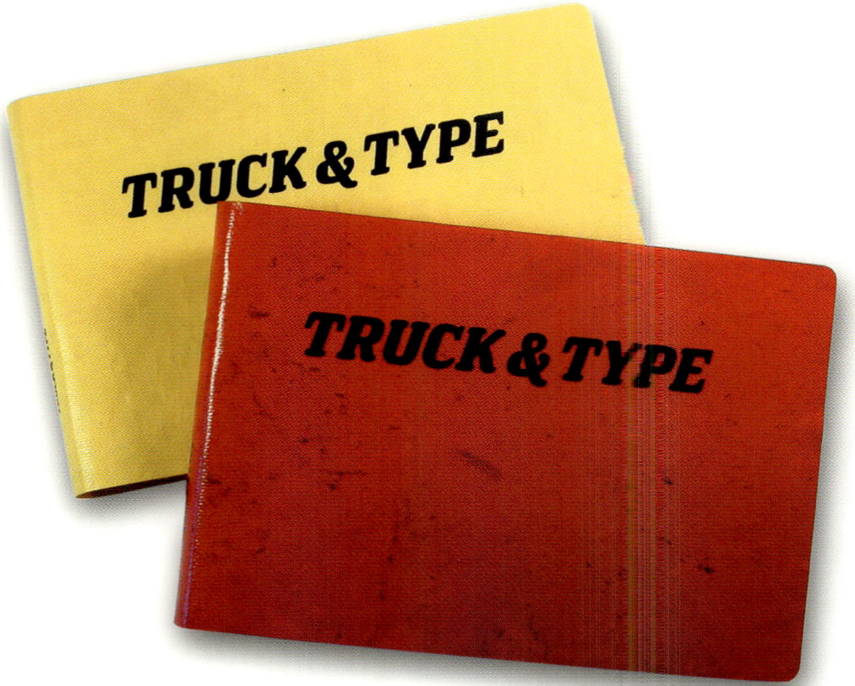

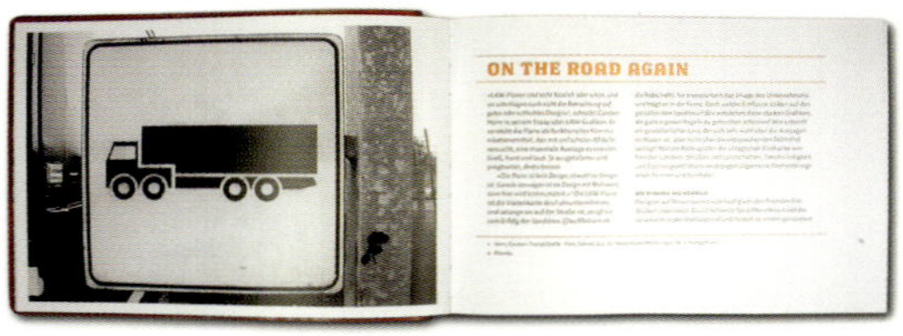

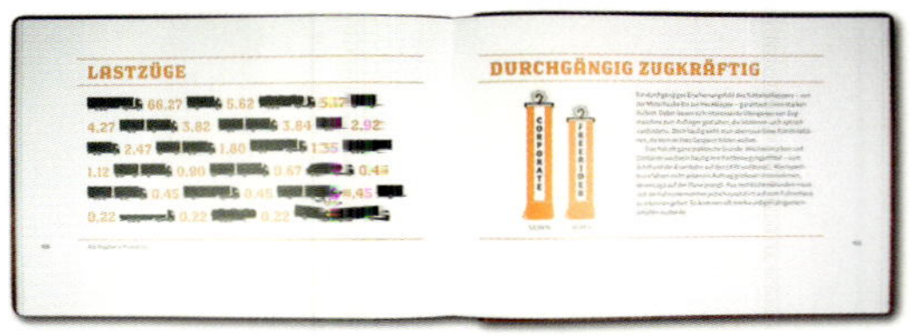

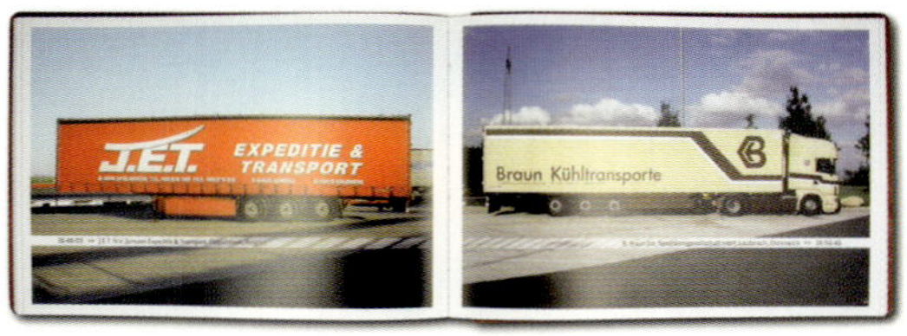

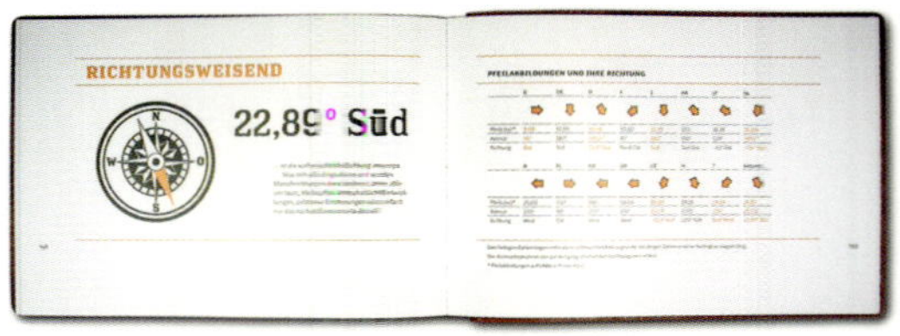

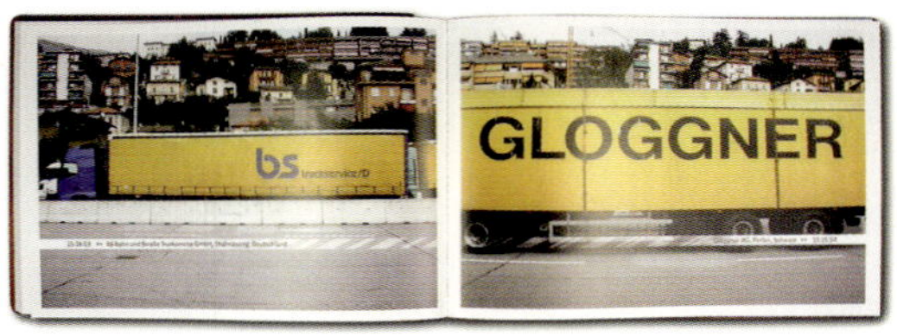

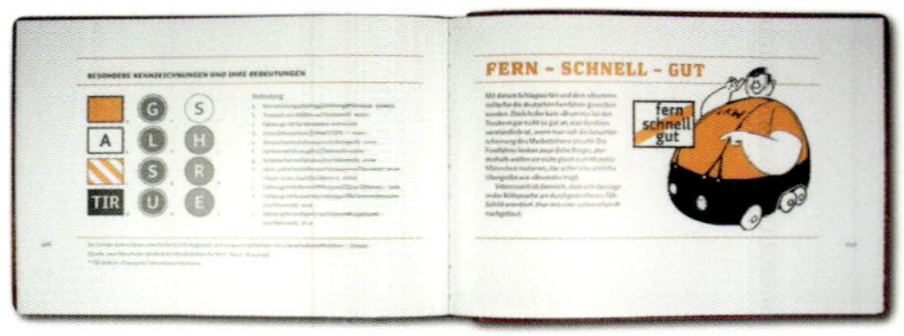

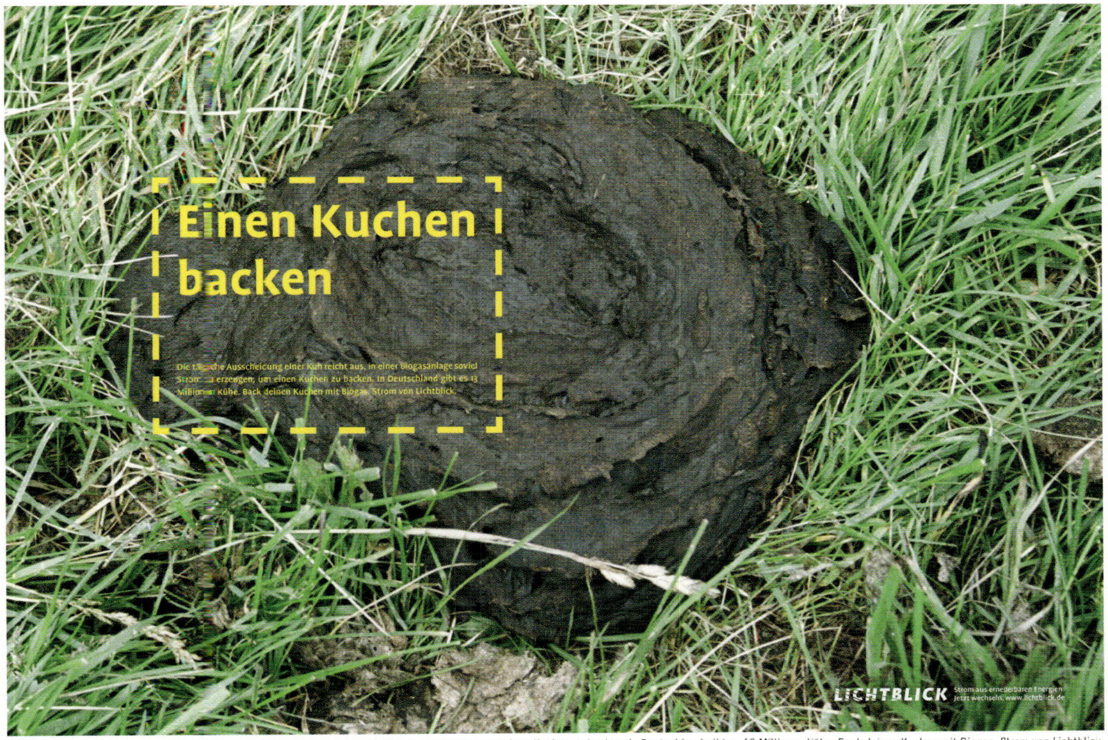

Die tägliche Ausscheidung einer Kuh reicht aus, in einer Biogasanlage soviel Strom zu erzeugen, um einen Kuchen zu backen. In Deutschland gibt es 13 Millionen Kühe. Back deinen Kuchen mit Biogas. Strom von Lichtblick.

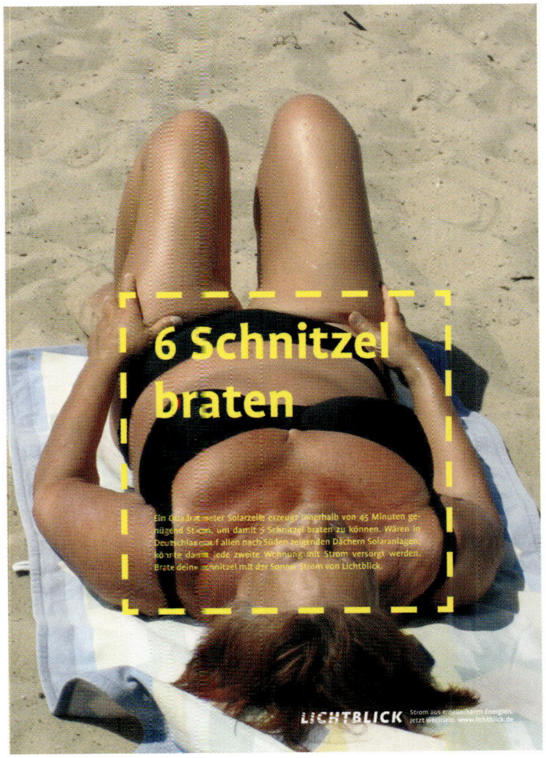

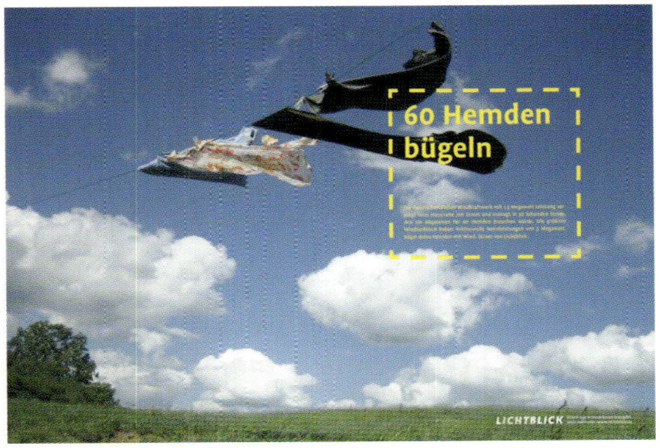

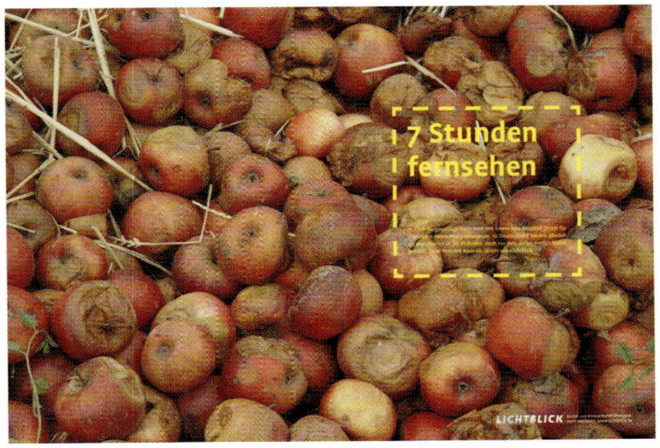

Auftraggeber Client
Bauhaus-Universität Weimar
Weimar, Germany

Projekt Project
Lichtblick
Anzeigen für den Ökostromanbieter Lichtblick
Ads for Lichtblick – a green power utility

Design
Marie Bauer
Berlin, Germany

Lichtblick ist der größte unabhängige Ökostromanbieter in Deutschland. Der Strom von Lichtblick wird zu 100% aus regenerativen Quellen wie Wasser, Biomasse, Sonnenenergie oder Windkraft erzeugt. Diese Energiequellen sind unendlich vorhanden und alltäglich; selbst der Kuhfladen auf der Weide kann in Strom umgewandelt werden.
Die Kampagne soll das Vertrauen in erneuerbare Energien stärken und Lichtblick bekannter machen, mit dem Ziel, die Nutzer zum Wechsel ihres Stromanbieters zu bewegen. Die Arbeit besteht aus neun Anzeigen, Postkarten, Informationsmaterial und Aufklebern.

Lichtblick is the largest independent ecological power utility in Germany. Its energy is generated entirely from regenerative sources like water, biomass, solar energy and wind power. These everyday energy sources are infinitely available – even a cowpat in a field can be turned into electricity.
The campaign is aimed at increasing the level of public confidence in renewable energies and raising awareness of Lichtblick to get users to change their electricity supplier. The project comprises nine ads as well as postcards, information material and stickers.

iF

iF Industrie Forum Design Hannover e. V.
Mitgliederliste ▪ list of members
Satzung des Vereins ▪ articles of association

■■ Vorstand / Board

01 Ernst Raue, Deutsche Messe AG, Hannover;
1. Vorsitzender / Chairman iF Hannover
02 Norbert H. Bargmann, Messe München GmbH, Munich
03 Christoph T. Böninger, brains4design GmbH, Munich
04 Jürgen Fitting, Andreas Stihl AG & Co. KG, Waiblingen
05 Fritz Frenkler, f / p design GmbH, Frankfurt
06 Prof. Harald Leschke, DaimlerChrysler AG, Sindelfingen
07 Ralph Wiegmann, iF Hannover;
Geschäftsführendes Vorstandsmitglied /
Executive Member of the Board

Geschäftsführung / Managing Director

Ralph Wiegmann, iF Hannover

Ehrenmitglieder / Honorary Members

1 Fritz Hahne, Hannover
2 Sepp D. Heckmann, Hannover
3 Karl-Heinz Krug, Düsseldorf
4 Prof. Herbert Lindinger, Hannover
5 Edwin A. Schricker, Rodach
6 Herbert H. Schultes, Fürstenfeldbruck
7 German A. Vorment, Hirschbach

	Members:		
1	Aha, Jürgen	Steinkleistraße 27	60435 Frankfurt
2	AL-KO Geräte GmbH	Ichenhauser Straße 14	89359 Kötz
3	Andreas Stihl AG & Co. KG	Badstraße 115	71336 Waiblingen
4	ARGUZ GmbH publishing	Mittelstraße 50	33602 Bielefeld
5	Ars Nova Collection	Salzstraße 10	49326 Melle
6	Artemide GmbH	Hans-Böckler-Straße 2	58730 Fröndenberg
7	Audi AG		85045 Ingolstadt
8	Bang & Olufsen Deutschland GmbH	Dr. Carl von Linde Straße 2	82049 Pullach
9	BASF AG	Carl-Bosch-Straße 38	67056 Ludwigshafen
10	BEGA Gantenbrink-Leuchten GmbH & Co.	Hennenbusch	58708 Menden
11	Blue C. Verlag GmbH, design report	Neckarstraße 121	70190 Stuttgart
12	brains4design GmbH	Böcklinstraße 54	80638 München
13	Braun GmbH Corporate Design	Frankfurter Straße 145	61476 Kronberg
14	Braun Wagner	Krefelder Straße 147	52070 Aachen
15	BREE Collection GmbH & Co. KG	Gerberstraße 3	30916 Isernhagen / Kirchhorst
16	B:SiGN Brodehl & Partner	Ellernstraße 36	30175 Hannover
17	Bulthaup GmbH & Co. KG	Werkstraße	84153 Aich
18	Carpet Concept Objekt-Teppichboden GmbH	Bunzlauer Straße 7	33719 Bielefeld
19	Crown Gabelstapler GmbH	Moosacher Straße 52	80809 München
20	DaimlerChrysler AG		71059 Sindelfingen
21	DDC Deutscher Designer Club e. V.	Hanauer Landstraße 139	60314 Frankfurt
22	designafairs GmbH	Tölzer Straße 2 c	81379 München
23	Designstudio Bliesener	Im Haspelfelde 31	30173 Hannover
24	design studio hartmut s. engel	Monrepos-Straße 7	71634 Ludwigsburg
25	Design Tech	Zeppelinstraße 53	72119 Ammerbuch
26	Deutsche Messe AG	Messegelände	30521 Hannover
27	Eschke, Christoph	Admiralitätsstraße 10	20459 Hamburg
28	EUROMEDIAHOUSE GmbH	Vahrenwalder Straße 263	30179 Hannover
29	Festo AG & Co. KG	Ruiter Straße 82	73734 Esslingen
30	Fleischer, Bianca	Sommerstraße 5	81543 München
31	fön,design_	Schiltachstraße 40	78713 Schramberg
32	FOLIA TEC Böhm GmbH + Co. Vertrieb KG	Neumeyerstraße 70	90411 Nürnberg
33	f/p design GmbH	Schweizer Straße 7	60594 Frankfurt
34	Gantenhammer, Andreas	Düsselstraße 21–25	40219 Düsseldorf
35	GDCdesign	Fraunhoferstraße 3a	90409 Nürnberg
36	GIRA Giersiepen GmbH & Co. KG	Dahlienstraße 12	42477 Radevormwald
37	Grohe Water Technology AG & Co. KG	Industriepark Edelburg	58675 Hemer
38	Grüter, Michael	Hainholzstraße 17	31558 Hagenburg
39	hannoverCare AG	Emil Meyer Straße 20	30165 Hannover
40	Hansgrohe AG	Auestraße 5–9	77761 Schiltach
41	Hasberg, Josef	Lokenbach 8–10	51491 Overath
42	HBK Braunschweig	Johannes-Selenka-Platz 1	38118 Braunschweig
43	Heidelberger Druckmaschinen AG	Kurfürsten-Anlage 52–60	69115 Heidelberg
44	IBM Deutschland GmbH	Pascalstraße 100	70548 Stuttgart
45	i/i/d Institut für Integriertes Design	Am Speicher XI	28217 Bremen
46	INOID DesignGroup	Reutlingerstraße 114	70597 Stuttgart
47	Interstuhl Büromöbel GmbH & Co. KG	Brühlstraße 21	72469 Meßstetten-Tieringen
48	Katsuta, Jiro	Elsternweg 46	47804 Krefeld
49	Kochan & Partner GmbH	Hirschgartenallee 25	80639 München

50	Köttermann GmbH & Co.	Industriestraße 2–10	31311 Uetze-Hänigsen
51	Kühne, Karl-Michael	Oelkersallee 31	22769 Hamburg
52	Le Peetz, Sebastian	Osterstraße 43 a	30159 Hannover
53	Lichtwark, Hildegund	Pfarrer-Linzbach-Str. 1	52388 Nörvenich
54	LOEWE OPTA GmbH	Industriestraße 11	96317 Kronach
55	Messe München GmbH	Messegelände	81823 München
56	MEYER-HAYOZ Design Engineering	Zollernstraße 26	78462 Konstanz
57	Moormann Möbel GmbH	An der Festhalle 2	83229 Aschau i. Chiemgau
58	Neumeister Design	Liebigstraße 8	80538 München
59	Nds. Ministerium für Wirtschaft, Technologie+Verkehr	Friedrichswall 1	30159 Hannover
60	OCO-Design O. K. Nüsse	An der Kleimannbrücke 79	48157 Münster
61	OCTANORM Vertriebs-GmbH	Raiffeisenstraße 39	70794 Filderstadt
62	Oventrop GmbH & Co. KG, F. W.	Paul-Oventrop-Straße 1	59939 Olsberg
63	Panasonic Design Company	Matsushita Electric Industrial Co., Ltd. 1-15 Matsuo-cho, Kadoma-city	Osaka 571-8504, Japan
64	Philips Design	Building HWD, Emmasingel 24	5600 MD Eindhoven, The Netherlands
65	Philips GmbH	Alexanderstraße 1	20099 Hamburg
66	Planlicht GmbH & Co. KG	Fiecht-Au 25	6130 Schwaz-Vomp, Austria
67	Poggenpohl Möbelwerke GmbH	Poggenpohl Straße 1	32051 Herford
68	Polvan Design Ltd.	Cemil Topuzlu cad. 79/2	34170 Istanbul, Turkey
69	PRODESIGN Bernd Brüssing	Turmstraße 39	89321 Neu-Ulm
70	rahe+rahe design	Konsul-Smidt-Straße 8C	28217 Bremen
71	Ramsden, Dale	Cradock Heights, Tyrwhite Avenues	Rosebank, Johannesburg, South Africa
72	Rat für Formgebung	Ludwig-Erhard-Anlage 1	60327 Frankfurt
73	RITTO GmbH & Co. KG	Rodenbacher Straße 15	35708 Haiger
74	Seibel Designpartner GmbH	Industriestraße 5	40822 Mettmann
75	Schlegel, Eberhard	Am Kapellenweg 4	88525 Dürmentingen
76	SCHOLZ & VOLKMER GmbH	Schwalbacher Straße 76	65183 Wiesbaden
77	Schwedux, J. + Filipschack, R.	Straße der Nationen 5,Expo-Park Ost	30539 Hannover
78	Sedus Stoll AG	Brückenstraße 15	79761 Waldshut
79	Sennheiser electronic GmbH & Co. KG	Am Labor 1	30900 Wedemark
80	Shacham, Avi	P.O. Box 512	Sede Warburg 44935, Israel
81	Siemens AG	Wittelsbacher Platz 2	80333 München
82	Spellmeyer, Prof. Gunnar	c/o FH Hannover Expo Plaza 2	30539 Hannover
83	Steckertpeter	Hohenzollernstraße 25	30161 Hannover
84	Steelcase Werndl AG	Georg-Aicher-Straße 7	83026 Rosenheim
85	Storck Bicycle GmbH	Carl-Zeiss-Straße 4	65520 Bad Camberg
86	Sun Same Enterprises Co., Ltd.	553, Chung Cheng S. Road	Tainan, 710 Taiwan
87	TRILUX-LENZE GmbH + Co KG	Heidestraße	59759 Arnsberg
88	Universität Wuppertal FB Industrial Design	Hofaue 35–39	42103 Wuppertal
89	VDID/DDV	Markgrafenstraße 15	10969 Berlin
90	Viessmann Werke	Industriestraße 5	35108 Allendorf
91	Volkswagen AG		38436 Wolfsburg
92	Weinberg & Ruf Produktgestaltung	Martinsstraße 5	70794 Filderstadt
93	wiege Entwicklungsgesellschaft mbH	Hauptstraße 81	31848 Bad Münder
94	Wilkhahn Wilkening + Hahne GmbH + Co.	Fritz-Hahne-Straße. 8	31848 Bad Münder
95	WINNER's Productdesigninnovation	Bergamtstraße 1	30890 Barsinghausen
96	Yokogawa Electric Corporation	2-9-32 naka-cho, Mushashino-shi	Tokyo 180-8750, Japan
97	ZUMTOBEL STAFF GmbH	Schweizerstraße 30	6850 Dornbirn, Austria
98	ZUMTOBEL STAFF GmbH & Co. KG	Grevenmarschstraße 74–78	32657 Lemgo

§ 1 Name und Sitz des Vereins

1. Der Verein trägt den Namen „iF – Industrie Forum Design Hannover e. V.".
2. iF – Industrie Forum Design Hannover ist im Vereinsregister eingetragen.
3. Der Verein hat seinen Sitz in Hannover.

§ 2 Zweck des Vereins

Der Verein verfolgt den Zweck der Förderung und Akzeptanz von Design als Teil der Wertschöpfungskette und als kulturelles Element der Gesellschaft. Die bewusste Gestaltung von Produkten und Lebensräumen für den privaten und öffentlichen Bereich sowie von benutzerfreundlichen Softwareanwendungen betrachtet der Verein als seine gesellschaftspolitische und kulturelle Aufgabe. Vereinsziele sind:

1. Die Anerkennung der Designleistung zur Erreichung von Unternehmenszielen und zur Sicherung von wirtschaftlichem Erfolg;
2. Die Organisation von Wettbewerben, Ausstellungen, Konferenzen, Vortragsveranstaltungen und weiteren Aktivitäten;
3. Die Veröffentlichung von Publikationen als Grundlage für Diskussionen;
4. Die Stärkung des Bewusstseins für Design in der Öffentlichkeit;
5. Das Angebot eines Forums zur Kommunikation auf neutraler Ebene;

Der Verein ist mit seinen Aktivitäten regional, national und international präsent.

§ 3 Geschäftsjahr

Das Geschäftsjahr ist das Kalenderjahr.

§ 4 Mitgliedschaft

1. Jede natürliche oder juristische Person kann Mitglied im iF e. V. werden. Der schriftliche Aufnahmeantrag wird von der Geschäftsführung geprüft und von ihr entschieden – in Zweifelsfällen beschließt der Vorstand mit einfacher Mehrheit über die Aufnahme des Mitglieds. Die Ablehnung der Aufnahme braucht nicht begründet zu werden.
2. Der Verein hat neben Mitgliedern auch Ehrenmitglieder. Ehrenmitglieder sind Persönlichkeiten, die sich im besonderen Maße um die Förderung und das Ansehen des „iF – Industrie Forum Design Hannover e. V." Verdienste erworben haben. Sie können durch einstimmigen Beschluss der Mitgliederversammlung zu Ehrenmitgliedern ernannt werden und haben in der Mitgliederversammlung kein Stimmrecht.
3. Die Mitgliedschaft endet durch

a) Austritt, der dem Verein gegenüber zu erklären ist. Der Austritt kann nur durch schriftliche Erklärung zum Ende eines jeden Geschäftsjahres mit dreimonatiger Frist erklärt werden,
b) Ausschluss aus dem Verein, über den der Vorstand einstimmig beschließt,
c) Tod eines Mitglieds.

§ 5 Organe

Die Organe des Vereins sind:

1. die Mitgliederversammlung
2. der Vorstand
3. die Geschäftsführung

§ 6 Mitgliederversammlung

1. Die Mitgliederversammlung ist zuständig für:

a) Entgegennahme des Jahresberichts des Vorstandes und des Berichts der Rechnungsprüfer,
b) Wahl des Vorstands,
c) Entlastung des Vorstands und der Geschäftsführung,
d) Satzungsänderungen,
e) Auflösung des Vereins,
f) Verwendung des Vermögens bei Auflösung des Vereins,
g) die Verwendung des jährlichen Gewinnvortrages bzw. die Behandlung eines Verlustes,
h) Entscheidung über neue Aktivitäten des Vereins,
i) Festsetzung des Jahresmitgliedsbeitrages,
j) Wahl der Ehrenmitglieder.

2. Jährlich ist eine ordentliche Mitgliederversammlung durchzuführen, zu der alle Mitglieder des Vereins eingeladen werden und in der insbesondere über das abgelaufene Geschäftsjahr, über den Rechnungsabschluss und das Ergebnis der Rechnungsprüfung zu berichten ist.
3. Die Mitgliederversammlung wird vom Vorsitzenden des Vorstandes mit einer Frist von mindestens einer Woche schriftlich unter Angabe der Tagesordnung einberufen. Die Tagesordnung der Mitgliederversammlung setzt der Vorsitzende des Vorstandes oder ein anderes von ihm benanntes Vorstandsmitglied fest. Anträge einzelner Mitglieder zur Tagesordnung können nachträglich auf die Tagesordnung gesetzt werden, wenn die Mitgliederversammlung es einstimmig beschließt.
4. Außerordentliche Mitgliederversammlungen können jederzeit vom Vorsitzenden des Vorstandes einberufen werden, wenn ein wichtiger Grund vorliegt. Die Einberufung muß erfolgen, wenn dies von mindestens 1/3 der Mitglieder schriftlich beantragt wird. Die Einberufungsfrist beträgt 2 Tage. In Eilfällen können die Einladungen telefonisch oder per Fax erfolgen.
5. Die Mitgliederversammlung wird von dem Vorsitzenden des Vorstandes, im Falle seiner Verhinderung durch ein von ihm bestimmtes anderes Vorstandsmitglied geleitet.
6. Die ordnungsgemäß einberufene Mitgliederversammlung ist beschlussfähig, wenn außer dem Versammlungsleiter mindestens drei weitere stimmberechtigte Mitglieder anwesend sind.
7. Jedes Mitglied hat in der Mitgliederversammlung eine Stimme. Beschlüsse der Mitgliederversammlung werden mit einfacher Mehrheit der bei der Abstimmung anwesenden stimmberechtigten Mitglieder gefaßt, soweit diese Satzung nicht etwas anderes bestimmt. Stimmberechtigte Mitglieder können sich in der Mitgliederversammlung vertreten lassen,

wobei auch die Übertragung des Stimmrechts auf den Vertreter zulässig ist. Bei Stimmengleichheit entscheidet die Stimme des Versammlungsleiters.
8. Eilbeschlüsse können im Umlaufverfahren schriftlich gefaßt werden.
9. Zu einem Beschluss über eine Änderung der Satzung ist eine Mehrheit von 3/4 der Stimmen der erschienenen stimmberechtigten Mitglieder erforderlich.
10. Die gefaßten Beschlüsse werden in einem Protokoll erfasst, das vom Protokollführer und dem Leiter der Mitgliederversammlung unterschrieben werden muß. Jedem Mitglied ist eine Niederschrift des Protokolls der Mitgliederversammlung zuzustellen.

§ 7 Vorstand

1. Der Vorstand des Vereins besteht aus dem 1. Vorsitzenden sowie maximal sechs weiteren Mitgliedern. Die Vorstandsmitglieder werden von den Mitgliedern auf die Dauer von vier Jahren gewählt. Sie verbleiben bis zur Neuwahl im Amt. Die Wahlen können aus wichtigem Grund auf der Mitgliederversammlung widerrufen werden.
2. Der Verein wird gerichtlich und außergerichtlich im Rahmen des Vereinszwecks durch den 1. Vorsitzenden vertreten. Der 1. Vorsitzende ist berechtigt, für Einzelaufgaben Dritten Vollmacht zu erteilen. Der Vorstand unterstützt die Geschäftsführung bei der Leitung des Vereins und beschließt über alle Vereinsangelegenheiten, soweit diese nicht der Mitgliederversammlung vorbehalten sind. Der Vorstand kann weitere zur Verwaltung des Vereins erforderlichen Personen wählen und entscheidet über sämtliche Einnahmen und Ausgaben des Vereins.
3. Über Aktivitäten besonderer Art und die Höhe der damit verbundenen Kosten beschließen nach Vorlage des Vorstandes die Mitglieder in der Mitgliederversammlung.
4. Die Mitgliederversammlung kann dem Vorstand eine Geschäftsordnung geben, nach der die Geschäfte des Vereins zu führen sind.

§ 8 Geschäftsführung

Der Vorstand bestellt eine Geschäftsführung. Diese führt die laufenden Aufgaben des Vereins nach Richtlinien des Vorstands durch. Sie ist auch zur Entscheidung über die Aufnahmeanträge der neuen Mitglieder bevollmächtigt. Die Geschäftsführung hat den Vorstand über alle Vereinsangelegenheiten von Bedeutung zu unterrichten. Sie ist an die Weisungen des Vorstandes gebunden und hat den Vorstand in allen wichtigen Angelegenheiten vorher zu konsultieren.

§ 9 Mitgliedsbeiträge

Der Jahresmitgliederbeitrag wird durch die Mitgliederversammlung bestimmt und staffelt sich zur Zeit wie folgt:
- Unternehmen: mindestens 600,– Euro
- Designer/Designbüros: mindestens 300,– Euro
- Privatpersonen: mindestens 150,– Euro

§ 10 Revision

Mit der Rechnungsprüfung wird die Revision der Deutschen Messe AG bzw. eine Wirtschaftsprüfungsgesellschaft beauftragt. Über das Revisionsergebnis ist in der ordentlichen Mitgliederversammlung zu berichten.

§ 11 Auflösung des Vereins

1. Ein Beschluss auf Auflösung des Vereins bedarf einer Mehrheit von 3/4 der stimmberechtigten Mitglieder, wobei die Mitgliederversammlung nur beschlussfähig ist, wenn mindestens 2/3 der stimmberechtigten Mitglieder anwesend sind.
2. Bei Auflösung oder Aufhebung des Vereins oder bei Wegfall des jetzigen Zwecks fällt das Vermögen, soweit es die eingezahlten Kapitalanteile der Mitglieder und den gemeinen Wert der von den Mitgliedern geleisteten Sacheinlagen übersteigt, an eine Körperschaft des öffentlichen Rechts zwecks Verwendung für kulturelle oder soziale Zwecke. Beschlüsse über die künftige Verwendung des Vermögens dürfen erst nach Einwilligung des Finanzamtes ausgeführt werden.
3. Die die Auflösung beschließende Mitgliederversammlung entscheidet über die Verwendung des Vereinsvermögens mit 3/4 Mehrheit.

§ 12 Überschüsse, Ausgaben, Buchführung

1. Etwaige Gewinne des Vereins dürfen nur für steuerbegünstigte Zwecke der Satzung verwendet werden. Die Mitglieder erhalten keine Gewinnanteile und in ihrer Eigenschaft als Mitglieder auch keine sonstigen Zuwendungen aus Mitteln des Vereins.
2. Es darf keine Person durch Verwaltungsaufgaben, die dem Zweck des Vereins fremd sind, mittels unverhältnismäßig hoher Vergütungen begünstigt werden.
3. Der Nachweis für die Mittelverwendung ist durch eine ordnungsgemäße Buchführung zu gewährleisten.

Hannover, 29. September 2005

Article 1. Name and registered office of the Association

1. The name of the Association is "iF – Industrie Forum Design Hannover."

2. iF – Industrie Forum Design Hannover is listed in the Register of Associations (Vereinsregister).

3. The Association's registered office is in Hannover, Germany.

Article 2. Object of the Association

The object of the Association is to support and gain acceptance for design as a link in the value chain and as a cultural component of society. The Association considers the targeted design of products, public and private living spaces, and user-friendly software applications as constituting its own civic and cultural mission. The Association aims to:

1. Recognize design achievements that help companies achieve their business goals and cement their economic success

2. Organize competitions, exhibitions, conferences, lectures, and other events

3. Issue publications to serve as the basis for discussions

4. Bolster the public's awareness of design

5. Provide a forum for objective dialogue on design-related issues.

The Association's activities give it a regional, national, and international presence.

Article 3. Financial year

The financial year corresponds to the calendar year.

Article 4. Membership

1. Any natural person or legal entity may become a member of the Association. Executive Management shall examine and decide on membership applications, which are to be submitted in writing. In cases where Executive Management is unsure, the Board of Management shall decide on whether to grant membership by means of a simple majority vote. The Association is not required to justify a negative decision on membership.

2. In addition to members, the Association also consists of honorary members. Honorary members are individuals who have been particularly meritorious in their support of "iF – Industrie Forum Design Hannover". They may be elected to honorary membership by way of a unanimous resolution passed at the General Meeting. They are not entitled to vote at the General Meeting.

3. Membership shall end

a) Through resignation, which must be declared to the Association in writing. This declaration requires a minimum of three months' advance notice before the end of the financial year, and takes effect at the end of the financial year.

b) By way of expulsion upon a unanimous resolution by the Board of Management.

c) At the time of a member's death.

Article 5. Official bodies

The Association's official bodies are:

1. The General Meeting

2. The Board of Management

3. Executive Management

Article 6. General Meeting

1. The General Meeting is responsible for:

a) Receiving the reports of the Board of Management and the independent auditors

b) Electing the Board of Management

c) Approving the actions of the Board of Management and the Executive Management

d) Changing the Articles of Association

e) Dissolving the Association

f) Using the Association's assets in the event that it should be dissolved

g) Using the annual profits or handling of any loss

h) Setting the annual membership fee

i) Electing honorary members

2. A regular General Meeting will be held annually. All members of the Association shall be invited to attend the General Meeting. The agenda of the regular General Meeting will, in particular, include reports on the preceding financial year, on the Association's financial statements, and on the results of the auditing of accounts.

3. General Meeting shall be called in writing by the Chairperson of the Board of Management, with a minimum of one week's advance notice and with such notice containing a meeting agenda. The agenda of the General Meeting is set by the Chairperson of the Board of Management or by another member of the Board of Management appointed by the Chairperson. Topics requested by individual members may be added to the agenda if the General Meeting gives its unanimous consent.

4. The Chairperson of the Board of Management may call an extraordinary General Meeting if an important reason exists for doing so. A meeting of this kind must be called if it has been requested in writing by at least one third of all members. Two day's advance notice is required. In urgent cases, invitations may be transmitted by telephone or fax.

5. General Meetings shall be presided over by the Chairperson of the Board of Management or, in the event that he or she should be prevented from doing so, by another member of the Board of Management appointed by the Chairperson.

6. A General Meeting that has been properly called has a quorum if, in addition to the individual presiding over the Meeting, at least three other members in possession of voting rights are present.

7. Each member has one vote at the General Meeting. Unless otherwise stipulated in these Articles, resolutions of the General Meeting shall be passed by way of a simple majority of those members present and entitled to vote. Members entitled to vote may appoint a proxy to attend the General Meeting in their place, including a proxy voting right. In the of a tie, the deciding vote shall be cast by the individual presiding over the Meeting.

8. Urgent resolutions may be passed by round-robin procedure.

9. Resolutions on amendments to the Articles of Association shall require a three-quarters majority vote by the attending members who are eligible to vote.

10. Approved resolutions must be recorded in writing and signed by both the individual presiding over the Meeting and the keeper of the minutes. All members shall receive a copy of the minutes of the General Meeting.

Article 7. Board of Management

1. The Association's Board of Management consists of the Chairperson and a maximum of six other members. The members of the Board of Management will be elected by the members of the Association for a term of four years. They shall remain in office until new elections take place. Elections may be revoked at General Meetings for good cause.

2. The Association will be represented in and out of court by the Chairperson of the Board of Management, who shall be permitted to delegate a power of attorney to third persons in individual cases.

3. The Board of Management shall support Executive Management in the running of the Association and shall decide on all matters concerning the Association, providing that the right to such decision-making is not reserved by the General Meeting. The Board of Management may elect other individuals required to assume administrative tasks and will decide on all of the Association's income and expenditures.

4. At the General Meeting, the members will decide on special activities and related expenditures as presented by the Board of Management.

5. The General Meeting may provide the Board of Management with rules of procedure governing the Association's operation.

Article 8. Executive Management

The Board of Management shall appoint the executive management, which will be charged with executing the ongoing tasks of the Association in accordance with the guidelines issued by the Board of Management. Executive Management is authorized to make decisions on membership applications for new members. Executive Management shall inform the Board of Management about all significant matters affecting the Association. Executive Management shall follow all instructions issued by the Board of Management und shall consult the Board of Management in advance regarding all important matters.

Article 9. Membership fees

The annual membership fee will be set at the General Meeting and is currently as follows:

- Companies: 600 minimum
- Designers and design studios: 300 minimum
- Individuals: 150 minimum

Article 10. Auditing

Accounting control will be assigned to Deutsche Messe AG's auditing department or to an independent auditor. The results of auditing will be reported at the regular General Meeting.

Article 11. Dissolution of the Association

1. A resolution to dissolve the Association shall require a three-fourths majority vote by the members who are entitled to vote, provided that at least two thirds of the members entitled to vote are present at the General Meeting.

2. Upon dissolution or annulment of the Association or in the event that the current object of the Association should cease to be applicable, any assets exceeding the capital shares deposited by the members and the fair market value of non-cash capital contributions made by members shall go to a public corporation that will use the assets for cultural or community welfare purposes. Resolutions on the future application of such assets may only be acted upon subsequent to consent thereto by the tax authorities.

3. The General Meeting which resolves to dissolve the Association shall pass a resolution on the application of the Association's assets by way of a three-fourths majority vote.

Article 12. Profits, expenditures, accounting

1. Any profits produced by the Association shall be used exclusively for the tax-privileged purposes of these Articles of Association. Members shall not receive shares of profit, nor shall they, in their capacity as members, receive any other forms of bestowal from Association funds.

2. No individual may be favored by receiving disproportionately high remuneration for administrative tasks not in keeping with the object of the Association.

3. Proof that the Association's funds are being used properly shall be guaranteed by way of adequate and orderly accounting procedures.

Hannover, September 29, 2005

PAGE, das führende Magazin für Design- und Publishing
PAGE, a leading design and publishing magazine

W&V Wochenmagazin für Marketing, Werbung, Medien und E-Business
W&V, Weekly Magazine for Marketing, Advertising, Media and Business
Germany

AGD
Allianz deutscher Designer
Germany

CIDA
China Industrial Designers Association
Taiwan

BCD
Barcelona Design Centre
Spain

ddc
Deutscher Designer Club
Germany

BNO
Beroepsorganisatie Nederlandse Ontwerpers
The Netherlands

Design Austria e.V.
Austria

CETRA/DPC
Design Promotion Center/China External Trade Development Council
Taiwan

BVDW
Bundesverband Digitaler Wirtschaft
Germany

AXIS, das führende japanische Designmagazin
AXIS, a leading Japanese design magazine
Japan

IdcN
International Design Center NAGOYA Inc.
Japan

IIDA
Israel Industrial Designers Association
Israel

MFT
Hungarian Design Council
Hungary

National Institute of Design India
India

sda
swiss design association
Switzerland

SGD+
Swiss Graphic Designers
Switzerland

Taipei Design Center Düsseldorf
Germany / Taiwan

VDID/DDV
Verband Deutscher Industrie Designer e.V.
Germany

index

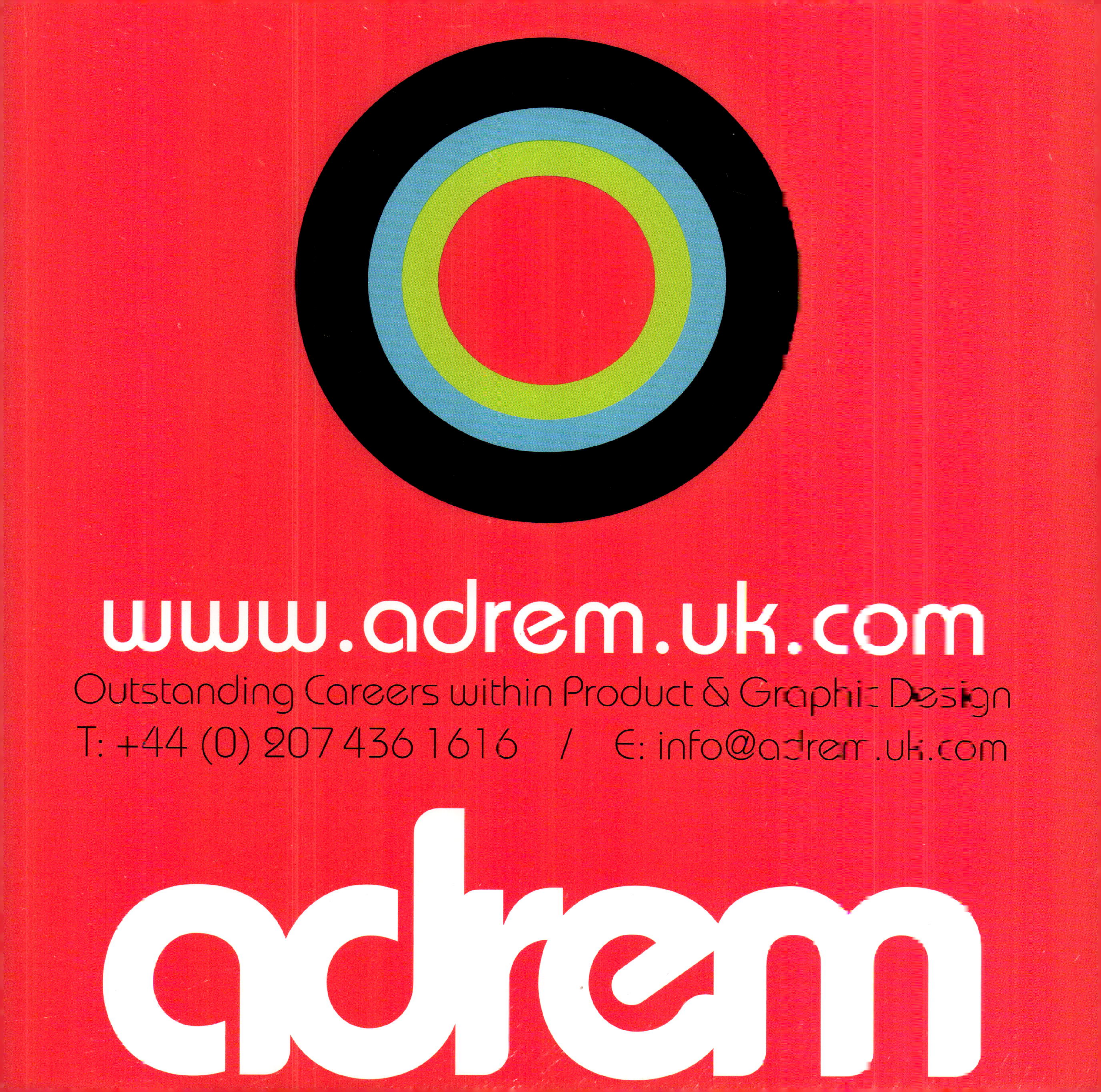

www.adrem.uk.com
Outstanding Careers within Product & Graphic Design
T: +44 (0) 207 436 1616 / E: info@adrem.uk.com
adrem

scrollan
Designbüro Peter Bünnagel,
Barbara Kotte, Anne-Lene Proff
Winsstraße 32
10405 Berlin, Germany
Phone +49(0)30-44318790
Fax +49(0)30-44318799
www.scrollan.de
▲ 93
● 93, 203

Hana Sedelmayer
Hasenböge 2
22159 Hamburg, Germany
● 313

Senckenbergmuseum
Senckenberganlage 25
60325 Frankfurt am Main, Germany
▲ 327

Miki Shimokawa
Schweizer Straße 7
60594 Frankfurt am Main, Germany
Phone +49(0)69-61991808
▲ 235

Siemens Audiologische Technik GmbH
Gebbertstraße 125
91058 Erlangen, Germany
Phone +49(0)9131-3080
▲ 112

Siemens Communication Networks
Ltd., Beijing (SCNB)
Jiu Xian Qiao Lu 14, Chaoyang District
100016 Beijing, China
Phone +86-10-64346688
▲ 71
● 71

SIGN Kommunikation GmbH
Oskar-von-Miller-Straße 14
60314 Frankfurt am Main, Germany
Phone +49(0)69-9443240
www.sign.de
● 133, 150, 151, 179

Silverstein Properties Inc.
530 5th Avenue
NY 10036 New York, United States
Phone +1-212-4900666
▲ 276

Simon & Goetz Design
GmbH & Co.KG
Westhafen Pier 1, Rotfeder-Ring 11
60327 Frankfurt am Main, Germany
Phone +49(0)69-96885547
● 209

Skidmore, Owings & Merrill LLP
14 Wall Street
NY 10005 New York, United States
● 276

Slade School of Fine Art
University College London
Gower Street
WC1E 6BT London, United Kingdom
Phone +44(0)20-76792313
www.ucl.ac.uk/slade/homepage.html
▲ 152

smart GmbH
Leibnizstraße 2–6
71032 Böblingen, Germany
Phone +49(0)7031-9071956
▲ 61

Catrin Sonnabend
Gustav-Adolf-Straße 38
63069 Offenbach, Germany
Phone +49(0)69-84848830
www.catrinsonnabend.de
● 321

Sony Deutschland GmbH
Hugo-Eckener-Straße 20
50829 Köln, Germany
Phone +49(0)221-5370
Fax +49(0)221-537349
www.sony.de
▲ 244–245

Staatskanzlei des Landes Nordrhein-Westfalen
Referat für Internationale
Angelegenheiten
Stadttor 1
40190 Düsseldorf, Germany
Phone +49(0)211-8371456
▲ 108

Staatstheater Mainz
Gutenbergplatz 7
55116 Mainz, Germany
Phone +49(0)6131-28510
▲ 214

Staatstheater Stuttgart
Schauspiel Stuttgart
Oberer Schlossgarten 6
70173 Stuttgart, Germany
Phone +49(0)711-2032262
www.staatstheater.stuttgart.de/schauspiel/index2.htm
▲ 185

Stadt Darmstadt
Luisenplatz 5
64283 Darmstadt, Germany
Phone +49(0)6151-132319
▲ 230

Stadtwerke Bochum GmbH
Ostring 28
44787 Bochum, Germany
Phone +49(0)234-9600
Fax +49(0)234-9604000
▲ 305

Strichpunkt
agentur für visuelle
kommunikation GmbH
Schönleinstraße 8a
70184 Stuttgart, Germany
Phone +49(0)711 62032732
www.strichpunkt-design.de
● 185, 232

Ströer Out-of-Home Media AG
Ströer Allee 1
50999 Köln, Germany
Phone +49(0)2236-96450
www.stroeer.de
▲ 113

Studio Urquiola srl
Via G.B. Morgagni 6
20129 Milan, Italy
Phone +39(0)2-2952271
● 253

Stylepark AG
Brönnerstraße 22
60313 Frankfurt am Main, Germany
Phone +49(0)69-29722222
Fax +49(0)69-29722223
www.stylepark.com
● 255

Georg Andreas Suhr Grafik Design
Leonhardtstraße 3
14057 Berlin, Germany
Phone +49(0)30-3246982
www.georg-andreas-suhr.de
● 139

Sulake Corporation Ltd
Eteläranta 12
130 Helsinki, Finland
Phone +358-9-6812270
www.sulake.com
● 53